The Atlas of Chick Development

The Atlas of Chick Development

Ruth Bellairs *and* **Mark Osmond**

Department of Anatomy and Developmental Biology, University College London, UK

ELSEVIER

Amsterdam • Boston • Heidelberg • London • New York • Oxford
Paris • San Diego • San Francisco • Singapore • Sydney • Tokyo

This book is printed on acid-free paper

Elsevier Academic Press
525 B Street, Suite 1900, San Diego, California 92101-4495, USA
http://www.elsevier.com

Elsevier Academic Press
84 Theobald's Road, London WC1X 8RR, UK
http://www.elsevier.com

British Library Cataloguing in Publication Data
A catalogue record for this book is available from the British Library

Library of Congress Catalog Number:

ISBN 0-12-084791-4

Printed and bound in Great Britian
05 06 07 08 9 8 7 6 5 4 3 2 1

Contents

Preface to First Edition

The study of the chick embryo reaches back into antiquity but continues to this day. Indeed, the chick embryo has for long been one of the most widely used laboratory animals for both teaching and research and the reasons are clear to see: fertilized eggs are cheap and available in large numbers all the year round; development is fast in comparison with most mammalian embryos and similar to that of the mouse, lasting a mere 21 days; perhaps more important, the physiology of the embryo is not complicated by the presence of a placenta nor, once the egg has been laid, by events taking place within the mother hen; the genetics are well known, both for the normal embryo and for many mutant forms; and lastly, being an amniote, the chick's development resembles that of a mammal in its broad outline. For these reasons it has found favour in many molecular biological and biochemical laboratories.

There has long been a need for an atlas of chick development, especially since a knowledge of developmental anatomy is less general in the scientific population than in former days. Most of the available textbooks of embryology pay little attention to stages of the chick embryo later than day 3, whilst the detailed accounts of Lillie (1952) and Romanoff (1960) give few illustrations of sectioned material.

In this atlas, we have provided labelled photographs of sections up to day 13, almost two-thirds of the incubation period. By this time, the major significant events have already taken place and the embryo has become so large that further sections through the whole body are inappropriate. In addition to the photographs, we have thought it valuable to include text on the development of the different systems, since many users of the book will be coming to the chick embryo for the first time and will probably be more familar with mammalian embryos. Not all the structures mentioned in the text can be found in the plates selected for this book, nor are all the structures that are illustrated discussed in the text. The plates follow a chronological sequence of sections through whole embryos whilst the text is based on the anatomical development of different organ systems. References to the plates made in the text are, therefore, not necessarily in sequential order. In identifying the structures in the later stages (days 9–13) we have had recourse to works on the anatomy of the adult fowl, particularly Ede (1964), King and McLelland (1979) and Baumel *et al.* (1993). Throughout this book we have used the Normal Table of Hamburger and Hamilton (1951) (see Appendix II) to indicate the developmental stage of an embryo during the first 3 days of incubation (i.e. up to stage 18) but have considered it more convenient to describe the older embryos in terms of days of incubation. We have tried to be consistent in terminology and where possible have favoured the anglicized rather than the latinized versions but, when in doubt, have followed the guidelines in Baumel *et al.* (1993).

We owe a particular debt of gratitude to Professor M. Kaufmann, who not only helped with his valuable advice on the preparation of the book, but generously gave us an essential supply of black and white label lines, otherwise unobtainable, to Dr Jeremy Cook who photographed the alizarin preparations and to A.J. Lee who prepared many of the text figures. We are deeply grateful to other colleagues and friends who have most kindly read through sections of the book: Dave Becker, Ann Burgess, Donald Ede, Marjorie England, Jay Lash, Pamela Lear, Paul Maderson, Drew Noden, Esmond Sanders and Willie Vorster. Their comments have helped us greatly, but we alone can be held responsible for any errors that may still be present.

We wish to acknowledge the generosity of the Leverhulme Trust, who provided a grant to R.B. to enable the expenses of microscopy and photography to be met. We would also like to thank the authors and publishers who have kindly given permission for reproduction of certain text figures.

Preface to Second Edition

In the second edition we have supplemented all sections of the original version. The text, as before, does not claim to give an all-embracing account of the development of the chick embryo but to provide an outline of the more important aspects. Some of these, such as the development of the coelomic cavities, have not attracted attention in recent years so that there has been little to add to the first edition. By contrast, others, such as the process of neurulation or the segmentation of somites, have provoked such extensive modern programmes of research that we have had to be more selective in our references than we would have liked. In mitigation, we can only plead that the text is intended as an introduction to the chick embryo and not as a total review of every known aspect of its development.

The number of plates has been increased, largely by the inclusion of many scanning electron micrographs. Some of the new plates have been generously supplied by other authors (the source duly acknowledged in each case) to whom we are most grateful. We are indebted to those authors and publishers who gave permission for the reprinting of text figures and plates, or advised us on particular points of fact, all of which have enabled us to extend the scope of the atlas. We are especially grateful to M. Bakst, M. Bancroft, A. Carretero, V.G. Djonov, D.A. Ede, H. Eyal-Giladi, R. Hirakow, T. Hiruma, M. Jacob, S. Matsuda, J. Ruberte, E.J. Sanders, C. Tickle, D.J. Wilson and G. Wishart. In addition, we thank those colleagues and friends who have given their time to help and advise us on technical aspects of the work, especially Alan Boyde, Claudio Stern and Ian Todd. We are grateful to the Samuel Sharpe Foundation for a grant to help in the preparation of the book.

Note on Plates

Fifty new plates have been introduced into the second edition. Most of these are scanning electron micrographs of either surface or internal views of the chick embryo although some additional light micrographs have also been included. The new plates are numbered 1–4 and 6–51. (Plate 1 from the first edition has become Plate 5 in the second.)

The remainder of the plates that were present in the first edition have been renumbered as Plates 52 *et seq.* and are photographs of serial sections taken by light microscopy. See the Glossary for definitions of longitudinal and transverse sections. The diagrams accompanying each of these plates provide an indication of the level and angle at which the sections are taken through the whole embryo but do not necessarily include all the structures shown in the photographs. In transverse sections through the embryos after the head has become bent ventral to the body (cervical flexure) the dorsal side of the head lies toward the bottom of the page, and the ventral side toward the top. As the series of sections progresses into the neck and trunk regions the dorsal side of the embryo lies to the top of the page (e.g. Plates 89, 90).

Plates 148–150 are photographs of whole embryos stained with alizarin red and alcian blue to show skeletal development.

Glossary

Achondroplasia: several mutant types of dwarfism in which ossification of bones is premature, reducing growth, e.g. *nanomelia* in birds (Greek: *a* = not; *chondro* = cartilage; *plasia* = plastic).

Acrosome: vesicular structure at the anterior end of the sperm containing hydrolytic enzymes (Greek: *akron* = extremity, top).

Activation of the egg: the stimulation of an egg cell, normally by fusion with a sperm, resulting in the onset of development (Latin: *activus* = active).

Activation of the sperm: the acquisition of motility by a sperm.

Agenesis: failure of part of the body to develop (Greek: *a* = not; *genesis* = to come into being).

Air space: the space between the two shell membranes at the blunt end of the egg. The chick pushes its beak into this space shortly before hatching and takes its first breath of air.

Albumen: the egg-white of a bird's egg (Latin: *albus* = white).

Allantoic stalk: the base of the allantois, which connects it to the hindgut.

Allantois: sac-like outgrowth from the hindgut consisting of endoderm covered with **splanchnic mesoderm**. It becomes highly vascularized and fuses with the **chorion** (*see* **Chorioallantois**). Excretory products are stored in its lumen (Greek: *allas* = sausage; *eidos* = form). N.B.: the chick allantois is balloon-like, not sausage-shaped.

Amniocardiac vesicle: bilaterally situated spaces forming in the mesoderm on either side of the head process. The dorsal wall of each cavity is composed of ectoderm and somatic mesoderm and gives rise to the **amnion**. The cavities become continuous medially with the cardiac cavity (becoming the pericardial cavity) and with the coelomic cavity laterally.

Amnion: the sac that envelops the embryo from about 3 days of incubation (Greek: *amnos* = a lamb).

Amniotes: the three classes of truly terrestrial vertebrates, reptiles, birds and mammals, all of which possess a true **amnion.**

Amniotic fluid: the proteinaceous fluid contained within the **amnion** and whose main function is to prevent dehydration and adhesions between the limbs and other regions of the body prior to cornification of the epidermis.

Amniotic folds: the folds of ectoderm, together with adherent somatic mesoderm, which fuse with one another to form the **amnion** and **chorion.**

Angiogenesis: development of certain parts of the vascular system (e.g. in the area vasculosa) by the formation of **blood islands,** which then join together by sprouting (but *see also* **Vasculogenesis**) (Greek: *angeion* = vessel; *genesis* = generation, development).

Animal pole: the uppermost surface of the fertilized egg, as apposed to the vegetal pole, the lowermost side.

Anlagen: (German) *See* **Primordium**.

Anterior: cranial, rostral: towards the head end (Latin: *ante* = in front of).

Anterior intestinal portal: the opening of the foregut in the early embryo into the lumen of the yolk sac (Latin: *ante* = before; *intus* = within; *porta* = door).

Anterior necrotic zone (ANZ): a patch of dead cells that appears on the anterior surface of the limb bud at stage 22 (about 3.5–4 days) (Latin: *ante* = before; Greek: *nekros* = dead body).

Anterior neuropore: the gap left between the unfused region of the neural folds at their most anterior end (*see* Plate 14). This pore becomes lost when the neural folds fuse at about stage 11 (*see also* **posterior neuropore**).

Apical ectodermal ridge: the thickened epithelium at the tip of a limb bud, which is essential for limb development (*see* Plate 26).

Apoptosis: *see* **Programmed cell death** (Latin: *ap* = before; Greek: *ptosis* from *piptein* = to fall).

Area centralis: the central area of a bird's blastoderm before the establishment of the area pellucida and area opaca.

Area opaca: the peripheral region of a young blastoderm, encircling the **area pellucida**. When removed from the yolk mass and viewed by transmitted light the tissue appears opaque because of the large amount of intracellular yolk droplets.

Area pellucida: the central area of a young blastoderm, which possesses little intracellular yolk and is therefore relatively transparent compared with the **area opaca**.

Area vasculosa: the proximal part of the **area opaca** after it has become vascularized.

Area vitellina: the distal part of the **area opaca** that remains unvascularized.

Atresia: the reduction or elimination of the lumen of an organ, or the failure of a duct to canalize (Greek: *a* = not; *tresio* = perforate, passage).

Atrio-ventricular canal: the constriction between the initial atrium and ventricle in the developing heart, which becomes divided into left and right sides by the growth of the atrio-ventricular cushions (Latin: *atrium* = vestibule; *ventriculus* = small belly; *canalis* = channel).

Atrophy: the wasting away of an organ or tissue. Although the term implies that this results from a lack of sufficient nourishment, this is not necessarily the case (Greek: *a* = not, *trophe* = food).

Auditory (otic) vesicle: *see* **Otic vesicle** (Latin: *auditus* = *hearing*).

Autopodium: the distal, flattened, end of a limb bud that will give rise to a foot or hand (Greek: *autos* = self; *podium* = foot).

Axial mesoderm: mesoderm lying in the midline (notochord) or immediately on either side of it (somitic mesoderm), which will take part in the formation of the axial structures, *i.e.* tissues arranged along the anterior–posterior axis of the body.

Axis: in a bilaterally symmetrical embryo there are three body axes, antero-posterior (head–tail, cranio-caudal), dorso-ventral (back–belly) and left–right (Greek: *axis* = axis).

Balbiani body: a region of the early oocyte rich in Golgi apparatus and mitochondria.

Bilaminar embryo: a blastoderm consisting of two layers only, **epiblast** and a **lower** layer.

Blastoderm, blastodisc, germinal disc, embryonic disc: the flattened disc of cells that forms during cleavage in the early embryo and persists until the end of gastrulation. By the time of laying two major regions are visible, the centrally situated **area pellucida** and the peripherally located **area opaca** (Greek: *blastos* = bud; *derma* = skin).

Blastodisc: *see* **Blastoderm.**

Blastomeres: cleavage stage cells, each completely enclosed by a cell membrane (*see also* **Open cells**) (Greek: *blastos* = bud; *meros* = part).

Blastula: an embryo during cleavage (*see also* **Blastoderm**).

Blood islands: isolated clusters of mesoderm cells in the **area vasculosa** (developing yolk sac), each consisting of a vesicle of endothelial cells surrounding the earliest red blood corpuscles.

Body folds: head, lateral and tail folds that raise the embryo above, and eventually separate it from, the extra-embryonic region of the blastoderm.

Bucco-pharyngeal membrane, oral plate, stomatodaeal plate: membrane formed by fused ectoderm of the mouth (buccal cavity) and the most anterior part of the foregut endoderm. This becomes broken down about stage 30, which results in the mouth and pharynx becoming in communication (Latin: *bucca* = cheek; *oris* = mouth; Greek: *stoma* = mouth).

Candling: technique for determining whether an incubated hen's egg contains a developing embryo. A bright light is shone through an egg of at least 9 days incubation and the embryo can be seen as a dark mass.

Cell cycle: the sequence of events through which a mitotic cell passes from its own formation until its division into two new cells.

Cell lineage: the history of a differentiated cell, with an emphasis on the cell types of its forbears.

Chalaza: one of the two spring-like structures projecting from the equatorial region of the vitelline membrane of the laid egg. The chalazae are thought to act as "balancers", helping the yolk to float in the albumen (Greek: *chalaza* = lump).

Chimera: an organism derived from two genetically different individuals other than by fertilization, e.g. by **xenoplastic grafting** (Latin from Greek: *chimaira* = she goat).

Chondrogenesis: the process of cartilage formation (Greek: *chondros* = cartilage).

Chorioallantois, allantochorion: the composite membrane formed by the fusion of the **allantois** with the **chorion**. Pressed beneath the shell membranes, it acts like a lung, exchanging oxygen and carbon dioxide through the pores in the shell.

Chorion (serosa): an extra-embryonic membrane formed, together with the **amnion**, from **ectoderm and somatic mesoderm**. After separating from the **amnion, the chorion comes to** lie close beneath the shell membranes and fuses with the **allantois** to form the **chorioallantois** (Greek: *chorion* = chorion).

Cleavage: the stage of development, after fertilization, during which many cells are formed by repeated cell division (German: *Kleben* = to adhere).

Cleidoic egg: the type of egg typical of birds and reptiles, possessing an almost impervious shell (Greek: *kleidoun* = to lock).

Cloacal bursa, bursa of Fabricius: a diverticulum on the dorsal side of the cloaca (Latin: *bursa* = purse).

Coelom: *see* **Lateral plate mesoderm**

Commitment (also known as **Determination**): a state of commitment (determination) is reached when a cell is no longer capable of changing its developmental path (i.e. no longer capable of **regulation**).

Compaction: the process by which a group of loosely associated cells becomes more compact, mainly as the result of increased cell adhesion. An example is the compaction of mesoderm cells to form a **somite**.

Congenital anomaly: an anomaly present in the individual from (and probably before) hatching (Latin: *con* = with; *gignere, genitum* = to beget).

***Coturnix coturnix* japonica:** the Japanese quail.

Cranial end: head end, anterior end (Greek: *cranium* = skull).

Creeper fowl: a mutant fowl with short limbs, a form of **achondroplasia**, which is lethal in the homozygous state.

Cuverian veins: another name for the Common Cardinal veins.

Cytodifferentiation: the changes taking place in a cell during differentiation.

Delamination: the splitting of a layer of cells into two layers.

Dermamyotome, dermomyotome: the dorsolateral side of a somite which will contribute to the dermis (**dermatome**) and muscle (**myotome**).

Dermis: the deeper layer of the skin, lying beneath the **epidermis** (Greek: *derma* = skin).

Determination: the process by which a cell or tissue becomes determined (**committed**) to follow a specific line of development rather than any other.

Developmental anatomy: the anatomical changes occurring during development.

Differential gene expression: *see* **Gene expression.**

Differentiation: the increasing complexity of the developing tissues in an embryo. Cells that are initially undifferentiated gradually acquire the molecular and histological characteristics of specialized cells.

Diploid: possessing the full complement of chromosomes as found in a somatic (body) cell (cf. **haploid**).

Distal: that part of an organ or tissue lying furthest from the midline (*see also* **Proximal**).

Diverticulum: a blind-ended extension from an organ, e.g. the caecum (Latin: *diverticulum* = byway).

Dorsal: the back aspect of the body (Latin: *dorsum* = back).

Dorsal mesentery: the mesentery that suspends organs in the body cavity from the dorsal body wall. In some regions the dorsal mesentery is renamed to indicate the organ it is serving, e.g. the dorsal mesentery that suspends the heart is known as the dorsal mesocardium.

Down feathers: the fluffy, earliest feathers of the newly hatched chick.

Ductus arteriosus: the remnant of the right sixth aortic arch artery which connects the ventral and dorsal aortae and enables blood to bi-pass the lungs during embryonic development. It closes on hatching (N.B. in mammals it is formed from the left aortic arch artery).

Ductus venosus: a complex series of venous channels in the embryonic liver through which blood from the yolk sac and the posterior part of the body pass before entering the posterior vena cava.

Ectoderm: the outermost (upper) layer of a blastoderm (Greek: *ektos* = outside; *derma* = skin).

Ecto-mesenchyme: obsolete term for the **neural crest**.

Egg tooth: a protruberance on the tip of the beak used for breaking through the shell membranes and shell at hatching. It is not a true tooth and does not possess dentine and enamel.

Embryonic disc: alternative term for **blastodisc/blastoderm**.

Endoblast: a term sometimes used for the lower layer of a bilaminar embryo, probably the equivalent of **hypoblast** (Greek: *endo* = within; *blastos* = bud).

Endoderm: the lowest layer of a trilaminar embryo (Greek: *endo* = within; *derma* = *skin*).

Endophyll: some of the earliest cells to form in the **lower layer** (Greek: *endo* = within; *phyllon* = leaf).

Ependyma: the epithelial layer of cells of the neural tube that surrounds the lumen.

Epiblast: the upper layer of a bilaminar embryo (Greek: *epi* = on, over; *blastos* = bud).

Epiboly: the flattening of an epithelial sheet of cells, and its consequent expansion. It is an integral component of the gastrulation movements.

Epithelial–mesenchymal interactions: the interactions that take place between adjacent epithelia and mesenchyme tissues during some types of **embryonic induction**, e.g. in the feather bud.

Epimere: the dorsal part of the myotome.

Epithelium: an histological term to describe a sheet of cells, in contrast to loosely arranged cells (**mesenchyme**). An epithelium may be of ectodermal, mesodermal or endodermal origin (Greek: *epi* = upon; *thele* = *nipple*).

Experimental embryology: analysis of an embryonic process by experimental means, e.g. extirpation or transplantation of part of the embryo to investigate the effect on development.

Extracellular material: material lying between cells consisting of fibres (e.g. collagen) and glycosamino-glycans, such as hyaluronic acid.

Extra-embryonic membranes: tissues formed along with the embryonic tissues. They play an essential role in supporting the embryo but do not form part of it (*see* **Allantois, Amnion, Chorion, Yolk sac**).

Fabricius, bursa of: *see* **Cloacal bursa**.

Falciform ligament: a remnant in the adult of the embryonic **ventral mesentery** (Latin: *falcis* = sickle).

Fate maps (prospective, presumptive fate maps): maps showing the expected developmental fate of the different regions in an early embryo where the tissues are not yet determined.

Feather buds/germs: the rudiment of a developing feather formed from a localized thickening of the epidermis underlain by a concentration of cells in the **dermis**, the dermal papilla.

Feather follicle: just before hatching the base of the **feather bud** sinks into a depression, the feather follicle.

Feather tract: the **feather buds** form in tracts (rows), the first appearing at about 5 days of incubation.

Fibroblast: connective tissue cell. Fibroblasts secrete fibronectin and pro-collagen. Confusingly, the term is used also for cells growing in **tissue culture** that have lost their specialized morphological characteristics and acquired a fibroblast-like appearance. Even cells of very different origins may be fibroblast-like and no longer distinguishable from one another but they do not necessarily correspond in behavioural and physiological aspects to true fibroblasts; care should therefore be taken in drawing conclusions about the behaviour of fibroblasts in culture from morphological characteristics alone (Latin: *fibra* = fibre; Greek: *blastos* = bud).

Fields, embryonic fields, individuation fields: field is an imprecise and almost obsolete term introduced in the early part of the twentieth century to indicate an area of an embryo that forms a particular organ or tissue, and it included the events involved in the formation of that tissue even if these were not known. The emphasis was on the co-ordination of many complex processes necessary for the differentiation of that organ or tissue. The term was also used to indicate the entire area throughout which the cells were capable of forming that organ or tissue, even though this ability was suppressed over much of it. This second usage of the term was, therefore, for a larger area than the first, e.g. the limb bud normally forms from a small, well-defined area on the flank of the embryo, but if that area is extirpated a much wider area of flank is capable of forming a limb, and this wider area was sometimes called the limb field.

Flexures: a vertebrate embryo becomes curved into a C-shape during development. The following bends (flexures) are recognised (*see also* Text-Figure 20):

a. **Cervical:** the flexure of the head on the trunk at the junction of the rhombencephalon and the spinal cord.
b. **Ventral mesencephalic;** a ventral bend in the midbrain.
c. **Pontine:** a ventral bend in the region of the rhombencephalon.
d. **Telencephalic:** a further bend in the forebrain.

Floor plate: a distinctive group of cells in the midline of the floor of the neural tube.

Follicle, ovarian: the epithelial layer (granulosa) surrounding each oocyte in the ovary.

Foramen: an aperture or hole.

Foregut: in the early embryo the anterior end acquires a floor and becomes a tube that is no longer open to the yolk sac except at its posterior end. This is the embryonic foregut. Its posterior end is marked by the **anterior intestinal portal**. The distinction between the foregut and the midgut is that the former possesses a floor whereas the midgut does not. Eventually, the entire gut acquires a floor and the term foregut is then confined to the region extending from the mouth to the umbilicus.

Forelimb bud, wing bud: a bulge of mesoderm covered with epithelium that will develop into a wing.

Frontal section, coronal: a longitudinal section that includes both left and right sides of the body.

Gamete: the **haploid** male or female **germ cell** (Greek: *gameein* = to marry).

Gastrula: an embryo undergoing **Gastrulation** (Latin: *gastrula* = little stomach).

Gastrulation: the period, after cleavage, when the cells that will form the future embryo migrate into their appropriate positions, forming **endoderm** and **mesoderm**, ready to start to **differentiate**. It results in the **bilaminar** embryo becoming **trilaminar**.

Genes: the hereditary elements, DNA, which pass from one generation to another (Greek: *genes* = born).

Gene expression: a gene acts (expresses itself) by producing **transcripts** which lead to the formation of specific proteins. Different genes act in different regions of the body and at different times (differential gene expression).

Genetic engineering: the production of a new genetic make-up by experimental means.

Genome: the entire complement of genes in any cell or organism.

Genotype: the genetic constitution of each cell in an organism. Not all genes are expressed in the **phenotype**.

Genital folds/urethral folds: paired swellings to the right and left of the cloaca, which will form part of the external genitalia.

Genital ridge: *see* **Nephrogenic ridge**.

Genital tubercle: protrusion that forms at the anterior end of the genital folds and gives rise to the penis.

Germ cells: the mature germ cells (**gametes**) are the ova and sperm, formed from the immature germ cells, the **oocytes** and **spermatocytes**, respectively, by the process of gametogenesis (Latin: *germen* = sprout, bud). *See also* **Primordial germ cells**.

Germ layers: ectoderm, mesoderm and **endoderm**.

Germ wall: the edge of a cleavage or gastrula stage embryo, which lies on the surface of a large yolk (as in a bird). The germ wall is the expanding edge of the **blastoderm**.

Germinal crescent: a region in the **area opaca** at the cranial (anterior) end of the avian **blastoderm** to which the primordial germ cells migrate prior to entering the blood stream and being transferred to the developing gonad.

Germinal disc, germ anlage: that part of a **blastoderm** of a yolky egg that contains the cytoplasm from which the embryo will develop.

Germinal epithelium: the mesothelial layer covering the gonad of the embryo, strands of which ingress to form the primary and secondary sex cords. The **primordial germ cells** become carried into the gonad during the ingression. Formerly, it was thought that cells of the germinal epithelium itself differentiated into germ cells, hence the terminology, but this idea is no longer accepted.

Germinal vesicle: the enlarged nucleus of the **oocyte** during **prophase**.

Gonadal ridge: *see* **Nephrogenic ridge**.

Granulosa: ovarian **follicle** cells (Latin: *granulum* = dim. Of *granum* = grain).

Haematopoiesis: the process of red blood cell formation (Greek: *haema* = blood; *poieen* = to make).

Haploid: a **germ cell** possessing half of the somatic complement of chromosomes (*see* **diploid**).

Head fold: a fold in the three germ layers that forms anterior to the head process (*see* **Body folds**).

Head mesenchyme: mesenchymal cells in the earliest stages of head formation. They may be of **mesodermal** or **neural crest** origin.

Head process: that part of the **notochord** that is formed by cells migrating anteriorly from **Hensen's node**.

Hensen's node: the anterior end of the **primitive streak**.

Heterogametic sex: the sex that produces two types of gametes, as opposed to the homogametic sex, which produces one type. The sex of the offspring is determined by the heterogametic sex, which is the male in mammals, and the female in birds (Greek: *heteros* = other).

Heteroplastic grafting: *see* **Quail**.

Hindgut: the simple gut tube at the posterior end of the early embryo. It extends into the tail region as the **tail gut**, and its anterior boundary is the **posterior intestinal portal** (*see also* **Anterior intestinal portal**).

***Hox* genes:** certain homeobox-containing genes important in establishing and maintaining the antero-posterior axis of the embryo.

Hypoblast: the lower layer of a **bilaminar embryo** (Greek: *hypo* = under; *blastos* = bud).

Hypomere: the ventro-lateral part of the somitic myotome, which gives rise to the muscles of the body wall and limbs.

Indifferent gonad: the gonad prior to its **differentiation** into ovary or testis.

Induction, embryonic: the interaction between two cells or tissues in which one, the inducer, affects the **differentiation** of the other.

Infundibulum: (a) **primordium** of the neurohypophysis (posterior pituitary), which arises from the floor of the diencephalon; (b) the funnel-shaped opening at the cranial end of the oviduct (Latin: a funnel, from *in* = in and *funder* = to pour).

Ingression: the process in bird embryos by which cells pass through the primitive **streak** to form the **endoderm** and **mesoderm** layers.

Intermediate cell mass, intermediate mesoderm: the **mesoderm** lying between the **somites** and the lateral plate and giving rise to most of the urino-genital system.

Invagination: the remoulding of a region of **epithelium** so that it sinks into the substance of the embryo, forming a cavity, e.g. the **optic cup** (Latin: *in* = in; *vagina* = sheath).

***In vitro*:** tissues removed from an organism and then grown away from the body are described as being grown *in vitro* or in culture (or **tissue culture**).

***In vivo*:** in living tissues or animal.

Koller's sickle (Rauber's sickle): crescentic region at the posterior end of the **area pellucida** that gives rise to the **primitive streak**.

Labile determination: *see* **Specification**.

Latebra: column of white yolk extending from the centre of the (yellow) yolk of a hen's egg to the periphery, just beneath the **ooplasm**.

Lateral: side, away from midline. (Latin: *lateris* = side).

Lateral plate mesoderm: the unsegmented **mesoderm** lying lateral to the **somites** in the early embryo. The upper (**somatic**) layer and the lower (**splanchnic**) layer are separated by a **coelom**.

Lecithin: a component of the yolk (Greek: *lekithos* = yolk).

Limb bud: the bulge on the side of the body that is the first stage in the formation of a limb.

Lipovitellins: lipoproteins forming the major component of the yolk granules (Greek: *lipos* = fat; Latin: *vitellus* = yolk).

Lower layer: the ventral layer of the **bilaminar embryo**, which consists of various populations of cells (*see* **Hypoblast**, **Endoblast**, **Endophyll**).

L.S.: longitudinal section; may be either **sagittal**, in which the plane of section passes from the ventral to the dorsal side, or frontal, in which the plane of section passes from left to right (Latin: *sagitta* = arrow).

Mammillary knobs: projections on the inner surface of the egg shell that act as anchor points for the shell membranes (Latin: *mammilla* = diminuitive of *mamma* = breast).

Margin of overgrowth: the periphery of the **blastoderm**, consisting of specialized area **opaca** cells.

Marginal layer: the major part of the wall of the neural tube in the early embryo.

Marginal sinus: *see* **Sinus terminalis**.

Medial: in or towards the midline, as opposed to **parietal**.

Medullary plate: *see* **Neural plate**.

Melanocyte: a cell of **neural crest** origin that produces melanin (Greek: *melas* = black; *kytos* = vessel, cell).

Mesenchyme: an histological term describing cells that are irregular in shape and arrangement (as opposed to **epithelial** cells, which are regular and arranged in sheets).

Mesendoderm: cells that become either **mesoderm** or **endoderm**.

Mesoderm: the middle layer of a **trilaminar embryo** (Greek: *mesos* = middle; *derma* = skin).

Mesonephros: the second kidney to develop (*see also* **Pronephros** and **Metanephros**).

Metamerism: repetition of a structure several times, e.g. the paired **somites** are repeated about 50 times in the chick (Greek: *meta* = among; *merism* = repetition).

Metanephros: the third and final kidney to develop (*see also* **Pronephros** and **Mesonephros**).

Midgut: in the early embryo, that part of the simple gut tube that lacks a floor and is, therefore, open to the yolk sac. *See also* **Foregut** and **Hindgut**.

Morphogenesis: the development of form (Greek: *morph* = form; *genesis* = birth).

Morphogens: chemical substances (e.g. retinoic acid) that may be necessary for the correct development of a tissue.

Mullerian duct: the developing oviduct.

Multipotentiality: the ability of most cells in an early embryo to develop along many different and alternative pathways. Most multipotent cells eventually become unipotent during development, becoming restricted to development into a single cell line, though others remain as **stem cells**.

Muscle mass: the early stage in the development of certain muscles.

Myogenesis: muscle formation (Greek: *myos* = muscle; *genesis* = birth).

Nasal (olfactory) placodes: thickened patches of **epithelium** that **invaginate** to form the nasal pits (Greek: *plax* = anything flat).

Nephric duct: the duct that serves the **pronephros** and, later, the **mesonephros**. It subsequently becomes the vas deferens in males.

Nephrogenic ridges: the paired ridges, left and right, formed on either side of the gut, which will give rise to the kidneys and their ducts, as well as to the gonads and genital ducts. (*See also* **Urogenital ridge**).

Neural arch: the dorsal part of a vertebra.

Neural crest: cells initially located at the dorsal side of the neural tube that migrate to form a range of tissues.

Neural plate: that portion of the **ectoderm** that will become the **neural tube**.

Neural tube: the precursor of most of the nervous system.

Node/primitive node: *see* **Hensen's node**.

Normal Table: *see* **Appendices**.

Notochord: a rod of cells formed from **Hensen's node** and which lies beneath the **neural tube**. One of the few truly midline structures in the body (Greek: *notos* = back; *chorde* = a cord).

Notochordal sheath: a thick band of connective tissue that wraps around the **notochord** after about stage 11.

Notoplate: the **notochord** together with the base of the **neural** tube.

Omphalo-mesenteric blood vessels: *see* **Vitelline blood vessels** (Greek: *omphalos* = navel; *enteron* = gut).

Oocyte: the female **germ cell** prior to **gamete** formation.

Ooplasm: egg cytoplasm.

Open cells: some of the earliest cells of the cleavage stage embryo are incompletely surrounded by a cell membrane and are continuous with the yolk. Ultimately, they become fully enclosed by cell membranes and are then known as **blastomeres**.

Optic cup: primordium of the eye formed as an outgrowth from the forebrain.

Organiser: the dominant region of the early embryo that controls the development of the body axis. The term was introduced to describe the dorsal lip of the amphibian blastopore, but is also applied to **Hensen's node** in birds.

Otic (auditory) placode: thickened region of **ectoderm** that invaginates to become the otic visicle.

Otic (auditory) vesicle: primordium of the inner ear.

Otocyst: *see* **Otic vesicle**.

Overgrowth, margin of: the periphery of the young **blastoderm** as it expands over the yolk by **epiboly**.

Ovulation: release of an ovum from the ovary.

Pander, nucleus of: white disc of yolk that lies beneath the early **blastoderm** at the surface of the **latebra**.

Paracrine factors: signalling molecules involved in cell–cell communication (as opposed to endocrine factors), e.g. TGFβ, FGF.

Paraxial: regions immediately to either side of the midline axis (i.e. of the **notochord**). The paraxial **mesoderm** gives rise to the **somites** (Greek: *para* = beside; *axis* = axis).

Parietal: the position in the body of an organ or tissue relative to the walls of the organism. This tends to be the furthest from the midline (Latin: *parietis* = wall).

Pattern formation: the formation of an orderly arrangement of tissues and organs in an embryo.

Periblast: an acellular region of mixed yolk and cytoplasm associated with the cleavage stage of large-yolked eggs (Greek: *peri* = around; *blastos* = bud).

Periderm: a temporary layer of squamous epithelial cells covering the simple epidermis in the early stages of skin development.

Perivitelline layer: *see* **Vitelline membrane**.

Pharyngeal (visceral) arches: segmental region of the pharynx corresponding to gill arches of fishes. They are separated from one another by **pharyngeal pouches** internally and by **pharyngeal grooves** externally.

Pharyngeal (visceral) grooves: *see* **Pharyngeal arches**.

Pharyngeal (visceral) pouches: *see* **Pharyngeal arches**.

Phenotype: the morphological expression of the **genotype** (Greek: *phainen* = to show).

Phosvitin: a component of yolk.

Placode: a thickened region of **ectoderm** that **invaginates** to form certain organs of special sense or certain cranial ganglia (Greek: *plax* = flat).

Polarity: the presence in an organism of a head and tail end, and dorsal and ventral sides (Greek: *polos* = axis).

Polarizing zone: the posterior region of the limb bud, which specifies position along the antero-posterior axis.

Positional information: a concept of development that supposes that the **differentiation** of a cell or group of cells is largely determined by their position in the embryo.

Posterior or caudal: tail end.

Posterior intestinal portal: the opening of the **hindgut** in the early embryo into the lumen of the yolk sac.

Posterior marginal zone: the posterior margin of the border between the **area pellucida** and **area opaca**. Cells migrate from here to form the **primitive streak.** *See also* **Koller's sickle**.

Posterior neuropore: an opening remaining at the posterior end of the neural tube in the final stages of neural tube closure. It becomes obliterated at about stage 11–12 (*see also* **Anterior neuropore**).

Prechordal mesoderm: mesoderm lying anterior to the **primitive streak** and head process.

Presumptive fate: *See* **Specification**.

Primitive foot plate: the flattened end of a leg bud that will become the foot.

Primitive hand plate: the flattened end of a wing bud that will become the avian equivalent of a hand.

Primitive node: *see* **Hensen's node**.

Primitive streak: the midline region of the early embryo where **ingression** of cells occurs during **gastrulation**.

Primordial germ cells: precursors of the **germ cells**, proper. In birds, they arise in the early **epiblast** and, later, journey to the gonad to become either **spermatocytes** or **oocytes**.

Primordium/Anlagen: the rudiment of an organ, e.g. the **optic cup** is the rudiment of the eye.

Proepicardial serosa: a specialized region of the heart that forms from the pericardium.

Programmed cell death: death of a group of cells that occurs as a normal event in the development of certain tissues in the embryo. *See also* **Apoptosis**.

Progress zone: cells at the tip of a limb that proliferate.

Proliferation: increase in the mass of a tissue by mitosis.

Pronephric duct: *see* **Nephric duct**.

Pronephros: the first kidney to develop (*see also* **Mesonephros** and **Metanephros**).

Prophase: the first phase of mitosis.

Prospective/presumptive fate: the fate that can be expected for a particular tissue, but is not yet fixed (**determined**) and may still change under experimental or adverse conditions.

Proximal: the part of an organ or tissue lying nearest the midline (*cf.* **Distal**).

Pyknosis: shrinkage and increased staining of the nuclei in dying cells; it is, therefore, a measure of cell death (Greek: *pyknos* = dense).

Quail: *Coturnix coturnix* japonica, the Japanese quail, whose nucleoli possess a distinctive appearance, which enables them to be recognized easily after **xenoplastic grafting**.

Rauber's sickle: *see* **Koller's sickle**.

Regression: the gradual disappearance of the **primitive streak**, from its anterior to its posterior end, as the cells **ingress** to become converted into axial tissues.

Regulation: the ability of a group of cells in the early embryo to switch from developing according to their prospective fate (as shown in the **fate maps**) and form into different tissues.

Roof plate: the mid-dorsal region of the **neural tube**.

Sagittal section: a **longitudinal** section passing through the body from dorsal to ventral sides (as opposed to **frontal section**) (Latin: *sagitta* = arrow).

Sclerotome: that part of a **somite** that will form vertebral cartilage (Greek: *skleros* = hard).

Segmental plate: the band of **mesoderm** posterior to the newly formed **somites** that will, itself, become somites.

Segmentation: the division of an initially continuous mass of tissue into segments (repeating units). **Somites** are the first segmental structures to form.

Serial sections: sections, whether transverse or longitudinal, passing systematically through an organism and mounted in correct sequence.

Sero-amniotic connections: the points at which the **amniotic folds** fuse over the embryo before separating to form the **chorion** (**serosa**) and **amnion**.

Serosa: *see* **Chorion**.

Shell membranes: two specialized membranes lining the shell.

Sinus terminalis/marginal sinus: a blood vessel that encircles the entire **area vasculosa**.

Somatic cells: body cells that are not **germ cells** (Greek: *soma* = body).

Somatic mesoderm: that part of the lateral plate **mesoderm** that lies beneath the **ectoderm** (*cf.* **Splanchnic mesoderm**).

Somatopleure: **ectoderm** together with the underlying **somatic mesoderm**.

Somites: paired blocks of **mesoderm** that form on either side of the neural tube and give rise to dermis, muscle and cartilage.

Somitomere: **primordium** of a **somite**, visible as a group of cells in the **segmental plate** that gradually become more and more adherent to one another. Best seen by scanning electron microscopy.

Specification: the first stage in the **differentiation** of an organ or tissue. Formerly known as **labile determination** to emphasize that the course of future development is not completely fixed (not yet **determined**); the term 'specification' is equivalent to **presumptive fate**.

Sperm nests: diverticula in the oviduct where sperm are stored awaiting the eggs to ovulate.

Spermatocyte: the male **germ cell** prior to **gamete** formation.

Splanchnic mesoderm: that part of the **lateral plate mesoderm** associated with **endoderm** (Greek: *splanchnic* = gut).

Splanchnopleure: the composite tissue formed by **splanchnic mesoderm** and **endoderm**.

Staging: *see* **Normal Tables** (Appendices I and II).

Stem cells: specialized undifferentiated cells that undergo repeated mitosis but remain able to differentiate if the conditions are favourable.

Stomatodaeal plate: the anterior tip of the early **foregut** where it is fused to the **ectoderm** (Greek: *stoma* = mouth).

Sub-blastodermal cavity: the fluid-filled space in the yolk immediately beneath the **blastoderm**.

Sub-germinal cavity: *see* **Sub-blastodermal cavity**.

Sub-germinal fluid: fluid in the subgerminal cavity.

Syndrome: multiple effects of a disorder.

Tail bud: the remnant of **Hensen's node**, after its **regression**, which forms the rudiment of the tail.

Tail fold: *see* **Body folds**.

Tail gut: extension of the hindgut into the tail.

***Talpid*:** a genetic mutant of the domestic fowl, whose main characteristic is short broad limbs that superficially resemble those of *Talpa*, the European mole.

Time-lapse cinematography/videography: a method of visualizing aspects of development, such as **gastrulation**, by taking a sequence of photographs of the same tissue but separated from one another by a selected interval of time. This has the effect of showing the process in a speeded-up form.

Tissue culture: *see* ***In vitro***.

Totipotency: the ability of a cell to **differentiate** into any cell type.

Trilaminar embryo: a three-layered embryo possessing **ectoderm, mesoderm** and **endoderm**.

Urogenital ridge: *see* **Nephrogenic ridge**.

Ureteric duct: also known as the metanephric duct. The future ureter.

T.S.: transverse section across the embryo.

Vasculogenesis: blood vessel formation (*See also* **Angiogenesis**).

Ventral: belly side, as opposed to dorsal side.

Ventral aorta: the main artery extending cranially from the heart in the early embryo.

Visceral arches: *see* **Pharyngeal arches.**

Vital dyes: dyes that are tolerated by living tissues with little or no toxicity.

Vitelline blood vessels: blood vessels of the yolk sac (Latin: *vitellus* = yolk).

Vitelline membrane/perivitelline layer: transparent elastic layer surrounding the yolk of a bird's egg.

Vitello-intestinal duct: *see* **Yolk sac**.

Wolffian duct: *see* **Nephric duct**.

Wolffian ridges: precursors of limb buds.

Xenoplastic grafting: grafting of tissues from one species to another e.g. quail to chick. (*see* **Quail**) (Greek: *xenos* = foreign; *plassein* = to mould).

Yolk granules, spheres, white: all components of the egg yolk.

Yolk sac: derived from the area opaca, the yolk sac absorbs yolk and transports it to the embryo via the **Vitelline blood vessels** in the yolk sac stalk.

Zone of polarizing activity (ZPA): *see* **Polarizing zone**.

Zygote: the cell that is formed after the fusion of the male and female **gametes**, i.e. a fertilized egg.

Some Highlights in the History of Chick Embryology

In many ways the history of chick embryology is the history of embryology itself because the great accessibility of the chick has always made it a favourite subject to study. All the great advances and discoveries in chick development have relevance for other vertebrates, including mammals, and some have brought about dramatic changes in the fundamental understanding of development itself. There are several excellent reviews on the subject (e.g. Needham, 1934; Oppenheimer, 1955; Mayer, 1939; Horder et al., 1986) and these deal with the growth in our general understanding of the processes involved, encompassing a range of species. This present account is restricted to the most significant events in the history of chick embryology.

The first clear-cut account of chick embryology was given by Hippocrates (460–377 B.C.). He is quoted by Needham as follows: 'Take 20 eggs or more and give them to 2 or 3 hens to incubate, then each day from the second onwards till the time of hatching take an egg, break it, and examine it. . .'. This orderly and systematic approach is meaningful even today. Aristotle (382–322 B.C.), who also followed this system, appears to have been the first to dissect the chick embryo.

During succeeding centuries many others were curious enough to open partially incubated eggs and often to record their findings. The seventeenth-century English physician and anatomist, William Harvey, based some of his discoveries and theories pertaining to the formation and function of the heart and circulation partly on his observations of chick development *in ovo*. He opened hens' eggs at various lengths of incubation and observed with magnifying lenses the different stages of heart formation and the commencement of beating. He also noted the formation of blood islands and their gradual incorporation into the circulation. Before this it was generally believed that the heart and circulation were idle until hatching, or birth in the case of mammals. Harvey observed the formation of other organs in the chick embryo and also described how the yolk was used as nourishment during development.

It was not until the invention of the microscope that it became possible to acquire some understanding of what was happening in the earliest stages. The Italian, Malpighi (1628–94), using the microscope, was the first to describe the blastoderm of the chick, the neural folds and groove, the optic vesicles, the early stages of heart development and the somites (see Needham, 1934).

More descriptions of chick embryos occurred during the seventeenth and eighteenth centuries but the most valuable achievement during this period was the resolving of one of the great controversies, that of **preformation** versus **epigenesis** and this was settled primarily by reference to chick development. The preformationists thought that the entire individual was present in miniature in the egg or, alternatively, within the sperm and that it simply grew until it hatched; there was thought to be no differentiation, merely growth. The epigeneticists considered that both growth and differentiation were involved and were necessary. Today, we are all epigeneticists. The argument was settled by Caspar Friedrich Wolff (1753–94), who was born in Berlin but spent most of his research life in St. Petersburg. He not only argued philosophically against preformation but clinched his arguments by providing two clear examples from the chick embryo. First, he pointed out that the blood vessels of the area opaca were not present from the beginning of incubation but that they developed from blood islands that joined up together. However, sceptics argued that perhaps the blood vessels were there all the time, although not visible to the human eye. Wolff countered that by searching for a tissue where this argument could not hold and, once again, found an excellent example in the chick. He pointed out that the gut is formed from a flat sheet of tissue that folds into a tube, the different stages being clearly visible. This cleverly chosen example was ultimately accepted and destroyed the case for preformation, not only in the chick but in all embryos.

The first description of the germ layers (ectoderm, mesoderm and endoderm) was made in the chick by the Russian embryologist, Pander (1794–1865), who was a follower of Wolff. However, it was the Estonian, Von Baer (1792–1876) who, with meticulous analysis of many embryos and the use of the light microscope, was able to show that the three layers were not merely a feature of the chick but were universal in vertebrates. Von Baer, himself, was opposed to the misleading ideas of recapitulation, current in the early nineteenth century, whereby the stages of development were thought to represent progressive stages of evolution. He discovered the notochord and studied the asymmetry in the chick, as well as the passage of materials from the yolk and the egg white into the embryo. Jane Oppenheimer (1955) described Von Baer as the 'most distinguished and influential of the *early nineteenth century* embryologists' and his work is said to have influenced Darwin.

The nineteenth century produced other great scientists and among these was Roux (1850–1924) who might be described as the most influential of the *late nineteenth century* embryologists, and whose name is commemorated in the journal Wilhelm Roux'Archives. He is often called the father of experimental embryology (*Entwickelungsmechanik*). Roux realized that the answers to many of the questions that were being asked about development could be obtained only by carrying out experiments directly on the embryo. He was also the first to pose many of those questions that still fascinate us today, such as the ways in which the planes of symmetry are established in the early embryo. His introduction of the experimental approach to embryology was a great step forward, both conceptually and technically, and during the first half of the twentieth century descriptive accounts of development became superseded by the experimental approach.

The effects of various changes in the physical or chemical environment on the chick embryo were investigated and simple experiments were performed on it *in ovo*. For example, Peebles (1898) inserted small pieces of sable hair into the anterior end of the primitive streak and found that they were carried posteriorly as Hensen's node regressed. More detailed studies on the morphogenetic movements were carried out by others, using spots of vital dyes to mark different areas whose shifting positions could then be followed closely. Gräper (1929) is especially noteworthy because he introduced the technique of recording the gastrulation movements of the chick by cinematography. Experiments of this type provided evidence toward the production of the early fate maps which, although they have been refined and modified in recent years by modern techniques, nevertheless were reliable enough to provide a basis for much of the experimental work that came afterwards.

During all these centuries, the chick was not the only embryo whose basic development had been studied and other species now proved to be easier subjects for some experimental studies. Many of these new experiments were carried out on amphibian embryos whose tissues were often easier to manipulate than those of the chick, so it is perhaps not surprising that the most famous experiment on embryos of the twentieth century was carried out on amphibians, resulting in the concept of the "organiser" as a major controller of development (reviewed by Spemann, 1938 and discussed in greater detail in Chapter 4). Inevitably, the question arose as to whether a similar "organiser" existed in other animals, and especially in amniotes. Experiments of this type on early mammalian embryos were technically not possible then and, consequently, attention became refocused on the chick. However, the large yolk and the fragility of the early blastoderm made the transplanting of tissues within the chick embryo extremely difficult. It was not until Waddington (1932) developed a procedure for removing the blastoderm from the egg and growing it *in vitro*, using a modification of a culture technique devised by bacteriologists, that it became possible. This technique was later modified and greatly improved by New (1955) and has since been the basis for much of the experimental work on the early blastoderm.

From our twenty-first century perspective, the latest highlight in the history of the development of the chick has been the introduction of molecular biology, so that we now understand so much more about the underlying forces controlling differentiation. The recent announcement of the completion of the sequencing of the chick genome has once again emphasized the critical place of the chick in embryology and developmental biology. It opens the way for the next step forward and the discovery of many of the controlling events in development. We can look forward to important new highlights in the near future.

Chapter 1

The Hen's Egg and its Formation

Birds of all species lay eggs and none is viviparous. Although this might seem to be a disadvantage, nevertheless birds have established themselves successfully in all types of terrain, due not only to their highly developed brooding behaviour and parental care, but also to the structure and composition of their eggs.

A bird's egg contains all the raw materials for the formation of the embryo apart from the oxygen, which enters through small pores in the shell. If these pores are experimentally clogged with wax, then no development occurs.

Text-Figure 1 shows the structure of an unincubated but fertilized egg of the domestic fowl. The yolk is about 2–3 cm in diameter and is enclosed by a thin transparent membrane, the vitelline membrane. The main components of the yolk are lipids and proteins and they are not alive. The embryo itself is a tiny blob of cytoplasm that floats to the uppermost surface of the yolk, coming to lie close beneath the vitelline membrane (which is sometimes called the perivitelline layer) and is the only part which is living. The egg white or albumen lies around the yolk and is in turn enclosed by two tough shell membranes that are fused together, except at the blunt end, where they are separated by an air space that is important in hatching. The inner surface of the calcareous shell is elaborated into little projections, the mammillary knobs, and strands from the shell membranes are attached to them (see p. 4).

Birds' eggs vary in size, shape and surface pigmentation according to the species, though there is considerable individual variation. Even in the same strain of domestic fowl some hens lay more elongated or more pointed eggs than others, and the pigment patterns of quail eggs vary greatly even within the same flock. However, the general morphology of the egg remains the same throughout the avian species and is related to the activities of the different regions of the reproductive tract in the hen.

THE REPRODUCTIVE TRACT OF THE HEN

The formation of a bird's egg takes place step by step along the oviduct in a way that is often compared to the assembly of man-made articles along a factory belt. Although both left and right ovaries and oviducts are formed in the embryo, the right degenerate and only the left remain.

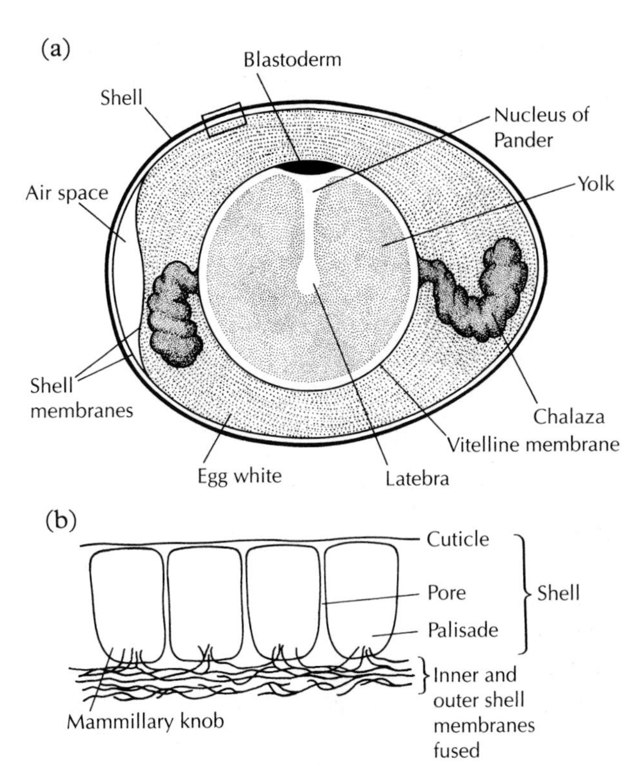

Text-Figure 1. *(a) Sagittal section through a hen's egg. The latebra and the Nucleus of Pander are regions of white yolk. The blastoderm lies over the Nucleus of Pander and below the vitelline membrane. The chalazae are balancers that support the yolk. The two shell membranes are separated by an air space at the blunt end. (b) Enlargement of the region of the shell shown in (a). (From Bellairs, 1971.)*

Many egg yolks are visible in the body of the laying hen, the largest being of comparable diameter to those in laid eggs. Inexperienced investigators tend to assume, wrongly, that these are ovulated (and so perhaps fertilized) eggs lying along the genital tract. Normally, however, only one egg is present at a time in the oviduct and the remainder of the yolks are still oocytes, each with its own follicle (*Plate 1a,b* and *Text-Figure 2*). The ovary of a laying hen contains many follicles and these vary in size according to whether they yet possess any yolk or to the amount of yolk already deposited in them. The largest oocyte usually measures about 3 cm in diameter and is almost ready to ovulate.

In a hen that has been laying regularly, several discharged follicles are visible in the ovary. These are the vascularized sacs that have been ruptured by the ovum during ovulation and they may vary in diameter from about 2 cm down to 2 mm or less. The size is an indication of the time that has elapsed since ovulation took place from that particular follicle, since they become progressively smaller as they are absorbed. In a hen that has gone 'off-lay', no large oocytes are visible, though small ones are still plentiful and measure up to about 5 mm in diameter.

The raw materials from which yolk is manufactured are formed in the liver of the laying hen and pass from there in the blood plasma to the ovary. Large quantities of lipids and proteins are present in the blood of the laying hen and these closely resemble and are often identical with similar substances in the yolk. They are not present in the blood plasma of male or juvenile birds nor of hens that have gone 'off-lay'. Labelled serum proteins injected into the blood stream of laying hens have been subsequently found in the yolk.

The substances that pass through the capsule of follicle cells into the oocyte are the definitive proteins and lipids from which the yolk is formed; there is no real synthesis of yolk in the oocyte but merely a rearrangement of materials. The yolk precursors are probably taken into the oocytes in the many coated vesicles which lie beneath the surface of the oocyte and possess receptors for the lipoproteins (Shen *et al.*, 1993). The smallest oocytes are probably incapable of utilizing these materials.

When the oocyte is about 2.5 mm in diameter the inner layer of the vitelline membrane (see below) becomes laid down, but the outer layer is not formed until after ovulation. By the time the oocyte is about 2.5 cm in diameter the inner layer has reached its maximum thickness of about 2.5 μm. In the final stage of oocyte growth, large vacuoles appear beneath its cell membrane (oolemma) and the number of villi at the surface decreases. When the oocyte has reached about 3 mm in diameter, yolk spheres become visible.

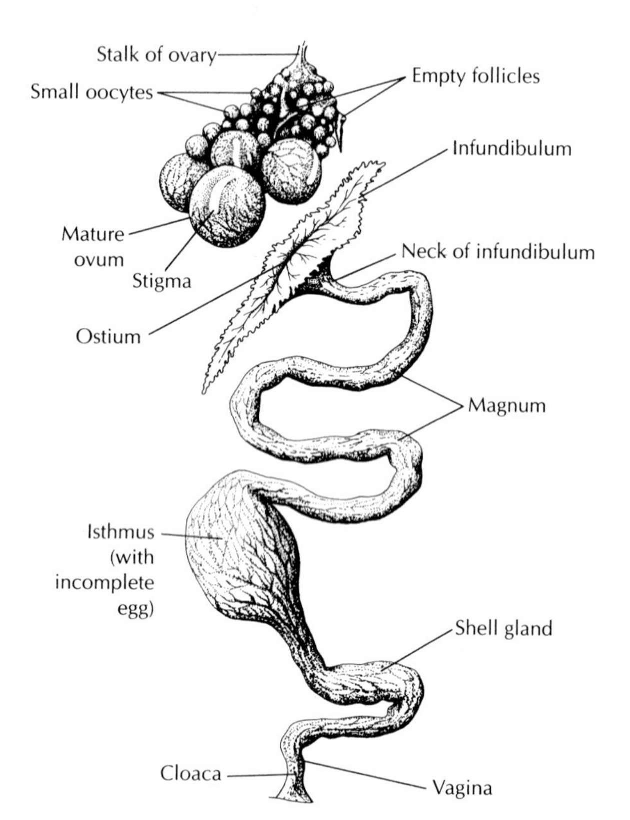

Text-Figure 2. *Ovary and left oviduct of the domestic fowl. (After Duval, 1889.)*

THE OVIDUCT

The regions of the oviduct are shown in *Text-Figure 2*. The most cranial region, the infundibulum, consists of a funnel about 9 cm wide at its opening immediately posterior to the ovary, leading into a narrower tubular region. The magnum follows and is the longest (about 30 cm) part of the oviduct. The isthmus is about 10 cm long and is succeeded by the shell gland (which, confusingly, is known also as the uterus) and the vagina, which leads into the cloaca. Each region is modified for a particular secretory function, which is best understood by following the egg down the oviduct. During this period, the embryo embarks on the earliest stages of development. The events are indicated in *Table 1*.

Table 1 Passage of the egg down the oviduct

Oviduct	Time spent	Event
Ovary	0 h	Ovulation
Infundibulum	15 min	Fertilization/thin albumen
Magnum	3 h	Albumen added
Isthmus	15 min	Shell membranes added
Shell gland	20 h	Shell added
Vagina	Negligible	Egg laid

Fertilization takes place as the egg is ovulated, in the funnel of the infundibulum, which contains no glandular tissue. It occurs before the laying down of the outer layer of the vitelline membrane, which happens at the junction of the funnel and tubular regions. The 'thick' albumen that immediately surrounds the yolk is secreted by the tubular region of the infundibulum and immediately blocks further sperm contact with the ovum (Wishart and Horrocks, 2000). The 'thin', watery, albumen is added in the magnum, which is rich in a mucus-secreting epithelium. The shell membranes are deposited in the isthmus.

Shortly after the egg enters the shell gland, fluid passes into the albumen so that it expands to its full size. This 'plumping' fluid, which is derived from the blood of the laying hen, carries with it the inorganic ions for the shell and deposits them on the outer side of the shell membrane.

The shell gland is about 5 cm long and highly muscular. Calcium ions pass from the blood of the hen to the shell membranes where they form columns of calcite (i.e. crystals of calcium carbonate, see p. 4) separated by pores (*Text-Figure 1b*) which will later allow the exchange of gases between the embryo and the outside environment. A small amount of organic material is secreted at the same time and forms a web that plays a role in the calcification. During its sojourn in the shell gland the egg slowly rotates; this probably helps to ensure an even distribution of the calcium ions but is also important in establishing the orientation of the embryo (see p. 17).

Detailed accounts of the egg and its components are given by Burley and Vadehra (1989) and Board and Sparks (1991).

THE OVULATION CYCLE IN THE LAYING HEN

Egg laying in wild birds is usually limited to one or two clutches a year, batches of eggs being laid one after the other over the space of a few days and then incubated together. The mature domestic fowl is capable of laying throughout the year. Nevertheless, the eggs are still laid in batches at the rate of about one per day (about every 26–28 h) for a period of between 4 and 6 days. There is then a rest, which may be as short as one day in a good layer, followed by another period of egg-laying days, and so on. The increased egg-laying in domestic birds has been brought about by a reduction in the length of time between clutches.

The process is under complex hormonal control from gonadotrophins produced by the anterior pituitary gland. An ovulation-inducing hormone (OIH), which is thought to correspond in many ways to the luteinizing hormone of mammals, is released by the pituitary and, if there is a sufficiently mature follicle available, it responds to the OIH by ovulating within about 4–8 h (Gilbert, 1971). If the hormone is injected into a laying hen it can provoke premature ovulation. Progesterone also plays a role, probably acting through the hypothalamus and thus controlling the release of the OIH by the pituitary. For a further account see King and McLelland (1979–89) and Etches (1996).

THE VITELLINE MEMBRANE (PERIVITELLINE LAYER)

The vitelline membrane is the transparent casing that encloses the yolk of the hen's egg and separates it from the albumen. It consists of two major layers, the inner layer, which is laid down in the ovary, and the outer layer, which is secreted in the oviduct. Bellairs *et al.* (1963) described the fine structure by transmission electron microscopy and their findings have since been confirmed by others. The fibres of both layers are rich in glycoproteins (reviewed by Wishart and Horrocks, 2000).

The inner layer is about 1–3.5 μm thick and consists of a meshwork of solid cylindrical fibres that mainly run parallel to the surface of the yolk and vary in thickness from about 0.2 to 0.6 μm (*Plate 2c*). The outer layer varies in thickness from about 0.3 to 9 μm and is composed of a variable number of sublayers lying one above the other. It also contains fibres, but these are thinner than those of the inner layer; strands of the outer fibres are extended out into the chalazae (*Text-Figure 1* and see below), where they are spirally arranged and coated with thick albumen. The material between the sublayers of the outer layer of vitelline membrane is probably albumen, and it is likely that there is a penetration of the outer layer by albumen, or that the secretion of the albumen starts to take place before the entire thickness of the outer layer has been laid down.

Each layer of the vitelline membrane consists of proteins, though the amino acid composition differs. The principal components of the outer layer are ovomucin, lysozyme and vitelline membrane outer proteins I and II (Burley and Vadehra, 1989). The ovomucin makes up the fibrous framework on which the outer layer is built, whilst lysozyme forms an electrostatic complex with the ovomucin that provides the bulk and strength.

The principal components of the inner layer are membrane glycoproteins I, II and III (see Burley and

Vadehra, 1989, for an account of their amino acid structure). It is thought that the glycoprotein II is concerned with the structural integrity of the inner layer. There is also a carbohydrate-containing fraction (Kido *et al.*, 1976) which includes lectins (Cook *et al.*, 1985). See Wishart and Horrocks (2000) for a fuller discussion of the biochemistry of the vitelline membrane.

TRANSPORT OF MATERIALS ACROSS THE VITELLINE MEMBRANE

During the first week of incubation, the albumen loses more water than could be attributed to evaporation and, at the same time, the yolk acquires more water (New, 1956). It is therefore assumed that water is passed through the vitelline membrane into the yolk and it has been calculated that about 16 g of fluid is drawn through the vitelline membrane during the first week of incubation (Romanoff and Romanoff, 1949) and collects as the sub-blastodermal fluid. When the chick blastoderm is grown in culture, fluid appears to be drawn through the membrane and to collect at the ventral side of the embryo. If the washed vitelline membrane is used like a 'dialysis' membrane, water-soluble proteins obtained from the egg are able to pass through it (Jordanov *et al.*, 1966).

Callebaut (1987) labelled albumen by injecting laying Japanese quail hens with tritiated tyrosine and harvesting the eggs. Non-radioactive yolks, still in their own vitelline membranes, were then obtained from non-injected laying hens. The labelled albumen and the unlabelled yolk, together with its vitelline membrane, were assembled in an unlabelled egg shell and incubated. Autoradiographs showed that during the first two days of incubation, albumen penetrated through the vitelline membrane and accumulated distal to the margin of overgrowth, the outer area of the blastoderm.

Sodium transport across the vitelline membrane from the albumen into the yolk appears to be controlled by the ectoderm of the embryo, since removal of the endoderm and mesoderm have been found to have little effect on it (Kucera *et al.*, 1994).

The possibility that the vitelline membrane itself selectively controls the passage of materials through it has frequently been investigated, but no clear evidence has been found (see discussion by Burley and Vadehra, 1989).

ALBUMEN AND THE CHALAZAE

The chalazae are a pair of spring-like structures that project from the equatorial region of the vitelline membrane into the albumen and are considered to act as balancers, maintaining the yolk in a steady position in the laid egg. Each chalaza contains fibres that appear to be identical with the fibres in the outer layer of the vitelline membrane, but are spirally coiled and embedded in albumen. Typically, they are illustrated as passing to the blunt and pointed ends of the egg (*Text-Figure 1*). If an egg is placed in an upright position, however, the yolk rotates so that the cytoplasm remains at the uppermost side, and the chalazae shift their position accordingly.

The albumen is not of the same consistency throughout (see above). This can be seen by emptying a fresh egg into a dish, the thinner, more watery, albumen spreading further than the thicker albumen which remains around the yolk. Within the unbroken egg, the same relationship occurs, the thicker albumen adhering to the vitelline membrane.

The albumen (egg-white), has been the subject of a considerable amount of research and its protein composition is well-documented (see Table 4.3 of Burley and Vadehra, 1989). The principal proteins are ovalbumin (54% of total protein), ovotransferrin (12%), ovomucoid (11%) and lysozyme (3.4%).

THE EGG SHELL AND ITS FORMATION

About 91% of the weight of the hen's egg shell consists of inorganic matter, and about 98% of this is calcite (Burley and Vadehra, 1989), the remaining 2% of inorganic material being composed of small amounts of magnesium carbonate and tricalcium phosphate. The remainder of the weight of the shell as a whole is made up of about 6% protein and some water.

The calcite crystals are arranged in columns, or palisades (*Text-Figure 1b*) which extend through almost the entire thickness of the shell and are closely packed together (Simkiss and Tyler, 1971). Morphologically, the shell is divided into two regions, the inner mammillary layer and the much thicker outer, spongy, layer. Gaps between the palisades act as pores and enable gases to be exchanged. Each crystal is rounded on the inside of the shell and attached to the outer surface of the shell membrane by a mammillary knob. These knobs mark the sites at which calcification begins, each starting as a small centre of crystallization on the outer shell membrane. Fibres from the shell membrane become encapsulated in the mammillary knobs. The outer shell membrane is so adherent to the knobs that in the unincubated egg it is impossible to separate them without damage. During the later stages of incubation, the embryo obtains calcium from the shell, which leads to a partial demineralization of the shell. It then becomes possible to separate the shell from the shell membranes (*Plate 2b*).

The outside of the egg shell of the domestic hen is covered by a thin cuticle (about 10 μm thick). Its role is not fully understood, especially as it is not present in all species of birds (discussed by Burley and Vadehra, 1989).

THE LAID EGG

The Egg Yolk

The yolk of the laid egg is enclosed in the vitelline membrane. Despite its apparently homogeneous appearance to the naked eye, the egg yolk possesses a well-defined structure (*Text-Figure 1a*). Its centre is marked by a ball of white yolk, the latebra, and from this a column of white yolk, the neck of the latebra, extends upwards to join with a disc of white yolk, the nucleus of Pander, which lies beneath the blastoderm. The latebra is surrounded by predominantly yellow yolk. When the blastoderm is dissected from the yolk, the nucleus of Pander is left behind and can be mistaken by the inexperienced worker for the blastoderm itself.

Sections through a hard-boiled egg often show the yolk to be arranged in a series of concentric rings around the latebra. Similar bands are visible in stained sections of yolks which have been fixed. It has sometimes been suggested that these alternating bands are of yellow and white yolk, the broader being laid down during the day and the narrower at night. However, in hens that have been fed on a uniform diet *ad libitum*, this banded pattern is not seen, so that it seems more likely that the pattern reflects the fact that the hen has eaten foods of differing pigmentation.

Physically, the yolk is an aqueous fluid in which float yolk spheres (also known as globules or yolk droplets) and lipid droplets (known also as yolk granules). Individual yolk spheres enclose a number of glistening lipid droplets, those taken from the yellow yolk possessing many small droplets, whereas those from the white yolk possess fewer but larger droplets (Bellairs, 1964).

Chemically, the yolk consists mainly of proteins and lipids in an aqueous solution. Methods for fractionation are discussed by Burley and Vadehra (1989).

The major chemical components of the yolk are:

- **α and β lipovitellins.** These are lipoproteins and the two forms differ from one another principally in the amount of protein-bound phosphorus they contain and in their amino acid composition. The relative proportion of α and β lipovitellins differs according to the breed of the hen. The lipovitellins constitute the main non-aqueous constituent of the yolk and are therefore an important component of the yolk granules.
- **Phosvitin.** This is the secondmost plentiful constituent of the yolk granules. It is a phosphoprotein and has many of the physical properties of a poly-electrolyte. It binds iron and so is used as an iron store for the developing embryo.
- **Low-density lipoproteins.** The low-density lipoprotein found in the yolk granules appears to be similar to that in the aqueous phase. This is sometimes called lipovitellenin (but incorrectly so, according to Burley and Vadehra, 1989). The main proteins in this class are the apovitellenins, AI and AII. These have been extensively studied and certain variations have been found in their composition between different species, e.g. between hens and geese. Apovitellenin I is as yet known only in the blood and yolk of laying birds; one of its characteristics is that it lacks the amino acid histidine. Apovitellenin II is present in the domestic hen and geese, but not in certain other species. Apovitellenins III, IV, V, VI have received less attention.
- **Livetins.** The α, β and τ livetins are a major component of the aqueous phase and appear to be identical to the serum globulin in the blood of the laying hen. Other livetins have also been reported as well as some vitamin-binding proteins.
- **Inorganic ions** and some **enzymes**.

Vitellogenin is the name given to a stable protein precursor, a phosphoglycolipoprotein that is synthesized in the liver of the laying hen and passes in the blood to the ovary. On entering the oocyte it becomes split to form phosvitin and the lipovitellin apoproteins (Burley and Vadehra, 1989).

Chapter 2
Techniques

This chapter is concerned with certain techniques that have been devised especially for avian embryos. It does not deal with methods that are standard for other biological tissues unless modifications are needed; many of these are described by Stern and Holland (1993) and Darnell and Schoenwolf (1997).

'MILKING' HENS

A traditional technique in the pigeon fancy has been that of removing an egg from the shell gland to reduce the weight of the bird before flight. The process, known as 'milking' the pigeon, involves firm stroking down the abdomen towards the cloaca. It is a useful means for obtaining cleavage stage embryos but is easier to perform with quail hens than domestic fowls.

STORAGE OF EGGS

Laboratories that do not have easy access to farms or hatcheries usually arrange to receive their eggs once a week. If carefully packed and handled the eggs should arrive in good condition whether they have travelled by road or rail. During transport, however, they may have been subjected to various hazards. If the boxes have been roughly handled, breakages may have occurred. If they have been subjected to abnormally low temperatures in winter, or abnormally high ones in summer, they may have suffered. The most common ill-effect is ageing, brought about by delay in reaching the laboratory. It is important to inform the supplier if this is the case as he may be unaware that it is important to you. If you then see no improvement, change your supplier.

If the eggs are not to be used immediately on arrival they should be stored in a cool room. The ideal temperature is about 12°C and if necessary this can be achieved by adapting a domestic refrigerator to this temperature. Under these conditions, eggs which were no more than 2–3 days old upon arrival will remain usable for up to a further week. It must be emphasized, however, that ageing eggs are unlikely to give such reliable results as fresher ones.

The changes that occur in eggs as they age include a decrease in the strength of the vitelline membrane, a higher albumen and yolk pH and decreased viscosity compared with a freshly laid egg (Kirunda and McKee, 2000). The strain of poultry also has an effect on the quality of the stored eggs (Lapao *et al.*, 1999). There is some indication that if eggs are to be stored for up to 14 days before use they are more likely to remain viable if they have been incubated briefly before storage, so as to have reached complete hypoblast formation (Fasenko *et al.*, 2001).

INCUBATION

The ideal temperature is about 37–38°C. There are several types of incubators available commercially. Ideally, an incubator should possess the following characteristics:

- The temperature should remain constant, not only throughout incubation, but also within the incubator, i.e. it should not be hotter at the top of the incubator than at the bottom. In a 'forced draft' incubator this is accomplished by an internal fan.
- There must be some arrangement to prevent dehydration of the eggs. This is usually a reservoir of water at the bottom of the incubator, which ensures that there is a high humidity.
- There should be a method for 'turning' the eggs, anything between about 30° and 180°. This is usually accomplished by a slow tipping of the racks containing the eggs, first one way then the other. The most critical period is from 4 to 7 days (New, 1966). In eggs that are not turned the young embryo tends to become stuck to the shell membranes, especially if there is some dehydration. One of the ill-effects of not turning is that

the albumen adjacent to the yolk sac becomes depleted of sodium (Latter and Baggott, 1996), which interferes with the passage of ion and water transport across the blastoderm into the yolk sac (see p. 4). Deeming (1989) reported that in unturned eggs there was a retardation in the formation of the allantoic and amniotic fluids, restricted albumen uptake and retarded growth.

Commercial incubators are usually designed to hold several hundred eggs, but some smaller models for about two dozen eggs are now available. These can often be fitted with special racks to make them suitable for other species, e.g. quail, duck or turkey eggs.

It is not always necessary to use commercial incubators, however, especially if the requirements for incubating eggs are for a limited period only. Tissue culture incubators may also be used, provided care is taken to maintain a reservoir of water in the base and either to turn the eggs manually by about 180° each day or, less satisfactorily, to stand them vertically with the blunt end uppermost. Eggs should not be introduced into an incubator that is regularly used for cell or tissue cultures, since the outside of an egg shell is non-sterile. It is not advisable to use histological slide incubators for incubating eggs, since they are often contaminated with xylene-based, or other toxic, mountants. Eggs should be removed from their packaging on arrival and should not be placed directly on top of one another in the incubator; it is important for them to have space for gaseous exchange.

CHICK–QUAIL CHIMERAS

Normal stages for the chick and the quail are given in Appendices II and I, respectively. Those for other avian species are listed in Appendix IV.

In carrying out transplants it is essential to be able to distinguish host from graft cells after development has taken place. This is possible by using two species, chick and quail, one for host and the other for graft. Grafting between two species is called xenoplastic grafting. The cells of the Japanese quail, *Coturnix coturnix japonica*, possess a large single nucleolus with dense heterochromatin, whereas those of the chick usually have two or more small nucleoli (Le Douarin, 1969). The difference is clearly shown by staining with Feulgen's technique or by a modified method of Harris' haematoxylin (Hutson and Donahoe, 1984). It is usual to use the chick as host and the quail as the donor of the graft, but ideally the converse experiment, with the quail as host and the chick as donor, should be performed also to test for the possibility of the species differences affecting the results. Quail embryos have a shorter period of incubation than the chick – 19 days as opposed to 21. Chick–quail chimeras may also be obtained by injecting dissociated cells of quail blastoderms into the subgerminal cavities of chicks (Naito *et al.*, 1991).

CULTURE TECHNIQUES

One of the major drawbacks of working with the chick embryo is that it rests on yolk so that even the slightest pressure causes it to shift around on its fluid bed. Injection into the embryo of carefully measured doses of drugs or other chemicals can result in erratic or misleading results if the substances leak into the yolk. A further problem is that by about 5 days of development the embryo has become so vascularized that some forms of microsurgery can lead to haemorrhage and death. Further difficulties are imposed by the shell, which makes access a problem.

To some extent these difficulties can be overcome by ingenious *in ovo* experiments but in many cases it may be necessary to carry out experiments on embryos that have been removed from the shell.

There are three main approaches:

- removing the early embryo from the yolk, after which it is grown in culture
- removing the early embryo from the yolk, carrying out microsurgery on it *in vitro*, and then grafting it back into another egg, or
- transferring the contents of the entire yolk into an artificial container.

The saline solution of Pannett and Compton (1924) has been used traditionally in preparing cultures of early chick embryos, but this is time-consuming to prepare and equally good results can be obtained by using commercially available saline buffer pellets (pH 7.0) dissolved in sterile water.

Explantation of the Blastoderm

The classical method was devised by New (1955) and procedural details are given by Stern (1993a). The technique is designed to maintain *in vitro* the same relationships between blastoderm, vitelline membrane and albumen as exist in the egg. A piece of vitelline membrane, with the blastoderm still attached to it, is removed from the yolk and placed embryo side upwards in a watch-glass. A glass ring is laid on the vitelline membrane so that it surrounds the embryo and the vitelline membrane is stretched taut by pulling its edges over the ring. Albumen is then introduced beneath the vitelline membrane and the

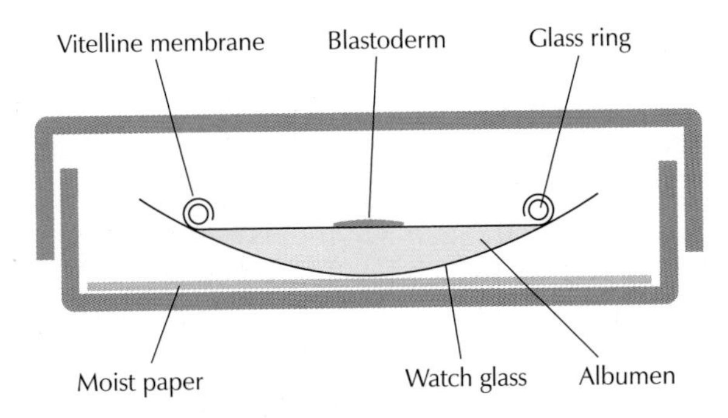

Text-Figure 3. *The culture technique of New (1955). (From Bellairs 1993 with permission of Poultry Science, http://www.tandf.co.uk.)*

whole is enclosed in a moist chamber (*Text-Figure 3*). Although the procedure is traditionally carried out using sterile glassware and instruments, sterility is not essential, since the bactericidal properties of albumen prevent infection except under grossly squalid laboratory conditions. Fungicidal infections are also unlikely to occur since these cultures are not normally maintained for more than 36 h. The technique is suitable for explanting embryos that have been previously incubated for periods up to about 48 h, and then for maintaining them in culture for up to about 36 h. It has been successfully used for filming morphogenetic movements (Bortier *et al.*, 1996). Development is usually slightly slower than in controls growing *in ovo*. Nevertheless, in successful explants the blastoderm expands to the glass ring and the embryo undergoes extensive differentiation. The success of development may be as high as 100% during the first 24 h for embryos explanted at stages between about 4 and 6 of Hamburger and Hamilton (Appendix II) but is likely to be much less for embryos explanted at younger stages. The success rate drops again with older embryos, partially because the diameter of the blastoderm is already so great at explantation that there is little space available for expansion. Furthermore, the vitelline membrane of the older embryos has less elasticity, though this difficulty can be circumvented to some extent by explanting the older embryos on vitelline membranes taken from unincubated eggs. It is usually possible to obtain some development during the second day after explantation, but it is seldom normal after the first 36 h.

Modifications of the New Technique

Albumen–agar technique

The purpose of this modification is to provide a semi-solid base beneath the vitelline membrane that is helpful when carrying out microsurgery. A few drops of Indian ink may be incorporated if a dark background is required. Chapman *et al.* (2001) recommend substituting a disc of filter paper for the glass ring so that the vitelline membrane adheres to the paper.

Inversion technique

In the traditional New technique the endodermal side of the embryo is uppermost, so that access to the ectoderm is difficult. Nicolet and Gallera (1963) solved the problem by adding an additional ring, which fits snugly inside the standard ring so that the edge of the vitelline membrane is gripped tightly between the two. It is then possible to invert the paired unit of rings so that the vitelline membrane is uppermost and the ectoderm may be reached by making a small slit in it.

Special culture chambers

Standard plastic 3.5 cm petri dishes may be substituted for the watch-glasses. Special moulded gelatine dishes have been devised by Kucera and Burnand (1987a) that incorporate a raised ring in the bottom of the dish and also allow for the possibility of gas exchange. They permit a rather longer period of incubation.

GRAFTING AND TRANSPLANTS

In Explanted Embryos

In embryos explanted in New or similar cultures the tissues are under tension and any small hole left in the blastoderm will become pulled into a larger one as the blastoderm expands.

- To remove a piece of tissue and replace it with a graft from a donor: First, prepare the graft by cutting through one layer of tissue only; this will ensure that clean outlines are obtained. Now cut through the remaining tissues and move the graft to the host using a broad-mouthed pipette. Using the same technique, cut a hole in the host slightly smaller than the graft so that a tight fit results when the graft is inserted. Draw off as much fluid as possible.
- To insert a graft between the germ layers (e.g. the classical organizer graft): Before stage 4, the germ layers are easily separated. A slit is made in the area opaca endoderm close to the border with the area pellucida. A blunt knife is inserted and pushed gently into the area pellucida to make a pocket in the mesoderm. A graft may now be introduced

into the pocket and positioned where required. Grafts may be introduced into the area opaca using the same procedure.

Shell-less Culture

Dunn (1991) reviewed methods for shell-less and semi-shell-less culture of avian embryos. In completely shell-less culture, the contents of the egg are tipped into a glass or plastic container, which is then incubated. The main advantage is that the embryo and its membranes are readily accessible for experiments. The main disadvantage is that the major source of calcium is lost to the embryo, which consequently suffers skeletal and other anomalies, reduced growth and failure to hatch. This 'disadvantage' is useful, however, for studying aspects of calcium transport and skeletal development (Tuan *et al.*, 1991).

The technique recommended by Dunn is as follows: After 3 days of normal incubation, the egg contents are tipped into hemispherical hammocks made from sheets of the plastic normally used for wrapping food. These hammocks are supported on tripods and are covered with a sterile Petri lid. The whole arrangement is then placed in a forced draft incubator in 1% carbon dioxide. Embryos younger than 3 days are apparently in greatest danger of desiccation and consequently show less satisfactory survival rates. An improvement in the technique would probably be to stand the tripod in sterile water in a plastic sandwich box which has been sterilized. The lid should then be placed on the box and sealed with masking tape. If two small holes have been made in the lid, the appropriate amount of carbon dioxide and air mixture may now be introduced by means of a hypodermic syringe and the holes sealed. In this way greater humidity as well as sterility will be retained.

In semi-shell-less cultures, the egg contents are transferred to a glass or plastic receptacle for carrying out the experiment. After this, they are transferred back into a true shell, which need not necessarily be from the same species. The chorioallantoic membrane develops in association with the foster shell, permitting resorption of calcium, though the embryos tend to be retarded in comparison with normal controls.

The semi-shell-less technique of Rowlett and Simkiss (1987) involves transferring the contents of an egg which has been incubated for 3 days to the empty shell of another fowl or even turkey egg. Approximately 55% of such embryos are likely to survive until day 17 of culture, but some have hatched and been reared to maturity. Unfortunately, this technique is not always suitable, since many important changes have occurred by 3 days of incubation. Perry (1988) and Naito *et al.* (1991), therefore, introduced an additional technique for raising newly fertilized eggs taken directly from the oviduct. The eggs are initially immersed in a solution of albumen and saline in a glass beaker for 24 h. During this period they absorb fluid and 'plump up' as they would in the oviduct. Afterwards they are transferred to foster egg shells (*Text-Figure 4*).

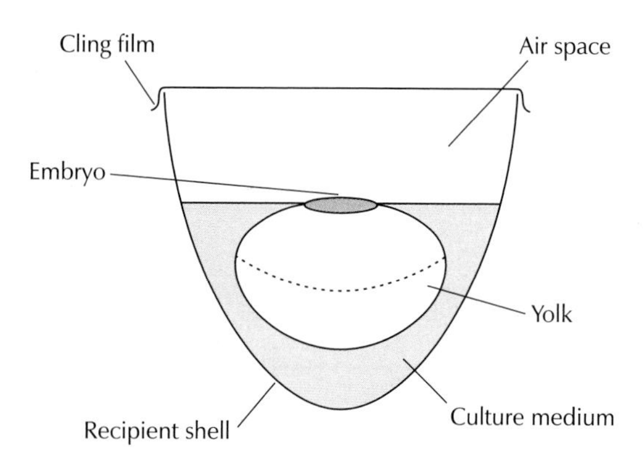

Text-Figure 4. *The final stage of the three-stage culture technique of Perry (1988).*

IN OVO TECHNIQUES

Preparation

The ideal approach must always be to disturb the environment as little as possible by carrying out experiments *in ovo* rather than *in vitro*. With young embryos the blastoderm rises to the top of the yolk, so the eggs should be allowed to settle on their sides for some minutes after removing them from the incubator.

Procedure

Wipe the shell with 70% methanol and using a vial saw or a sharp scalpel outline an area about 2–3 cm. square by scratching the upper surface of the shell. Now gently saw or cut through the shell but leave the shell membranes intact until the shell has been removed. The shell membranes may then be picked away with fine forceps. The embryo should be visible but, if not, pick away more shell and shell membranes with the forceps. Sometimes it is advantageous to lower the yolk away from the shell at this stage. Using a sharp knife or a mounted needle, make a small incision through the blunt end of the shell into the outer air sac. The air is then released and the yolk drops down. After the operation has been performed, both holes are sealed with clear plastic tape, care being taken that it is pressed down firmly so that no air can enter, or albumen escape.

From about 8 days of incubation the position of the embryo in the shell can be established by candling the unopened egg, which enables the experimenter to position the window in the most suitable site. The egg is held against a bright light; a 60 W bulb is adequate but it is important to cover the sides of the egg to shut out excess light. The embryo can be seen as a dark patch with the highly pigmented eyes showing most strongly. With unfertilized eggs, no patch is visible.

Grafting *in ovo*

Small pieces of tissue (e.g. somites) may be extirpated and replaced with grafts of similar size, although if the experiment is carried out after about 3 days of incubation the embryo is so vascularized that heavy bleeding may prevent success. It is often helpful to inject an Indian ink solution beneath the embryo to give a high contrast background, although this may have mild toxic effects. Bagnall (1988) recommended dilution of the Indian ink with yolk spun on a bench centrifuge. To operate, a small hole is made in the vitelline membrane over the appropriate region of the embryo, the ectoderm above the selected tissue (e.g. somite) is peeled away and the tissue removed with a fine needle, taking care not to damage the underlying endoderm as this can result in yolk flooding through the gap. A graft may now be inserted into the space; a tight fit is desirable. As much fluid as possible is removed from above the embryo so that the graft does not float out of position. The shell is then sealed with adhesive tape and incubation is resumed with the window uppermost.

Intracoelomic grafting is a method for maintaining tissues within the coelomic cavity of a host embryo. A graft is inserted into the extra-embryonic coelom at about stage 6 and pushed gently through into the embryonic coelom. After the body becomes folded off from the yolk sac, communication between the extra-embryonic and embryonic coeloms is lost but the graft may survive not only until hatching but throughout the life of the adult bird. This useful technique is seldom utilized these days. For a full description consult Hamburger (1947) or Wilt and Wessels (1967).

The **technique of Martin (1990)** is not easy but can provide excellent results. The entire area pellucida of the experimental embryo, together with a surrounding circle of area opaca (but not the entire area opaca), is removed from its own yolk, operated upon in a Petri dish and then regrafted back into another egg. The host egg having been incubated for the same length of time, a slit is made in the vitelline membrane over its embryo and the vitelline membrane is trimmed away until a large part of the area opaca is exposed but the periphery remains intact and still adherent to the vitelline membrane. The area pellucida is cut out of the host and discarded. The operated embryo is now laid over the hole and its area opaca fused onto that of the host. This is done using fine ophthalmological scissors, which slowly pinch the host and graft tissues together in such a way that the cut edges become fused and the discarded portions can be removed. The endoderm of the area opaca of the host embryo becomes fused to the endoderm of the area opaca of the grafted embryo. The secret of this technique is to proceed slowly, taking very small 'bites' with the scissors and allowing the blades to remain pressed together for several seconds after each cut. Gradually, by progressing around the circumference of the join the two tissues become 'stitched' together.

An additional technique for transplanting tissues *in ovo* is described by Stern (1993a,b) and techniques for grafting pieces of limb buds by Tickle (1993).

Injection Techniques

Avian chimeras can be formed by injecting quail cells at stage X (Etches *et al.*, 1993) and XI–XIII (Watanabe *et al.*, 1992) into chicks of a similar age. Petite *et al.* (1990) described a technique for producing chimeras by injecting cells from stage X embryos into the subgerminal cavity of a different strain.

A localized electroporation method developed for use with avian embryos is described by Atkins *et al.* (2000).

Techniques for making corrosion casts of blood vessels are described by Carretero *et al.* (1993) and Hiruma and Hirakow (1995). Examples supplied by these authors are shown in *Plates 40–43.*

PREPARATION OF SERIAL SECTIONS

Older embryos (from about 3 days of incubation) are prepared using standard histological methods. Difficulties may arise with the younger stages, however, because of the delicacy of the material and the need to prevent the blastoderm from curling at the time of fixation. Removal of the blastoderm from the yolk may be accomplished in at least three ways.

- A thin paper disc is laid over the blastoderm so that it adheres to the vitelline membrane. Using a circular cut through the vitelline membrane, the disc is removed, together with the embryo, and washed briefly in saline to remove any yolk adhering to it. It may then be placed in the fixative and the paper be removed later.

- One end of a piece of nickel–chrome wire (about 22 gauge) is bent into a ring about 15 mm in diameter and the other end is fashioned into a handle at right angles to it. The ring is heated to red heat in a flame and quickly applied to the vitelline membrane and the area opaca. The vitelline membrane and area opaca stick to the ring and if a fine sharp blade is passed around the outer side of the ring, they may be lifted cleanly from the yolk (Daniel, 1955). The blastoderm is still fixed to the wire to keep it flat.
- The procedure used for preparing New cultures may also be employed. The blastoderm, together with the vitelline membrane, is peeled off the yolk in a saline bath. The vitelline membrane is then stretched around a ring and fixative is dropped gently on to the embryo.

A variety of fixatives may be used when preparing material for light microscopy, but that of Schoenwolf and Watterson (1989) is recommended. Carnoy's, Bouin's, formal buffered saline and Zenker's are all satisfactory. For embryos up to about 4 days fixation need not last longer than 2 h, though it can be extended overnight provided that the pots are covered to guard against evaporation. After fixation the specimens are dehydrated in a series of increasing concentrations of alcohol. The time spent in each grade of alcohol can be as short as 10 min for young embryos. Transferring embryos from pot to pot is to be avoided because of their fragility. When passing through the final stage of dehydration, it is advantageous to add a few drops of a stain, such as 1% light green in absolute ethanol, so that the specimens will subsequently be visible in the wax block. After a quick rinse in xylene, the embryos are transferred to the paraffin wax at about 50°C. A common fault is to leave these delicate young specimens for too long in the wax; two consecutive periods of 30 min each are adequate. Longer periods are needed for older embryos.

Sections illustrated in this book were cut at 10 μm. The staining procedures for avian embryos were standard. Most of the sections illustrated in this book were stained with haematoxylin and eosin or with Masson's trichrome. The whole mount embryos in which both cartilaginous and ossified tissues were stained blue and red, respectively (*Plates 148–150*), were prepared according to the method of Dingerkus and Uhler (1977).

Preparation for Scanning Electron Microscopy

The techniques are basically the same as those used for preparing adult tissues for scanning electron microscopy. The major difficulties lie in the fragility of the early embryo so that extreme care must be taken, especially during the mounting of the specimens.

INSTRUMENTS

No instruments peculiar to the chick are necessary. Operations are generally carried out using tungsten needles (Stern, 1993a,b), ophthalmological knives and scissors, sharpened steel sewing needles or sharpened Borradaile knives.

COMMON ABNORMALITIES

Abnormal embryos occur infrequently among commercially supplied eggs, the most common being Siamese twins formed from a duplication of the primitive streak (see Newman, 1917) followed by an interaction between the two axes. Similar Siamese twins can be produed experimentally (*Plate 3*). Micropthalmia of one or both eyes, cyclopia and anopthalmia present infrequently. A well-documented range of mutations exists, maintained in various research centres (e.g. Storrs Agricultural Experiment Station, The University of Storrs, Connecticut, USA and the Department of Avian Sciences, University of California, Davis, California, USA).

The individual names of the mutants have usually been chosen because of their most obvious anomaly, though frequently there are additional malformations, e.g. the **Talpid** mutants (*Plate 25*) are so-called because of their short, broad mole-like limbs which superficially resemble the limbs of the mole, *Talpa*, although they have, in addition, abnormalities of the face and a short vertebral column. These anomalies are all brought about by a single recessive gene (Landauer, 1967). Talpid mutants have appeared spontaneously several times and all are not necessarily the result of the same mutation. The **Creeper** mutant is characterized by shortness of the extremities, especially the legs, and is an achondroplastic, lethal mutation (Landauer, 1967).

The study of the development of mutants has proved valuable in understanding aspects of normal development, e.g. the discovery of the essential role played by the apical ectodermal ridge in normal limb development was made by Zwilling (1949) when studying the early embryology of the **Wingless** mutant. These results stimulated an extensive series of experiments in which the apical ectodermal ridge of a normal embryo was separated from its underlying mesenchyme and then recombined with tissues from the Creeper mutant. More recently, the study of the molecular basis of mutations has opened up new approaches (see Abbott and Pisenti, 1993).

Chapter 3
Early Stages

FERTILIZATION AND PRE-LAYING

There is a tendency to think of the development of the chick embryo as starting at incubation and lasting 21 days, but not only does fertilization take place before the egg is laid, but some of the most critical events occur during this period. They have attracted comparatively little attention compared with those of the older stages, largely because of the practical difficulties involved in their study.

The structure of the **avian sperm** corresponds to that of other vertebrate species in that there is a head, consisting mainly of the nucleus and acrosome, a middle piece with mitochondria, and a long flagellum. In the domestic fowl, however, the nucleus is so long and slender that the head does not bulge out (*Text-Figure 5*; *Plate 1*) in the way characteristic of most vertebrate sperm. Sperm of this type are typical of non-passerine birds and are described as fusiform or vermiform because of their slender shape. The fowl sperm nucleus is about 12 μm long and 0.6 μm in diameter (Wishart and Horrocks, 2000). For a full description by transmission electron microscopy see Bakst and Howarth (1975) and Thurston and Hess (1987). Avian sperm, unlike those of mammals, develop their fertilizing ability whilst still in the testes and, as there are no accessory glands corresponding to those in mammals, the semen passes directly into the genital tract of the hen (Wishart and Horrocks, 2000). Fertilization takes place in the infundibulum of the oviduct (*Text-Figure 2*).

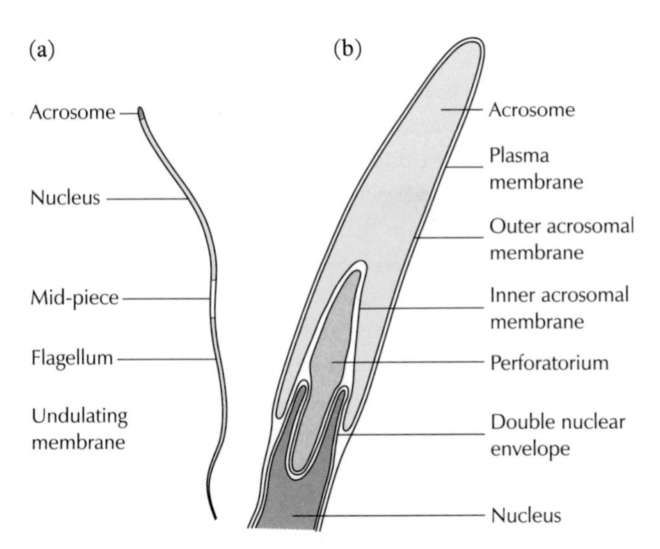

Text-Figure 5. *Morphological features of the sperm of the domestic fowl. (a) the entire spematozoan; (b) the acrosomal region (see also Plate 1c). (After Wishart and Horrocks, 2000, with permission of Springer-Verlag, Heidelberg.)*

After mating some, but not all, of the sperm are transported to the (lower) sperm storage tubules that are situated at the junction of the vagina and the shell gland (uterus). It is not clear what factors determine which sperm are transported and which are not, though differences in sperm motility or in sperm selection by the vaginal environment have both been suggested. There are additional sperm storage tubules at the upper end of the infundibulum. The sperm storage tubules do not appear from their histology to be secretory and probably act merely as receptacles, the sperm lining up in an orderly manner with their heads to the epithelium of the tube and their tails hanging in the lumen. As many as 100 sperm may be stored in one tubule (Bakst *et al.*, 1994). The main significance of the sperm storage tubules is that it enables fertilization of a clutch of eggs to take place without the need for synchronized copulation. This is important, since fertilization must occur immediately after ovulation in the brief period, perhaps only 15 min (Wishart and Horrocks, 2000), before the outer layer of the vitelline membrane is laid down and blocks sperm entry. The sperm are probably metabolically quiescent whilst in the tubes, the concentration of calcium in the tubular fluid apparently inhibiting sperm motility (Holm *et al.*, 2000), but become activated upon leaving. There is some evidence that a neural stimulus may bring about the release of the sperm from the storage tubules, since individual tubules are separately innervated (Freedman *et al.*, 2001). The sperm appear to be released continuously and not necessarily as the result of any event in the oviduct. Once the egg enters the infundibulum, many sperm rapidly contact the surface of the inner layer of the vitelline membrane and pass through it, apparently dissolving the granular material lying between the fibres with hydrolytic enzymes released from the acrosome (Wishart and Horrocks, 2000).

In penetrating through the inner layer, each sperm first binds to receptors on it, which appear to be carbohydrate based (Robertson *et al.*, 2000). There is evidence that the breakdown of the acrosome membrane (acrosome reaction) at this time is induced by an *N*-linked oligosaccharide with terminal *N*-acetyl-glucosamine residues (Horrocks *et al.*, 2000). This appears to trigger the rupture of the membrane around the acrosome, resulting in the release of hydrolytic enzymes, which digest away a small area of the inner layer of the vitelline membrane so that the sperm is able to pass through it.

One of the problems of large-yolked eggs is that of how the sperm nucleus finds and fuses with the female nucleus. In some small-yolked eggs (e.g. sea urchins and mammals) there are mechanisms to prevent the entry of more than one sperm, but in birds many sperm pass into the egg. However, only one successfully fertilizes it, the male and female haploid nuclei fusing to form a zygote possessing the diploid number of chromosomes about 3 h after ovulation (Waddington *et al.*, 1998). The first mitotic division occurs after about a further hour (Wishart and Horrocks, 2000). The remaining (supernumerary) sperm may undergo division, before breaking down.

Spontaneous parthenogenesis occurs in some birds. Olsen (1962), using an inbred strain of Beltsville White turkeys, obtained an incidence of nearly 40% parthenogenetic embryos, of which some survived into adulthood; they are thought to have doubled their chromosomes by retention of the second polar body. Such a biological solution to the problems of fertilization is, however, an abnormal one.

Even before ovulation the egg cytoplasm has almost separated from the yolk and, because it is lighter, has come to lie on the uppermost side of the egg, the animal pole, and is known as the germinal disc. Components of the germinal disc include the metaphase II meiotic nucleus and discarded first polar body, mitochondria and glycogen granules (see Bakst and Howarth, 1977, for a full description). As the egg yolk, enclosed in its vitelline membrane, passes down the oviduct it becomes covered first by the albumen, then by the shell membranes and finally by the shell (see Chapter 1). During this time the embryo undergoes cleavage, this process differing greatly in birds from that found in mammals. Cleavage furrows are visible in fertilized eggs taken from the oviduct (*Text-Figure 6*; *Plate 5*), although the first 16 cleavage cells are not completely enclosed in

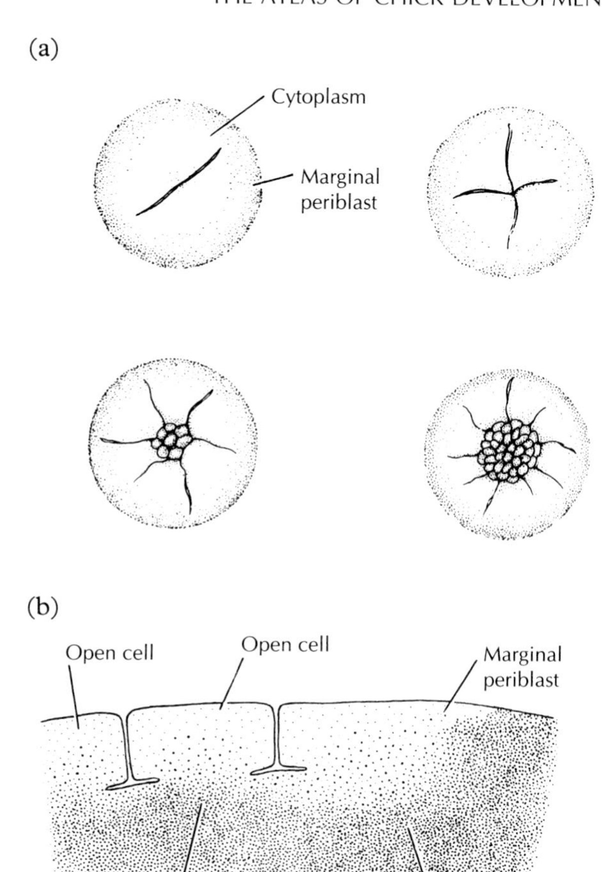

Text-Figure 6. *The structure of the open cells in early cleavage. (a) four stages seen from dorsal side of embryo; the yolk is not illustrated but the marginal periblast is visible; (b) vertical section through the edge of an early cleavage stage embryo. (From Bellairs et al., 1978, with permission of The Company of Biologists Ltd.)*

their own cell membranes but are open ventrally to the yolk mass; they are known as 'open cells'. With each mitotic division the cell membrane becomes completed around one of the daughter 'cells', whilst the other migrates into the adjacent yolky cytoplasm, carrying some cytoplasm with it, so that a new open cell is formed. By the 64-cell stage the most centrally situated cells have each become completely surrounded by a cell membrane, but there is still a region of mixed yolk and cytoplasm (ooplasm) beneath and around the embryo known as the subgerminal and marginal periblast, respectively (*Text-Figure 6*). During this time fluid is drawn from the albumen through the vitelline membrane and accumulates beneath the embryo, forming the fluid in the subgerminal cavity (*Plate 5b*).

The most important event for the cleaving embryo is the establishment of the axes of polarity. First, the dorsal and ventral sides become apparent (the dorso-ventral axis) and then the antero-posterior (the cranio-caudal axis). With the interaction of these two axes the embryo acquires a left and a right

(a)

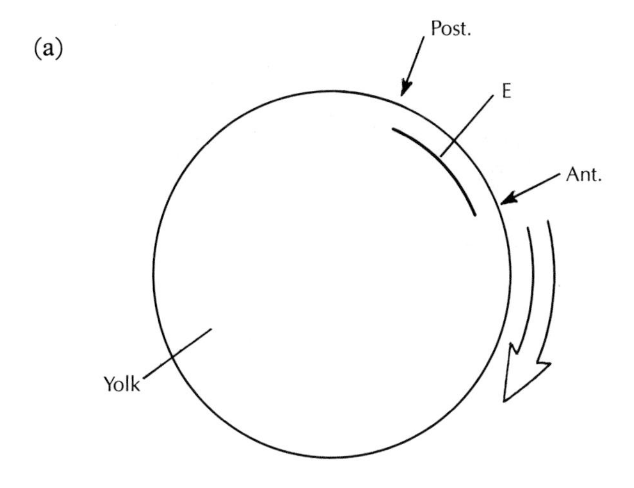

(b)

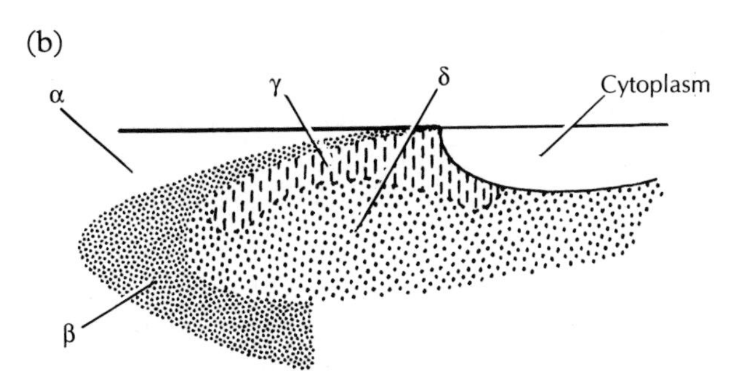

(c)

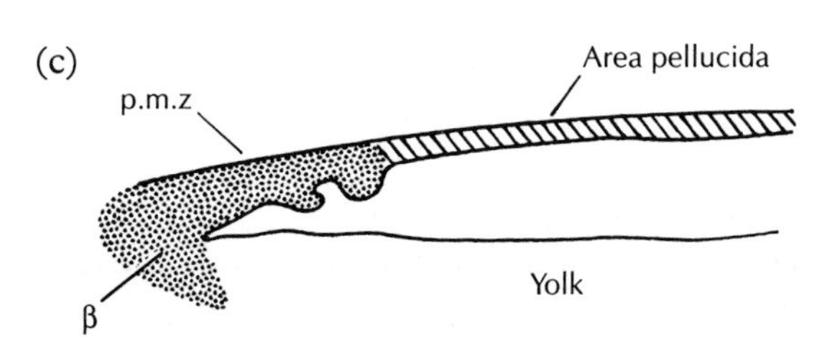

Text-Figure 7. *(a) The yolk during rotation in the shell gland. The direction of rotation is indicated by the broad arrow. The embryo no longer lies uppermost on the yolk but has been tipped to one side. Ant = future anterior end of embryo, Post = future posterior end. (b) Vertical section through ooplasm of a quail oocyte shortly before ovulation. α, β, γ and δ = ooplasmic components. (c) Vertical section through the posterior region of a late cleavage stage embryo. The β-ooplasm has become located in the posterior marginal zone (p. m. z.). (After Callebaut, 1983.)*

side. If these differences between the different regions were not established the embryo would continue to grow in size but lack any form or pattern to its body. The establishment of the dorso-ventral axis is directly associated with the relationship between the yolk and the cytoplasm.

During its early stages the chick embryo is a flat disc, the blastoderm. Its lower surface rests on the yolk and becomes the ventral side of the embryo, while the cells at the upper side lie immediately beneath the vitelline membrane and become the dorsal side. There is a potential difference of 25 mV across the blastoderm between the acidic ventral side and the basic dorsal side (Stern and Canning, 1988). It seems probable, therefore, that the differing environments at the two sides of the blastoderm play a role in bringing about the dorso-ventral polarity.

The establishment of the antero-posterior axis results from a shifting of the ooplasmic components by gravity. This takes place during the 20 or so hours spent in the shell gland (uterus) when the entire egg rotates at about 10–12 turns per hour as the shell is laid down and in so doing tips the blastoderm from its original position on top of the yolk. The future anterior (head) end forms in the region that has been tipped lowest (*Text-Figure 7*), whereas the future posterior (tail) end forms from the highest region (Eyal-Giladi and Fabian, 1980). This tilting is accompanied by a shifting of the yolky ooplasm, including the nucleus of Pander (see Chapter 1) beneath the oocyte so that the more lightweight of its components, especially the so-called β-ooplasm, come to lie beneath one side of the blastoderm, which, probably as a result, beomes the future posterior end (Callebaut, 1987). The nature of the supposed interaction between the periblast and the blastoderm is not understood, but it is clear that the antero-posterior axis is not yet completely determined at this stage, since it can be experimentally changed even after laying; if the blastoderm of a laid egg is bisected before incubation it may form twinned axes (*Plate 3b*) and these do not necessarily have the same orientation as one another. Normally, however, a range of *Homeobox* and other genes that are associated with the establishment and maintenance of antero-posterior polarity begins to be expressed soon after laying (see Chapter 5).

Ooplasmic Segregation

By the time of laying the unincubated blastoderm already consists of several regions. *Text-Figure 8a* is a diagram of the ventral side of an unincubated quail embryo which lies over the nucleus of Pander. Koller's sickle (see below) is visible at the posterior end as a crescent and this is packed with ooplasmic particles, the whole being encircled by the area opaca. A region called the marginal zone is apparent, lying between the area opaca and area pellucida, which is thickest posteriorly (the posterior marginal zone). Koller's sickle lies immediately anterior to the posterior marginal zone and extends out as two lateral horns. The endophyll area (see below) is situated in the central region of the area pellucida. Callebaut *et al.* (2001) describe a further region, the anti-sickle, which lies at the extreme anterior border of the area pellucida.

During the formation of the oocyte four ooplasms (**α**, **β**, **δ** and **γ**) have been distinguished in association

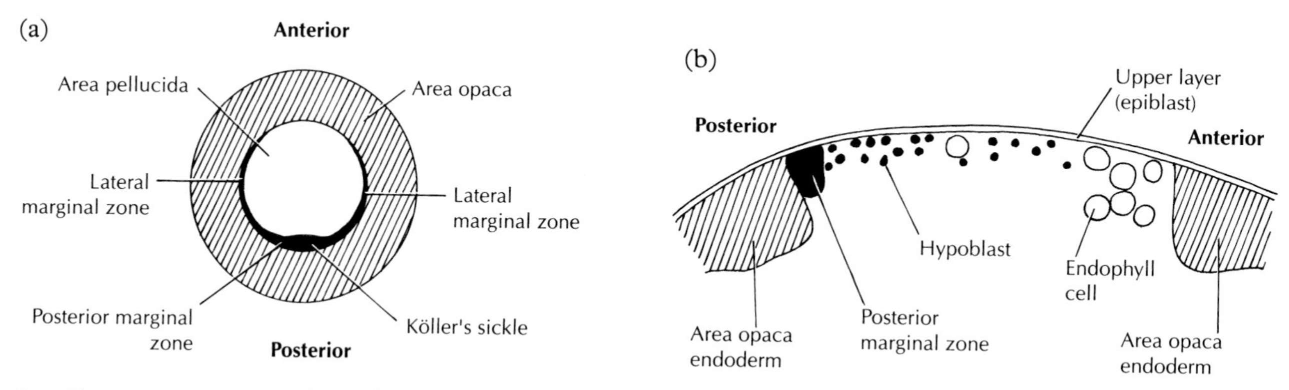

Text-Figure 8. *(a) Lower surface of an unincubated chick embryo; (b) section through the same embryo passing along the anterior–posterior axis of the area pellucida.*

with the germinal disc of the quail, each possessing its own pattern of organelles (Callebaut, 1987). During the tilting process the **β** and **δ** ooplasms shift and become distributed to different regions of the embryo, each then playing a significant role in the development of its region. The **α** ooplasm segregates very early from the other ooplasmic regions and disappears before laying. The **β** ooplasm forms Koller's sickle. The **γ** ooplasm becomes incorporated into the central region of the area pellucida, whilst the **δ** also contributes to the central region and segregates later into the primordial germ cells (Callebaut, 1987). It appears therefore that there is a relationship between the tilting of the egg, the redistribution of the four types of ooplasm and the establishment of the antero-posterior axis, though the nature of the supposed interaction between the ooplasms and the blastoderm has yet to be uncovered.

Repeated cell divisions occur during cleavage so that by the time the egg is laid the chick embryo possesses several thousand cells. Estimates range from 32 000–42 000 cells (Stepinska and Olszanska, 1983) to as many as 60 000 (Spratt, 1963).

THE EARLY POST-LAYING STAGES

The transformation of a group of unspecialized cells into a basic embryonic body occurs during the first 2 days of incubation and during this time the embryo is at its most vulnerable to disturbances. Many of the abnormalities seen in late embryos or hatchlings have their origins in mishaps at these early stages.

When the egg is laid its temperature falls from that of the hen's body (about 40°C) to that of its surroundings, which is normally well below the incubation temperature of 37–38°C. The result is that the development of the embryo slows dramatically and virtually ceases below about 25°C, remaining arrested until incubation begins. In the wild, a hen may take several days to complete the laying of a clutch of eggs and only then begin to incubate them, the result being that all the chicks hatch at much the same time (but see also Chapter 13).

The morphological changes occurring during the first 24 h are best seen by dissecting the embryo from the surface of the yolk and examining it by transmitted light. The most striking feature is the division into the **area pellucida**, the almost transparent central region, and the **area opaca**, the more opaque peripheral ring (*Text-Figure 8*). *Plate 10a* illustrates these regions by scanning electron microscopy, though the technique has made them no longer transparent. Sections show that there are two layers, the upper (epiblast) and the lower, throughout the area opaca and in the posterior part of the area pellucida (*Text-Figure 8b* and *Plate 5*). The opacity of the area opaca is due to the presence of large numbers of intracellular yolk droplets in the lower layer. Intracellular yolk droplets are present in the area pellucida also, but they are much smaller and fewer than those in the area opaca so that the tissue is relatively transparent. The area pellucida gives rise mainly to embryonic tissues, whereas the lateral borders of the area pellucida together with the entire area opaca form extra-embryonic tissues only (see Chapter 13 for more details of the area opaca).

The developing primitive streak starts to be visible in the area pellucida at about 10 h of incubation (stage 2 of Hamburger and Hamilton, 1951; stage XIV of Eyal-Giladi and Kochav, 1975). (For staging of embryos see Appendices I and II.) It appears as a dark triangular region in the upper layer, its apex lying in the area pellucida and its base along the border with the area opaca. By about 18 h of incubation (stage 4 of Hamburger and Hamilton) it has reached its full length, extending about two-thirds of the way across the area pellucida and its most important region, Hensen's node, is visible as the swollen tip at the anterior end (*Plate 7*).

Two key structures involved in the formation of the primitive streak are the posterior marginal zone and Koller's sickle.

The Posterior Marginal Zone and Koller's Sickle

The marginal zone lies at the border between the area opaca and the area pellucida and is already visible at the time of laying, It is widest at the posterior end of the area pellucida, the multilayered posterior marginal zone, and becomes progressively thinner in its lateral and anterior regions. Koller's sickle is a crescent-shaped area at the anterior edge of the posterior marginal zone (*Text-Figure 8*) and is also known as Rauber's sickle, since it was described initially by Rauber in 1876 (Callebaut *et al.*, 2000b). The significance of the posterior marginal zone and Koller's sickle lies in the role they play in the formation of the primitive streak.

The cells of the posterior marginal zone and of Koller's sickle are in direct contact with the underlying yolk, whilst the area pellucida, which is anterior to them, lies over the fluid-filled subgerminal cavity. This close association with the yolk is of great significance, since there is evidence that the ooplasm contributes to the posterior marginal zone and Koller's sickle (see p. 17). There has been some disagreement about the number of cell layers present in this region and the precise relationship of Koller's sickle to the posterior marginal zone. This is not surprising since it is difficult to distinguish the different types of cells by traditional staining, though the situation is easier in the quail than in the chick. It should be noted that some authors (e.g. Eyal-Giladi, 1991) regard Koller's sickle and the posterior marginal zone as being one tissue. According to Zehavi *et al.* (1998) there is a high level of cell proliferation in both the posterior marginal zone and Koller's sickle.

The Lower Layer

Four types of cells contribute ultimately to the lower layer. At the time of laying, usually about stage X of Eyal-Giladi and Kochav (1975), the area pellucida consists of an upper layer, the epiblast, underlain by scattered clumps of cells each about 30–50 μm in diameter (*Plate 5c*). These clumps are the first indication of a developing lower layer and were named **endophyll** by Vakaet (1962). Alternative names are polyinvaginated cells (Eyal-Giladi and Kochav, 1975) and primary hypoblast (Stern, 1990). They are situated in the anterior (cranial) extension of Koller's sickle (Callebaut *et al.*, 1998) and are thought to arise by ingression from the upper layer, though the evidence is based mainly on the staining of fixed material (Fabian and Eyal-Giladi, 1981). Most, if not all, the endophyll cells are thought to give rise to the primordial germ cells (see p. 66) and they may also contribute to the yolk sac stalk (Stern, 1990).

Shortly after the start of incubation the second component of the lower layer, the **hypoblast** proper, arises at the posterior end of the area pellucida. It appears to form principally by migration from Koller's sickle, and by about 12 h of incubation (stages XIII–IV) has come to lie beneath the entire area pellucida. As it spreads from the posterior to the anterior end of the area pellucida it forms a loose sheet in which some of the endophyll cells are caught and carried anteriorly. Ultimately, they become located in the germinal crescent (see p. 66–7). The hypoblast cells, individually about 15–20 μm in diameter and yolky, are attached to one another loosely. There is no basal lamina associated with them and no apparent dorso-ventral specialization, so the sheet cannot be said to be a true epithelium. Hypoblast cells have been studied by time-lapse cinematography in tissue culture (Sanders *et al.*, 1978) and have been found to settle and migrate readily, even on untreated glass, laying down a thick layer of fibronectin. According to Rosenquist (1971) the hypoblast cells give rise to the endoderm of the yolk sac stalk (but see below). Several markers of the hypoblast are now available, e.g. *Otx2*, HNF3β, *gsc* and cCer/caronte (discussed by Foley *et al.*, 2000).

At about stage 2–3 the posterior hypoblast becomes replaced by the third component of the lower layer, the **endoblast**, which is sometimes considered to consist of two types of cells, the **sickle endoblast** (Vakaet, 1970) lying beneath Koller's sickle, and the **junctional** endoblast situated further laterally (Callebaut *et al.*, 2001). The endoblast is derived partially from Koller's sickle (Callebaut and Van Neuten, 1994) and partially from the lateral posterior marginal zone (Stern and Ireland, 1981). The sickle endoblast has been found to form the yolk sac endoderm and to influence the formation of the yolk sac blood vessels (Callebaut *et al.*, 2000a); this finding does not necessarily contradict that of Rosenquist (mentioned above) but reflects the fact that Rosenquist did not distinguish between hypoblast and sickle endoblast.

Bachvarova *et al.* (1998) produced a fate map (see p. 22) for discussion of fate maps) by staining selected groups of cells in and around Koller's sickle with DiI/DiO (*Text-Figure 9*). Using goosecoid (*gsc*) as a marker, which is expressed in the hypoblast but not the endoblast, they concluded that the two tissues are derived from different locations in the deep layers of the sickle endoblast.

The fourth and final component of the lower layer is the **definitive endoderm**, which consists of cells that have ingressed through the primitive streak after it has fully formed and it becomes the endoderm of the developing gut. These four components are illustrated in *Text-Figure 10*.

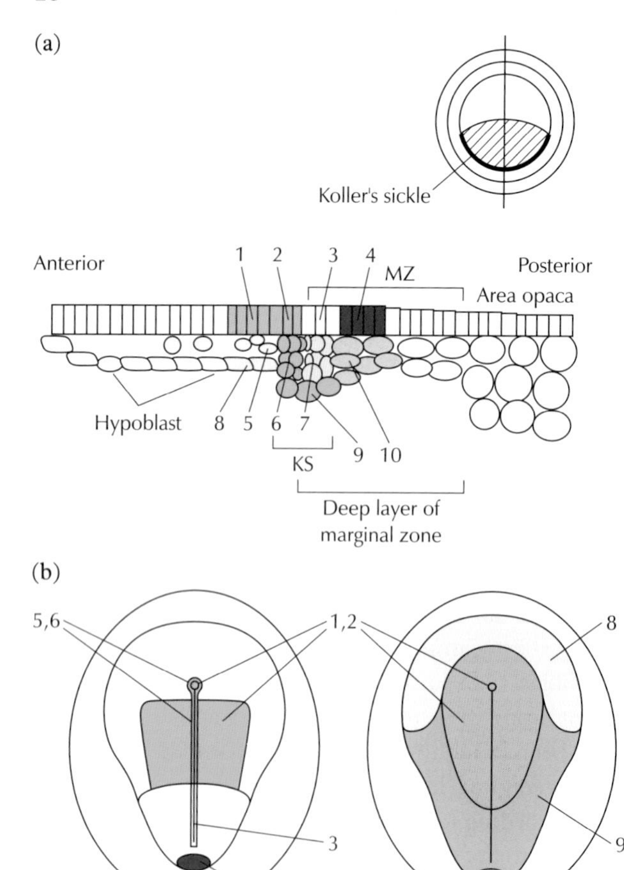

Text-Figure 9. *(a) Midsagittal section of a prestreak embryo at about stage XII (see Appendix I) showing the various structures at sites of labelling with DiI/DiO. (b) Positions of cells at stage 4 (Appendix II) derived from the labelled sites. The fates of the upper layers (ectoderm and mesoderm) are shown in the left diagram, and of the lower layers in the right diagram. Upper: Sites 1, 2, 5 and 6 contribute to Hensen's node, the anterior streak and the embryonic mesoderm. Site 3 contributes to the posterior streak and the embryonic mesoderm. Site 4 remains in the posterior area opaca. Lower: Sites 1 and 2 contribute to the embryonic endoderm. Site 8 is found in the hypoblast of the germinal crescent. Site 9 is found in the endoblast, whilst sites 7 and 10 remain posterior to the primitive streak. (After Bachvarova et al., 1998, with permission of the Company of Biologists Ltd.)*

The Upper Layer (Epiblast)

During this period when activity is taking place in the lower layer, the embryo is expanding over the yolk and the upper layer becomes thinned from about four to six cells to one cell deep, probably by intercalation and changes in cell shape. Canning and Stern (1988) showed by staining with the monoclonal antibody HNK-1 that there is a pepper-and-salt pattern in chick embryos at stages XII and XIII, indicating that the cells are not identical throughout. Later, Stern and Canning (1990) found that the HNK-1 positive cells are precursors of the mesoderm and definitive endoderm of the primitive streak. A similar conclusion was previously drawn by Bancroft and Bellairs (1974) who showed that the pattern and distribution of the microvilli on the dorsal surface of the epiblast varied according to the region (*Plate 6*).

Between stages X and 3 there is a convergence of cells to the posterior midline, followed by anteriorward movements along the midline (Vakaet, 1984; Hatada and Stern, 1994) (*Text-Figure 11*). Modern maps indicate that there is considerable overlap of the prospective fates in these early stages. This is possible because most, if not all, of the cells are not completely determined as yet. They are merely **specified** (labilely determined) as can be shown by transferring small pieces of tissue from different regions to a foreign environment (e.g. by grafting to the chorioallantoic membrane, or explanting into tissue culture). For example, gut endoderm is capable of developing from almost the entire area pellucida of a stage XI to XIII embryo, though normally it forms only from cells of the posterior third of the area pellucida. Hatada and Stern concluded that the cells in the anterior region of the area pellucida are normally prevented from developing into gut endoderm because of influences from neighbouring cells.

The Primitive Streak

The primitive streak plays a critical role in the development of the embryo. It is first visible at about 6–7 h of incubation as a triangular-shaped dark shadow at the posterior end of the area pellucida (Stage 2 of Normal Table of Hamburger and Hamilton; see Appendix II). It gradually extends further anteriorly until about 13 h of incubation (stage 3+) when it reaches the centre of the area pellucida and during the next 2 h it achieves its final length. (This is not shown in the Hamburger and Hamilton table in Appendix II, their stage 4 already illustrating the beginning of the next, the head process, stage). The primitive streak is so-called because it appears as a dark double line across the relatively translucent area pellucida. The darkness is due to the large number of cells which have become piled up there, waiting to ingress (*Plates 4b, 7a–d*).

In a classical experiment, Waddington (1932) showed that if the upper layer (epiblast) of a prestreak embryo was explanted in culture on its own it failed to form a primitive streak and consequently an embryo. If the lower layer (hypoblast) was present then a primitive streak could form. The traditional view, therefore, was that the primitive streak was induced in the epiblast by the underlying hypoblast. Eyal-Giladi (1991) and her colleagues (e.g. Khaner, 1995), who carried out a series of extirpation and transplantation experiments to test this concept, considered that the hypoblast was derived, at least in part, from the posterior marginal zone (*Text-Figure 8*) and so investigated which regions of the posterior

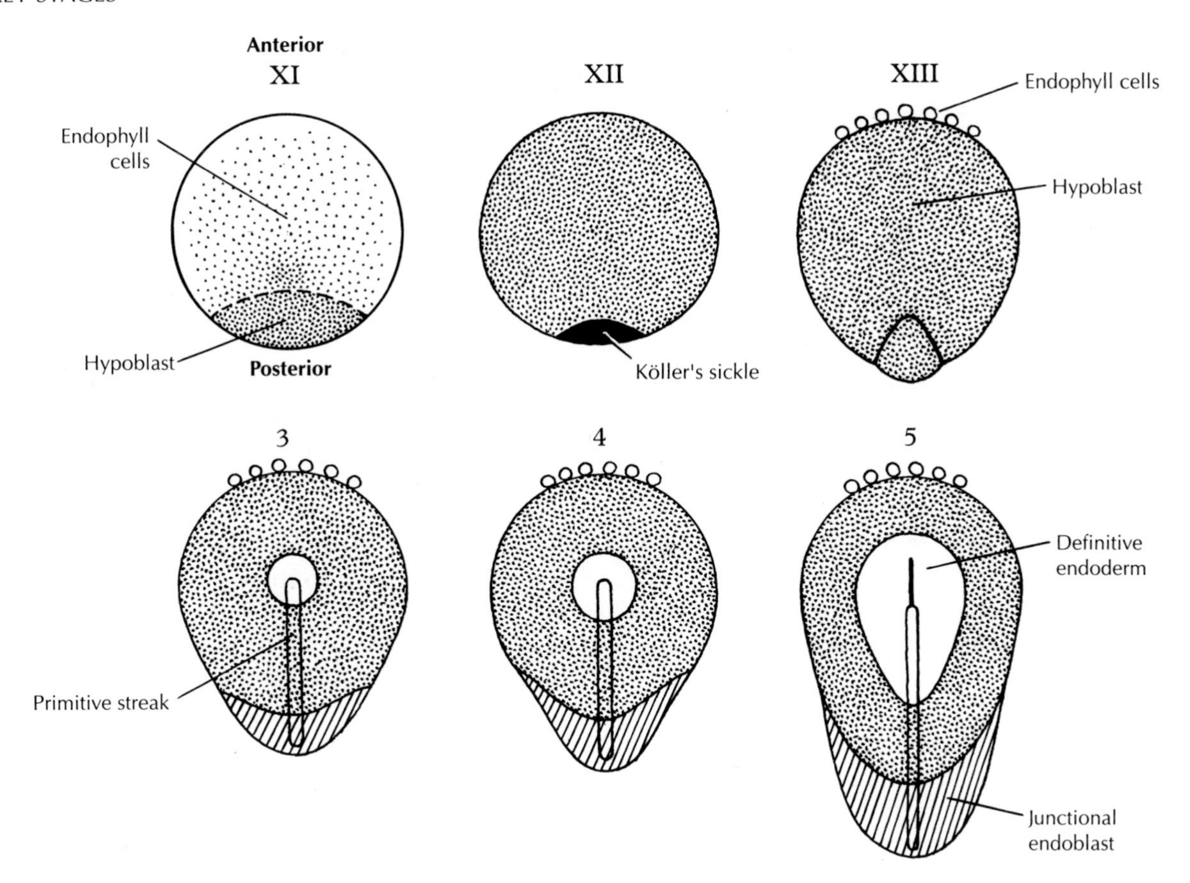

Text-Figure 10. *The relations of the tissues that form the lower layer of the area pellucida. Initially the hypoblast spreads from the posterior end and forms a continuous sheet (stages XI, XII and XIII of Eyal-Giladi and Kochav). The junctional endoblast then spreads from the posterior end and the definitive endoderm ingresses through the primitive streak (stages 3, 4 and 5 of Hamburger and Hamilton). Another component of the lower layer is the endophyll cells (future primordial germ cells), which move to the anterior border of the area pellucida. (After Sanders et al., 1978, with permission of the Company of Biologists Ltd.)*

marginal zone were capable of inducing a primitive streak. They found that there was a gradient of potential along the marginal zone in its ability to induce a primitive streak, the posterior part having the highest ability. If this region is extirpated then the two lateral marginal regions on either side act as high points and two primitive streaks form. Eyal-Giladi suggested that the formation of a single primitive streak in normal development depends on an interaction between inductive action from the medial posterior marginal zone and inhibitory effects from the lateral marginal zone.

Many investigators have sought to discover the precise source of the inductive region, but the hypothesis that the hypoblast itself is the inductor has now been abandoned. Experiments (Khaner, 1998) have implicated Koller's sickle in addition to the posterior marginal zone. This is of particular interest because Koller's sickle in birds is thought to be comparable to the Nieuwkoop centre in amphibians. The Nieuwkoop centre is a region of vegetal cells in the amphibian blastula, which lies beneath the mesoderm of the dorsal lip of the blastopore and induces it to become the amphibian organizer of Spemann.

The posterior marginal zone of the chick has many similarities to the Nieuwkoop centre in that it is associated with the induction of the principal organizer region (i.e. Hensen's node), and it expresses genes homologous to those implicated in the Nieuwkoop centre (viz. *Vg1* and *Wn8c*; Bachvarova *et al.*, 1998) and can induce a complete axis from the upper layer (Callebaut *et al.*, 1998; Bachvarova *et al.*, 1998). The ability to carry out this induction is greatest in the region of Koller's sickle, but can occur in the region just posterior to the sickle (Bachavarova *et al.*, 1998; Zehavi *et al.*, 1998).

cVg1 (the chick version of the amphibian *Vg1* gene) is a member of the transforming growth factor (TGFβ) family of signalling molecules. An ectopic primitive streak can be induced in the marginal zone by *cVg1* (Shah *et al.*, 1997) and this is related to the fact that the marginal zone expresses *Wnt8c* (Hume and Dodd, 1993). Skromme and Stern (2001) showed that the area pellucida can also form an ectopic primitive streak in response to *cVg1* provided that *Wnt* is also supplied. Moreover, inhibition of endogenous *Wnt* activity with *Crescent* or *Dkk1* in the marginal zone blocks the axis forming ability of the *cVg1*.

Other genes that are expressed in the region of the posterior marginal zone and/or Koller's sickle include *goosecoid* (*gsc*), *GSX*, *CNOT1*, *CNOT2*, *HNF3β*, *Otx2* and *chordin*, all of which originally have sickle-shaped domains but, as the primitive streak forms, become elongated along it.

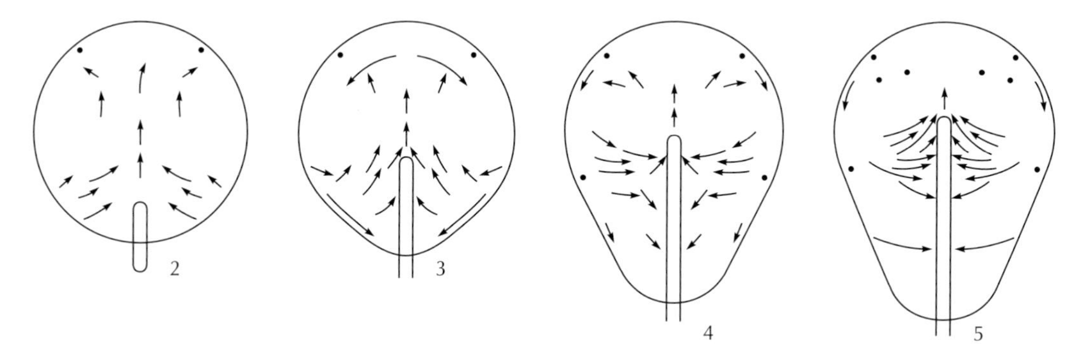

Text-Figure 11. *Gastrulation movements in the upper layer of the area pellucida during formation of the primitive streak and the onset of regression (stages 2–5). (After Vakaet et al., 1980, with permission of Academic Press (Elsevier).)*

Koller's sickle, therefore, contains precursor cells of Hensen's node and of its derivative, the prechordal plate (see p. 29), which express characteristic organizer markers. The posterior marginal zone and Koller's sickle do not, however, possess the same ability as the amphibian organizer to induce a complete axis, although they are able to induce a primitive streak.

Formation of the Primitive Streak

Where do the cells of the primitive streak itself come from? Principally, they arise from a population of epiblast cells in the posterior midline (Lawson and Schoenwolf, 2001). Time-lapse cinematography shows that there is a convergence of epiblast cells at the posterior end of the area pellucida towards the midline and then a forward migration (Vakaet, 1984; *Text-Figure 12*) and this is supported by transplant studies (Eyal-Giladi, 1991). Further evidence comes from DiI cell labelling experiments (e.g. Hatada and Stern, 1994) between stages X and 3.

GASTRULATION MOVEMENTS

Gastrulation is the period during which mass migrations of the cells take place so that they come to lie in the appropriate positions ready to start differentiation. The formation of the primitive streak is the first indication that gastrulation has begun. In amniotes (reptiles, birds and mammals) gastrulation continues at the posterior end of the embryonic axis when differentiation has already begun at the anterior end, and does not cease as abruptly as it does in amphibians.

Fate maps are charts of the early embryo showing the location of groups of cells that will form specific tissues at later stages (e.g. neural tissue, somites, heart), provided that development proceeds normally. Many experiments have shown that most cells at the primitive streak stage are still able to change their fate and form different tissues and, for this reason, their fates are more properly referred to as presumptive or prospective fates. As the mass migrations of the cells take place the arrangements of the presumptive tissues also change. Investigations, therefore, which tell us about the fate maps also reveal the changing patterns of cell migrations. Techniques for labelling (see Chapter 2) have greatly improved in recent years. Early investigators dabbed on spots of vital dyes (i.e. dyes that can be used on living tissues without any obvious damage to them, such as Nile blue sulphate or Bismark brown) or inserted particles of carbon into the cells, but fading of the dyes or shifting of the particles sometimes affected the results. More reliable approaches involved the replacement of a specific region of the embryo by an identical but labelled one using modern agents such as the carbocyanine dyes, DiI and DiO.

The routes taken by the migrating cells as the primitive streak forms were analysed by Vakaet (1984). With the onset of gastrulation the convergence movements (*Text-Figure 12*) lead to a build up of cells which form the primitive streak. Initially, there is no clear groove in the primitive streak, though, as the migrating cells continue to converge and pile up, the developing streak becomes narrower and the groove deeper (*Plate 7a,b*). According to Hatada and Stern (1994) the convergence begins between stages XI and XIII in the epiblast and most of the cells that will contribute to Hensen's node have reached the centre of the blastoderm before the primitive streak starts to form. A second migration of cells is then initiated at stage 2 when the lower layer moves forward. The first population of cells is destined to give rise to the prechordal mesendoderm, the notochord and some somite cells, whereas the second gives rise mainly to definitive endoderm. By stages 3+/4 most cells of Hensen's node express the organizer markers, *chordin*, HNF3β and *goosecoid*.

The primitive streak elongates by extending not only anteriorly, but also posteriorly (Harrisson *et al.*, 1988) so that the posterior end eventually moves into the area opaca, causing the area pellucida as a whole to become pear-shaped. The fully formed primitive streak (about stage 4 of Hamburger and Hamilton, 1951) reaches approximately to the centre of the original circle of the area pellucida. The result

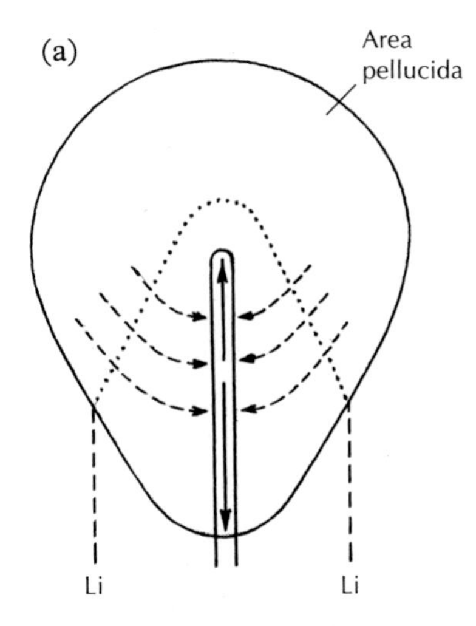

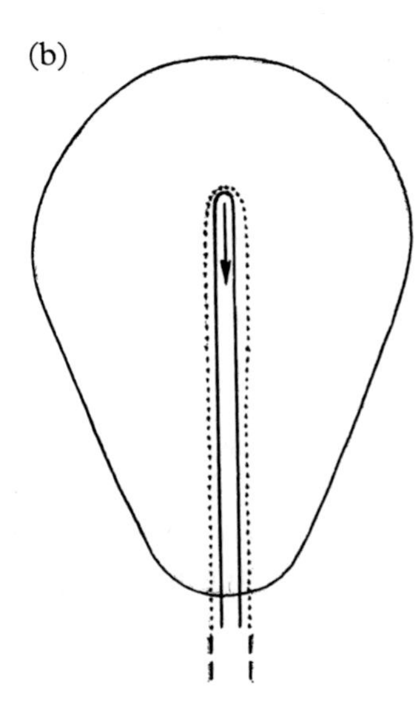

Text-Figure 12. *Gastrulation movements in the area pellucida of the chick embryo. (a) During formation of the primitive streak: convergence movements in the upper layer are shown as curved, broken arrows; Li = limit of the area that will ingress. The area pellucida has become pear-shaped because of the stretching of the primitive streak along its length (shown by arrow on the primitive streak). (b) During regression: the major movement is from anterior to posterior. (After Vakaet, 1984, with permission of Academic Press (Elsevier).)*

of all these movements is that the presumptive areas come to lie around the anterior end of the primitive streak, especially around the extreme end, which is then known as Hensen's node (sometimes called the primitive node). Although most of these movements appear to be taking place in the posterior half of the area pellucida, there is also a migration of cells away from the midline in its anterior half (*Text-Figure 11*).

Vakaet's findings were based on the careful analysis of xenoplastic grafts of quail cells (see Chapter 2) followed by time-lapse cinematography and a histological analysis of the fate of the quail cells. The maps obtained were very similar to those of earlier authors (Pasteels, 1937) who used vital dyes to label their cells. A criticism of the technique, however, is that Vakaet's grafts may have included cells from several layers and so the movements may have reflected more than just the superficial layers. Nevertheless, they provide us with a general picture of the pattern of the cell migrations at this time.

The more recent labelling of individual cells or small groups of cells with dyes such as DiI (see Chapter 2) have provided evidence that many of the cells do not remain in the same relationship to one another during gastrulation. Moreover, the clear-cut boundaries between the different presumptive areas illustrated in the fate maps of earlier investigators are no longer acceptable. Many experiments have shown that, with the exception of the presumptive notochord, there is still considerable flexibility of fate (labile determination) in these tissues, at least until the primitive streak is fully formed at stage 4.

With the development of the primitive streak the left and right sides become visible, though they are not yet fully determined until Hensen's node is formed and a sequence of molecular differences becomes apparent on the two sides. These appear to be set in train as Hensen's node forms. Up to this point the *sonic hedgehog* gene has been active on both sides but now it ceases to be expressed on the right side. The basic features are that the left side expresses *sonic hedgehog* and *lefty-1*, but the right side is prevented from doing so by activin signalling. A cascade of molecular events on the left side results in the activation of the *pitx2* gene and the suppression of *snail*, whilst a different cascade on the right side results in the activation of *snail*. For a summary of these and other events see Gilbert (2000) and Dathe *et al.* (2001). Recently, however, Dathe *et al.* (2003) have reported that by the time that the asymmetric expression of these molecules is apparent, Hensen's node already shows some morphological differences between the left and right sides. It seems likely, therefore, that the critical events occur before Hensen's node is visible.

INGRESSION AND CELL MIGRATION AWAY FROM THE PRIMITIVE STREAK

We can now consider the morphogenetic events in the primitive streak itself. All the presumptive areas, apart from the definitive endoderm, are arranged on the upper surface and they then ingress through the primitive streak. **Ingression** is the process by which the cells leave the epiblast and become part of the primitive streak before migrating away from it to give rise to the definitive endoderm and mesoderm. (The term 'invagination' is no longer considered appropriate to describe the passage of cells through the primitive streak, since it implies the formation of a cavity, such as the archenteron of the amphibian embryo. No cavity is associated with the primitive streak.) The cells lose their epithelial character and become mesenchymal as they enter the primitive streak and this appears to be largely due to a change in cell surface carbohydrates (Sanders, 1991). *Plate 7c,d* shows sections through the primitive streak viewed by scanning electron microscopy, and *Plate 7e* shows a section just to the left of the primitive streak. Subsequently most of the cells become arranged in epithelia again as they become transformed into specific tissues. These tissues are: (1) the definitive **endoderm**, which will form the lining of the gut and the respiratory system; (2) the head process and **notochord**; (3) the paraxial **mesoderm** (somites), the intermediate mesoderm (which forms the urino-genital system) and the lateral plate mesoderm (which lines the coelom and contributes to the extra-embryonic membranes. Thus, there is a sequence of morphological changes in most cells as they pass through the primitive streak from epithelial to mesenchymal and back to epithelial. A further derivative of the ectoderm is the neural crest (see p. 35 *et seq.*).

One of the immediate consequences of gastrulation is to convert the two-layered embryo (consisting of the upper (epiblast) and lower layer) into a three-layered one, composed of ectoderm, mesoderm and endoderm. This is the result of cell migration and there is no evidence to suggest that it is due to any high level of cell proliferation in the primitive streak or in the cells that migrate from it (Sanders *et al.*, 1993).

There is some disagreement as to whether the paths taken after ingression are controlled by the cells themselves or by their environment. According to Garcia-Martinez and Schoenwolf (1992), cells migrating out from pieces of anterior primitive streak which have been grafted into a posterior position follow a more lateral pathway than if they had been left *in situ*, whereas cells from grafts that had been implanted into a more anterior position follow a more medial pathway. They concluded, therefore, that the extent of migration depended on the position of the cells within the primitive streak. By contrast, Psychoyos and Stern (1996) found that the migrating cells followed pathways that depended more on the tissue that they would form rather than on their position within the streak; they plotted the migration routes for the major tissues.

Much of the recent work on gastrulation has centred on the mechanisms by which the orderly sequences of cell migration are controlled. With the expansion of our knowledge of the genes that are active during this period there has been an extensive reworking of many of the older experiments in an attempt to relate molecular changes to distribution and differentiation of individual tissues. For example, the gene *chordin* is associated only with the notochord and may therefore be used as a marker for it. Some genes are expressed widely throughout the early blastoderm but then 'disappear' or become restricted to a specific region; (e.g. *cGata2* is strongly expressed in the epiblast at stages X–XIII but disappears abruptly at stage 3+ and a new domain subsequently appears at stage 4 in the prospective non-neural ectoderm (Sheng and Stern, 1999). Others appear faintly at first in the region that will become a specific tissue and then slowly increase in intensity as the tissue differentiates. In such cases it may be difficult to determine whether the gene is responsible for the differentiation of the tissue or is merely expressed as the tissue differentiates, and it is important under these circumstances to try to remove the gene product to see if the tissue is able to differentiate without it.

Now that the whole chick genome has been sequenced, many genes are being described and it is not always clear what role they play in development. There can be little doubt that few of them act alone. Most, if not all, probably have antagonists that suppress them and it seems likely that many need to act synergistically with other genes to bring about differentiation.

There is much evidence that changes in **cell surface receptors** and their relationships to the extracellular materials play an essential role in cell migration. The most important glycosaminoglycan (GAG) in the extracellular material is hyaluronic acid, accounting for 80–90% of the total GAG (Sanders, 1989). Collagen type IV is present in the basement membrane of the epiblast, but the interstitial collagens (II, III and IV) do not appear to play a role at this time. It is likely that after leaving the primitive streak at least some of the migrating cells use the basal lamina beneath the ectoderm as a guide. Other cells, however, appear to crawl over one another and it is not surprising therefore that the extracellular matrix glycoprotein, fibronectin, plays an important role. The peptide of fibronectin that is most concerned with cell adhesion is the RGD sequence, which binds to receptors on the cell surface, and will do this even if the remainder of the fibronectin molecule is absent, but not if the sequence of the amino acids is changed (e.g. to GRD). When GRD or antibodies to the fibronectin receptor were applied to the gastrulating embryo, the adhesive qualities of the fibronectin were seriously decreased or eliminated and the migration of the mesoderm cells was greatly reduced (Brown and Sanders, 1991).

Growth factors are also involved in the transition from epithelial to mesenchymal morphologies. TGFβ1 becomes increasingly expressed in the mesoderm cells as they migrate laterally away from the primitive streak and appears to reinforce the factors responsible for the conversion from epithelium to mesenchyme (Sanders and Prasad, 1991).

Expansion of the Blastoderm

At the time of laying, the chick blastoderm measures about 5–8 mm in diameter and has embarked on the process of **epiboly**, the spreading of an epithelial sheet in which the cells become flattened and increase their surface area. In birds, the blastoderm eventually comes to cover almost the entire yolk and most of the active migration is by cells at the edge of the blastoderm.

In the unincubated egg the blastoderm lies freely on the surface of the yolk, but after about 12 h of incubation it has become attached to the inner surface of the vitelline membrane. Under normal circumstances, however, the attachment is limited to the periphery of the blastoderm, and a special relationship appears to exist between the edge cells (also known as the margin of overgrowth) and the inner surface of the vitelline membrane (*Plate 2*). New (1959) showed that the young blastoderm is unable to expand if its edge is detached from the vitelline membrane, although if it is allowed to reattach, migration continues. Some migration can take place if the cells are explanted onto other substrates, such as glass, but expansion is slower. The edge cells are largely responsible for exercising a degree of tension throughout the blastoderm and if this is not maintained, abnormal or stunted embryos develop (Bellairs *et al.*, 1967). Kucera and Burnand (1987b) have calculated that the tension is of $1–1.2 \times 10^4$ dyn/cm^2 and is generated by radial contraction of the cells of the area opaca.

The edge cells are a specialized extension of the area opaca ectoderm and are therefore epithelial. The extreme tips of these cells consist of filopodia and, when studied by time-lapse filming, they can be seen to undergo ruffled membrane activity as they move across the inner surface of the vitelline membrane (Bellairs *et al.*, 1969). They are rich in microtubules and if treated with Nocadazole, a microtubule inhibitor, these become disassembled and migration ceases (Mareel *et al.*, 1984). Microfilaments also play a role since, if these are disrupted by treatment with cytochalasin B, migration ceases (Chernoff and Overton, 1979). Fibronectin too is important in the relationship between the edge cells and the vitelline membrane (Lash *et al.*, 1990). Although the diameter of the blastoderm increases as it expands over the upper half of the yolk, the edge cells do not undergo cell division (Monnet-Tschudi and Kucera, 1988), but there is a band of rapidly mitosing cells lying proximally to them (Flamme, 1987) which may contribute to the population of edge cells.

Chapter 4
Establishment of the Embryonic Body

When the primitive streak is fully formed (stage 4) the first stages of gastrulation are complete and the tissues that will develop into the body are arranged around its anterior end. During the next 12 h they will be transformed into a basic embryonic axis consisting of ectoderm, including neural plate and neural crest, notochord, somites, lateral plate and intermediate mesoderm, and endoderm. These changes are illustrated in:

- Normal Tables in Appendix II
- selected stages of whole-mounted embryos (*Plates 8, 9*).
- selected stages of whole-mounted embryos seen by scanning electron microscopy (*Plates 10–12*).
- histological sections (*Plates 53–56*).

Apart from those destined to form the notochord and the base of the neural plate, none of these cells is yet 'committed' ('determined') and it is possible to alter its fate experimentally. For example, a patch of presumptive epidermis may still be induced to form neural plate if a graft of primitive streak is inserted beneath it at this stage.

The key structure in the primitive streak stages is Hensen's node (alternative names are primitive node or primitive knot and, more recently, chordoneural hinge by Charrier *et al.*, 1999), the anterior tip of the primitive streak, where the primitive groove ends in the primitive pit and is enclosed anteriorly and laterally by the swollen primitive folds (*Plate 11c*). It corresponds in many ways to the dorsal lip of the blastopore in amphibians. One of the most famous and influential experiments ever performed on an embryo was that of Spemann (1938) who grafted a piece of tissue from the dorsal lip of an amphibian blastopore into the flank of a similar embryo of the same stage of development, which resulted in a secondary axis. It became apparent that the graft had influenced the host tissues adjacent to it to form an embryo, and Spemann concluded that the dorsal lip of the blastopore possessed special properties which enabled it to 'organize' the other tissues around it to form an embryonic axis. The dorsal lip of the gastrula therefore became known as the 'organizer'. Subsequently, it was shown by Waddington (1932 and later, summarized by Waddington, 1952) that the same sort of activity occurred in birds. He demonstrated that if Hensen's node was extirpated, stripped free of any adherent endoderm and implanted beneath the ectoderm at the edge of the area pellucida of another embryo, it was able to induce a secondary axis consisting of neural tube, gut and somites. It soon became apparent, moreover, that the ability to induce a secondary axis was not restricted to Hensen's node but was present also in at least the anterior third of the primitive streak (Waddington and Schmidt, 1933), although the power to bring about an induction of neural tube was found to be greater in Hensen's node than in any other part of the primitive streak (see below for neural induction).

A major characteristic of Hensen's node is that it expresses the chick homeobox-containing gene, *goosecoid* (*gsc*) (Izpisúa-Belmonte *et al.*, 1993). This gene was initially derived from a *Xenopus* blastopore dorsal lip cDNA library and is, therefore, considered as a marker for the organizer region. Its presence in Hensen's node emphasizes the analogous organizer roles of the amphibian dorsal lip and chick Hensen's node. An important finding of Izpisúa-Belmonte *et al.* was that *goosecoid* is first expressed in a restricted region of middle layer cells in Koller's sickle, implying that these cells are the precursors of Hensen's node.

By the time the primitive streak is fully formed the main presumptive areas lie clustered around its anterior end and it now expresses a range of markers. For example, in addition to *goosecoid*, it expresses HNF3β, *Otx2*, ADMP, FGF8 and *chordin*, though some, e.g. *goosecoid*, *Otx2*, ADMP and FGF8, are no longer expressed after stage 4. At later stages Hensen's node expresses other genes, such as *sonic hedgehog*, *noggin* and *nodal*.

Detailed fate maps for stages 3–4 have been published by Selleck and Stern (1991) and

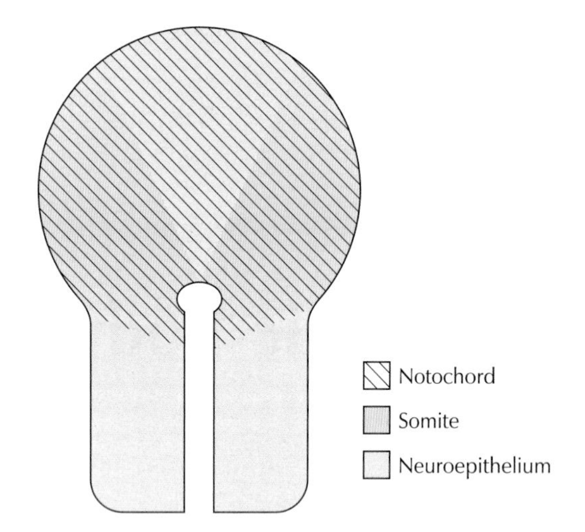

Text-Figure 13. *Fate map of the tissues in Hensen's node at the fully grown primitive streak stage (stage 4). (After Selleck and Stern (1991), with permission of the Company of Biologists Ltd.)*

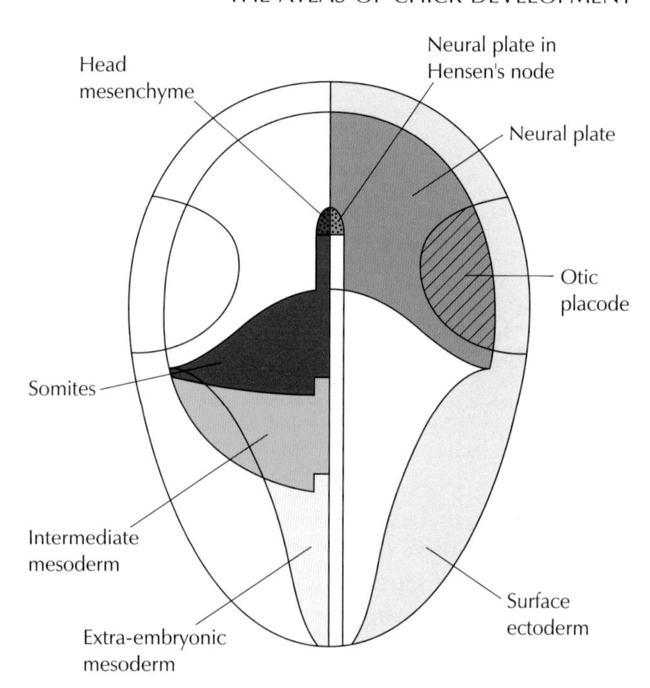

Text-Figure 14. *Fate map of the ectoderm and primitive streak at about stage 4 to show the layout of the presumptive areas. (After Garcia-Martinez et al., 1993, reprinted by permission of Wiley-Liss, Inc., a subsidiary of John Wiley and Sons, Inc.)*

Garcia-Martinez *et al.* (1993). Selleck and Stern's (1991) fate maps of Hensen's node were based on DiI labelling of small groups of cells and lysine-rhodamine-dextran labelling of single cells. They concluded that the node consists of several clearly defined regions (*Text-Figure 13*), each possessing its own presumptive fate. By stage 4 all the definitive endoderm has ingressed and the node contains only the mesoderm and that part of the presumptive neural plate that will form the floor plate. By stage 5 the presumptive notochord region has become narrower in the midline, since many cells have ingressed to form the head process and the notochord.

The lateral folds of Hensen's node give rise to the medial halves of the somites, whilst a separate region of primitive folds posterior to the node forms the lateral half of each somite. The fate map of Selleck and Stern was limited to the primitive streak, but Garcia-Martinez *et al.* (1993) have published a map of the entire area pellucida ectoderm at stage 4, based on the results of chick–quail transplantation experiments (*Text-Figure 14*).

Other maps include:

- the location of the presumptive heart rudiments in the primitive streak at stages 3–4 (Garcia-Martinez and Schoenwolf, 1993)
- the early neural plate (Bortier and Vakaet, 1992)
- the primitive streak (Psychoyos and Stern, 1996) (*Text-Figure 15*) and the mesoderm (Sawada and Aoyama, 1999 (*Text-Figure 16*); Lopez-Sanchez *et al.*, (2001)
- a spinal cord fate map at regression stages (Catala *et al.*, 1996)
- tail bud (Le Douarin, 2001) (*Text-Figure 17*).

Until about stage 3+ most of the mesoderm cells are as yet uncommitted to their individual fates, but the presumptive notochord cells, situated in Hensen's node and the anterior end of the primitive streak, appear to be totally committed to form notochord. The evidence comes from a wealth of experiments in which tissues are transplanted to other regions of the embryo (e.g. Selleck and Stern, 1992; Garcia-Martinez and Schoenwolf, 1993). The cells in most of the grafts were able to ignore their presumptive fate and develop according to their new position, but the presumptive notochord cells appeared to be able to form only notochord.

HEAD PROCESS AND REGRESSION

In the fully formed primitive streak the presumptive areas are grouped around its cranial end (see above). As regression occurs, they become rearranged to lie along the future body axis. The presumptive notochord cells in Hensen's node and the anterior primitive streak give rise to the **notochord** proper, a rod of cells running along the length of the embryo. The most anterior part of this rod is the **head process** (*Plate 10b*); it is formed by the anteriorward migration of cells from Hensen's node, whereas the trunk notochord is formed by the caudalward movement of cells during the process of regression. The first appearance of the head process is as a small dark triangle anterior to Hensen's node at stage 4 (Normal Table, Appendix II), which marks the completion

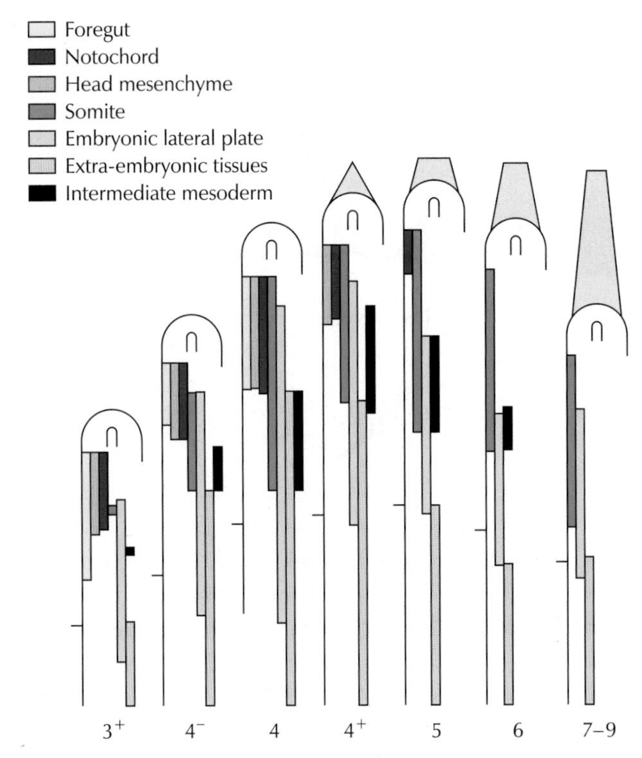

Text-Figure 15. *Fate maps of the primitive streak summarizing the relative antero-posterior positions of the precursors of the main tissue types along the axis. The prospective cell types are shown side-by-side for clarity, but this should not be taken to imply any defined medio-lateral position in the diagram. (After Psychoyos and Stern, 1996, which should be consulted for further details. With permission of the Company of Biologists Ltd.)*

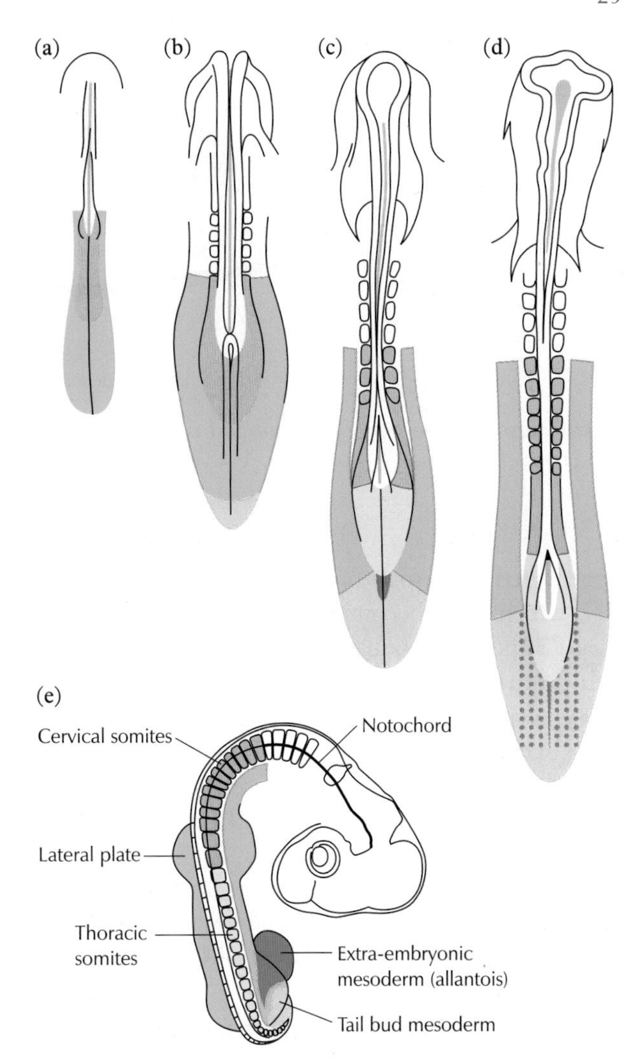

Text-Figure 16. *Fate maps of the mesodermal cells of the primitive streak between stages 6 and 20. (a) Stage 6; (b) Stage 8; (c) stage 9; (d) stage 10; (e) stage 20 (After Sawada and Aoyama, 1999, with permission of UBC Press.)*

of the formation of the primitive streak. The onset of regression initiates the start of the disappearance of the primitive streak. Differentiation of the tissues takes place gradually from anterior to posterior as regression occurs, so that the development of the anterior tissues is always in advance of the posterior ones.

The head process must be distinguished from the **prechordal mesoderm**, which lies anterior to it and does not form notochord. Like the head process it is derived from Hensen's node, but there is evidence to suggest that TGFβs from the underlying endoderm repress the notochord characteristics (Vesque *et al.*, 2000).

There has been much confusion over the term **regression**. Essentially, it describes the gradual disappearance of the primitive streak, which shortens from anterior to posterior as the presumptive mesodermal tissues ingress and leave it, but it is also used to describe a series of morphogenetic movements that take place at this time. These are shown in *Text-Figure 11*. The main feature is that most of the presumptive notochord cells in Hensen's node and the anterior primitive streak, together with the associated floor-plate cells, move toward the posterior end of the embryo and during this process become deposited as a trail of cells which differentiates mainly into the notochord (*Plates 15 and 16*). There is recent evidence to indicate that the floor plate of the neural tube and a longitudinal strip of endoderm are deposited at the same time (Le Douarin, 2001) (*Text-Figure 17*). The individual cells of the notochord become greatly elongated by the stretching and extension of the area pellucida, which occurs at this time. It is, however, not only the presumptive notochord and floor plate tissues that are affected by the regression movements. Other presumptive tissues that are still within the primitive streak (*Text-Figure 15*) also become extended. The morphogenetic movements involved in node regression also take place to a lesser extent in the regions lateral to the primitive streak, so that all the presumptive areas become elongated along the body axis. By about stage 11 the remnants of Hensen's node and the primitive streak lie at the posterior end of the area pellucida and have become

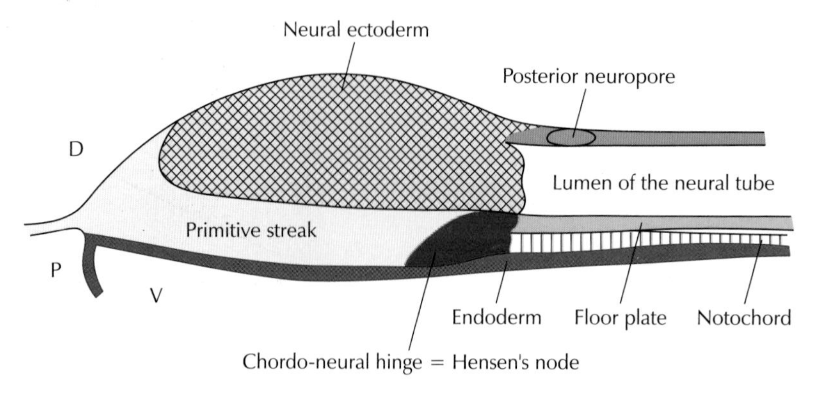

Text-Figure 17. *Fate map of the tail bud of the quail at a stage of about 26 somites (corresponding to about stages 15–16 of the chick). D= dorsal; P= posterior; V= ventral. (From Le Douarin, 2001, with permission of UBC Press.)*

aggregated into a mass of cells, the tail bud (*Plates 9, 12, 72*), which then becomes undercut by the tail fold (see Chapter 5). About this time there is a extensive cell death in the tail bud. According to Le Douarin (2001) this is due to the absence of adjacent structures secreting *sonic hedgehog* protein, and the tissues can be rescued if *sonic-hedgehog*-secreting tissues are implanted. A fate map of the tail bud is illustrated in *Text-Figure 17.*

NEURAL INDUCTION

The ability of the ectoderm to form neural plate extends beyond the presumptive neural plate region. There is ample evidence that if a small piece of ectoderm, together with its underlying mesoderm, is taken from almost anywhere in the anterior half of the area pellucida at stage 3 and isolated (even on the chorioallantoic membrane: see p. 117), it can give rise to a neural tube.

Most fate maps for stages 3–4, whether produced by vital staining (e.g. Pasteels, 1937), by the use of grafts labelled with tritiated thymidine (Rosenquist, 1966), or by the use of chick–quail transplant chimeras (e.g. Garcia-Martinez *et al.*, 1993), show that the region which normally gives rise to the neural tissue (neural plate) is an extensive area of epiblast lying over Hensen's node and extending anteriorly, laterally and posteriorly. According to Garcia-Martinez *et al.* (1993) it lies about 0.5 mm behind the node, although other investigators (e.g. Bortier and Vakaet, 1992) have found that the presumptive neural plate does not extend further posteriorly than Hensen's node.

For many years after the organizer region was discovered in amphibians, attempts were made to identify the supposed chemical nature of a signalling molecule, but it was not until about 70 years later that molecular biological techniques made this possible. Current opinion (reviewed by Streit and Stern, 1999) is that in *Xenopus* the ectoderm in general is inhibited from forming a neural plate by cell-to-cell signalling of the bone morphogenetic proteins, BMP2, 4 and 7, but that in the midline the organizer secretes antagonists to the BMPs so that the ectoderm is able to develop into neural plate. The antagonists include *chordin*, *noggin*, *follistatin*, *Xnr3*, *cerebus* and *gremlin* and are thought to initiate neural plate of an anterior (rostral) character. The organizer is then thought to convert some at least of these cells into more posterior (caudal) neural plate by emitting further signals, such as retinoids. Streit and Stern (1999), however, pointed out that in the chick the process is more complex. The region where BMP4 and two of its antagonists (*chordin* and *noggin*) are active appears to be the boundary between neural and non-neural ectoderm. They suggest that the main functions of *noggin* at this stage may be in somite formation, by protecting the prospective somite cells from the inhibitory actions of BMPs, whilst *chordin* is primarily involved in axis formation.

Two markers, *ERNI* (early response gene for neural induction) and *Sox3* (a similar early response gene) are induced by Hensen's node (as well as other neural inducters such as head process mesoderm), and each may, therefore, be used as an early marker for neuralizing activity. Streit and Stern found that when cells from the middle layer of Koller's sickle were grafted into the area opaca of a primitive streak stage embryo, which is a region that would not normally form neural tissue, both *ERNI* and *Sox3* were induced. Markers for later development of the neural tissue were, however, not induced unless epiblast from the centre of the area pellucida was also included. Streit and Stern, therefore, suggested that neural induction is initiated before the start of gastrulation by signals (FGF signals) from a small population of organizer precursor cells in Koller's sickle and that when these cells spread anteriorly and make contact with a second precursor population centre (in the middle of the area pellucida epiblast) the two groups of cells interact with one another to produce

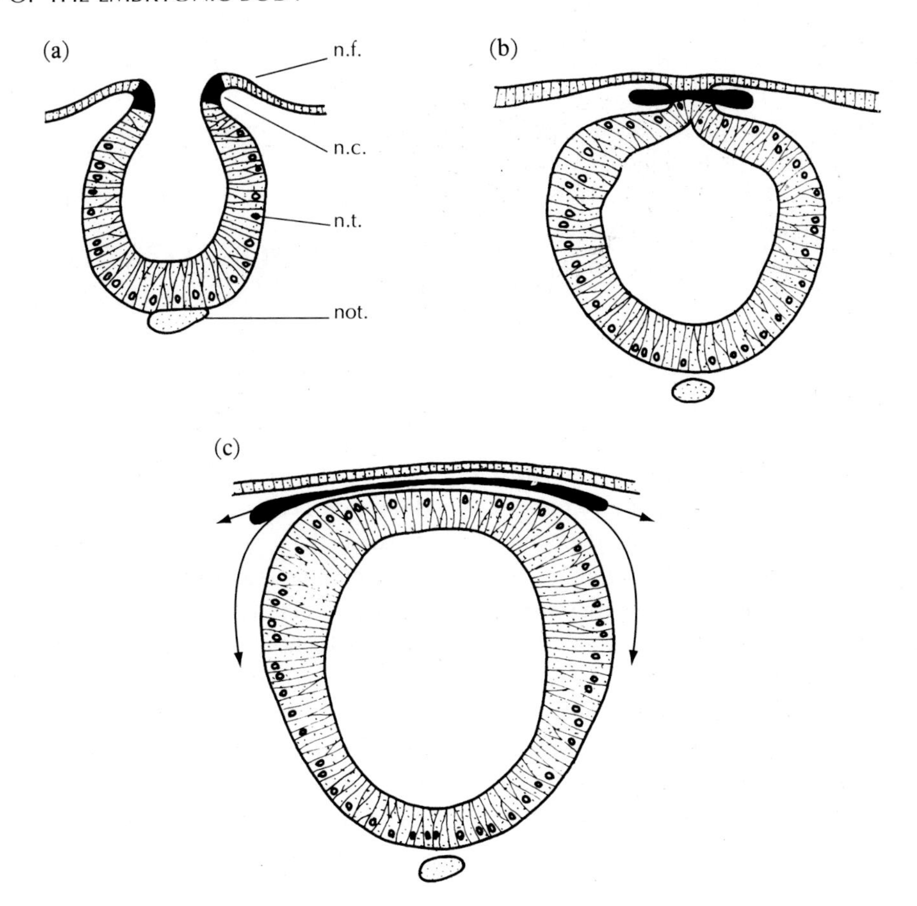

Text-Figure 18. *Section through the neural tube at three consecutive stages to illustrate the origin of the neural crest. Arrows show the paths taken by the migrating neural crest cells. n.c. = neural crest; n. f. = neural fold; n.t. = neural tube; not = notochord.*

a functional organizer; this then leads to a cascade of reactions including the sensitization of the epiblast to the BMP antagonists. Interestingly, transplantation experiments indicate that the inducing ability is restricted to the medial region of Hensen's node, and is not present in the deep portion of the posterior-lateral part (Storey *et al.*, 1995).

By the time that the neural tube has formed it already possesses a dorso-ventral polarity (see below).

FORMATION OF THE NEURAL TUBE

Initially there is no sharp morphological border between the cells of the neural plate and the adjacent ectoderm, the height of the epithelium changing gradually in the transitional zone (*Plate 54*), but if this transitional zone is removed, the neural plate is unable to form neural folds (*Plates 10, 13*) and remains flat (Moury and Schoenwolf, 1995). The traditional view is that the neural plate is induced by Hensen's node, even though some neural plate formation can take place in its absence. The inducing ability of the various subpopulations of the node was investigated by Storey *et al.* (1995). After grafting carefully selected small populations of cells into an extra-embryonic site beneath ectoderm that would not normally form neural plate, they concluded that the neural inducing ability was associated not only with specific regions of Hensen's node but also with certain types of cells.

Initially, the neural plate is a pear-shaped area that subsequently becomes folded into a neural tube (*Text-Figure 18*) and then differentiates along its entire length into specialized regions of brain and spinal cord. Transplant experiments have shown that the age of the node determines the regions of the central nervous system that are induced by it. Young nodes (stages 2–4) can induce both anterior and posterior central nervous system, but older nodes can induce only posterior central nervous system (Storey *et al.*, 1992).

After the neural plate has formed, the morphogenetic movements of regression lead to a narrowing and posterior extension of the trunk region and a widening of the brain region (*Plate 12*). These regression movements are accompanied by dramatic changes in the shapes and arrangements of the cells, the epidermal cells becoming flatter and the neural cells deeper. The notochord is attached to the floor plate (the floor of the neural plate) (*Plate 71*) and acts as a hinge about which the neural plate folds and bends to form a tube (Schoenwolf and Smith, 1990). The floor plate expresses HNF3β and is probably induced by notochord, since grafts of notochord can induce an ectopic floor plate which expresses HNF3β (Ruiz I Altaba *et al.*, 1995). According to Catala *et al.* (1995), notochord and floor plate express so many genes in common that they should be considered as the same tissue.

Theories about the possible mechanisms involved in the process of neural folding are discussed by Moury and Schoenwolf (1995) who emphasize that neurulation, especially neural plate formation, is a multifactorial process resulting from forces both intrinsic and extrinsic to the neural plate. Changes in the tenascin component of the cells are associated with changes in cell shape, whereas changes in the cell membranes are involved in the increasing adhesion of the cells to one another.

Fusion of the neural folds begins at about stage 8 at the level of the midbrain and is rapidly followed by fusion throughout the entire brain and the anterior regions of the trunk. The final point of fusion in the brain is the anterior neuropore (*Plate 14*), which is obliterated at about stage 12. The neural tube closes progressively down the trunk until by stage 13 the process is almost complete and an opening remains only at the posterior end as an aperture, the posterior neuropore (*Plate 9*). As the neural folds fuse, the cells of the neural crest (see below) come to lie at its dorsal side. The ectoderm that is not enclosed within the neural tube subsequently becomes the epidermis.

The dorsoventral patterning of the neural tube is brought about by the interaction of at least two major gene products. There is now extensive evidence that the ventralization is brought about by *sonic hedgehog* protein from the notochord, whilst the dorsalizing influence results from BMPs probably secreted by the ectoderm (Patten and Placzek, 2002).

The way that the neural tube forms in the tail bud (see p. 30) is designated **secondary neurulation** to distinguish it from the rolling up of the neural plate described above, which is sometimes called **primary neurulation**. In secondary neurulation, mesenchyme cells on the dorsal side of the tail bud become condensed into an epithelial-like rod, which then separates off from the remaining tissues and becomes hollowed out. A region of overlap occurs where the so-called **primary neural tube** of the trunk gives way to the secondary neural tube of the tail (*Plate 72*).

Transverse sections through the early neural tube of the trunk region (*Plate 15*) show that it possesses thick lateral walls (*Plates 15, 61*); these will give rise to both the neurones and glia. The dorsal and ventral regions of the tube, the roof plate and floor plate respectively, are narrower with wedge-shaped cells, and form neuroglia only.

REGIONALIZATION OF THE NEURAL TISSUE

We have seen (p. 28) that in addition to the cells forming the notochord and head process, several other types of cell originate in Hensen's node (Selleck and Stern, 1991). These are the **definitive endoderm**, which gives rise to the gut, the **mesendoderm** cells, which pass anteriorly to lie beneath the anteriormost end of the node and contribute to the gut and prechordal plate mesoderm, and the **lateral cells**, which contribute to the somites. Initially, it is difficult to identify the types of cells as they leave the node since, morphologically, they are indistinguishable from one another. For this reason they are often collectively, but confusingly, called mesendoderm cells. The ability of Hensen's node to induce neural tissue is lost when the mesendoderm cells have left it (Storey *et al.*, 1992).

Markers for the forebrain are GANF, CNOT1 or Nkx2.1, which are normally expressed in the neural plate as it forms (Boettger *et al.*, 2001) though they can be induced experimentally by grafts of the mesendoderm. Grafts taken from the region of the primitive streak just posterior to Hensen's node are still able to induce neural plate, though not forebrain. It appears therefore that regionalization is not established at the time of neural induction but that the forebrain is induced at the anterior end of the neural plate by the underlying mesendoderm.

Foley *et al.* (2000) have put forward a model to explain forebrain patterning. They found that the migrating hypoblast induced the expression of early forebrain markers (*Sox3* and *Otx2*) but was unable to induce forebrain. They suggested that a role of the hypoblast was to direct the cell movements in the overlying epiblast so that the prospective forebrain was removed from the caudalizing effects of the node.

There are probably a large number of other genes involved in regionalization of the neural tube. For example, the homeobox gene *cDix* is restricted to the presumptive ventral forebrain region of the neural plate that ultimately gives rise to the hypothalamus and adenohypophysis (Borghjid and Siddiqui, 2000). By the time that the neural tube has formed it already possesses a dorso-ventral polarity. *Sonic hedgehog* secreted by the notochord and floor plate is important in specifying ventral telencephalic cells (Gunhaga *et al.*, 2000). The response of the neural cells to *sonic hedgehog* can be altered *in vitro* by BMP signals causing cells that would have become ventral to change into dorsal type cells (Liem *et al.*, 2000).

FORMATION OF THE NOTOCHORD AND SOMITES

The notochord and somites (*Plates 15, 59*) are structures that are found only in the early embryo and are not present at later stages.

The **notochord** (*Plate 16*) plays a critical role in development. As we have seen above, it is derived from Hensen's node and is formed as regression takes place. It occupies a central position running down the length of the body and during its early stages is closely adherent to both the floor plate of the neural tube and to the underlying endoderm. The notochord plays an important role in the dorso-ventral patterning of the neural tube. We have seen above that the notochord expresses *sonic hedgehog* protein, which has a ventralizing effect on the neural tube, but it also has a ventralizing effect on the somites. Subsequently, the neural tube and notochord become encased in the vertebrae (see Chapter 10). Remnants of the notochord survive as the nucleus pulposus in the intervertebral discs.

The *somites* (*Plates 8, 9, 62, 68, 71*) are paired blocks of mesoderm that form in a row on either side of the notochord. They are the first segmented structures to develop in the embryo, and their layout influences that of all the other segmental structures that form subsequently; these are the vertebrae and ribs, the cranial and spinal nerves, the vertebral arteries and the skeletal muscles and ligaments.

The mesoderm that will give rise to the somites is located around the anterior end of the primitive streak at about stages 3–4 (*Text-Figures 13–15*), including the lateral edges of Hensen's node, and probably starts to ingress soon afterwards. Cells pass into and through the primitive streak and then migrate laterally, passing between the ectoderm and the endoderm, usually keeping to the same side of the embryo, although occasionally individual cells may leave the primitive streak on one side and then move to the contra-lateral side. Some of the ingressing cells become lateral plate mesoderm, others intermediate mesoderm and the remainder (nearest the neural plate) give rise to the segmental plate (sometimes called pre-segmental plate) mesoderm (*Plates 8, 17, 62*) from which the somites will form. The fate of any cell is determined soon after leaving the primitive streak and appears to be brought about by the relative concentration of several signals. There is a high concentration of TGFβ-like molecules, especially BMP4, in the lateral plate. (This region corresponds to the ventral region of amphibian embryos and so the effect of BMP4 is said to be a ventralizing one.) The node, meanwhile, produces *noggin* protein, which is an antagonist of BMP4 and so plays a role in preventing the pre-somitic mesoderm from developing into lateral plate (Streit and Stern, 1999). The interplay of these and other signals is discussed by Pourquié (2001).

As the cells leave the primitive streak two major morphogenetic movements are occurring in the mesoderm. These are the medio-lateral migration of the ingressing cells and the antero-posterior movements associated with regression, and it seems likely that these two streams of migrating cells interact with one another (discussed by Ooi *et al.*, 1986).

The first pair of somites appears at about stage 7 and succeeding pairs are laid down sequentially further and further posteriorly (*Plate 8*). The cells that will become somitic mesoderm enter the segmental plate as they leave the primitive streak. The left and right segmental plates lie on either side of the neural tube (*Plate 72*) and each consists of a strip of condensed mesoderm. They are easy to dissect from other tissues, especially if the embryo is treated for about 15 min with 1% trypsin in calcium- and magnesium-free saline. The cells at the posterior end of the segmental plate are mesenchymal, whereas those at its anterior end are arranged as an epithelial ball. Groups of cells at the anterior end of the two segmental plates become separated off simultaneously to form the left and the right of a pair of somites. Each newly formed somite is a ball of columnar epithelium, the walls consisting of a single layer of cells (*Plates 18, 62*) surrounding a cavity, the **somatocoele**, which contains mesenchyme cells.

A new pair of somites forms from the anterior end of each segmental plate about every 90–100 min, but meanwhile a steady stream of mesenchyme cells is being added to the posterior end of the segmental plate. As a result the segmental plate remains visible, although there is a continual turnover of the individual cells which constitute its population. Between stages 10 and 18 it possesses the equivalent of about 10 pairs of somites (Packard and Meier, 1983). Once the cells have entered the segmental plate they are committed to becoming part of a somite. Differentiation into somites takes place gradually along the segmental plate as the original mesenchyme cells become arranged into an epithelium and, in appropriate preparations, it is possible to see part-formed somites (named **somitomeres** by Meier, 1979).

Although best seen by scanning electron microscopy, the somitomeres are also visible in segmental plates that have been dissected from the embryo and viewed by light microscopy.

Traditionally, the somites are numbered sequentially from the anterior to the posterior end of the embryo, which corresponds to the sequence in which they form. An alternative scheme for the numbering of somitomeres and somites was devised by Ordahl (1993), the aim being to provide a standard description of the developmental state of each somitomere and somite when comparing different stages. For example, the most recently formed somite of a 13-somite embryo (stage 11) and that of a 26-somite embryo may be considered comparable. Although there are certain advantages to this scheme, they are probably overshadowed by the possibilities for confusion generated, especially as it involves the numbering of the somites from posterior to anterior. In general, therefore, it is probably best to use this scheme only under very specialized circumstances.

Two genes that have been implicated in initiating epithelialization of the somites are *paraxis* (Sosic *et al.*, 1997) and *Epha-4* (Schmidt *et al.*, 2001), the latter gene, at least, being probably induced by signals from the adjacent ectoderm.

As the cells gradually become arranged into epithelia, their cell to cell adhesiveness increases, the most anterior somite compacting immediately before segmenting off from the segmental plate. This increased adhesiveness appears to be brought about, at least in part, by the action of the glycoprotein, fibronectin (Lash *et al.*, 1984), which increases in concentration as the somitomeres mature along the segmental plate. This can also be brought about experimentally by the synthetic peptide, GRGDS, which corresponds to the adhesive segment of fibronectin (Lash *et al.*, 1987). Each newly segmented somite contains about 2500 cells (Christ and Ordahl, 1995).

One of the greatest problems in embryology has been that of explaining the underlying causes of segmentation. What causes a strip of tissue, the segmental plate, to become broken up into a repeating series of somites? Another way of looking at the same problem is to ask, 'What causes the boundaries between somites to be established?' Although a number of alternative theories have been proposed in the past, the concept that oscillating genes control the process is now widely accepted. This idea was first put forward in a theoretical paper by Cooke and Zeeman (1976) and is now supported by modern molecular evidence. Members of the *Notch* signalling cascade, especially *lunatic fringe*, appear to be essential components in setting up the oscillatory clock (Pourquié, 2001). For recent reviews of other molecules involved in segmentation see Stockdale *et al.* (2000) and Pourquié (2001).

Individual somites appear to be rectangular (*Plate 17*) or globular (*Plate 18*) in shape when seen from the dorsal or the ventral side, but transverse sections show them to be wedge-shaped, the medial wall being curved to the shape of the neural tube. Their shape is largely dependent on the tension exerted by the collagen fibrils in the extracellular matrix that surrounds them and attaches them to the adjacent epithelia, the ectoderm, endoderm, notochord and neural tube (Veini and Bellairs, 1990; Chernoff *et al.*, 2001). The earliest somites are formed before the neural tube has rolled up completely and have a flattened shape (*Text-Figure 33a*). As the neural tube closes the somites apparently become drawn upwards and acquire the more familiar shape seen in transverse sections. Longitudinal sections show each somite as an epithelial ball of cells with walls one cell thick and centred around a lumen (*Plate 18a,b*). Mesenchymal cells are present in the lumen.

Each newly formed somite consists of four main regions, the medial, lateral, anterior and posterior (*Text-Figure 59*). Experiments in which regions of the chick segmental plate have been replaced with the corresponding tissue from quail embryos have shown that these regions have differing fates.

One of the first visible changes in the mature somite is that the basement membrane in the medio-ventral region breaks down and the cells lose their epithelial character and become mesenchymal (*Plates 19b, 71*). They are the **sclerotome** cells. The breakdown of the ventral wall of the somite releases the cells within the somitocoele and these also become part of the sclerotome.

A further change follows as the epithelium at the dorso-lateral side increases in height and is then known as the **derma-myotome** (often spelled dermo-myotome). In longitudinal sections it can be seen as a cap with its edges curved inwards (*Plates 19, 71*); it gives rise to the dermatome and myotome. The sclerotome and the derma-myotome are formed almost exclusively from the cells of the medial half of each new somite, the derma-myotome from the dorsal part, the sclerotome from the ventral region. The cells of the lateral part of the somite migrate away to form the muscles of the limbs and probably of the ventral body wall (see Chapter 10). There is evidence that this medio-lateral patterning of the somites is the result of a cascade of molecular signals. The neural tube produces *Wnt1*, which promotes *noggin* expression in the medial part of each somite, and this antagonizes the BMP4 which is produced by the lateral plate mesoderm (Hirsinger *et al.*, 1998). The patterning of the somites as they segregate into different regions involves many more molecular changes (reviewed by Stockdale *et al.*, 2000). The dorso-ventral patterning appears to be brought about principally by *sonic hedgehog* and *Wnt* signalling, whilst antero-posterior polarity is due to *notch*.

The anterior and posterior regions of the somite are separated from one another by a distinct gap, **von Ebner's fissure** (*Text-Figure 59*). The spinal nerves are able to enter the anterior half of each somite but not the posterior half.

The first myotome cells form in the anterior-medial region of the somite from the edge cells of the derma-myotome. By the time they have extended into the posterior half of the somite they have already begun to produce muscle proteins and are, therefore, probably already determined, though experiments have shown that the sclerotome cells are not (Dockter and Ordahl, 2000). Once the myotome has formed it becomes separated from the dermatomal component of the somite by a basement membrane. Later in development, dermatome cells migrate dorsally to become the dermis.

THE LATERAL PLATE MESODERM AND THE INTERMEDIATE MESODERM

The **lateral plate mesoderm** arises from the posterior region of the area pellucida and ingresses through the primitive streak before the paraxial (somitic) mesoderm, spreading out to reach the edge of the area opaca and then continuing into this region as extra-embryonic mesoderm. Unlike the somitic mesoderm, the lateral plate is not segmented but by stage 4 is present as two sheets, the somatic layer, which lies beneath the ectoderm (together forming the bilaminar sheet, the somatopleure) and the splanchnic layer, which lies over the endoderm (together forming the bilaminar layer, the splanchnopleure) (*Text-Figure 24*; *Plate 72*). They are separated by a space, the future coelom (see Chapter 8). The development of the lateral plate appears to be linked in some ways to the somites. For example, *Epha-4* is expressed both in the lateral edge of the somites and in the lateral plate mesoderm and, as discussed above, the BMP4 produced by the lateral plate affects somite development. The lateral plate mesoderm subsequently forms the mesenteries, the lining of the pleural, cardiac and abdominal cavities, and the major substance of the heart as well as contributing to the extra-embryonic membranes.

The **intermediate mesoderm** lies between the somites and the lateral plate mesoderm, to both of which it is initially attached (*Plate 56*). It appears at the same time as the somites and develops simultaneously with them from anterior to posterior down the trunk, but it differs from the somites in that it is not overtly segmented and does not extend into the postcloacal tail. It later forms most of the urino-genital system (see Chapter 7).

THE NEURAL CREST

Like the notochord and somites the neural crest is found only in the early embryo and is not present at later stages. It is, as its name implies, a strip of cells situated along the dorsal side of the early neural tube, forming a 'crest' down its entire length (*Plate 20a,b*). Its former name, ectomesenchyme, is descriptive of its origin and fate rather than of its position, in that it is an ectodermal derivative that gives rise to a wide range of mesenchymal structures. There is an extensive literature on the neural crest; For a recent summary see Le Douarin and Kalcheim (1999). It originates from the ectoderm at the junction of the non-neural ectoderm and the presumptive neural ectoderm (Selleck and Bronner-Fraser, 1995) and there is evidence that the two tissues interact to induce it. Dickinson *et al.* (1995) found that neural crest could be induced from neural plate in culture either by ectoderm or by two proteins secreted by the ectoderm (BMP4 and 7). These two proteins induce the expression of *slug* protein and *Rhob* protein in the cells which become neural crest and in the absence of either one of them the neural crest cells fail to leave the neural tube. Even the extra-embryonic ectoderm can be induced to form neural crest (Ruffins and Bronner-Fraser, 2000). *Rhob* is involved in the production of cytoskeletal elements that are needed for migration (Hall, 1999). The non-neural ectoderm remains able to induce neural crest from the neural ectoderm until about stage 10, though the neural plate loses its ability to respond after about this time (Basch *et al.*, 2000). These authors found that the marker, *slug*, was present in the neural folds even by the end of gastrulation, and suggested, therefore, that neural crest induction involves a series of stages. The gene, *lunatic fringe*, which is expressed in the neural tube but not in the neural crest cells, also plays a role, since excess *lunatic fringe* results in excess proliferation of neural crest cells at least in the cranial region (Nellemann *et al.*, 2001).

The neural crest is first visible as a thickened region just before the neural tube closes (*Text-Figure 18*, *Plate 53*). Its formation and subsequent development follow the usual anterior to posterior sequence. As the neural tube closes, the neural crest becomes lodged between it and the overlying ectoderm

(*Text-Figure 18*), and as the three tissues separate from one another the neural crest cells settle down onto the dorsal surface of the neural tube (*Plates 20, 21*). Shortly before they leave the neural tube they flatten and reorientate so that the longest axis of each cell is arranged at right angles to that of the embryo and there is a reduction in the intercellular spaces (Bancroft and Bellairs, 1976). The emigration from the neural tube takes place in a wave from anterior to posterior down the body. An important aspect is that it follows on closely after the wave of somite segmentation and there is evidence that each pair of somites influences the release of the neural crest in its vicinity by affecting the BMP4 and *noggin* interrelationship. BMP4 appears to play a major role in releasing the neural crest cells from the neural tube. Sela-Donenfeld and Kalcheim (2000) found that in regions of the embryo where *noggin* expression is high in the neural tube, BMP4 is inactive and neural crest cells fail to emigrate, but progressive reduction in *noggin* activity coincides with activation of BMP4 and the emigration of neural crest cells. The authors suggested, as a result of a series of experimental ablations, that an inhibition of *noggin* was produced by the dorso-medial part of the epithelial somites. This would explain why neural crest delamination follows somite segmentation.

Once the neural crest cells have left the neural tube they follow well-defined pathways through the body until they reach their target tissues. The major cues that direct their migration are provided by the surrounding environment: e.g. the neural crest cells pass through the anterior half of each somite but never through the posterior half. This rule is followed even in experimental situations where the somites are rotated or inverted so that their relationship to the emigrating neural crest cells is altered. *F-spondin*, which is expressed in the posterior region only of the somites, appears to play a role in inhibiting the crest cells from entering this region

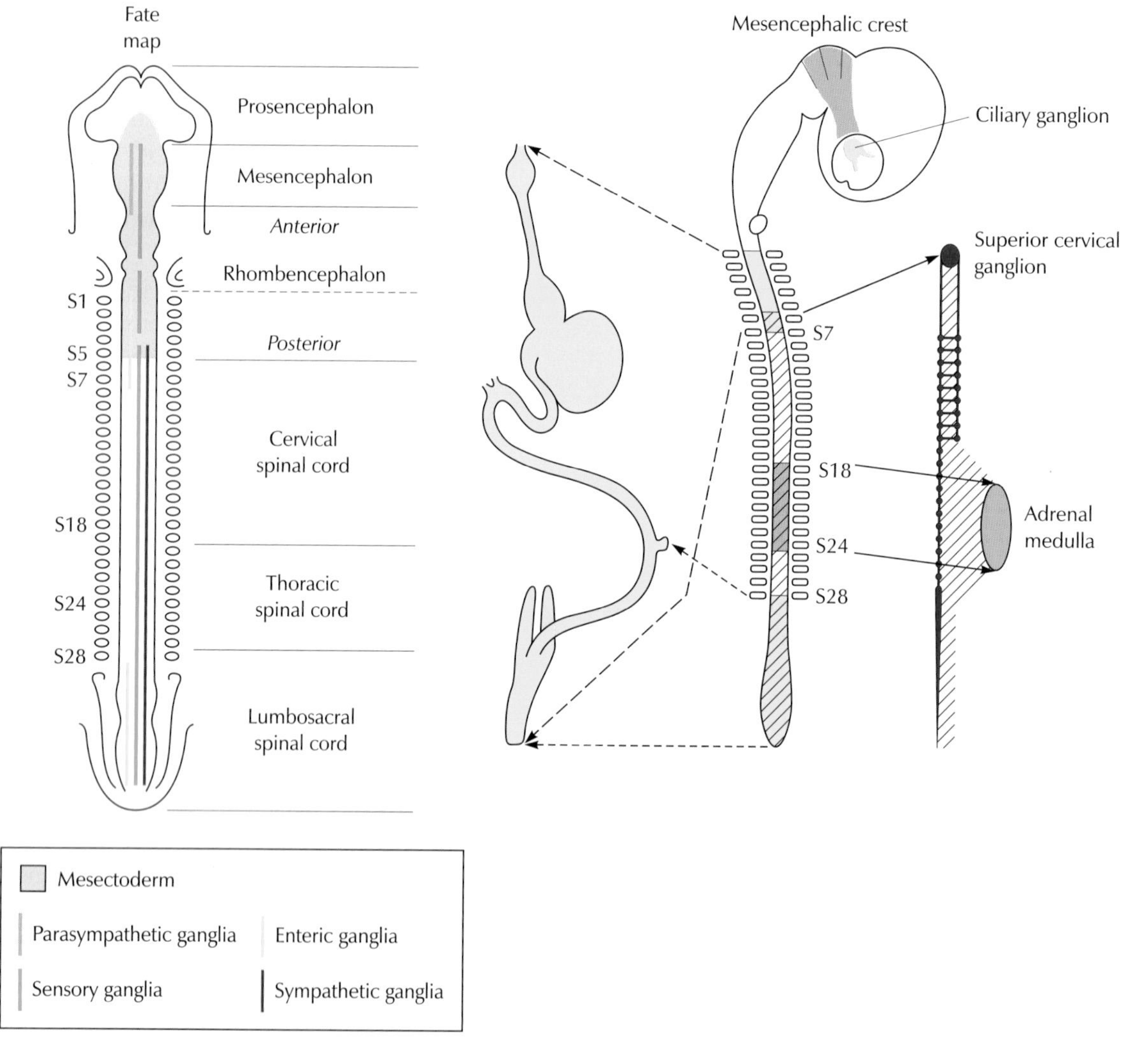

Text-Figure 19. *Regions of the neural crest and their contribution to the autonomic nervous system: the* ***cranial*** *neural crest gives rise to the cranial nerves and ganglia; the* ***vagal*** *neural crest (level somites 1-7) and the* ***sacral*** *neural crest (posterior to somite 28) form the nerves of the gut; the* ***cardiac*** *neural crest (level somites 1-3) contributes to the heart; the* ***trunk*** *neural crest cells give rise to the sympathetic neurons, those between somite level 18-24 forming the adrenal medulla. (After Le Douarin and Kalcheim, 1999, with permission of Cambridge University Press.)*

(Debby-Brafman *et al.*, 1999). Similarly, *ephrin* proteins inhibit the passage of crest cells through this region (Newgreen *et al.*, 1986). The neural tube also influences the direction in which the neural crest cells migrate, since if it is inverted in its dorso-ventral axis the crest cells that were destined to migrate down the side of the neural tube continue to do so, even though that leads them toward the ectoderm instead of the endoderm.

The neighbouring tissues exert much of their influence on the neural crest through the medium of the extracellular matrix which surrounds them, the most important components being fibronectin, laminin, tenascin, collagen and various proteoglycans. Integrins and other adhesion molecules enable the neural crest cells to interact with the extracellular matrix. It is likely that the differing concentrations in which these materials are present in different regions of the body play a role in directing migration. The environment plays a role also in the final differentiation of the neural crest cells. Most neural crest cells appear to be pluripotent at the time they leave the neural tube in that they can often be made to abandon their presumptive fate and differentiate according to a new environment if transplanted. The major exception is that the trunk neural crest, which does not form cartilage or bone in the trunk, appears to be unable to do so even when transplanted into the head.

Neural crest which arises in the head region (cephalic neural crest), and forms cartilage and bone of much of the head and face skeleton, also gives rise to connective tissue as well as Schwann cells and cranial sensory ganglia (see Chapter 9). Neural crest in the vagal (just posterior to the head) and sacral regions forms the entire enteric (gut) nervous system. The neural crest in the trunk migrates as two separate streams; the dorsal one passes laterally beneath the ectoderm until it reaches the mid-ventral body wall. Eventually, it migrates into the substance of the dermis and forms the pigment cells. The ventro-lateral stream passes through the anterior halves of the somites, some cells remaining there and differentiating into dorsal root ganglia, whilst others give rise to sympathetic ganglia, Schwann cells and adrenomedullary cells. The cardiac neural crest extends from the first to the third somites, between the cranial and the trunk neural crests (*Text-Figure 19*).

Some interesting parallels exist between the neural crest and the primitive streak (see Bellairs, 1987). In each case cells leave an epithelium and migrate to another part of the body; they move from the outer layer to the inside of the embryo and upon reaching their destination become arranged into new structures. The cells that leave the primitive streak inevitably differ from those of the neural crest in that they give rise to a different range of tissues. There is, however, an additional difference. The cells that arise from the primitive streak all give rise to epithelial tissues on reaching their destination (somites, lateral plate, pronephric duct, endoderm), but in birds the cells that arise from the neural crest seldom become rearranged as epithelia.

Chapter 5

External Appearance and Polarity

We have seen in previous chapters that in the early stages the chick embryo is a flat disc, the blastoderm, the centre of which is destined to form the embryonic body, whilst the periphery will give rise to the yolk sac and amnion. The first indication of the body proper is at about stage 6 (about 24 h of incubation) when the head fold develops (*Text-Figures 23, 39*; *Plates 10, 53*). Anterior to the tip of the head process a crescent-shaped groove appears in the entire thickness of the blastoderm. The result is that the region of the future head, together with the developing foregut, becomes raised above the flat blastoderm (*Plate 53*). Sections show that the head is covered with ectoderm and is separated at its tip from the blastoderm. The folds continue to expand posteriorly and at the same time the growing head extends further anteriorly, becoming more and more defined. The lateral body folds (*Text-Figure 24*, which start to form as longitudinal grooves of the ectoderm and somatic mesoderm on either side of the embryo, are continuous with the posterior extensions of the head fold. There is a high level of cell division in the ectoderm of the folds (Miller, 1982); by stage 15 the lateral folds extend to the anterior wing level (somites 15–17) and by stage 16 have reached the level of somites 17–20. Meanwhile, the tail fold has begun to form at the posterior end (*Text-Figure 40*). By stage 17 the head fold (*Text-Figure 39*), lateral body folds (*Text-Figure 24*) and the tail fold have met to form a continuous groove around the periphery of the body, lifting it up above the surface of the yolk sac. By stage 20 the body is no longer continuous with the extra-embryonic tissues except in the region of the umbilicus where the head, tail and lateral body folds converge.

Although the foldings are most conspicuous in the ectoderm and somatic mesoderm, they also include the endoderm and the splanchnic mesoderm, so that they lead to the early stages in the formation of the gut. The foregut (*Plate 53*) forms as the head fold develops and the hindgut is initiated by the tail fold (see Chapter 8).

Once the body starts to become raised above the blastoderm, changes in the external appearance begin (*Plate 9*). The cranial flexure (*Text-Figure 20*; *Plates 9c,d, 12a*) is visible from about stage 12, though measurements of the cranial angle show that it begins as early as stage 10 (Goodrum and Jacobson, 1981; Pikalow *et al.*, 1994). It consists of the head bending ventrally toward the yolk sac and there is evidence that it is brought about, at least partially, by the ventral bulging and elongation of the prosencephalon (Goodrum and Jacobson, 1981). The infundibulum at the anterior end of the neural tube is linked to the foregut by Rathke's pouch (*Plate 80*); as the brain elongates, the growing neural tube has to bend around the foregut. By stages 12–13 the head has begun to rotate also so that its left side comes to lie against the yolk sac and its right side to lie uppermost (see normal table, Appendix II; *Plate 9c,d*). At this stage the trunk region has not yet turned and its ventral side still lies against the yolk sac with its dorsal side uppermost, but gradually the rotation spreads down the body until at about stage 20 the entire embryo has rotated and lies on its left side.

Two other flexures occur that affect the shape of the body. In the cervical flexure (*Text-Figure 20b*; *Plates 9 and 12*) which forms between about stages 14 and 23 (2–4 days), the head bends round in the neck region so that it comes to lie at right angles to the trunk, its ventral side lying against the ventral side of the pharynx. Experimental evidence suggests that the cervical flexure is related to looping of the heart (Flynn *et al.*, 1991), though whether the contraction of the heart from a straight tube to a looped one is responsible for pulling the head around in an arc (Flynn *et al.*, 1991) or, conversely, whether the head flexures are responsible, at least in part, for cardiac looping (Männer *et al.*, 1993) is not yet clear. Meanwhile the trunk itself becomes curled around so that the tail bud comes to lie close to the head (*Text-Figure 20c*; *Plate 12d*).

From this stage onwards the body shape is an indicator of many of the events taking place beneath

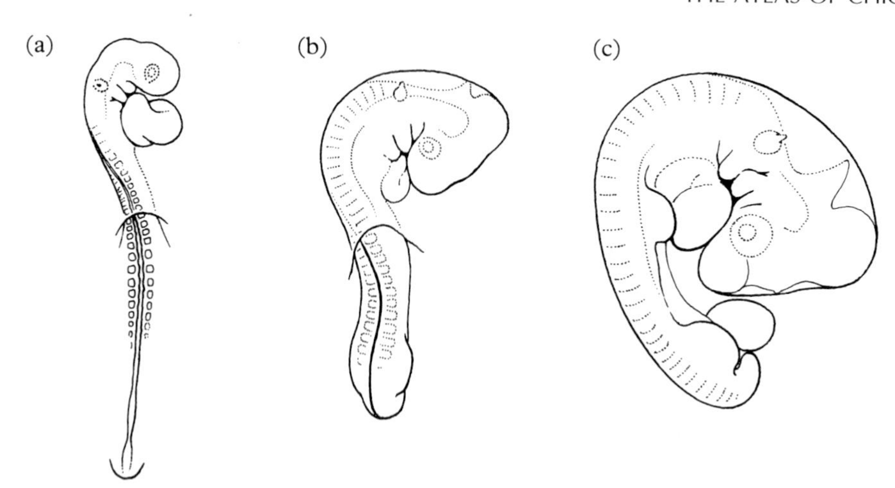

Text-Figure 20. *(a) Cranial flexure; the head has turned to the right but the trunk has not (about stage 14). (b) Cervical flexure: the head now lies at right angles to the trunk (about stage 18). (c) Trunk curvature: the trunk has now turned so that the whole embryo lies on its side and the tail bud is close to the head (about stage 22).*

the ectoderm. These include the formation of the limb buds (*Plate 12b,d*), starting at about stage 17, with their enlargement and elaboration until the major features of the wing or leg have formed at about 10 days. The tail bud (*Plate 27*), the developing eyes, the beak and the pharyngeal pouches are all conspicuous aspects of the developing embryo, as is the bulge caused by the increasing size of the heart. These and other external features are categorized in the Normal Tables of Hamburger and Hamilton (1951; see Appendix II) and the wing shapes by Murray and Wilson (1994) (see Appendix III).

FACE

The face is formed mainly from mesenchyme cells of the neural crest. Before the face starts to form, the anterior part of the head consists of an ectodermal coating over the forebrain with a few scattered mesenchyme cells. The facial structures are derived from cells which migrate between the forebrain and the ectoderm during day 4 of incubation, after the stomatodaeal plate has ruptured (see Chapter 8) and form a group of outgrowths. These are the left and right maxillary and mandibular processes of the first pair of pharyngeal arches which lie on either side of the stomatodaeal opening, and the medial fronto-nasal process (*Text-Figure 65a*) which lies above it. The bone morphogenetic proteins BMP2 and BMP4 appear to play a role in the outgrowth of these structures (Francis-West *et al.*, 1994). The nasal placodes (*Plate 23*) become visible as shallow pits at stages 15–16 and indent further by stage 18 on either side of the fronto-nasal process. The maxillary processes grow towards the midline of the face during days 4 and 5 and fuse with the sides of the fronto-nasal process. The fronto-nasal process itself becomes extended forward to form the upper jaw, whilst the two mandibular processes also fuse in the midline and become extended forward to form the lower jaw (*Text-Figure 65*). The facial outgrowths form as a result of an interaction between the overlying epithelium and the neural crest mesenchyme. The presence of the epithelium is essential, since without it the neural crest mesenchyme cells will not form the primordia (Saber *et al.*, 1989) but the type of facial prominence that does develop is determined by the mesenchyme. The evidence comes from recombinant experiments in which the epithelium has been separated from the mesoderm of one primordium and been replaced by the epithelium of another (Richman and Tickle, 1989).

The site at which the particular primordium forms, however, appears to be dependent on events in the ectoderm. For example, before the cranial neural crest cells have arrived in the facial region the ventral head ectoderm expresses FGF8 in two domains that correspond to the future mandibular processes, and BMP4 at the site of the future maxillary processes. The spatial relationship between the forebrain and the facial features appears to be coordinated by local retinoid signalling that maintains the expression of FGF8 and *sonic hedgehog* (Schneider *et al.*, 2001; Le *et al.*, 2001).

The maxillary processes meet in the midline and form the palate and the upper lip (beak). High levels of *Msx* gene transcripts have been correlated with regions of outgrowth (Brown *et al.*, 1997). Apoptosis takes place in the **periderm** (see p. 105) but the epithelial cells proper break down to form mesenchyme at the region of junction (Sun *et al.*, 2000).

A morphometric analysis of changes within the fronto-nasal process was published by Patterson and Minkoff (1985). The development of the primary palate was described by Yee and Abbott (1978). Extirpation of the fronto-nasal process leads to the reduction of the upper beak, agenesis of the primary palate and impaired development of the maxillary processes and the palatal shelves (McCann *et al.*, 1991).

ORIGIN OF THE LIMBS

The potential limb regions become visible from about stage 15 (50–55 h) (*Plate 12* and Appendix III) as slightly thickened ridges of the somatic lateral plate mesoderm, though, according to Stephens *et al.* (1992) a limb-forming region can be recognized as early as stage 11. The wing buds (*Plates 73, 82*) form at the level of somites 15–20 and the leg buds (*Plates 73, 82*) at the level of somites 26–32, these levels apparently being determined by the distribution expression of various *Hox* genes. For example, the forelimb buds form at the most anterior level of expression of *Hox-6* (Burke *et al.*, 1995; Burke, 2000) As the lateral body folds form, the limbs come to lie along the sides of the body wall. At 3 days of incubation each limb bud is about 1 mm in length by about 1 mm in width (*Plate 5* of Appendix II). Each consists of an envelope of ectoderm enclosing a core of mesoderm; the ectoderm is derived from the ectoderm of the lateral body wall and the mesoderm is formed from the somatic lateral plate, although this subsequently becomes supplemented by cells migrating in from the somites. The somatic mesoderm gives rise to the tendons, skeleton, dermis and connective tissues of the limbs, whilst the somitic cells form the muscles.

The first morphological step in the formation of the limb bud is the proliferation of the lateral plate mesoderm cells and this appears to be brought about by the production of the paracrine factor FGF10 by the lateral plate mesoderm itself. When beads soaked in FGF10 were inserted under ectopic ectoderm supplementary limb buds were induced (Sekine *et al.*, 1999). The FGFs themelves appear to be induced by members of the *Wnt* family of growth factors. In particular, *Wnt2b* is expressed in the wing region and *Wnt8c* in the leg (discussed by Tickle and Munsterberg, 2001).

In their early stages of development the wing and leg buds are similar to one another morphologically, but by about stage 24 they have begun to acquire their individual characteristics. These events are controlled, at least in part, by the transcription factors TBX5, which is present in the wing buds, and TBX4, which is found in the leg buds. Experimental introduction of these factors into a developing limb bud has been shown to influence its future development into either a wing bud or a leg bud.

At the tip of the limb the ectoderm becomes thickened and is then known as the apical ectodermal ridge (*Plates 85, 94*). Important interactions take place between the ridge and the underlying mesoderm. The ridge itself is induced to form by the mesenchyme cells, largely through the secretion of FGF10. But the mesenchyme is subsequently dependent on the presence of the ridge. If the ridge is removed the limb fails to differentiate further, whereas if an additional apical ectodermal ridge is grafted onto the limb bud a supernumerary limb structure develops (Saunders *et al.*, 1957).

The mesenchyme immediately beneath the **apical ectodermal ridge** is a region of high mitotic activity, being the source of the additional cells that are needed as the limb bud elongates. It is known as the **progress zone** and consists entirely of cells that are as yet undifferentiated. It appears to be maintained by FGF8 secreted by the apical ectodermal ridge. There is also evidence that *Wnt* signalling is involved (Kawakami *et al.*, 2001). Gradually the most proximal cells in the progress zone leave it, being replaced by more cells distally as mitosis continues. The first cells to leave the progress zone form the most proximal structures of the limb, and successively, as more and more cells leave the progress zone they become destined to form progressively more distal structures. The specification of the tissues that form along the limb bud is due to the activity of a sequence of *Hox* genes. In the wing bud the humerus forms first, apparently under the influence of *Hox-d9* and *10*, which is expressed at the time throughout the limb bud. This first phase is followed by the formation of the radius and ulna (middle phase) and then by the metacarpals and digits (third phase), and during these periods there is a rearrangement of the expression of these genes and the expression of additional *Hox* genes. These skeletal structures are formed initially in cartilage and become ossified later (see Chapter 10). Once the regions of the limb have become laid down the apical ectodermal ridge disappears (Wolpert, 2002). During this period when the proximo-distal axis is already present, two further axes become established. The first is that of the anterior–posterior axis. This has been shown by a series of experiments to be controlled by the so-called ZPA (zone of polarizing activity), a region of mesenchyme at the posterior border of the early limb bud near its junction with the body wall (*Text-Figure 21*). If this region is transplanted to the anterior border of another, intact limb bud it leads to the formation of additional, duplicated limb skeletal elements with a polarity defined by the graft usually in mirror image to that of the host limb. Similar duplications were seen in which beads soaked in retinoic acid were implanted in the same region, the degree of duplication being dose dependent (Tickle, 1991).

Riddle *et al.* (1993) demonstrated that the ZPA was characterized by the gene *sonic hedgehog* and that

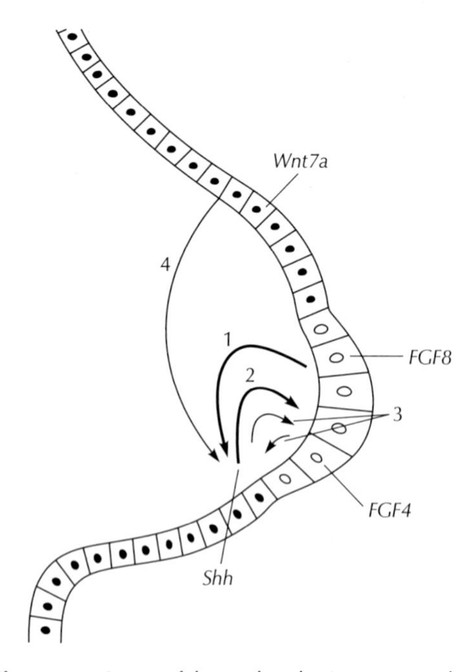

Text-Figure 21. *Some of the molecular interactions by which limb bud formation is initiated and maintained. Stage 1: the apical ectodermal ridge (AER) expresses FGF8, which induces sonic hedgehog (Shh) in the zone of polarizing activity (ZPA). Stage 2: ZPA expresses Shh, which induces FGF4 in the posterior AER. Stage 3: Shh and FGF4 reciprocally maintain each other. Stage 4: the dorsal ectoderm expresses Wnt7a, which helps maintain Shh in the ZPA.*

similar duplications could be obtained when cells that were capable of secreting *sonic hedgehog* protein were grafted into a chick limb bud. In the normal limb bud the *sonic hedgehog* also controls the level of FGF4 in the posterior region of the apical ectodermal ridge. The role of *sonic hedgehog* in controlling the antero-posterior axis is not, however, a straightforward one, as BMP2 and BMP4, as well as retinoic acid, play a part.

Each limb develops also a dorso-ventral axis so that, for example, the dorsal (upper) side of the foot differs from the ventral (plantar) side. This appears to be due mainly to the activity of events in the ectoderm. In particular, *Wnt7a* is active in the dorsal ectoderm and is responsible for a cascade of events (*Text-Figure 21*).

The autopod (hand or foot) is the final region of each limb to be laid down. TGFβ2 and activin signalling are involved in the skeletogenesis of the digits (Merino *et al.*, 1999a). (For a recent review of digital development see Sanz-Ezquerro and Tickle, 2003). The 'hand' of the wing is highly modified in birds, but the foot retains the characteristic vertebrate plan of five digits. The foot initially forms as a flattened paddle-like structure (Appendix III) in which the cartilaginous elements are clearly distinguishable yet still connected by mesenchyme (*Plate 116*). In the duck and other web-footed birds this interdigital region remains as a web, but in the chick it is removed by apoptosis except at the base, so that the toes become separated.

Other specialized areas of the limb bud, in addition to the ZPA, are the **anterior necrotic zone** (ANZ) and the **posterior necrotic zone** (PNZ), which not only play a role in the shaping of the limbs but are centres from which morphogens appear to diffuse and interact with other tissues, playing an essential role in the patterning of the limbs. There is evidence that retinoic acid is an active principle secreted by these zones (reviewed by Paulsen, 1994).

These changes in the external appearance of the limbs are illustrated and described in the Normal Table of Hamburger and Hamilton 1951 (see Appendix II) and by scanning electron microscopy by Murray and Wilson (1994) (see Appendix III).

The tail bud is shown in *Plate 27*.

GROWTH OF THE EMBRYO

Growth, a process usually defined as the increase in mass of an organism, is brought about mainly by the continual rise in the population of cells present in the embryo.

Changes in cell shape and size, as well as the accumulation of extracellular materials, also affect the shape and size of developing organs. Some regions undergo a more rapid rate of cell division than their neighbours, leading to a build-up of tissue. Growth factors, such as FGF, play an important controlling role.

APOPTOSIS

As well as regions of high proliferation there are also areas that have a high rate of cell death. Probably in all embryonic tissues there is a continuous loss of cells through death, but in certain parts of the developing body the death rate outstrips the proliferative rate. This means that just as some regions become enlarged by rapid cell division, others become eroded away. The dying cells undergo a process of breakdown and gradually become phagocytosed by other cells. Apoptosis is also known as 'programmed cell death' because in many cases the patches of cells die in a particular location of the embryo at a specific time in development and play an important role in morphogenesis. Two classical examples in the chick occur in the shaping of the wing bud and of the toes. Apoptotic cells are, however, found in normal embryos even as early as gastrulation (Bellairs, 1961) and after that in many well-defined sites in the differentiating tissues, e.g. in the mesonephros (Chapter 7), in the heart

(Chapter 6), in the sclerotome (Sanders, 1997), in the nervous system (Oppenheim *et al.*, 1999), in the neural crest (Jeffs and Osmond, 1992) and in the tail bud (Sanders *et al.*, 1986), branchial arches and lateral body wall (Hirata and Hall, 2000). In most of these examples cell death is focused on a highly localized region and occurs within a restricted period of time, e.g. cell death within the endocardial cushions of the developing heart occurs only between about days 5.5–7.5 (Keyes and Sanders, 1999).

Regions of cell death play a specific role in the shaping and patterning of organs, though they can sometimes be 'rescued' from death if treated with an appropriate growth or differentiation factor. For example, regions of programmed cell death in the early nervous system of the chick embryo can be counteracted by the application of *sonic hedgehog* protein (Charrier *et al.*, 2001). Hirata and Hall (2000), who have reviewed the temporo-spatial patterns of cell death from stages 1 to 25 have concluded that cell death is a feature of development at all these stages but that there are changing patterns, depending on the specific stages of growth, differentiation and morphogenesis.

SYMMETRY AND ASYMMETRY

We have already seen that the embryonic axes (dorso-ventral and antero-posterior) begin to be established even before the egg is laid, but the finalization of this polarity does not take place until later. Although the antero-posterior polarity is initiated by the effects of gravity during cleavage, it is still possible to overcome it until the primitive streak has begun to develop. At that time a series of *Hox* genes comes into play which interact with one another so that each level of the body becomes categorized in a unique way. It acquires its own 'positional information' (Wolpert, 2002). There are a number of markers of antero-posterior polarity, e.g. *Gata2* protein is expressed in the area opaca epiblast prior to primitive streak formation as a gradient along the antero-posterior axis, being highest anteriorly (Sheng and Stern, 1999) and disappears at stage 3+ (though it reappears later in other regions, such as the lateral part of the somites).

With the establishment of antero-posterior polarity the embryo acquires a left and a right side. Initially, the two sides are symmetrical, being mirror images of one another, e.g. a left wing and a right wing, which essentially remain so throughout development. With the differentiation of the internal organs, however, the two sides diverge and certain regions become highly asymmetrical, as shown by the heart, liver and gut.

The first signs of left–right asymmetry appear to be associated with Hensen's node and involve a molecular cascade. It is possible to disrupt the left–right arrangement of the body by interfering with the cascade. For example, if cells secreting *sonic hedgehog* are transplanted to the right of the node, nodal is induced symmetrically and the looping of the heart may occur randomly to the left or right.

Chapter 6

Heart, Blood Vessels and Lymphatics

HEART

The heart proper begins to develop at about stage 9 or 10 and is initially present as a simple tube which then loops into an S-shape (*Plates 28–30*). Even as it forms, the basic regions become apparent, first the truncus and ventricle, then the atrium and lastly the sinus venosus. The first beats are at about stage 10 and the circulation has become well established by about stage 16. Division into left and right sides takes place during days 3–5.

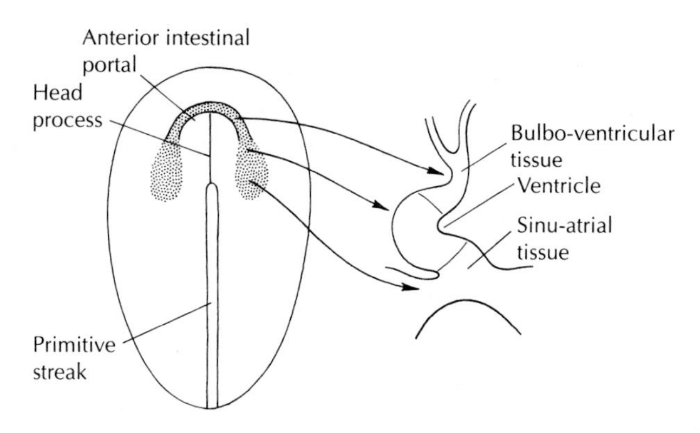

Text-Figure 22. *Chick embryo at stage 5 to show the location of the precardiac mesoderm and the fates of the different regions in the newly formed heart at about stage 10. (After DeHaan, 1965.)*

Origin of the Heart

The heart is formed from two symmetrical areas of the early embryo. The primary heart field has long been recognized as two patches of splanchnic mesoderm lying one on either side of the head process at stages 5–6 after ingression (Rawles, 1943) (*Text-Figure 22*). It is derived from cells that are present in the primitive streak at stage 4 (Selleck and Stern, 1991), though transplantation experiments show that they are still capable of developing into other tissues (Inagaki *et al.*, 1993). These two areas migrate cranially and join together anterior to the head process to form a crescent, the cardiac crescent (also known as the cardiogenic or crescentic plate).

A secondary heart field has been identified more recently as a result of modern labelling experiments (e.g. Mjaatvedt *et al.*, 2001; Waldo *et al.*, 2001; Kelly and Buckingham, 2002). It forms from splanchnic mesoderm which lies in the midline anterior to the simple heart tube at about stage 10 and its cells subsequently become incorporated into the myocardium of the truncus (see below).

Single-cell analysis of isolated presumptive heart cells taken from stage 4 embryos and grown *in vitro* has established that they are capable of forming muscle-specific myosin heavy chains, though they do not appear to be able to express chains specific for atrial or ventricular muscle unless grown at high density (Gonzalez-Sanchez and Bader, 1990). The cells are, however, not yet determined and if transplanted elsewhere can still give rise to other tissues (Veini and Bellairs, 1990; Climent *et al.*, 1995); conversely, tissues taken from outside the primary heart field area are still capable of becoming incorporated into the heart (Orts-Llorca and Collado, 1970). Nevertheless, there is evidence that the primary heart field mesoderm generates a subpopulation of cells, cardiomyocytes and endocardial endothelial cells, and that their precursors are already established by stage 3 (Wei and Mikawa, 2000).

After the precardiac cells have migrated into the anterior lateral plate they come under the influence of signals from the underlying endoderm (reviewed by Lough and Sugi, 2000). The paracrine factors required to allow cardiac myocytes to form include growth factors (FGFs), especially FGF8, and bone morphogenetic proteins (BMPs). Repressors of cardiogenesis, which include *Wnt8c*, are expressed in the posterior lateral mesoderm but not in the primary cardiac fields (Marvin *et al.*, 2001; Yutzey and Kirby, 2002).

When the lateral plate mesoderm splits into two layers, somatic and splanchnic (*Plate 53*), the splanchnic layer becomes closely associated with the endoderm. Initially, the space between the two layers, which will become the coelomic cavity, is present only in the future trunk region of the embryo but intercellular spaces gradually join up together and extend the coelom cranially. By stages 7–8 this newly formed coelomic cavity is now arranged as paired vesicles, one on either side of the head process, known as the amnio-cardiac vesicles (*Text-Figure 23*) because they take part in the formation of the amnion (see p. 116) as well as the pericardial cavity. They show a tendency to swell up in embryos under unsuitable osmotic conditions. By the time the amniocardiac vesicles have formed, the precardiac mesoderm of the primary heart field has become localized in the splanchnic mesoderm. Organized myofibrils are already present in the precardiac mesoderm of the primary heart field as the sheet of cells migrates anteriorly, and rhythmic contractions occur at the time of fusion (Colas *et al.*, 2000).

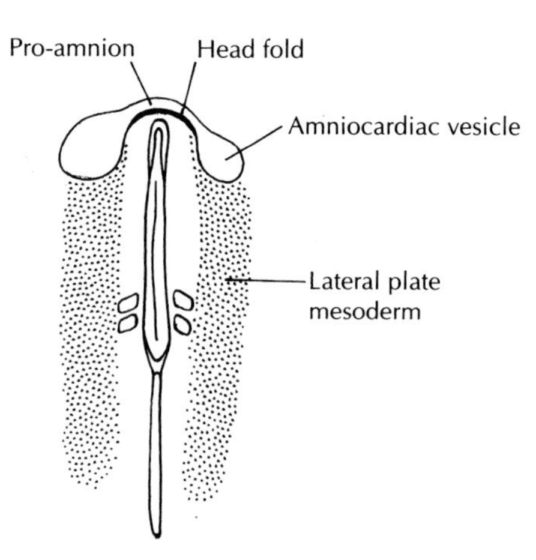

Text-Figure 23. *An embryo at stage 7, showing the head fold and amniocardiac vesicles. The lateral plate mesoderm lies over the body coelom.*

The Migration of the Precardiac Mesoderm

The precardiac mesoderm adheres so closely to the underlying endoderm that it is difficult to separate the two layers cleanly without the use of enzymes. This is because the interface between the endoderm and mesoderm is richly coated with fibronectin. If the endoderm is removed, the precardiac mesoderm does not migrate. Formerly, it was thought that the mesoderm cells were carried passively into the midline by the folding of the foregut endoderm, but time-lapse studies have shown that they migrate independently of it (DeHaan, 1965). Linask and Lash (1986) and Easton *et al.* (1990) concluded that the precardiac mesoderm selectively migrated up a gradient of fibronectin and, with the use of synthetic peptides, were able to show that the fibronectin was an essential component of the endoderm substratum. If migration of the precardiac mesoderm on the substratum is disrupted, either surgically or chemically, the resulting heart will be deficient in its anterior portion, or even divided into two separate hearts (Osmond *et al.*, 1991).

As the foregut forms, the precardiac cells are moved into the midline and begin to condense into bilateral tubular structures. As the developing foregut loses contact with the yolk sac the two endodocardial tubes of the heart fuse in the midline to form a simple tube, starting anteriorly and progressing posteriorly. The process can best be seen by studying a sequence of sections from progressively older embryos (*Text-Figure 24*). Fusion of the two endocardial tubes begins, anterior and ventral to the anterior intestinal portal, at about stage 8 (Coffin and Poole, 1988). At the same time, the two sides of the foregut join medially (*Plate 28*), forming the characteristically V-shaped pharynx (*Plate 55*). The left and right sheets of splanchnic mesoderm ventral to the pharynx combine to form the dorsal and ventral mesocardia (*Plate 56*). The latter is so transient, however, as to be seldom visible, rupturing at about stage 10. The simple tube is initially suspended into the coelom beneath the pharynx by the dorsal mesocardium but this structure is also transient, persisting only until about the stages 12–13. As the lateral body folds form (see p. 39) the coelom surrounding the heart becomes isolated from the more posterior coelom and is henceforth known as the pericardial coelom or cavity (*Plates 31, 66*).

The foregut forms in an antero-posterior sequence and its development is paralleled by the anterior to posterior formation of the heart. The most anterior part of the heart, the future truncus arteriosus (*Plates 29a, 66*), forms first and is then succeeded by the ventricle (*Plates 29, 65, 66*) the atrium (*Plate 29*) and the sinus venosus (*Text-Figure 25*), in that order. The region where the ventricle narrows to form the truncus is known as the conus (bulbus) arteriosus.

Development of the heart and foregut is closely associated with the head fold formation (*Text-Figure 39*) and with the lateral body folds (*Text-Figure 24*). These folds can be considered separately in longitudinal and transverse sections, but they are continuous with one another, and the effect is that a horseshoe-shaped trench undercutting the future head region (*Text-Figure 39*).

The posterior extension of the foregut can be prevented experimentally by cutting through the anterior intestinal portal (DeHaan, 1959) or by inserting a mechanical obstacle such as a lump of thick

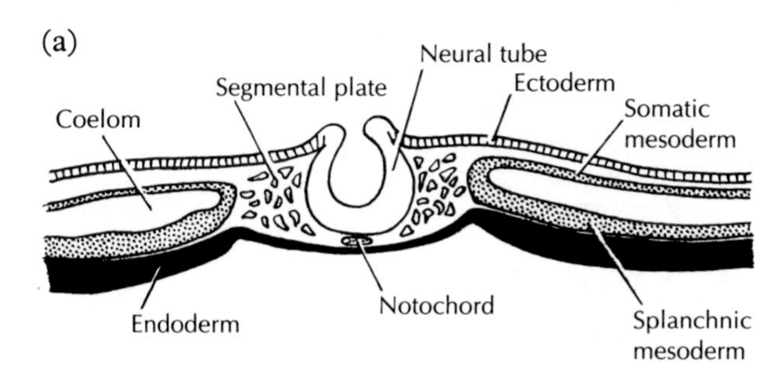

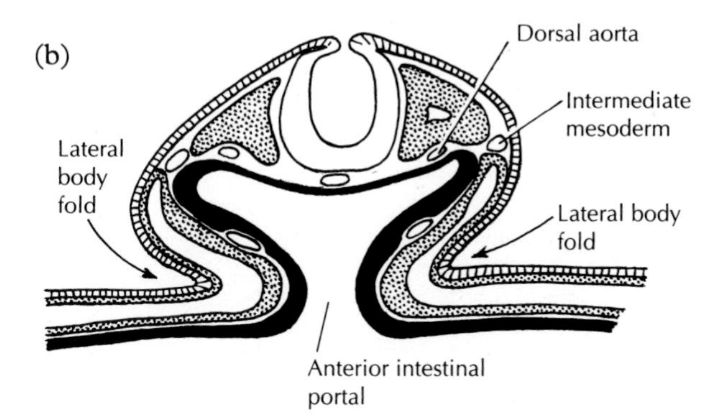

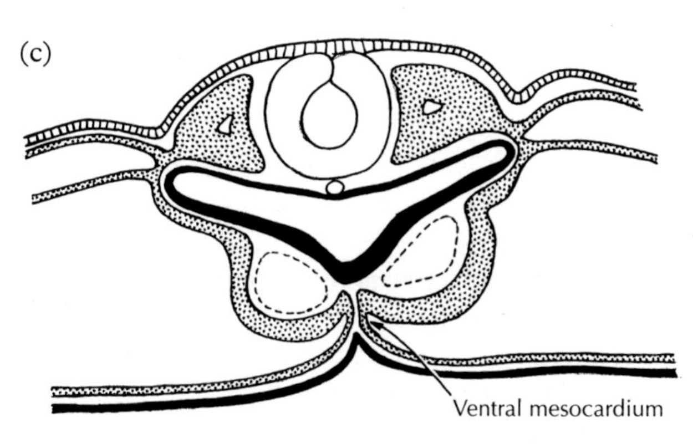

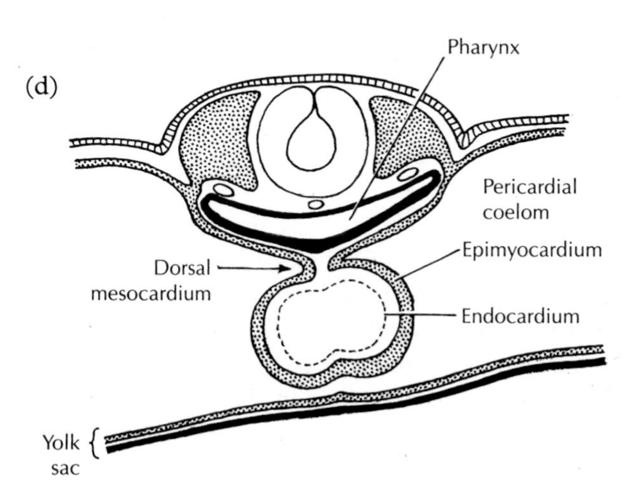

Text-Figure 24. *(a–c) A series of transverse sections through an embryo at stage 8 to illustrate the lateral body folds and early stages of heart formation, (a) being the most posterior and (c) the most anterior. (d) Transverse section across the heart at stage 10 to illustrate the loss of the ventral mesocardium.*

albumen. In either case the result is that the two sides fail to fuse and diplocardia results. Each side then forms a small but perfect heart which begins to beat independently and at the appropriate time.

Once the heart has formed into a simple tube (*Plates 28, 29*), two distinct layers are visible (*Plate 70*): the thin inner layer, the endocardium, and the thicker outer layer, commonly known as the epimyocardium. Subsequently, the heart becomes three layered, these layers being the **endocardium**, the **myocardium** and the thin enveloping **epicardium**. The traditional view has been that the precardiac mesoderm gives rise to all three layers of the heart, but several authors using transmission or scanning electron microscopy (e.g. Shimade and Ho, 1980), or experimental techniques (Männer, 1993) have concluded that the epicardium arises from the pericardium (i.e. from the splanchnic mesoderm of the pericardial wall) and the traditional view has now been modified. The relevant region of the pericardial wall, now known as the **pro-epicardial serosa** (*Plate 32*), surrounds the entire region near the sinus venosus, and contributes primarily extracardiac cells to the developing heart (Männer *et al.*, 2001). As the pro-epicardial serosa starts to develop at stages 13–14, bleb-like protrusions form at the ventral wall of the left and right horns of the sinus venosus. Cells from the pro-epicardial serosa are then transferred across the pericardial cavity to the surface of the developing ventricles, either as free-floating aggregates or by the formation of bridges across the lumen. The primordial epicardium then spreads as a continuous epithelial sheet over the myocardium (Männer *et al.*, 2001). For more information on the origin of the pericardial serosa in the quail see Virágh *et al.* (1993) or Männer *et al.* (2001).

Looping

Looping begins at about stage 10 and normally occurs to the right (*Plates 29, 30, 34*). The right ventricular wall bulges outwards and becomes convex at the same time as the left wall becomes concave (*Text-Figure 25*). Coincidentally, the heart rotates to the right. There is no well-proven explanation as to how this is brought about, nor as to why the initial loop is to the right though, as we have seen (p. 43), molecular differences between the right and left sides of the embryo have become established before the heart begins to form. The homeobox gene, *Pitx2*, has been found to play a crucial role, *Pitx2c* being present in the left lateral plate as well as in the left side of the heart tube and head mesoderm. Ectopic expression of this gene resulted in random looping (Yu *et al.*, 2001). Nowadays, most authorities believe that many factors are involved and the idea that the heart is forced into a loop because it is in a confined space has long been abandoned. There is some evidence that heart looping and cranial flexure

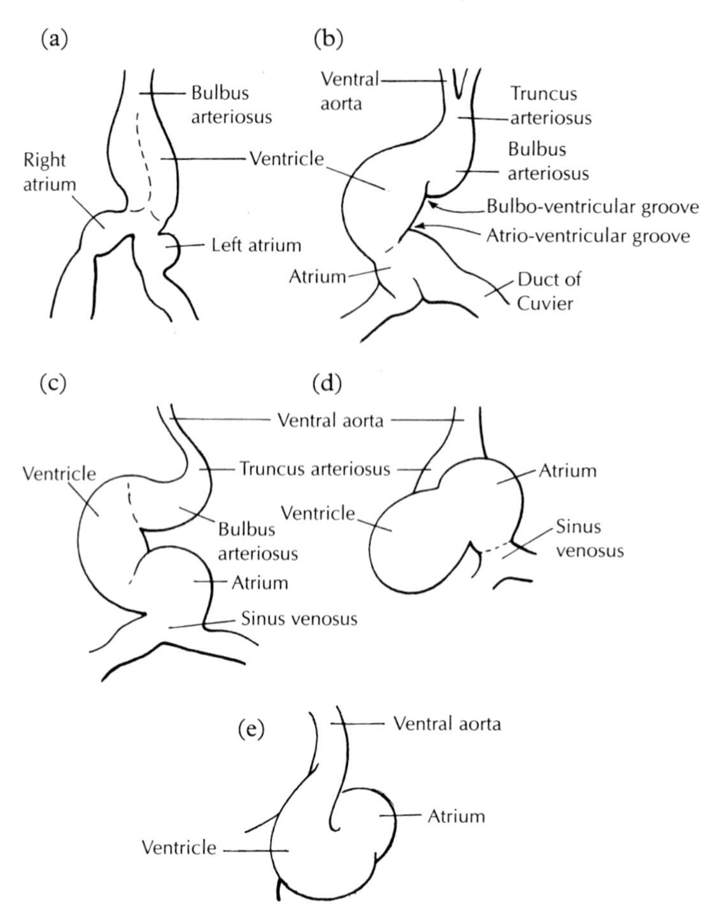

Text-Figure 25. *Looping of the heart seen from the ventral aspect. (a) stage 10; (b) stage 12; (c) stage 13. The fully looped heart at stage 16, seen from the left side (d) and the right side (e). (Partially after Patten, 1950.)*

are related (see Chapter 5), but although the cranial flexure may depend on heart looping, heart looping can occur in the absence of cranial flexure (Männer *et al.*, 1989).

The simple tube is capable of looping as soon as it has formed. The work of Butler (1952) is much quoted though not easy of access. Butler isolated newly formed chick hearts *in vitro* and found that although they bent they did not rotate to the right. Castro-Quezada *et al.* (1972) showed that if parts of the simple heart tube were removed, looping occurred in the remaining sections. It seems likely that in both these experiments the initiation of the process had begun at the molecular level before extirpation of the tissue. Indeed the pre-loop heart is already slightly asymmetrical (Stalsberg, 1969).

There is evidence that an important role in looping is played by the actin bundles. These become arranged in a circumferential pattern as the tube forms but are more prominent on the right than on the left side. When small crystals of cytochalasin B were inserted into the right side of the heart, bundles at that side were disrupted and looping took place to the left. By contrast, when crystals were inserted into the left side, looping took place as normally, to the right (Itasaki *et al.*, 1991). These authors have suggested therefore that the bundles in the right side of the heart normally generate tension and lead to dextro-looping. This explanation, however, fails to account for the difference in distribution of the actin bundles on the two sides of the heart, and for the reversal of looping in the right-hand side in diplocardia.

There is a possibility that the actin bundles are influenced by the cardiac jelly, which fills the lumen of the early heart and consists of a complex matrix of glycosaminoglycans, collagen and non-collagenous glycoproteins, with hyaluronic acid as the principal constituent. It has been suggested that the cardiac jelly produces an outward pressure so that the walls in the heart are under tension and become deformed (Nakamura *et al.*, 1980). This concept was questioned, however, by Baldwin and Solursh (1989) who cultured rat embryos in the presence of hyaluronidase but found that normal looping took place. Thus, although the cardiac jelly may play a role in looping, it does not appear to be an essential one.

After the dorsal mesocardium breaks down, the heart is anchored only at its anterior end (to the developing aorta) and at its posterior end (where it is still attached to the endoderm at the anterior intestinal portal) (*Plate 29a*). It therefore lies freely in the future pericardial cavity (*Text-Figure 24*). Further refinements in the looping and rotation of the heart have been described in detail by Patten (1922), Romanoff (1960) and Coffin and Poole (1988). The result is that the ventricular region moves posteriorly, whilst the anteriormost region moves into a more ventral position. During the time that the looping is taking place, cells from the secondary heart field become added to the ventricular outflow tract, forming the conus (bulbus) arteriosus, between stages 12 and 22 (2–4 days) (*Plate 30*). FGFs and BMPs are expressed in the pharyngeal endoderm at this time and appear to be necessary for the differentiation of the muscle and the conotruncal cells from the secondary field.

Septation of the Heart

The divisions of the heart into the major regions, conus, ventricle, atrium and sinus venosus, are enhanced by the development of grooves, the bulbo-ventricular and atrio-ventricular grooves, between stages 16 and 20, (*Text-Figure 25*). The simple tube is subsequently divided into left and right channels by the growth of a series of septa, so that after hatching, the two sides of the heart are completely separate. The septa are derived from the endothelial cells that line the heart, under the influence of TGFβ (Darland and D'Amore, 2001). The process of septation is comparable to that in the human heart, which is extensively described in text books of human embryology, though there are certain differences.

At stages 16–17 two apposed ridges, the atrio-ventricular cushions, project into the lumen of the atrio-ventricular canal from its dorsal and ventral walls. The cushions are initially formed from acellular cardiac jelly which then becomes invaded by cells from the endocardium (see Icardo, 1989, for a morphological description of the cushions). The transcription factor, *slug*, is expressed in this region and is necessary for the initiation and conversion of epithelium to mesenchyme (Romano and Runyan, 2000). The cushions meet and fuse in the mid-line toward the end of day 5, leaving a channel, the atrio-ventricular canal, on either side (*Plate 90*). Many authors have now shown that BMPs and TGFβs play a critical role in the epithelio-mesenchymal transformation of the endothelial cells during the formation of the atrio-ventricular cushions (e.g. Boyer and Runyon, 2001; Keyes *et al.*, 2003). Meanwhile a septum, the interatrial septum (*Plate 90*), which begins to form at about stage 14 as a small ridge at the cranio-dorsal end of the single atrium, grows down to meet the atrio-ventricular cushions. By 4 days (about stage 24) the septum has begun to fuse with the cushions at about the same time as the cushions are fusing with one another. The result is to divide the atrium into left and right compartments, but the two sides do not become completely separated because many small holes, the interatrial foramina, break through the septum and allow communication between the left and right atria. This is important because most of the venous blood returns to the right side of the heart, only a trickle from the pulmonary vein returning to the left. The passage of blood through the foramina into the left atrium enables the pressure in the two sides of the heart to be balanced, which is necessary for their equivalent growth. Strands of tissue overlap the holes so that they open only when the right atrium becomes expanded during diastole and blood then passes from the right to the left atrium. When the atrial wall contracts during systole equal pressure is re-established in the two atria and the foramina close.

In birds, there is development of neither a single large foramen corresponding to the foramen ovale of mammalian embryos, nor of a secondary septum corresponding to the septum secundum of the mammalian heart.

At the time of hatching there are important changes in the circulation that result in an equalization of pressure in the two sides of the heart and which in turn results in the closure of the foramina. In the domestic fowl, though apparently not in all birds, both the pulmonary vein and part of the left atrial tissue become absorbed into the interatrial septum, and in this respect also the development of the chick heart differs from that of mammalian hearts (Quiring, 1933).

The conversion of the single ventricle into right and left components starts during day 3 as the left ventricle bulges and during day 4 as the right ventricle bulges. These bulges are thought to be due to the force of the blood streams. Part of the original wall of the single ventricle remains between the two bulges and is the source of the interventricular septum. This septum (*Plates 36, 91, 92, 105*), which lies initially at the apex of the ventricles, extends towards the atrio-ventricular cushions and fuses with them by day 6, leaving only a small gap, the interventricular canal.

During day 5, the proximal part of the truncus arteriosus becomes incorporated into the right atrium, whilst the distal part becomes divided into the aorta and the pulmonary arteries by the growth of the aortico-pulmonary septum (*Plates 31, 105*). This forms from two endocardial ridges which arise from opposite sides of the truncus and meet in the middle. The position in which they arise is apparently determined by the spiral course taken by the two blood streams leaving the ventricles. The result is that the aortico-pulmonary septum (*Plate 105*) develops in a spiral manner and separates the truncus into its two channels, the aorta and the pulmonary artery. (For a full discussion of the process see Icardo, 1990). Between days 6 and 7 (stages 30, 31) extensive apoptosis occurs in the myocardium of the outflow tract and, since this coincides with extensive remodelling of this region, its function is thought to enable the great vessels to align properly over their respective ventricles (Rothenberg *et al.*, 2003). By the end of day 7, the septum has extended back into the interventricular canal and become continuous with the interventricular septum, so that the right ventricle now opens directly into the pulmonary artery whilst the left ventricle opens into the aorta. (For a discussion of some of the differences in this aspect of development between chick and mouse embryos, see Icardo, 1990.) The aortico-pulmonary septa are patches of condensed mesenchyme formed principally by neural crest cells (see Chapter 4) which arise from between the region of the otic placode and the posterior level of somite 3. The migration of these neural crest cells into the developing outflow tract depends on the action of BMP2-4 and if this is inhibited by *noggin* before septation, the endocardial cushions fail to form properly (Allen *et al.*, 2001). Extirpation of this region of neural crest results in failure of the trunco-bulbus septum to form, whilst partial removal leads to defects such as tetralogy of Fallot (Kirby, 1990); replacement with neural crest from other regions fails to repair the defects. It appears, therefore, that the 'cardiac' neural crest cells are already committed to form cardiac cells (Kuratani and Kirby, 1991). Removal of the nodose ganglion at the same time as these 'cardiac' neural crest cells results in even greater anomalies.

During days 4–5 the valvulae venosae and the sinu-atrial valves (*Plates 33, 90*) develop from the wall of the right atrium at its junction with the sinus venosus. The atrio-ventricular valves (*Plate 136*) are formed partially from the endocardial cushions and partially from the atrial and ventricular muscles. The ventricular walls become greatly thickened (*Plates 36, 37, 136*) and muscular with extensive trabeculae carnae. The sinus venosus and the sinu-atrial valves are retained as definitive structures in the adult bird heart (Quiring, 1933), whereas in mammals the sinus venosus is absorbed into the right atrium.

During the period of separation of left and right sides of the heart changes occur in their histology. By 5 days of incubation, bands of muscles have begun to form in the atria and between days 7 and 11 these become the 'muscular arches' which are a characteristic of the avian atria and appear to play an important role in systole.

Change in Relative Position of the Heart

In its initial stages as a simple tube, the heart lies at the level of the rhombencephalon, but by day 4 of incubation when all the main chambers have been formed, it is situated at the level of the wing buds. The 'descent' of the heart is largely the result of an increase in the length of the neck and brain, which therefore shift forward in relation to the heart.

The Heart Beat

The hearts of birds differ from those of mammals in that they have heavier workloads because of the requirements for flight, very large volumes of blood passing to the wings and flight muscles. A review of the adult heart and circulation is given by West *et al.* (1981).

If isolated cells from the precardiac mesoderm at stage 7 are maintained *in vitro*, they will begin to pulsate, though each at its own rhythm. If they come into contact with one another then they beat in synchrony. According to DeHaan (1990), who exchanged pieces of tissue within the precardiac mesoderm area, the heart mesoderm is not yet stably coded for its future intrinsic beat rate, and local cues from the surrounding tissues can alter it as it differentiates.

Action potentials in the cardiac mesoderm can be recorded at stage 9, first appearing in the left caudal region of the ventricle (Hirota *et al.*, 1987) before either the atrium or the sinus venosus have formed. This indicates the first sign of the pacemaker. Contractility is visible in the heart from about stage 10. By about stage 12, waves of excitation sweep forward from the caudal pacemaker and lead to a rhythmical contraction of the muscles, an event which causes the blood to start circulating (DeHaan, 1990). The beat rate is higher at the posterior (sinu-atrial) end than at the anterior (truncal) end. As with other tissues, gap junctions play an essential role in the electrical conductivity of the heart. Overall junctional resistance depends on the number of channels in a junction, their single channel conductance and the probability that the channels are open. Interestingly, although the size of the heart increases by more than a thousand times during the embryonic period, the time for the spread of excitation remains relatively constant because the conduction velocity increases rapidly in the ventricles (DeHaan, 1990). The atrio-ventricular node and the upper part of the atrio-ventricular bundle (conducting tissue) develop in the interatrial septum at 5.5 to 6 days (Argüello *et al.*, 1988).

BLOOD VESSELS

The first blood vessels are extra-embryonic and are formed by the process of angiogenesis (sprouting). Blood islands become visible in the proximal region of the yolk sac, the area vasculosa, at about stage 8. They develop from isolated clumps of cells in the splanchnic mesoderm. The peripheral cells of each blood island become the endothelium, whilst the remaining cells become the haematopoietic (primitive) blood cells. Columns of cells sprout from each blood island and become canalized and fuse with those from adjacent blood islands to form angioblastic plexuses. In this way a network of branching and intercommunicating vessels is formed (*Plate 38*).

By contrast, most of the embryonic blood vessels are formed *in situ* by the development of endothelial precursor cells, angioblasts, which then fuse together to form vesicles, but without the associated development of haematopoietic cells, except in a small region of the aorta (Risau and Flamme, 1995). This process is often called vasculogenesis to distinguish it from angiogenesis, though in each case an angioblastic plexus is formed. The endothelial precursor cells are mainly derived from the splanchnic mesoderm by undergoing a transition from epithelium to mesenchyme. This is due to an induction from the underlying endoderm, though this can be experimentally initiated by treatment with FGF2 (Poole *et al.*, 2001) so that it is thought that the angioblasts are induced by FGF2. The vascular endothelial growth factor, VEGFR2, does not appear to be involved in angioblast induction, though it may play a role in their growth and arrangement into a vascular pattern (Poole *et al.*, 2001). Even somatic mesoderm is able to form angioblasts if cultured in the presence of endoderm (Pardenaud and Dieterlen-Lièvre, 1999). The vesicles accumulate along the future pathways of some of the earliest blood vessels, such as the dorsal aorta, and may fuse either to each other or to existing vessels. Experiments in which labelled tissues have been transplanted to other regions of the early embryo (Pardenaud and Dieterlen-Lièvre, 1993) have shown that most mesodermal populations, but especially those of the splanchnopleure, contain cells that have the ability to form the endothelium of blood vessels. Not all the embryonic blood vessels form *in situ*, however; e.g. the intersomitic arteries are formed at stage 9 by sprouting from the dorsal aorta (Coffin and Poole, 1988).

Once the main vessels have been laid down, refinements of the pattern take place with the formation of extra vessels by angiogenesis (Pardenaud and Dieterlen-Lièvre (1999). The question as to why some vessels become arteries and others veins has not yet been elucidated. Some role appears to be played by the circulation at later stages but there is evidence of differences in expression of neuropilins, which are receptors for class 3 semaphorins. Both np-1 and np-2 are expressed in the blood islands of 24 h-old chick embryos, but by 48–72 h np-1 is confined mainly to the arteries and np-2 to the veins (Herzog *et al.*, 2001).

The Extra-embryonic Blood Vessels

These are the vessels that develop in the extra-embryonic membranes; they are known as the omphalo-mesenteric (vitelline) vessels. These carry blood from the embryo to the yolk sac and back again and their function is to transport partly digested yolk to the embryo. The sequencing of the pattern of the vessels of the yolk sac is a complex one and a detailed account is given by Romanoff (1960).

The pattern begins from about stage 9 with the blood islands in the splanchnic mesoderm of the proximal part of the area opaca joining those in the distal part of the area pellucida to form a meshwork of capillaries. By the time that circulation begins (about stage 12), a series of channels of varying size has formed. By about stage 16 the vascular area of the yolk sac has become surrounded by the sinus terminalis (*Text-Figure 26*), a vein that sharply delineates the area vasculosa from the non-vascular area vitellina, which lies more peripherally. The capillaries that form near the body are relatively large and gradually become transformed into a major pair of vessels, the left and right **omphalo-mesenteric (vitelline) arteries**. Blood passes from the dorsal aorta at about the level of the 18th pair of somites, into the right and left omphalo-mesenteric arteries and then into the area vasculosa. It enters a series of smaller and smaller branching arteries, eventually moving through a meshwork of fine capillaries that extends over the proximal part of the area vasculosa (the developing yolk sac). It is then collected up into the anterior or posterior **omphalo-mesenteric (vitelline) veins**, either directly or by way of the sinus terminalis. Blood returns from these veins to the heart. Initially, a left and right pair of anterior omphalo-mesenteric veins develop (*Text-Figure 26*), but by about 3.5 days the two begin to anastomose and by the end of the day 4 only one remains, the unfused remnant of

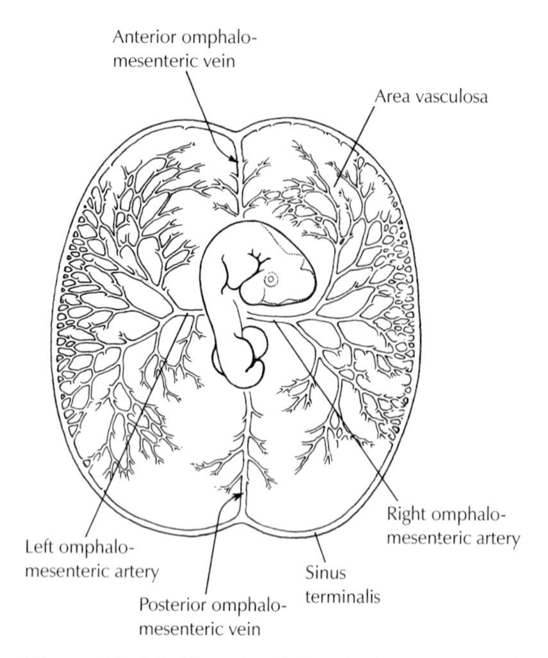

Text-Figure 26. *Vitelline circulation in the area vasculosa at 4 days, seen from the dorsal aspect. The vitelline vessels become the omphalo-mesenteric vessels in the embryo.*

the other one having atrophied. There is some disagreement in the literature as to whether it is the left or the right which survives, but it is possible that there is some individual variation. The posterior omphalo-mesenteric vein also appears to be paired in some cases but eventually becomes single.

Although the general pattern of the omphalo-mesenteric vessels is similar in different individuals, the details of the network show many variations. Those capillaries through which blood flows most rapidly become the largest vessels (Hughes, 1937). According to Romanoff (1960), changes in the circulation during day 3 cause some of the capillaries that were originally arterial to become transformed into veins, and the sinus terminalis to begin to regress from about stage 20 (4 days). By the end of day 5 however, a dense network of blood vessels has formed. By stage 16 the two omphalo-mesenteric arteries have fused into a single vessel which lies between the fifth and tenth somites. The omphalo-mesenteric vessels continue to branch and extend into the endodermal folds on the inner side of the yolk sac through most of the incubation period. The yolk sac is withdrawn into the body at hatching and the omphalo-mesenteric vessels continue to absorb yolk for the first day, after which they begin to degenerate.

The Allantoic and Chorioallantoic Circulations

The allantois starts to develop from about stage 18 (3 days) though in some individuals it may not appear till 4 days. Shortly after its appearance it forms a ramifying network of capillaries that receive blood via the allantoic (umbilical) arteries and return it to the body via the allantoic (umbilical) veins.

Fusion of the allantois with the chorion to form the chorioallantois begins during days 6–7. The pattern of vascularity in the chorioallantois begins now to differ from that in the inner layer of the allantois. The capillaries of the outer layer of the allantois proper become highly modified (*Plate 39*) and penetrate not only into the chorion but even pass through it and come to lie between the shell and the shell membranes. Here they are well placed to take part in the exchange of gases. Romanoff (1960) reported that the capillary vessels are of such a narrow bore that blood corpuscles can pass through only in single file.

The Dorsal Aortae

Paired dorsal aortae are visible beneath the somites (*Plates 13, 61*) by about stage 8. At about stage 15 the right and left dorsal aortae begin to fuse together. They are brought into proximity by the enlargement of the coelom and the formation of the dorsal mesentery. This process is related to the folding of the body wall (*Text-Figure 24*). The fusion remains incomplete in the extensions of the aortae anterior to the heart, which subsequently become the internal carotids (*Text-Figure 27; Plate 75*). At the most posterior end the dorsal aorta extends into the tail as a small vessel, the caudal artery (*Text-Figure 31; Plates 42, 137, 138*).

The pattern of development of the embryonic arterial system is comparable to that of mammalian embryos, the major difference being that in adult birds the aorta loops to the right, whereas in mammals it loops to the left. In order to understand how these differences are brought about, we must consider the development of the aortic arches.

The Aortic Arches

The aortic arches (*Plates 40, 41, 80–84, 87*) are the blood vessels that supply the pharyngeal arches (Chapter 8) and they serve as a communication between the ventral and dorsal aortae. They are formed mainly from neural crest cells (Brockman *et al.*, 1990). A splanchnic plexus forms around the foregut and gives rise to the pharyngeal arch arteries as well as the pulmonary and bronchial vessels

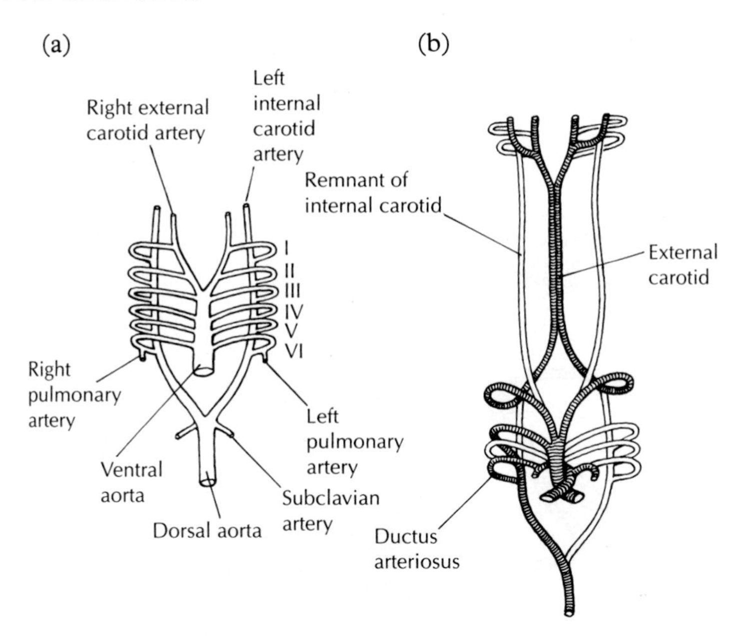

Text-Figure 27. *The relationship between dorsal aorta, ventral aorta and aortic arches. (a) Basic pattern in vertebrate embryos; not all vessels are present simultaneously. (b) Pattern in late bird embryo after modification. (After Kerr, 1919.)*

(DeRuiter *et al.*, 1993). The ventral aorta is the main artery into which the truncus arteriosus leads (*Text-Figure 27, Plates 28, 40, 65, 84*). It bifurcates into left and right vessels which extend forward as the paired external carotids (*Text-Figure 28*; *Plates 40, 41*), whilst the paired dorsal aortae extend forward as the internal carotids. The aortic arches do not all develop at once and some are of a transitory nature, disappearing before others have even begun to form. They are paired, serving the left and right pharyngeal regions. A morphological study of their formation was published by Hiruma and Hirakow (1995) using corrosion casts analysed by scanning electron microscopy (*Plates 40, 41*). The first pair forms at about stage 9 (Coffin and Poole, 1988) and passes through the left and right first pharyngeal (mandibular) arches, respectively. The remaining arterial arches form sequentially over the next 3 days of incubation (*Table 2*, p. 72). The basic pattern is characteristic of all amniotes and is indicated diagrammatically in *Text-Figure 27*. The aortic arches are formed, like the dorsal and ventral aortae, from solid angioblastic cords which develop *in situ* from mesoderm and subsequently become canalized. As the pharyngeal arches differentiate into other structures, the individual arterial arches either disappear or become converted. The cranial extensions of the dorsal aortae become the internal carotids in all young amniote embryos, whilst the cranial extensions of the ventral aortae become the external carotids. They are retained as such in mammals, but in birds the internal carotids in the head anastomose with the external carotids at about 6.5 days (Hughes, 1934), and the remnants of the internal carotids in the neck region disappear. The paired external carotids (*Plates 40, 41*) come to lie in close proximity to one another, fusing into a single vessel in some species (e.g. herons), though not in the fowl. The carotids branch to supply all the arteries of the head; the main branches have formed by 7.5 days and are illustrated in *Text-Figure 28*. A number of other anastomoses occur between the arteries of the head, forming structures that are sometimes compared to the circle of Willis of mammals. Some anastomoses take place also between arteries and veins.

Another important way in which birds differ from mammals is that in birds, (*Text-Figure 27*) the communication of the fourth arterial arch with the dorsal aorta is retained on the right and lost on the left (about days 6–7), whereas in mammals the the left is retained and the right lost. In amphibians and reptiles both sides are retained, resulting in a double 'systemic' arch.

In the chick it has been possible to induce experimentally the mammalian pattern by ligaturing the fourth arch on the right side (Stéphan, 1949), so that the blood is then diverted through the left aortic arch, which, therefore, fails to become obliterated. Experiments of this sort provide further evidence of the importance of the blood flow in modifying the vascular pattern but have little to tell us about the factors that provoke the normal closure and disappearance of blood vessels during embryogenesis. Hughes (1937) suggested that the 'posteriorward shift' in the position of the chick heart and aortic arches as the heart 'descends', together with the torsion and elongation of the truncus arteriosus,

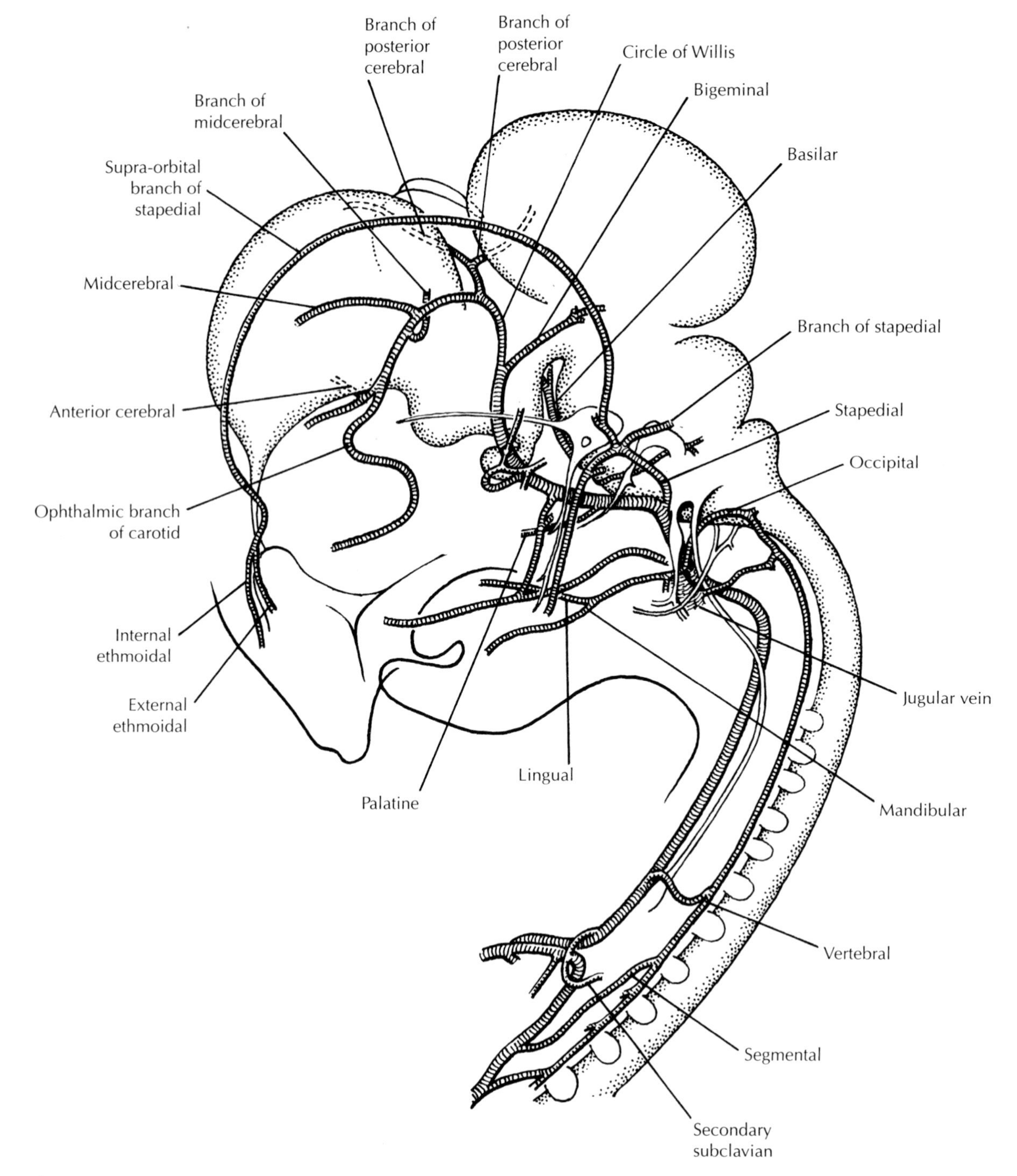

Text-Figure 28. *Arteries of the head and neck at 7.5 days. (After Hughes, 1934.)*

lead to mechanical tension which obliterates the lumina of the aortic arches (but see p. 50).

The fifth arch is transitory and disappears on both sides soon after formation (*Plate 40*). Although it is traditionally considered that the pulmonary arteries in amniotes are derived from branches of the sixth aortic arches (as in *Text-Figure 28*), there is now evidence that they develop as branches of the fourth aortic arch which connect with the endothelial vesicles in the pulmonary mesenchyme (Noden, 1990), and that only later does a communication arise with the sixth aortic arch.

On the right side the connection with the aorta remains as the ductus arteriosus. This becomes occluded around the time of hatching though it may remain patent for several days.

The two dorsal aortae unite posterior to the heart (*Plates 42, 90*) to form a single median vessel which gives rise to:

- The arteries supplying the limbs.
- The segmental (intersegmental) arteries (*Text-Figure 28; Plate 43*), which are branches of the dorsal aorta and extend out between the somites. They contribute to the formation of the vertebral and subclavian arteries (*Plates 41, 133, 134*), and by stage 11 have formed between all existing somites.

- The omphalo-mesenteric (vitelline) arteries (*Plates 4, 88*).
- The primary mesonephric arteries (*Plate 43*) from which the secondary and tertiary mesonephric arteries are formed, the latter becoming the afferent glomerular arteries (Carretero *et al.*, 1995).
- The allantoic (umbilical) arteries (*Plate 107*).
- The caudal artery (*Plates 42, 137*), which is the most posterior end of the dorsal aorta.

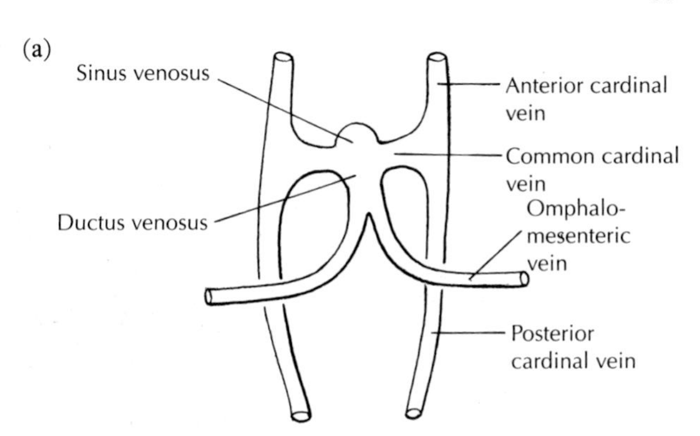

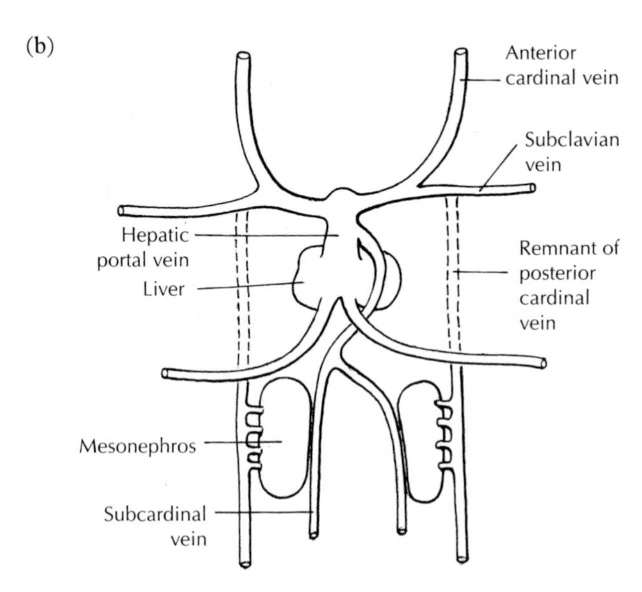

Text-Figure 29. *(a) The cardinal venous system at 3 days; (b) modifications that have occurred by 4 days.*

THE VENOUS SYSTEM

The embryonic venous system of birds is similar to that of mammals. It forms as an H-shaped structure which subsequently undergoes extensive modification. The sides of the H are derived from the left anterior and posterior cardinal veins and from the right anterior and posterior cardinal veins, whilst the crossbar is formed by the left and right common cardinals (ducts of Cuvier) which bring blood to the sinus venosus (*Text-Figure 29* is a simplified diagram, but *Text-Figure 30* indicates the curving of the sides of the H). The common cardinals begin to form at about stage 11 and by about stage 12 have fused with the omphalo-mesenteric (vitelline) veins, and the circulation has started. The cardinal veins arise *in situ* by the segregation of endothelial cells from the mesoderm (Coffin and Poole, 1988).

The anterior cardinal vessels (*Plates 73, 74*) begin to develop at about stage 8 and are formed by the fusion of a number of vascular spaces which include the left and right head veins (vena capitis), which lie on either side of the rhombencephalon.

The anterior cardinal veins (*Plate 74*) and the **vena capitis medialis** (median head vein) subsequently give rise to the main venous vessels of the head and neck. A network of vessels, the primary capillary plexus of the head, develops simultaneously. It becomes supplied with blood by the aorta in the forebrain and midbrain and is drained by the vena capitis medialis in the hindbrain region. Further veins, the left and right **vena capitis lateralis** (lateral head vein), have formed by about stage 17 and taken over the drainage of the forebrain. There are many changes in the patterning of the blood vessels of the head during the first half of incubation, vessels forming, anastomosing and often degenerating. They are described by Hughes (1934) and will not be considered here.

The anterior cardinal veins give rise to the jugular veins and the ventral veins of the neck region. Birds differ from mammals in that they do not form a true innominate vein, but instead retain both anterior cardinals, although a small cross-connection, the **interjugular anastomosis** (anastomotic vein), forms at about 7 days (*Plate 123*) and the right anterior vena cava becomes longer than the left. The pulmonary veins are formed from the blood vessels that develop in the splanchnic mesoderm surrounding the foregut (DeRuiter *et al.*, 1993) and they extend into the heart via the remnant of the dorsal mesocardium (Webb *et al.*, 2000). The left and right subcardinal veins (*Plates 102, 106*) are derived as branches from the posterior cardinals (*Plates 85, 86*). By day 6, however, much of the posterior cardinal system has disappeared and the subclavians have formed connections with the anterior cardinals (*Text-Figures 29, 31*). The jugular veins on each side join the vertebral and subclavian veins near the heart to form the **anterior vena cava**.

The Omphalo-Mesenteric (Vitelline) Veins

These bring blood back to the body from the yolk sac (*Text-Figure 26, 30,* Plates *4, 62, 65*), entering the body at the level of the anterior intestinal portal and

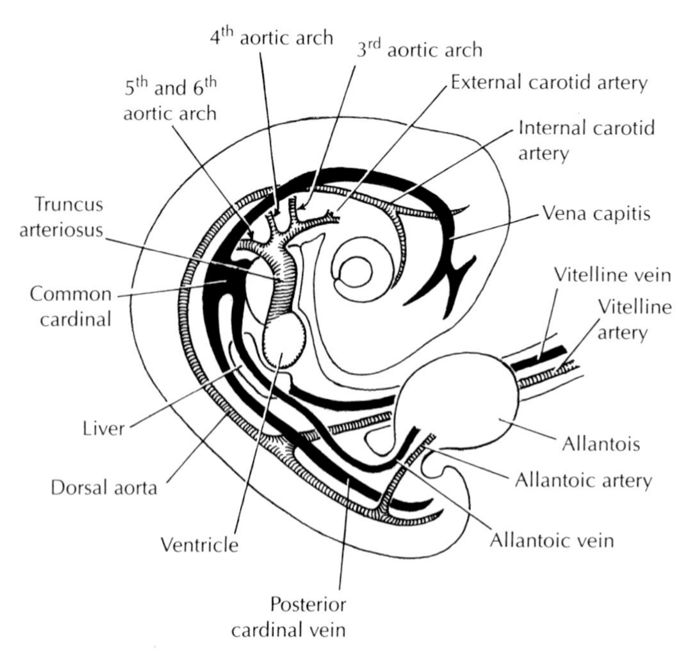

Text-Figure 30. *The main blood vessels of a 4-day-old chick prior to modification of the carotid arteries. (After Patten, 1950.)*

merging into a single channel, the **ductus venosus** (*Plate 91*), which empties into the sinus venosus (*Text-Figure 29, Plate 90*). During subsequent development, the ductus venosus becomes enveloped by the developing liver and breaks up into many channels, though its anterior end remains intact and contributes to the **posterior vena cava** (*Text-Figure 31, Plate 118*). The situation is similar to that in mammalian embryos where the ductus venosus is usually described as a shunt passing through the liver. By day 5 the hepatic portal vein has formed from the omphalomesenteric vein (*Text-Figure 29*).

The Posterior Vena Cava (*Plates 94, 110, 122*)

The formation of the posterior vena cava involves the formation and destruction of a series of vessels. The posterior cardinals (*Text-Figures 29, 31*) are fully formed by about stage 17, sprouts from them becoming the segmental veins which pass into the intersomitic spaces along with the segmental arteries. Additional sprouts extend from the posterior cardinals and anastomose to form a new series of veins, the left and right subcardinals (*Text-Figures 29, 31, Plate 43*), which lie close to the left and right posterior cardinals respectively, on the medial side of the mesonephroi and lateral to the aortae. Branches from the subcardinals collect blood from the hindlimbs and tail as well as from the intestines, and pass it via the mesonephric veins on to the posterior cardinals. The mesonephric veins are therefore renal portal vessels (see Chapter 7). (See Arcalis *et al.*, 2002, for a detailed description of the vasculogenesis and angiogenesis of the quail mesonephros.) The posterior vena cava (*Plate 94, 118*) begins to form at about 3.5 to 4 days. Both the posterior cardinals (*Plate 86*) and the subcardinals (*Plates 43, 88*) contribute to it. The subcardinals fuse just caudal to the omphalomesenteric artery by day 5 (*Text-Figure 31b*), and the anterior part of the posterior vena cava subsequently fuses with the right subcardinal vein (*Text-Figure 31*). By day 6 the anterior ends of the posterior cardinals have disappeared so that all the blood from the posterior end of the body must now pass through the mesonephric kidneys (*Plates 42, 43, 94, 96*) on its way to the subcardinals. The subcardinals, therefore, become the efferent vessels of the mesonephros. Using scanning electron microscopy of vascular erosion casts, Carretero *et al.* (1997) discovered that an additional blood supply reaches the anterior part of the mesonephros by a hitherto undescribed vessel, the anterior mesonephric portal vein (*Plate 43*). Later, the mesonephric kidneys become replaced by the metanephric kidneys (*Plates 43, 122*) (see Chapter 7) and the posterior parts of the subcardinals disappear and are replaced by the renal veins. The metanephros possesses both an anterior and a posterior portal vein. The posterior metanephric portal vein forms from the mesonephric portal vein, itself a derivative of the posterior cardinal vein, whilst the anterior metanephric portal vein is formed from a new vessel unconnected with the mesonephros. It connects with the vertebral venous sinus. The posterior part of the posterior cardinals fuse together on day 6 (*Text-Figure 31c, Plate 42*) and send a branch, the **coccygeal vein**, into the tail. Further anteriorly, the posterior cardinals give rise to the **sciatic** and **iliac veins**.

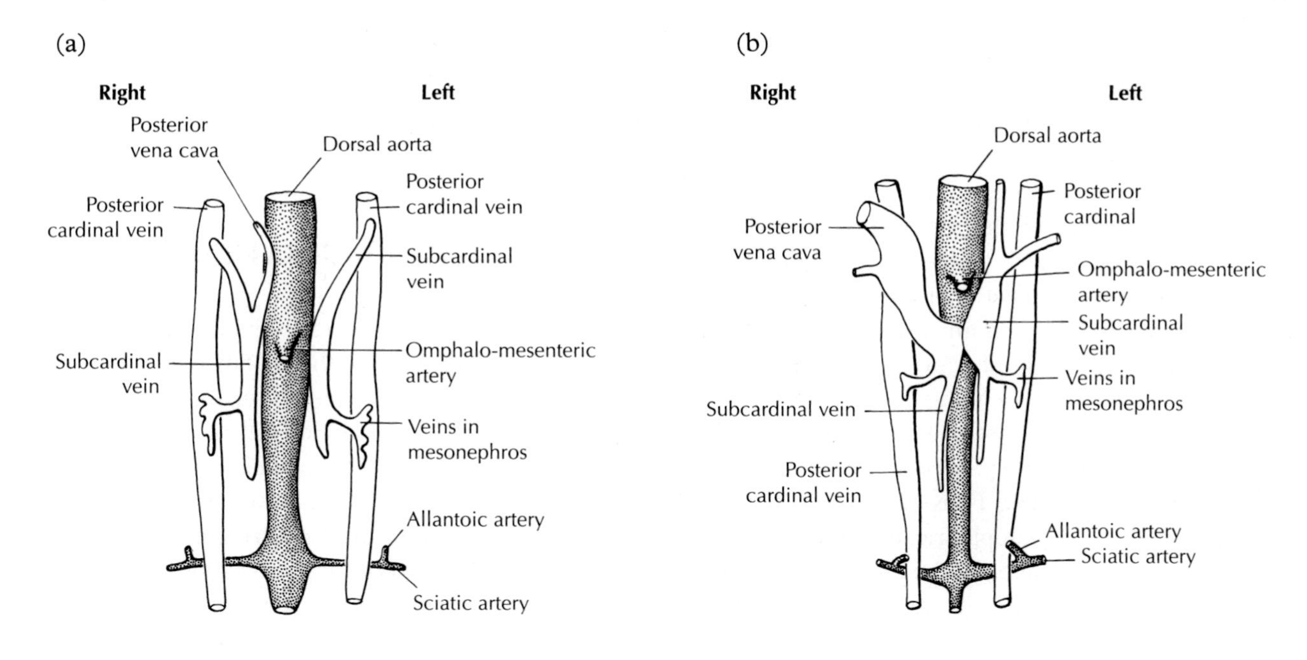

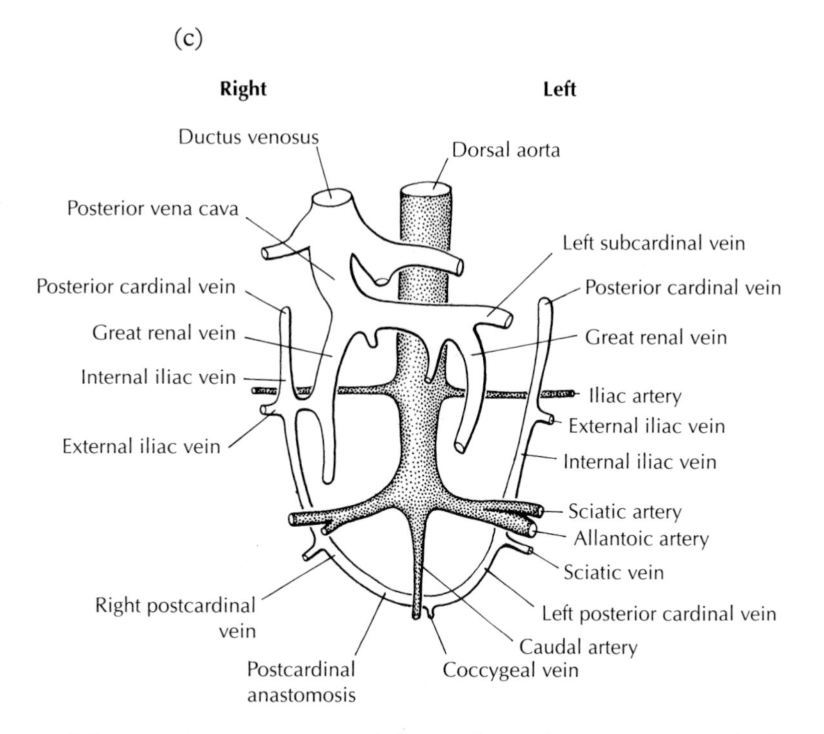

Text-Figure 31. *Development of the posterior vena cava and the renal portal system at: (a) 60 h; (b) 5 days; (c) 14 days. (After Lillie, 1952.)*

The internal iliac is also known as the posterior mesonephric portal vein (Carretero *et al.*, 1997).

THE ENDOTHELIAL AND HAEMOPOIETIC CELLS

The formation of the blood cells is closely linked with the development of the endothelium.

The first **erythrocytes** (the primary lineage) in amniotes arise from stem cells in the yolk sac and possess a specific haemoglobin (Dieterlen-Lièvre, 1984), though eventually these stem cells die out (Dieterlen-Lièvre, 1997). A second set of stem cells arises from day 5 in the embryonic organs and gives rise to the secondary lineage of erythrocytes, which possess different haemoglobin patterns. Experiments utilizing chick–quail chimeras show that it is the stem cells from the intra-embryonic tissues that normally colonize the thymus and suggest that these embryonic stem cells arise in the lateral plate mesoderm during days 2–4 of incubation

(Dieterlen-Lièvre, 1984). They are then identifiable in the mesenchyme near the thoracic and abdominal aorta and subsequently colonize the thymus and bone marrow where haemopoiesis begins. Subsequent grafting experiments have shown that the angioblastic capacity of the mesoderm depends on whether it is associated with ectoderm or endoderm (Pardenaud and Dieterlen-Lièvre, 1993). These authors suggest that there is an early haemangioblastic rudiment consisting of common ancestors for endothelial cells and haemangiopoietic cells, and that for these cells to become committed to the haemangioblastic line they need to be induced by the underlying endoderm. Thus, mesoderm that lies against ectoderm (somatopleuric mesoderm) never becomes vascularized and, as the authors point out, this finding correlates with the lack of vascularity of the amnion, which is composed of ectoderm and somatic mesoderm. By contrast, the splanchnopleural mesoderm does become vascularized, and this includes the rich vascularization of the allantois (endoderm plus splanchnic mesoderm). They propose that the endodermal factor is likely to be a growth factor.

Haematopoiesis begins in the spleen at about 10 days, though haemopoietic cells are present from day 4. Lymphocytic homing in the spleen begins just before hatching (Yassine *et al.*, 1989).

THE LYMPHATICS

Until recently, the lymphatic system was thought to be formed by sprouting from the venous system but new, highly specific, markers for lymphatic endothelium have shown that lymph angioblasts are present in the mesoderm a considerable time before the lymph sacs develop. Lymphangioblasts migrate from the somites into the somatopleure and contribute to the lymphatics of the limb (Wilting *et al.*, 2001). A series of lymphatic capillaries arises from a plexus. Lymph 'hearts' are muscular swellings which pulse and help the lymph to flow and are thought to be a modification of the lymph vessels. It is difficult to see the lymph vessels in histological sections because of their small size in the embryo; even in the adult they seldom exceed 1 mm in diameter. A detailed account of the adult lymphatic system is given by Rautenfeld (1993).

Chapter 7
Urino-Genital System

The development of the urino-genital system follows the usual amniotic pattern, but it includes certain differences that are related to the production of large-yolked eggs, one of the most important being the reduction and loss of function of the right ovary and oviduct. The kidneys of birds differ from those of mammals in certain respects and birds possess a renal–portal system (see Chapter 6) that is not present in mammals.

THE URINARY SYSTEM

With the exception of the cloaca and the primordial germ cells, the urino-genital system is derived from the intermediate mesoderm which lies between the somites and the lateral plate, to both of which it is initially attached (*Plate 56*). It appears at the same time as the somites and develops simultaneously with them, from anterior to posterior down the trunk, but it differs from the somites in that it is not overtly segmented and does not extend into the region of the post-cloacal tail. The intermediate mesoderm becomes a thick band, the so-called **nephrogenous mesenchyme** or nephrogenic cord (*Text-Figure 32; Plate 78*) extending down the trunk on either side of the gut. By 3 days of incubation they have formed a ridge, the **urinogenital (nephric) ridge**. It gives rise initially to the anteriorly situated **pronephric kidneys**, which are generally considered to be non-functional, then to the more posteriorly located **mesonephric kidneys** (*Plate 43a,b*), which are active throughout much of embryonic life, and finally to the **metanephric kidneys** (*Plate 43c*) which is fully functional by the day 15 of incubation. These kidneys are arranged in pairs, left and right. The pro- and mesonephric kidneys are continuous with one another (*Text-Figure 32*) and initially there is not always a clear division between where the one ends and the next begins so that they are sometimes called together the nephric kidney to distinguish them from the metanephric kidney. Each kidney is supplied with an excretory duct. The duct that serves the pronephros and subsequently the mesonephros has commonly been known in its early stages as the pronephric duct and in its later stages as the mesonephric duct. However, many authorities now prefer to call it the nephric duct throughout its entire period as a urinary structure. When the nephric duct is no longer needed as a urinary duct it becomes the **Wolffian duct**, and ultimately the **vas deferens** in males. Prior to the development of the metanephros a new duct forms, the **metanephric (ureteric) duct**, which later becomes the ureter.

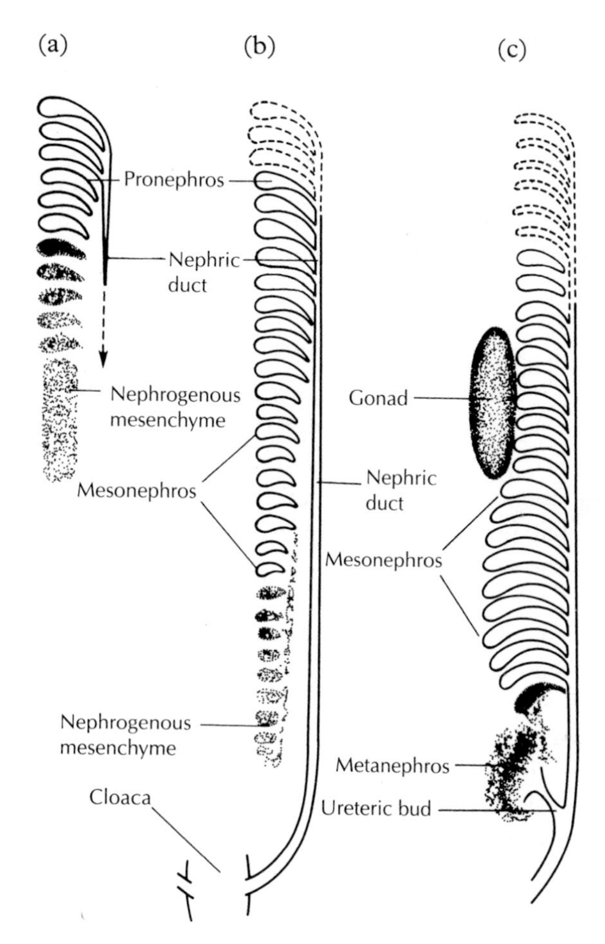

Text-Figure 32. *Plan of the relationship of (a) the pronephros, (b) the mesonephros and (c) the metanephros. (After Burns, 1956, with permission from Elsevier Press.)*

The pronephric region of the nephric kidney begins to form at about 36 h of incubation (about stage 10) (Abdel-Malek, 1950) and develops gradually from about the level of the 6th to that of the 13th somites. It becomes separated into a dorsal portion which gives rise to the nephric duct, and a ventral part which forms the tubules (Hiruma and Nakamura, 2003). Beginning at about stage 13, it degenerates from anterior to posterior, and has completely disappeared by day 8 of incubation (Romanoff, 1960). The mesonephros (*Plates 43a, 44–46*) overlaps the pronephric region and extends from about the level of the 9th to the 15th somites and, like the pronephric kidney, it forms sequentially from anterior to posterior. It functions from about day 5 to day 11 of incubation. A vascular cast of the mesonephros of a 6-day quail embryo is shown in *Plate 43*. There is some discrepancy in the literature as to when the mesonephros starts to degenerate, but it has probably ceased to function in the chick by about day 15 of incubation. For a description of mesonephric degeneration see Carretero *et al.* (1995).

The first-formed (pronephric) region of the nephric kidney appears to be induced before stage 8 by signals from the adjacent paraxial segmental plate mesoderm. Mauch *et al.* (2000), using *Pax-2* mRNA expression and Lim1/2 antibody staining as markers for the conversion of intermediate mesoderm into pronephric tissue, showed that if the intermediate mesoderm was surgically separated from the somitic mesoderm the pronephros failed to develop. By contrast, the presence or absence of the neural tube, or the notochord or the lateral plate, did not affect the development of the pronephric region.

The gene *Frizzled-4* (*cFz-4*), which is characteristic of the metanephric kidney, is first expressed in the pronephric region posterior to the third somite at stage 10, its expression increasing with development though gradually becoming restricted to the newly formed glomeruli and tubules of the mesonephros and metanephros (Stark *et al.*, 2000).

The mesonephric nephrons are probably induced by the nephric duct. After experiments in which parts of the nephric duct were ablated, or in which migration of the duct was interrupted (e.g. Gruenwald, 1941; Bishop-Calame, 1965), no further differentiation of the mesonephros took place. Not all experimental evidence supports the idea of induction of the mesonephric kidney by the nephric duct, however, some favouring the possibility of induction by other tissues (e.g. by the gut endoderm, Croisille *et al.*, 1976).

The histological structure of a vertebrate nephron is shown in *Text-Figure 34*. A major component is the glomerulus, a tuft of capillaries that carries afferent blood and is formed as an outgrowth from the dorsal aorta. Vascular casts of encapsulated **glomeruli** from mesonephric and metanephric kidneys are illustrated in *Plate 43b,c*, respectively, and sections through mesonephric glomeruli are shown in *Plate 46*. The proximal and distal tubules of each nephron unite to form a collecting duct that communicates with the nephric (or metanephric) duct. This basic structure undergoes modification in later stages. The glomeruli of the pronephric region are first visible at about stage 17 (Abdel-Malek, 1950) but the nephric tubules in this region do not become fully patent and the blood vessels in the glomeruli are not continuous with the aorta. It is for these reasons that the pronephros is considered to be non-functional. The nephric tubule of the mesonephros, which extends from the glomerulus to the nephric duct, elongates to form a highly coiled structure consisting of proximal and distal regions, together with several intermediate (or transitional) 'segments'. Zemanova and Ujec (2002), who examined the potential difference across the tubular epithelium in embryos of 7 days incubation (stage, 30–31), concluded that the distal nephron, unlike the proximal, is capable of active transport at this stage. Excretory products from the blood are filtered through the glomerulus and pass into the nephric duct.

By 8 days of incubation and until day 20, transforming growth factor α (TGFα) becomes expressed in the distal nephric tubules, and from day 14 of incubation, and continuously thereafter into the adult, is found in the distal nephric tubules of the metanephros (Diaz-Ruiz *et al.*, 1993).

A distinction between internal and external glomeruli is sometimes made (*Text-Figure 35*). External glomeruli, which are the first to form, are nephrogenic masses hanging freely in the body cavity, forming the nephrogenic ridge (*Text-Figure 33*). Internal glomeruli, which develop later, lie within the nephrogenic ridge and are individually enclosed within a capsule, an internal glomerulus together with its capsule constituting a Malpighian corpuscle. External glomeruli are characteristic of the most anterior (pronephric) region, whereas internal glomeruli are characteristic of the mesonephric and metanephric regions.

Each nephric duct (*Text-Figure 32; Plate 45*), like its kidneys, develops from the intermediate mesoderm and is visible at about 30 h of incubation (stage 9). During the next 30 h it lengthens steadily and fuses with the cloaca (*Plate 79*) at about stage 17.

There are several possible, though not mutually exclusive, ways in which the duct might lengthen. They include differentiation of intermediate mesoderm *in situ*, addition of cells from the lateral plate, a high level of mitosis and/or cell elongation, and finally the migration of mesodermal cells at the caudal end. The evidence strongly supports the idea

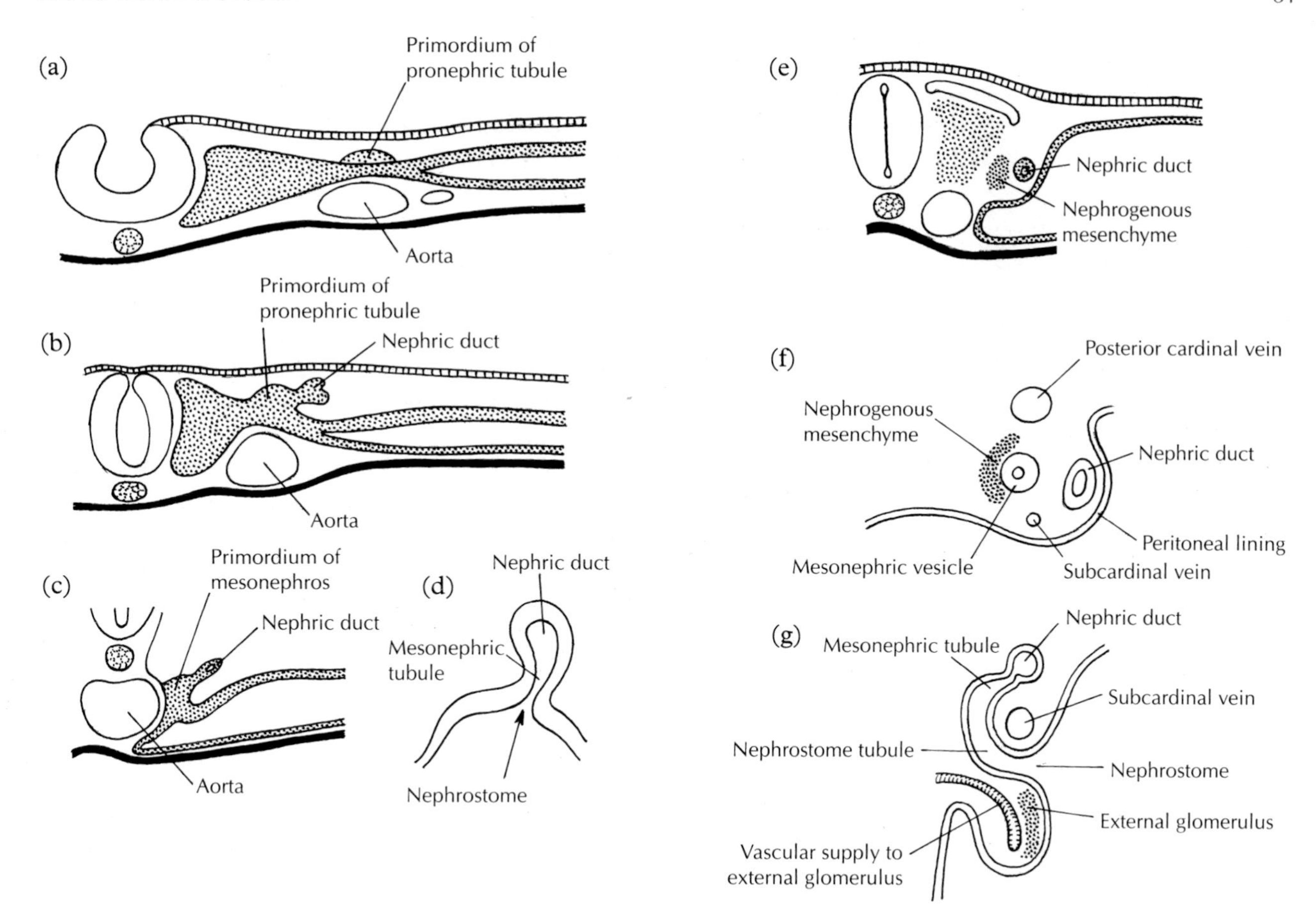

Text-Figure 33. *Transverse sections to show stages in the formation of the nephric duct and kidney. (a, b) The primordium of the nephric duct and a pronephric tubule at about stage 10, at about the level of somite 11 or 12. (c, d). The primordium of the mesonephros at about stage 11 at the level of somite 11. (e) At about stage 16 at the level of somite 22. (f) At about stage 17 at the level of somite 15, showing the nephric duct alongside nephrogenous mesenchyme. The increased mass of tissue now bulges into the coelomic cavity and is known as the nephrogenic ridge. (g) At about stage 18 at the level of somite 12, showing the mesonephric tubule and an external glomerulus with its vascular supply. (All adapted from Abdel-Malek, 1950, reprinted by permission of Wiley-Liss. Inc., a subsidiary of John Wiley and Sons, Inc.)*

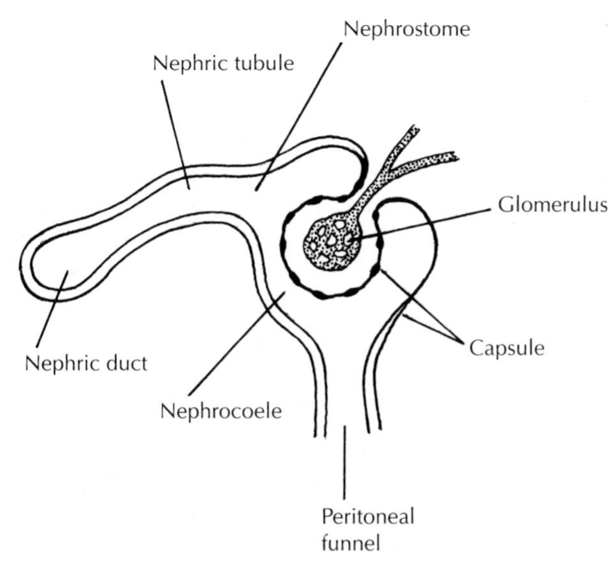

Text-Figure 34. *A vertebrate nephron. (After Torrey, 1965.)*

that migration plays a significant role (discussed by Bellairs *et al.*, 1995) and is based mainly on the results of transection experiments (Bishop-Calame, 1965), of the effects of destruction of the caudal end (Boyden, 1927; Jacob and Christ, 1978), and of the composition of the duct in experimentally produced chick–quail chimeras (Martin, 1971). As it lengthens, the duct becomes epithelial but at its posterior (i.e. undifferentiated) end it remains mesenchymal (*Plate 45*). In amphibian embryos, the nephric duct has been shown experimentally to migrate actively toward the cloaca, apparently up an adhesive (haptotactic) substrate gradient (Poole and Steinberg, 1982), though the nature of the gradient material is unknown, and to recruit cells *en route* (Cornish and Etkin, 1993). The substrate probably includes fibronectin (Bellairs *et al.*, 1995).

An important controversy, extending back for more than a hundred years, surrounds the early stages of development, that of whether the first nephric ducts form before (e.g. Jarzem and Meier, 1987) or after (e.g. Abdel-Malek, 1950) the first nephrons in the pronephric kidney. Related to this is another question, that of whether the ducts form from the nephrons or the nephrons from the ducts. The evidence is largely morphological. Lear (1993)

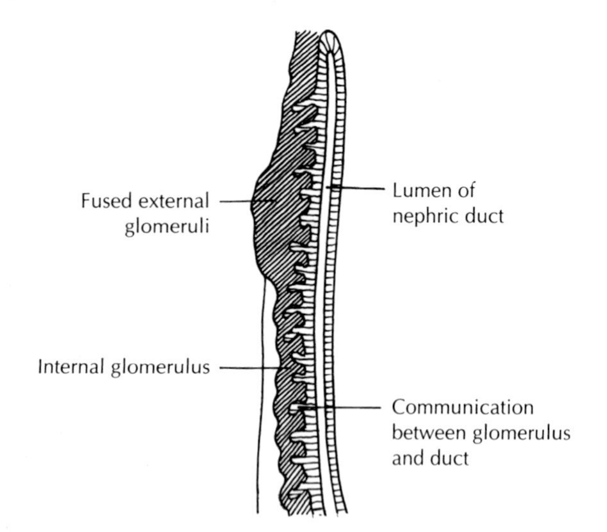

Text-Figure 35. *Longitudinal section to illustrate the relationship between the external and internal glomeruli. (After Abdel-Malek, 1950, reprinted by permission of Wiley-Liss, Inc., a subsidiary of John Wiley and Sons, Inc.)*

who used both scanning electron microscopy and immunocytochemistry, concluded that the duct begins to form before the nephrons.

The metanephric diverticulum (the future ureter) arises from the posterior end of the nephric duct just short of its junction with the cloaca and begins to form at the end of day 4. It extends first dorsally and then anteriorly, passing close to and parallel with the nephric duct (*Text-Figure 36*), though separated from it by the posterior cardinal vein. The question of how the metanephric duct is guided to its destination is an intriguing one, since unlike the nephric duct, which travels towards the posterior end, the metanephric travels away from it. It seems likely therefore that the nephric and metanephric ducts are either responding to different cues in the environment, or that there is a significant change in the distribution of environmental cues in the time interval between the completion of migration of the nephric duct and the start of that of the metanephric duct.

The metanephric kidney (*Plate 43c*) begins to develop after the metanephric duct has contacted the most posterior region of the nephrogenous mesoderm, which has not been incorporated into the mesonephros. The procedure appears to be similar to that in mammals; if the nephric duct is experimentally prevented from making contact with the cloaca then no metanephric duct develops and no metanephric kidney is formed. This, and many other experiments, provide compelling evidence that the metanephros is induced by the metanephric (ureteric) diverticulum.

Although the nephrogenic tissue of the metanephros is originally continuous with that of the mesonephros, the two sets of kidneys become separated by the degeneration of the junctional region on days 4–5. Some of the mesonephric veins that supply the mesonephros degenerate, along with the mesonephros itself, whilst others colonize the developing metanephros (Carretero *et al.*, 1997).

The metanephric kidney (*Plates 94, 96*) is formed partly from the most caudal (metanephric) end of the nephrogenous mesenchyme and partly from the metanephric duct. The mesenchyme gives rise to the renal corpuscles and the cortical tubules as well as the capsule and stroma of the kidney; the metanephric duct forms the medullary (collecting) tubules and ureter. It is vascularized by branches of the dorsal aorta. Our understanding of the interactions between the epithelial metanephric duct and the metanephric mesenchyme is based mainly on the mouse and is reviewed by Gilbert (2000). By the time it reaches the metanephric mesenchyme, the metanephric duct has begun to branch and, contrary to the usual pattern of organ development, the posterior branches form first, followed progressively by more and more anterior ones; this may be related to the fact that the duct itself contacts the posterior region first. The metanephric mesenchyme consists of two zones, the inner lying between the outer zone and the

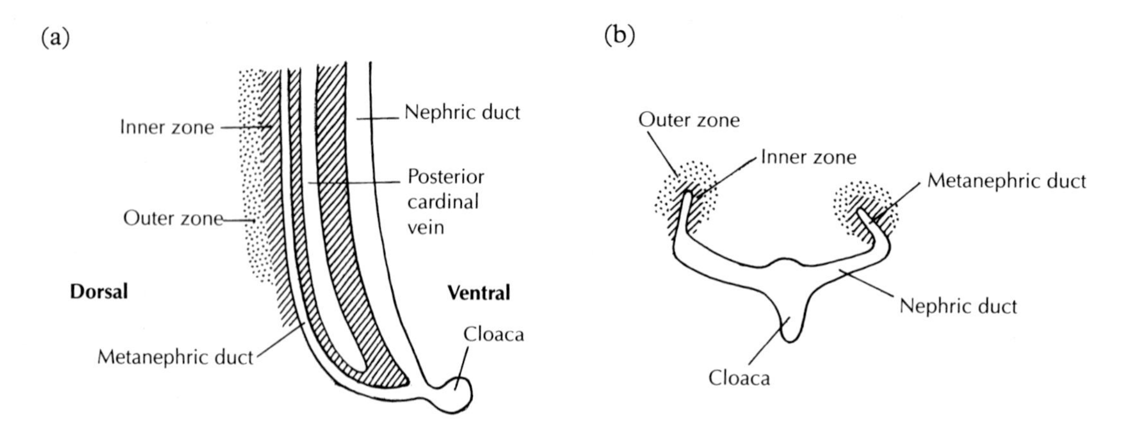

Text-Figure 36. *Relationship of the nephric (Wolffian) duct and metanephric duct at about 4 days. The metanephric mesoderm consists of an inner and outer zone. (a) lateral view; (b) ventral view. (After Abdel-Malek, 1950, reprinted by permission of Wiley-Liss, Inc., a subsidiary of John Wiley and Sons, Inc.)*

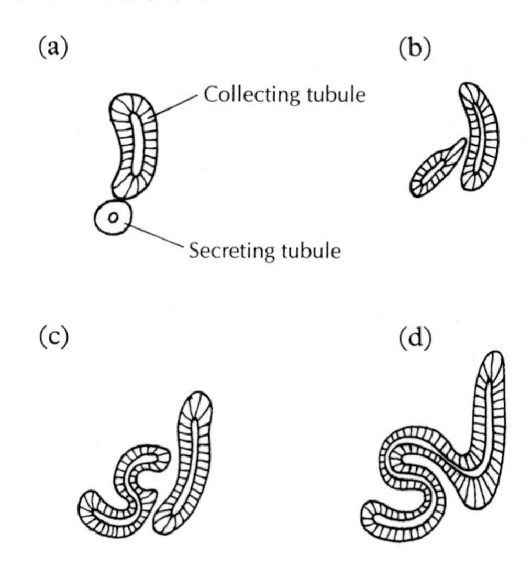

Text-Figure 37. Sequence to show development of S-shaped metanephric tubule. (a) Secreting tubule lying close to collecting tubule. (b, c) The secreting tubule has elongated and become S-shaped. (d) The S-shaped secreting tubule now opens into the collecting tubule. (a–c) about 9 days, (d) about 11 days. (Modified after Schreiner, 1902.)

metanephric duct (*Text-Figure 36*). The primary capsule of the kidney is formed from the outer zone. At about 7 days, each branch of the metanephric duct induces adjacent to it the formation of a cluster of cells formed from the inner region. A lumen develops within each cluster, which then transforms into an epithelial vesicle. Each vesicle elongates into an S-shape, a secretory tubule, the proximal portion of which fuses with the collecting tubule, whilst the distal portion becomes the renal corpuscle (*Text-Figure 37*). Further elaboration takes place and by day 11 the entire metanephric kidney has formed secreting tubules, and has begun to excrete. The glomeruli are supplied by segmental branches of the dorsal aorta.

About days 5–6 of incubation, the cloacal wall expands to engulf the posterior end of the nephric duct so that its junction with the metanephric duct becomes absorbed, the result being that the metanephric duct acquires a separate opening into the cloaca from the nephric duct. The metanephric duct will become the ureter. The process is similar to that in mammals except that birds retain the cloaca and do not form a bladder.

The metanephric kidney of adult birds is three-lobed and elongated and does not possess the 'kidney-shape' characteristic of mammals. The left and right kidneys lie at the same antero-posterior level and are not staggered as in mammals. The loop of Henle is less well developed than that of mammals.

In the adult kidney there are three types of nephrons, mammalian type I (juxtamedullary) with long loops of Henle, mammalian type II with short loops of Henle, and reptilian type. These differentiate after day 14 but maturation is not complete until after hatching (Gambaryan, 1992). In the mammalian kidney the cortex and medulla are clearly defined, but in the avian metanephric kidney the two are intermingled.

THE GENITAL SYSTEM

Like the urinary system, the genital system forms from the intermediate mesoderm, the two sets of organs being closely linked. By day 3, the nephrogenous mesenchyme has enlarged and given rise to a pair of ridges that hang down from the body wall into the peritoneal cavity on either side of the gut (*Plate 44b*). As the genital components develop, each ridge becomes known as a uro-genital, nephrogenic or **genital ridge** (*Plate 77*). It is covered by the peritoneal epithelium derived from the splanchnic mesoderm. Although an inspection of the chromosomes will disclose the sex of the embryo, it is not possible in the early stages of gonad development to determine from its histology or gross morphology whether it will become a testis or an ovary. This is, therefore, called the **indifferent stage**. There is a large literature to show that further differentiation is affected by hormonal activity. For example, female embryos develop testes if implanted at 3 days of incubation with testes from 13-day-old male embryos (Stoll *et al.*, 1993) and early treatment of males with oestradiol leads to feminization of the gonads and retention of the Mullerian duct (Faucounau *et al.*, 1995). In birds the **heterogametic sex** (ZW) is the female, whereas the homogametic sex (ZZ) is the male. This is the reverse of the situation in mammals, where the heterogametic sex (XY) is the male and the homogametic (XX) is the female.

By 3.5 days the primordial germ cells have begun to colonize the gonad and initially lie in the peritoneal epithelium. They can be distinguished morphologically from the peritoneal cells by their size, about 10 μm in diameter, as opposed to the peritoneal cells, which are about 5 μm, but more reliably by specific staining (see below). The primordial germ cells enter both the right and left gonads and undergo proliferation from 10 days, but the degree of cell death is greater in the right than in the left (Ukeshima and Fujimoto, 1991), and the right ovary has itself undergone considerable atrophy by the time of hatching.

There are two further components of the indifferent gonad, the **rete cords** and the **primary sex cords**. The rete cords are solid columns of epithelial cells that appear on day 5 among the mesenchymal stroma, and form a meshwork in the anterior region of the gonad by branching and anastomosing. At a later stage they become canalized in the male and form a connection between the primary sex cords and the modified Wolffian tubules, thus forming the rete testis. In the female this connection is not made.

The primary sex cords appear toward the end of day 5. They differentiate into the seminal tubules in the male and the medullary cords in the female. They form by proliferation of the epithelial cells that invade the mesenchymal stroma and they carry the primordial germ cells with them into the substance of the gonad. As both the rete cords and the sex cords proliferate, the indifferent gonad increases greatly in size, causing the genital ridge to bulge further into the coelomic cavity. In both sexes the left gonad is larger than the right and contains more primordial germ cells. The oestrogen receptor mRNA is present in both males and females at stage 26 before any morphological differentiation has taken place between males and females (Andrews *et al.*, 1997). Once the sex differentiation of the gonads has begun, however, the expression of the *aromatose* gene (*cAROM*), which is essential for the synthesis of oestrogen, is restricted to the female. At the time that the morphological differences between the left and the right ovary start to become apparent the expression of the *aromatose* gene in the left ovary becomes much greater than that in the right (Villalpando *et al.*, 2000). Administration of an *aromatose* inhibitor induces testis development in a genetic female (Nakabayashi *et al.*, 1998) and masculine type genitalia (Burke and Henry, 1999). Conversely, the administration of oestrogen induced an ovo-testis in genetically male embryos (Nakabayashi *et al.*, 1998). The normal development of the male is characterized by the production of anti-Mullerian hormone (Oreal *et al.*, 1998) and the expression of Sox9 in the genital ridges (Kent *et al.*, 1996; Morais da Silva *et al.*, 1996).

Sexual Differentiation

The **Testis** *(Plate 128)*

By the end of day 7 the sex cords have proliferated so much that the rete cords have become confined to the hilum. The primary sex cords, which will differentiate into the seminiferous tubules, continue to branch and proliferate and anastomose but do not become canalized until about day 20. The **primordial germ cells** begin to divide and differentiate about day 13 into spermatogonia. The Sertoli cells are thought to be derived from the so-called germinal epithelium (that part of the peritoneal epithelium that encapsulates the gonad) which becomes thinner from about day 11.

We have seen that as the rete cords become canalized they provide a passage for the sperm from the seminiferous tubules to the mesonephros. These mesonephric tubules become the vasa efferentia and their glomeruli disappear. The nephric (Wolffian) duct is no longer an excretory duct and becomes the vas deferens, subsequently carrying the sperm to the cloaca.

The **Ovary** *(Plate 136)*

Only the left ovary is functional in adult birds. The indifferent gonad becomes recognizable as an ovary at about 7–8 days by the expansion of the primary sex cords, which individually increase in cross-section and length at about 9–11 days. Subsequently, a second set of cords, the **secondary sex cords**, are formed from the epithelium. The primary sex cords become the medulla of the ovary, whilst the secondary sex cords become the cortex.

The primordial germ cells lie in the secondary sex cords and differentiate into oogonia during day 7. The right gonad fails to form a cortex and remains rudimentary unless the left ovary is removed, when the right undergoes compensatory growth (Groenendijk-Huijbers, 1967). It is capable of differentiation, however, if the left ovary is extirpated, though it forms a testis and not an ovary. It is usually considered that the gonad is incapable of becoming a true ovary without the cortex. Such an experiment does not, however, affect the genetic make-up of the individual.

Unlike the situation in the testis, no communication is made between the ovary and the mesonephros, and the latter then degenerates, apart from the region that corresponds to the rete testis in the male. This is found as a rudiment, the epio-ophorum, in the female.

The Sex Ducts

The potential to form either male or female structures is not confined to the gonads but is also a feature of the sex ducts and the external genitalia. Each individual initially possesses paired male (**Wolffian, nephric**) and paired female (**Mullerian** or **paramesonephric**) ducts (*Plate 44*), but by day 8 of incubation there is a higher density of progesterone-receptor-containing cells in the epithelium of the urino-genital sinus of the female than of the male (Gasc, 1991).

We have seen that the male (Wolffian) duct is used in the early stages for excretion by the mesonephros (and is then known as the mesonephric or nephric duct) but is subsequently converted into the male genital duct. This duct undergoes degeneration in the female though remnants may persist in the adult.

The female (Mullerian) ducts (*Plates 47, 136*) develop adjacent to the Wolffian ducts but at a later stage, and according to Gruenwald (1941) are induced by them. Both of the Mullerian ducts of the male and the right Mullerian duct of the female begin to degenerate about day 8. In the male, the degeneration involves extensive cell death and there is some evidence that this is related to the secretion of the male hormones by the testis (Wolff, 1953). The Mullerian ducts have usually disappeared in the male by about day 12, whereas in the female the right Mullerian duct persists until about day 18.

Our understanding of the way in which the Mullerian duct forms has recently been revised. The traditional concept has been that the Mullerian duct develops first as a groove that appears on the dorso-lateral surface of the nephrogenic ridge and rolls up into a tube, which then sinks beneath the surface of the genital ridge. Jacob *et al.* (1999), however, using scanning electron microscopy, have shown that it begins as a placode-like thickening in close contact with the Wolffian duct at stage 19, and subsequently becomes canalized, the process being almost completed by about stage 28. The Mullerian duct extends by means of groups of cells which detach from the solid cord and move posteriorly, using the basal lamina of the Wolffian duct as a substrate (Gruenwald, 1941). There is a high level of cell proliferation at this stage (Jacob *et al.*, 1999). The blind posterior end meanwhile extends by cell elongation and rearrangement and partially by active migration. According to Gruenwald (1941), if the nephric duct is interrupted, the Mullerian duct does not migrate beyond the break and it seems likely, therefore, that the Mullerian duct uses the Wolffian (nephric) duct as a guide. The ducts in the female fuse with the cloaca about day 11. The left duct then undergoes modification into ostium, glandular regions and the shell gland (*Text-Figure 2*). The opening of the duct into the cloaca does not, however, occur until after hatching.

The External Genitalia

In both sexes genital folds and swellings (*Plates 48, 49, 137*) arise at about 6 days of incubation around the cloaca, and a genital tubercle forms at the anterior end. In the mouse, the development of the external genitalia is, in part at least, dependent on *sonic hedgehog* signalling from the urethral epithelium (Perriton *et al.*, 2002) but it is not clear whether the same processes are involved in birds. The appearance of the external genitalia varies considerably in different species (discussed by Romanoff, 1960). In the goose and the male mallard duck the genital tubercle is highly developed even in the embryo, whereas in the domestic fowl it is rudimentary until after hatching. The appearance of the external genitalia is affected by the hormonal activity of the embryo. Treatment with 17β-oestradiol leads to an increase in the proportion of embryos exhibiting phenotypically female external genitalia (Coco *et al.*, 1992).

The development of the external genitalia has been described by Bakst (1986) (*Plates 48, 49*). Paired 'genital eminences' have formed in the cloacal membrane by day 5 (*Plate 48a*) and have fused together to form the genital tubercle by day 5.5 (*Plate 48b*). Meanwhile, paired 'genital swellings' have formed at the base of the tail, and by day 6 they have not only merged with each other but have grown to overhang the genital tubercle (*Plate 48c*). By day 7 the enlarged genital tubercle has acquired a bi-lobed appearance (*Plate 48d*) and by day 8 the genital swellings have fused with each other and extended like a collar around the genital tubercle (*Plate 48e,f*). The phallus is formed from the genital tubercle (now the median phallic body) with contributions from two (left and right) lateral phallic bodies. The phallus enlarges and rotates towards the floor of the proctodaeum so that it is visible only after dissection. Meanwhile, radial folds arise around the periphery of the cloaca, and by day 16 the cloacal aperture has become transversely oriented with the smaller, dorsal lip overhanging the ventral one (*Plate 49c,d*).

The genital tubercle is initially a solid mass of mesenchyme covered with epithelium, but in the male chick a groove develops along its length, the edges of which come together and convert it into a tube, the external seminiferous duct. This gives rise to the penis, which shows remarkable morphological variation in different species, e.g. in the mallard duck the penis is a conical organ with a triple spiral. The male tubercle in the 1-day-old hatched chick is slightly bigger than that of the female and can be used by experienced operators for sexing. Commercially, however, this procedure has been largely superseded by the introduction of breeds of fowl in which the colour or patterning of the down is sex-linked.

The Primordial Germ Cells

The primordial germ cells are the cells that give rise to the germ cells in the gonads from which the gametes are ultimately derived. They begin to differentiate in the early embryo long before the gonads, or indeed any other organs, have started to form and subsequently migrate through the body to the germinal ridge to colonize the developing gonads.

One of the classical problems of embryology has been that of the origin of the primordial germ cells. Weismann (1885) proposed that a specific region of cytoplasm, the so-called **germ plasm**, was set aside in the fertilized egg and that this was the forerunner of the entire lineage of the germ line. This theory, usually called the 'continuity of the germ plasm' supposed that the cytoplasm of the germ line was always kept separate from that of the somatic cells from generation to generation. The alternative concept is that the germ cells arise *de novo* in each individual from somatic cells. Nieuwkoop and Satasurya (1979), who reviewed the situation throughout the animal kingdom, concluded that the idea of a continuity of the germ plasm is an out-of-date and unneccessary concept and that the differentiation of germ cells and of somatic cells is much the same. Similarly, McLaren (1981) assembled evidence to suggest that all cells are initially totipotent and that their development into germ cells or somatic cells is dependent on the position in which they find themselves in the embryo during the crucial early stages.

Nevertheless, a germ plasm of sorts does appear to exist in most species of animals. For example, at the posterior pole of the *Drosophila* egg an assembly of mitochondria, fibrils and polar granules is present which is essential for the formation of primordial germ cells (Mahowald, 1971) and a similar region is present in the amphibian egg. Recently, a germ plasm has been identified in the chick embryo using an antibody to the chick vasa homologue (CVH) protein (Tsunekawa *et al.*, 2000). The vasa gene is expressed in the germ line of a number of other species and is therefore regarded as a characteristic of the germ plasm. Tsuenakawa *et al.* (2000) traced the *vasa* protein from the first cleavage stage until it eventually became part of the germ cells. It was associated with the Balbiani body (mitochondrial cloud) of the oocyte and subsequently with the primordial germ cells, and then with the germ cells proper until the formation of the **spermatids** and oocytes in the adult. The function of the *vasa* protein is not yet clear. The fact that it is possible to trace components of this 'germ plasm' first to the primordial germ cells, then to the gonadal germ cells, does not, however, prove that it is totally committed. Indeed, there are indications that the germ plasm is capable of altering its fate under suitable experimental conditions, e.g. Kagami *et al.* (1995) found that genetically male primordial germ cells could differentiate into functional ova, and genetically female germ cells into male sperm.

Interestingly, a chick gene, crescent, which is first found at stage X (see Appendix I) in the centre of the area pellucida (where the primordial germ cells arise) subsequently becomes localized to the **germinal crescent** (Pfeffer *et al.*, 1997), though it is not clear whether it is associated with the primordial germ cells.

The main reason why this has been such a difficult problem is a technical one. Until modern markers became available it was not possible to identify the primordial germ cells of birds with confidence before about stage 4 when they started to arrive in the anterior germinal crescent (*Text-Figure 38*) which is located at the anterior end of the area pellucida at its junction with the area opaca. They did not appear to arise in that region but it was not known how they came to be there. Modern work, which has provided answers to this problem, falls into two categories, specific labelling and microsurgical experiments.

The primordial germ cells from the primitive streak stages onwards stain strongly with periodic Schiff stain (Meyer, 1964) but cannot be stained in this way at earlier stages. A number of specific markers have been reported that not only pick out the primordial germ cells in the germinal crescent but also identify them at earlier stages. Their major characteristic is that they indicate a strong carbohydrate component at the surface of the primordial germ cells (e.g. Pardenaud *et al.*, 1987; Loveless *et al.*, 1990; Didier *et al.*, 1990). For a review see D'Costa *et al.* (2001).

Although some differences exist between the results of different workers, nevertheless they are in agreement that primordial germ cells are present from the time of laying (unincubated stage) and are initially in the epiblast. Evidence from a series of experiments (Eyal-Giladi, 1991; Karagenc *et al.*, 1996) in which pieces of the early blastoderm were isolated and allowed to develop *in vitro* suggests that the primordial germ cells arise from the central (i.e. embryonic) region of the area pellucida. These cells are also known as the **endophyll cells** (Vakaet, 1970), or the primary hypoblast by others (e.g. Stern, 1990) (see

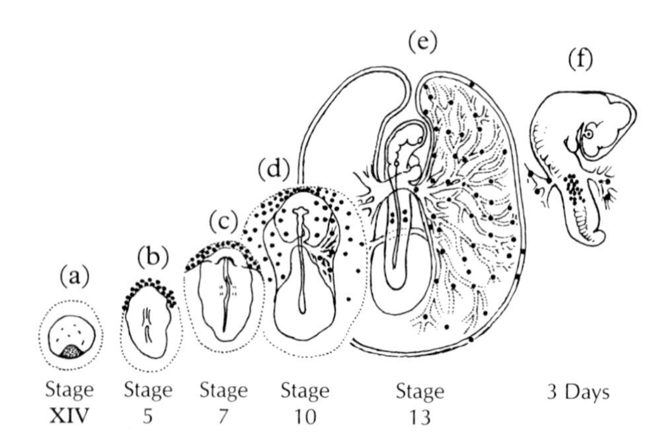

Text-Figure 38. *The primordial germ cells, shown as black dots, arise in the early area pellucida from endophyll cells (a) and migrate to the germinal crescent at the anterior end of the area pellucida (b–d). They then enter the blood stream and are dispersed (e), eventually passing into the gonads (f). (After Nieuwkoop and Satasurya, 1979.)*

Chapter 3). With the forward extension of the main hypoblastic sheet from posterior to the anterior end of the area pellucida, the primordial germ cells become displaced anteriorly, though it is not known whether they are carried passively by the hypoblast or whether they migrate actively along with it. By stage 5, the primordial germ cells which leave the hypoblast land among the mesoderm cells, which have spread out from the primitive streak and then probably migrate actively toward the **germinal crescent**. The numbers of primordial germ cells at stages IX to XIII were found to be about 150 (Karagenc *et al.*, 1996). In the early pre-primitive streak stage in the quail Pardenaud *et al.* (1987) reported a mean number of 19 ± 7 at stages XII–XIII, rising to 135 ± 40 by stages 4–5. It is not clear whether these differences in numbers are due to species variation or to the differences in counting. When chick embryos were subjected to simulated microgravity in a clinostat, there was a significant reduction in the numbers of primordial germ cells (Li *et al.*, 2002). Further experiments will be needed, however, to decide whether gravity plays a role in the formation or of mitosis of primordial germ cells.

If the germinal crescent is extirpated, the gonads that subsequently develop possess no germ cells and are, therefore, sterile (Willier, 1937). The size of the germinal crescent varies not only in individual chick embryos but also among different species of birds (Nieuwkoop and Satasurya, 1979). In the chick and turkey it extends around the anterior half of the area pellucida, whereas in the quail it reaches as far posteriorly as Hensen's node and even further posteriorly in the duck (D'Costa *et al.*, 2001).

The primordial germ cells arrive in the germinal crescent at stages 4–5 and remain there until the region becomes vascularized at about stage 10. They then enter some of the blood vessels, probably by an active migration, and are carried through the body. Vascular dissemination is found in birds and some reptiles but not in mammals. The greatest number of primordial germ cells in the circulating blood was reported as about 13 per 10 μl at stage 14, though this was followed by a reduction in succeeding stages as the cells entered the gonad (Tajima *et al.*, 1999).

There have been many attempts to explain how the primordial germ cells leave the blood stream and enter the gonad. The possibility that they become trapped by narrow-bore capillaries in the developing gonad now seems unlikely (Nieuwkoop and Satasurya, 1979). There is some evidence that primordial germ cells may be specifically attracted to the genital ridges, perhaps by some chemotactic effect (Kuwana *et al.*, 1986). It appears that they leave the blood stream at about stage 16–17 in a region just posterior to the omphalo-mesenteric (vitelline) artery and enter the splanchnopleure. The splanchnic epithelium shifts with the body folding (see p. 40) to become part of the coelomic lining at the site of the future gonad (Ukeshima *et al.*, 1987). Initially, the numbers of primordial germ cells entering the left and right gonads are similar but by stage 24 greater numbers are present in the left ovary than the right, and by stage 27 more are present in the left testis than in the right (Zaccanti *et al.*, 1990). (For reviews of primordial germ cells see Gomperts *et al.*, 1994; Ginsburg, 1997.)

Chapter 8
Gut, Coelom and Respiratory System

EARLY STAGES

The gut of adult birds differs from that of mammals in some important ways. Although birds possess salivary glands, they lack teeth and so are unable to crush and break up food in the mouth. The food passes directly down the oesophagus and collects at its lower end, the **crop**. From here it enters the anterior part of the stomach, the **proventriculus**, where it is subjected to digestive enzymes. It then moves into the gizzard, the highly muscular part of the stomach, where it is ground up. At its posterior end too the gut of birds differs from that of mammals in that it is not separate from the urinary and genital ducts, but shares with them a common cloaca and cloacal aperture. As in all vertebrates, the entire gut, apart from the stomatodael and proctodael regions, is formed from a lining of endoderm and a covering of mesoderm. The histology of the stomach and gastro-intestinal tract have been compared in chick and mouse embryos by Smith *et al.* (2000a) who concluded that the anterior portion of the stomach possessed a similar glandular histology as well as a similar expression of the secreted factors *Wnt5a* and BMP4. The posterior stomach in both species expressed *Six2*, BMPR1B and *Barx1*.

We have seen (Chapter 2) that the definitive endoderm, which becomes the gut endoderm, arises by ingressing from the upper layer through the anterior end of the primitive streak and then spreading to form a disc which becomes drawn posteriorly into a pearshape. The simple tubular embryonic gut is formed by the folding of the body (see Chapter 6; *Text-Figures 24, 39*) and by day 3 consists of the

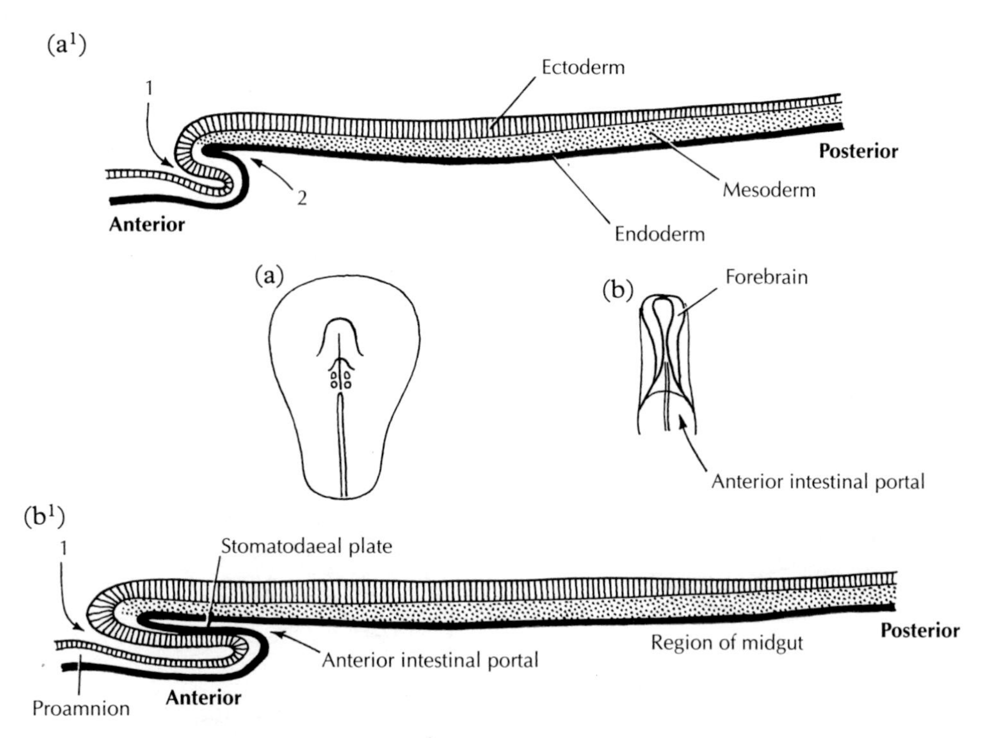

Text-Figure 39. *Formation of head fold and foregut. (a, a1) embryo at stage 7 in surface view (a) and sagittal section (a1). Arrow 1: head fold; arrow 2: foregut. (b, b1) embryo at stage 8, showing surface view of anterior end (b) and sagittal section (b1). The head fold has deepened and, as the foregut has lengthened, the posterior end of the foregut, the anterior intestinal portal, has shifted posteriorly.*

foregut, **midgut** and **hindgut**. The foregut, which starts to appear at about stage 8, is composed of endoderm with closely associated mesoderm. The first part of it to form lies anteriorly, ending blindly at the stomatodaeal plate where it is fused to the ectoderm (*Plates 53a, 55*). In the head the gut is surrounded by head mesenchyme but throughout the greater part of its length the gut endoderm is associated with splanchnic mesoderm.

The endoderm forms the epithelial lining of the gut and the ducts of the mucous glands, whilst the mesoderm gives rise to the muscular wall and associated structures. The shape of the walls and the lumen vary greatly along the gut. In the oesophageal region the foregut has a uniform structure with a round lumen and thick walls (*Plates 74, 75, 90*) but in the more posterior regions the shape of the lumen and the thickness of the walls vary. For example, see sections of the gizzard (*Plates 36b, 94*) and colon (*Plates 102, 108*). The open, posterior end of the foregut diverticulum is the anterior intestinal portal (*Text-Figure 41, Plates 9, 53, 57*); it is not a fixed point but an opening into the foregut. Matsushita (1995, 1996) has traced the fates of different regions of the endoderm and the splanchnic mesoderm of embryos of 1.5 days incubation (approximately stages 9–11) using DiI labelling. Although there is some overlap between areas, nevertheless, the cells for each presumptive organ are already 'roughly' segregated. Those endoderm cells located anteriorly and medially contribute to the anterior end of the gut, whilst cells located posteriorly become part of the posterior regions of the gut, though they are situated more laterally at this stage than the anterior cells. As the gut tube continues to form, these lateral cells become drawn medially. The endodermal cells between the first and tenth somites contribute to the gut from the level of the posterior pharynx to the jejunum, and also to the yolk sac.

The morphogenetic movements that bring about the formation of the foregut extend further and further posteriorly so that there is a gradual movement of cells toward the intestinal portal and through it. As a consequence, the foregut increases in length and the anterior intestinal portal is displaced further and further posteriorly, the population of the cells in its walls continually changing as the cells become incorporated into the foregut. Between stages 7 and 10, the anterior intestinal portal is displaced posteriorly at the rate of about 0.1 mm h^{-1}, but by stage 12 has slowed to less than 0.02 mm h^{-1} (Seidl and Steding, 1978).

Endo-mesodermal interactions play an essential role in the development of the gut. Interestingly, the presumptive endoderm cells for the individual regions of the gut were found by Matsushita (1996) to lie anterior to the corresponding mesoderm cells. He suggested, therefore, that the epithelial–mesenchymal interactions that take place as the organs develop may not begin until the two tissues become aligned with each other as the result of morphogenetic movements. Nevertheless, each of these regions of presumptive gut endoderm of stage 10 embryos appears to be already regionally committed, since if each region is individually cultured with flank somite mesoderm, with which

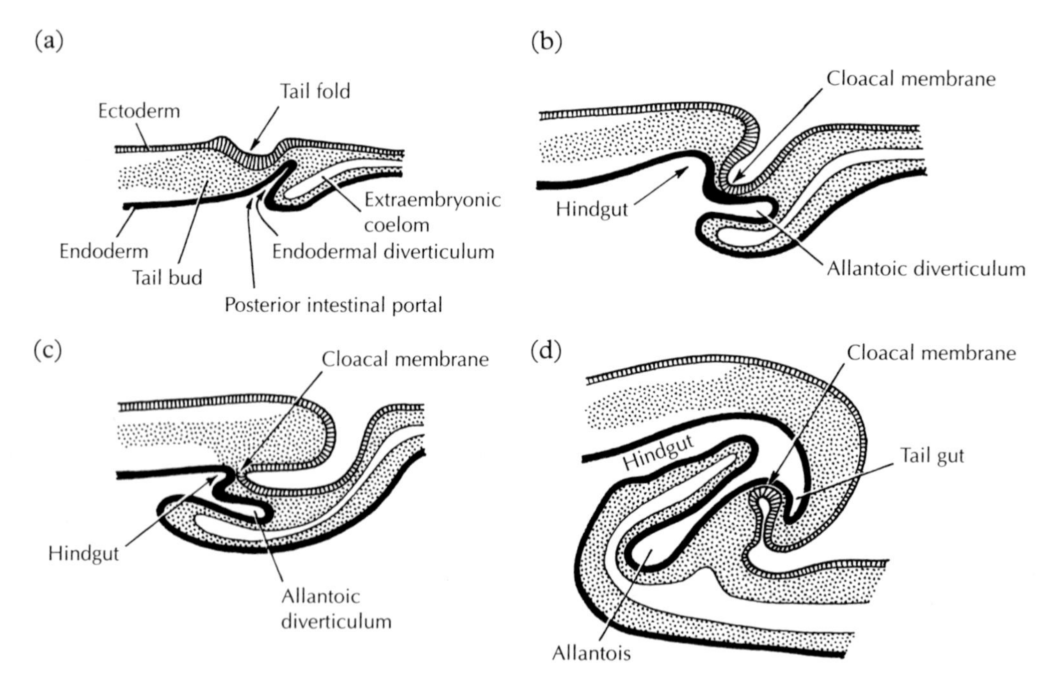

Text-Figure 40. *Formation of tail fold, allantoic diverticulum and hindgut shown in sagittal section at stages 12–13 (a); 13–14 (b); 18 (c); and 20 (d). With the formation of the tail fold, the cloacal membrane becomes drawn ventrally.*

it would not normally come into contact, it forms tissues expressing its own appropriate genes (Matsushita *et al.*, 2002). There is evidence that the signals include bone morphogenetic proteins (BMPs) that are expressed by the mesoderm. In particular, BMP-2, 4 and 7 are expressed by the mesenchyme of the proventriculus prior to the formation of the proventricular glands (Narita *et al.*, 2000).

The hindgut, which forms in a comparable manner to the foregut, is present as a simple tube by about stage 14, its anterior border being the posterior intestinal portal (*Plate 68*). A small section of the posterior gut extends into the tail region (*Text-Figure 40*) and is known as the tailgut.

The **allantoic diverticulum** (see Chapter 13) starts to arise from the hindgut in embryos at about stage 18, though there is considerable individual variation. The midgut differs from the foregut and hindgut in that, although it possesses lateral body walls, it lacks a ventral wall and is in communication with the yolk sac (*Text-Figure 39, Plate 65*). By the end of day 5 the foregut and hindgut have extended greatly, whilst the midgut has been reduced to a narrow region, only about 150 μm in diameter, the **yolk sac stalk** (*Plate 121*). Both the foregut and hindgut meanwhile extend in length. (It should be noted here that in the later embryo and adult the term foregut refers only to that region which lies anterior to the liver, and hindgut to the region posterior to the caecum; the midgut is represented by the intervening portion of gut.)

The differentiation of the simple gut tube into specific regions along its length appears to be associated with localized differences in gene expression. An example given by Smith *et al.* (2000b) is that BMP4 is expressed throughout the hindgut and midgut but not in the early gizzard and this is correlated with differences in the expression of BMP receptors. *Hox* genes are also associated with differentiation of the various regions of the gut (Sakiyama *et al.*, 2000). Both BMP4 and the specific *Hox* genes appear to be induced by the endodermal signal *sonic hedgehog* (Roberts *et al.*, 1995).

Sonic hedgehog gene also acts as a morphogen, being involved in the morphogenesis and cytodifferentiation of the gut (Fukuda and Yasugi, 2002). Smith and Tabin (2000), who have analysed the lineage relationships of cells derived from the original gut have found that clonally related cells have become restricted to a single organ by stage 12. They reported that from day 4 to at least day 6 the proventriculus was characterized by *Wnt5a*, the gizzard by *BMPR1B*, the pyloric sphincter by *Nkx2.5* and the small intestine by *Nkx2.5*.

The Beak

See Chapter 11.

The Oral Cavity

The mouth is formed partly from the stomatodaeum (*Plates 55, 60*), an ectodermal depression that appears beneath the forebrain at the end of day 2, and partly from the most anterior part of the foregut itself. The stomatodaeum is specified by signals from the endoderm (Withington *et al.*, 2001). **The stomatodaeal (oral) plate** is the region where the endoderm of the foregut fuses with an ectodermal depression of the head, no mesoderm intervening (*Text-Figure 41*). The stomatodaeal plate breaks down in the chick at the end of day 3 so that the anterior (ectodermal) and posterior (endodermal) parts of the oral cavity are now continuous with one another, and the lumen of the foregut comes into communication with the amniotic cavity. The former position of the stomatodaeal plate is visible in later embryos by the position of Rathke's pouch (*Plate 73*), which arises immediately anterior to it. Further development of the mouth includes the formation of the jaws from the first pair of pharyngeal arches (see Chapter 10).

The tongue (*Plates 93, 128, 132*) develops principally, as in mammals, from two swellings in the floor of the pharynx in the region of the foramen caecum (the point of origin of the thyroglossal duct) (*Plates 73, 74*) (see Chapter 12 for thyroid development), the **tuberculum impar**, which forms during day 4 at the junction of the first and second pharyngeal arches, and the **copula** (hypobranchial swelling), formed from the basihyal and basibranchial cartilages. Further contributions to the tongue are made by the lateral lingual ridges after the fusion of the main rudiments (see also Chapter 10 for tongue muscles, which are derived from the dermomyotomes). The entire organ is covered with an epithelium. Huang *et al.* (2001) found that the

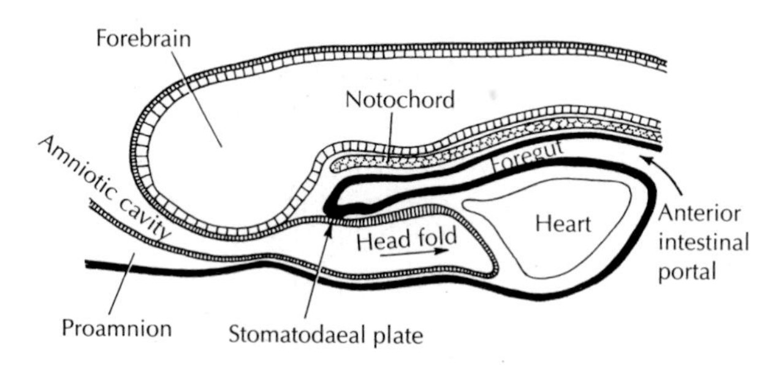

Text-Figure 41. *Sagittal section to show head of embryo at about stage 9. The anterior end lies to the left and the posterior to the right.*

growth of the tongue occurred throughout the tissue and was not associated with any distally located growth zone of the type found in limb bud development. They reported also that the epithelial and mesenchymal components of the tongue were controlled by different genes. There was a high expression of *sonic hedgehog* and *patched* in the dorsal epidermis until about 8 days of incubation, after which the intensity declined and cornification began. These genes were not expressed in the mesenchyme. (See also *Text-Figure 69*, which shows the skeleton of the tongue.)

The Pharynx

The first-formed and most anterior part of the foregut tube is the pharynx, which lies immediately posterior to the mouth. Its central lumen is initially round in transverse section, but a series of paired pouches extends laterally, the pharyngeal (visceral) pouches (*Plate 74*), so that it becomes trough-shaped in transverse section. The dorsal wall, which lies beneath the notochord and somites, is thinner than the ventral one, which is underlain by splanchnic lateral plate mesoderm (*Plate 53*). In the chick there are four well-developed pairs of pharyngeal pouches, together with a fifth which is smaller. Remnants of a sixth have been described in some birds and may be present in the chick. These pouches form sequentially between stages 10 and 22 and coincide with a series of grooves, the **ectodermal grooves**, that indent the ectoderm. At these points the ectoderm and endoderm fuse and in the case of the first three pouches a temporary perforation is formed (*Text-Figure 42*). This communication of the pouches with the outside

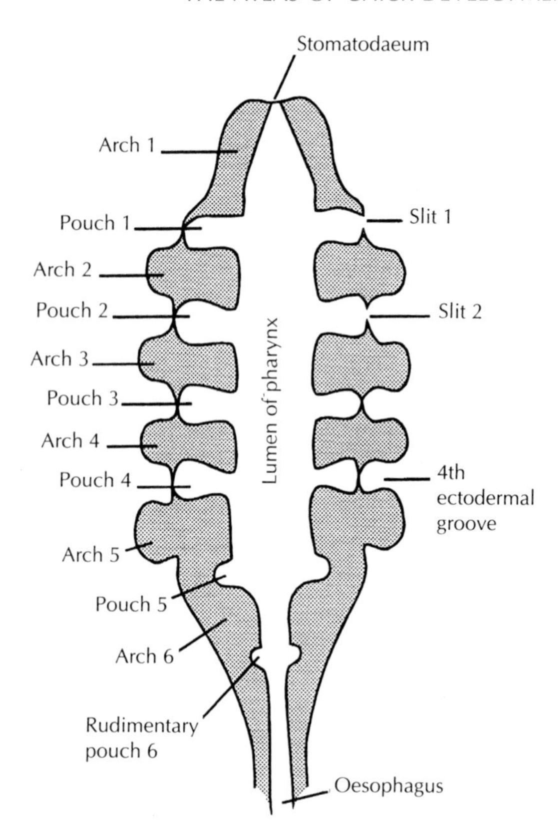

Text-Figure 42. *Stylized diagram of a section along the pharynx to show the relationship between the arches and pouches. On the right side the first and second pouches have perforated where they have contacted the associated ectodermal groove.*

(i.e. with the amniotic cavity) resembles the gill slits of fishes. Similar perforations are unusual in mammals but can occur as an abnormality, e.g. as branchial fistulae in humans.

The slits corresponding to pouches 1 and 2 are visible at stage 14, and to pouch 3 at stage 15. The fusion of the endoderm of the pouch and the ectoderm of the groove is possible because the mesoderm is lacking in those regions, although it is present in the remainder of the pharyngeal wall and forms the pharyngeal arches (*Plate 74*). Each arch consists of a band of mesoderm that passes ventral to the pharynx and each becomes continuous with the mesoderm of the arch on the other side of the body and is covered externally by ectoderm and internally by endoderm. Pouches and arches alternate (*Text-Figure 42*), the nomenclature being that arch 1 lies anterior to pouch 1, which is followed by arch 2 and then by pouch 2 and so on.

The arches and pouches give rise to much the same structures as in mammals (see Table 2). The changes begin to take place during the fourth day, and the perforations of the first three pouches close during days 4–5. Each of the three pouches is subdivided into a dorsal and a ventral diverticulum, and these give rise to different tissues.

Table 2 Pharyngeal derivatives

Arches	Pouches	Derivatives
I Maxilla		Palato-quadrate
I Mandibular		Meckel's cartilage
	I	Part of tubo-tympanic recess
II Hyoid		Median: copula I and basihyals
		Lateral: columella and ceratohyals
	II	Thymus II from wall
III		Median: copula II; lateral:
		Ceratohyal II and epibranchials
	III	Dorsal: thymus III
		Ventral: parathyroid III
IV		No skeletal contribution?
	IV	Dorsal: thymus IV
		Ventral: parathyroid IV
V		No skeletal contribution?
	V	Ultimo-branchial body

The **thymus** forms principally from the dorsal wall of the third pharyngeal pouch and lengthens along the jugular vein. The dorsal wall of the second pouch and of the fourth pouch, together with mesoderm of the pharyngeal wall, contribute to it so that it becomes a series of glandular patches running down either side of the neck (*Plates 115, 132, 141*). Experiments involving chick–quail chimeras have shown that in the chick the **thymus rudiment** is colonized by waves of lymphocyte precursors during a period of about 36 h, starting at about 6.5 days (*Le Douarin et al., 1984a*).

The **ultimo-branchial bodies** develop from the fifth pouches and become the ultimo-branchial glands of the adult (Dudley, 1942; Kameda, 1984, 1993), which are involved in calcium metabolism (Simkiss, 1967). Their endocrine cells, the C-cells, are responsible for the production of calcitonin.

The **pharyngeal arches** give rise primarily to cartilages and muscles (see Table 2 and Chapter 10). Each arch is served by an aortic arch artery (*plates 40, 41*) which acts as a link between the ventral and dorsal aortae (see Chapter 6), and by a cranial nerve (see Chapter 9).

The primordium of the thyroid, the **thyroglossal duct**, develops as an invagination in the floor of the pharynx at the level of the second pair of pouches (*Plates 73, 74*).

The larynx, trachea and lungs form from an invagination in the floor of the posterior part of the pharynx (see below).

The Oesophagus and Crop

The oesophagus of birds differs from that of mammals in that ridges (plicae) run along its length and protrude into the lumen. They are larger in the domestic fowl, which tends to eat hard food, than in birds which feed on small insects. In the early embryo, the oesophageal region of the foregut has a more uniform structure than the pharynx with a round lumen and thick walls (*Plates 68, 84*). The lumen becomes reduced from about day 5 to day 6 (*Plate 90*), proceeding from anterior to posterior, ultimately becoming obliterated, apparently due to the rapid growth of the epithelial lining. It starts to re-open during days 6–8, but now proceeding from posterior to anterior. A morphometric analysis was published by Gheri-Bryk *et al.* (1993).

The oesophagus becomes innervated about day 5 by the invasion of fibres from the vagus nerve, and by day 8 **Auerbach's plexus**; by day 13 the **Meissner's plexus** is visible in suitably stained specimens. Solid epithelial buds grow into the tunica propria of the associated mesoderm from day 16, giving rise to the mucous glands.

The **crop** (*Plates 99, 100*) begins to form during days 5 and 6 by a swelling at the posterior end of the oesophagus at about the level of the seventh vertebra. By the end of day 7 it bends to the right and has acquired a lumen about 0.25 mm wide, as opposed to that of the oesophagus proper, which is about 0.16 mm. The crop continues to grow in size as a diverticulum at the right side of the oesophagus (*Plate 115*), its walls and plicae (*Plate 142*) being thinner than those of the main part of the oesophagus.

The Stomach

The region of the foregut which will form the stomach has begun to swell by the end of day 3 and to shift toward the left of the midline. Shortly afterwards the division into **proventriculus** (*Plates 110, 118, 119*) and **gizzard** (*Plates 35, 36*) begins. The shift of the stomach leftwards continues and by day 7 the gizzard has swollen into a thick-walled sac that points posteriorly (*Plate 106*).

The glands of the proventriculus have begun to form in the chick by about day 7. Endodermal invaginations invade the mesoderm layers and undergo budding and by day 11 have become multilobed (*Plate 136*). Further modifications take place during the remainder of the incubation period. Simple glands also form in the proventriculus (*Plate 139*).

The gizzard of the adult is highly muscular and is innervated by Auerbach's plexus, formed partially from neural crest cells (Le Douarin and Kalcheim, 1999) and partially by derivatives of the vagus nerve (Zimmerman *et al.*, 1995). The development of this thickened region is apparent in gross sections from about day 6 (*Plates 36, 106*) but becomes more marked as development proceeds, with the formation of circular muscles. The smooth muscle regulatory proteins caldesmon and myosin are, however, expressed before this, as early as day 5 (Paul *et al.*, 1995). Although the secretory glands begin to form about day 12, mucosubstances have been reported to be present as early as day 3 by Matsushita (1998), who examined the gizzard by means of antimucus antibodies. Stimulation of the nerve supply to the gizzard elicited little response, whether intramural or vagal, until after hatching (Ohashi *et al.*, 1993).

The pyloric sphincter appears to be induced at the caudal end of the developing gizzard by BMP4 signalling from mesoderm of the small intestine from about day 2.5 (Smith and Tabin, 1999). *Nkx2.5* is a marker for the sphincter (Smith *et al.*, 2000b).

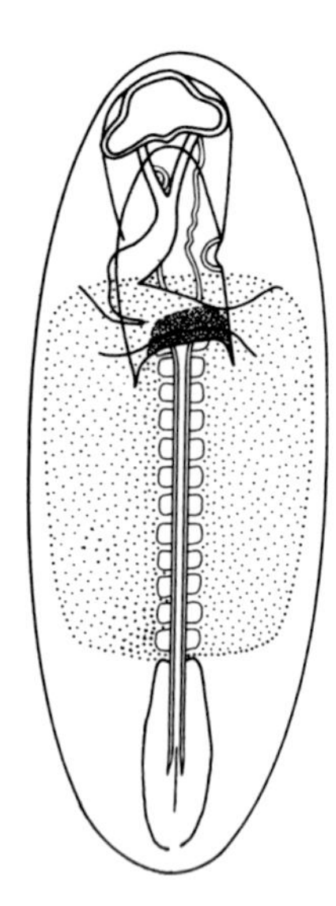

Text-Figure 43. *Distribution of hepatic endoderm (dark shading) and mesodermal areas (light shading) in a stage 12 embryo. (From Croisille and Le Douarin, 1964.)*

The Liver and Gall Bladder

The presumptive liver areas are closely associated with those of the heart and together are known as the cardio-hepatic regions. They lie on either side of Hensen's node at stages 4–5. But whereas the heart is an entirely mesodermal structure, the liver is formed from both mesoderm and endoderm (*Text-Figure 43*). At stage 12 the endodermal component is located in the floor of the anterior intestinal portal and is derived from a diverticulum of the gut and by stage 14 it invades the mesoderm that will form the matrix of the liver. The mesodermal component is part of the ventral mesentery, itself derived from the splanchnic mesoderm, and is present as two bilateral patches extending to about the level of the 15th somite (*Text-Figure 43*). The endodermal component becomes split into two diverticula around the ductus venosus, an anterior and a posterior part (sometimes called the left and right diverticula, respectively). By stages 15–16 the anterior diverticulum has extended over the ductus venosus as far as the common cardinal veins. By stage 17 the anterior diverticulum lies to the left of the ductus venosus and the posterior to the right. The gall bladder and bile duct (*Plate 119*), the **ductus choledochus**, form at about stage 18 as an appendage of the right diverticulum. From about 11 days, the lining cells of the gall bladder undergo dramatic changes (Gheri-Bryk *et al.*, 1990). Experiments in which the endodermal and mesodermal components have been separated from one another or recombined with other tissues have shown that until about stage 8 the endoderm is not determined to form liver. After this, the endoderm is induced by the mesoderm of the hepato-cardiac area. A secondary induction, now

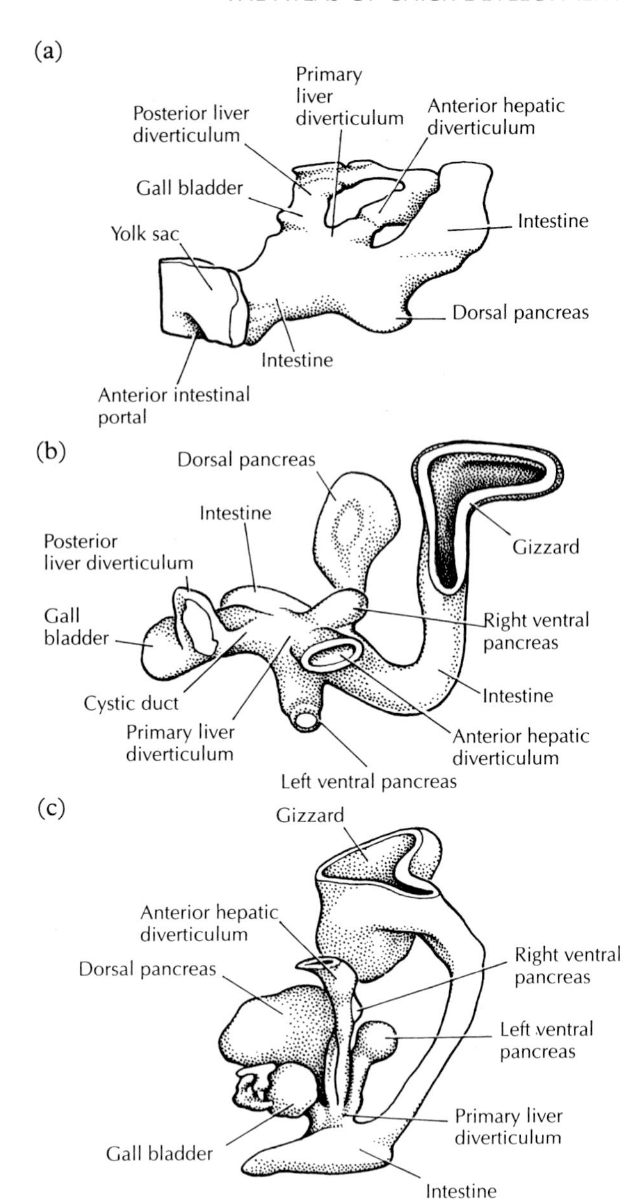

Text-Figure 44. *Development of the biliary and pancreatic ducts and of the gall bladder. (a) stage 18 (3 days); (b) 4 days; (c) 5 days. (After Romanoff, 1960.)*

from the hepatic mesenchyme, leads to the formation of hepatic cords (Croisille and Le Douarin, 1964).

The liver primordium is visible at the end of day 2. As it grows, it comes into contact with the body wall (*Plate 119*). The veins that pass through the liver are discussed in Chapter 6 (e.g. allantoic vein, hepatic vein, subcardinal vein).

The Pancreas

The pancreas of birds develops from three rudiments, one dorsal and two ventral (*Text-Figure 44*). The dorsal pancreas (*Plate 119*) forms at about stages 17–18 as an outgrowth from the dorsal wall of the gut, and its formation is associated with signals from the notochord which repress the expression of *sonic hedgehog* in that region (Hebrock *et al.*, 1998). The left and right ventral pancreatic rudiments evaginate from the base of the hepatic diverticulum and are clearly visible toward the end of day 4 (Patten, 1950). The initial budding of the pancreatic rudiments from the gut epithelium is associated with the transcription factors *Pdx-1* and *ngn3* (Grapin-Botton *et al.*, 2001).

The three rudiments of the pancreas start to fuse together during day 6, although the process is not completed until about day 9. The number of pancreatic ducts that are retained depends on the species of bird, but usually all three are present in the adult domestic fowl and open near the end of the duodenum, close to the two bile ducts (for a discussion of the variation in species, see McLelland, 1979). Cell differentiation of the pancreas begins on day 6.

THE INTESTINES

The Small Intestine

The future divisions of the intestine can be recognized by day 6 (*Plate 94*). A small duodenal loop, which lies beneath the right lobe of the liver, is followed by the duodenal–jejunal flexure (*Text-Figure 45*) and a further loop connecting it to the yolk sac stalk. The caecum, which is present as a small bulge, lies at the junction of the small and large intestines. In later stages the pancreas forms within the duodenal loop. The small intestine, particularly the ileum, grows extensively and, as in other amniotes, is herniated into the umbilical cord (*Text-Figures 45, 46, Plate 127*). This is possible because both the internal and external coeloms, which were in communication in the early embryo (*Plates 77, 128*). remain so through the umbilical cord (*Plate 128*). The apex of

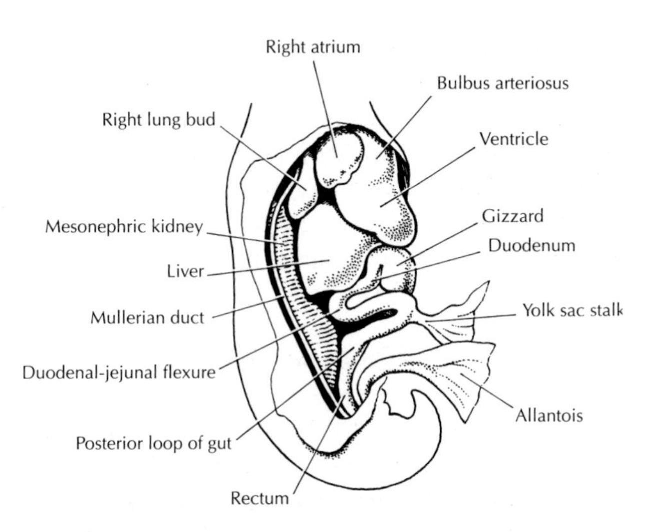

Text-Figure 45. *View of the dissected trunk of an embryo of 6 days to show the gut and other internal organs. (After Duval, 1889.)*

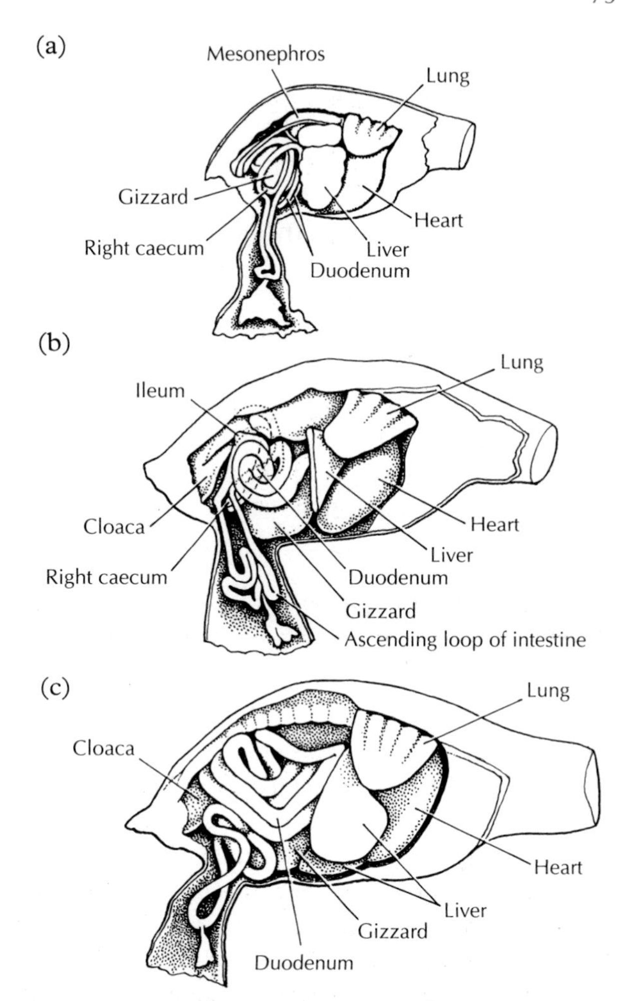

Text-Figure 46. *Development of the intestinal loops herniated into the umbilical cord, seen from right side after removal of body wall. (a) 11 days, 21 h; (b) 12 days, 22 h; (c) 15 days, 22 h. (After Romanoff, 1960.)*

the loop is continuous with the yolk sac (*Text-Figures 45, 46; Plate 137*). By day 6, the intestinal loop has begun to rotate through 90°, the anterior section moving to the right and the posterior section to the left, a process that is completed by day 11. Further convolutions take place (see Romanoff, 1960) and the gut is gradually drawn back into the abdomen from about day 17, though the process is not completed until about day 19 when the remnant of the yolk sac stalk is also withdrawn. A morphometric analysis of the ileum was carried out by Gheri and Gheri-Bryk (1987). The utilization of the yolk in the newly hatched chick is described by Noy and Sklan (1998).

The Caeca

Although most mammals have only one caecum, birds have two. They begin to form during day 4 as bilateral swellings approximately midway between the umbilical wall and the base of the allantois. By day 7 they have grown to 2–4 mm in length and by the time of hatching to about 30 mm. A morphometric analysis was carried out by Gheri *et al.* (1992).

The Large Intestine

The large intestines (*Plate 107*) lie posterior to the caeca and begin to elongate after day 4, though they are still only about 20 mm at the time of hatching. At about day 6 the lumen of the large intestine occludes, but has recanalized by about day 12. A morphometric analysis of the ileum was carried out by Gheri and Gheri-Bryk (1987).

The Tail Bud, Hindgut and Tail Gut

The endoderm that will form the hindgut lies posterior to the last somite at 1.5 days of incubation (about stage 10) and becomes greatly elongated antero-posteriorly during gut formation (Matsushita, 1999).

The hindgut begins to appear at about stage 13 and its development is closely associated with that of the tail bud (*Plates 68, 79*). (For a review of the tail bud see Griffith *et al.*, 1992).

The **tail bud**, which starts to form at about stage 11, is derived from the remains of Hensen's node after it has regressed down the primitive streak. As the tail fold (posterior body fold) forms, it undercuts the node, which, therefore, becomes wrapped dorsally, ventrally and posteriorly in ectoderm (*Text-Figure 40*). The **hindgut** begins to form at about stage 14 and its most posterior section becomes an integral part of the tail and is there known as the tail gut (*Plate 80a*). Both the mesoderm and the endoderm of the tail gut express *Hox-a13*, which appears to be involved in the epithelial–mesenchymal interactions that take place during tail growth and patterning (De Santa Barbara and Roberts, 2002). The tail gut becomes obliterated, largely as a result of cell death (Miller and Briglin, 1996) and has usually disappeared by about 3.5 days (stage 21).

The development of three important structures is closely associated with that of the hindgut, viz. the **allantois** (*Plate 86*), the **cloacal membrane** (also known as the anal plate, **cloacal plate** or proctodaeal membrane) (*Plate 122*) and the cloacal bursa (**bursa of Fabricius**) (*Plate 138*).

The Allantois

Like the gut, the allantois is formed essentially from endodermal epithelium with associated mesoderm. It begins to arise about stage 15 and is usually considered to form as an outgrowth from the hindgut, though some investigators have maintained that it develops before the hindgut. According to Matsushita (1999), pre-allantoic endoderm cells are confined to a small area at the posterior end of the primitive streak. It is only as the tail fold forms that it becomes relocated as a diverticulum of the hindgut. For further development of the allantois see Chapter 13.

The cloacal (proctodaeal) membrane becomes visible at about stages 12–14 as a region just posterior to the tail bud, where an indentation of the ectoderm fuses with the posterior end of the gut endoderm (*Text-Figure 40*). As the tail fold forms, the cloacal membrane becomes shifted relatively anteriorly. Its position marks the junction between the trunk and tail. The cloacal membrane perforates just before hatching and the ectodermally derived part becomes incorporated as the proctodaeum into the cloaca, as opposed to the endodermally derived components of the cloaca, known as the coprodaeum and the urodaeum, respectively (*Plates 139, 140*).

The Cloacal Bursa (Bursa of Fabricius)

During day 4 the bursa of Fabricius (cloacal bursa) (*Plates 138, 140*) forms as a diverticulum on the dorsal side of the cloaca. Its role is related to that of the thymus (Schoenwolf and Singh, 1981) and the spleen (Glick and Olah, 1993). It develops as a result of an interaction between the endoderm and its surrounding mesoderm (Le Douarin *et al.*, 1984a) and soon becomes highly plicated. Lymphoid tissue develops on day 12–13 (Goldschneider and Barton, 1976) and the synthesis of immunoglobulins begins soon after.

THE COELOM, MESENTERIES AND RESPIRATORY SYSTEMS

The coelom, or body cavity, begins to develop as the lateral plate mesoderm forms (*Plate 53*). We have seen (Chapter 4) that by stage 8 the mesoderm on either side of the midline posterior to the head is starting to form into somites and lateral plate (*Plate 61*). The lateral plate mesoderm remains unsegmented but becomes arranged as two separate sheets, the upper, somatic, layer lying close beneath the ectoderm, and the lower, splanchnic, layer, which adheres to the endoderm (Chapter 5). The space between the two sheets is the coelom or body cavity. Before the embryonic body begins to fold the coelom is arranged as two separate regions, the left and the right coeloms, and each of these extends distally beyond the tissues that will contribute to the embryonic body and into the extra-embryonic region. The embryonic and extra-embryonic coeloms are, therefore, in communication, a situation that enables intracoelomic grafting to be carried out at this time (see Chapter 2). As the head fold forms, each lateral coelom extends anteriorly, forming the amnio-cardiac vesicles (*Text-Figure 23; Plate 52*), and the two sides unite anterior to the head at about stage 7. The lateral regions subsequently join to form the heart, whilst the more anterior part, the proamnion, gives rise only to the amnion. By stage 8 the lateral body folds have begun to form, which will eventually lead to the closing off of the ventral body wall, not only bringing the two lateral coeloms into communication with one another, but separating off the embryonic from the extra-embryonic coeloms, as well as playing a role in the formation of the gut and the mesenteries (*Text-Figures 24, 47*). The gut endoderm becomes folded initially into a trough that is open ventrally

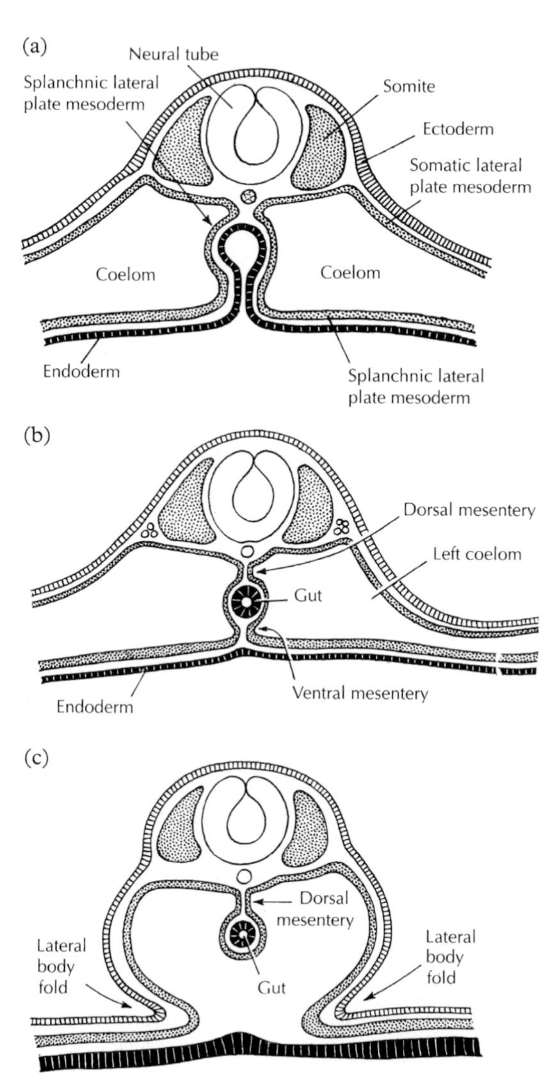

Text-Figure 47. Transverse sections to show right and left coeloms (a, b) fusing to form a single coelom (c). The dorsal and ventral mesenteries (a, b) form from splanchnic mesoderm, the ventral mesentery breaking down (c) over much of the length of the gut, whilst the dorsal mesentery persists.

but gradually separates from the extra-embryonic endoderm of the yolk sac, except in the region of the midgut, the splanchnic mesoderm of the lateral plate remaining closely adherent to the gut endoderm during this process (*Plate 53*). The left and right sheets of splanchnic mesoderm meet and fuse with one another both dorsally and ventrally to the post-pharyngeal gut as it forms (*Text-Figures 47a,b*). The dorsal fused sheet is the **dorsal mesentery** (*Plate 35*) and the ventral fused sheet is the **ventral mesentery** (*Plate 75*). The left and right coelomic cavities thus come into communication with one another and the gut hangs down in the dorsal mesentery. The ventral mesentery survives only in the region of the heart, liver and cloaca.

Throughout its length the coelomic cavity is lined by an epithelial sheet, the **peritoneum**, derived from the splanchnic mesoderm, and any organ or tissue which may appear to protrude into the coelomic cavity does not normally penetrate this sheet but merely invaginates it.

In adult birds, as in mammals, the coelom is divided into three basic compartments, pericardial (*Plate 90*), pleural (*Plate 117*) and peritoneal (*Plate 91*), but there are important differences between the two classes of vertebrates. The first stages are comparable but differences develop in the relative position of the organs, since in birds the heart 'shifts' (but see p. 50) further caudally than it does in mammals. A major difference, however, is that the air sacs, developing from the lungs, penetrate between the mesenteries, further dividing the coelom. McLelland and King (1970, 1975) identified eight coelomic cavities in adult birds. They are the pericardial cavity, the left and right pleural cavities, and five peritoneal cavities (the left and right dorsal hepatic cavities, the left and right ventral hepatic cavities and the intestinal cavity; *Text-Figures 48, 49*). These are all present by 5 days. No true diaphragm is present in birds.

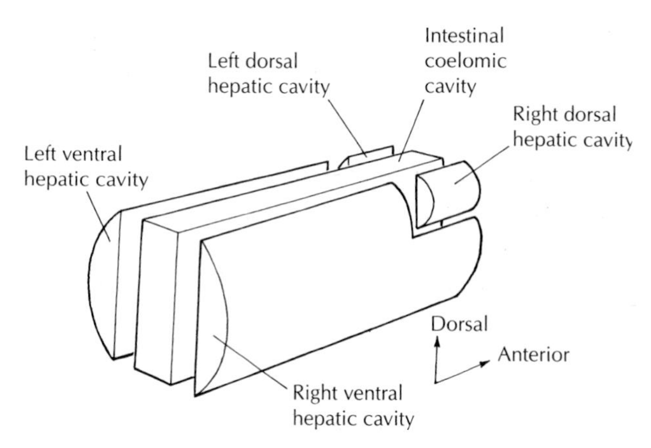

Text-Figure 48. Highly schematic diagram of the five peritoneal cavities in an adult fowl. (After McLelland and King, 1970, with permission of Elsevier GmbH.)

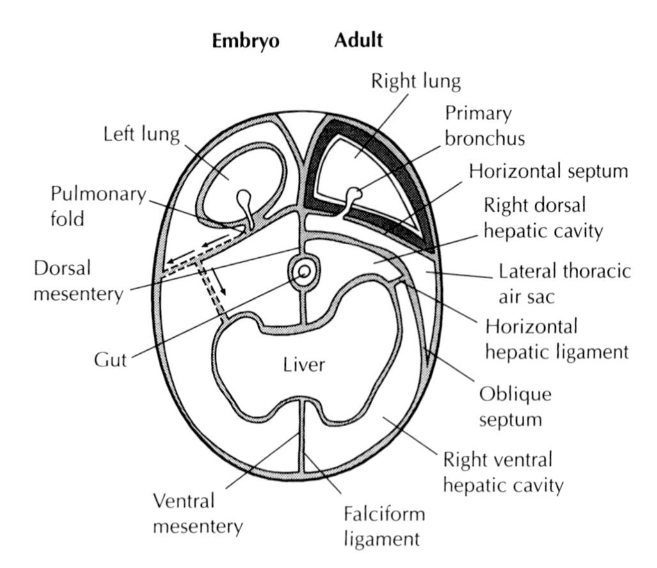

Text-Figure 49. *Transverse section through the pleural region of an embryo (left) and an adult (right). Each pulmonary fold grows ventro-laterally to fuse with the peritoneal lining of the body wall and the liver (indicated on the left by dotted lines and arrows). Later, the lateral thoracic air sac penetrates between the two layers of the pulmonary fold, separating it into the horizontal septum and the oblique septum (shown on right). (After McLelland and King, 1970, with permission of Elsevier GmbH.)*

Division of the Initial Coelom

The **pericardial cavity** (*Plates 31, 34, 90*), or pericardial coelom, is formed from the most anterior part of the coelom (see amnio-cardiac vesicles in *Text-Figure 23*) and is visible at about 22 h in whole-mounted embryos. As we have seen, the heart develops from thickened splanchnic mesoderm and is initially supported by dorsal and ventral mesenteries (also known as the dorsal and ventral mesocardia) (*Plate 56*), though these break down later, leaving the tubular heart attached to the walls of the pericardial cavity at its anterior and posterior ends only (*Plate 62a*). Shortly after its formation, the pericardial cavity (*Plate 56*) is still in communication with the left and right coelomic regions (*Text-Figure 24*), which themselves extend further posteriorly. It then becomes separated from them by the formation of the lateral mesocardial folds. These arise dorso-laterally from the splanchnic epithelium over the **omphalo-mesenteric veins.** Searls (1986) ascribed the relationship between the pericardial and pleural coeloms to caudal movement of the aortic arches.

The **pleural cavities** become divided from the abdominal cavities by the pulmonary folds (*Plate 91*) each of which arises medially from a lung bud and extend laterally, left or right (*Text-Figure 49*). (In mammals the corresponding folds arise laterally and extend medially.) Each pulmonary fold fuses with the peritoneal lining of the body wall to form the **pleuro-peritoneal membrane**, with the pericardium to form the **pleuro-pericardial membrane**, and with the peritoneum covering the liver to form eventually, the **lateral hepatic ligament** (*Plate 136*). During development the lateral thoracic air sacs on each side invade the pulmonary fold (*Text-Figure 49; Plate 136*), splitting apart the two layers so that two separate septa are formed, the horizontal septum (also known as pulmonary aponeurosis, sacco-pleural membrane (*Plate 136*), or pulmonary diaphragm) and the oblique septum (also known as thoraco-abdominal diaphragm or sacco-peritoneal membrane). (Note: these are not true diaphragms, as found in mammals, so this term is best avoided.) The ventral mesentery persists as the falciform ligament (*Plate 136*), a transverse sheet connecting the ventral body wall (inner face of the sternum) with the ventral surface of the liver. It thus corresponds with the falciform ligament of mammals. The post-hepatic septum is an extension of the mesentery arising at the level of the gizzard.

The **dorsal** and **ventral hepatic cavities** (abdominal coeloms) (*Plate 136*) (see *Text-Figure 48*) lie on either side of the midline. Their mode of formation is not clearly understood and no recent investigation appears to have been carried out.

The **intestinal coelomic cavity** (*Text-Figure 48; Plate 136, 137*) contains all of the intestinal tract from the proventriculus to the rectum, including the pancreas and liver, as well as the gonads, the spleen and the abdominal air sacs.

THE RESPIRATORY TRACT

The lungs of birds differ markedly from those of mammals and other vertebrates, their unique modifications being related to the higher metabolic rate of birds. These modifications include the absence of alveoli and the anastomosis of branches of the bronchi so that the lung resembles a sponge in structure (*Plates 50, 117*). The air is thus able to pass quickly throughout the organ. The avian lung differs also from the mammalian one in that it is not elastic and expandable and that air sacs extend posteriorly from it. These air sacs, which are poorly vascularized, expand during inspiration and contract during expiration, like bellows, so that the air is continually passing through the lungs, where blood gas exchange occurs. The motive force for this activity is provided by the abdominal muscles.

In the adult bird the air is taken into the respiratory system through the **external nares** and nasal sacs and enters the mouth through a V-shaped slit, the **internal naris** (*Plate 114*). The glottis (*Plate 131*), the opening into the trachea, lies posterior to the tongue. The tracheal wall is modified to become the larynx

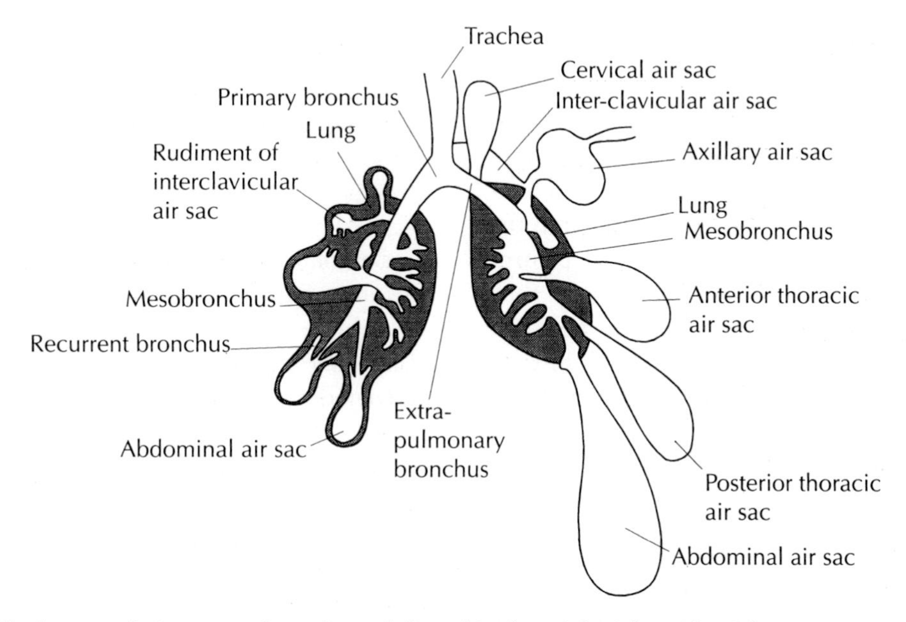

Text-Figure 50. *The lungs and air sacs at about day 4 (left) and in the adult (right). (After Ede, 1964.)*

(*Plate 142*) and is supported by tracheal cartilages (*Plate 131*). The trachea bifurcates into two primitive bronchi, one passing to each lung, giving rise not only to the air passages within the lung, but also to the air sacs which extend to different parts of the body (*Text-Figure 50; Plate 144*).

The larynx, trachea and lungs are derivatives of the gut and, like the gut, are formed from a thin lining of endoderm and a thicker covering of mesoderm. As with the gut, epithelial–mesenchymal interactions play an important role in the differentiation of the various regions. The pharyngeal endoderm gives rise to the epithelium of the bronchial tree, whilst the mesenchyme forms the muscles and connective tissue of the lungs as well as the blood vessels and the lymphatic vessels. The pulmonary and bronchial vasculature are both derived from a splanchnic plexus that forms around the foregut before the lung buds develop (DeRuiter *et al.*, 1993). As the two lung buds grow they extend in a posterior direction. *Hox-b* genes appear to play a critical role in demarcating different regions. *Hox-b5* and *Hox-b6* expression characterizes the trachea, bronchial tree and airsacs (Sakiyama *et al.*, 2000).

The first morphological sign of the development of the respiratory tract is the laryngo-tracheal groove (*Plate 84*), which becomes visible in the midline of the floor of the pharynx during days 3 or 4 of incubation arising posterior to the fourth pharyngeal pouches. As it extends posteriorly, it becomes closed off from the pharynx along most of its length, forming a tube, the trachea, although it remains open to the pharynx at its most anterior end, this opening constituting the glottis (*Plate 131*). There is a temporary closure of the lumen of the trachea during day 4 (Romanoff, 1960). By the day 6 the trachea has lengthened and its anterior end has expanded to form the larynx, the lumen of which becomes temporarily obliterated between days 8 and 11 of incubation. The development of the larynx and trachea was studied by transmission electron microscopy between days 10 and 21 of incubation (Walsh and McLelland, 1978). The **syrinx** (*Plate 116*), the vocal organ of birds, forms at the junction of the trachea and bronchi. Unlike mammals, birds possess no vocal cords in the larynx.

During day 3, **cartilaginous rings** begin to form in the mesoderm surrounding the trachea and bronchi and, during day 9, syringeal cartilages and muscles start to develop. By day 11, the syrinx of the male is larger than that of the female. (For a fuller description of the development of the trachea and syrinx, see Romanoff, 1960.)

The **lungs** (*Plate 50*) start to arise during day 3 as paired evaginations from the laryngo-tracheal groove. The distal part of each evagination forms the lung and the proximal part of the bronchus (*Plate 116*). Each bronchus may be further subdivided into two regions: the mesobronchus, which is the part that becomes partly enclosed in the lung, and the primary (or extrapulmonary) bronchus, the part outside the lung (*Text-Figure 50*). Epithelial–mesenchymal interactions are involved in lung branching (Stabellini *et al.*, 2001). By about the end of day 4 the lungs have become small, smooth-surfaced, sacs lying on either side of the oesophagus (*Plate 34*). By day 5 the distal end of the mesobronchus has swollen to become the abdominal sac. The pulmonary veins and pulmonary arteries (*Plate 116*) become visible about days 5–6. Surfactant protein mRNA has been detected in the lungs as early as day 14 (Johnston *et al.*, 2000) and appears to be formed by similar cellular pathways to those used by mammalian embryos, despite the differences in lung structure and function (Sullivan and Orgeig, 2001).

The air sacs, which have all formed by outpushings of the mesobronchi, develop their own bronchial tubes after hatching. Apart from those in the cervical air sac, these bronchi pass back into the lung proper and are known as the recurrent bronchi. The abdominal air sacs (*Plates 136, 138*) start to migrate into the abdomen about day 9, penetrating the pleuroperitoneal septum *Plate 117*) and from day 10 they expand, reaching the posterior end of the coelom by day 15. The other air sacs possess a similar history, forming as dilatations of the bronchi and then penetrating to other regions of the body (Locy and Larsell, 1916). Considerable development takes place after hatching. For a recent study of the post-hatching lungs by scanning electron microscopy see Maina (2003).

Respiration begins via the chorioallantoic membrane about 1 day before hatching. For an account of the respiratory rhythms just before and after hatching see Chiba *et al.* (2002). The growth rate of heart and lungs from 15 days of incubation to 5 days post-hatching are considered by Yang and Siegel (1997).

Hatching

Gas exchange takes place via both the chorioallantoic membrane and the lungs (Menna and Mortola, 2002) though, as the circulation in the chorioallantoic membrane gradually ceases, the exchange becomes restricted to the lungs. About a day before hatching the chick pierces the inner shell membrane with the egg tooth (see p. 108) and begins to draw air into its lungs from the air space (*Text-Figure 1*). After a further 8–9 h the beak breaks through the shell over the air space (pipping) and, after a further 3–4 h and a complex series of hatching movements, the chick emerges from the shell. Cheeping begins while the chick is still in the egg and there is some evidence that it can play a role in the synchronization of hatching in a batch of eggs (Freeman and Vince, 1974).

Chapter 9

Nervous System

The nervous system is formed from the neural plate and the neural crest (see Chapter 4). Ectodermal placodes contribute to the organs of special sense.

THE BRAIN

The anterior end of the neural tube enlarges to form the brain. *Text-Figure 51* shows a fate map of the most anterior region of the neural plate based on the results of chick–quail transplants. By stage 9 (*Plate 55*) the primary optic vesicles (see below) have begun to form, and by stage 10 the brain has become divided into three primary regions, the **prosencephalon** (forebrain), the **mesencephalon** (midbrain) and the **rhombencephalon** (hindbrain) (*Plate 8*). By stage 11 the rhombencephalon has become subdivided into the **metencephalon** (cerebellum and pons) and the **myelencephalon** (medulla oblongata) (*Plate 9*).

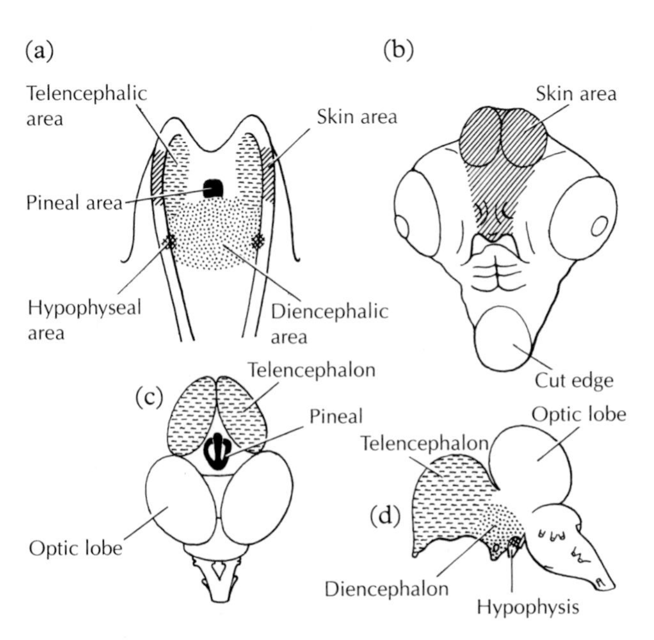

Text-Figure 51. *The origin of the different tissues of the head. (a) Fate map of the anterior end of the neural tube at stage 7 (24 h). (b) Skin area of the face at 8 days (stage 34), which has originated from the lateral neural folds. (c,d) the telencephalic and diencephalic regions at 8 days (stage 34), including the hypophysis. (After Couly and Le Douarin, 1988, with permission of the Company of Biologists Ltd.)*

The division of the prosencephalon into telencephalon and diencephalon (thalamus) has begun to take place by stages 12–13. Paired swellings, appear just anterior to the optic vesicles and form the telencephalic vesicles, each of which communicates with the median telocoele (the central lumen of the brain in that region) by the foramen of Monro. At about the same time the cranial flexure has begun to appear as a bend at the anterior end of the mesencephalon, and the cervical flexure begins at about stage 18 (day 3) (*Text-Figure 20*; Chapter 5). The general topography of the brain at 4 days (stage 23) is shown in *Text-Figure 52*. By 10 days (stage 36) the cranial flexure has almost gone and the cervical flexure has become reduced, so that the neck is less convex and the head no longer tucked against the thorax (see Normal Table in Appendix II). The walls of the telencephalic vesicles begin to thicken at about stage 13 (2 days) and will form the cerebral hemispheres, their lumina becoming the lateral ventricles. The cerebral hemispheres have expanded so much that by 11 days (stage 37) they overlap the diencephalon and by 18 days (stage 44) the mesencephalon (*Text-Figure 53*).

The communication between the optic vesicles and the anterior end of the diencephalon has become reduced to a stalk by about stages 13–14 (*Plate 66*). The wall of the optic recess (*Plate 73*), a depression in the floor of the diencephalon between the optic vesicles, becomes very thin by stage 18 (3 days), but a thickened region immediately posterior to it marks the development of the optic chiasma (*Text-Figure 52a*; *Plate 124*). The pineal organ begins to form at about stage 19 as an evagination from the median part of the roof of the diencephalon (*Text-Figure 52a, Plate 82*). A description of its development is given by Calvo and Boya (1979). The infundibulum (see Chapter 12 and *Plates 81, 97*) develops posterior to the chiasma and combines with Rathke's pouch (*Plate 73*) to form

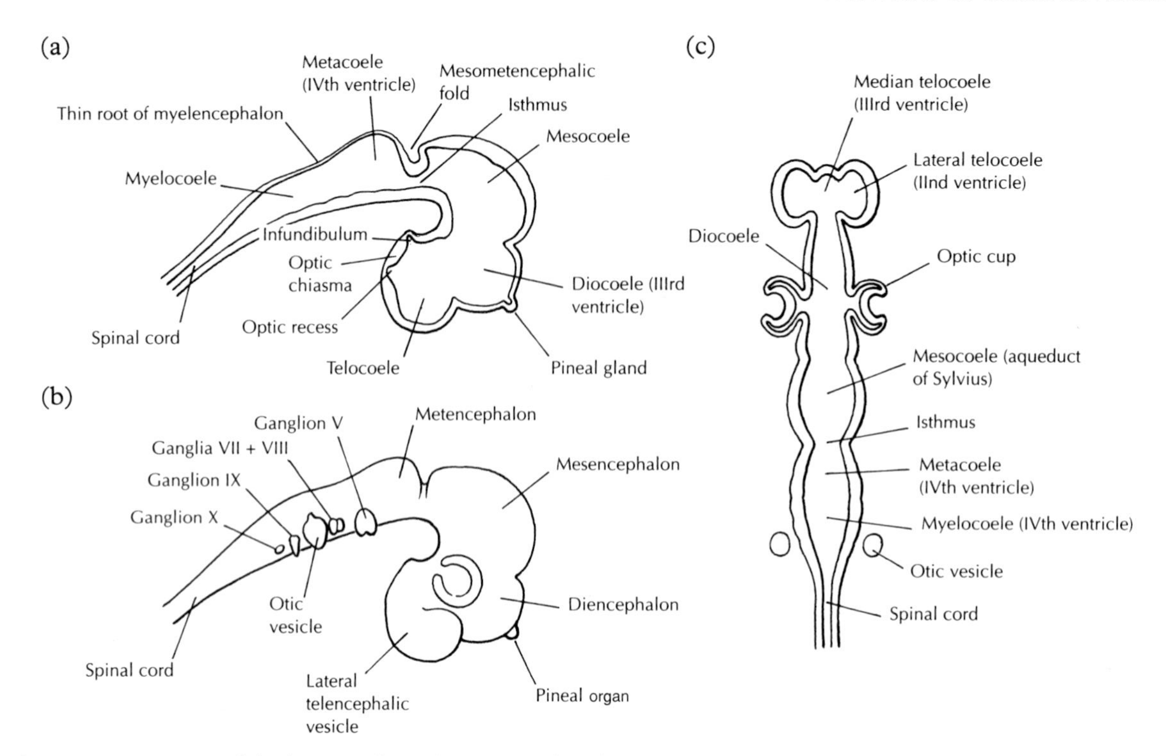

Text-Figure 52. *Diagrams of the brain to show the topography of a 4-day (about stage 23) embryo. (a) Sagittal section. (b) View of brain from right side showing some of the cranial nerve ganglia. (c) Schematic frontal section plan, with flexures straightened out. (After Patten, 1950.)*

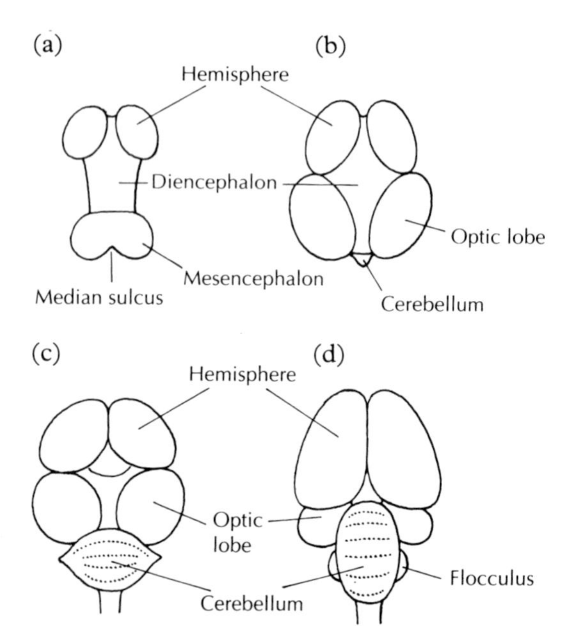

Text-Figure 53. *Dorsal view of the brain to show how the cerebral hemispheres, formed from the telencephalic vesicles, expand and overlap the diencephalon and mesencephalon. (a) 3 days (stage 18); (b) 7–8 days (stage 33); (c) 12 days (stage 38); (d) 18 days (stage 44). (After Romanoff, 1960.)*

the pituitary. A useful fate map of the anterior forebrain at stage 8 is based on the analysis of chick–quail chimeras (Cobos *et al.*, 2001).

The mesencephalon has increased greatly in size by about stage 18 (3 days) and become clearly marked off posteriorly from the rhombencephalon by a narrower region, the isthmus (*Text-Figure 52*) located in the meso-metencephalic fold (*Plates 73, 82*). By about stage 22 (3.5–4 days) the mesencephalon has grown so large that it projects over the metencephalon. The differentiation of the optic lobes from the mesencephalon begins to occur at about 5–5.5 days (stage 27) but even as early as 3 days the surface of the mesencephalon is covered with fibres, foreshadowing the enormous size of the optic lobes in the adult, which itself is correlated with the importance of vision in birds.

The rhombencephalon is characterized by a thin-walled roof in the early stages, though by about 4 days this becomes thicker in the anterior region as the metencephalon develops (*Plate 82*). The walls of the metencephalon become thicker as the two halves of the cerebellum begin to form. During the following days they enlarge and by 9–10 days (about stage 36) they have fused together in the midline. During the remainder of the incubation period the cerebellum increases in size and complexity and by day 16 almost abuts the cerebral hemispheres (day 18 shown in *Text-Figure 53*). The myelencephalon has acquired the characteristic shape of the medulla oblongata by about 9–10 days (about stage 36; *Plates 113, 124*). Many fibre tracts have developed in the brain by 5 days (stage 26) (see Romanoff, 1960). These are best seen in silver-stained preparations but are often visible in sections stained with haematoxylin and eosin.

We have seen (Chapter 4) that a range of genes play a role in patterning of the brain and spinal cord. Expression of *Wnt8b* and *Wnt7b* are of importance in the patterning of the forebrain (Garda *et al.*, 2002), while *Wnt1* and *Wnt3a* may be implicated in the specification of the dorsal spinal cord (Muroyama *et al.*, 2002). The dorso-ventral patterning of the spinal cord is also dependent on *sonic hedgehog*; cells in chick neural plate explants differentiate into ventral cell types under the influence of *sonic hedgehog* (Robertson *et al.*, 2001).

HISTOLOGY OF THE SPINAL CORD AND DEVELOPMENT OF THE FIBRE TRACTS

A fate map of the spinal cord in the thoracic, lumbar and caudal regions was published by Catala *et al.* (1996). At the time of its formation from the ectoderm the neural tube consists of a thickened columnar epithelium, but changes soon take place in its shape and histology. All the cells of the neural tube are elongated with their long axes radiating out from the ependyma to the periphery. By stage 16 (2.5 days) the neural tube possesses two layers, the **ependyma**, which lines the neural canal and contains a large number of mitotic cells, and the **marginal layer** (*Text-Figure 54*), and by stage 18 (3 days) the **mantle layer** is also recognizable. Neuroblasts are visible from about stage 15 (2 days) in the ventro-lateral part of the tube; this region is about six cells deep, whereas in the floor plate (the keel) there is only a single layer of cells.

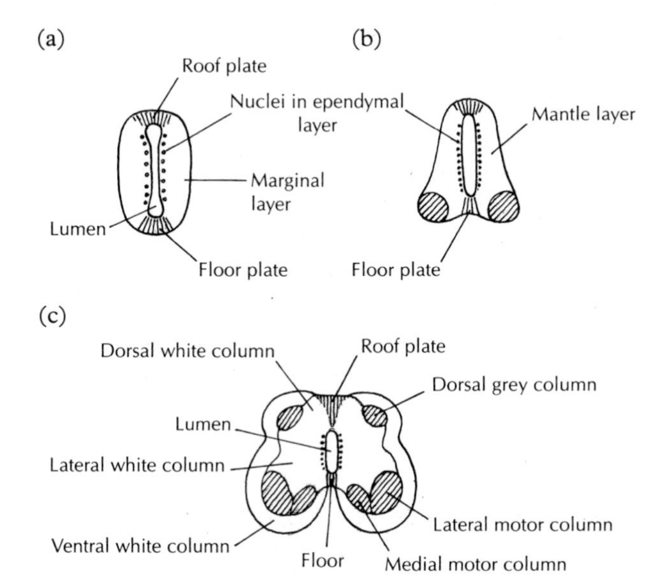

Text-Figure 54. *Development of the spinal cord at (a) 2.5 days (stage 16); (b) 4.5 days (stage 24); (c) 8 days (stage 34). (After Romanoff, 1960.)*

By stage 16 spinal nerves have developed and by stage 22 (3.5–4 days) regions of 'grey' and 'white' matter are recognizable. Nissl substance is visible in embryos of this stage stained with basic dyes. Dorsal and ventral horns can be seen in the grey matter from day 7, (about stage 31) and glial cells in the white matter. During the following days the spinal cord becomes larger in transverse section and there is a change in shape of the lumen from a longitudinal slit to an almost square or round shape. This is accompanied by the development of dorsal and ventral fissures.

SPINAL NERVES

Each spinal nerve has both a somatic and a splanchnic component, the former serving the somatopleure and the body axis and the latter the viscera, and each part has both motor and sensory components. By 8 days (stage 34) there are 38 pairs of spinal nerves. The spinal ganglia are of neural crest origin.

The segmentation of the spinal nerves is associated with that of the adjacent somites, so that each spinal nerve emerges between two vertebrae. After leaving the neural tube each axon passes through the anterior half of the sclerotome and is inhibited from going through the posterior half. The neural crest cells are similarly constrained to migrate through the anterior and not the posterior half (Keynes and Stern, 1984).

The ventral roots of the spinal nerves have emerged in the cervical region by about stage 13 and appear progressively in the more posterior regions during the following day. The spinal ganglia begin to appear early in day 3 and the dorsal roots that arise from them develop during days 3–4 (*Plate 84*). The neurons in the dorsal root ganglia vary in structure. The size of the dorsal root ganglia also varies, being especially large for the nerves innervating the limbs. By stage 20 (3–3.5 days), the ganglia of the brachial segments 14–15 are more than 80% larger than those in the cervical segments 5 and 6, and this has been correlated with the colonization of the brachial region by a larger number of neural crest cells (Goldstein *et al.*, 1995).

The motor axons of the spinal nerves innervate the myotomes to which they attach and become transported by the developing muscle. The sensory fibres subsequently migrate along the motor nerves. Lance-Jones (1988), who studied the somitic level of origin of the hindlimb muscles, found that the motor neurons supplying them arose from the same somitic level and, subsequently (Lance-Jones, 1990), that the axons were guided by the limb somatopleure.

Spinal nerve plexuses develop in the brachial and lumbo-sacral regions from the 13th to 16th and the 23rd to 29th spinal nerve, respectively. Chick spinal nerves have been found to undergo a transient period of spontaneous activity during their development, with regularly occurring episodes interspersed with quiescent periods (Wenner and O'Donovan, 2001).

The meningeal layers of the spinal cord, and the endoneurium of the peripheral nerves are all formed from mesenchyme cells of non-neural crest origin, whereas the Schwann cells are of neural crest origin (Halata *et al.*, 1990).

Oligodendrocytes first appear in the ventral region of both the spinal cord and the metencephalon (Davies and Miller, 2001) under the influence of *sonic hedgehog*. Conversely, the specification of oligodendrocyte precursors is inhibited by bone morphogenetic proteins (BMPs) (Mekki-Dauriac *et al.*, 2002).

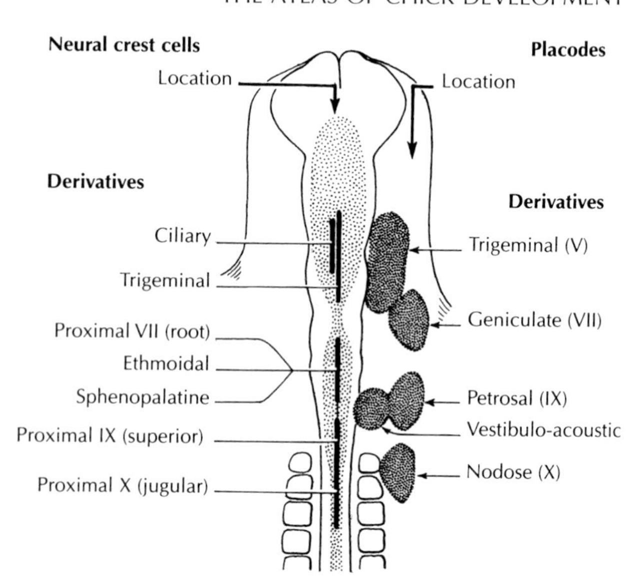

Text-Figure 55. *Schematic drawing of a stage 9.5 chick embryo indicating the position of the neural crest and placodal anlagen (precursors) for cranial, sensory and autonomic ganglia. (After Noden, 1988, with permission of the Company of Biologists Ltd.)*

THE CRANIAL NERVES AND GANGLIA

The development of each of the cranial nerves is closely associated with that of fibre tracts and nuclei in the brain. The following cranial sensory nerve ganglia are partially derived from the neural crest and partially from epidermal placodes: the trigeminal (V), geniculate (VII), vestibular acoustic (VIII), and the proximal ganglionic components of the petrosal (IX) and the nodose (X). The head neural crest cells migrate between stages 9 and 11 and some condense lateral to the brain and join with cells from the epidermal placodes which will become neuroblasts. D'Amico-Martel and Noden (1983) have mapped the position of the neural crest and placodal anlagen (*Text-Figure 55*) at stage 9.5 and the position of the cranial ganglia at 12 days.

Cranial Nerve I (Olfactory Nerve)

(*Plates 114, 131*)

The sense of smell is poor in birds and this is reflected in the small size of the olfactory nerve. It forms by the growth of cells from the nasal placode (*Plate 23c,d*) to the olfactory lobes, which begins about stages 17–18 (3 days). Fibres become visible at about 3–4 days.

Cranial Nerve II (Optic Nerve)

Vision is highly developed in birds and this is indicated by the large optic nerves and optic lobes (*Plate 97*). The optic nerves grow from the retina toward the ventral region of the diencephalon, where they cross to form the optic chiasma (*Plates 124, 131*) and terminate on the contralateral side of the brain. The retinal fibres enter the optic stalk at about stage 18 (3 days) and cross through the optic chiasma at about the beginning of day 4, reaching the surface of the optic lobes on day 5. By day 7 many fibres have spread over the surface of the optic lobes.

Cranial Nerve III (Oculomotor Nerve)

(*Plates 84, 124*)

This is the main nerve to the extra-ocular muscles and is especially large in birds because of the great size of the eyes. The left and right oculomotor nerves grow from the ventral side of the midbrain immediately posterior to the hypophysis and innervate the superior and inferior rectus muscles of the eye. The ciliary ganglia appear at about stage 10 and each is situated two-thirds of the distance from the proximal to the distal end of the oculomotor nerve. The ciliary nerve, which runs from the ganglion to the iris, is visible according to some authors at 4 days (stage 23).

Cranial Nerve IV (Trochlear Nerve)

This special somatic nerve arises at about 3–3.5 days (stage 20) at the dorsal surface of the midbrain posterior to the oculomotor nerve and innervates the superior oblique muscle of the contralateral eye.

The left and right nerves cross at about 3.5 days (stage 21) at the dorsal surface of the midbrain and then start to grow ventrally over the lateral face of the mesencephalon.

Cranial Nerve V (Trigeminal Nerve)

The various branches of the trigeminal nerve appear at about stages 13–14 together with the neural crest-derived trigeminal (Gasserian/semilunar) ganglion, which is easily recognizable (*Plates 82, 83, 87*) by its large size and its situation at the widest part of the mesencephalon. The sensory parts of the nerve are formed from the ganglion, and the motor components are derived from the rhombencephalon. The major branches of the trigeminal nerve are the ophthalmic (*Plate 81*) and the maxillary-mandibular, the latter subdividing into its maxillary and mandibular branches (*Plates 81, 83*) during day 4.

Cranial Nerve VI (Abducens Nerve)

This cranial nerve (*Plate 131*) arises during day 3 in the midline from the ventral side of the myelencephalon and has migrated to the primordium of the lateral rectus muscle by stage 24 (4.5 days). There is no ganglion associated with it. The main nucleus, which is situated ventral to the floor of the fourth ventricle, appears at 5 days. The accessory nucleus, which lies between the descending root of the trigeminal and superior olive nerve, is derived from cells which have left the main nucleus.

Cranial Nerve VII (Facial) and Cranial Nerve VIII (Acoustic or Vestibulocochlearis Nerve)

The development of these two nerves is closely associated. The facial nerve (*Plate 123*), which becomes recognizable at about stage 11, is formed partially from neural crest and partially from the epibranchial placode of the first pharyngeal groove. A branch from the facial ganglion has reached the upper end of the hyoid arch by stage 13, and has extended into the third rhombomere by stage 15. Cells leave the auditory epithelium at about stage 18 to form the acoustic ganglion (shown at 11 days in *Plate 131*) and at about the same time the facial and acoustic ganglion columns fuse with one another (*Plate 83*). Meanwhile cells bud from the placode over the hyoid arch and fuse with the facial and acoustic columns to form the geniculate ganglion (*Plates 82, 87*). By about 3.5–4 days (stage 22) the geniculate ganglion is attached to the rhombencephalon.

Early on day 4 the acoustic ganglion is present as two parts and has become independent of the facial nerve, but by day 5 it is a large bilobed structure, closely applied to the auditory epithelium. The vestibular ganglion is derived almost entirely from the epibranchial placode. At 5 days the motor nucleus of the facial nerve is a single group of cells but by day 7 this has become divided into a dorso-medial and a ventro-lateral group.

Cranial Nerves IX and X (Glossopharyngeal and Vagus Nerves)

The glossopharyngeal nerve (*Plate 83*) innervates the pharyngeal muscles and the salivary glands. The vagus nerve (*Plate 87*) supplies the cervical and thoracic regions and much of the abdominal viscera. The peripheral differentiation of the vagus nerve was described by Kuratani and Tanaka (1990). Sato *et al.* (2002) found that electrical stimulation of these nerves between days 4 and 8 of incubation elicited a pattern of optical recordings which changed with advancing development.

A large ganglionic complex begins to form at about stage 13 at the root of the two nerves, the most rostral part being the superior ganglion of the glossopharyngeal nerve (*Plate 82*), and the more caudal part the jugular ganglion of the vagus (*Plate 84*). In addition a ganglion develops on the trunk of each nerve a short distance from the brain, the **petrosal ganglion** (*Plates 124, 125*) on the glossopharyngeal and the **nodose** on the vagus. The nodose ganglion provides sensory innervation to the heart and other viscera (Harrison *et al.*, 1995).

Both the petrosal and the nodose ganglia receive a contribution of cells from the ectoderm (the nodose from the ectoderm dorsal to the fourth pharyngeal cleft and the petrosal from a more anterior region), in addition to that from the neural crest. The nodose ganglion begins to appear at stage 18 (3 days) as a group of cells lateral to the dorsal aorta, where it is joined by the anterior cardinal vein (D'Amico-Martel and Noden, 1983), but it subsequently shifts posteriorly.

By stage 13 the cells of the glossopharyngeal–vagus part of the neural crest have already migrated under the ectoderm and over the first and second somites. By stage 15 the distal part of the migrating column of cells has started to bifurcate, the anterior part going to the third visceral arch and the posterior part to the fourth visceral arch.

The ganglia of the IXth and Xth cranial nerves form just posterior to the auditory vesicle at the dorsal side of the second and third visceral furrows.

Cranial Nerve XI (Spinal Accessory Nerve)

This nerve supplies the pharynx and shoulder muscles. In the 4–5 day embryo the strand runs from the ganglion and dorsal root of the third cervical nerve to the root of the vagus.

Cranial Nerve XII (Hypoglossal Nerve)

This nerve supplies the muscles at the base of the tongue. It leaves the ventral side of the rhombencephalon by several roots opposite the third and fourth somites, which unite to form a single trunk. The first axons appear at about stages 17–18 (3 days). During days 4–5 the nerves extend posteriorly and enter the pharynx.

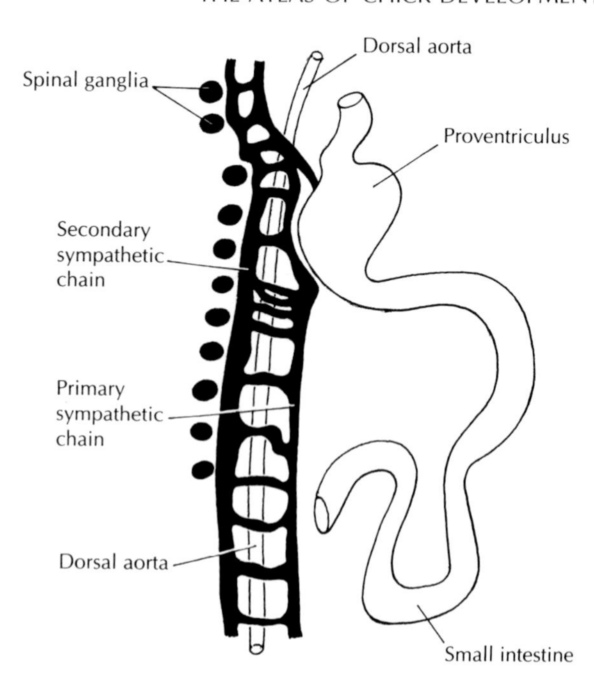

Text-Figure 56. *Autonomic nervous system showing primary and secondary sympathetic chains at 10 days. (After Romanoff, 1960.)*

THE AUTONOMIC NERVOUS SYSTEM

This is of neural crest origin. The levels of the neural crest that give rise to the various autonomic ganglia as well as to the adreno-medullary cells were determined experimentally by exchanging strips of neural tube, together with the associated neural crest, between chick and quail embryos (Le Douarin *et al.*, 1984a,b). All the sympathetic ganglia arise from the neural crest posterior to the level of the fifth somite. The adreno-medullary cells form from the neural crest between levels of the 18th and 24th somites. All the parasympathetic enteric ganglia receive neural crest cells from somite levels 1–7, but those parasympathetic ganglia that lie posterior to the yolk sac stalk receive additional neural crest cells from the sacro-lumbar region. The ganglion of Remak is derived from the lumbro-sacral crest posterior to the level of the 28th somite.

The first signs of the primary sympathetic chain are visible at the end of day 3 as two rows of cells lying one on either side of the neural tube, ventral to the somites and immediately lateral to the aorta. Ganglia form on the chains about days 4–5. The secondary sympathetic chains, which start to appear early on day 5, grow out as sprouts from the primary chain and then become connected with one another to form the secondary chain, which runs close to the spinal ganglia (*Text-Figure 56*). The primary chain then becomes reduced in size and it eventually disappears. There are about 38–39 secondary sympathetic ganglia (*Plate 115*), consisting of 13 cervical, 7 thoracic, 14 lumbro-sacral and 4–5 coccygeal. The rami communicantes are visible in silver preparations by 5–7 days and pass from the secondary sympathetic ganglia to the roots of the spinal nerves.

The secondary chain (*Text-Figure 56*) becomes a continuous nerve cord (the paravertebral) between days 4 and 8, running from the superior cervical ganglion to the coccyx. It bifurcates around each rib but remains as a single structure between the ribs. It connects with the plexuses in the following regions: the aortic plexus (which becomes modified further into the coeliac plexus around the coeliac artery, as well as the pelvic and hypogastric plexuses), the medullary plexus of the adrenal gland, the splanchnic plexus and Remak's ganglion in the rectal region.

The times at which enteric neurons first appear in different regions of the gut have been plotted by Fairman *et al.* (1995). Most of the enteric nervous system is formed from neural crest cells of the vagal region, though there is a contribution also from sacral neural crest (Burns and Le Douarin, 2001).

ORGANS OF SPECIAL SENSE

The Eye

The first sign of the development of the eyes is a bulging at the lateral sides of the prosencephalon at about stage 9 (*Plate 55*). These are the rudiments

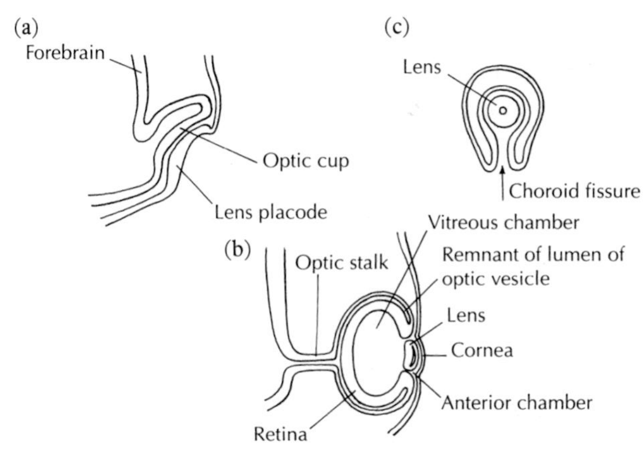

Text-Figure 57. *Development of the optic cup and lens. (a) Stage 14, section through developing retina and lens placode. (b) Stage 18, transverse section through the eye to show retina and developing lens. (c) Stage 18, frontal section through the eye to show choroid fissure.*

of the optic vesicles which lie beneath the head ectoderm. By stage 10 each vesicle has begun to constrict at its base (*Plate 58*) until by stage 12 its connection to the brain has been reduced to a narrow stalk (*Plates 62, 66*), but the lumen of the optic vesicle is still in communication with the prosencephalon via the stalk.

Meanwhile, the distal part of each optic vesicle (the future sensory layer), invaginates and presses against the proximal part (the future pigment layer of the retina, iris and ciliary body). This results in the formation of the **optic cup** (*Text-Figure 57, Plate 66*), the elimination of the original lumen of the optic vesicle and the formation of a new lumen, the future **vitreous chamber**. The invagination of the optic cup does not start at the extreme lateral border of the optic vesicle but from a more ventral position. The wall of the cup is therefore lacking on the ventral side and the invagination is thus able to continue along the inferior (ventral) side of the optic stalk, so that a groove forms, along which the optic nerves and blood vessels subsequently pass. The gap in the ventral wall of the optic cup is the **choroid fissure** (*Text-Figure 57, Plate 81*). By the time the optic cup has formed regional differences are already marked within it.

The lens is formed from the lens placode (*Plate 24*), a thickening of the ectoderm formed in response to an inductive signal from the optic cup. It is first visible at about stage 12 (Bancroft and Bellairs, 1977) and invaginates at about stage 15 to form the lens vesicle (*Plates 66, 76*), which sinks beneath the surface of the ectoderm, the latter becoming the cornea. By about stage 18 (3 days) the wall of the proximal side of the lens vesicle has greatly thickened, whereas the wall at the distal side has become thinner. The thickening is brought about by an increase in the length of the individual cells that stretch across the entire thickness of the walls, a band of nuclei being visible within the thickening (seen as a dark band in *Plate 98*). A high mitotic rate at the periphery of the lens results in its continuing growth. By day 4 the lens cavity has been obliterated and at the same time the lens has begun to acquire its characteristic, lentoid, shape.

As the lens continues to grow, the cells in the thickened region lose their ability to divide and become converted into fibres that will become the core of the adult lens. Fibroblast growth factor (FGF) signals appear to play a role in the conversion to fibres (Le and Musil, 2001). New fibres are formed from the cells at the periphery of the lens which divide rapidly and become arranged in concentric circles around the original core. By the time of hatching there are three concentric layers of fibres, the core (0.8 mm in diameter), the intermediate layer of irregularly arranged fibres, and the radial layers which continue to grow after hatching. The major lens protein is Δ-crystallin (not δ-crystallins as in mammals), though α- and β-crystallins can be detected when the lens placode has only just begun to form (Zwaan and Ikeda, 1966). The lens capsule, which is an extracellular material with a high collagenous component, starts to form about the day 7.

As the lens loses contact with the ectoderm a space is formed, the **anterior chamber of the eye** (*Text-Figure 57*). Neural crest cells, which have reached the margin of the optic cup by stage 18, have migrated into the anterior chamber by stage 23 (Beebe and Coates, 2000). The corneal epithelium develops from the ectoderm covering the anterior chamber, whilst the corneal stroma forms from the mesenchyme and becomes visible on day 4 as a thin layer beneath the epithelium. It becomes thicker as mesenchyme cells migrate into it during day 7, and by the formation of Bowman's membrane from 11 days, followed by Decemet's membrane from about 13 days.

The **iris** arises from cells at the margin of the anterior chamber at about day 7. Removal of the lens results in disorganization of the components of the anterior chamber (Beebe and Coates, 2000).

The **retina** is formed from the optic cup. Its inner layer becomes the neural retina and its outer layer the pigmented retina. FGF8 is associated with the differentiation of the neural retina and BMP7 with that of the pigmented retina (Vogel-Hopker *et al.*, 2000).

The **choroid** and **sclera** differentiate from the mesenchyme around the optic cup, forming the inner pigmented vascular layer, and the outer, fibrous layer, respectively. The melanophores of the choroid are derived from cells of the neural crest that reach the eye during day 2 and develop pigment on day 7. Cartilage starts to form in the sclera on day

8. On day 9, 14 papillae become visible on the conjunctival sclera; each is composed of a solid cord of epithelium and becomes associated with a condensation in the underlying mesenchyme. By 13 days the papillae have disappeared and the condensations have become 14 scleral ossicles (*Plate 150*).

The **eyelids** start to form at about 7 days (stage 31) from a circular fold of skin surrounding the eye which becomes modified to form the upper and lower eyelids. A semicircular fold within this circular fold becomes the nictitating membrane. The Harderian gland begins to develop on day 11 (stage 37) from epithelial cones on the conjunctiva (Niedorf and Wolters, 1978).

The **choroid fissure** usually begins to close in the region near the lens about day 4 (stage 23) though accounts vary. At this time a ridge of mesoderm, carrying with it a blood vessel, migrates along the choroid fissure into the posterior chamber of the eye and enlarges during day 5 (about stage 26) to form the pecten (*Plate 123*). Subsequently it becomes wrapped around by the ridges of the choroid fissure in the optic stalk and is totally covered by day 8. The pigment cells of the pecten are derived from the pigmented retina (Yew, 1978). The pecten is a structure characteristic of birds and it is thought that it acts not only by bringing oxygen and nutritive materials to the eye but it may also play a role in vision.

The development of the pecten, the neural retina and the retinal pigment epithelium are all influenced by retinoic acid and *sonic hedgehog*. BMPs also play a role in the development of these structures. Adler and Belecky-Adams (2002) showed that overexpression of *noggin*, which binds several BMPs at the optic vesicle stages, led to micropthalmia, together with abnormalities of the retinal pigment epithelium and lens. Similar treatment at the optic cup stages resulted in extensive anomalies including coloboma and pecten agenesis. BMP4 is expressed in the developing neural retina where it appears to be responsible for apoptosis (Trousse *et al.*, 2001). The **optic nerve** (*Plate 123*) is derived principally from the retina, its axons growing along the choroid fissure on their way to the brain (see p. 87).

The vitreous humour is secreted by the cells of the optic cup.

The Ear

The **inner ear** is derived from the otic placode (*Plates 22, 23, 70*) and the middle and outer ear are formed from the region of the first ectodermal groove.

The left and right **otic placodes** (*Plates 22, 23*) are visible in embryos at about stage 10 (Bancroft and Bellairs, 1977) as thickened regions on either side of the head just anterior to the somites. By stage 12 each has invaginated to form an otic pit, probably as a result of induction by FGF3 (Vendrell *et al.*, 2000), and by stage 14 the opening of the pit has become greatly restricted and the whole structure now becomes known as the **otic vesicle** or otocyst (*Plates 65, 73*). If the basal lamina associated with the otic placode is enzymatically disrupted, the placode fails to invaginate properly (Moro-Balbás *et al.*, 2000). The otic vesicle loses its communication with the outside at about stage 18. The endolymphatic duct is a blind-ended outgrowth from the dorsal wall of the otic vesicle.

During days 5–7 the distal end of the endolymphatic duct (*Plate 88*) enlarges to form the **endolymphatic sac** (*Plate 110*) which grows above the brain and comes to lie along the dorso-lateral side of the myelencephalon. The remainder of the otic vesicle meanwhile has become loosely divided (*Plate 103*) into a **superior chamber**, which will form the semicircular canals and utricle, and an **inferior chamber**, which will give rise to the saccule and cochlea.

The **semicircular canals** start to develop on days 4–5 as three grooves in the wall of the superior chamber, which each invaginate and close over to form a tube. The lumen of each tube is continuous with that of the superior chamber of the otic vesicle, which itself will become the utriculus. The anterior semicircular canal (which lies in the sagittal plane) forms first, followed by the external semicircular canal (which lies in the horiziontal plane) and finally by the posterior canal (which lies in the frontal plane). The canals then move out from the main body of the otic vesicle, each carrying with it a thin fold of otic wall. By days 7–8 this thin sheet has perforated so that the canals remain attached to the otocyst only at the ends. Swellings which form along the canals during days 5–6 form the ampullae. There is some evidence that BMP2, secreted by the epithelium of the otocyst, is involved in the early stages of the formation of the semicircular canals and may affect the chondrogenesis of the otic capsule around the canals (Chang *et al.*, 2002). BMP4 is essential for normal development of the semicircular canals (Gerlach *et al.*, 2000).

The **saccule** begins to appear on day 7 as a swelling on the dorso-medial wall of the inferior chamber. The cochlea and the cochlear duct start to form during day 6 by growing out from the ventral region of the inferior chamber.

The **middle and outer ear** are formed from the first pharyngeal pouch and the surrounding wall of the pharynx (see Jaskoll and Maderson, 1978). The dorsal part of the first pharyngeal pouch gives

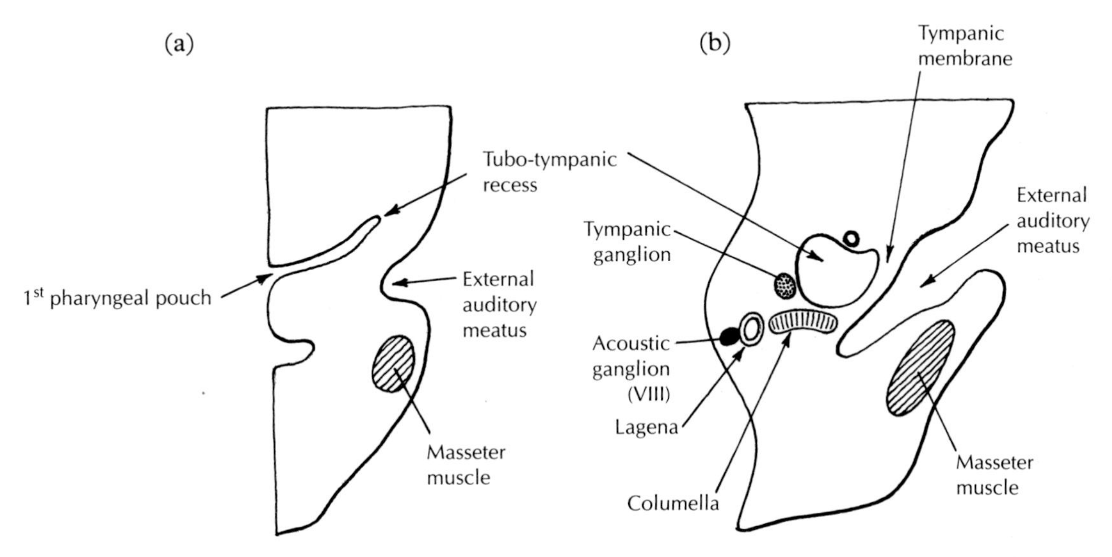

Text-Figure 58. *Transverse section through the right middle and outer ear at (a) 4 days; (b) 12 days. (After Romanoff, 1960.)*

rise to the tubo-tympanic recess which later becomes the auditory (Eustachian) tube (*Text-Figure 58*). The **external auditory meatus**, which forms from the first ectodermal groove (*Plate 114*), is apparent by day 6) and further elaborated by day 12 when it then lies in juxtaposition to the tympanic cavity. The thickened region of the wall which separates them is the primordium of the tympanic membrane. Birds possess a single auditory ossicle and not three as in mammals. The cartilaginous predecessor of this bone, the **columella**, is present by 6 days and lies between the tympanic membrane and the cochlea and the tubo-tympanic recess is directed towards it. For a full description of the development of the middle ear and tympanum see Jaskoll and Maderson (1978).

In a more recent study, the maturation of the tympanum layers is discussed by Chin *et al.* (1997). For a report on the changing sizes of the individual components of the inner ear between 10 days of incubation and adulthood see Cohen *et al.* (1992), and on patterns of chondrification and ossification see Cohen and Hersing (1993).

For an atlas on the gross anatomy of the chick inner ear see Bissonette and Fekete (1996). The neural relationships in the vestibular region of the 11-day-old chick embryo are described by Diaz *et al.* (2003).

The Nose

The nasal (olfactory) placode (*Plates 23, 73*) becomes visible at stage 16 (Bancroft and Bellairs, 1977) as two thickened regions anterior to the eyes. Like the optic and otic placodes each nasal placode invaginates, and it forms the **nasal pit** (*Plate 85*) on days 3–4. A slight depression on the medial wall of each pit at about 4–5 days is thought to be the **organ of Jacobson**. Each nasal pit remains open to the surface of the head during its earliest stages of development, this opening eventually becoming the **external naris** (see Chapter 5). It becomes closed from about day 6, however, by the proliferation of epithelial cells forming a plug and does not reopen until hatching, by which time the epithelial plug has degenerated.

The pits deepen, partly due to the bulging up of the ectoderm around the opening, forming the lateral and median fronto-nasal processes. The cavities of the nasal pits initially open into the oral cavity, but when the palate forms between days 5 and 7, the upper (nasal) part of the oral cavity becomes partly separated off from the lower part and as a consequence the internal nares (*Plate 104*) become displaced to the posterior end of the mouth. In birds, however, the palatine processes do not fuse completely along the length of the **median palatine fissure** (*Plate 132*), even though they meet in the midline on day 4. Birds are therefore said to have a split palate.

Three **conchae** (turbinals) (*Plate 132*), the superior, middle and inferior, develop from the lateral wall of the nasal cavity and project into the lumen, the middle appearing on day 5, the superior at days 5–6 and the inferior (or vestibular) during day 7. The inferior turbinal is found only in birds, although not in all birds.

The lateral nasal glands form between days 8 and 14, starting as a solid mass in the septal wall of the nasal cavity.

Chapter 10
Skeleton and Muscles

SKELETON

Birds resemble other amniotes in that most of the skeleton is initially laid down in cartilage which subsequently becomes ossified. Bones formed in this way are known as **cartilaginous bones**, as opposed to **membrane bones**, which are ossified directly from mesodermal tissues and do not go through a cartilaginous phase. Birds differ from other amniotes, however, in possessing air-sacs, which form as outgrowths from the lungs (see Chapter 8) extend into the bones and even replace the marrow of the long bones. This process, known as **pneumatization**, leads to a reduction in weight and is helpful for flight, but does not take place until after hatching (Schepelmann, 1990). Fusion of some of the vertebrae to form the **synsacrum** (see below) occurs during embryonic development and is an adaptation to reduce the shock of landing after flight.

The Vertebral Column

The vertebrae develop from the sclerotomes and the notochord. The first pair of somites forms at stage 7 (about 24 h of incubation) and succeeding pairs continue to be laid down until about 50 pairs have formed, by stage 22 (3.5–4 days). As we have seen (Chapter 4), each somite gives rise to three major components, the dermatome, the myotome and the ventrally located sclerotome. In the same way as the segmentation of the somites proceeds from anterior to posterior, so the differentiation of individual somites into the three components takes place in a chronological sequence down the body and this is followed by the formation of the vertebrae in the same order. A useful consequence is that the maturation of the somitic tissues may be studied not only by comparing embryos of different ages, but also by comparing somites in different positions along the body of the same embryo. *Plates 71* and *72* illustrate a series of transverse sections through the somites of the same embryo at the levels indicated.

The origin of the **notochord** has been described on p. 33. It extends from the level of Rathke's pouch to the tip of the tail bud by about stages 20–21 and is one of the few medial structures in the body. It is round or oval in transverse section and until about stage 16 it measures about 30 µm in diameter. After that its diameter increases, reaching about 115–160 µm by about stage 23 (4 days). This increase is due to the development of vacuoles which subsequently secrete material that supplements the notochordal sheath (peri-notochordal tube) and are visible from about stage 11 (*Plate 16*).

We have discussed in Chapter 4 that the secretion of *sonic hedgehog* protein by the notochord and floor plate leads to the formation of sclerotomal cells (which are mesenchymal) from the medio-ventral part of the somite which, until then, has been epithelial. Some, though not all, of the sclerotomal cells now express *Pax-1* and *Pax-9* and migrate toward the notochord to form the notochordal sheath (Ebensperger *et al.*, 1995). *Pax-1* and *Pax-9* are considered to be among the markers of sclerotome cells, others being *twist* and *scleraxis* (Christ *et al.*, 2000). *Sonic hedgehog* (*Shh*), and possibly *noggin*, are essential for the survival and maintenance of these *Pax-1* expressing cells (Dockter, 2000). *Sonic hedgehog* is considered to act as a ventralizing signal, since it is present in the notochord and the ventral region of the neural tube (floor plate), whereas its antagonist, BMP4, has a dorsalizing effect, as do members of the *Wnt* family (produced by the ectoderm and neural tube). It appears that *Shh* signalling from the notochordal sheath actively attracts the migrating sclerotome cells. About day 10 of incubation the diameter of the notochord starts to be reduced and by day 14 the vacuoles have disappeared. The size of the sclerotome depends, at least in part, on the balance between the dorsal and ventral signalling, but may also be affected by adjacent tissues (reviewed by Christ *et al.*, 2000).

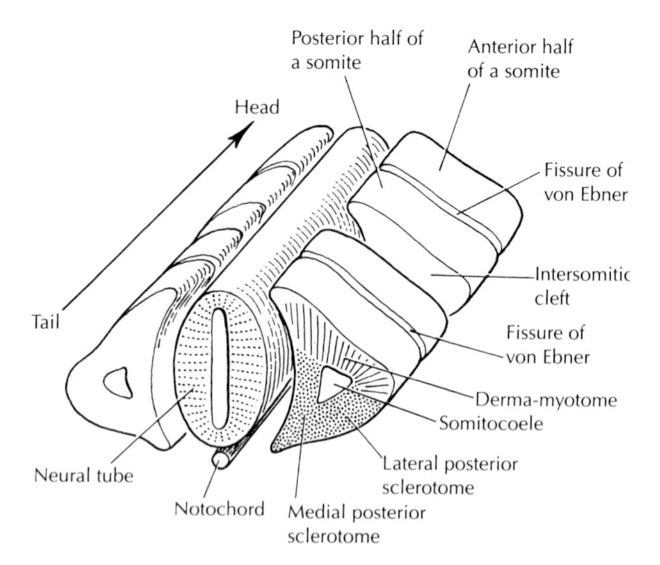

Text-Figure 59 *Two pairs of somites, illustrating regions as designated by Christ and Ordahl (1995).*

The anterior half of each sclerotome becomes partially separated from the posterior half by a discontinuity, the **fissure of von Ebner** (*Text-Figure 59*), the cells being more densely packed in the posterior half. The fates of the different regions have been traced mainly by immunostaining (Christ and Ordahl, 1995; Ebensperger *et al.*, 1995). (The mesenchyme cells, which initially lie in the somitocoele, contribute to the sclerotome). Most of the skeleton is formed by the sclerotome cells, though the clavicle, coracoid, sternum and pelvic girdle are derived from somatic mesoderm. The lateral part of the posterior region of each sclerotome (*Text-Figure 59*) gives rise to the neural arches and pedicles of the vertebrae and, in the thorax, to the ribs. The medial part of the sclerotome forms the perinotochordal sheath and the vertebral body, and contributes to the vertebral disc. In addition, the sclerotome cells form the ligaments and the periosteum as well as contributing to the endothelium of certain blood vessels (Huang *et al.*, 1994).

The vertebrae are cartilaginous at 5 days and ossification begins at about 11–13 days (*Plates 148–149*). Until recently, a controversy has existed as to whether the segmentation of the vertebrae is 'intersegmental' or not. The traditional concept holds that, as the sclerotomal component of each vertebra is derived from the posterior half of one pair of somites and the anterior half of the succeeding pair, the position of each vertebra is not identical with that of either of the two pairs of somites from which it arose but with the intersomitic clefts that separate them. Traditionally, therefore, the vertebrae are considered to be 'intersegmental', in that they alternate in position with the somites.

This concept was challenged by Verbout (1985) who suggested that the sclerotomal components of each vertebra were derived exclusively from one pair of somites only. Attempts to settle the problem by extirpating either one or two somites and studying the effect on the relevant vertebrae (e.g. Bagnall *et al.*, 1988) did not give clear-cut results. More recent work, based on immunohistochemical studies (Christ and Wilting, 1992) and on the grafting of one and a half somites, to eliminate the possibility of a wrong orientation of the graft (Huang *et al.*, 2000a), concluded that the resegmentation theory was in essence correct. Interestingly, the spinous process of each vertebra has been found to have a different origin from that of the rest of the vertebra (Monsoro-Burq *et al.*, 1994; Aoyama and Asamoto, 2000) and appears to be formed as a result of the influence of BMP4 secreted by the dorsal side of the neural tube (Dockter, 2000).

The individual vertebrae of birds possess the typical structure found in all vertebrates: ventral to the spinal cord is the vertebral body or centrum; dorsal and lateral to the spinal cord are the neural arches. Each adult vertebra possesses an articulating surface at the anterior and posterior aspect of each vertebral body, and paired articulating processes, the prezygapophyses at the anterior end and the postzygapophyses at the posterior end. It is thus able to articulate with the preceding and succeeding vertebrae. There are over 40 vertebrae in the hatched domestic fowl, of which 14 are cervical, 7 thoracic, 5 lumbar, 2 sacral and 10 caudal, whilst several fused vertebrae form the pygostyle (*Text-Figure 60*). These figures do not apply to all birds, variations existing between species not only in the numbers in each region but also in the total number present. The atlas and axis are formed from somites 6–8. Each thoracic vertebra articulates with a pair of dorsal ribs (dorsal costa). Except for the first two pairs, each dorsal rib articulates also with a ventral (or sternal) rib, and each bears a posteriorly projecting process, the uncinate process (*Text-Figure 60; Plate 149*), to strengthen the rib cage. The most anterior four of the five ventral ribs articulate also with the sternum.

After the sclerotome cells have migrated around the neural tube and notochord they form membranous structures, the **arcualia**, the forerunners of the cartilaginous (later bony) vertebrae and the ligaments and periosteum. Chondrification begins in each vertebra at certain well-defined centres. The vertebral bodies are cartilaginous at 5 days and the dorsal spines at 6.5 days (Shapiro, 1992), and by day 8 almost every part of the vertebra has become cartilaginous. As the membranous vertebrae form, some mesenchyme remains between them, and this, together with remnants of the notochord, forms the intervertebral discs. The three primary centres of ossification in each vertebra are located in the

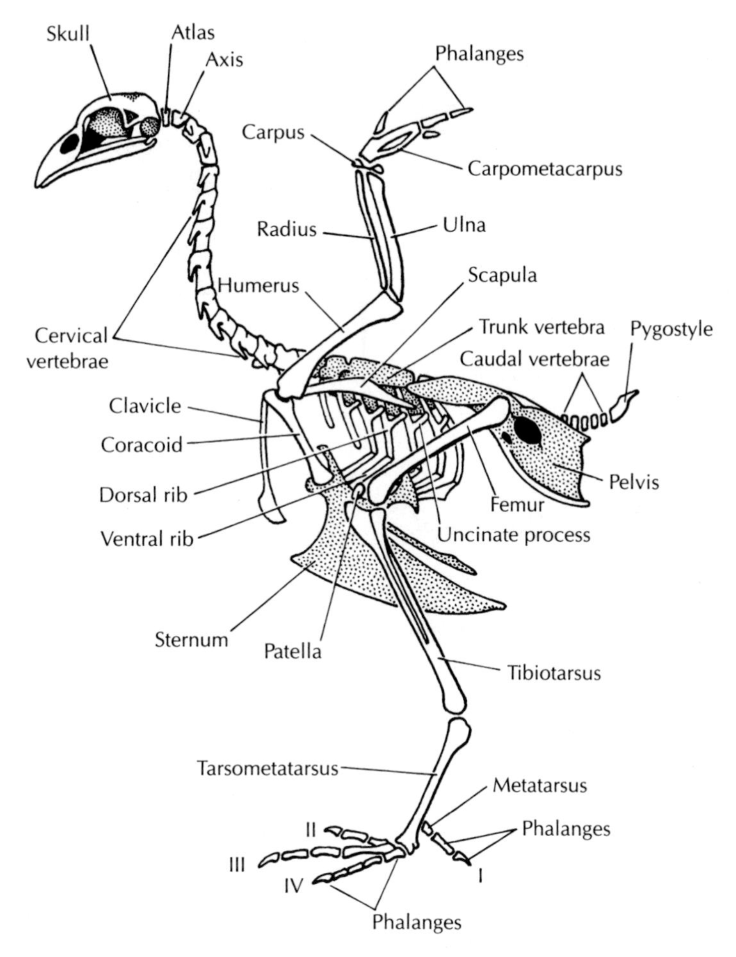

Text-Figure 60 *Components of the skeleton of an adult fowl. (After Ede, 1964.)*

vertebral body and the two sides of the neural arch respectively. Ossification begins in the vertebral body about days 12–13 but is not completed, even in the cervical vertebrae, by the time of hatching.

The type of vertebra which develops (i.e. cervical, thoracic, lumbar, sacral or coccygeal) is determined by the position of the cells, even before segmentation, along the body axis. This was shown by transplanting regions of segmental plate mesoderm to other locations in the chick embryo by Kieny *et al.* (1972), who found that the graft formed a vertebra corresponding to its original position and not to its new one. The different regions of the spine appear to be controlled by the activation of specific *Hox* genes (Burke, 2000).

A characteristic of the bird's skeleton is that extensive fusion occurs between certain vertebrae. The synsacrum (*Plates 110, 138*) is formed by the fusion of the most posterior (seventh) thoracic, the five lumbar, the two sacral and the five most anterior of the caudal vertebrae. This rigid structure is itself fused to the ilia of the pelvic girdle, the entire lot being immovable. Ossification in the synsacral vertebrae is not completed until just before hatching. Ossification of the last caudal vertebrae begins on day 19 and has been completed by hatching.

The intervertebral discs are formed from the densely packed fibrous tissue which surrounds the remnant of the notochord, the nucleus pulposus (Christ and Wilting, 1992).

Ribs

The components of the avian ribs are illustrated in *Text-Figure 60*. Vertebral (dorsal) and sternal (ventral) ribs, together with the trunk and intercostal muscles, are derived from somites, the ribs from the sclerotomes and the intercostal muscles from the derma-myotomes. The evidence comes from the results of experimental ablation and transplantation experiments (Chevalier, 1975; Huang *et al.*, 2000d). The proximal part of each rib forms from the medial part of the thoracic sclerotome, whilst the distal part forms from the lateral region (Olivera-Martinez *et al.*, 2000). The development of the proximal region depends on signals from the notochord, whereas that of the distal ribs is dependent on signals from the somatopleure, of which bone morphogenstic protein 4 (BMP4) is an important component (Sudo *et al.*, 2001). Suggestions that the derma-myotome might contribute cells to the ribs (Kato and Aoyama,

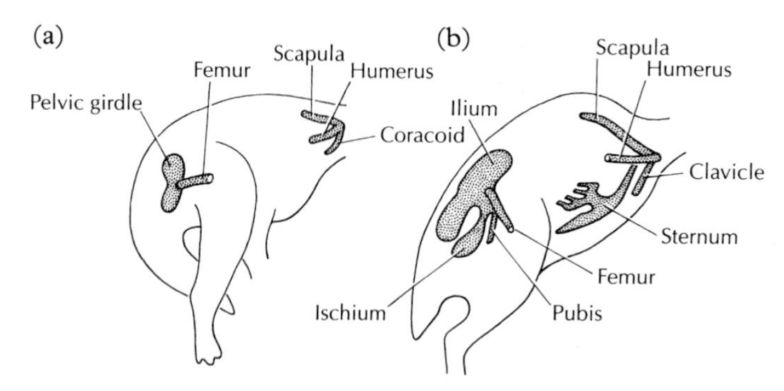

Text-Figure 61 *Developing pectoral and pelvic girdles at 7 days (a) and 9 days (b) seen from the right side. (After Chevallier, 1977, with permission of the Company of Biologists Ltd.)*

1998) have been discounted by others (Huang *et al.*, 2000d; Sudo *et al.*, 2001).

The only somites capable of forming ribs are those of the thoracic region and they are already specified to do so by the time they have segmented. Even that part of the segmental plate from which they form is capable of developing ribs if explanted to a region of the body where ribs do not normally form (Nowicki and Burke, 1999). The specificity for ribs in this region appears to be due to the action of the *Hox* genes at that level.

Outgrowth of sclerotome and somitocoele cells begins at about stages 14–15 (Pinot, 1969). Initially, all the neck and trunk vertebrae bear rudiments of ribs and these are derived as outgrowths from the vertebrae. The cervical ribs of the adult are small, apart from the last two pairs, and each chondrifies from a single centre. In the thorax the rudiments of the ribs migrate ventrally and contact the sternum. They chondrify from two centres, one close to the vertebra and the other near the sternum. The ribs begin to appear in membrane from about the end of day 6, become cartilaginous about 7–7.5 days, and ossify at 10–12 days (*Plate 149*).

Although processes comparable to ribs form from the lumbar and sacral vertebrae they do not separate from the vertebrae. In the chick, no ribs develop in association with the atlas vertebra.

Sternum

The sternum and the ventral muscles are derived from the somatopleural mesoderm (Chevallier, 1975, 1977) at the level of somites 12–26. The sternum begins to form at about 8 days as a bilateral pair of mesodermal condensations lying in the dorso-lateral wall of the thorax immediately posterior to the pericardial cavity. These '**sternal plates**' migrate ventrally and fuse together progressively from the anterior to the posterior end on day 9 (*Text-Figure 61*), forming not only the main body of the sternum but also the keel. Chondrification is completed by 9 days and ossification begins in the chick at about the time of hatching. There are five centres of ossification.

The Pectoral Girdle

The pectoral girdle consists of three components, the **scapula**, the **coracoid** (alternatively called the procoracoid, see Huang *et al.*, 2000) and the **clavicle**, each of which is present as a pair of bones on the left and right, respectively (*Text-Figure 62*). In the adult bird the scapula is firmly attached to the rib cage by muscles and ligaments, whilst the coracoid and clavicle act as supports for the wing (*Text-Figure 60*). According to Chevallier (1977), using chick–quail grafting experiments, the scapula is derived from somitic mesoderm of levels 5–24. However, Huang *et al.* (2000c), also using chick–quail grafts, found that although the scapula blade was formed from cells derived from somites 17–24, its neck and head developed from somatopleural mesoderm; whatever their origin, the cells that formed

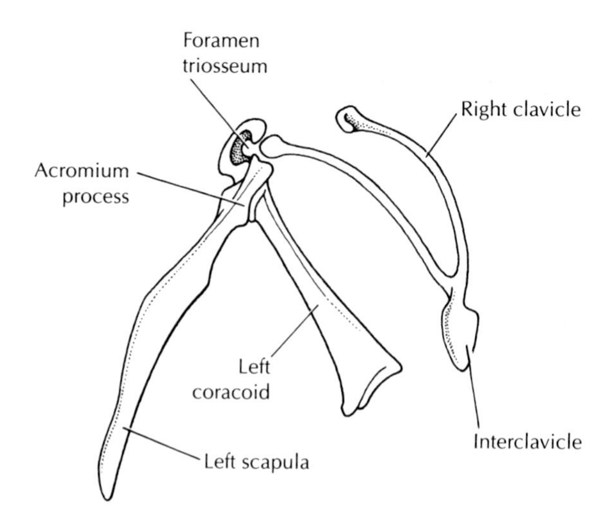

Text-Figure 62 *Components of the left half of the pectoral girdle of an adult fowl. (After Ede, 1964.)*

the scapula all expressed *Pax-1*. The coracoid and clavicle are formed from the somatopleural mesoderm at levels 10–15 and 15–17, respectively (Gumpel-Pinot, 1984). However, long-term fate mapping shows that the clavicle receives also some cells migrating from the cranial neural crest (McGonnell *et al.*, 2001).

Both the scapula and the coracoid have become established in membranous bone by day 5 and chondrification begins at about the end of day 6. Ossification starts on day 12, although the junctional region of the two bones remains permanently cartilaginous.

The clavicle (furculum, wishbone) is essentially a membrane bone and ossifies directly from membrane. It begins as a condensation of mesenchyme at stages 31–32 (7.5 days) and starts to become ossified at stage 33 (8 days). There is then a brief appearance of secondary cartilage, though this lasts merely from about stages 35–36 (Hall, 2001). The left and right clavicles meet and fuse about days 12–13, forming the wishbone (furculum or merrythought). Defects of the shoulder girdle have been produced by treating embryos with signalling molecules that lead to the expression of BMPs, which in turn suppress *Pax-1* in the limb (Hofmann *et al.*, 1998).

The Pelvic Girdle

The pelvic girdle (*Text-Figure 61*) is formed from somatopleural mesoderm (Chevallier, 1977). On day 4 a mesenchymal condensation begins to separate into the three components of the pelvic girdle, the dorsally situated **ilium**, the laterally situated **ischium** and the latero-ventrally situated **pubis**. Chondrogenesis begins on day 6. The ilium begins to ossify on day 12, the pubis about day 13 and the ischium about day 14. The ilium fuses with the synsacrum, the ischium fuses with the ventral part of the ilium and the pubis fuses with the border of the ischium. The acetabular region is still chondrogenic at hatching and fusion of the pelvic girdle with the sacral vertebrae has still not occurred.

The Wing and Leg Skeletons

See Chapter 5 for early stages in limb development. Like all pentadactyl limbs, the **wing** of the fowl consists of proximally, the **humerus**, followed by the **radius** and **ulna** and distally by the carpals, metacarpals and phalanges (*Text-Figure 63; Plate 26*), the digits being the last part of the wing to form. These structures are clearly visible in the young embryo but the distal ones have undergone considerable

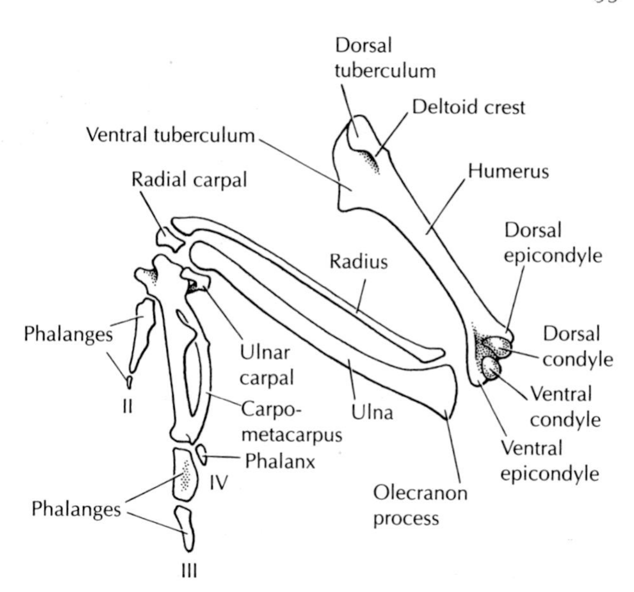

Text-Figure 63 *Components of wing skeleton of adult fowl. (After Ede, 1964.)*

modification by hatching. After the humerus, radius and ulna have formed in membrane from the mesenchyme of the wing bud, chondrogenesis takes place at about 6 days and ossification starts at about 7 days. The carpal and metacarpal bones begin to appear between 6 and 8 days (*Plate 116*). Thirteen separate carpal bones have been described but by day 13 most have lost their individuality, having fused either with each other or with the metacarpals, thus forming the carpo-metacarpus. By day 8 the metacarpals have been reduced to four cartilages, the first having disappeared, and metacarpals 3 and 4 become fused distally (*Text-Figure 63*). The first of the phalanges does not extend out as a digit but is thought to be one of the 13 carpal bones. Digits II–V become modified, II and III being the longest, whereas IV remains short and fuses with V (*Plate 26*). As the wing develops, its position relative to the pectoral girdle changes (Yander and Searle, 1980).

The hindlimb skeleton develops from the same mesoderm as the pelvic girdle. The precartilaginous precursors of the **femur**, **tibia** and **fibula** are laid down by day 5, whereas those of the **tibiotarsus** (*Plate 107*) and **metatarsus** (*Plate 137*) appear during the early part of day 6. The femur becomes chondrified during day 5 and ossification starts on day 6. The bone marrow cavity begins to form on day 8.

The tibia and fibula become chondrified during day 6 and ossification begins shortly afterwards. The patella forms in membrane between days 8 and 9 and starts to chondrify on day 10; ossification does not take place until about 11 weeks after hatching.

The foot and ankle region of birds is greatly modified from the basic pentadactyl plan. The embryo possesses a group of cartilaginous structures which fuse with either the distal end of the tibia or with the second, third or fourth metatarsal and so are lost as separate entities. Of the five digits that form, the fifth disappears, the second, third and fourth extend and after being ossified they fuse at their proximal ends, not only with each other, but also with the distal tarsals, to form the tarso-metatarsus. The first digit remains short, and the number of phalanges differs in different toes, e.g. toe 1 has two phalanges, whereas toe 4 has five (see *Plate 25*).

The cartilaginous elements of both the wing and the leg appear about stages 27–28 (5 days). They initially form as a number of chondrogenic rays in the flat **handplate** or **footplate**. The number of digits, as well as the type of each digit (e.g. second or fourth toe) is determined by signalling from the **polarizing zone** (Panman and Zeller, 2003). Experiments in which grafts of polarizing tissue have been inserted into abnormal places in a host embryo have produced additional digits, the type of each digit depending on its distance from the polarizing region. For example, the digit that forms nearest to the polarizing region differentiates into the most posterior digit, toe 4, whereas the digit furthest from the polarizing region becomes toe 1.

Chondrogenesis appears to be associated with transforming growth factor beta (TGFβ), and apoptosis between the digits is associated with signals of BMP4. For a review of digit formation see Sanz-Ezquerro and Tickle (2003).

The Skull

The skull is formed from three tissues whose respective origins in the early embryo are discussed in Chapter 4: the cranial neural crest, the paraxial head mesenchyme and the first five pairs of somites (*Table 3*). The evidence comes principally from transplantation experiments involving chick–quail chimeras. The paraxial head mesoderm has formed by stages 5–7 and the first five pairs of somites by stage 8, by which time the head has started to lift from the surface of the yolk sac.

The cranial neural crest does not form over the brain until about stage 11. The most anterior cells of the neural crest migrate over the prosencephalon and around the optic vesicles; they form the maxillary and fronto-nasal processes and contribute to the cornea (Noden, 1983b) (*Text-Figure 65*; see Face, p. 40). The neural crest cells arising posterior to the optic vesicle move ventrally to form the pharyngeal arches (*Text-Figure 64*). The differentiation of the cranial neural crest into skeletal structures frequently depends on an interaction with adjacent tissues,

Table 3 Origin of head skeleton

I. BONES OF NEURAL CREST ORIGIN (based on data from Couly *et al.*, 1993)

Membrane bones

of face:	*of skull*:
Nasal	Frontal
Maxilla	Parietal
Vomer	Squamosal
Palatine	Columella
Quadrato-jugal	
Mandible	

Cartilaginous bones

Interorbital septum
Basipresphenoid
Scleral ossicles
Ethmoid, pterygoid
Meckel's
Quadrato-articular
Hyoid
Pars cochlearis of otic capsule (partly)

II. BONES OF PARAXIAL HEAD MESODERM ORIGIN

Cartilaginous bones

Supra-occipital
Sphenoid
Pars canalicularis and cochlearis of otic capsule

III. BONE OF SOMITIC ORIGIN

Cartilaginous bones

Basi- and exo-occipital
Pars canalicularis of otic capsule (partly)

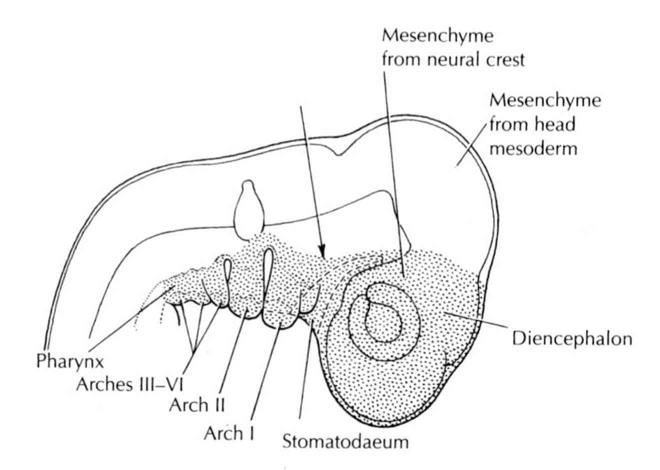

Text-Figure 64 *Distribution of the cells derived from the head mesenchyme and neural crest, respectively. Arrow shows interface between the two regions. (After Noden, 1987, reprinted by permission of John Wiley and Sons. Inc.)*

though the precise elements which form are probably inherent in the cranial neural crest cells themselves.

The paraxial head mesenchyme migrates anteriorly to form bones associated with the most anterior regions of the skull, the supra-occipital, sphenoid, pars canicularis and cochlearis of the

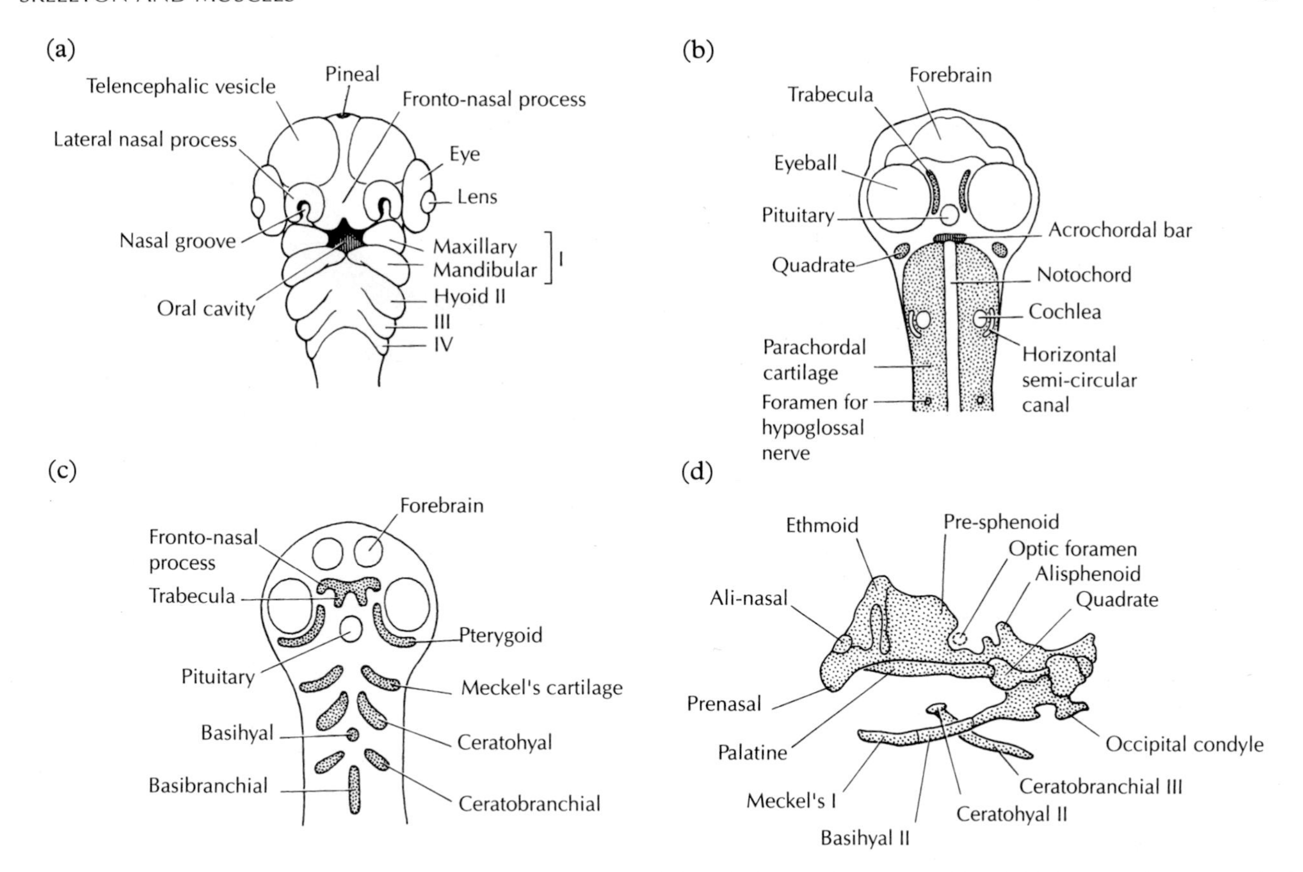

Text-Figure 65 *Developing head structures: (a) Day 4, viewed from below. (After Patten, 1950.) (b) Day 5, after removal of dorsal structures, viewed from above. (After Parker and Bettany, 1877.) (c) Day 5, after removal of lower jaw, viewed from below. (After Parker and Bettany, 1877.) (d) Day 7, lateral view of chondrocranium. (After Parker and Bettany, 1877.) Stippled regions represent cartilages. I–IV = pharyngeal arch origins.*

otic capsule (Couly *et al.*, 1993; Noden, 1983b). The contribution of the anterior somites to the skull is shown in *Text-Figure 66*, which is based on the chick–quail grafts of Huang *et al.* (2000c). These authors point out that the border between head and neck is in the centre of somite 5, and this corresponds with the expression boundary of *Choxb-3*.

The controversial idea that the vertebrate skull has been derived during evolution from segments corresponding to the somites of the trunk and tail received some support from the claim of Meier (1981) that somitomeres, which are a feature of the trunk (see p. 33), were also visible in the head mesoderm. Computer-assisted reconstructions of the head mesenchyme, however, failed to substantiate this claim (Freund *et al.*, 1996) and the concept is probably best avoided.

The skull is formed from both cartilage and membrane bones. The cartilages form most of the base of the skull, the capsules of the ear and nose, the inner wall of the orbit, the jaw articulation and the hyoid skeleton, whilst the membrane bones provide most of the vault and sides of the skull, the jaws and palate. For a detailed description of the development of the chondrocranium see Vorster (1989).

The development of the skull inevitably reflects that of the underlying brain. By day 5 the cranial flexure (see p. 39) is well defined and the cerebral hemispheres and eyes are prominent. The nasal pits are visible, whilst the oral cavity, which lies beneath the forebrain, is bordered above by the fronto-nasal process (which separates the two nasal pits), and laterally by the maxillary processes. The pharyngeal arches are well defined. The anterior end of the notochord lies just posterior to the developing pituitary (*Text-Figure 65b*) and the **parachordal** cartilages lie on either side of it, whilst the acrochordal cartilage has formed at about 4.5 days as a small transverse bar anterior to it. Each parachordal lies beneath the midbrain and hindbrain and is continuous laterally with the chondrified wall of the otic vesicle, the **pars cochlearis**. The cochlea and semicircular canals (see p. 88) are already present within the otic vesicle. The paired **trabeculae** develop anterior to the notochord (*Text-Figure 65c*) and lie beneath the diencephalon and cerebral hemispheres, extending through the interorbital region to the nasal sacs.

Posterior to the parachordals the notochord becomes surrounded by three cartilaginous rings, which correspond with the first three occipital

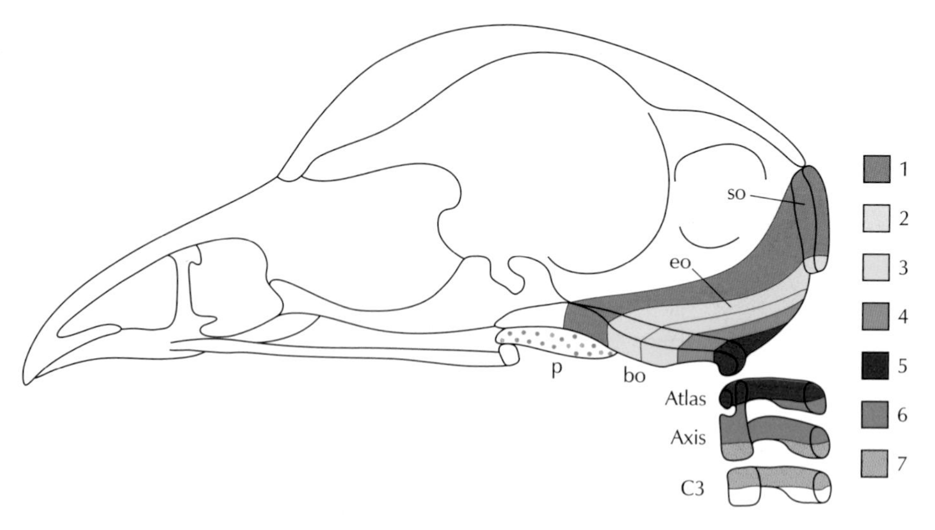

Text-Figure 66 *Distribution of somite-derived cells to the neurocranium. Somites are labelled consecutively from cranial to caudal. C3 = third cervical vertebra; bo = basioccipital bone; eo = exoccipital bone; p = parasphenoid bone; so = supraoccipital bone. (From Huang et al., 2000c.)*

vertebrae. The first two fuse with the parachordals, whilst the third becomes the atlas vertebra (*Plates 124, 125*). The term **basal plate** (*Plate 129*) is applied to the collection of cartilages lying beneath the brain, extending from the trabeculae to the occipital vertebrae. Its posterior border is the ventral edge of the foramen magnum.

By days 6 and 7 of incubation the trabecular region has become longer than the parachordal region. The trabeculae fuse together during day 7 and are then considered as two regions, the **interorbital** (*Plate 114*) and the **ethmoidal** regions, the whole structure extending from the pituitary fossa to the tip of the head. The ethmoidal region itself is divided into the **pre-nasal cartilage** from which the beak will form, and the **ali-nasal cartilage**. The preotic region of the basal plate grows both dorsally and laterally and forms the **alisphenoid** which extends from the posterior edge of the orbit to the anterior border of the cranial cavity (*Text-Figure 65d*).

The Pharyngeal Arch Cartilages

The formation and development of the pharyngeal arches is discussed in Chapter 8. The fate of the cartilages only is considered here (*Text-Figure 65*).

By the end of day 2, three to four pairs of pharyngeal arches are present and by day 4 the first and second have become much bigger than the rest. During day 6 the first arch (the maxillary-mandibular) gives rise to the palato-quadrate in the upper jaw and to Meckel's cartilage in the lower jaw (*Plate 103*), which articulates with the articular process of the palato-quadrate. Meckel's cartilage elongates greatly and its tips fuse at about 16 days. Meanwhile the palate has begun to form as a roof to the oral cavity (*Plates 101, 132*). Its development resembles that of the mammalian palate in that a shelf, the **palatine process** (*Plate 101*), grows medially from the lingual surface of each maxillary process, but differs from that of mammals in that the two shelves do not meet in the midline but remain separated by a slit that is the **internal naris** (*Plate 114*).

The second arch, the **hyoid**, forms a median ventral structure and paired lateral cartilages. The median structure gives rise to the **basihyal** (or copula I), which forms part of the **basi-branchial** (see below). The lateral cartilages develop into the **columella** of the ear (*Plates 124, 126, 130*) and the **ceratohyals**, which give rise to the **entoglossal cartilage** and part of the **cornua** of the hyoid.

The third pharyngeal arch gives rise centrally to the basibranchial (or copula II) and laterally to the paired **ceratobranchial** and **epibranchial cartilages**.

Ossification of the Skull

The membrane bones begin to be ossified from about day 9, though there are discrepancies in the exact times given by different authors. Most of the skull bones have undergone at least some ossification by day 14 (see Table 22 of Romanoff, 1960).

The **scleral ossicles** (*Plate 150*) are a ring of 14 overlapping bony plates in the sclera around the margin of the cornea. Their future position is indicated on day 8 by a series of conjunctival papillae which disappear on day 12 as ossification occurs

(Coulombre *et al.*, 1962). Each scleral papilla induces a scleral ossicle between 6.5 and 10 days (Hall, 1981).

THE MUSCLES

Most of the muscles in the body are formed from somites, particularly the skeletal muscles of the dorsal and dorso-lateral regions of the trunk and of the limbs. The lateral plate mesoderm, on the other hand, gives rise to the muscles of the lateral and ventral body wall and to the tendons of the limbs. The muscle sheaths are also derived from the lateral plate mesoderm (Ordahl and Le Douarin, 1992), while the extra-embryonic lateral plate mesoderm gives rise to the muscles of the amnion and allantois. The neural crest, prechordal mesoderm, paraxial head mesenchyme and the most anterior somites all contribute to the muscles of the head (Noden, 1983a). The precursors of these muscles are arranged in an antero-posterior sequence which corresponds with the future location of the motor nerves.

As discussed in Chapter 4, the dorsal region of each somite develops into the derma-myotome, which then gives rise to both dermis and myotome. The skeletal muscles form from the myotome. The most important part of the derma-myotome is the **dorsal medial lip** (*Text-Figure 59*), since if this extirpated no further myotome cells develop (Ordahl *et al.*, 2001). Cells from this region have, traditionally, been thought to migrate around the curved epithelium of the derma-myotome. However, the results of recent labelling experiments using DiI suggest that they do not do this, but rather join the other cells of the myotome and migrate directly (Denetclaw *et al.*, 2001). The cells of the myotome are orientated longitudinally along the body axis, in contrast to those of the derma-myotome, whose orientation is at right angles to it.

The most medial cells of the myotome form the **epaxial** muscles, while the lateral edge of the myotome forms the **hypaxial** muscles. These lateral cells lose their epithelial nature and become mesenchymal, migrating to form the abdominal body wall and the limb muscles (Ordahl and Le Douarin, 1993). The connective tissue of the epaxial muscles is formed from the somites, but that of the hypaxial muscles is derived from the somatic lateral plate mesoderm (Christ and Ordahl, 1995).

We have seen in Chapter 4 that the newly formed somite is not yet determined as to the future of its cells. If the somite is experimentally inverted, the myotomes develop from cells that would normally have given rise to sclerotome, while the sclerotome develops from cells that would have formed myotome. See Borycki and Emerson (2000) for a review.

The factors controlling the development of muscles from the somites and other muscle-forming cells include a large group of genes, of which the **myogenic regulatory factor** (*MRF*) genes are the most important (Brand-Saberi and Chirst, 1999). The timing of expression for these different genes was found to be similar in different mesoderm poulations, i.e. *Myf4* was expressed first, followed by *MyoD* and then *myogenin* (Hacker and Guthrie, 1998). Two of these genes, *MyoD* and *Myf5*, which are expressed in both somitic epaxial (back) and hypaxial (intercostal) muscle progenitor cells, are activated simultaneously in the dorsomedial cells of the epithelial somite (reviewed by Borycki and Emerson, 2000).

The activation of *MyoD* and *Myf5* in the epaxial cells is brought about by influences from the notochord and neural tube on the newly formed somites (Pownall *et al.*, 1996). If these tissues are separated from the somites at this time, the epaxial muscles fail to form and the somite cells become apoptotic (Sanders and Parker, 2001).

There are other genes involved in the development of these tissues. For example, *sonic hedgehog* (*Shh*) is needed for development of both the myogenic and chondrogenic lines (Teillet *et al.*, 1998). *Wnt* signalling molecules, which are expressed by the dorsal neural tube and the ectoderm, appear to be involved in the formation of the myogenic cells in the somite (Wagner *et al.*, 2000). The gene *Lbx1* is expressed in the migrating hypaxial muscle precursors at occipital, cervical and limb levels from stage 14 (Dietrich *et al.*, 1998).

General Morphology and Nomenclature of Muscles

No single monograph appears to exist on the development of muscles in the domestic fowl, though accounts of specialized regions are available (e.g. limb muscles). Most of the body muscles are laid down between about days 5 and 9 and the adult pattern is well established by day 10. Identification of muscles from about 9 days of incubation is therefore usually based on the pattern of the muscles in the adult, though lack of uniformity in the terminology can be confusing. Muscles have been given different names by various investigators and although some of these names correspond to those used for mammalian muscles they are not necessarily identical in function. The names given in this book are based principally on those used by Berger (1960) and Vanden Berge and Zweers (1993). Where suitable, alternatives are shown in parentheses.

Limb Muscles and Tendons

Limb muscles form from somites (*Table 4*), and their associated tendons and connective tissue develop from somatic mesoderm (Chevallier *et al.*, 1978).

For detailed morphogenesis of the wing muscles see Sullivan (1962), Beresford (1983) and Zhi *et al.* (1996), of the thigh muscles see Romer (1927) and Drushel and Caplan (1991), and of the lower leg muscles see Wortham (1948). The wing muscles are formed from somites 16–21, those of 18–20 giving rise to most of the muscles of the hand (Zhi *et al.*, 1996). These authors found that anteriorly located somites contributed to muscles on the ulnar side, but the hand muscles were derived from both anterior and posterior somites. Cells from these somites actively migrate into the wing bud between stages 15 and 18 (Schramm and Solursh, 1990). The leg muscles form from somites 26–32 (Gumpel-Pinot, 1984; Lance-Jones, 1988); cells from these somites migrate into the leg bud mainly during stages 17 and 18 (Jacob *et al.*, 1979). The somites which give rise to limb muscles also supply endothelial cells for the limb buds, though the migrating cells do not follow the same route (Huang *et al.*, 2003).

The first indication of muscle development in a limb bud is the formation of two pre-muscle masses (*Table 5*) which form close to the girdle. They become visible in the wing bud at about stages 24–26 (4.5–5 days) lying dorsal and ventral to the condensation of cartilage that will form the wing skeleton (*Plate 99*). Lobulations in the ventral mass indicate some of the future muscles (Sullivan, 1962). Each dorsal and ventral mass undergoes a sequence of divisions, giving rise to the series of muscles shown in *Table 5*.

Differentiation takes place in a proximo-distal direction. *Text-Figure 67* shows a reconstruction of the wing at 6–6.5 days (stage 29). A scanning electron microscopic study of normal wing development has been prepared by Murray and Wilson (1994) as a supplement to the table of Hamburger and Hamilton (1951) (see Appendix III).

Table 4 Somitic level of muscles formed from somites (based on data from Zhi *et al.*, 1996)

Wing and scapula muscles	16–24
Pectoralis muscles	13–22
Intercostal muscles	19–26
Abdominal wall muscles	27–29
Leg and pelvic girdle muscles	26–32

Table 5 Differentiation of pre-muscle masses of the wing bud after Sullivan (1962)

DORSAL MASS

Tensor propatagialis and deltoid
Scapulohumeralis, cranialis and caudalis
  Subscapularis and subcoracoideus
  Coracobrachialis caudalis
Latissimus dorsi
Triceps brachii

VENTRAL MASS

Pectoralis
Supracoracoideus
Coracobrachialis cranialis
Coracobrachialis caudalis
Biceps brachii
Brachialis

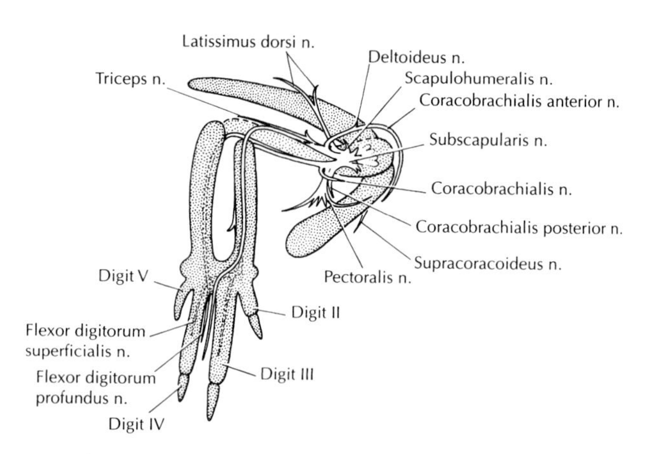

Text-Figure 67 *Pre-muscle masses (stippled) of right wing bud together with nerves at about 6–6.5 days. (After Sullivan, 1962, with permission from CSIRO Publishing, http://www.publish.csiro.au/journal/ajz.)*

Development of the leg muscles takes place between days 6 and 8. The dorsal and ventral pre-muscle masses (*Table 6*) are well defined at 6 days. Each then undergoes a series of subdivisions comparable to those of the pre-muscle masses of the wing bud. The dorsal mass of the thigh region forms a superficial group and a deeper group of muscles. Both the dorsal and the ventral pre-muscle masses divide at the knee into proximal and distal components, the proximal ultimately forming the muscles of the lower leg and the distal the muscles of the foot (*Table 7* and *Text-Figure 68*).

The tendons of the limbs also form from the dorsal and ventral pre-muscle masses (*Plates 99, 100*), though they develop from different populations with different cell lineages. The tendons are derived from the somatopleure, whereas, we have seen, the muscles are formed from cells that have migrated

Table 6 Differentiation of pre-muscle masses of the thigh (after Romer, 1927)

DORSAL MASS

Superficial dorsal mass
- Triceps femoris
- Iliotibialis
- Ambiens
- Femorotibialis
- Iliofibularis

Deep dorsal mass
- Iliofemoralis
- Iliotrochanterici

Ventral mass
- Obturator
- Puboischiofemoralis
- Ischiofemoralis
- Ischioflexorius
- Caudilioflexoralius

Table 7 Differentiation of pre-muscle masses of the lower leg (modified after Wortham, 1948)

DORSAL MASS

Superficial
- Peroneus longus
- Tibialis cranialis

Deep
- Peroneus brevis
- Extensor digitorum longus

Foot
- Extensor hallucis longus
- Abductor digiti II
- Extensor proprius digiti III
- Extensorbrevis digiti IV

VENTRAL MASS

Superficial
- Flexor perforans digiti III
- Flexor perforans digiti II
- Gastrocnemius

Deep
- Flexor perforatus digiti II, III, IV
- Flexor digitorum longus
- Popliteus

Foot
- Flexor hallucis brevis
- Adductors for digiti III, IV

from the somites (Christ *et al.*, 1977; Chevallier *et al.*, 1978). However, absence of a specific muscle is accompanied by absence of the corresponding tendon (Chevallier *et al.*, 1978). The tendons begin to form in the lower leg at the end of day 6 or the start of day 7 (Wortham, 1948). Schweitzer *et al.* (2001), using a specific marker, *Scleraxis*, showed that they are derived from a population of cells in the mesenchyme induced by ectodermal signals. As each muscle forms from the pre-muscle mass, the tendon grows forward from it to its place of insertion. Each tendon grows along a digital ray, which is essentially a band of collagen around which the tendon condenses (Ros *et al.*, 1995). All the leg tendons are defined by 9 days, apart from those of the toes, which are unable to complete their insertions until day 11 when the final skeletal cartilages are formed. Like the muscles, the tendons continue to enlarge and differentiate during the succeeding days of incubation. For further information on tendon development in the avian limb see Oldfield and Evans (2003).

Muscles of the Thorax and Abdomen

The thoracic and abdominal muscles are formed from the somitic myotomes, which become divided into a dorsal part (the **epimere**, which gives rise to the epaxial muscles) and a ventral part (the **hypomere**, which forms the hypaxial muscles). The myoblasts of these layers migrate laterally and ventrally, passing through the substance of the somatic mesoderm immediately beneath the ectoderm. The epimeres form the extensor muscles of the vertebral column, whilst the hypomeres form the flexor muscles of the vertebrae, the limb muscles and the intercostal muscles. The hypomeres give rise also to the muscles of the abdominal wall.

Muscles of the Tongue

The deep muscles of the tongue are formed from anterior somites, though the precise somites and their individual contributions remain controversial, despite modern chick–quail grafting experiments. For example, the glossal muscles have been reported to receive cells from somites 2–5 (Noden, 1983a), 1–5 (Couly *et al.*, 1992) or 2–6 (Huang *et al.*, 1999, 2001). The skeleton of the tongue consists of the **basibranchial** cartilage, which supports the base of the tongue, and the **paraglossum** (entoglossum), which supports the body of the tongue. In addition, paired lateral wings are formed by the **ceratobranchials** and the **epibranchials** (*Text-Figure 69*). Huang *et al.* (1999, 2000a) have classified the tongue muscles according to their origins and insertions into **glossal** (attached to the entoglossal cartilage), **suprahyoid** or hypobranchial (extending from the mandible to the hypobranchial skeleton) and **infrahyoid** muscles (which originate from the larynx, tracheal cartilage, sternum and clavicle and are attached to the hypobranchial skeleton). They conclude from their experiments that somites 2–6 give rise to all

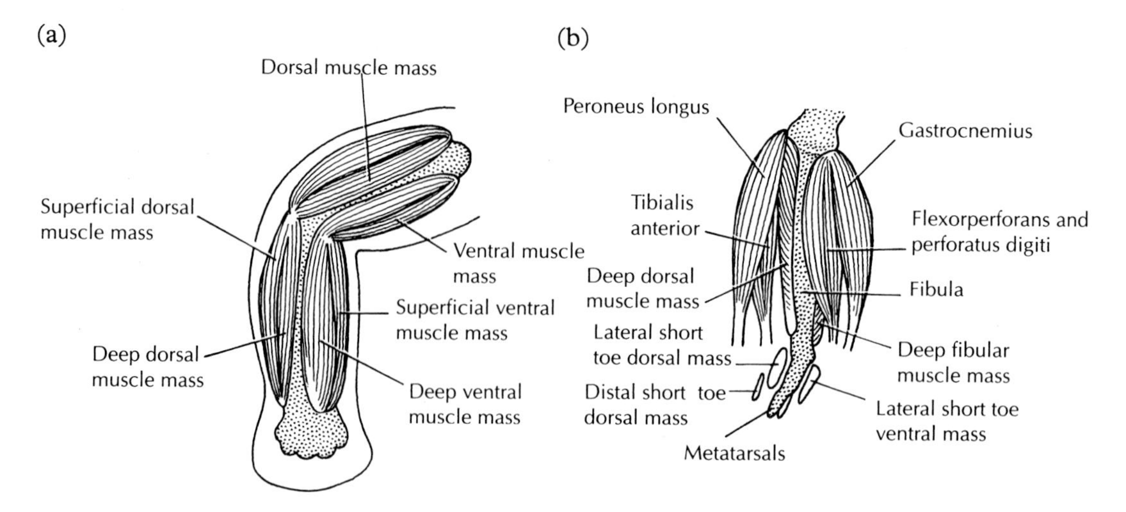

Text-Figure 68 *(a) Longitudinal division of pre-muscle masses of the thigh at 6.5 days into superficial and deep layers, with incomplete separation in the region of the knee. (b) Differentiation of peroneus longus and tibialis anterior muscles from the superficial dorsal pre-muscle mass, at about 6.75 days. There is incomplete separation of the flexor perforans et perforatus digiti group of muscles from the lateral portion of the superficial ventral muscle mass and complete separation of the short toe muscles masses from the leg muscle masses. (After Wortham, 1948.)*

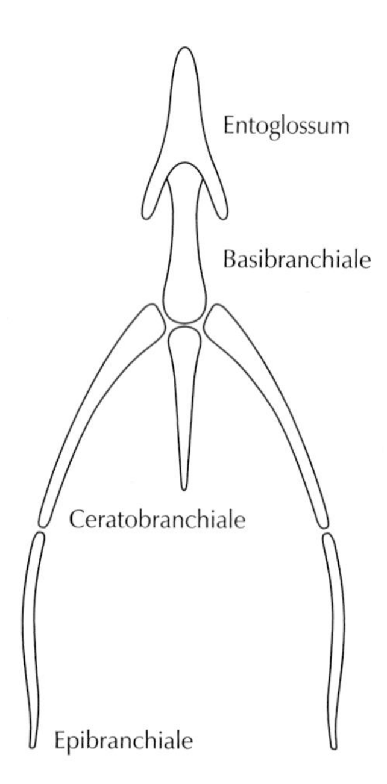

Text-Figure 69 *Schematic drawing of the tongue skeleton. (From Huang et al., 1998, with permission from the Company of Biologists Ltd.)*

the glossal and infrahyoid muscles but not the suprahyoid muscles. Huang *et al.* (1999, 2000a) found that myoblasts from different somites intermingled during their migration to the tongue, apparently streaming to a focal point and then migrating in a stream. The migrating muscle cells carry with them the hypoglossal nerve. The signalling cascades involved in the development of the tongue are described by Huang *et al.* (2001). The more superficial muscles are derived from the mesoderm of the first, second and third pharyngeal arches and are innervated by the trigeminal, facial and glossopharyngeal cranial nerves, respectively.

Head and Neck Muscles and Connective Tissue

Those muscles formed from the fronto-nasal process have been traced back to the neural crest. The remainder of the head muscles, including the external eye muscles, as well as all the voluntary muscles of the face, are formed from the paraxial head mesenchyme (Noden, 1983; Hacker and Guthrie, 1998), which is the unsegmented mesoderm situated anterior to the somites in the early embryo and lying on either side of the neural tube (*Table 8*). Its cells migrate to the first, second and third branchial arches and colonize the core of each arch, forming a muscle mass (*Text-Figure 70*). At much the same time neural crest streams to line the periphery of each of these branchial arches.

The prechordal mesoderm, which lies anterior to the notochord and medial to the paraxial mesenchyme is found subsequently beneath the anterior region of the midbrain and the prosencephalon, and gives rise to the oculorotatory muscles innervated by the oculomotor nerve (Jacob *et al.*, 1984). The tip of the notochord lies beneath the caudal midbrain and the hindbrain (Noden, 1992).

According to Noden (1983a) somites 3–7 only, contribute to in the formation of the cranio-cervical muscles, but Huang *et al.* (2000c) found that somites 2–8 are involved. The connective tissues of the

Table 8 Embryonic origin of cephalic and cervical muscles (modified after Noden, 1983a, by permission of Wiley-Liss Inc., a subsidiary of John Wiley and Sons Inc.)

Muscle group	Myoblast origin	Connective tissue origin
Occipito-cervical, epaxial, hypaxial	Somites 3–7	Somitic mesoderm
Caudal external laryngeal	Somites 3–6	Lateral mesoderm
Tongue	Somites 2–5	Neural crest
Intrinsic laryngeal	Somites 1–2	Lateral mesoderm
Jaw opening and hypobranchial	Paraxial head	Neural crest
Jaw closing	Paraxial head	Neural crest
Dorsal oblique	Paraxial head	Neural crest
Other extrinsic ocular	Paraxial head	Neural crest

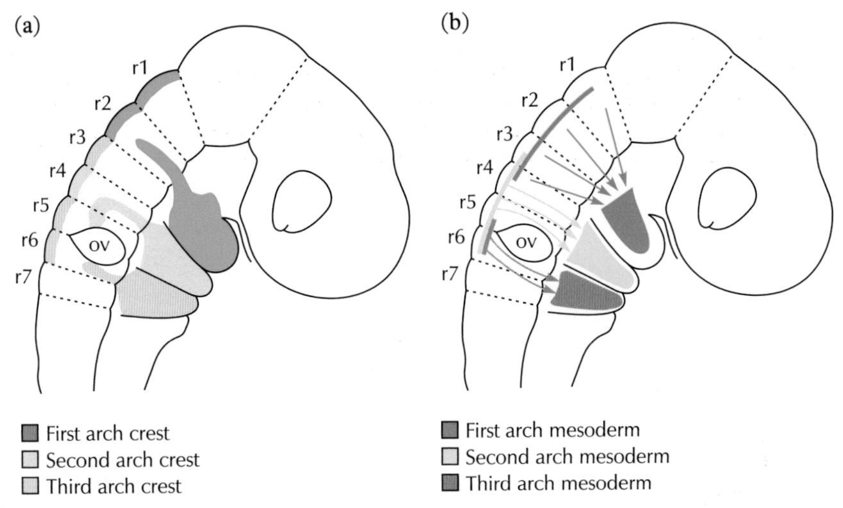

Text-Figure 70 *Migration of (a) the cranial neural crest and (b) the cranial paraxial mesoderm. r1–r7 = rhombomere position of populations of cells; ov = otic vesicle. (From Hacker and Guthrie, 1998, with permission from the Company of Biologists Ltd.)*

head also have differing origins. Those of the mid-facial, peri-ocular, glossal and visceral arch regions are derived from neural crest. The laryngeal and tracheal connective tissues develop from the lateral plate mesoderm of the level of the first two pairs of somites (Noden, 1986). The connective tissues for the head elevator and depressor muscles are formed from somites (see Noden and DeLahunta, 1985).

Heart Muscle

See Chapter 6.

Chapter 11

The Integument

The term integument includes not only the skin and feathers but also the beak, claws, scales, comb and wattle and the uropygial gland. Its morphology has been studied since the nineteenth century (Lucas and Stettenheim, 1972; Bereiter-Hahn *et al.*, 1986) and, because of its diversity, ease of access and amenability to experimental manipulation, it has been important in the study of cell and tissue interactions. Each specialized structure is formed from an interaction between its two components, the dermis and the epidermis. Much research has been focused on analysing these interactions by experimentally separating and then recombining them (reviewed by Sengel, 1976) and, more recently, on the cytological and molecular changes involved (reviewed by Chuong, 1993, Noveen *et al.*, 1995).

The **dermis** is formed mainly from neural crest mesenchyme in the facial region, from prechordal mesoderm over the cranium and from the dermatome or the lateral plate in the trunk region. The dermis of the epaxial (dorsal) region is derived from the dermatome, but of the hypaxial (lateral) region is from the somatic layer of the lateral plate mesoderm (Christ *et al.*, 1983). In the early stages the dermis is mesenchymal and is not present as a distinctive tissue until about 5 days of incubation (Sengel, 1976). Olivera-Martinez *et al.* (2001), who carried out a series of grafting experiments, demonstrated that the specification of the dermis to produce feathers was the result of a signal from the neural tube, probably *Wnt1.*

The **epidermis** forms from the ectoderm, which is initially a single layer but starts to become two layered at about stage 12. The outer layer is the **periderm**, a temporary layer of squamous epithelial cells that forms at about the time that the neural tube closes and acts as a barrier between the internal tissues and the amniotic fluid (Sengel, 1976). It becomes sloughed off at about 18 days of incubation.

Classical studies have shown that the rate of integumentary differentiation varies significantly in different regions of the body. Mayerson and Fallon (1985) reported that not only can early feather primordia be recognized at any stage between early 31 and 39 (7–13 days) but that there are important differences between the pterylae. Studies of scale differentiation in the legs (Sawyer *et al.*, 1986) provide a similar picture. Consequently, it can be misleading when discussing the differentiation of avian integument to give precise timing for specific developmental events.

THE FEATHERS

Adult feathers do not cover the body uniformly but are arranged in tracts (**pterylae**) separated by regions where few, or no, feathers occur (**apterylae**). The feathers, as well as the scales, are formed from β-keratins, and the apteria express α-keratins (Prin and Dhouailly, 2004). There is considerable morphological variation between species, and the domestic fowl is exceptional in two respects (Lucas and Stettenheim, 1972). First, the embryonic apterylae are still recognizable in the adult. Second, whilst the major body feathers (including those of the wings and the tail) consist of **contour** feathers, the **plumules** (downy feathers) and **filoplumes** (fine, thread-like structures) that lie beneath them are relatively sparse in comparison with the situation in a duck or a goose.

The hatching plumage, consisting of **down feathers**, is soon replaced by the juvenile plumage and this in turn gives way to the adult plumage, which is characterized by various forms of contour feather. The precise relationship between the widely studied 'embryonic feather germs', which form the hatching plumage, and the juvenile and adult plumage does not appear to have been investigated and it is not known if the feather follicles of the adult bird are formed before hatching.

The first visible sign of an individual feather rudiment is a white spot on the surface of the body; this is an epidermal placode, a localized thickening in the epidermis which then becomes underlain by

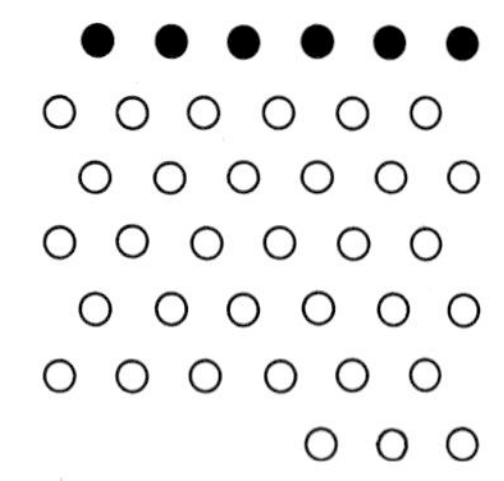

Text-Figure 71. *The sequence of patterning of the feather rudiments on the right side of the lumbar region of the back. Row 1 (black) develops first and succeeding rows (white) are arranged so that the feathers come to lie in a pattern of oblique and longitudinal rows.*

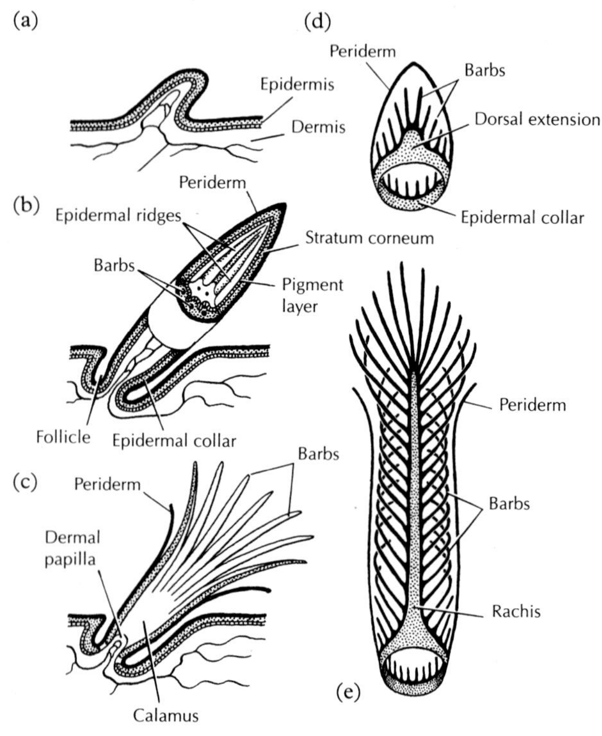

Text-Figure 72. *Stages in the formation of a down feather (a–c) and an adult contour feather (d,e). (After Ede, 1964.)*

a condensation of cells in the dermis, the **dermal papilla**. The first feather rudiments appear at about 5 days as a single row in the lumbar region, separated by spaces where no feathers develop. As additional rows are laid down, the new rudiments are always opposite the spaces between the rudiments in the preceding row. The sequence of the patterning was described by Mayerson and Fallon (1985), each row, or tract, being established in precise relation to the preceding row so that the feathers become arranged in a hexagonal pattern of oblique and longitudinal rows (*Text-Figure 71*). This is brought about by the interaction of BMPs, Follistatin and the receptor tyrosine kinase Eph-A4 (Patel *et al.*, 1999).

Each dermal papilla and the ectoderm above it proliferate and the entire structure becomes extended into an elevated epidermal cylinder with a dermal core (*Text-Figure 72*), the feather bud. The two sides grow unequally so that the apex moves posteriorly and the entire structure lies almost flat on the surface of the skin. There is a region at the base of the feather bud, the **epidermal collar**, which has a high mitotic rate. The melanophores become arranged in rows along the feather. By day 11 a series of longitudinal ridges have appeared in the stratum corneum of the epidermal walls and between 12 and 18 days they form the barbs and their associated barbules. The **calamus** is a hollow region at the base of the feather where the barbed ridges remain fused. About day 20 the base of the feather germ sinks into a depression, the **feather follicle**. A thin sheath of epidermal cells, the **periderm**, remains around the feather germ until after hatching but splits as the chick dries and so releases the barbs and barbules. The down feathers are pushed out of the follicle by the apices of the **juvenile feathers**, which are similar in structure to the adult feathers. For a morphological description of the embryonic down feathers, see Matulionis (1970).

Adult feathers are of three types:

- the **contour feathers**, which are the main body feathers including a wide variety of wing and tail feathers
- the **plumules** or downy feathers, which lie beneath the contour feathers
- the **filoplumes**, which are fine thread-like feathers closely associated with contour feathers.

All these definitive feathers are formed from the dermal papillae laid down in the embryo. All parts of each feather shaft are derived from the embryonic epidermal cells, the collar, which surrounds the dermal papilla. The pulp is formed entirely from the papilla.

The development of the feathers is an example of epithelial–mesenchymal interactions. The ectodermal epithelium is incapable of forming a feather unless induced by the underlying mesoderm; if the papilla is removed, no more feathers develop. Nevertheless, it is the epidermis that determines where the feather will eventually develop. A series of thickenings (placodes) appears in the epidermis and each of these promotes a condensation of the mesoderm immediately beneath it. These condensations do not form in the absence of the epidermis (Sengel, 1976) and they do not form in the mutant *Scaleless*, which does not develop feathers (Goetinck and Sekellic, 1972). Each dermal condensation then induces the epidermis to form a feather bud. Interestingly, although the feathers are formed almost entirely from the epidermis, their future adult form is determined by the mesoderm. This has been shown by experiments in which epidermis from one region of the body has been stripped of its associated dermis, recombined with dermis from another part of the body and then grown in tissue culture. For example, if the dermis beneath wing bud epidermis is removed and replaced with dermis from the back of the embryo, the feathers that develop are characteristic of the back and not of the wing (Sengel, 1986). The inductive signals involved in these interactions are complex and not yet fully understood. Among the factors that have been implicated are: from epidermis to dermis, BMPs (Scaal *et al.*, 2002); and from dermis to epidermis, FGF10 (Tao *et al.*, 2002), *cDermo-1* (Scaal *et al.*, 2002), *Wnt6* (Chodankar *et al.*, 2003), *sonic hedgehog* and BMP2 (Harris *et al.*, 2002), and *Sox 18* (Olsson *et al.*, 2001). For a recent series of reviews on further development see Yu *et al.* (2004).

The regular hexagonal arrangement of the feather buds (see above) appears to be the result of an individual action of each developing epidermal placode which inhibits the formation of a similar placode in its immediate vicinity. The homeobox gene, *HB9*, is expressed in the epidermal placodes and feathers as well as in the dermal condensations, but not in the interplacode regions (Kosaka *et al.*, 2000). Conversely, the *cDeltaa-1* gene is widely expressed in the dermis of the Scaleless mutant and its experimental overexpression in dermis from a non-mutant embryo inhibits feather formation (Viallet *et al.*, 1998). The morphogenesis of the induced feathers has been analysed by transgenic experiments which have shown that BMP enhances the size of the rachis, *noggin* increases branching and *sonic hedgehog* (*Shh*) causes webbing of the branches (Yu *et al.*, 2004). A fast, and inexpensive technique for commercial sex-typing of ostrich chicks has been devised by Malago *et al.* (2002), using DNA extracted from feathers.

Pigmentation of the Feathers

See Pigmentation, below.

Feather Muscles

Four small smooth muscles form in the dermis around the feather follicle and these are used for erecting and depressing the formed feather.

THE PERIDERM

The diversity of form of the integument covering the avian leg and foot, and the regional distribution in keratins (only some of which closely resemble the feather keratins) have stimulated the study of epithelial–mesenchymal interactions that produce patterns and control differential gene expression (Sawyer *et al.*, 1986; Barnes and Sawyer, 1995). The broad, overlapping scutate and scutellate **scales** on the anterior and posterior surfaces of the tarsometatarsus are best known for their similarity to scales of archosaurian reptiles (Maderson, 1985). They arise as a series of placodes during days 10–11, form scale ridges by day 12, and elongate distally until the end of day 17, by which time the typical epidermal proteins are detectable. All types of avian scale lack the pronounced dermal condensations characteristic of feather primordia, although a transient increase in mesenchymal density is detectable at the scale ridge stage. Epidermal placodes are absent from the tubular, reticular scales on the plantar surface.

The ptilopodous ('feather-feet') phenotype seen in various breeds of the domestic fowl, as well as in some other galliformes and owls (Lucas and Stettenheim, 1972), and the possibility of its experimental induction, is often cited as supporting evidence for the evolutionary origin of feathers from scales. But, since small feathers can be induced to form on all types of scales by *in ovo* application of retinoids (Sengel, 1986; Sawyer *et al.*, 1986), it is more likely that these feathers result from the upsetting of regulatory genes (Michaille *et al.*, 1995; Noveen *et al.*, 1995) that are usually suppressed over the legs and feet.

The first rudiments of the scutellate scales (*Text-Figure. 73*), which appear on the dorsal side of the foot, have formed placodes by day 11 and scale ridges by day 12; distal outgrowth of the scale ridges takes place between 12 and 17 days (Sawyer, 1972). Each tiny scale increases in size and becomes keratinized and overlaps the neighbouring scale, and blood vessels penetrate into the papilla.

BEAK AND CLAWS

The ectoderm of the beak region has started to thicken by day 6 along the upper and lower jaws. By day 7 the egg tooth has begun to develop as a projection from the tip of the upper jaw (*Text-Figure 74*); it grows rapidly from about 3 mm in length at 10 days to about 3.5 mm at 14 days. Its role is to break through the shell at hatching, after which it is shed.

The beak starts to become cornified on day 10 and has hardened by day 14. The periderm of the beak is not sloughed off until about days 19–20.

Claws, which are modified scales, are visible at about 10 days (stage 36) on the toes and have started

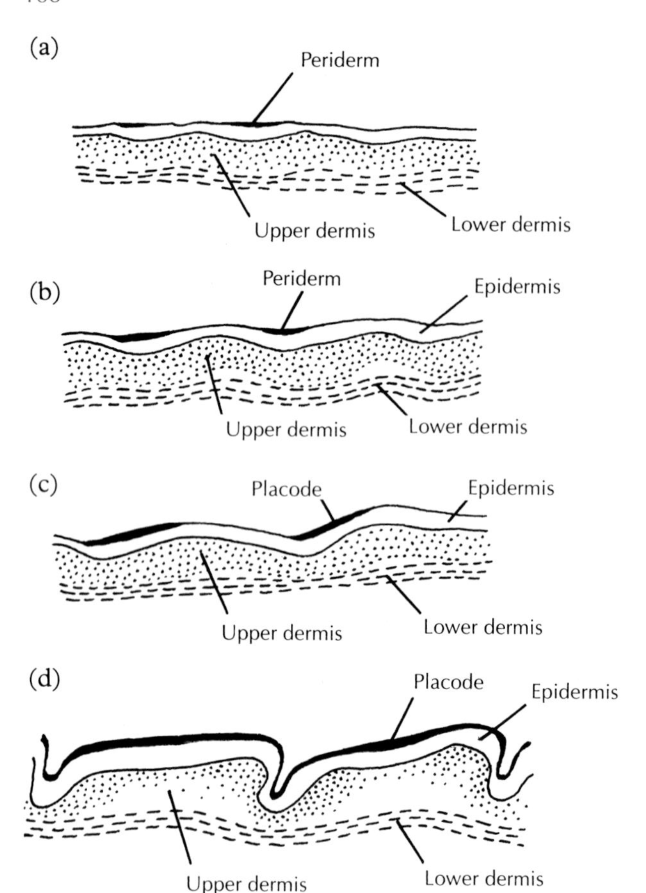

Text-Figure 73. *Development of two anterior metatarsal scales. (a) Late stage 35: symmetrical placodes. (b) Stage 36: the placodes are becoming asymmetrical. (c) stage 37: 'hump' stage. (d) Stage 38: two scale ridges now formed. (After Sawyer, 1972, with permission of Wiley-Liss, Inc., a subsidiary of John Wiley and Sons Inc.)*

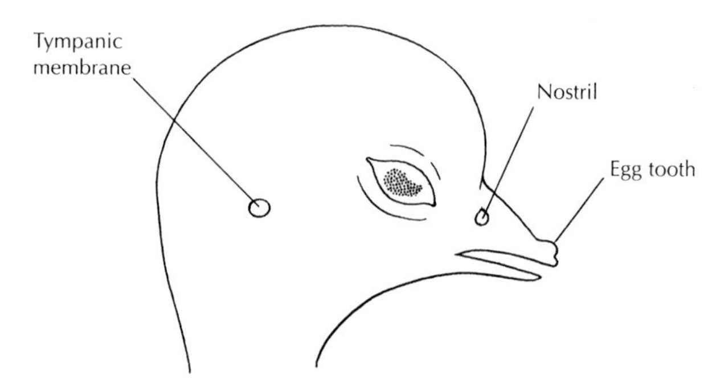

Text-Figure 74. *The beak of an embryo at 12 days to show the egg tooth.*

to become cornified by 14 days (stage 40, see normal tables, Appendices II, III). A transitory thumb-like claw is visible on the first digit of the wing at 10 days.

THE COMB

The comb is first visible as a ridge along the head at about days 6–7. It originates as a thickened and vascularized dermal thickening beneath the ectoderm (Menon *et al.*, 1981). Small papillae develop along the posterior two-thirds of the region, and are the forerunners of the adult serrated comb. As it enlarges during the following days, it becomes highly vascularized. Wattles start to appear at 11 days and develop in a similar way to the comb.

THE UROPYGIAL GLAND

Birds possess no integumentary glands apart from the uropygial (preen) gland, which forms on the dorsal side of the body at the base of the tail. It starts as a bilobed structure that arises on day 10 from a pair of ectodermal invaginations, one on either side of the vertebrae. These become paired ducts and each buds off further sacs. The two structures unite on day 14 but secretion does not begin until after hatching. For further details see Lucas and Stettenheim (1972), Quay (1986) and Fukui (1988, 1989).

PIGMENTATION

The yellow of the beak, skin and legs is due to lipochromes, which are carotenoid pigments. They appear at the time of keratinization and are derived from the carotenoids of the yolk. The melanin pigments are all derived from those neural crest cells that migrate beneath the ectoderm and above the somites. These cells are already specified to become melanocytes by the time they begin to migrate and there is evidence to suggest that this is due, at least in part, to *Wnt* signalling from the dorsal neural tube (Jin *et al.*, 2001). These authors found that as the *Wnt* signalling increased, BMP4 decreased, suggesting that *Wnt* and BMP4 signalling had antagonistic functions, *Wnt* promoting melanocytes and BMP4 neuronal and glial cell lineages.

The premigratory neural crest cells express the endothelin receptor B (ETRB), and if they migrate along the dorsal pathway they continue to do so, whereas those that pass along the ventro-medial pathway express a different endothelial receptor (Dupin *et al.*, 2000). Only those neural crest cells that are specified as melanocytes are able to take the dorsal path (Erickson and Goins, 1995). In the normal embryo these dorsally migrating cells are unable to enter the ventral pathway. By contrast, in the development of the Silkie fowl they pass not only along the dorsal route but also along the ventral one, so that eventually most of the connective tissues become pigmented. Faraco *et al.* (2001) have provided experimental evidence that in the normal embryo environmental barriers, such as lectins, prevent the migrating melanoblasts from taking the ventral route, but that these barriers are not present in the Silkie embryo.

Those melanocytes that enter the wing bud of the normal embryo migrate in a proximo-distal direction under the influence of fibroblast growth factors FGF2 and FGF4, produced by the apical ectodermal ectodermal ridge (Schöfer *et al.*, 2001).

The melanocytes that lodge in the feather buds undergo extensive mitosis and enlarge and begin to secrete melanin from about days 8–9 (for a discussion see Le Douarin and Kalcheim, 1999). In breeds such as the White Leghorn the melanophores differentiate as in the darker-feathered breeds but die before depositing pigment, so that the feathers are non-pigmented.

INNERVATION AND SENSORY RECEPTORS IN THE INTEGUMENT

The innervation of the skin is restricted to the dermis and to a ring of skin around the base of each feather.

Merkel Cells

Halata *et al.* (1990), using chick–quail chimeras, showed that the Merkel cells are of neural crest origin and that in the limb they migrate along with the Schwann cells. They are found not only in the dermis (Grim and Halata, 2000) but also in other epithelial-derived structures such as the hard palate (Nafstad, 1986), the tongue (Toyoshima *et al.*, 1993) and the beak (Saxod, 1980; Halata and Grim, 1993). Their function is unknown.

Herbst Corpuscles

Herbst corpuscles are found in the skin and beak and consist of an inner core of cells derived from the neural crest and an outer capsule from the adjacent mesoderm (Halata *et al.*, 1990). Temperature receptors have been described in adult skin, but their development does not appear to have been described.

Chapter 12
Endocrine Glands

The endocrine glands are the pituitary, thyroid, parathyroids, adrenals, pancreas and gonads. All become fully formed and functional, as endocrine glands, in the embryo. General reviews and their physiological activity in birds are given by Höhn (1961), whilst certain aspects of endocrinological activity are covered by Scanes *et al.* (1987). Here we deal mainly with their morphological development.

THE PITUITARY GLAND

The pituitary (hypophysis) is formed partly from the **infundibulum** (*Plate 73*), a depression in the floor of the diencephalon, and partly from **Rathke's pouch** (*Plate 73*), an ectodermal invagination from the roof of the mouth. It is possible that the ventral diencephalon induces the formation of Rathke's pouch and this supposition is supported by transplantation experiments (Gleiberman *et al.*, 1999). The infundibulum, which forms the neural or posterior lobe of the pituitary (the neurohypophysis) starts to appear at about stage 14, according to Jacobson *et al.* (1979) who distinguish between this structure and a 'false' infundibulum that arises earlier and does not contribute to the pituitary. During the following 2 days the true infundibulum increases and extends both anteriorly and posteriorly. Rathke's pouch, which is destined to become the anterior lobe of the hypophysis and the pars tuberalis, arises at about stage 17 and its anterior end fuses with the infundibulum. Its posterior end breaks away from the roof of the mouth at about 7 days of incubation. Birds differ from mammals in that they do not form a pars intermedia.

The term **adenohypohysis** is used for the pituitary lobes derived from Rathke's pouch. They are the pars distalis or larger lobe, and the pars tuberalis or smaller lobe, which is formed by two outgrowths from the pars distalis that fuse together. The pituitary is encapsulated by the dura mater.

The ectoderm which is destined to form Rathke's pouch is situated at stages 4–5 in the midline anterior to Hensen's node. With the formation of the head fold and head it becomes shifted so that by stage 14 it lies in the roof of the mouth, just anterior to the stomatodeal membrane. (For a detailed account see Jacobson *et al.*, 1979). Even as early as stage 20 (day 3 of incubation) distinct areas can be recognized histologically in Rathke's pouch that will give rise to the distinct regions of the pars distalis and pars tuberalis (Sasaki *et al.*, 2003). By 4.5 days the α-subunit mRNA of glycoprotein hormones can be found, and luteinizing hormone may be detected by immunoreactivity in the cephalic lobe by day 6 (Kameda *et al.*, 2000).

The pituitary growth hormone plays an essential role in post hatching growth and development, but the pituitary itself does not begin to secrete it until about half way through incubation, despite the fact that the growth rate of this organ is highest in the early embryo. It appears therefore that the early stages are independent of pituitary growth hormone, although there is accumulating evidence that immunoreactive growth hormone is present in many early embryonic tissues as early as day 3 of incubation, especially the neural tube, notochord, limb bud, somites, heart, stomach, liver and kidney. The problem is reviewed by Harvey *et al.* (2000) who conclude that the hormone acts locally at this time in a paracrine or autocrine fashion

The pars distalis has appeared by day 5 (stage 26) at the tip of Rathke's pouch (Jacobson *et al.*, 1979). A capsule of connective tissue starts to form around the pars distalis at 13 days and has enveloped the whole pituitary by 16 days. The lumen of the pituitary is apparent at 6 days but has disappeared by 11 days. Investigations involving immunocytochemistry and immunoassays detected cells producing the following hormones: ACTH at 7 days, growth hormones at 12 days, prolactin at 6 days, LH and FSH at 4.5 days, thyrotropin at 6.5 days, and mesotropin and vasotocin at 8.5 days (Scanes *et al.*, 1987).

THE THYROID GLAND

The thyroid forms during days 2–3 from a median invagination from the floor of the pharynx into the underlying mesoderm at the level of the second pair of pharyngeal pouches (*Plate 74*). During day 3 it becomes a sac still open to the lumen of the pharynx, but on day 4 this connection is reduced to a solid stalk, the **thyroglossal duct** (*Plate 92*), and the organ itself becomes bilobed. According to Knospe *et al.* (1991) this is due to the enlarging oesophagus pressing the trachea on to the rudiment of the thyroid. During day 6, the thyroglossal duct degenerates and each lobe becomes further lobulated and extensively vascularized by vessels from the subclavian arteries (derivatives from the third aortic arches). The thyroid increases in size throughout this period from a length of 167 μm and a depth of 83 μm to a length of 209 μm and a depth of 250 μm by day 5 (Romanoff, 1960).

The thyroid consists entirely of an epithelium of endodermal origin until day 6, when it becomes invaded by mesenchyme. The epithelial cells now become arranged into cords, and sinusoids develop from the mesenchyme. By day 8 the organ measures about 0.4 mm and by day 12 a lumen has formed in each cord, which then dilates. The first appearance of colloid droplets is on day 7 (Hilfer, 1964) but it is not until about days 10–13 that the adenohypophysis–thyroid interactions become established (reviewed by Thommes, 1987). Before this, the function of the thyroid is apparently independent of the adeno-hypophyis, but after this it becomes dependent on it (Thommes, 1987). The growth of the embryo, as measured by body weight, skeletal size, muscle growth and growth of cartilages and bone, is greatly influenced by thyroid hormones in late embryonic development (King and May, 1984).

THE PARATHYROID GLANDS

The parathyroids are traditionally considered to be formed from the ventral parts of the third and fourth pharyngeal pouches (i.e. endoderm), as in mammals, though Merido-Velasco *et al.* (1996) have concluded from chick–quail grafting experiments that parathyroid III is of ectodermal origin. The parathyroids become innervated by branches of the vagus nerve (Egawa and Kameda, 1995). By 9.5 days the anterior lobe has become semicircular and contains epithelial cords separated by mesenchyme in which capillaries run. The parathyroids, which are supplied by branches of the carotid body artery (Kameda, 1990) become vascularized during the second half of incubation and have become fully differentiated by day 17 (Simkiss, 1967). It is thought that the parathyroid hormones are probably not necessary for calcium uptake during the greater part of embryonic life (Simkiss, 1991) as they are later on.

THE ADRENAL GLANDS

As in other amniotes the adrenal glands are formed from two sources, the cortex and the medulla. The adrenals of birds differ from those of mammals in that the cortex does not encapsulate the medulla but is intermingled with it and in this respect resembles the adrenals of amphibians and reptiles.

The cells that will form the cortex (or inter-renal tissue) arise from the peritoneal epithelium of the nephrogenic (genital) ridge (see p. 63). During day 4 a groove becomes visible immediately posterior to the pronephric glomeruli, extending from about the level of the 17th to the 22nd somites. Cells that leave the epithelium of this groove, proliferate rapidly and invade the underlying mesoderm. By the end of day 4 they have formed solid nodular structures lying between the mesonephros and the dorsal aorta and are no longer attached to the peritoneal epithelium.

During the following days there is extensive proliferation and by day 7 the cells have become arranged in cords. Each adrenal mass is in contact with the ventral side of the mesonephros. By day 8 the size of each adrenal mass, as seen in transverse sections, is approximately twice that of the dorsal aorta.

The adrenal medulla (or chromaffin tissues) is derived from cells of the sympathetic chain, which in turn have been formed from the neural crest. These cells migrate individually during days 4 and 5 from the sympathetic chain toward the mesentery and pass between the aorta and the cortical nodules. They then cluster around and penetrate into the cortical nodules. By day 7, the chromaffin cells have started to differentiate and become arranged in cords between the cortical cords. During the following 2 days the tissues continue to intermingle and by day 8 a meshwork of nerve fibres has formed. After day 9 development consists principally of further growth. (*Plate 136*).

THE PANCREAS

The development of the pancreas has been described in Chapter 8. The endocrine cells are formed from the endoderm (Le Douarin, 1988). Insulin starts to be secreted by the β-cells of the pancreas on day 4 but is not produced in large amounts until about day 12. Glucagon begins to be secreted from day 3 (see Freeman and Vince, 1974).

The insulin and glucagon islets are anatomically separate in birds and production of the enzymes responsible for processing the prohormones occurs soon after the islet cells first appear (Rawdon, 1998; Rawdon and Larsson, 2000).

THE GONADS

The development of the gonads has been described in Chapter 7. Both female and male hormones are produced by the indifferent gonad. Oestrogenic hormones are secreted by the interstitial cells of the medulla from about day 4, whereas testosterone is produced by the cord cells. The pituitary–gonadal axis is established from about day 13 (see discussion by Freeman and Vince, 1974). Luteinizing hormone, (LH) is present in the blood early in development but does not reach a high enough level to stimulate steroidogenesis in the gonads until 13.5–14.5 days (Woods, 1987). Nevertheless, steroidogenic factor-1, which regulates steroidogenic enzyme expression, was detected by *in-situ* hybridization in the undifferentiated genital ridge of both sexes as early as stage 21–22 (3.5 days). By stages 30–35 it had become higher in the ovaries than in the testes and eventually was highest of all in the left (functional) ovary (C.A. Smith *et al.*, 1999). There was a reduced response in the right ovary as it regressed, and a comparable reduction was noted by Pedernera *et al.* (1999) in gonads treated with follicle-stimulating hormone (FSH). Both the left and right ovaries of 8-day embryos responded to FSH by secreting steroids, but in embryos older than 13 days the right ovary failed to respond.

GUT ENDOCRINE CELLS

There are many endocrine cell types in the gut that produce serotonin and a range of regulatory peptides. Most appear between 10 and 12 days of incubation and have been shown by chick–quail grafting experiments to be derived from the endoderm and not, as previously thought from the neural crest. The gut mesenchyme plays a significant role in promoting their development. For a review of gut endocrine cells in birds see Rawdon and Andrew (1999).

Chapter 13
Extra-Embryonic Membranes

The embryo not only grows and differentiates but undergoes those physiological processes that are necessary for life, some of which are possible only with the aid of the extra-embryonic membranes. These structures are continuous with the embryonic tissues but are not part of the embryonic body proper and are discarded at hatching. They are:

- the **yolk sac**, which encloses the yolk and is so constructed that the partially digested yolk is passed into the embryonic blood stream
- the **chorioallantois**, which has three functions, respiratory exchange, calcium transport and storage of nitrogenous wastes produced by the embryo
- the **amnion**, which is a sac surrounding the embryo and which secretes fluid that cushions the embryo and prevents it from becoming dehydrated.

THE YOLK SAC

The yolk sac (*Plates 4, 65*) is formed from the area opaca (see Chapter 3) and consists initially of a thin epithelial ectoderm underlain by a layer of tall endodermal cells containing many intracellular yolk droplets, these layers being continuous with the area pellucida upper and lower layers, respectively (*Text-Figure 8*). By about 12 h of incubation the periphery of the area opaca has become attached to the inner surface of the vitelline membrane (*Plate 2c*). At this stage the vitelline membrane is taut around the yolk and is used as a substratum by the peripheral cells of the area opaca migrating under it. The result is that the blastoderm itself is under tension and if that tension is released abnormal development follows (Bellairs *et al.*, 1967). The migration takes place all round the periphery of the area opaca, the so-called 'margin of overgrowth', so that the entire blastoderm gradually expands, the region showing the greatest increase in area being the area opaca. Mesoderm cells, which have ingressed through the primitive streak and migrated laterally, colonize the area opaca at about stage 5 (Kessel and Fabian, 1985). They are continuous with the lateral plate mesoderm of the embryo and are similarly arranged into somatic and splanchnic layers. Blood islands begin to form in the proximal part of the splanchnic mesoderm, and by about 48 h they have joined up to form a series of blood vessels that is in communication with the vascular system of the embryo. About this time it is usual to substitute the term yolk sac for that of area opaca. The vascularized region of the yolk sac is known as the **area vasculosa** and is to be distinguished from the distal part, which is known as the **area vitellina**. The two regions are clearly demarcated by a blood vessel at the periphery of the area vasculosa, the **sinus terminalis**, which also indicates the lateral extent of the splanchnic mesoderm (*Text-Figure 26*). It is likely that the mesoderm continues to invade the area vitellina. The vitelline membrane breaks down over the embryo during day 3 and its value as a substratum is lost. The edge cells of the yolk sac pass over the equator of the yolk between 5 and 7 days.

Before the area opaca has become vascularized, the embryo is nourished by intracellular yolk droplets. They are most conspicuous in the large yolky endoderm cells of the area opaca, but smaller ones are present in the ectoderm, mesoderm and endoderm cells of the embryo itself. The intracellular yolk drops in the embryo have become used up by about 48 h, the embryonic circulation has become established and the intracellular yolk of the area opaca has become available. By the time these new stores have become depleted, the yolk sac circulation has developed further and is so modified that the embryo is able to tap the entire contents of the yolk sac (*Plate 79*). The development of the yolk sac circulation is considered in Chapter 6 and begins at about stage 9.

The extra-embryonic coelom (*Plate 65*), which is the space between the somatic and splanchnic lateral plate mesoderm, is continuous with the embryonic coelom for about 3 days, the communication being lost as the body folds (see Chapter 5). The blood islands form in the splanchnic mesoderm (see Chapter 6). Mesoderm cells surrounding them become arranged as vesicles which then fuse to form vessels (Kessel and Fabian, 1985). No mesoderm is present in the area vitellina. Meanwhile the chorion (see below), begins to form from the somatopleure.

The endodermal layer of the yolk sac undergoes extreme elaboration. By day 4 its inner surface consists of ridges and folds of endoderm which become highly vascularized and hang loosely into the yolk. The ridges become more and more elaborate as incubation proceeds. Digestive enzymes are secreted by the endoderm into the yolk and the partially digested material is then absorbed into the cells (Mobbs and McMillan, 1981; Lambson, 1970) and passes from there into the blood vessels and back to the embryo (see Chapter 6).

The embryo becomes lifted above the yolk sac by the formation of the head, tail and lateral body folds (see Chapter 5) but remains connected to it by the yolk sac stalk, which becomes narrower and narrower during incubation. By day 19 most of the yolk has been absorbed and the remnants of the yolk sac and its contents start to be drawn up into the body by contraction of the abdominal muscles and serve to nourish the hatchling during its first day. Day-old chicks may therefore be transported without needing to be fed.

THE AMNION AND CHORION

These two tissues are formed together. The first indication is a crescentic fold of ectoderm, together with its underlying somatic mesoderm, which starts to overlap the forebrain at about stage 12. The inner part of the fold will become the amnion, whilst the outer will form the chorion (*Text-Figure 23; Plates 68, 69*). The arms of the fold extend further and further posteriorly so that by stage 16 they have reached as far as somites 10–18 (see Normal Table, Appendix II). The growth of the lateral amniotic folds is characterized by increased cell proliferation (Miller *et al.*, 1994). As the lateral amniotic folds rise up they meet in the midline and fuse, the inner and outer layers becoming separated from each other, (Overton, 1989). A similar crescentic fold begins to form immediately posterior to the tail bud at about stage 17, though there is great variation among embryos in the exact timing. This fold extends anteriorly. By about stage 18, the anterior, lateral and posterior regions of the amniotic fold have united above the region of the hind limbs, forming initially a small oval opening, the **amniotic umbilicus**. Although this soon closes, the inner and outer layers remain attached to one another at this point, the so-called **sero-amniotic connection**, though elsewhere the two membranes, amnion and chorion, are separate. The cells of the amnion now secrete fluid into the amniotic cavity around the embryo.

A characteristic of the amnion is that it lacks blood vessels. According to Pardenaud and Dieterlen-Lièvre (1993) this may be due to the fact that it has no contact with endoderm when it is forming, unlike the yolk sac and allantois, which are both vascularized. Another important feature of the amnion is that the muscular component is not innervated. Despite this, vigorous muscular contractions take place between days 5 and 9 (Müller *et al.*, 1983) and these help to prevent adhesions of the amnion to the embryo until day 12, or between different parts of the embryo. After this period, the skin has undergone differentiation and the danger is lessened. The contractions have been found to last from 15 to 60 s with a frequency af about 11–18 per 10 min, depending on the temperature (Nechaeva and Turpaev, 2002).

THE CHORIOALLANTOIC MEMBRANE

The chorioallantoic membrane is formed by the fusion of the chorion (see above) and the allantois. The chorion, which consists of ectoderm underlain by splanchnic mesoderm, remains continuous at its lateral edges with the somatopleure covering the yolk sac, but elsewhere it becomes associated with the allantois.

The allantois has begun to form from the hindgut by about stage 18 as an endodermal diverticulum with a covering of mesoderm (see Chapter 8). Although its speed of development varies between individuals it has usually become a balloon-like structure outside the body by 4 days (*Text-Figure 30; Plate 82*) and has begun to fuse with the inner layer of the chorion by about 6–7 days, forming the chorioallantoic membrane (also known as the allantochorion). When this has happened, the chorioallantoic membrane consists of three layers, **ectoderm** (from the chorion), **mesoderm** (fused somatic mesoderm from the chorion and splanchnic mesoderm from the allantois) and **endoderm** (from the allantois). The chorioallantoic membrane has a rich vascular system that develops within the mesodermal layer and is served by paired allantoic (umbilical) arteries and paired allantoic (umbilical) veins (*Plates 106, 107*) (see Chapter 6). The pattern of the blood vessels in the allantois and chorioallantois is the subject of many studies (e.g. DeFouw *et al.*, 1989).

The network is extended and elaborated by continual intussusceptive growth. Djonov *et al.* (2000) using methyl methacryalate casting (*Plate 39*) found the intussusceptive growth was also significant in the formation of vascular trees, especially between days 10 and 11. It is, perhaps, not surprising that the allantois has been found to be a site capable of producing both haemopoietic and endothelial cells (Caprioli *et al.*, 2001). By 16 days the chorio-allantoic membrane has become so large that it covers most of the yolk sac and adheres to it in places, although the two circulations remain separate (Romanoff, 1960).

The chorioallantoic membrane becomes closely pressed against the shell membranes, which, together with its extensive system of blood vessels, enables it to act as a gas-exchange organ (Freeman and Vince, 1974; Paganelli, 1991) receiving oxygen and eliminating carbon dioxide through the pores in the shell. This proximity has the additional benefit of enabling calcium ions to be absorbed from the shell and transported in the blood stream to the embryo. During the first 10 days, the embryo obtains its calcium from the yolk sac, but the require-ments are greatly increased during the last 11 days as the bones begin to ossify, and calcium is then obtained mainly from the shell. The physi-ology of calcium mobilization has been studied extensively, using culture methods that eliminate the shell and so provide for calcium deficiency (Tuan *et al.*, 1991).

The third role of the chorioallantoic membrane is to store the excretory products. The mesonephros begins to excrete into the allantois during day 4, mainly as urea, ammonia and uric acid (Fisher and Eakin, 1957; Freeman and Vince, 1974). At hatching, the chorioallantoic membrane is left attached to the inside of the discarded shell.

Plate Section

All staging of embryos, unless stated otherwise, is according to the table of Hamburger and Hamilton (see Appendix II). Magnifications are shown as bars for most of the scanning electron micrographs and whole mounts, and as direct magnifications for sectioned material.

PLATE 1

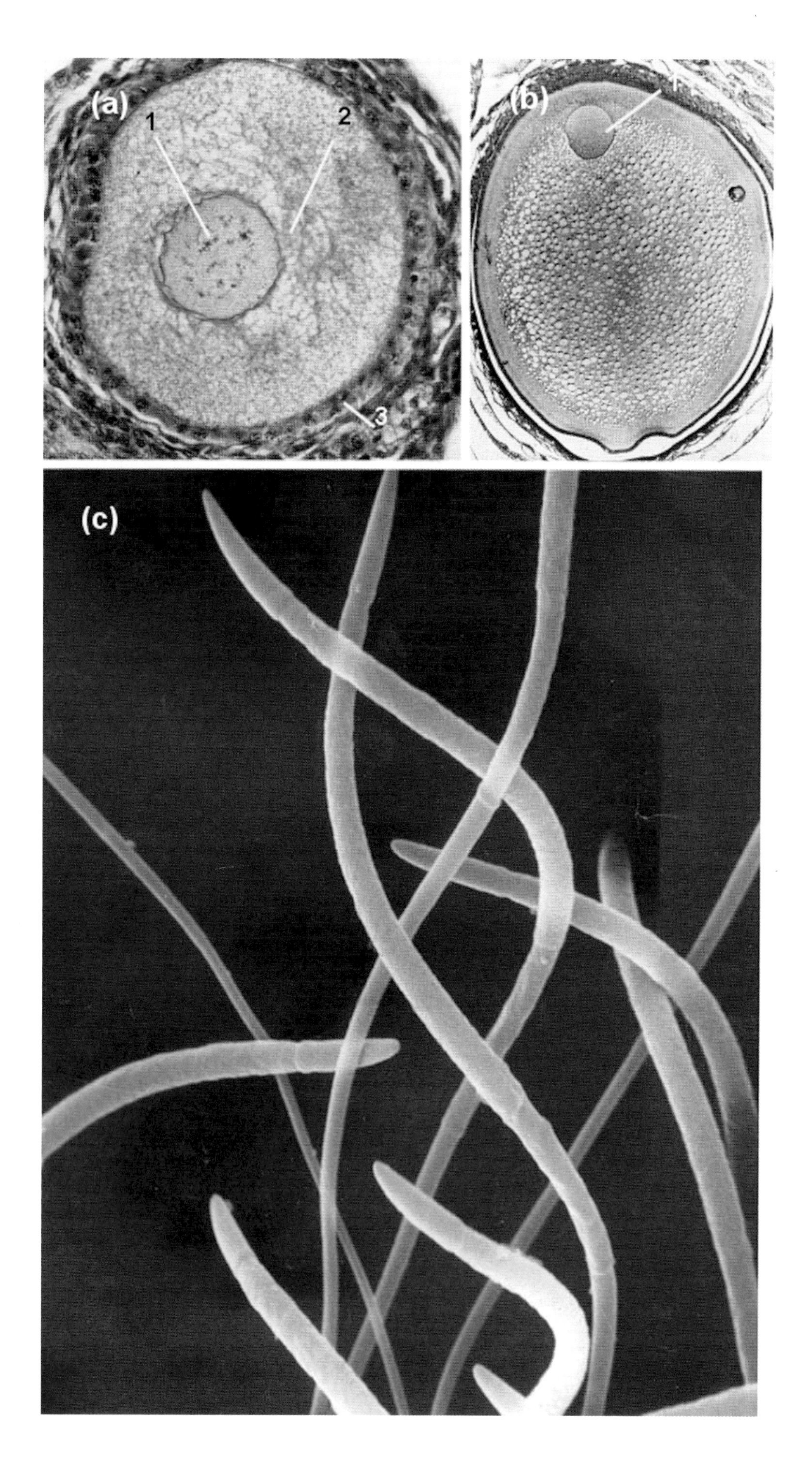

Plate 1

(a) Light micrograph of an oocyte at the Balbiani stage with the nucleus lying centrally. Diameter of oocyte = 100 μm.
(b) Light micrograph of an oocyte at a later stage in which the cytoplasm has become vacuolated prior to yolk deposition. The Balbiani body has disappeared and its components now lie in the periphery. The nucleus has shifted to one side, the future animal (dorsal) pole. Diameter of oocyte = 2 mm.
(c) Scanning electron micrograph of sperm from a domestic fowl. Note the small diameter (10 μm) of the head of each sperm.

1. nucleus
2. Balbiani body
3. follicle cells

(b) Reprinted from Bellairs (1967), with permission of the Company of Biologists Ltd.
(c) Kindly supplied by M. Bakst.

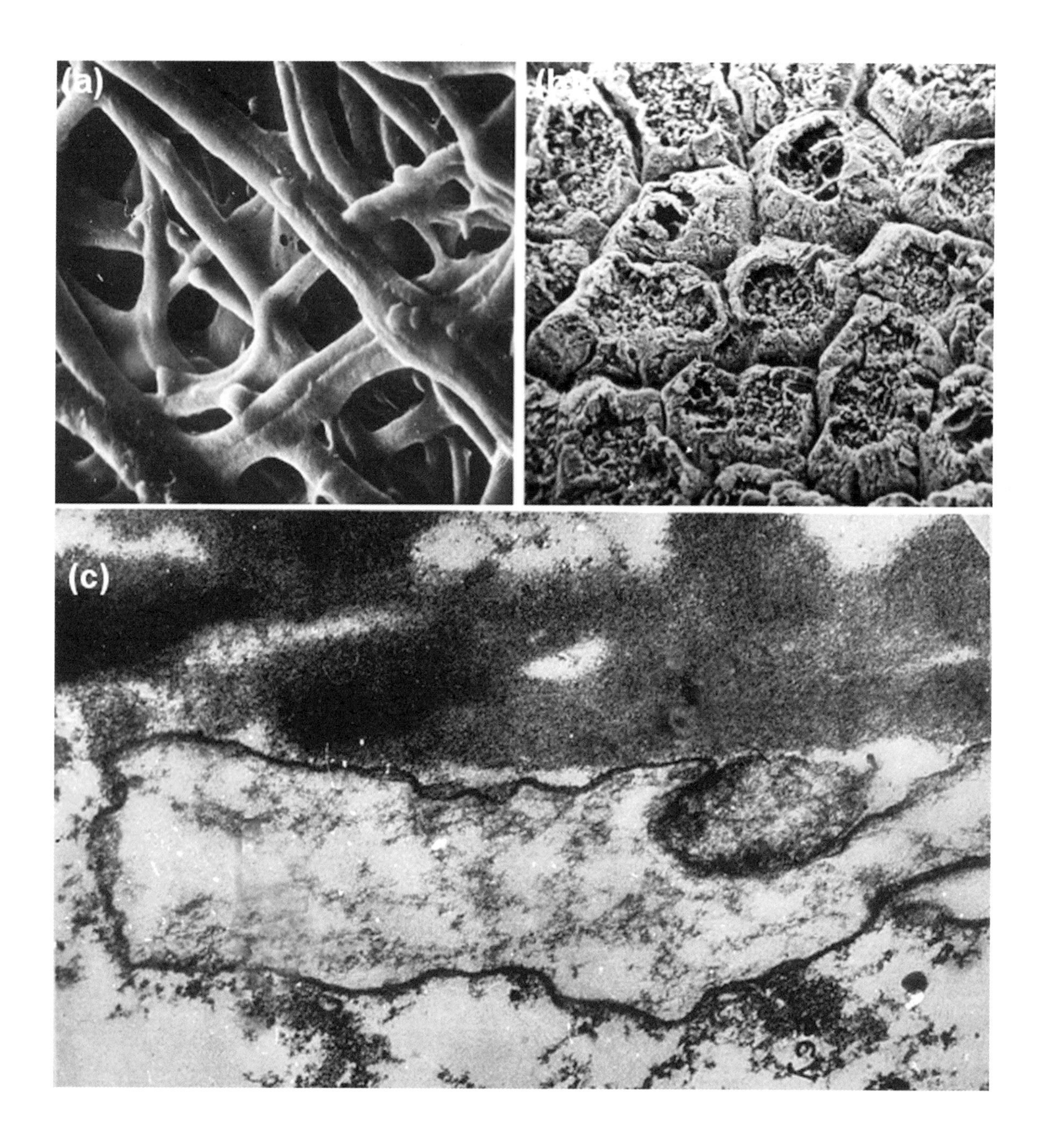
(a)
(b)
(c)

Plate 2 Electron micrographs

(a) Scanning electron micrograph of the outer shell membrane from an egg incubated for 18 days. ×3000.
(b) Scanning electron micrograph of the inner surface of a shell from an egg incubated for 18 days. The mammillary knobs have been eroded away as the calcium has been removed during incubation. ×2000.
(c) Transmission electron micrograph through the area opaca of a stage 4 embryo. The leading cell process is firmly attached to the inner surface of the vitelline membrane (top of picture). ×60 000.

(a) Reprinted from Bellairs and Boyde (1969), with permission of Springer-Verlag.

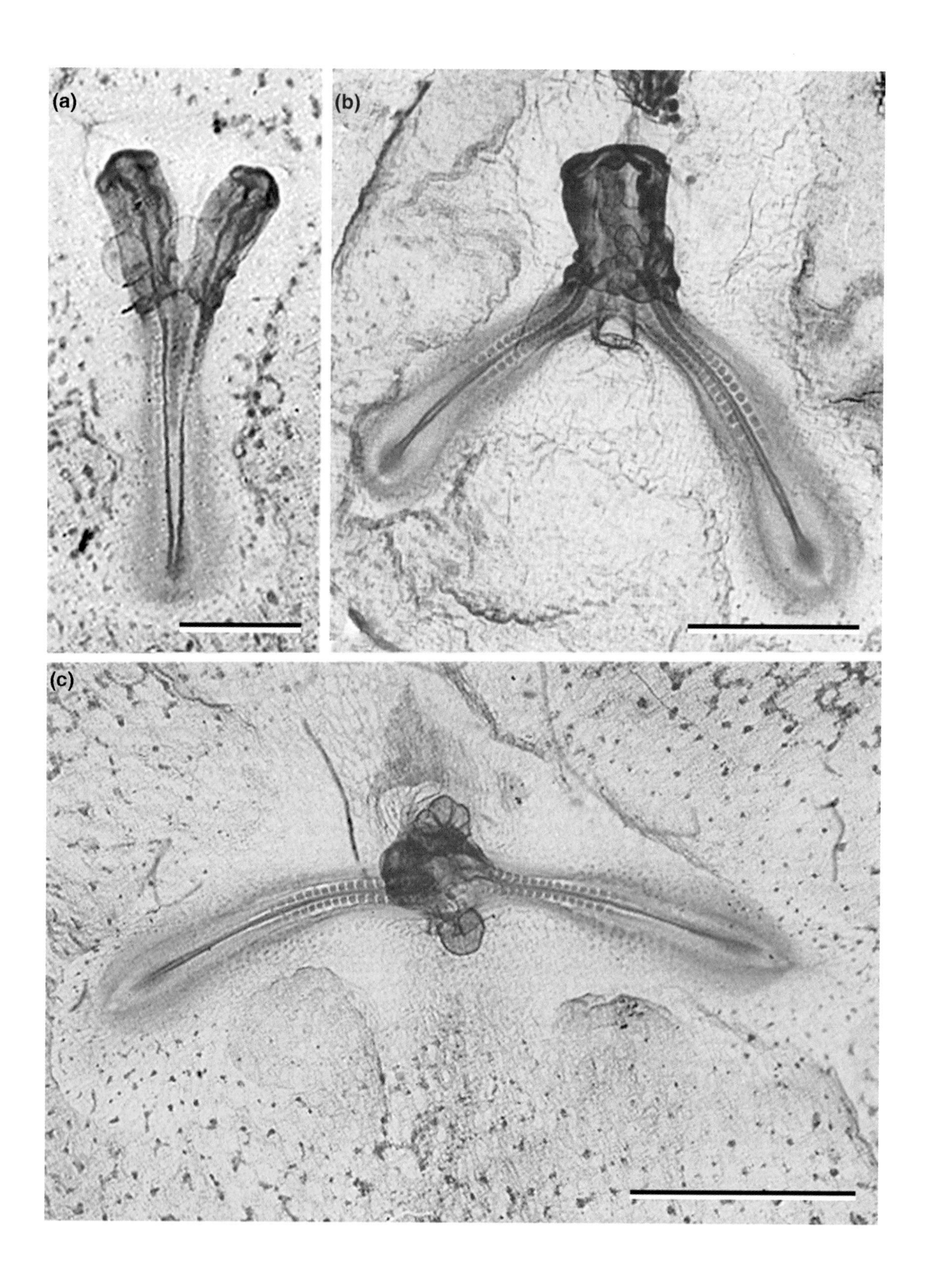
(a)
(b)
(c)

Plate 3 Light micrographs of whole-mounted embryos

(a) Naturally occurring *duplicitas anterior.* This anomaly appears to have developed from a single primitive streak that split into two at its anterior end, so that two Hensen's nodes were formed. As they regressed they were brought into contact by the gastrulation movements. Bar = 1 mm.

(b) Experimentally produced *duplicitas posterior,* formed by bisecting a pre-streak quail blastoderm. Two primitive streaks developed and their anterior ends collided, as in (c). Bar = 2 mm.

(c) Naturally occurring *duplicitas posterior.* This appears to have resulted from two primitive streaks having developed on the same blastoderm. The head processes probably collided at about stage 5. Regression of Hensen's node appears to have occurred normally since two segmented trunks have formed. Bar = 3 mm.

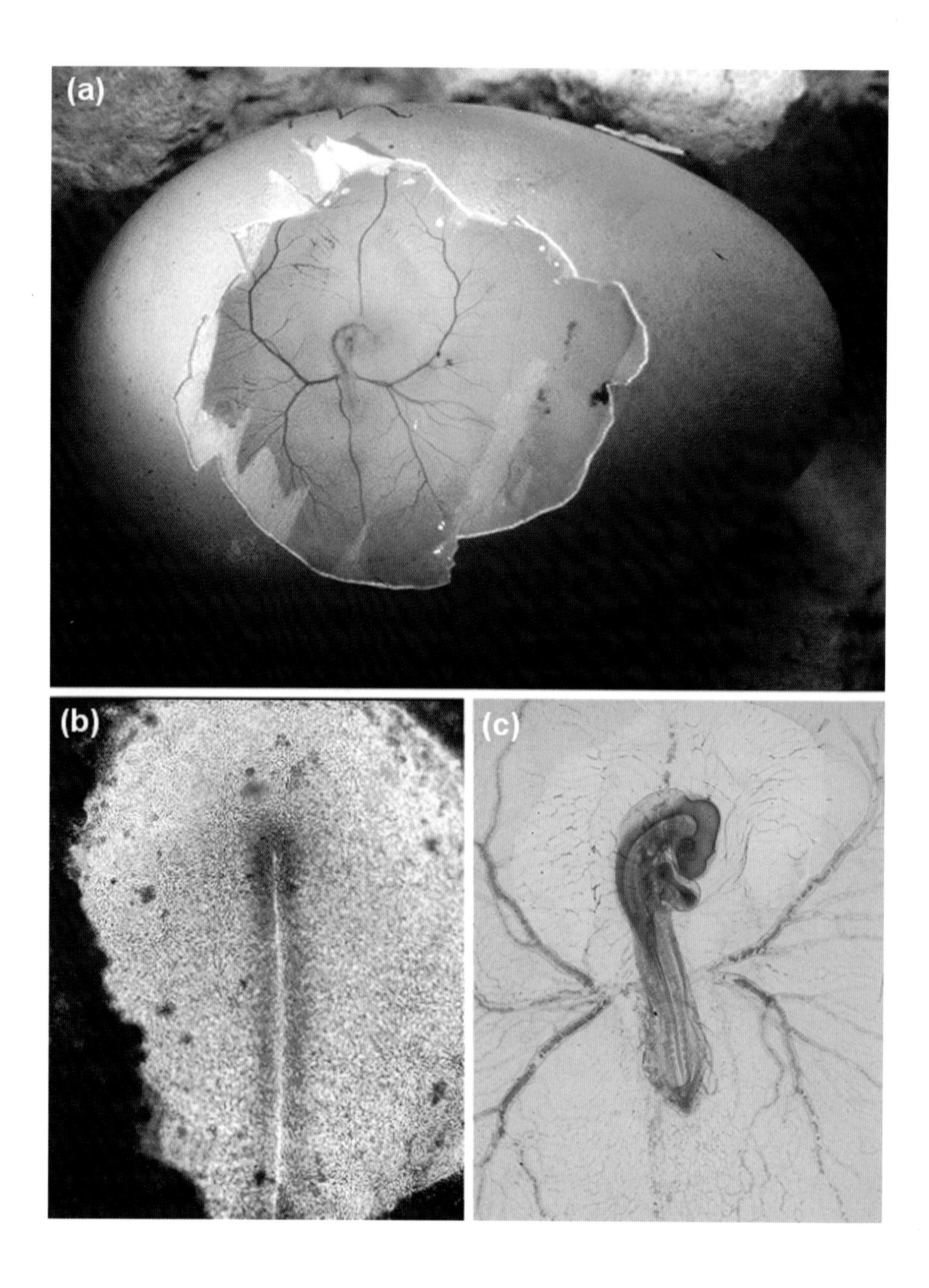
(a)
(b)
(c)

Plate 4 Light micrographs

(a) Hen's egg incubated for 58 h and opened to show the embryo at stage 16 together with the omphalo-mesenteric (vitelline) vessels spreading over the yolk. For labels see *Text-Figure 26*. ×3.

(b) A full-length primitive streak. Note: this important stage was omitted from the table of Hamburger and Hamilton (Appendix II) and is intermediate between their stages 3+ and 4. The head process has not yet begun to form in this embryo but is already present in Hamburger and Hamilton's figure 4. For convenience, this stage is sometimes called stage 3++. ×30.

(c) Embryo at stage 17 to show the omphalo-mesenteric vessels (see *Text-Figure 26*). ×8.

PLATE 5

(a)

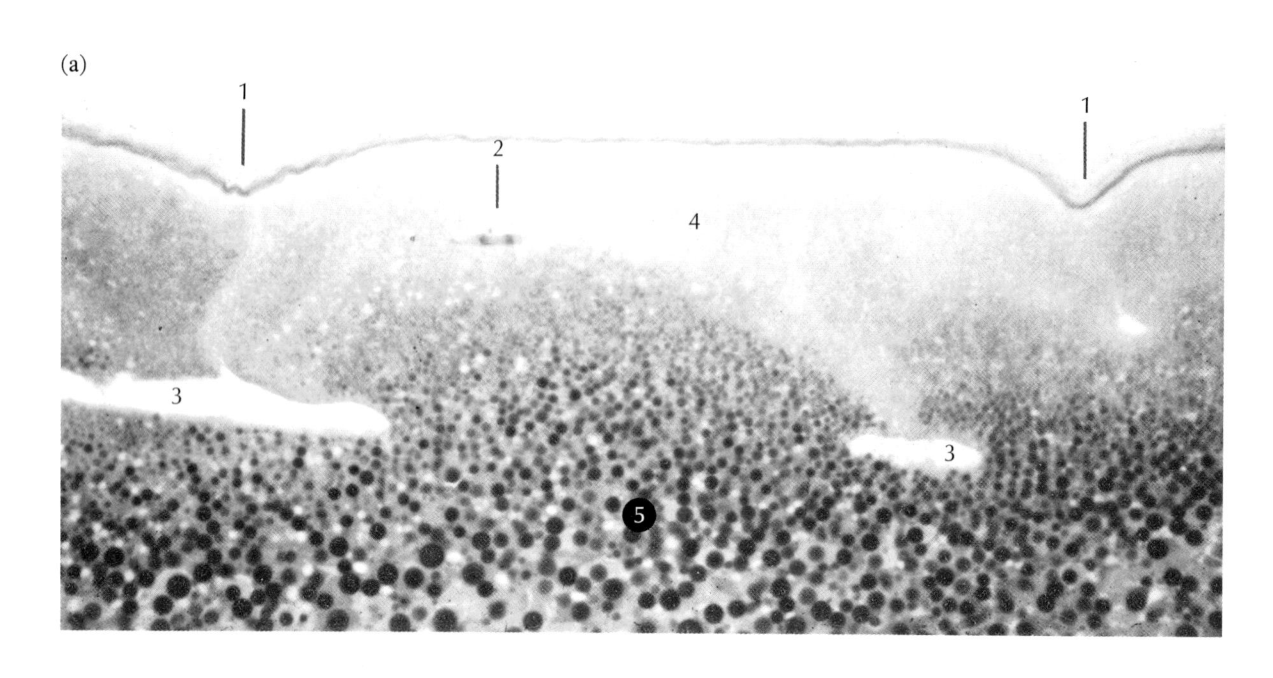

(b)

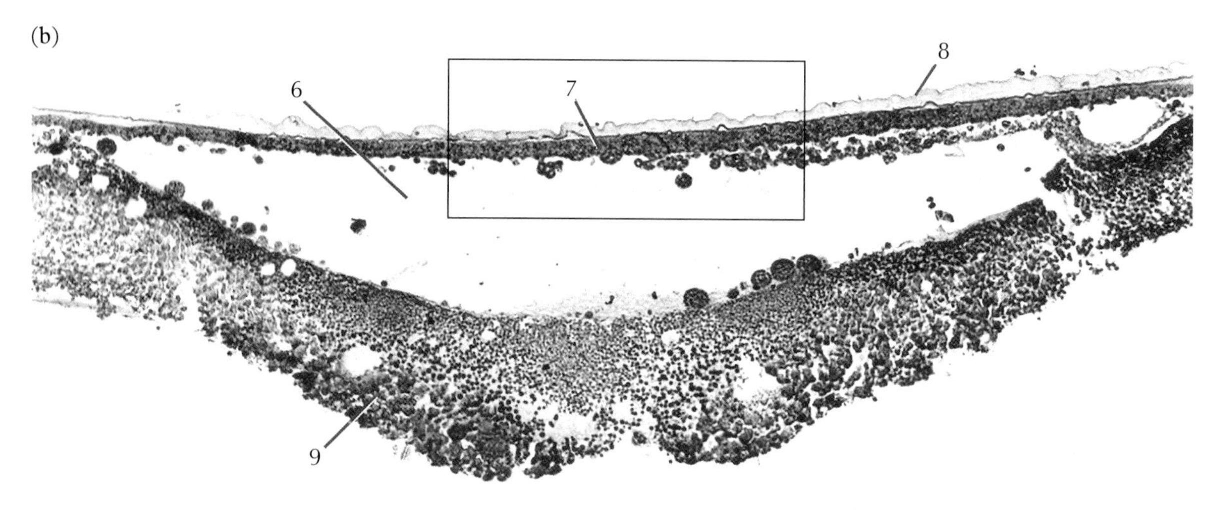

(c)

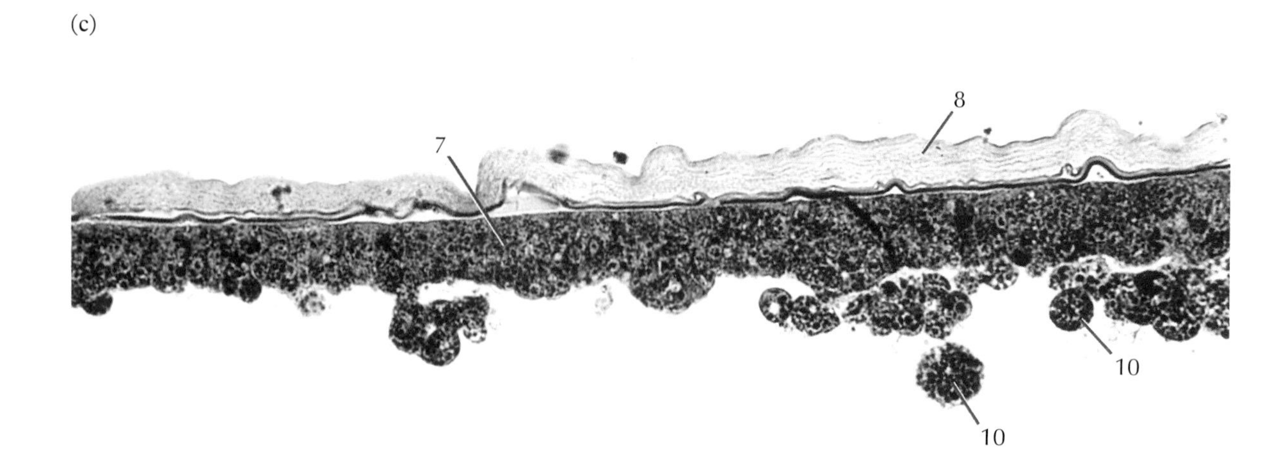

Plate 5 Light micrographs

(a) Vertical section through a 12 cell stage embryo taken from the oviduct, showing open cells. ×200.
(b) Vertical section through a stage X embryo (see Appendix 1), after laying but prior to incubation. ×60.
(c) Enlargement of region shown in (b). ×200.

1. cleavage furrows
2. mitotic spindle
3. base of cleavage furrow
4. cytoplasm
5. periblast
6. subgerminal cavity
7. upper cell layer
8. vitelline membrane
9. nucleus of Pander
10. endophyll cells of lower layer

Reprinted from Bellairs *et al.* (1978), with permission of the Company of Biologists Ltd.

Plate 6

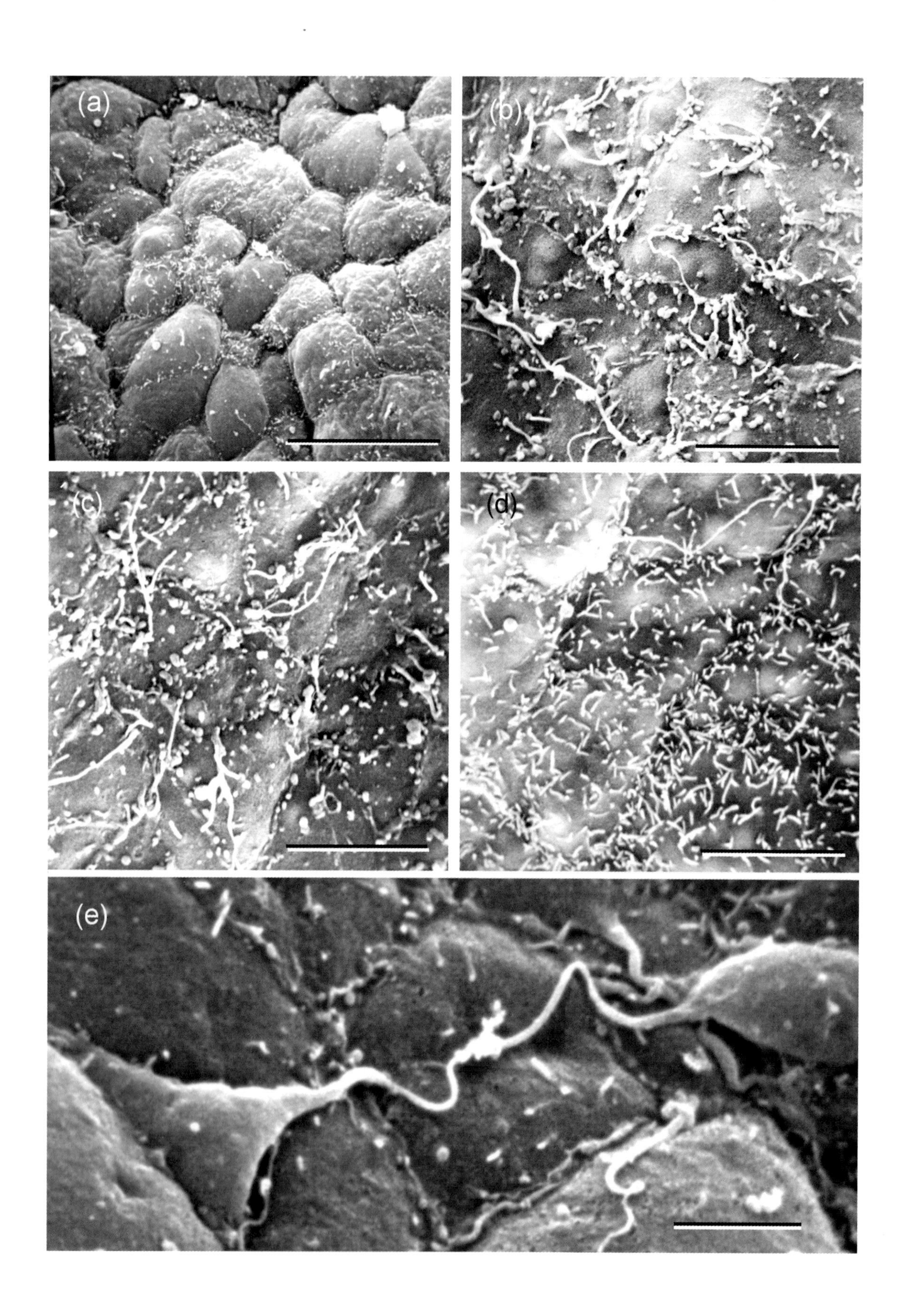

Plate 6 Scanning electron micrographs of the dorsal side of early embryos

The dorsal surface of the epiblast (upper layer) is covered with microvilli, though the pattern and distribution vary according to the region.

(a) Epiblast cells at the lateral side of the area pellucida of an unincubated embryo (stage XII: see Appendix I). Small globular microvilli and crypts between the cells are frequent. Bar = 100 μm. (b–e) are all taken from a single embryo at stage 5, about 20 h of incubation.
(b) Lateral to Hensen's node. Bar = 5 μm.
(c) Lateral to the primitive streak, about half way along its length. Bar = 5 μm.
(d) Posterior end of the area pellucida, especially rich in microvilli. Bar = 5 μm.
(e) Telophase bridge. Bar = 2.5 μm.

Plate 7

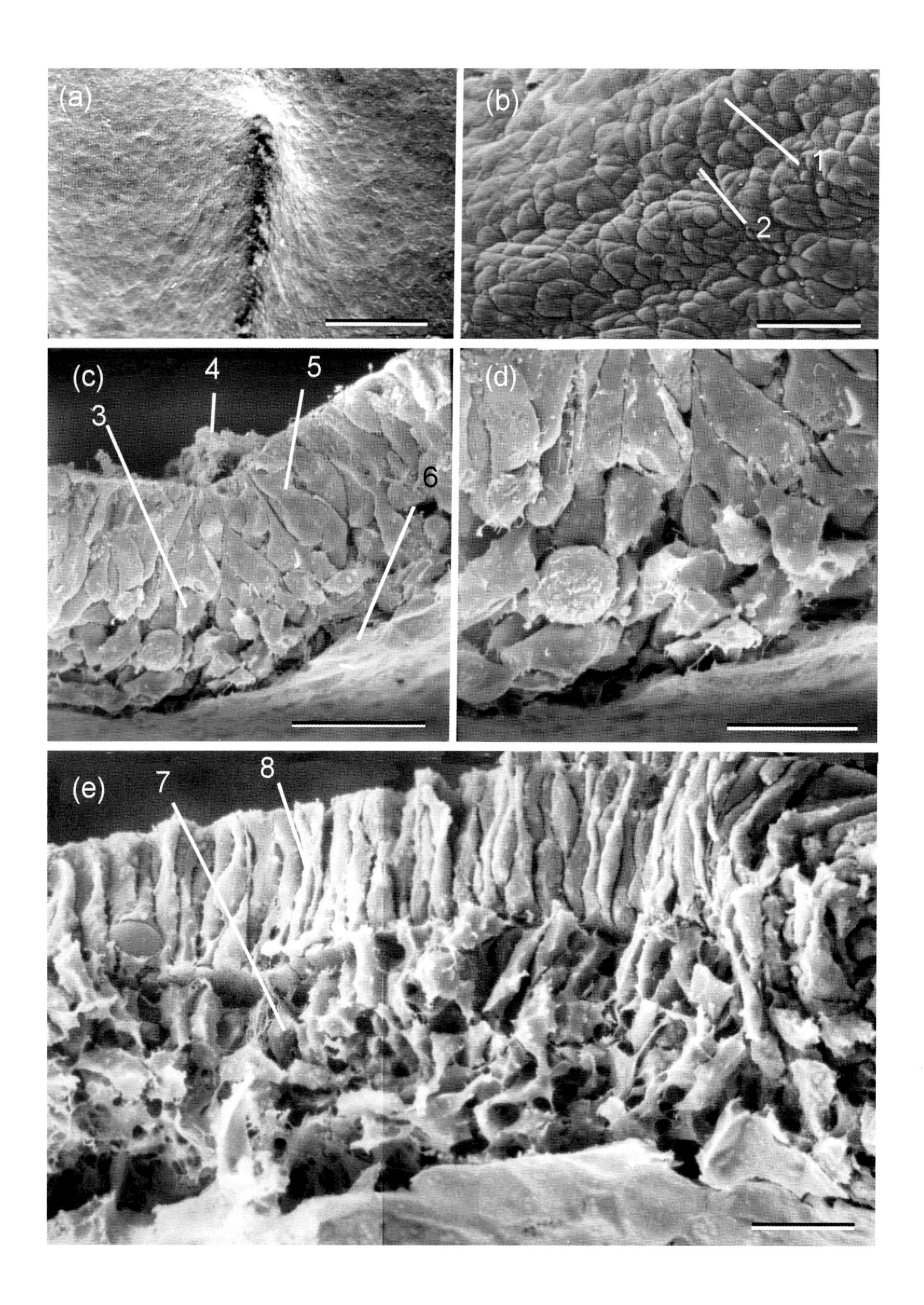

Plate 7 Scanning electron micrographs of the primitive streak

(a) Anterior end of the primitive streak (Hensen's node) of a stage 3+ embryo, 16 h of incubation. Bar = 500 μm.
(b) Detail of the primitive streak (running from bottom left to top right) of a stage 4 embryo, 18 h of incubation. Bar = 40 μm.
(c) Transverse section through the primitive streak of a stage 4 embryo. Bar = 20 μm.
(d) Enlargement of part of (c) above. Bar = 10 μm.
(e) Transverse section lateral to the primitive streak of a stage 4 embryo. Bar = 15 μm.

1. primitive fold
2. primitive groove
3. mesoderm cells emerging from primitive streak
4. cellular debris in primitive streak
5. bottle-shaped cell in primitive streak
6. endoderm
7. mesoderm cell
8. ectoderm

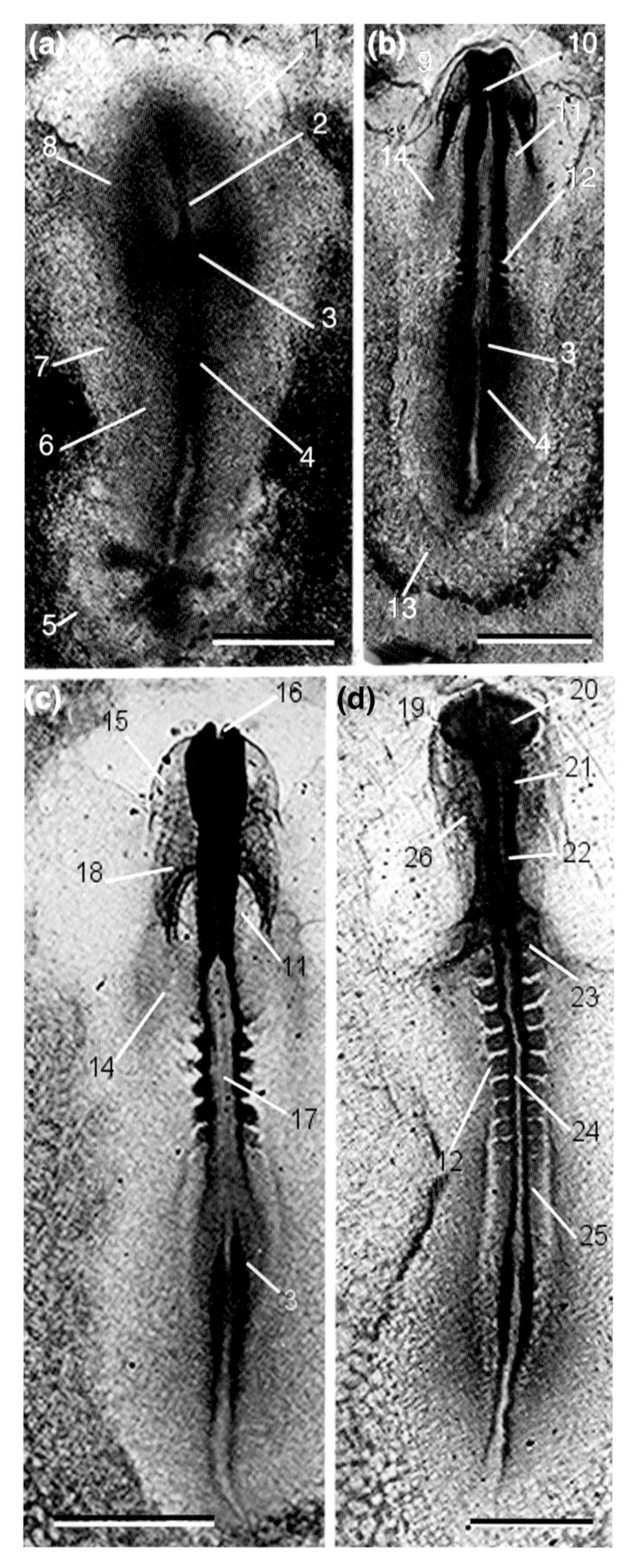
(a)
1
2
8
3
7
6
4
5
(b)
9
10
11
14
12
3
4
13
(c)
16
15
18
11
14
17
3
(d)
20
19
21
26
22
23
24
12
25

Plate 8 Light micrographs of whole-mounted embryos at early stages

(a) Stage 5. Bar = 1 mm.
(b) Stage 8. Bar = 1 mm
(c) Stage 9. Bar = 1 mm.
(d) Stage 10. Bar = 0.5 mm.

1. amniocardiac vesicle
2. head process
3. Hensen's node
4. primitive streak
5. border of area opaca
6. invaginating mesoderm
7. area pellucida
8. border of invaginating anterior mesoderm
9. head fold
10. neural folds fusing to form prosencephalon (forebrain)
11. anterior intestinal portal
12. somite
13. area vasculosa
14. heart-forming area
15. amnion
16. anterior neuropore
17. open neural plate
18. foregut
19. optic vesicle
20. prosencephalon (forebrain)
21. mesencephalon (midbrain)
22. rhombencephalon (hindbrain)
23. opening of foregut into anterior intestinal portal
24. neural tube
25. segmental plate
26. wall of foregut

PLATE 9

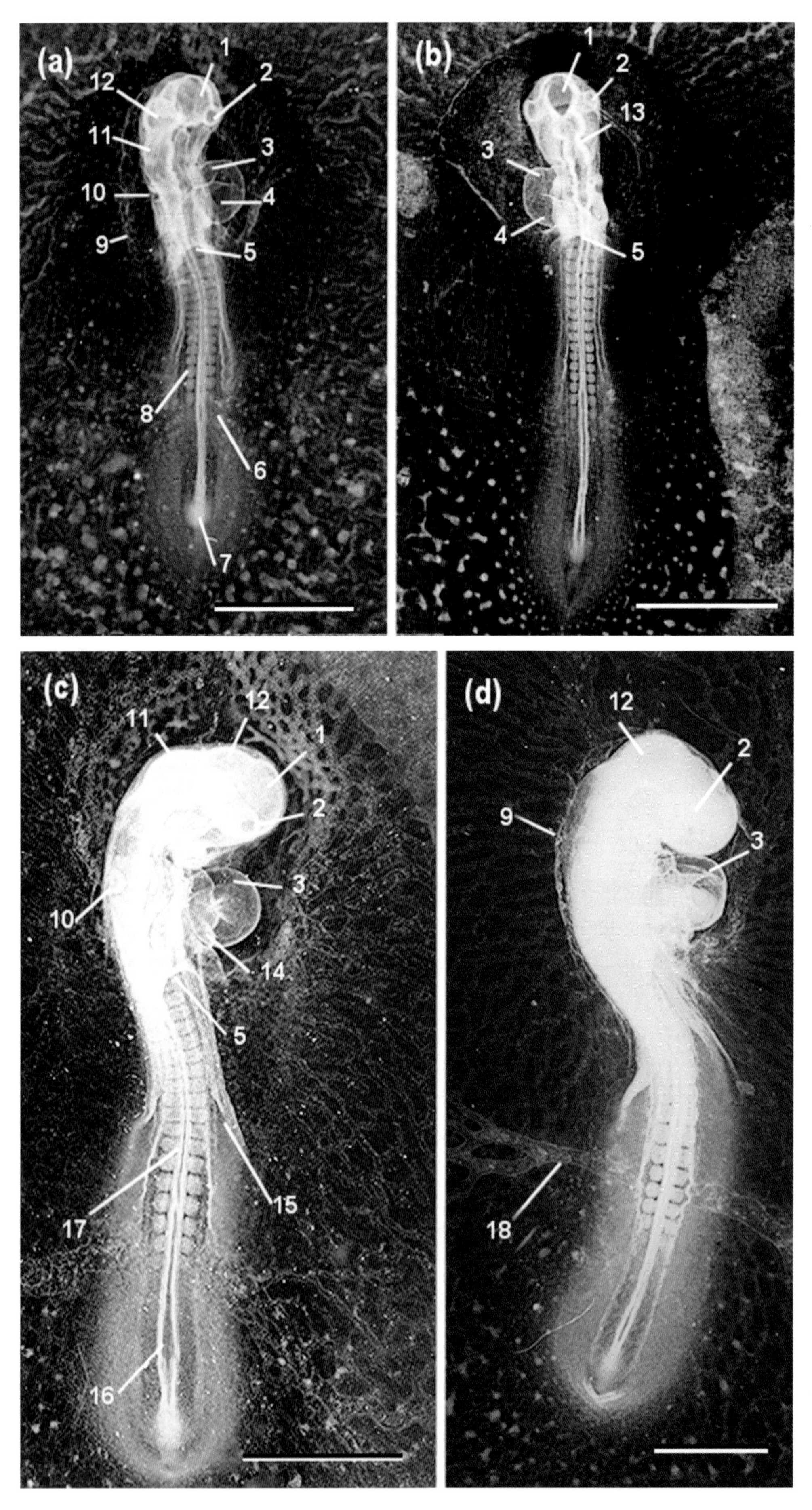

(a)
1
2
12
11
3
10
4
9
5
8
6
7
(b)
1
2
13
3
4
5
(c)
11
12
1
2
3
10
14
5
17
15
16
(d)
12
2
9
3
18

Plate 9 Light micrographs of whole-mounted embryos showing body flexures

(a) Stage 13, illustrating the beginning of the cranial flexure as seen from the dorsal side. The head is turned to the right. Bar = 1 mm.
(b) Stage 13, a different embryo. Seen from the ventral side. Bar = 1 mm.
(c) Stage 14, illustrating cervical flexure. The head now lies at right angles to the trunk. Bar = 1 mm.
(d) Stage 15, showing an increased cervical flexure. Bar = 1 mm.

1. prosencephalon (forebrain)
2. optic vesicle
3. atrium
4. ventricle
5. anterior intestinal portal
6. segmental plate mesoderm
7. remnant of Hensen's node (tail bud)
8. somites
9. amnion
10. otic vesicle
11. metencephalon (cerebellum)
12. mesencephalon
13. myelencephalon (medulla oblongata)
14. sinus venosus
15. lateral body fold
16. posterior neuropore
17. neural tube
18. vitelline (omphalo-mesenteric) blood vessels

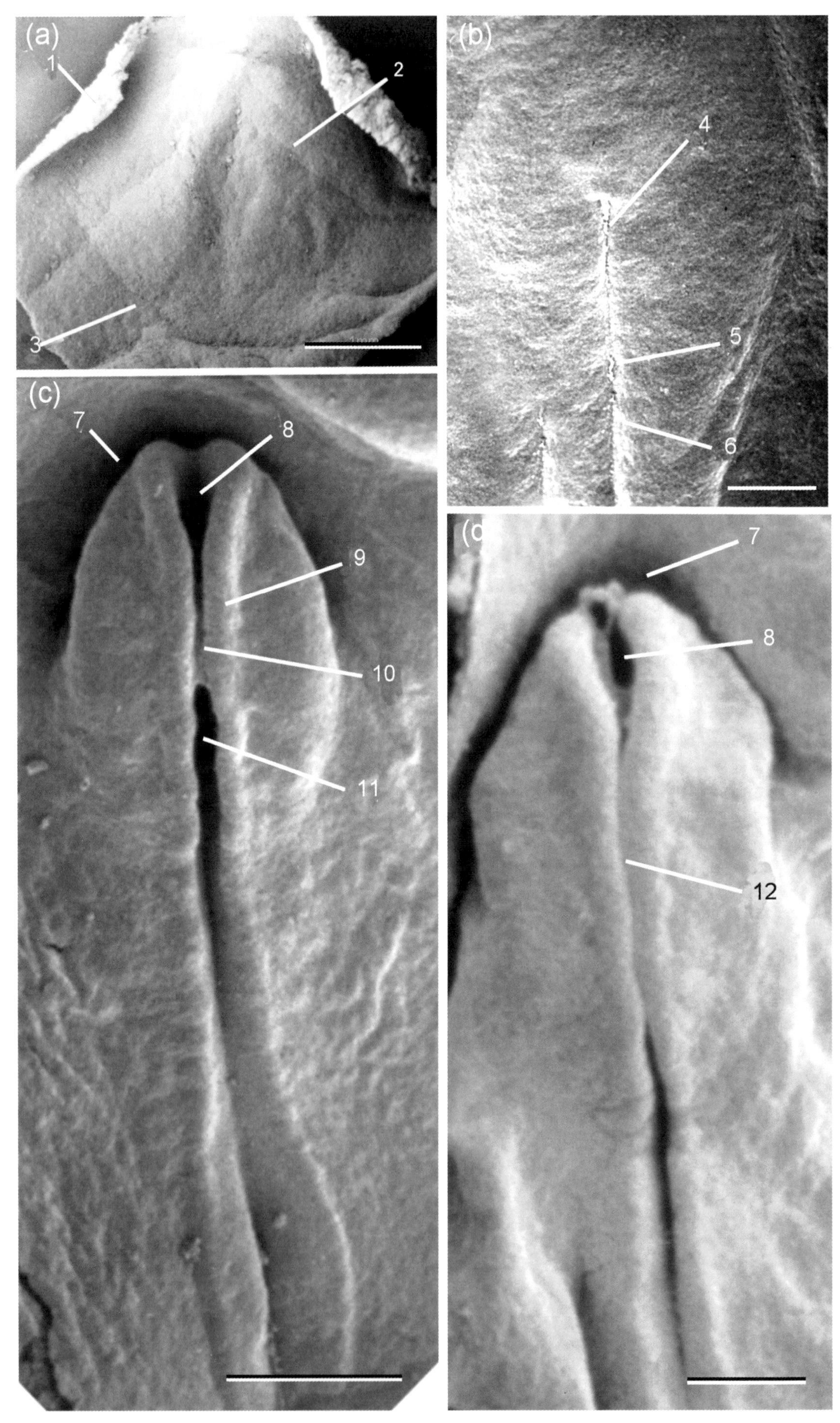
(a)
1
2
3
(b)
4
5
6
(c)
7
8
9
10
11
(d)
7
8
12

Plate 10 Scanning electron micrographs of early embryos from the dorsal side

(a) Stage XII (see Appendix I). This embryo is about 6 h of incubation. The region of Koller's sickle is not visible. The rolled back edge of the area opaca shows the yolky cells of the ventral side. The fold in the centre is artefactual. Bar = 0.6 mm.
(b) Stage 5. This embryo is after about 20 h of incubation and ingression has begun. The area pellucida is now pear-shaped and bisected by the primitive streak. The area opaca is not shown. Bar = 350 μm.
(c) Stage 8. This embryo is after about 26 h of incubation. The neural folds have almost met in the midbrain region and appear to have begun to fuse. The head fold extends beneath the forebrain. Bar = 300 μm.
(d) Stage 8. The embryo is after about 30 h of incubation and the undercutting of the head by the head fold has become more pronounced. The neural folds have now met and fused over a longer distance but the brain remains slightly open at the anterior neuropore. Bar = 110 μm.

1. endodermal side of area opaca
2. dosal side of area pellucida
3. border between area opaca and area pellucida
4. head process
5. Hensen's node
6. primitive streak
7. head fold
8. anterior neuropore
9. neural fold
10. neural folds at region of first fusion
11. unfused neural folds
12. fused neural folds

(b) Reprinted from Bancroft and Bellairs (1974), with permission of Springer-Verlag.
(c) Reprinted from Bancroft and Bellairs (1975), with permission of Springer-Verlag.

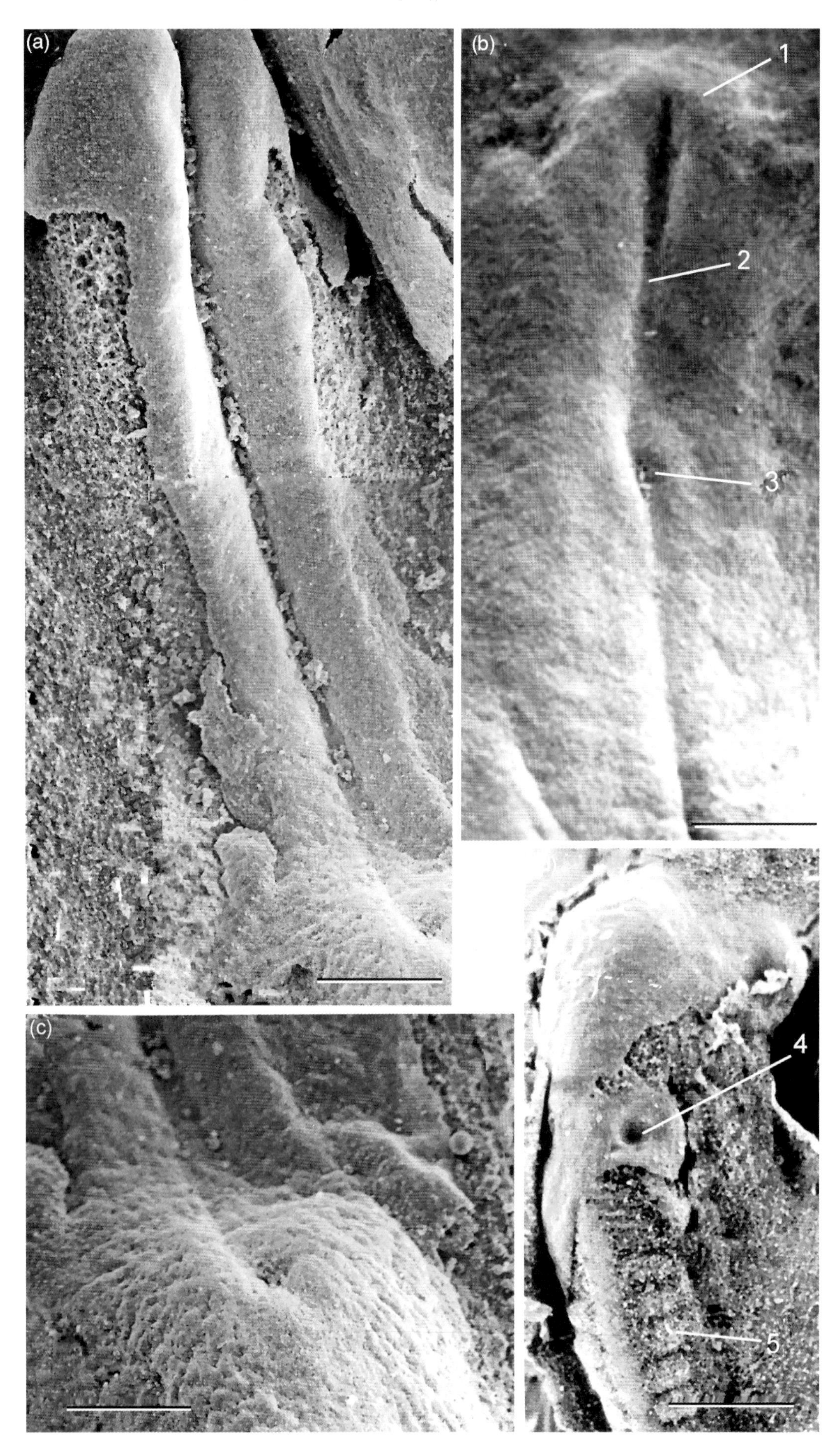
(a)
(b)
1
2
3
(c)
4
5

Plate 11 Scanning electron micrographs showing dorsal views of head region

(a, b) Embryos at stage 6, about 24 h of incubation. The head fold has begun to develop and the anterior end of the head process and the neural folds are now visible. Hensen's node has retreated some distance along the primitive streak. No somites have yet formed. In (a) the ectoderm has been dissected from the regions lateral to the neural folds. Bar in (a) = 300 μm; bar in (b) = 500 μm.

(c) An enlargement of Hensen's node shown from the lower right corner of (a). Bar = 110 μm.

(d) Stage 13+. This embryo is after about 50 h of incubation. The ectoderm has been dissected from the anterior end of the trunk (posterior end not shown) revealing six pairs of somites. The head has turned to the right and is partially submerged. Bar = 200 μm.

1. head fold
2. head process
3. Hensen's node
4. otic pit (see *Plate 22*)
5. somites

Plate 12

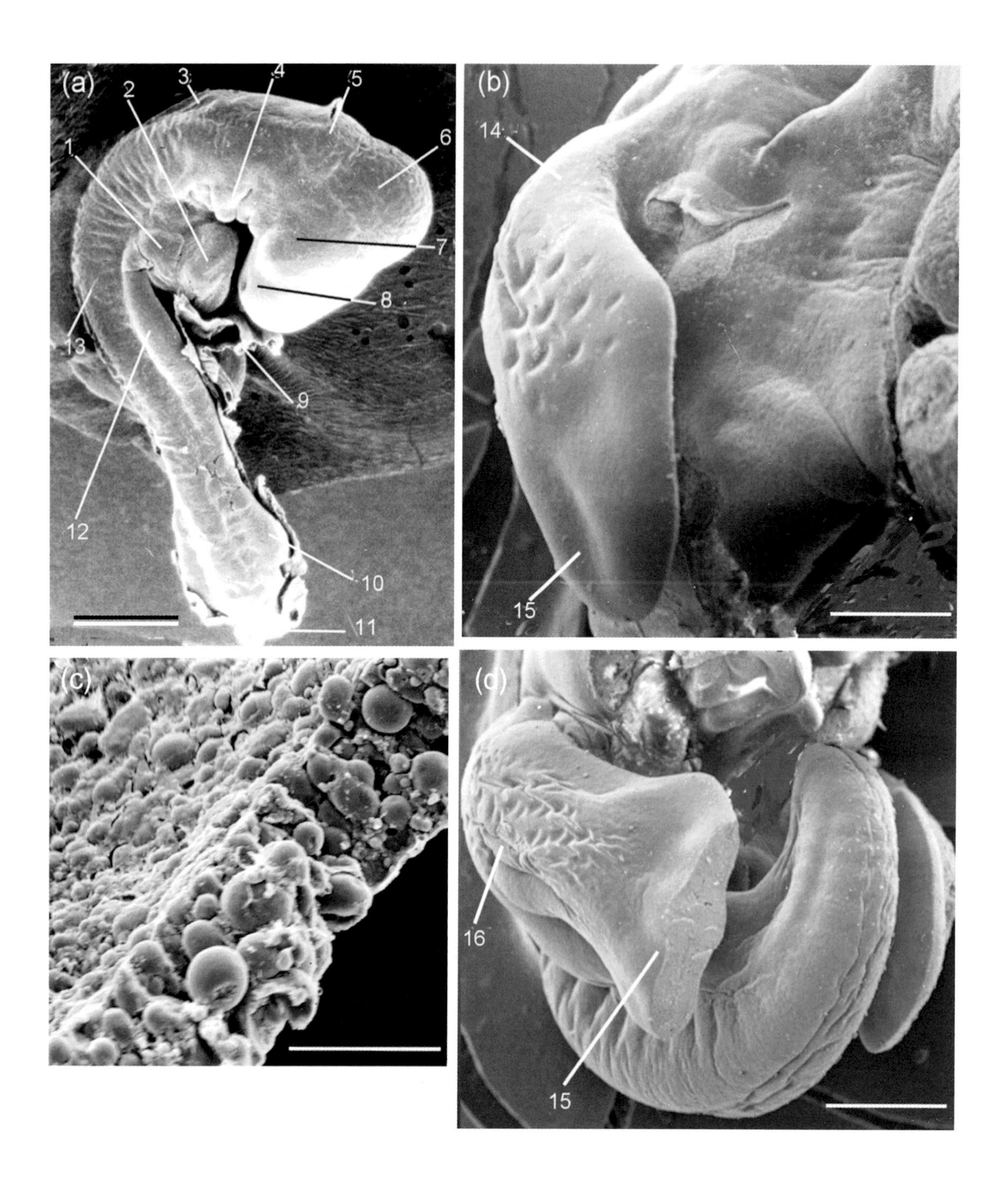

Plate 12 Scanning electron micrographs showing body flexures and area opaca

(a) Stage 18, about 68 h of incubation. The embryo lies on its left side on top of the yolk and has undergone both cranial and cervical flexure. The amnion has been dissected away from over the embryo but remnants are visible. Bar = 1 mm.
(b) Stage 27–28, 5.5 days of incubation. The right wing bud is seen from the ventro-lateral aspect showing the interdigital grooves. Bar = 250 μm.
(c) Area opaca from a stage 7 embryo, about 24 h. The specimen has been broken across the tissue to show the large yolky cells, the dorsal (ectodermal) surface being in the upper left of the figure. Bar = 150 μm.
(d) Stage 27–28, 5.5 days. Hindlimb buds and tail bud of specimen shown in (b) above. Interdigital grooves are visible. The distal edge of the right leg bud has been dissected to show the underlying mesoderm. Bar = 250 μm.

1. atrium
2. ventricle
3. myelencephalon
4. hyoid arch
5. metencephalon
6. mesencephalon
7. eye
8. nasal pit
9. remnant of amnion
10. right leg bud
11. tail bud
12. right forelimb bud
13. somites
14. right elbow
15. interdigital groove
16. right knee

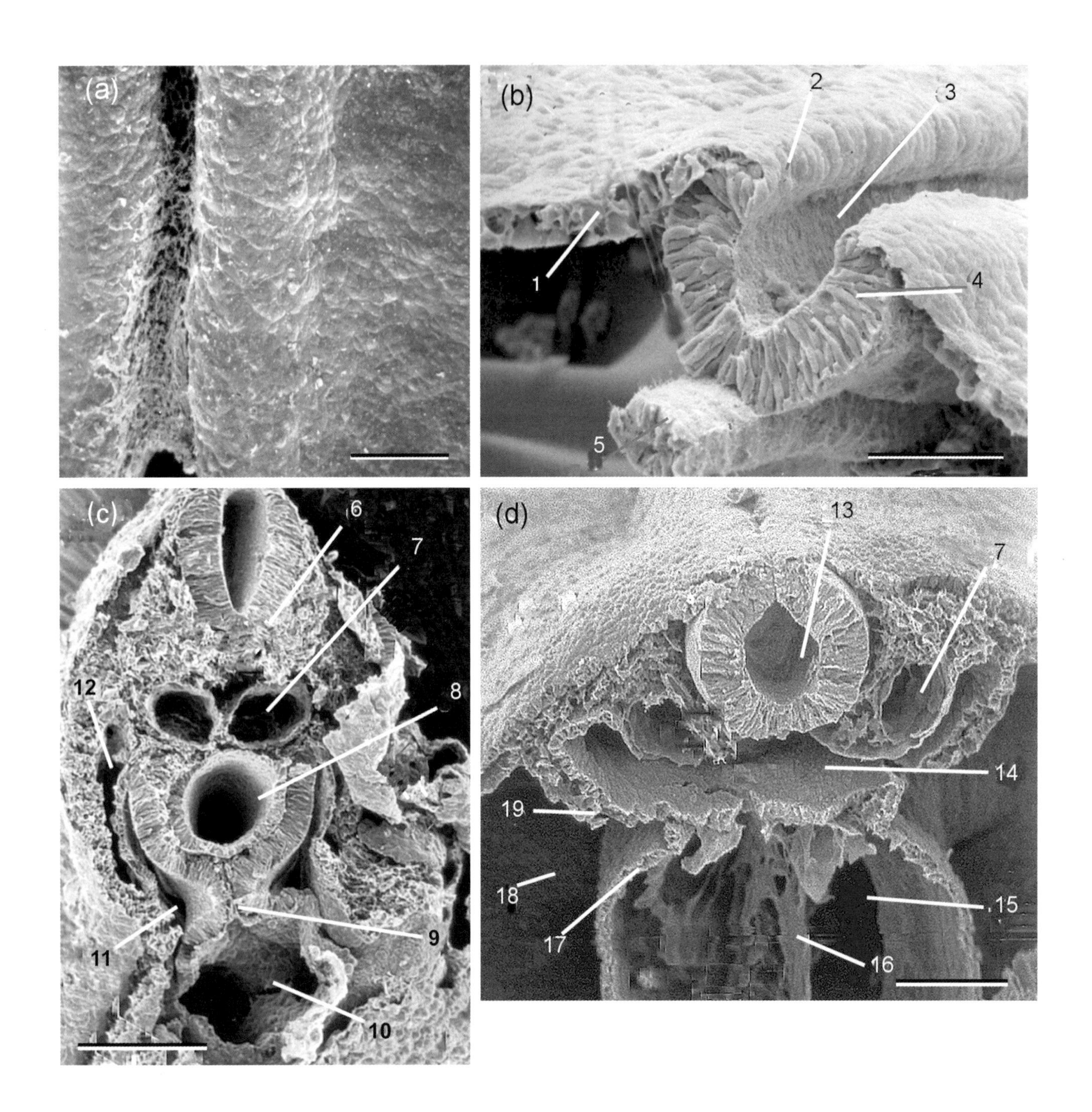
(a)
(b)
1
2
3
4
5
(c)
6
7
8
9
10
11
12
(d)
13
7
14
15
16
17
18
19

Plate 13 Scanning electron micrographs of neural folds and neural tube

(a) Neural folds seen from the dorsal side. Stage 8, about 28 h incubation. Bar = 100 μm.
(b) Transverse section across the trunk region to show the open neural plate. Stage 9 embryo, about 30 h of incubation. Bar = 50 μm.
(c) Transverse section across the anterior trunk. Stage 14, about 52 h of incubation. Bar = 100 μm.
(d) Transverse section at the level of the heart. Stage 10, about 35 h of incubation. Bar = 100 μm.

1. ectoderm
2. neural fold
3. neural groove
4. neural plate
5. notochord
6. neural tube
7. left dorsal aorta
8. gut
9. ventral mesentery
10. omphalo-mesenteric vein
11. coelom
12. anterior cardinal vein
13. lumen of neural tube
14. pharynx
15. left atrium
16. interatrial septum
17. wall of atrium
18. pericardial cavity
19. splanchnic mesoderm

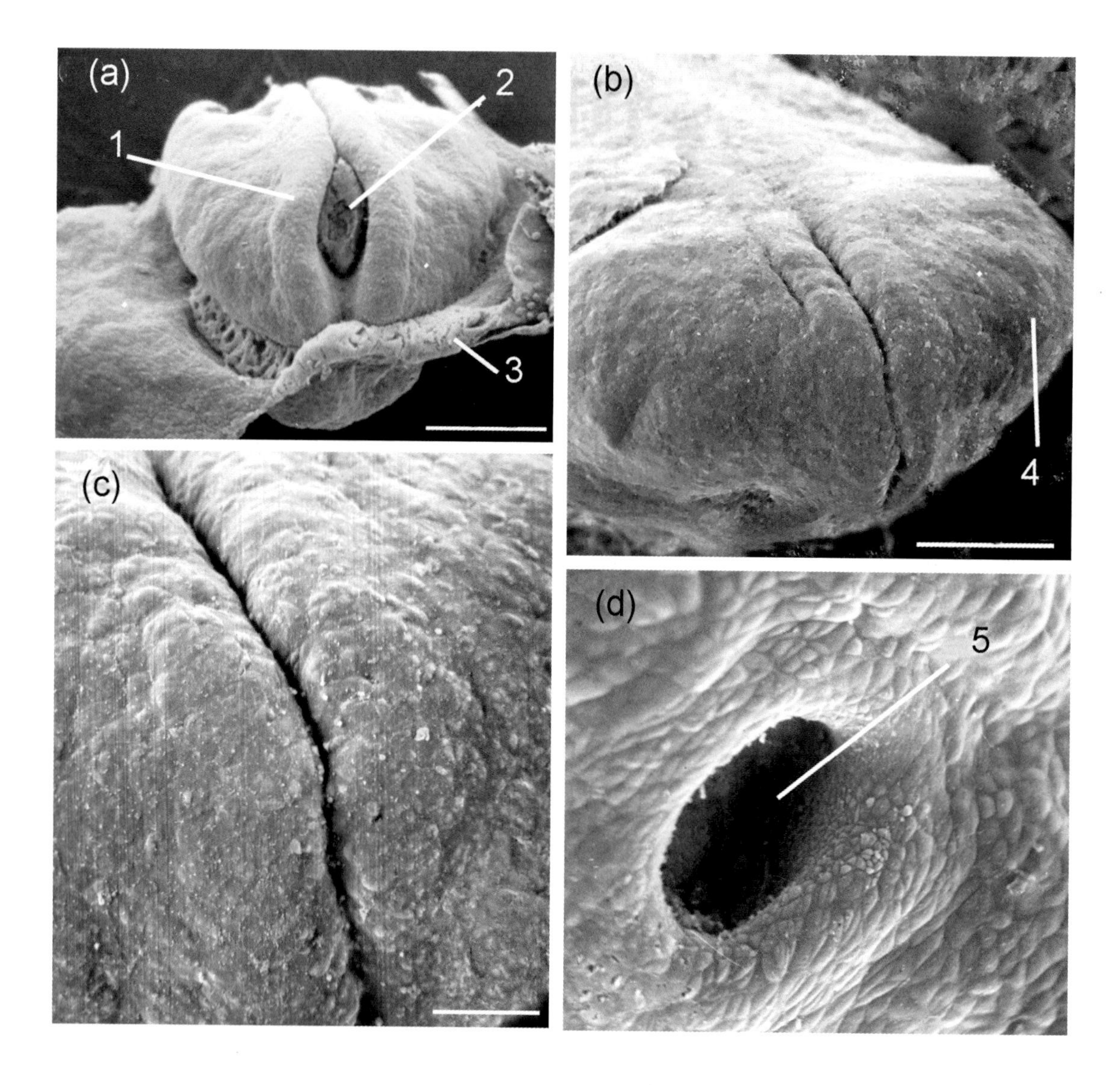

Plate 14 Scanning electron micrographs showing closure of the anterior neuropore

(a) Open anterior neuropore at the tip of the forebrain. Stage 8 embryo, about 28 h of incubation. Bar = 1 mm.
(b) Dorsal side of the forebrain to show the neural folds almost closed. Stage 10 embryo, about 34 h of incubation. Bar = 200 μm.
(c) Enlargement of part of (b). Bar = 50 μm.
(d) Final point of closure of the anterior neuropore, much enlarged, of a stage 10 embryo, 35 h of incubation.

1. neural fold
2. anterior neuropore
3. amniotic ectoderm
4. ectoderm covering optic vesicle
5. final remnant of anterior neuropore

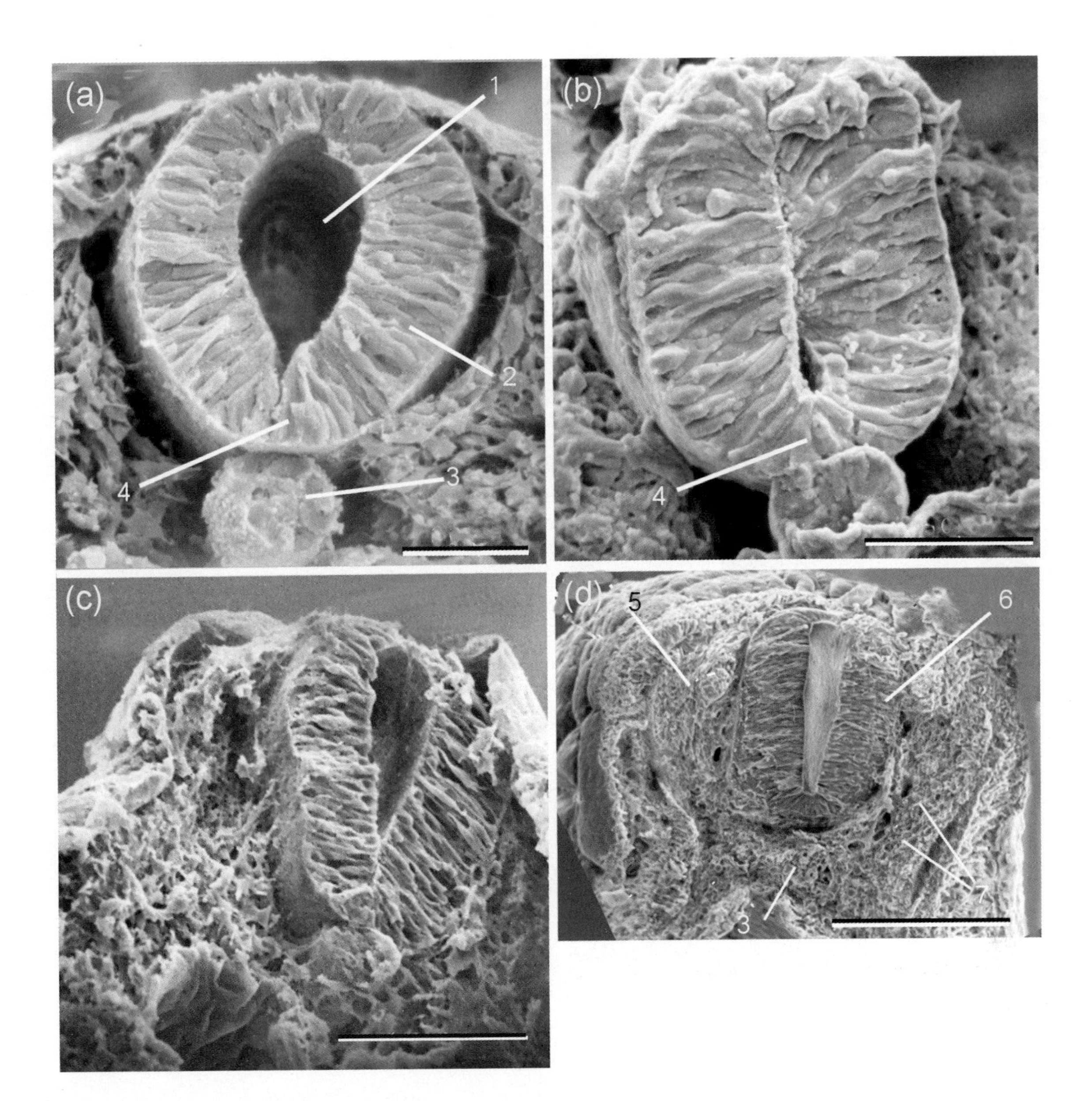

Plate 15 Scanning electron micrographs of the neural tubes of four embryos in the anterior trunk region

(a) Transverse section across the neural tube of a stage 10 embryo. The lumen is widely open. Bar = 30 μm.

(b) Transverse section across the neural tube of a stage 12 embryo. The lumen is almost closed and the lateral walls have become thicker, whilst the basal plate is more distinct. Bar = 50 μm.

(c) Transverse section across the neural tube of a stage 14 embryo. Bar = 30 μm.

(d) Transverse section across the neural tube of a stage 15 embryo showing the spread of the mesoderm around the notochord. Bar = 25 μm.

1. lumen of neural tube
2. neural tube
3. notochord
4. basal plate
5. sclerotome
6. wall of neural tube, now thicker
7. blood vessels

Plate 16

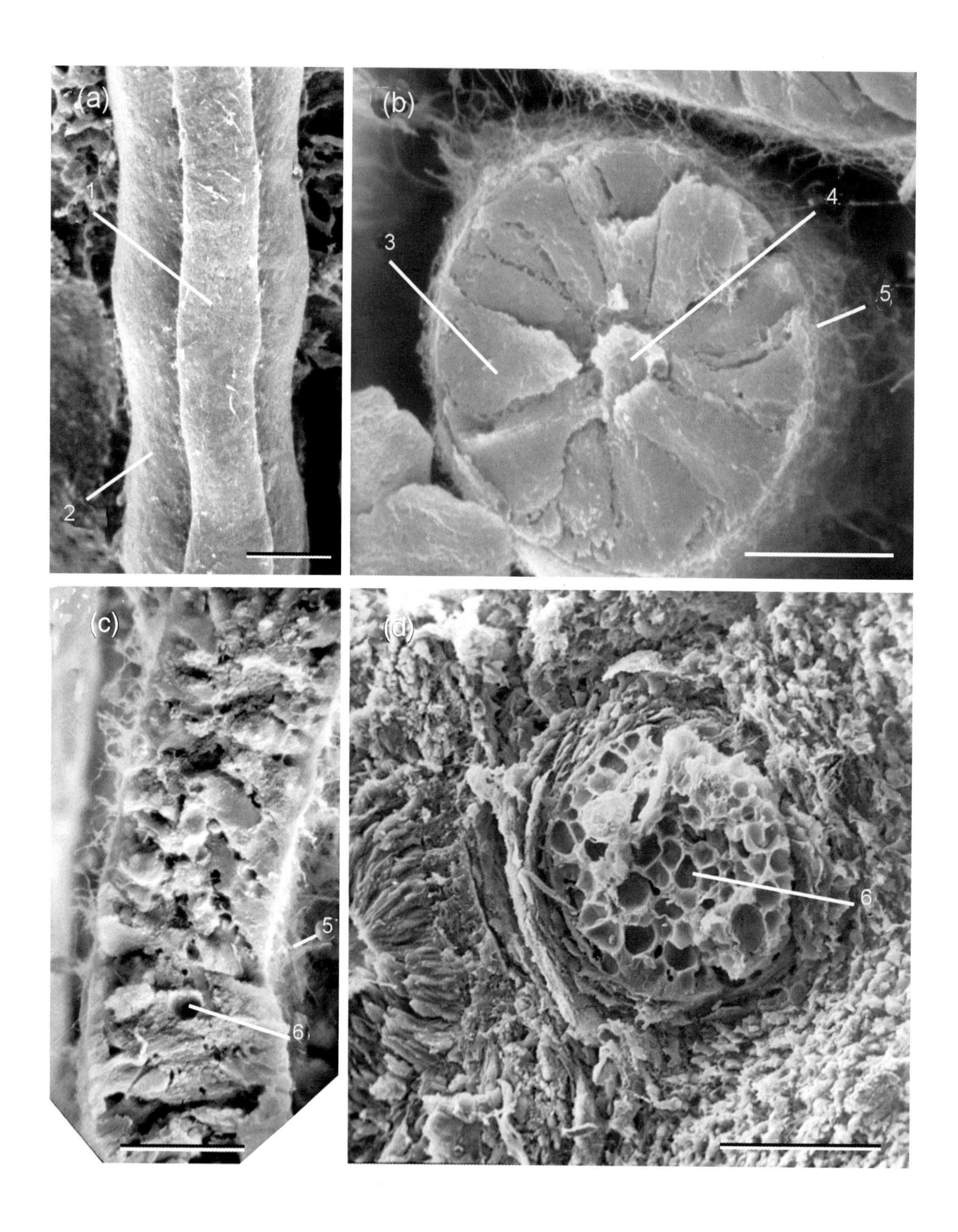

Plate 16 Scanning electron micrographs of notochord

(a) Ventral view of the notochord of a stage 11 embryo (about 44 h incubation) after removal of endoderm and somites. Bar = 30 μm.
(b) Transverse section across the notochord of a stage 6 embryo (about 24 h incubation). Each cell is wedge-shaped with its widest (basal) end peripherally, giving a radial arrangement to the tissue. The apical ends of the cells surround a small lumen containing cells. Bar = 5 μm.
(c) Longitudinally broken section through the notochord of a stage 10 embryo. The extracellular materials are apparent on both dorsal and ventral surfaces. Bar = 10 μm.
(d) Transverse section across the notochord of a stage 23 embryo. Each cell is highly vacuolated. Bar = 7.5 μm.

1. notochord
2. neural tube
3. wedge-shaped cell
4. lumen with cells
5. extracellular material
6. vacuolated cell

(b, d) Reprinted from Bancroft and Bellairs (1976), with permission of the Company of Biologists Ltd.

PLATE 17

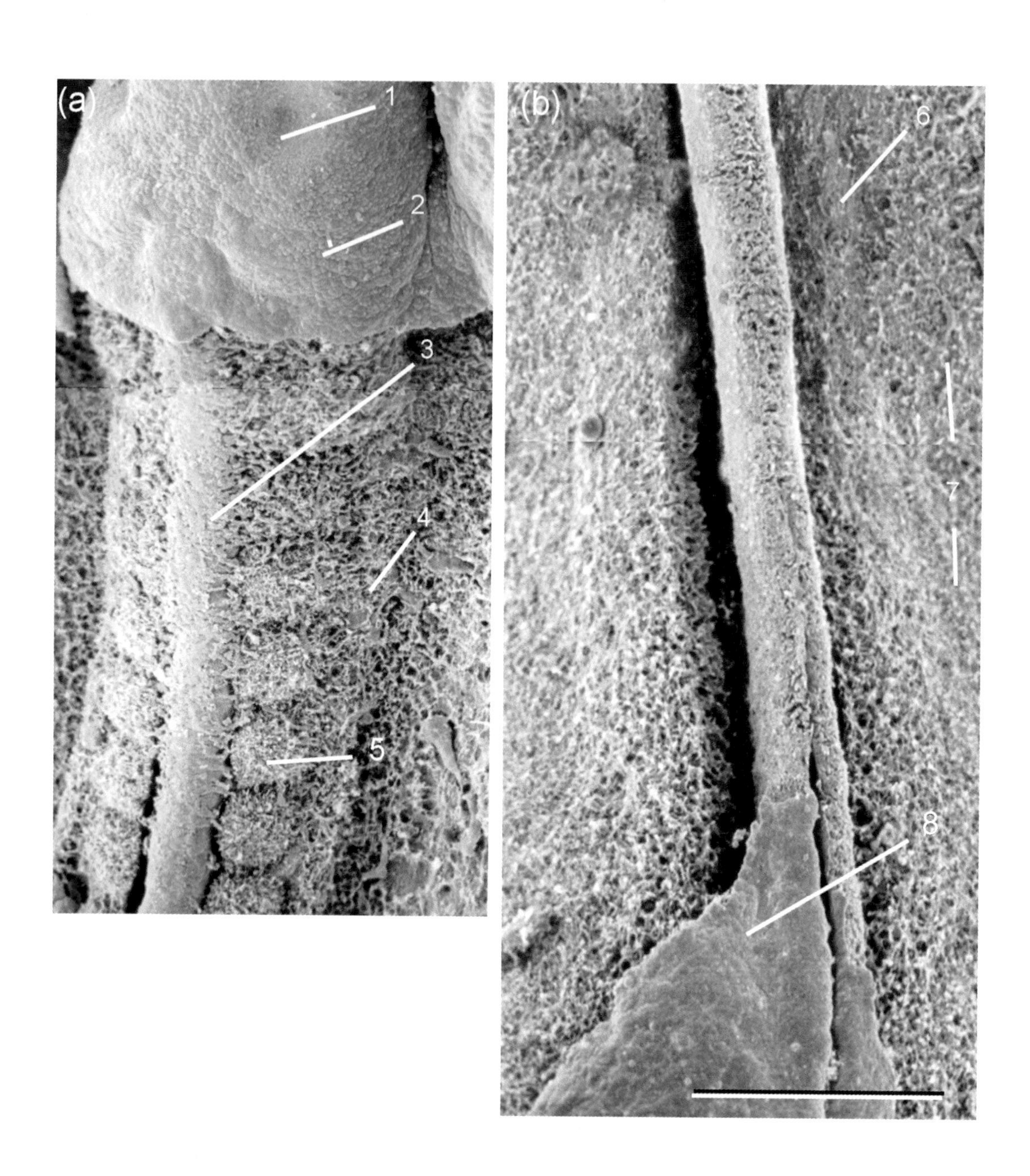

Plate 17 Scanning electron micrographs of segmented and unsegmented mesoderm

(a) Anterior end of a stage 11 embryo after removal of much of the ectoderm.
(b) Posterior end of the same embryo at the same magnification. Bar = 200 μm.

1. otic placode (see also *Plate 22*)
2. ectoderm covering head
3. neural tube
4. intermediate mesoderm
5. somite
6. segmental plate mesoderm
7. lateral plate mesoderm
8. ectoderm

Plate 18 Scanning electron micrographs of somites and nephric duct

(a) Somite from a stage 11 embryo, broken open in a sagittal plane, lying against the notochord (left) and the neural tube (right). The walls of the somite are composed of a single layer of epithelial cells surrounding a central cavity, the somitocoele. The adjacent anterior and posterior somites have been removed. Bar = 75 μm.

(b) Enlargement of central region of (a) showing the cells in the lumen and the apical borders of the cells bordering it. Bar = 10 μm.

(c) Somite, nephric duct and lateral plate of a stage 14 embryo seen from the dorsal side after removal of the ectoderm and neural tube. Bar = 100 μm.

(d) Nephric duct of a stage 14 embryo, showing the fine fibrils and filopodia passing over it between the adjacent somite and lateral plate. Bar = 50 μm.

1. notochord
2. intersegmental angioblastic cords which will contribute to the right cardinal vein
3. right nephric duct
4. left nephric duct
5. enlargement of a nephric duct to show filopodia

(a, b) Reprinted from Bellairs (1979), by permission of the Company of Biologists Ltd.

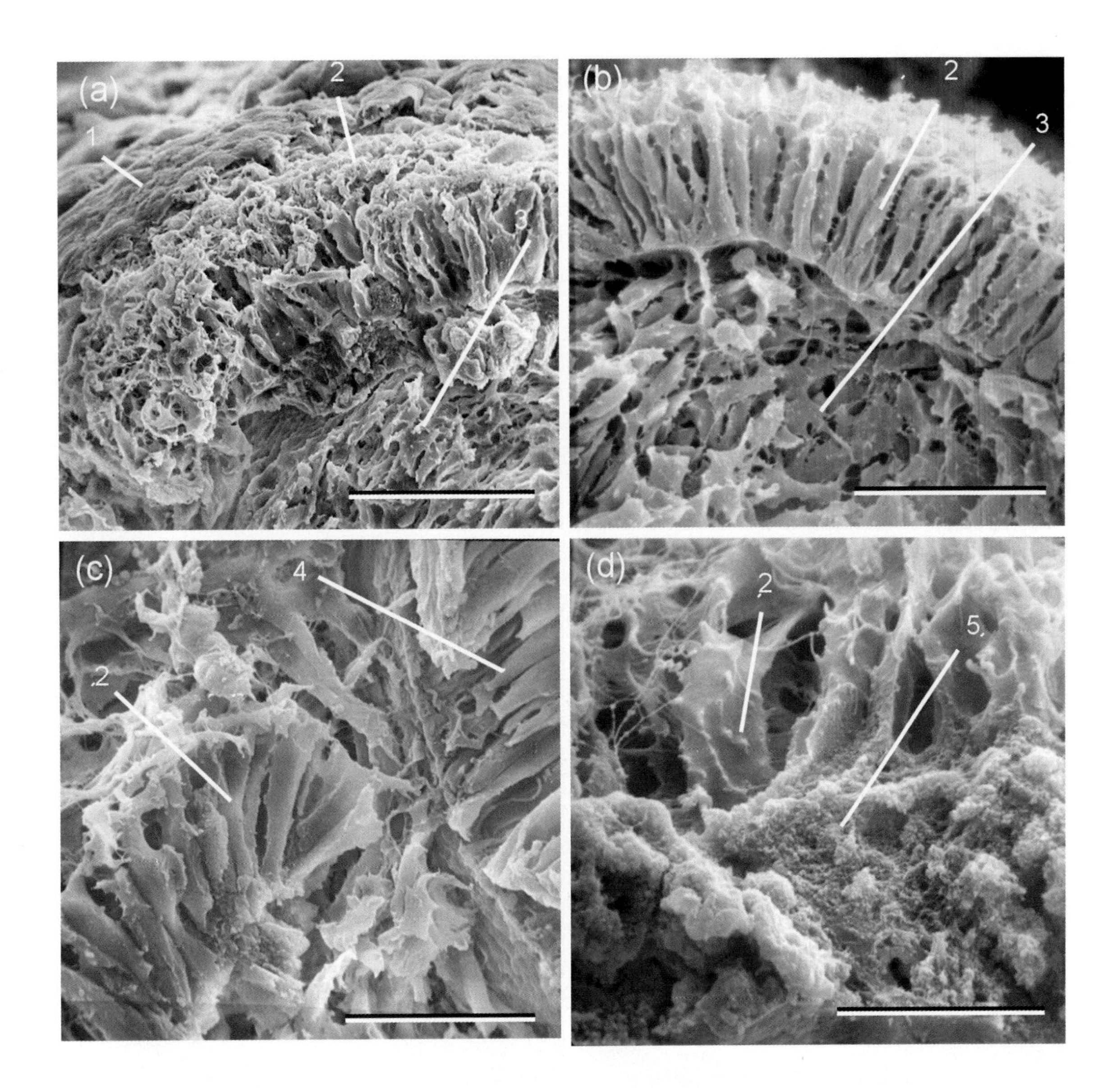

Plate 19 Scanning electron micrographs of the derma-myotome

(a) Part of the stage 15 specimen shown in *Plate 15*. The ectoderm has been partially removed from over the dermatome. Bar = 28 μm.

(b) Section through the differentiating somite of a stage 14 embryo showing the simple columnar epithelium of the dermatome and the mesenchymatous cells of the sclerotome. Bar = 24 μm.

(c) Dermatome of a stage 16 embryo to show its relationship to the neural tube. Bar = 40 μm.

(d) Dermatome and myotome at stage 17, showing the many contacts between the dermatome cells. Bar = 10 μm.

1. ectoderm
2. dermatome
3. sclerotome
4. neural tube
5. fractured myotome cells

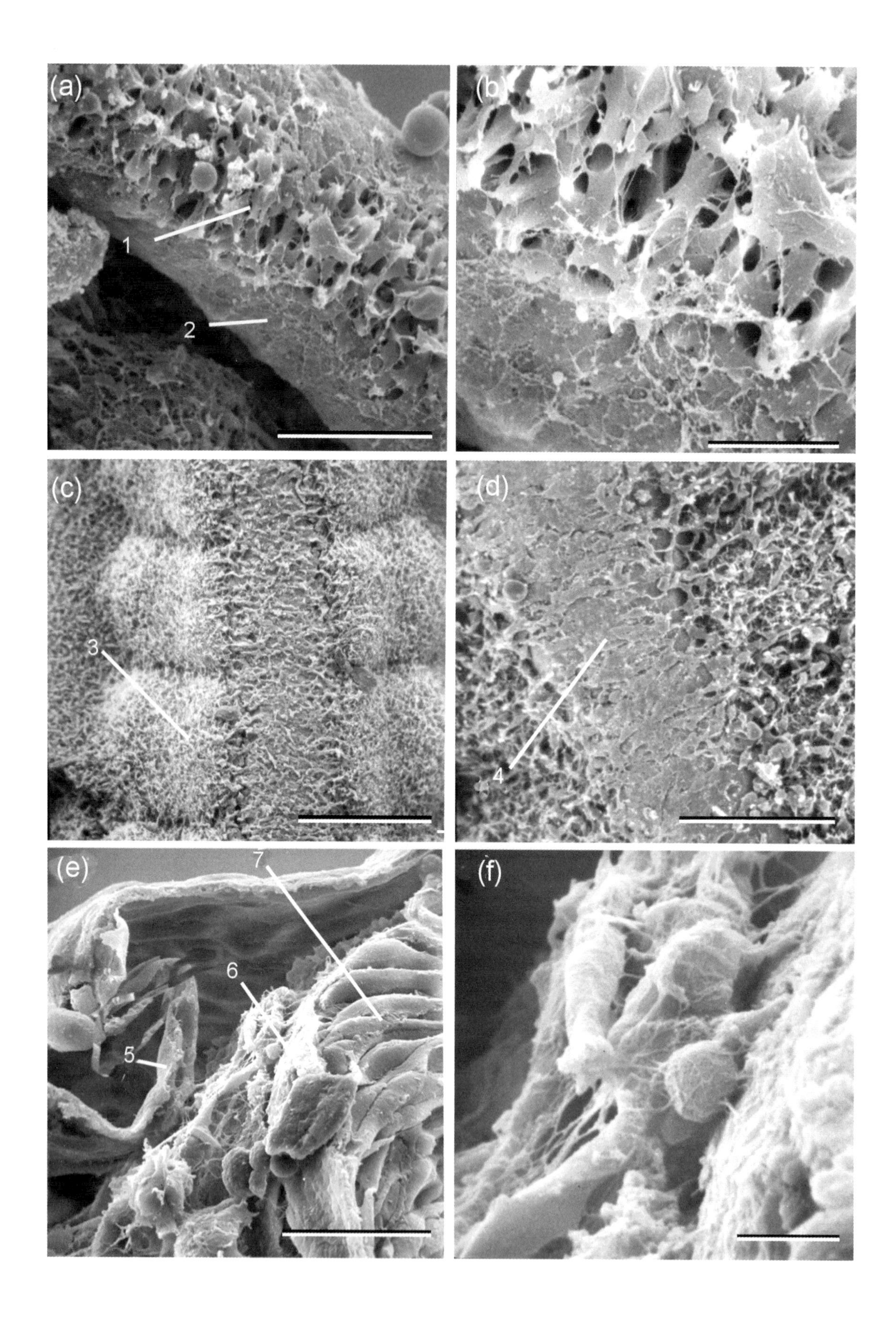
(a)
1
2
(b)
(c)
3
(d)
4
(e)
7
6
5
(f)

Plate 20 Scanning electron micrographs showing neural crest I

(a–d) After removal of ectoderm.

(a) Surface view of the neural crest in the unsegmented region of a stage 11 embryo (about 43 h) after removal of overlying ectoderm. Bar = 30 μm.

(b) Enlargement of part of (a). The cells are not yet flattened and the intercellular spaces are conspicuous. Note the filament-like strands of extracellular materials. Bar = 10 μm.

(c) Surface view of the neural crest overlying the dorsal side of the neural tube and somites at the level of the 13th somite of a stage 14 embryo. Bar = 100 μm.

(d) Surface view of the neural crest at a level anterior to (c). The cells are flatter than in (c). Bar = 50 μm.

(e) Transverse section across the edge of a neural tube showing the migrating neural crest cells of a stage 14 embryo. Bar = 3 μm.

(f) Enlargement of part of (e). Group of neural crest cells caught in the act of migration down the lateral edge of the neural tube. Bar = 1.5 μm.

1. neural crest cell
2. neural tube
3. somite
4. flattened neural crest cells on the surface of the neural tube
5. ectoderm
6. neural crest cell fixed in transit
7. neural tube cells.

(a, b) Reprinted from Bancroft and Bellairs (1976), with permission of Blackwell Publishing, Ltd.

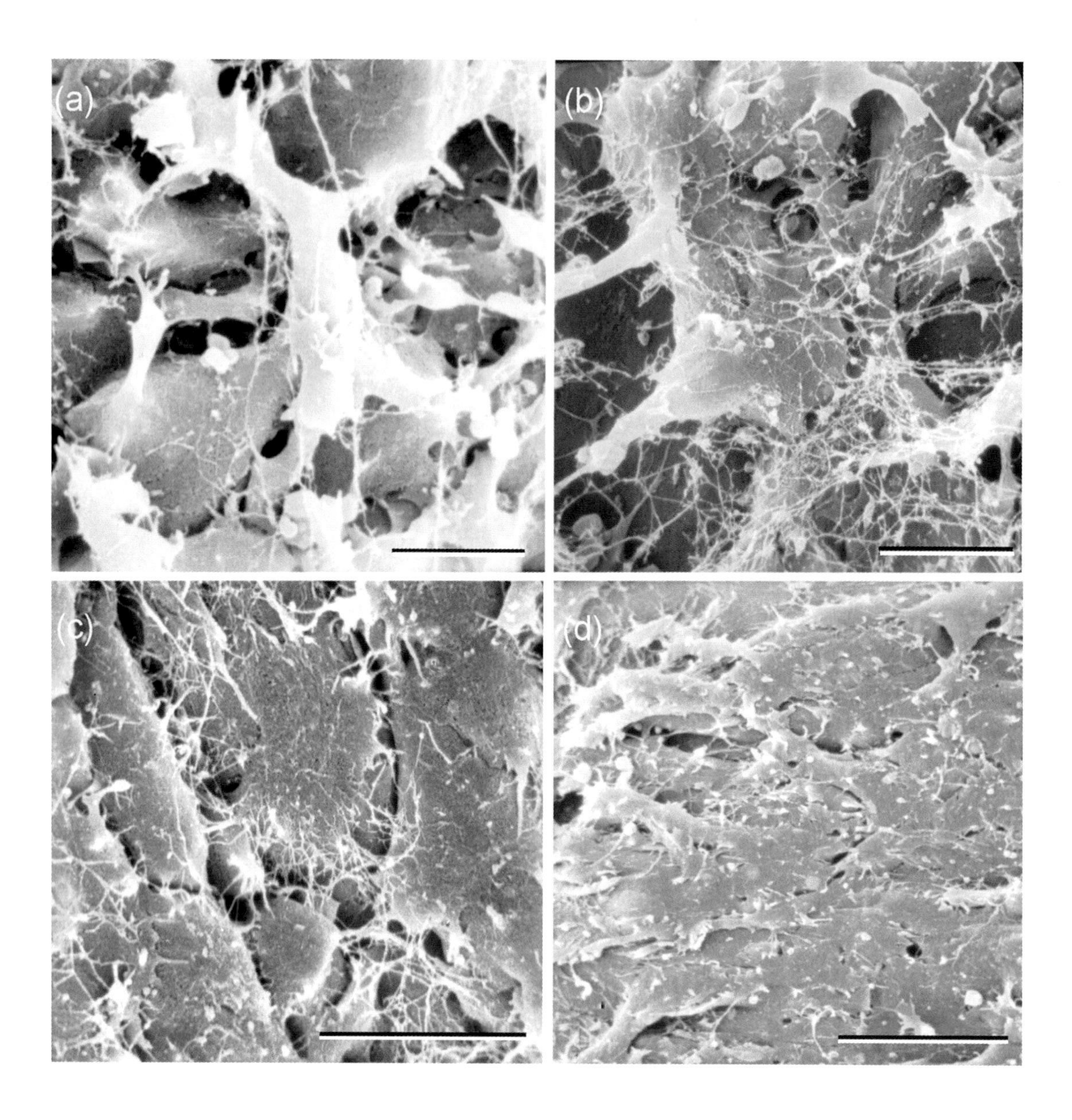

Plate 21 Scanning electron micrographs showing neural crest II

(a–c) Dorsal views of neural crest cells all taken from the same stage 14 embryo after removal of the ectoderm. They illustrate stages in the maturation of the neural crest cells. In each case, the anterior end of the embryo is at the top of the photograph.

(a) The least mature neural crest, from the posterior end of the embryo. Bar = 20 μm.

(b) Neural crest from the middle of the trunk at the level of the 13th somite. Bar = 20 μm.

(c) The most mature neural crest from the anterior end of the trunk at the level of the first somite. With increasing maturity the cells become flattened onto the surface of the neural tube and reorientated from an antero-posterior direction to a medio-lateral one. Bar = 15 μm.

(d) Fully flattened and reorientated neural crest from the level of the first somite of a different stage 14 embryo. Bar = 10 μm.

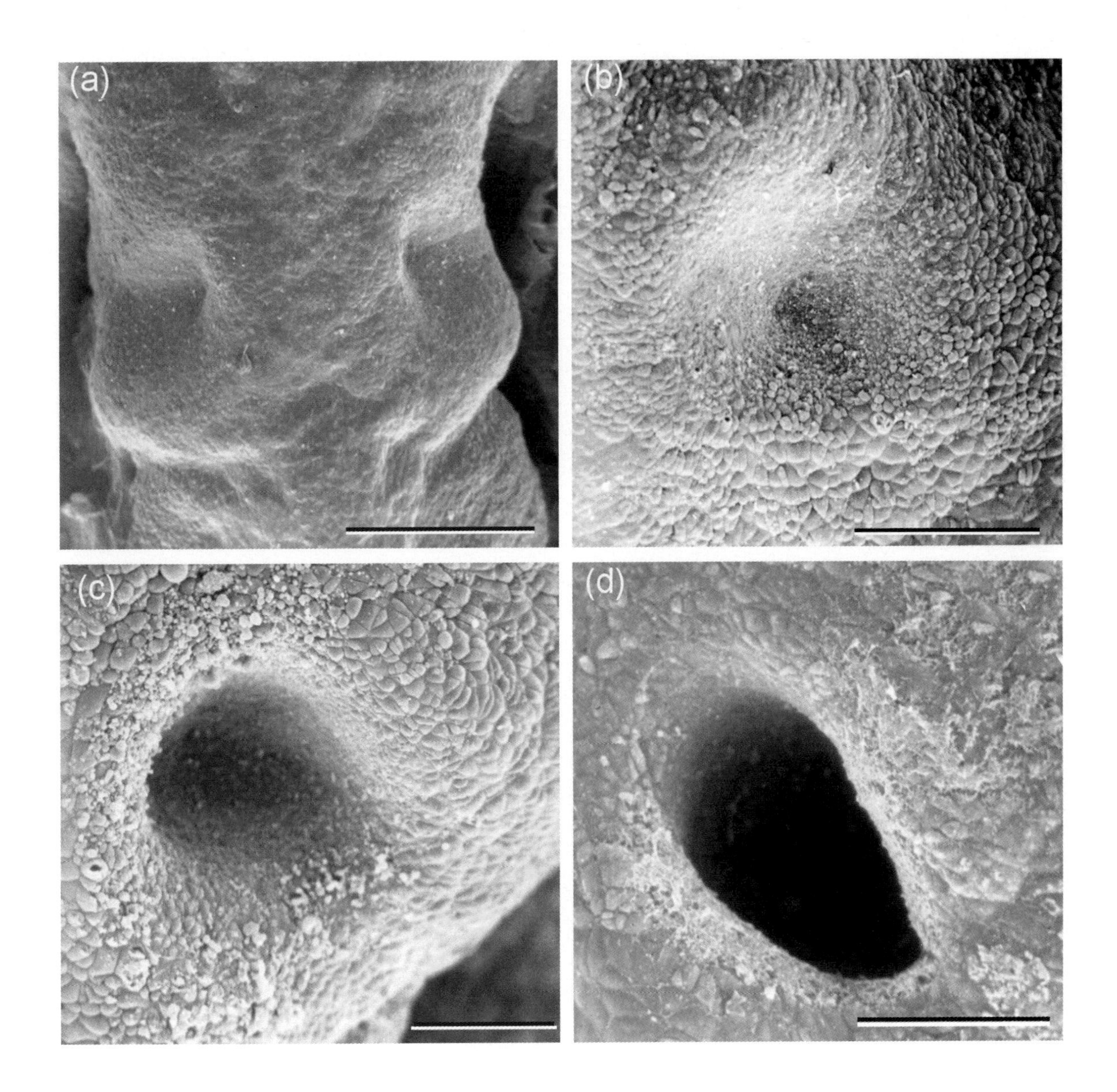

Plate 22 Scanning electron micrographs of otic placodes and pits

(a) Otic placodes of a stage 11 embryo, seen from the dorsal side (see also *Plate 17a*). Bar = 140 μm.
(b) Otic placode at stage 12. Bar = 30 μm.
(c) Otic pit at stage 13 (see also *Plate 11d*). Bar = 30 μm.
(d) Otic pit at stage 14. Bar = 50 μm.

All figures reprinted from Bancroft and Bellairs (1977), with permission of Springer-Verlag.

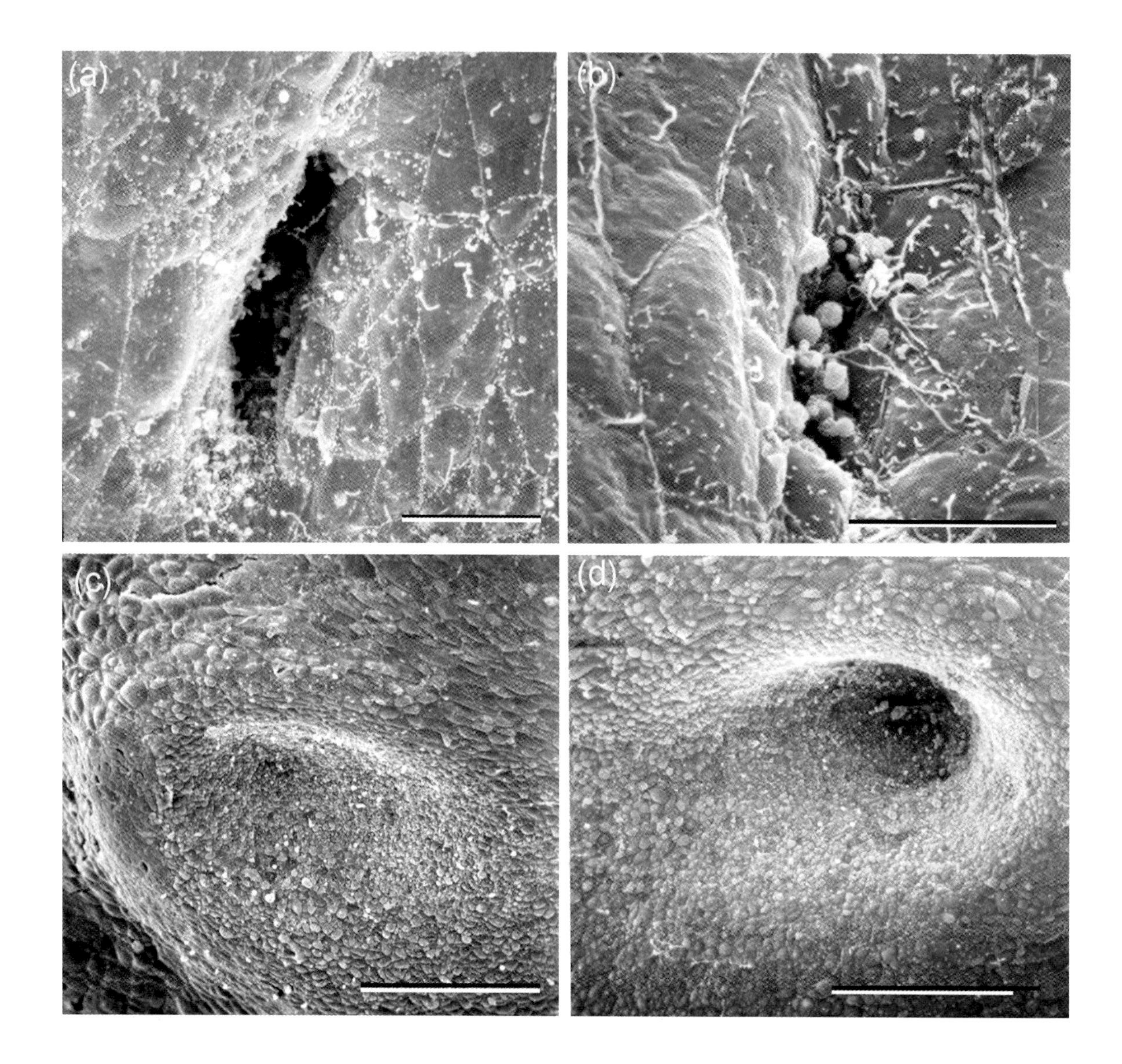

Plate 23 Scanning electron micrographs of otic pits and nasal placodes

(a) Otic pit at stage 16. Bar = 20 μm.
(b) Otic pit at stage 17. Bar = 10 μm.
(c) Nasal placode at stage 16. Bar = 85 μm.
(d) Nasal pit at stage 18. Bar = 85 μm.

All figures reprinted from Bancroft and Bellairs (1977), with permission of Springer-Verlag.

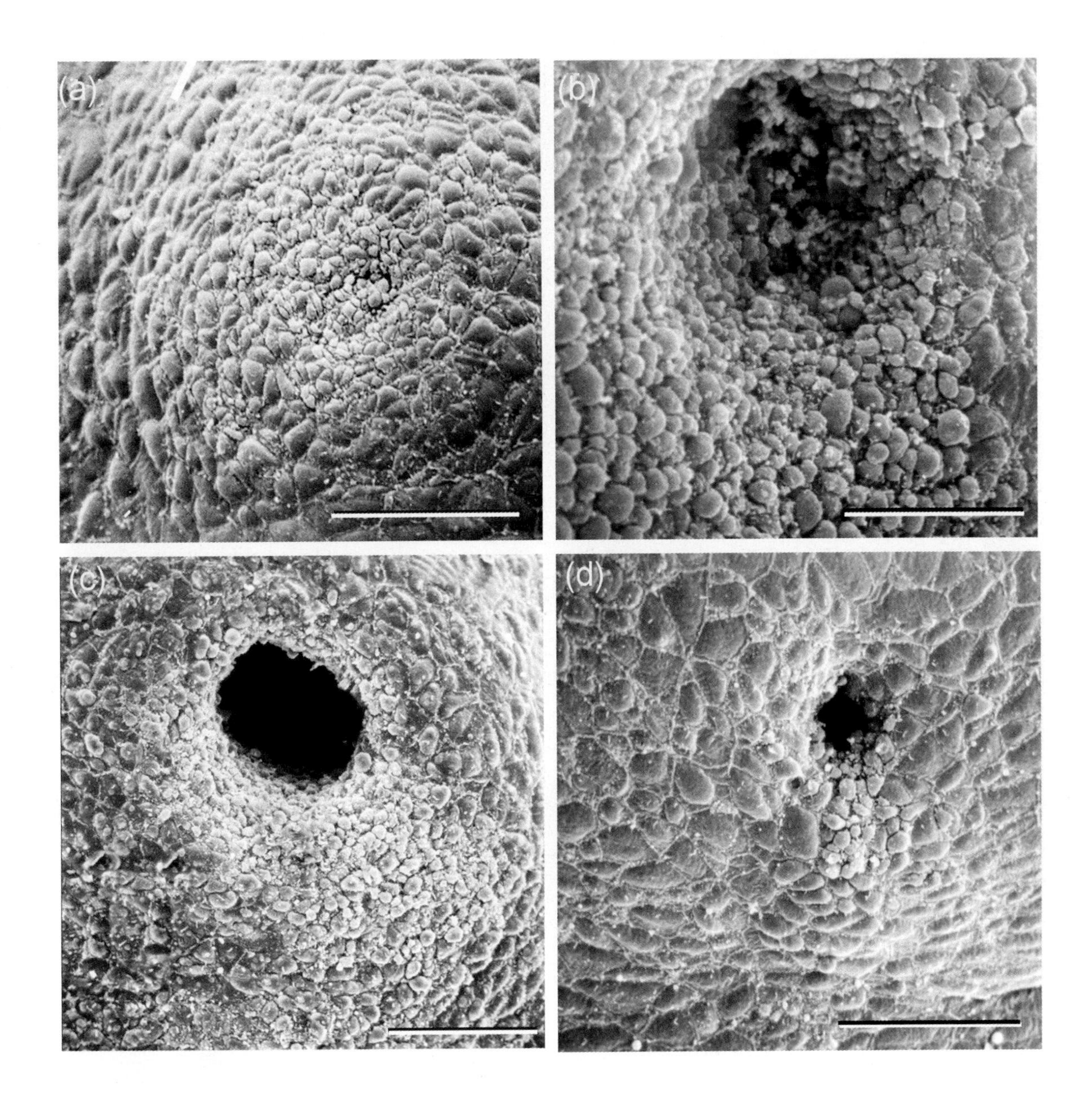

Plate 24 Scanning electron micrographs of the lens placode

(a) Lens placode at stage 13 (about 50 h incubation). Bar = 50 μm.
(b) Lens placode at stage 14, beginning to invaginate (about 52 h incubation). Bar = 25 μm.
(c) Lens placode at stage 15 (about 54 h incubation) almost completely invaginated. Bar = 100 μm.
(d) Remnant of invaginating lens placode at stage 17 (about 58 h incubation), only a small cavity now remaining. Bar = 25 μm.

(a) Reprinted from Bancroft and Bellairs (1977), with permission of Springer-Verlag.

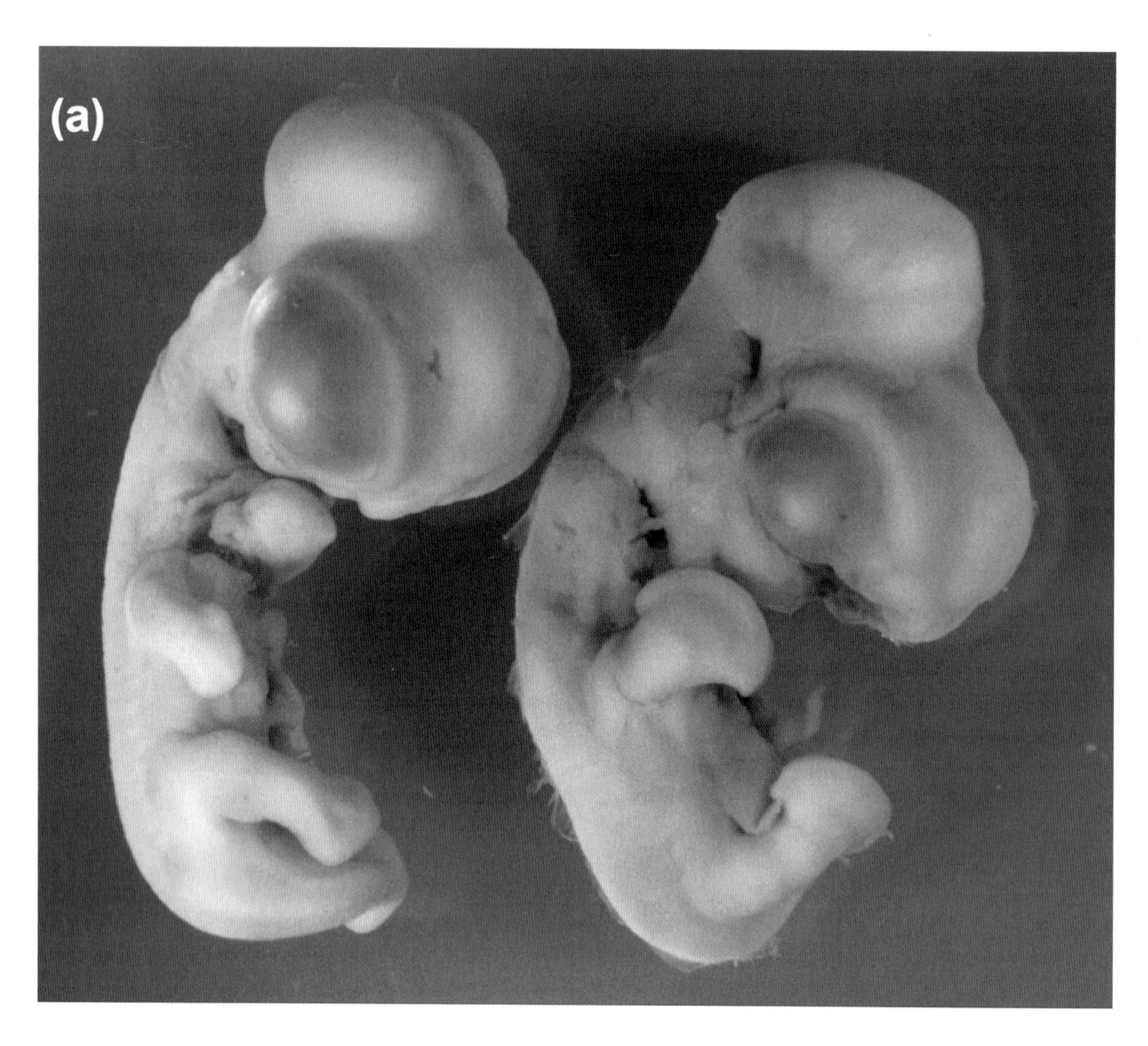
(a)

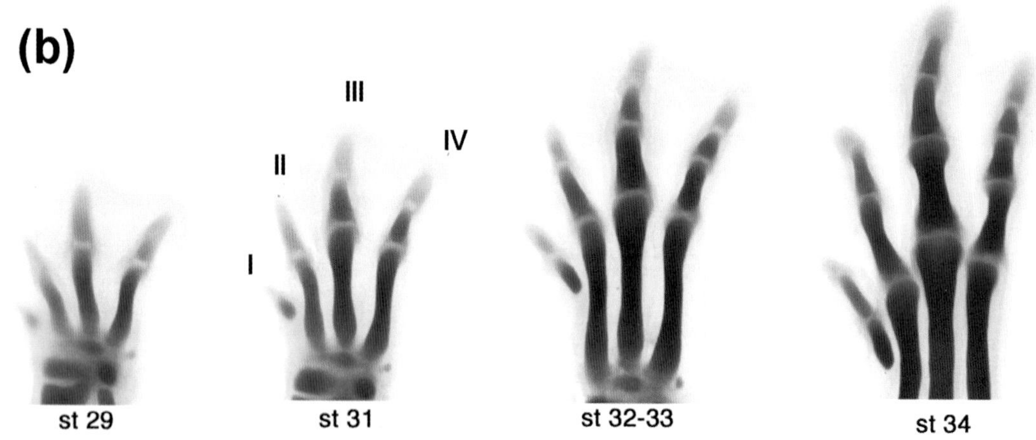
(b)
III
IV
II
I
st 29
st 31
st 32-33
st 34

Plate 25 Light micrographs of whole embryos and skeletally stained feet

(a) Normal (left) and talpid mutant (right) to show differences in the limb buds and the head.
(b) Right feet of normal embryos between stages 29 and 34.

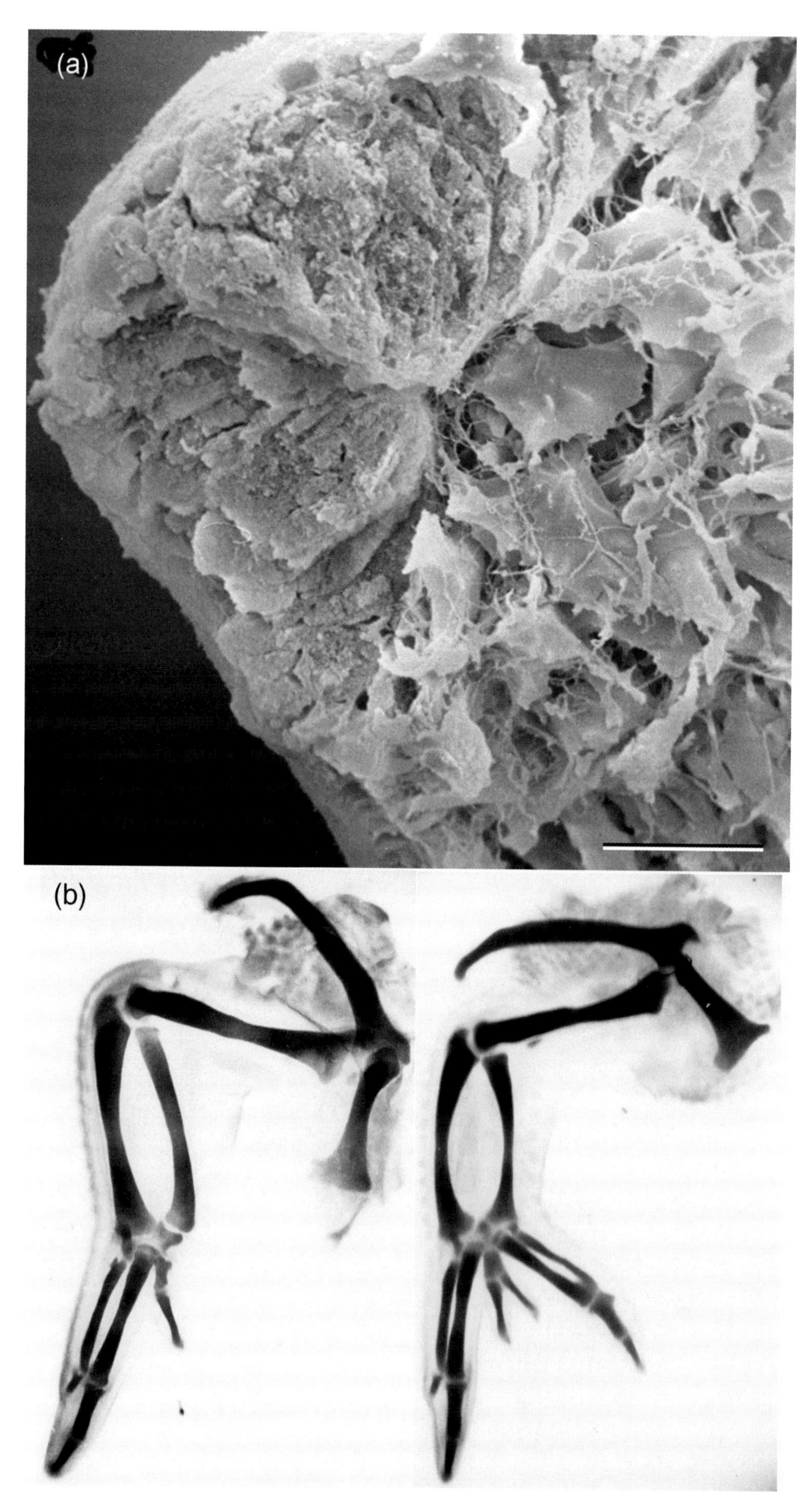
(a)
(b)

Plate 26 Scanning electron and light micrographs of limbs

(a) Scanning electron micrograph showing a section through the apical ectodermal ridge of the wing bud of a stage 26 embryo (5 days of incubation). Bar = 17 μm.

(b) Light micrographs of a skeletally stained normal wing skeleton (left), and duplicated wing skeleton (right) produced by grafting a polarizing region to the anterior margin of a stage 20 wing bud.

(b) Kindly supplied by C. Tickle.

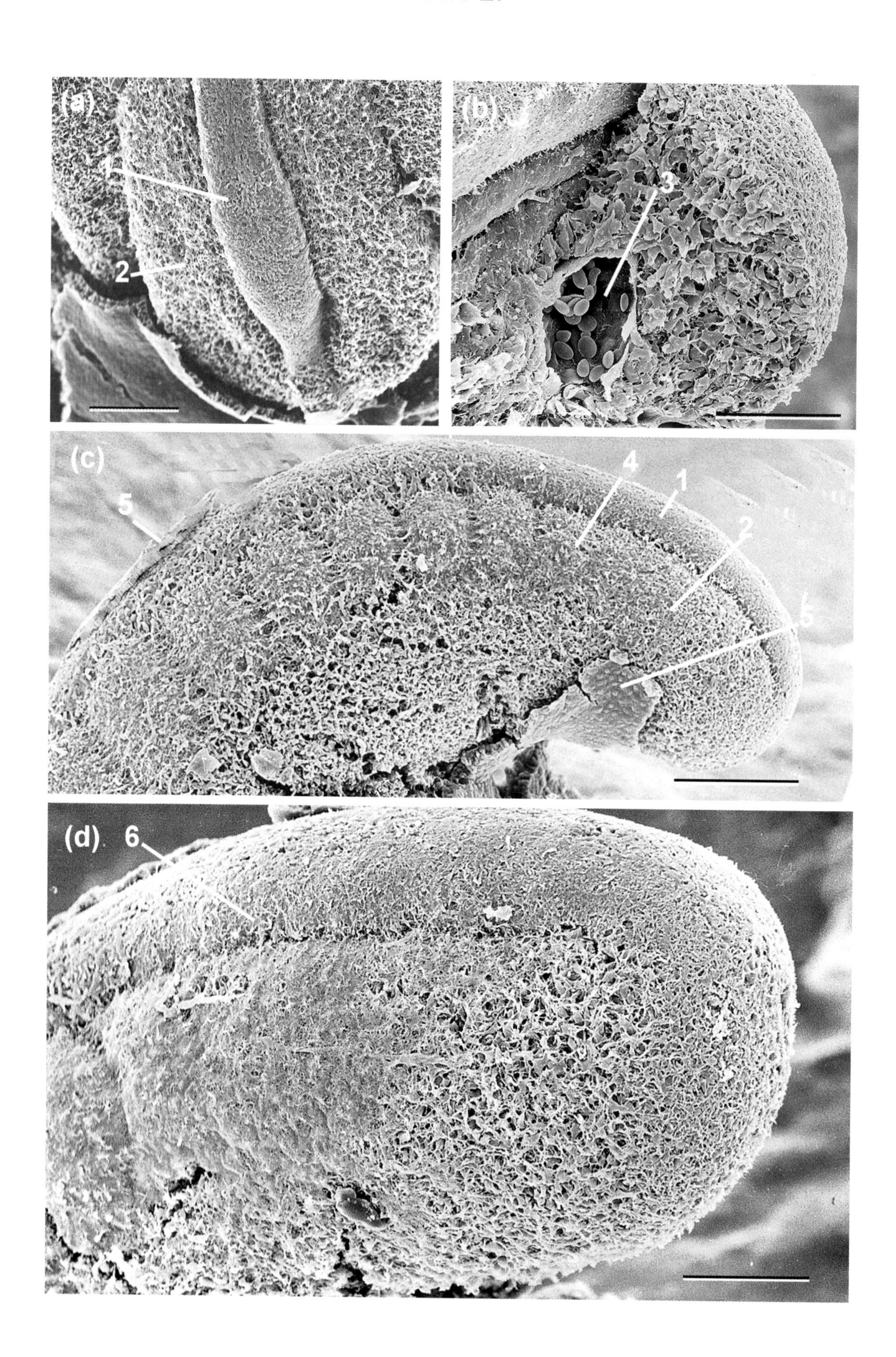
(a)
1
2
(b)
3
(c)
5
4
1
2
5
(d)
6

Plate 27 Scanning electron micrographs of the tail bud

(a) Dorsal view of the tail bud of a stage 16 (52 h of incubation) embryo after removal of the ectoderm. Bar = 50 μm.
(b) Tail bud of a stage 22 (3.5 days of incubation) embryo. A block of mesoderm has been dissected from the left side. Bar = 20 μm.
(c) Tail bud of a stage 21 (3 days of incubation) embryo after removal of the ectoderm. Bar = 20 μm.
(d) Tail bud of an embryo at stage 22 after removal of the ectoderm. Bar = 11 μm.

1. neural tube
2. unsegmented mesoderm
3. blood vessel
4. somite
5. ectoderm
6. neural crest cells on surface of neural tube.

Reprinted from Bellairs and Sanders (1986), with permission of Springer-Verlag.

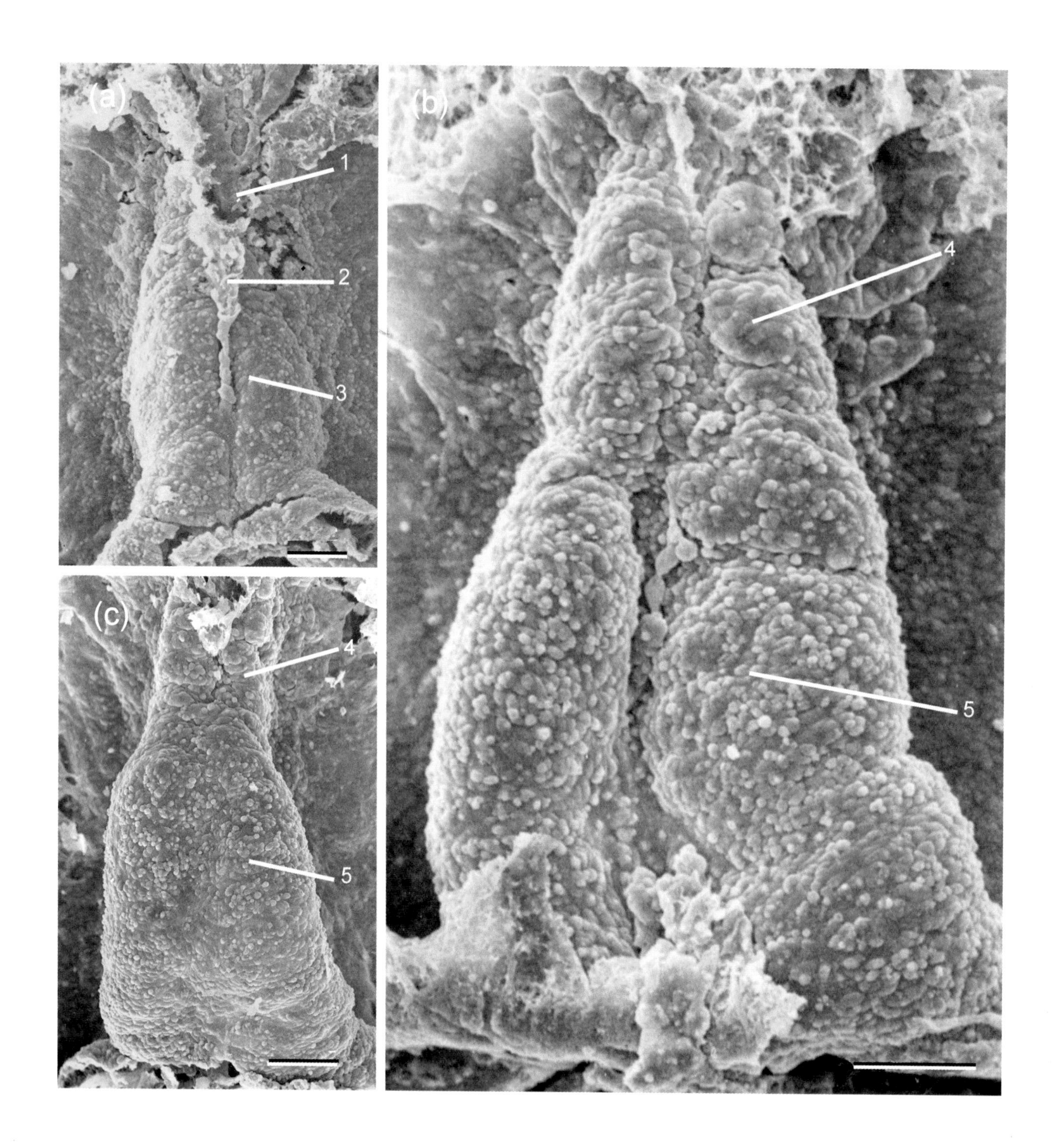
(a)
1
2
3
(b)
4
5
(c)
4
5

Plate 28 Scanning electron micrographs of early stages in heart development I

Ventral views of the early heart tube between stages 8 and 10 (29–38 h). In (a) and (b) the left and right sides are not yet completely fused.

(a) 8 somite stage. Bar = 50 μm.
(b) 9 somite stage. Bar = 50 μm.
(c) 10 somite stage. Bar = 50 μm.

1. ventral aorta
2. ventral mesocardium
3. cardiac tube
4. developing bulbus cordis
5. developing ventricle

Reprinted from Hiruma and Hirakow (1985), with permission of Springer-Verlag.

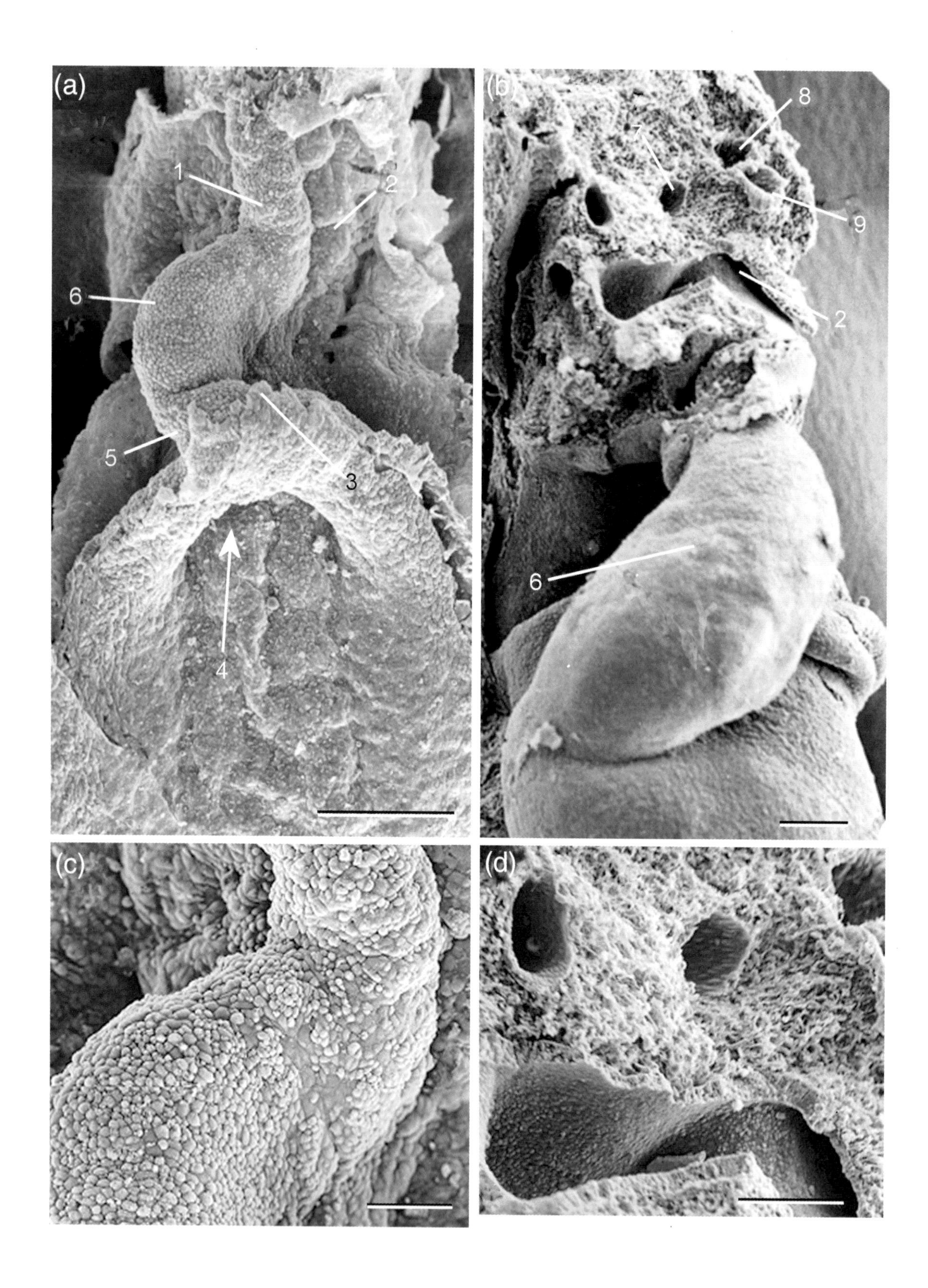
(a)
1
2
6
5
3
4
(b)
7
8
9
2
6
(c)
(d)

Plate 29 Scanning electron micrographs of early stages in heart development II

(a) Ventral view of the developing heart of a stage 12 embryo (about 48 h incubation) from which most of the endoderm has been removed. Bar = 200 μm.
(b) Ventral view of the heart of a stage 22 embryo (about 3.5 days incubation). Bar = 200 μm.
(c) Enlargement of part of (a). Bar = 40 μm.
(d) Enlargement of part of (b). Bar = 100 μm.

1. truncus arteriosus
2. pharynx
3. cut edge of splanchnopleure
4. anterior intestinal portal
5. atrium
6. ventricle
7. left dorsal aorta
8. left anterior cardinal vein
9. left aortic arch

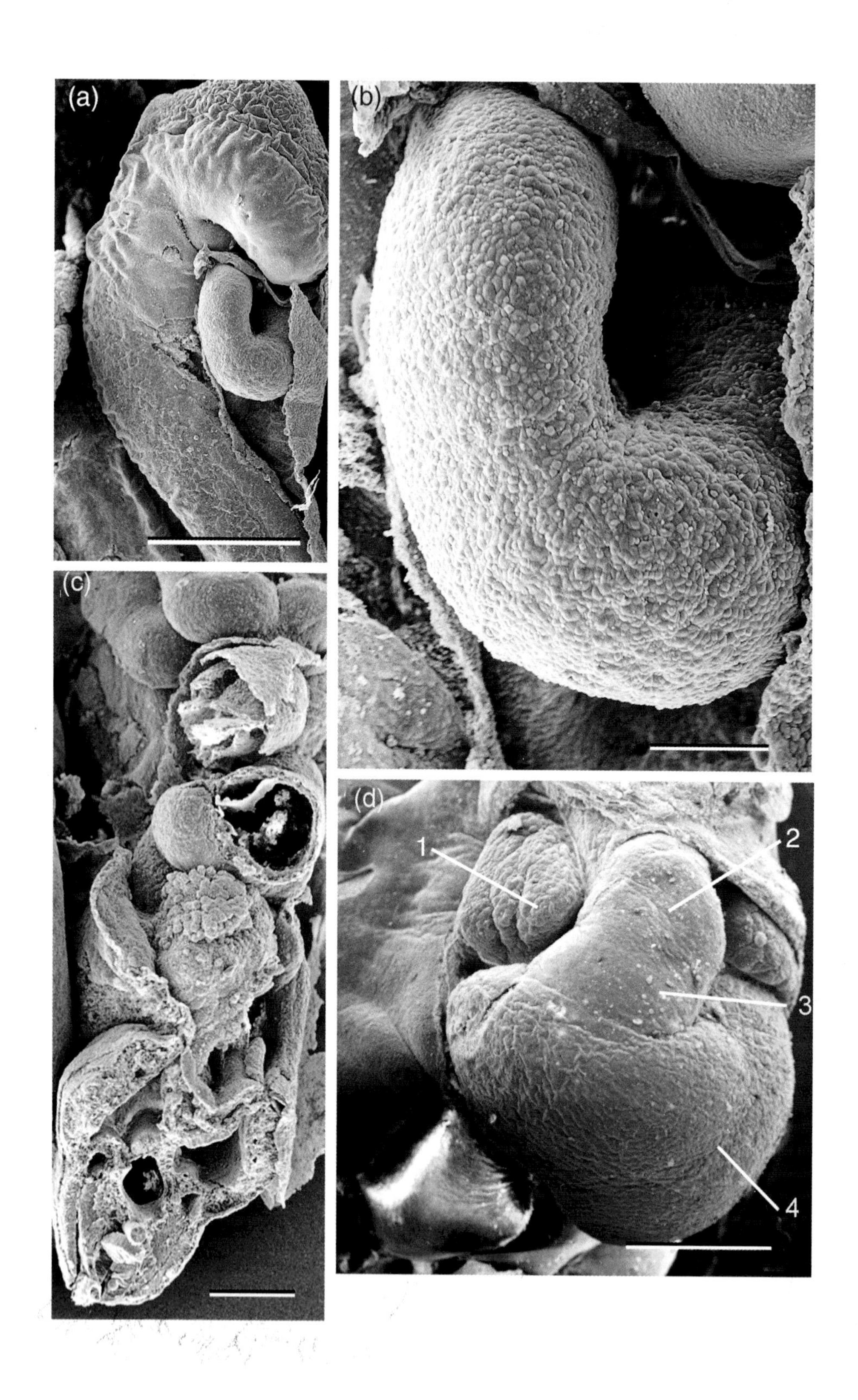
(a)
(b)
(c)
(d)
1
2
3
4

Plate 30 Scanning electron micrographs of heart III

(a) Stage 18 embryo (about 3 days) showing the head and trunk with heart exposed. Bar = 400 μm.
(b) Enlargement of heart of (a). Bar = 100 μm.
(c) Low-power view of a dissected stage 18 embryo (see *Plates 31* and *32*). Bar = 300 μm.
(d) Heart of an embryo at stage 27 (6 days of incubation). Bar = 400 μm.

1. atrium
2. truncus arteriosus
3. conus (bulbus) arteriosus
4. ventricle

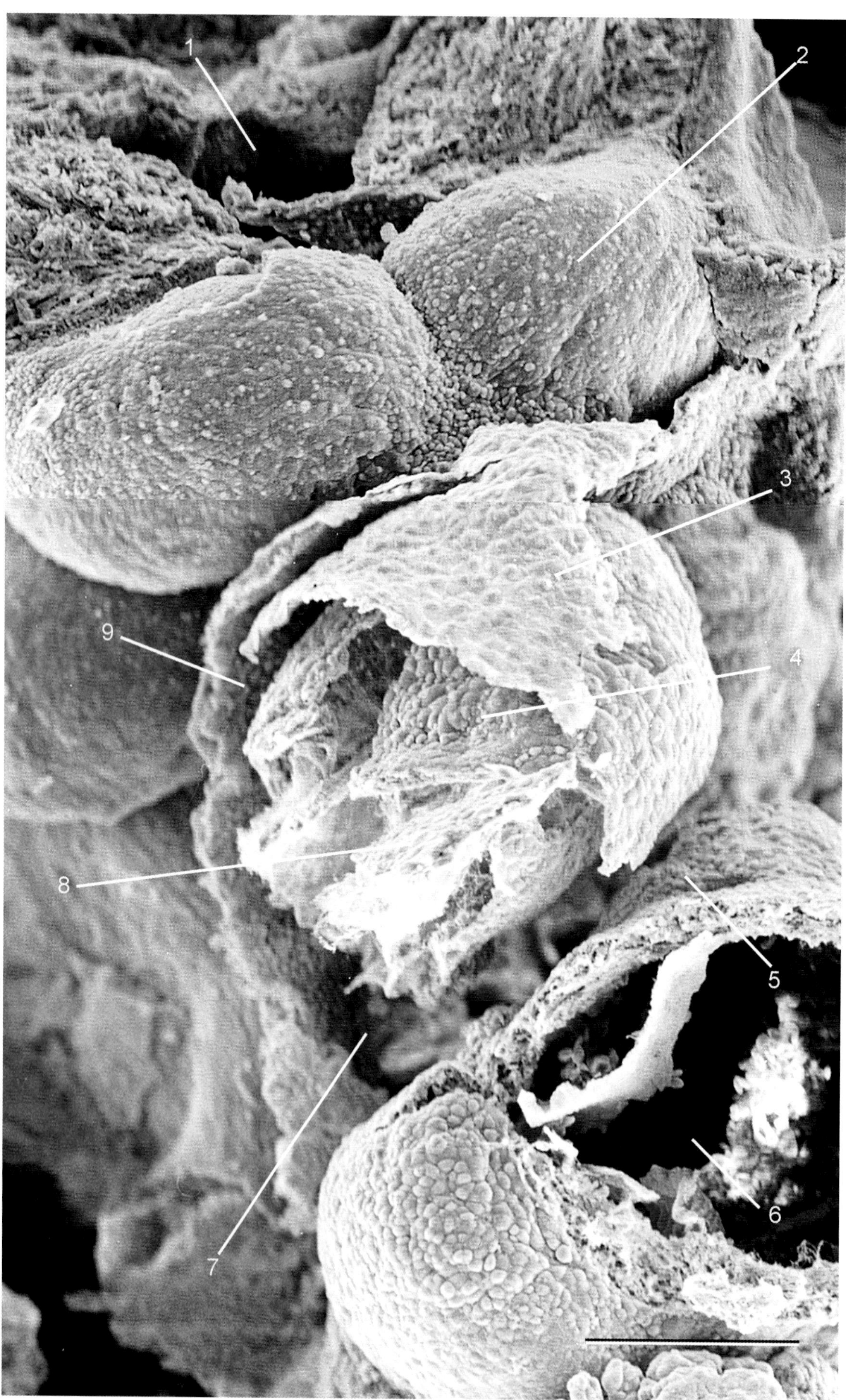
1
2
3
4
5
6
7
8
9

Plate 31 Scanning electron micrographs of internal organs I

Enlargement of portion of *Plate 30c*. The walls of the atrium and ventricle have been partially removed. Bar = 200 μm.

1. pharynx
2. mandibular arch
3. ectoderm of ventral body wall
4. wall of truncus arteriosus
5. wall of ventricle
6. lumen of ventricle
7. pericardial coelom
8. aortico-pulmonary septum
9. epicardium

Plate 32

Plate 32 Scanning electron micrographs of internal organs II

Enlargement of part of *Plate 30c*. Region adjacent to *Plate 31*. Bar = 150 μm.

1. ventricle
2. pro-epicardial serosa (also known as pericardial villi) covering sinus venosus
3. atrium
4. sinus venosus
5. posterior vena cava
6. hindlimb bud

Plate 33

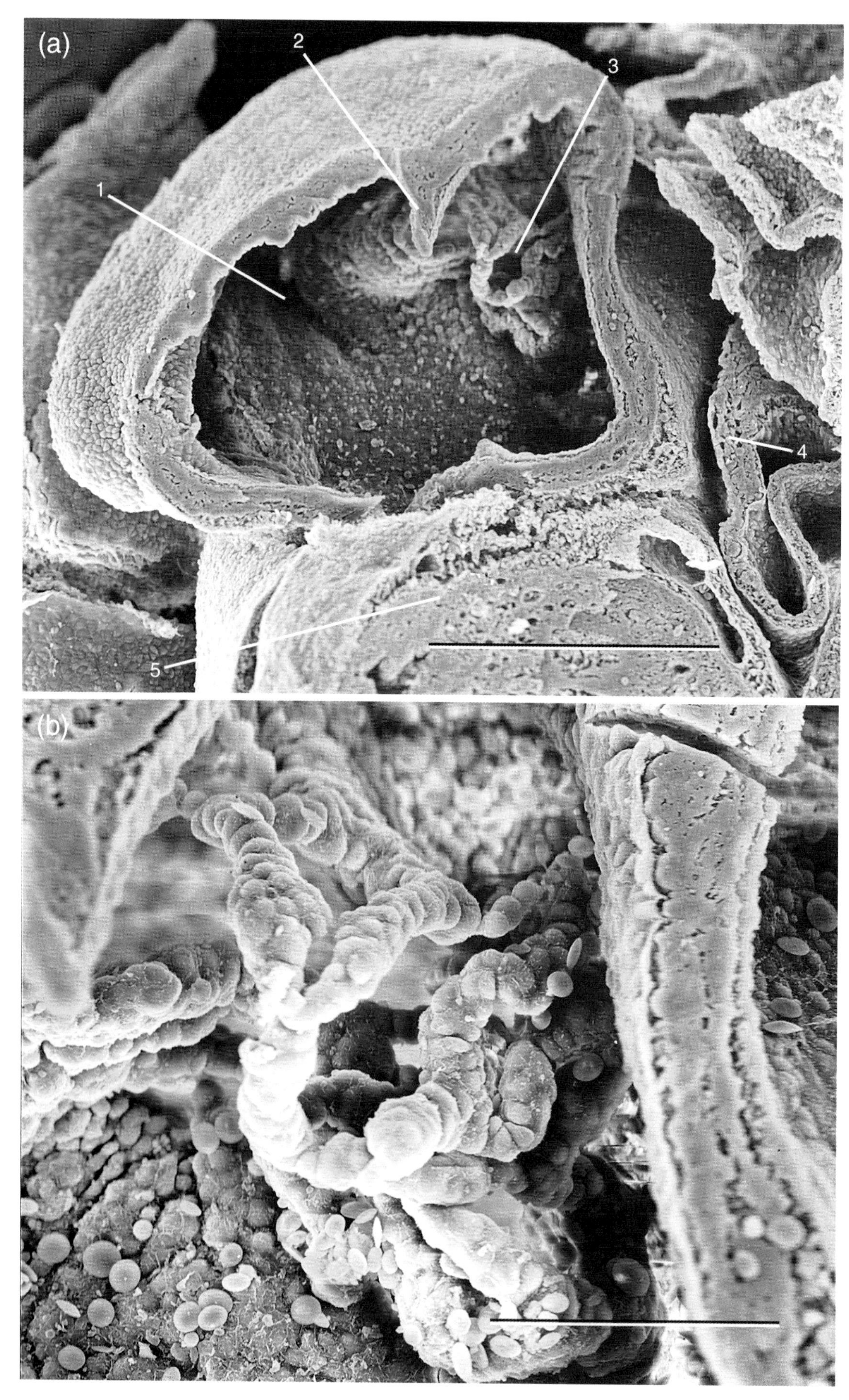

Plate 33 Scanning electron micrographs of internal organs III. Atrium

(a) Section across atrium at 5 days of incubation. Bar = 500 μm.
(b) Enlargement of part of (a) showing the developing sinu-atrial valves. Bar = 130 μm.

1. lumen of atrium
2. interatrial septum
3. sinu-atrial opening
4. wall of sinus venosus
5. ventricle

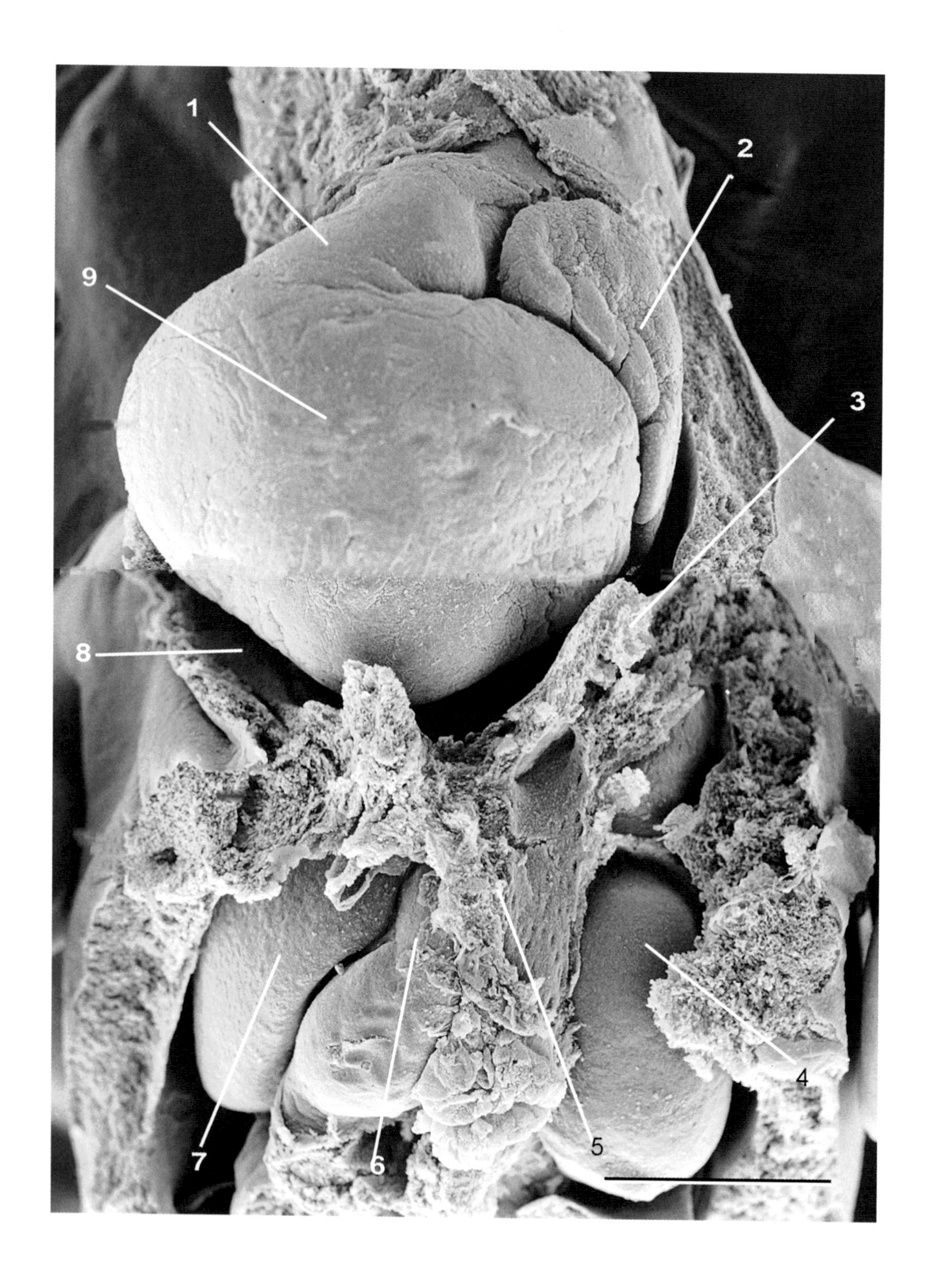
1
2
9
3
8
4
7
6
5

Plate 34 Scanning electron micrographs of internal organs IV. Heart and lungs

Embryo at stage 29 (6 days of incubation) dissected to show heart and lungs. Bar = 500 μm.

1. truncus arteriosus
2. left atrium
3. remnant of left subclavian vein
4. left lung
5. mesentery
6. gizzard
7. right lung
8. pericardial cavity
9. ventricle

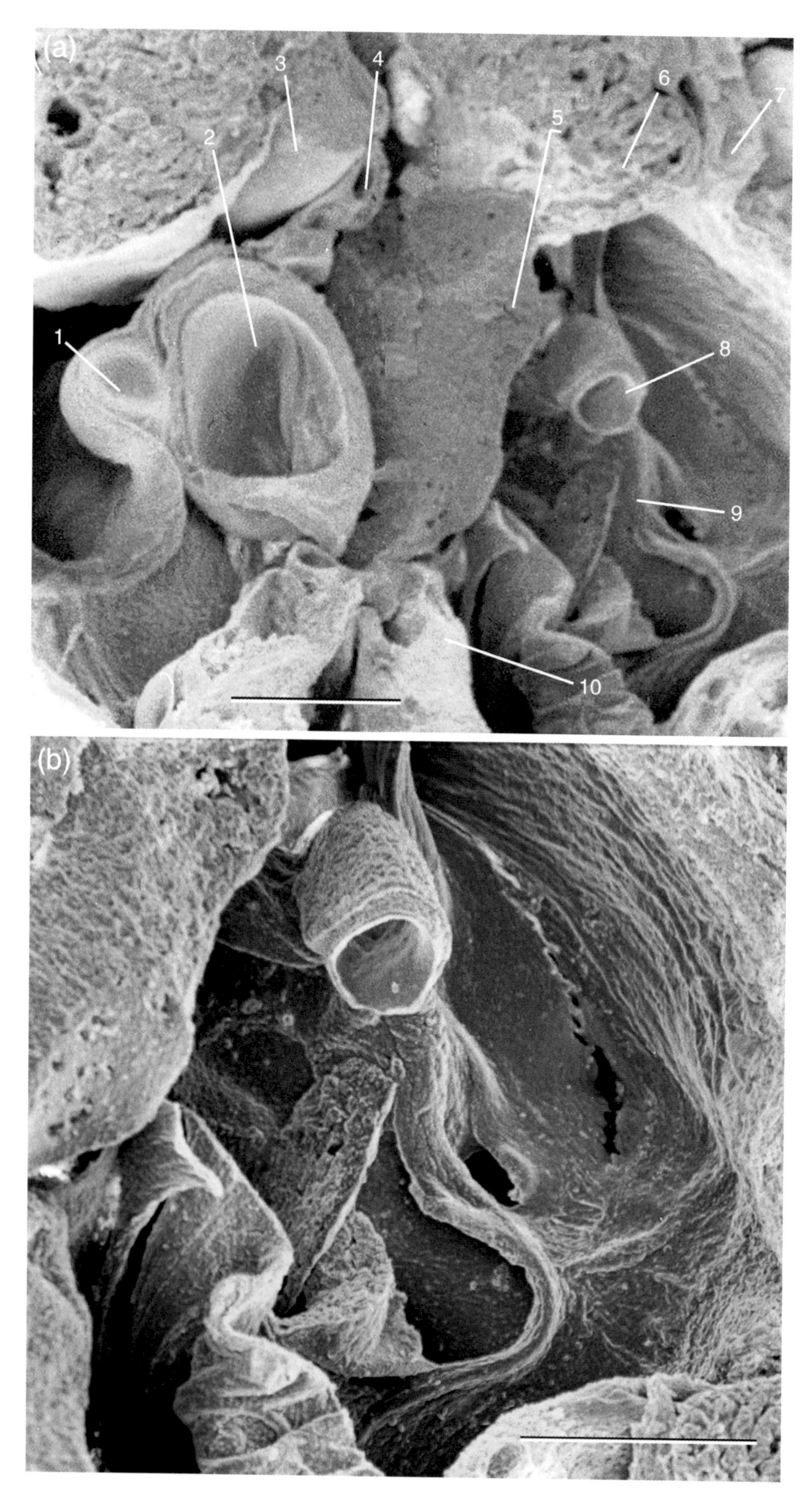
(a)
1
2
3
4
5
6
7
8
9
10
(b)

Plate 35 Scanning electron micrographs of internal organs V

(a) Abdominal cavity of an embryo at stage 35 (9 days). Bar = 1 mm.
(b) Enlargement of colon. Bar = 500 μm.

1. duodenum
2. gizzard
3. right ovary
4. metanephric duct
5. dorsal mesentery
6. mesonephros
7. left Mullerian duct
8. colon
9. mesentery
10. rectum

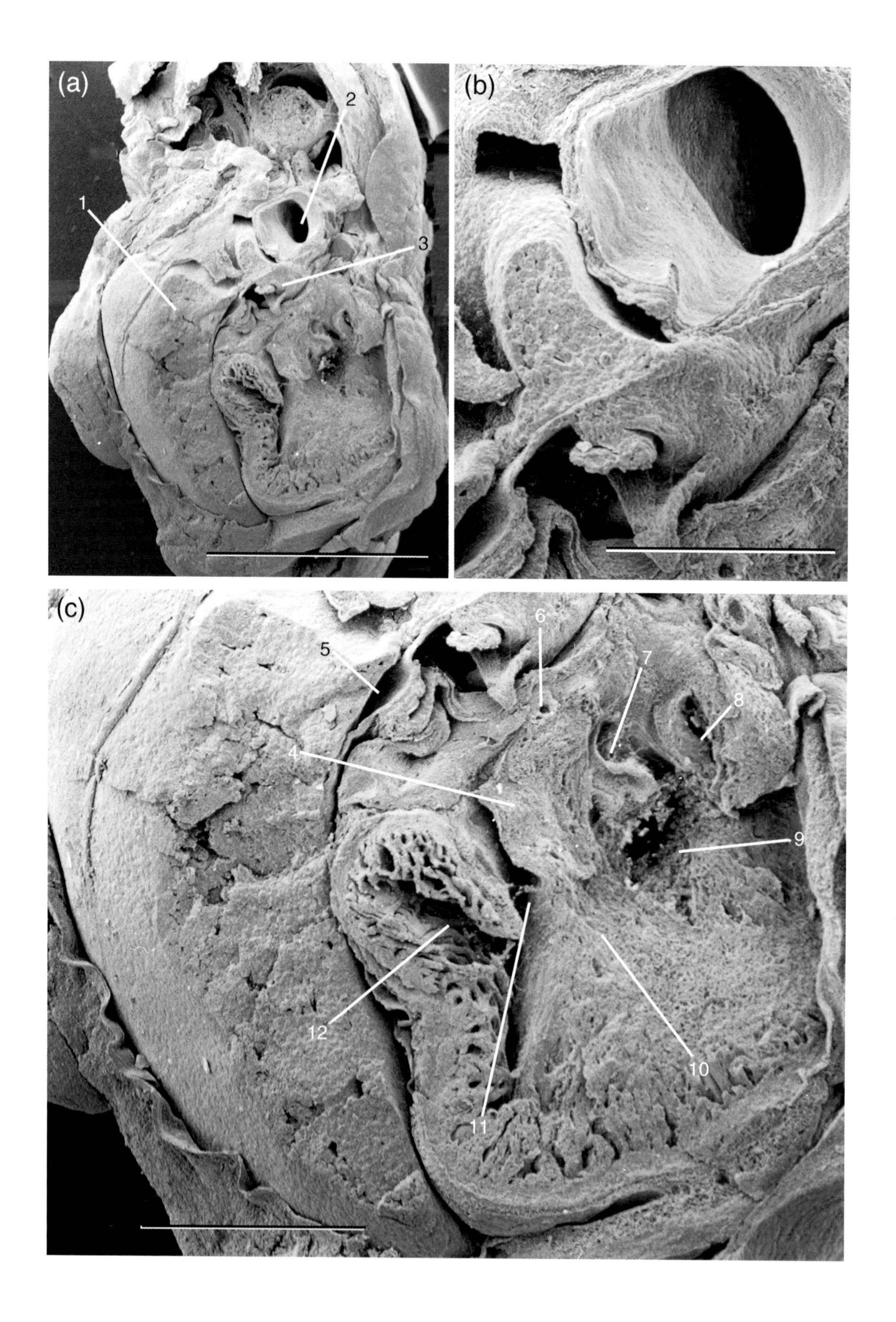
(a)
1
2
3
(b)
(c)
4
5
6
7
8
9
10
11
12

Plate 36 Scanning electron micrographs of internal organs VI. Heart and gizzard

(a) Thoracic region of a stage 39 embryo (13 days incubation) after removal of the ventral body wall. Bar = 3 mm.
(b) Enlargement of the gizzard in (a). Bar = 750 μm.
(c) Enlargement of the heart in (a). Owing to the angle of section, both the interatrial and interventricular septa are visible as frontal sections. Bar = 500 μm.

1. liver
2. gizzard
3. sinus venosus
4. interatrial septum
5. pericardial cavity
6. coronary artery
7. left atrium
8. sinus venosus
9. left atrioventricular canal
10. interventricular septum
11. right atrio-ventricular canal
12. lumen of right ventricle

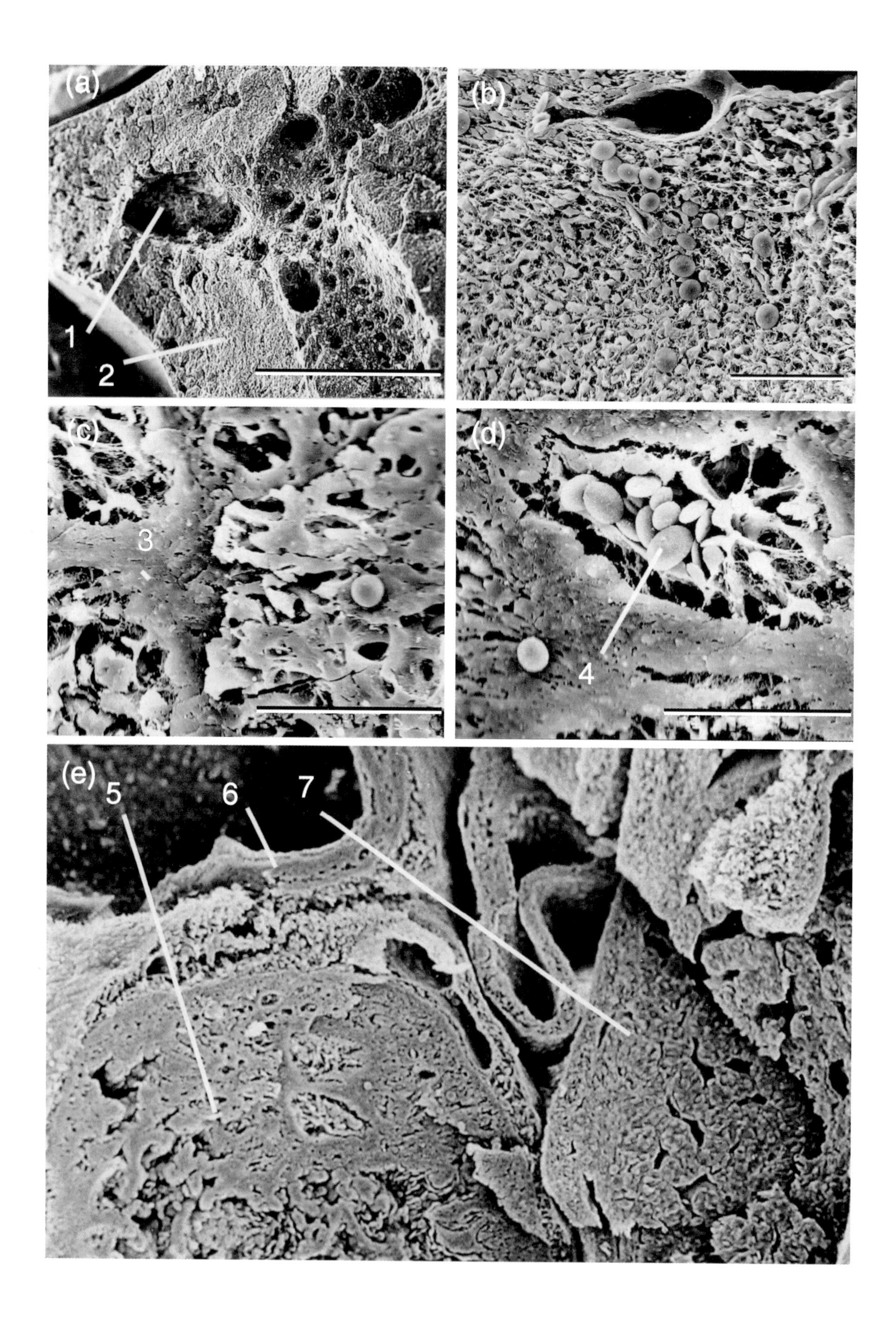
(a)
1
2
(b)
(c)
3
(d)
4
(e)
5
6
7

Plate 37 Scanning electron micrographs of liver and ventricle

(a) Section through liver in (e) below. Bar = 500 μm.
(b) Enlargement of part of (e). Bar = 50 μm.
(c) Section through ventricle in (e). Bar = 500 μm.
(d) Section adjacent to that in (c).
(e) Section through liver and ventricle at 5 days.

1. liver sinus
2. parenchyma
3. myocardium
4. red blood corpuscles (avian red blood corpuscles retain their nuclei and so are not biconcave like those of mammals)
5. ventricle
6. wall of atrium
7. liver

PLATE 38

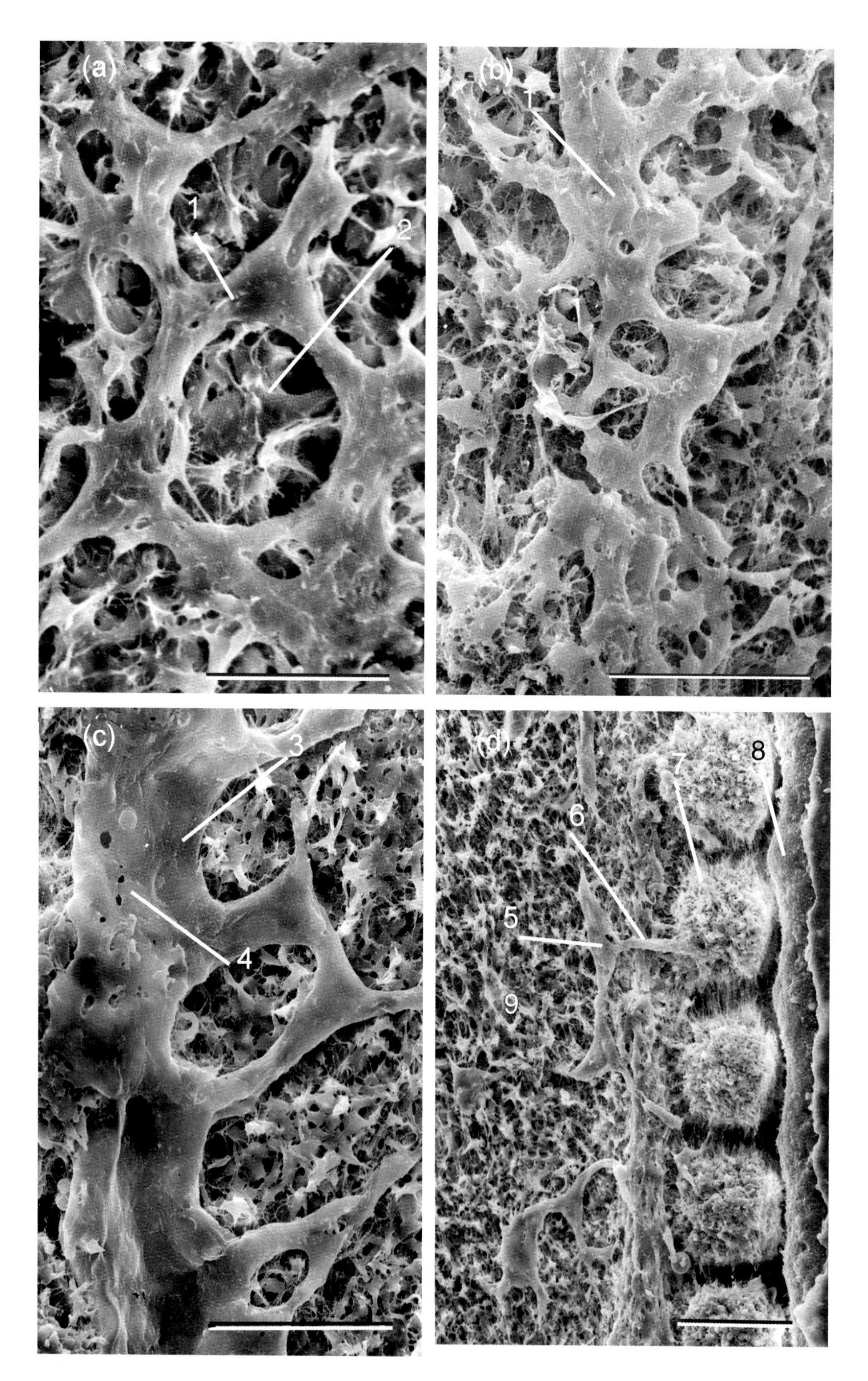

Plate 38 Scanning electron micrographs showing early stages of blood vessel formation

All specimens are viewed from the ventral side after removal of the endoderm.

(a) Angioblastic plexus in the area vasculosa of a stage 8 embryo. Bar = 50 μm.
(b) Angioblastic plexus in a stage 9 embryo in the process of forming the dorsal aorta. Bar = 50 μm.
(c) Angioblastic plexus at a slightly more advanced stage (stages 9–10) during the formation of the dorsal aorta. Bar = 50 μm.
(d) Developing left vena cava and intersegmental veins of a stage 10 embryo. Bar = 50 μm.

1. angioblastic plexus
2. mesoderm cell
3. dorsal aorta
4. pores between as yet incompletely fused angioblasts
5. left anterior cardinal vein forming
6. developing intersegmental artery
7. somite
8. neural tube
9. lateral plate mesoderm

Reprinted from Hirakow and Hiruma (1981), with permission from Springer-Verlag.

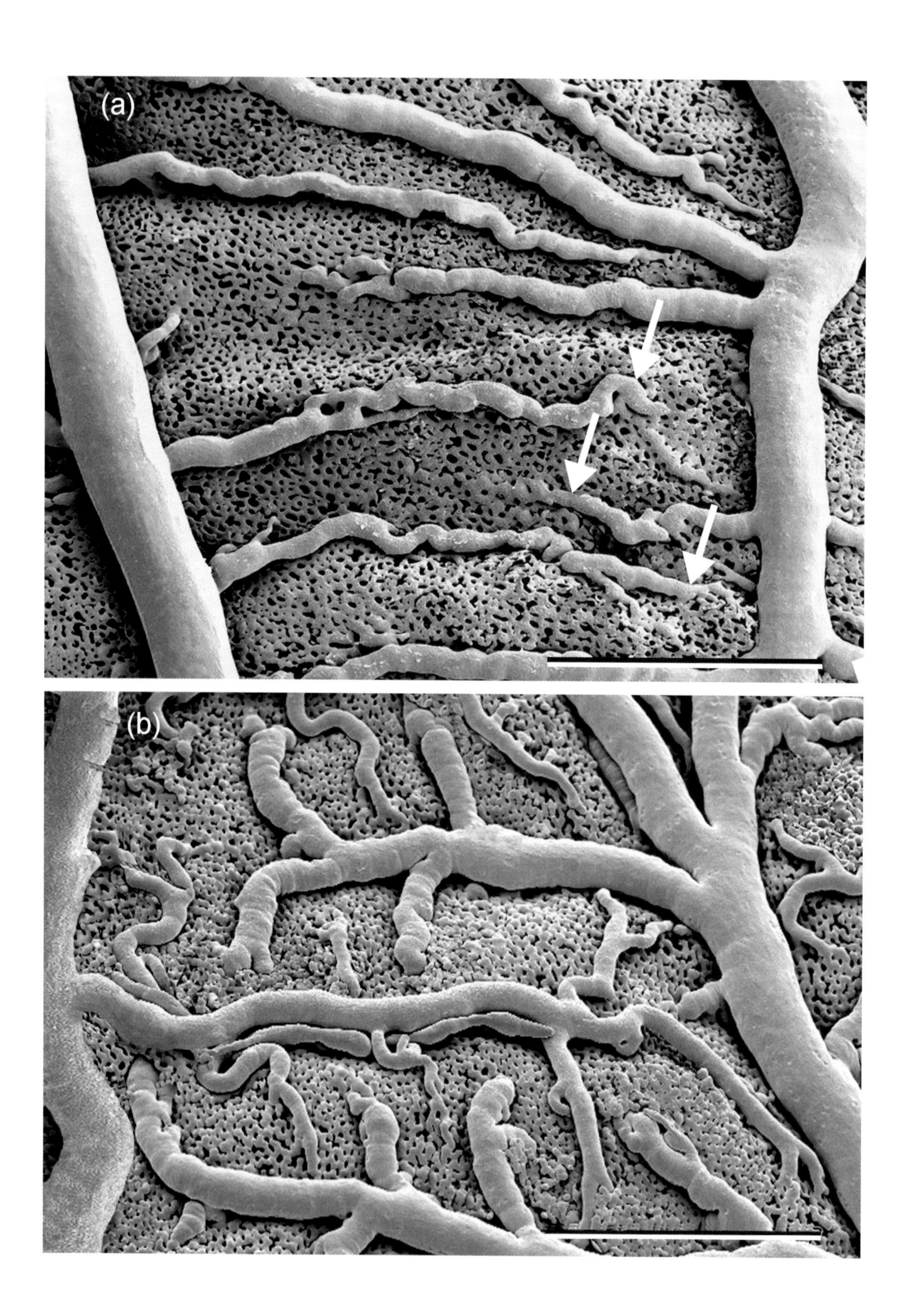
(a)
(b)

Plate 39 Scanning electron micrographs showing vascularization of the chorioallantoic membrane

(a) 11 days of incubation. Bar = 300 μm.
(b) 14 days of incubation. There is a marked increase in the number of branches in the capillary network. Bar = 300 μm.

Arrows indicate where 'feed vessels' merge with the capillary network.

Reprinted from Djonov *et al.* (2000), with permission of Springer-Verlag.

PLATE 40

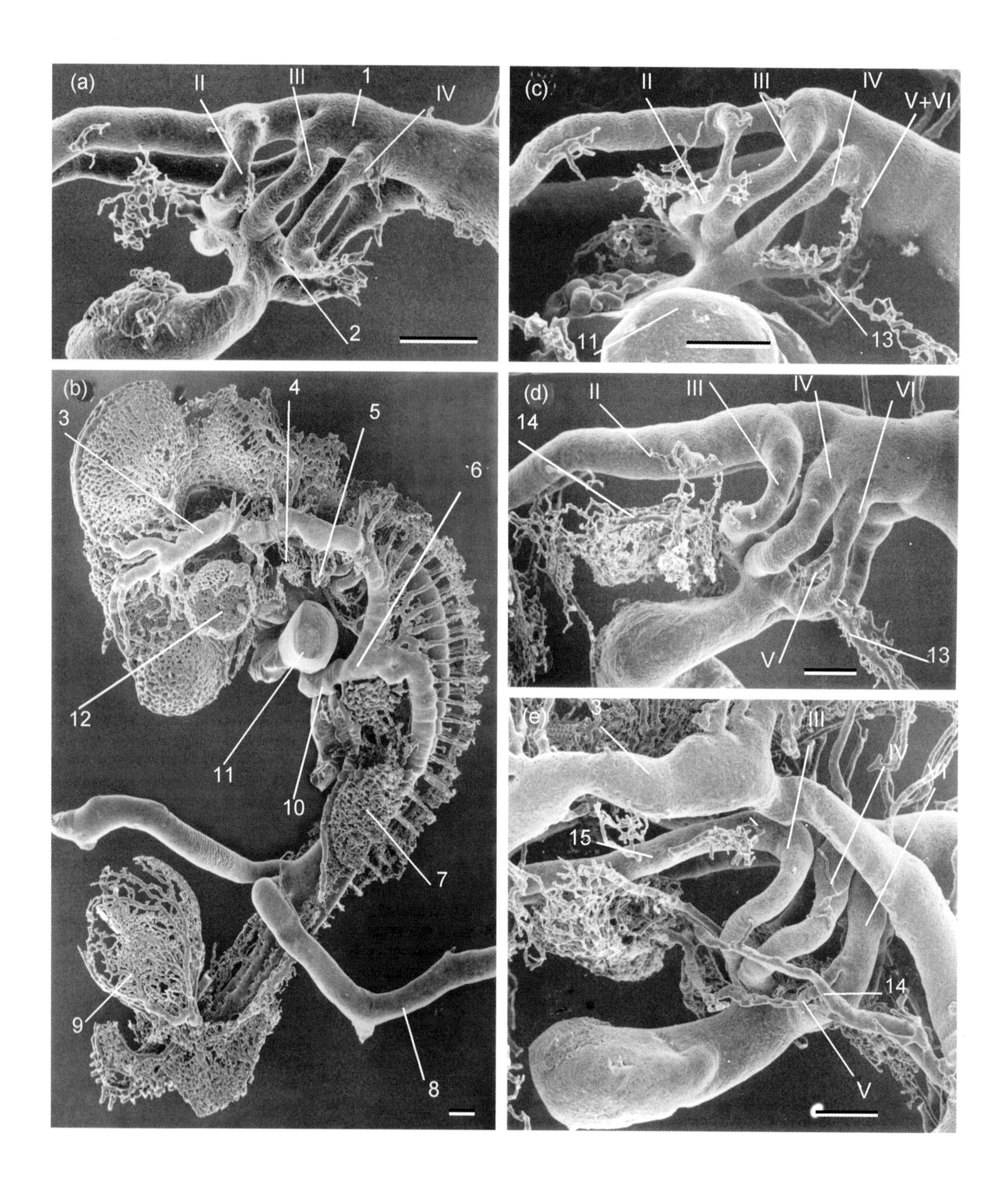

(a)
II
III
1
IV
2
(b)
3
4
5
6
12
11
10
7
9
8
(c)
II
III
IV
V+VI
11
13
(d)
II
III
IV
VI
14
V
13
(e)
3
III
IV
VI
15
14
V

Plate 40 Scanning electron micrographs of casts of pharyngeal arch arteries. Left side view

Casts of pharyngeal region.

(a) Stage 19. Bar = 250 μm.
(b) Stage 20, general view. Bar = 250 μm.
(c) Stage 20. Bar = 250 μm.
(d) Stage 23. Bar = 250 μm.
(e) Stage 24. Bar = 250 μm.

The individual pharyngeal arteries are labeled I–VI. The leg plexus, the tail bud plexus and the intersegmental arteries are also visible in (b) (not labelled).

1. dorsal aorta
2. ventral aorta
3. anterior cardinal vein
4. superficial capillary plexus of the mandibular arch
5. superficial capillary plexus of the hyoid arch
6. aortic root
7. plexus of wing bud
8. omphalo-mesenteric (vitelline) artery
9. blood vessels in allantois
10. truncus arteriosus
11. heart
12. eye
13. pulmonary artery
14. primary external carotid artery
15. stapedial artery

Reprinted from Hiruma and Hirakow (1995), with permission of Springer-Verlag.

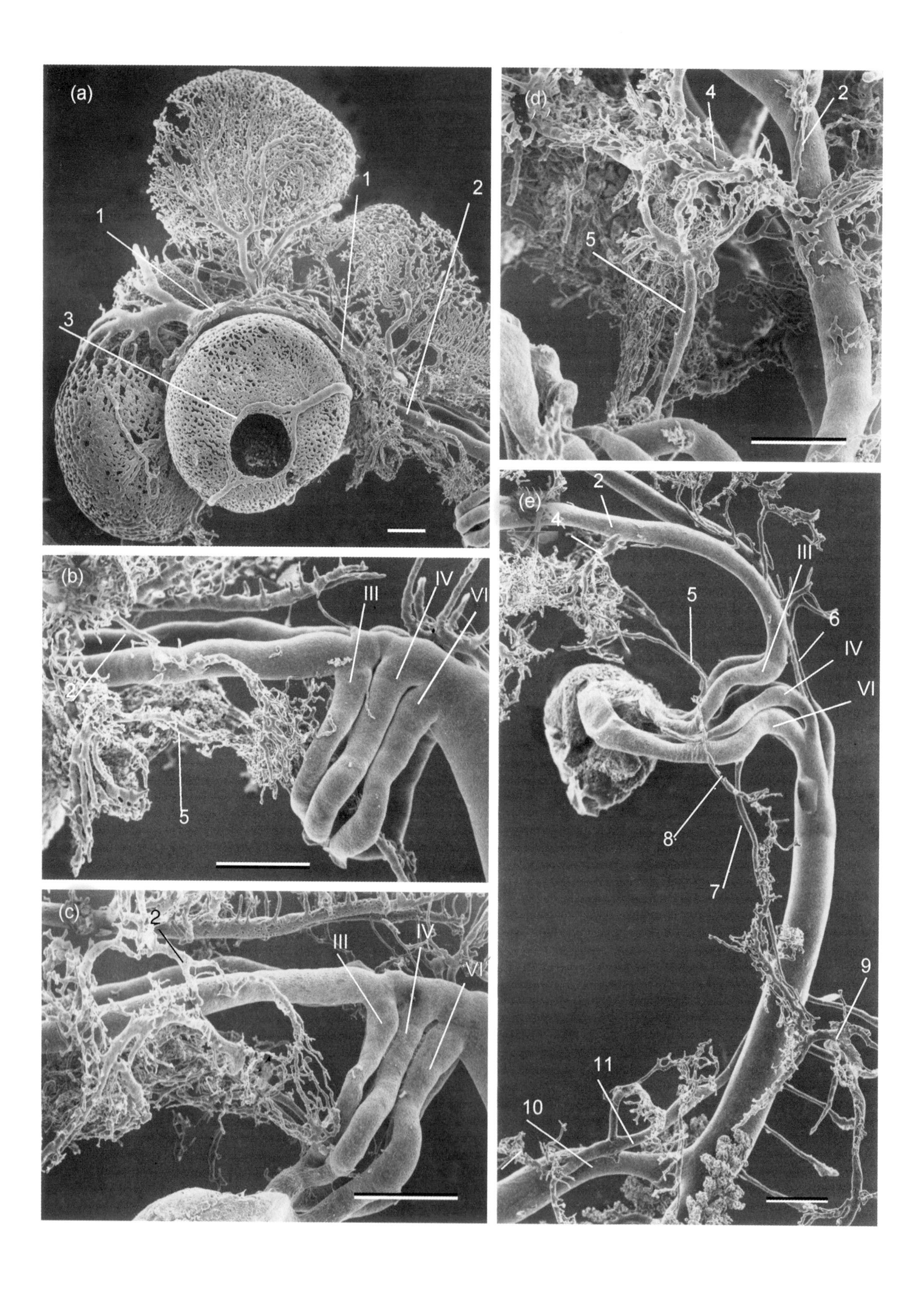
(a)
1
1
2
3
(b)
2
III
IV
VI
5
(c)
2
III
IV
VI
(d)
4
2
5
(e)
2
4
5
III
6
IV
VI
8
7
9
11
10

Plate 41 Scanning electron micrographs of casts of pharyngeal arch arteries (continued)

(a) Stage 25 (4.5–5 days). Bar = 0.5 mm.
(b) Stage 25. Bar = 0.5 mm.
(c) Stage 26 (5 days). Bar = 0.7 mm.
(d) Stage 27 (5–5.5 days). Bar = 0.5 mm.
(e) Stage 28 (5.5–6 days). Bar = 0.7 mm.

The individual arch arteries are labeled I–VI.

1. supraorbital branch of stapedial artery
2. stapedial artery
3. capillary plexus draining into supraorbital vein
4. secondary external carotid artery
5. primary external carotid artery
6. carotid duct
7. pulmonary artery
8. secondary subclavian artery
9. primary subclavian artery
10. anterior mesenteric artery
11. coeliac artery

Reprinted from Hiruma and Hiakow (1995), with permission of Springer-Verlag.

PLATE 42

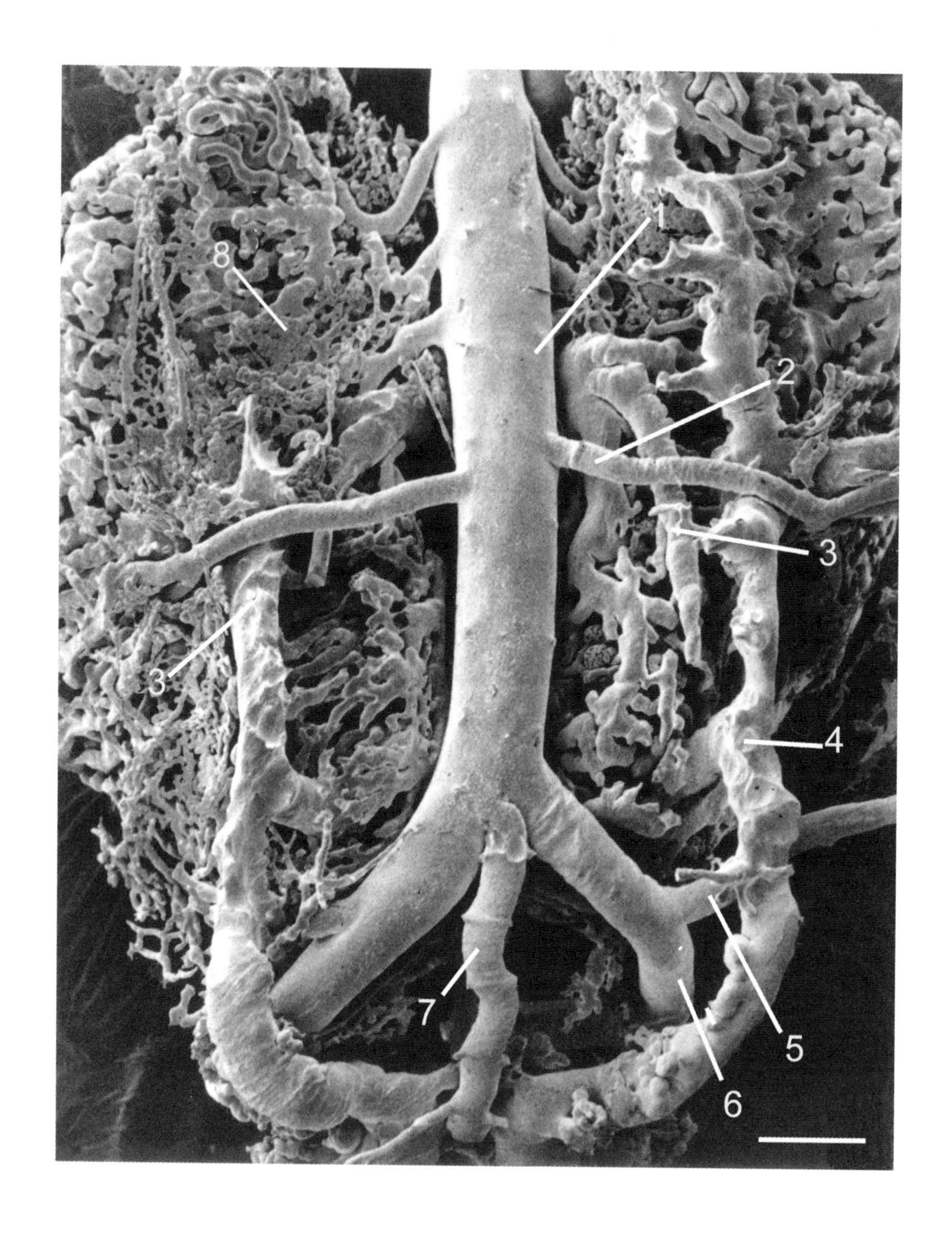

Plate 42 Scanning electron micrographs of a vascular cast of the trunk

Dorsal view of the mesonephric region of an embryo at stage 32 (7.5 days). Bar = 400 μm.

1. dorsal aorta
2. external iliac artery
3. great renal vein (future metanephric venous system)
4. posterior mesonephric portal vein (formed from posterior cardinal vein)
5. sciatic artery
6. allantoic artery
7. caudal artery
8. mesonephros

Modified from Carretero *et al.* (1993), with permission of *Scanning Microscopy*.

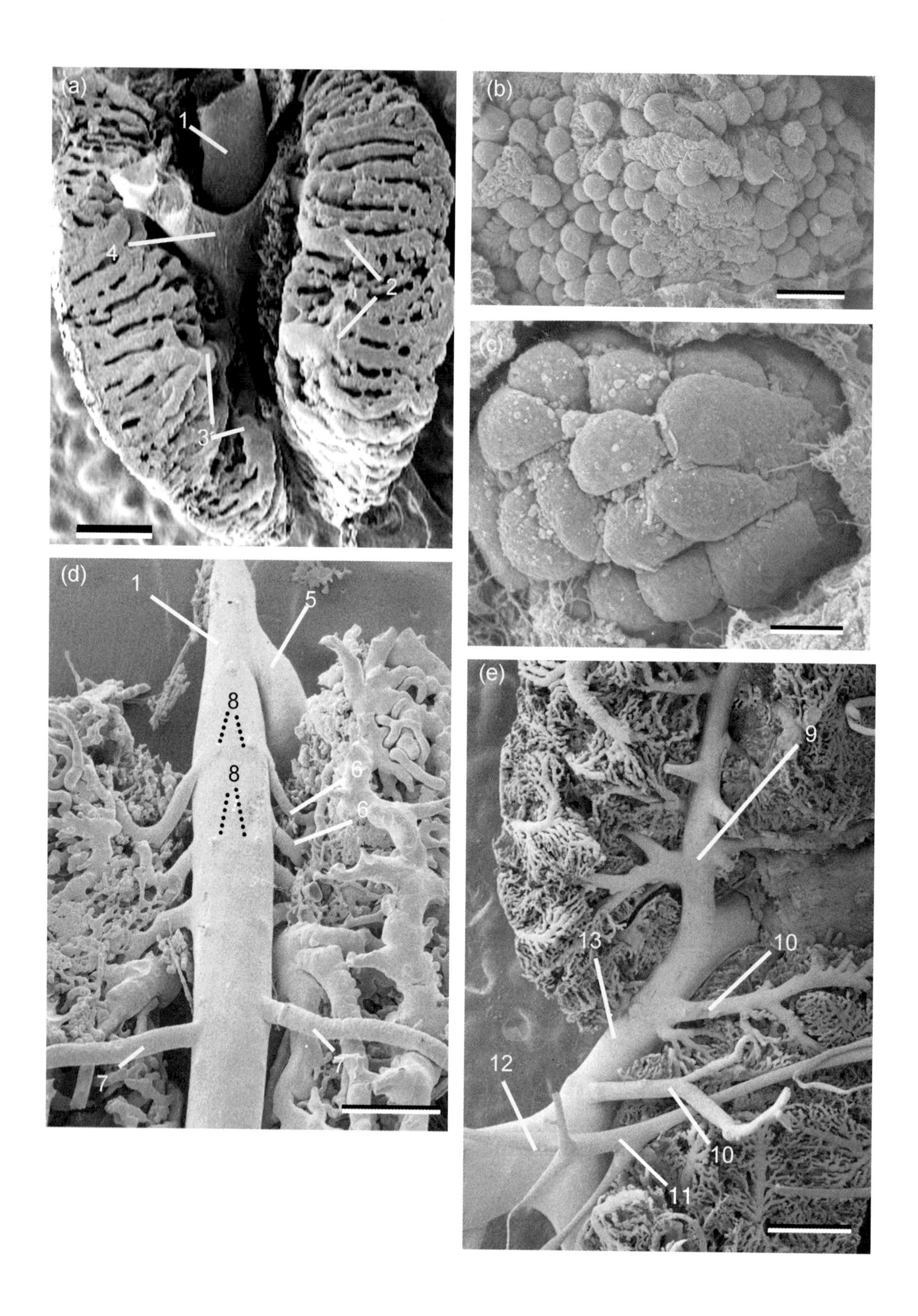
(a)
1
2
2
3
3
4
(b)
(c)
(d)
1
5
8
8
6
6
7
7
(e)
9
10
13
12
10
11

Plate 43 Scanning electron micrographs of vascular casts of the trunk (continued)

(a) Ventral aspect of a vascular cast of the mesonephros of a quail embryo incubated for 6 days. Bar = 0.5 mm.
(b) Mesonephric glomerulus of a chick embryo at stage 22. Bar = 12 μm.
(c) Metanephric glomerulus of a chick embryo at stage 37. Bar = 3 μm.
(d) Dorsal aspect of the dorsal aorta of a chick embryo at stage 32 to show the origin of the primary mesonephric arteries. Bar = 400 μm.
(e) Dorsal view of a vascular cast of a stage 40 chick embryo. Part of the metanephros has been removed to allow observation of the anterior metanephric portal vein. Bar = 0.67 mm.

1. dorsal aorta
2. subcardinal venous plexus
3. subcardinal veins
4. posterior vena cava
5. anterior mesenteric artery
6. primary mesonephric arteries
7. external iliac artery
8. origins of intersegmental arteries
9. anterior metanephric portal vein
10. branches to middle renal division of the metanephros
11. posterior metanephric portal vein
12. external iliac vein
13. common iliac vein

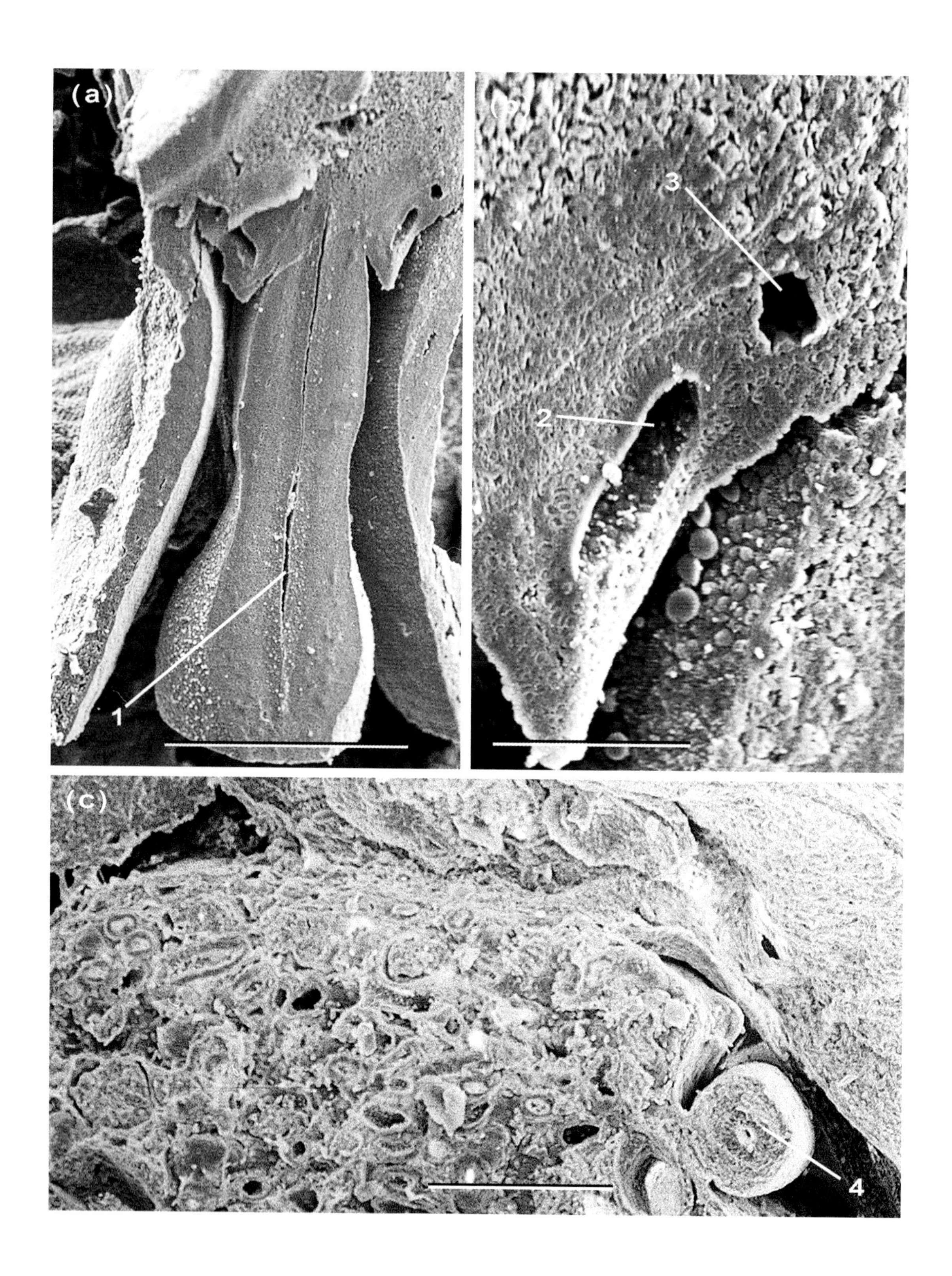
(a)
1
3
2
(c)
4

Plate 44 Scanning electron micrographs of the abdomen

(a) Small intestine of a stage 31 embryo (7 days). Bar = 200 μm.
(b) Enlargement of (a) to show left nephric duct and left subclavian vein in genital (nephrogenic) ridge. Bar = 100 μm.
(c) Mesonephros and oviduct of a stage 35 embryo (9 days). Bar = 130 μm.

1. lumen of intestine
2. subclavian vein
3. nephric duct
4. Mullerian duct (oviduct)

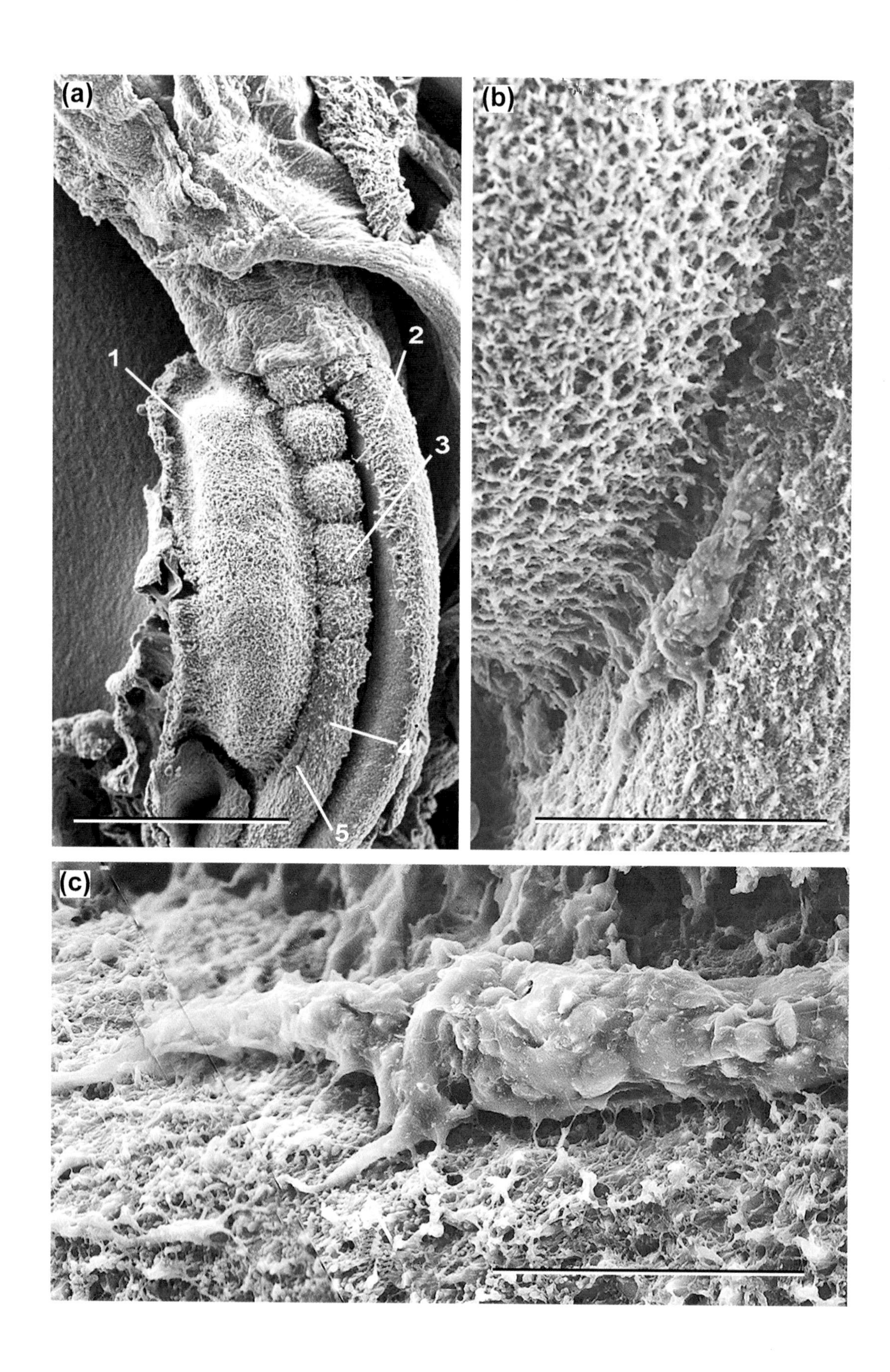
(a)
1
2
3
4
5
(b)
(c)

Plate 45 Scanning electron micrographs of the nephric duct

(a) Left lateral view of a stage 14 embryo after removal of the ectoderm. Bar = 0.5 mm.

(b, c) Enlargements at the posterior end of the mesonephros to show the posterior tip of the nephric duct. Bar of (b) = 130 μm. Bar of (c) = 50 μm.

1. mesonephros
2. neural tube
3. somite
4. unsegmented mesoderm
5. tip of nephric duct

(c) Reprinted from Bellairs *et al.* (1995), with permission of Wiley-Liss, Inc., a subsidiary of John Wiley and Sons, Inc.

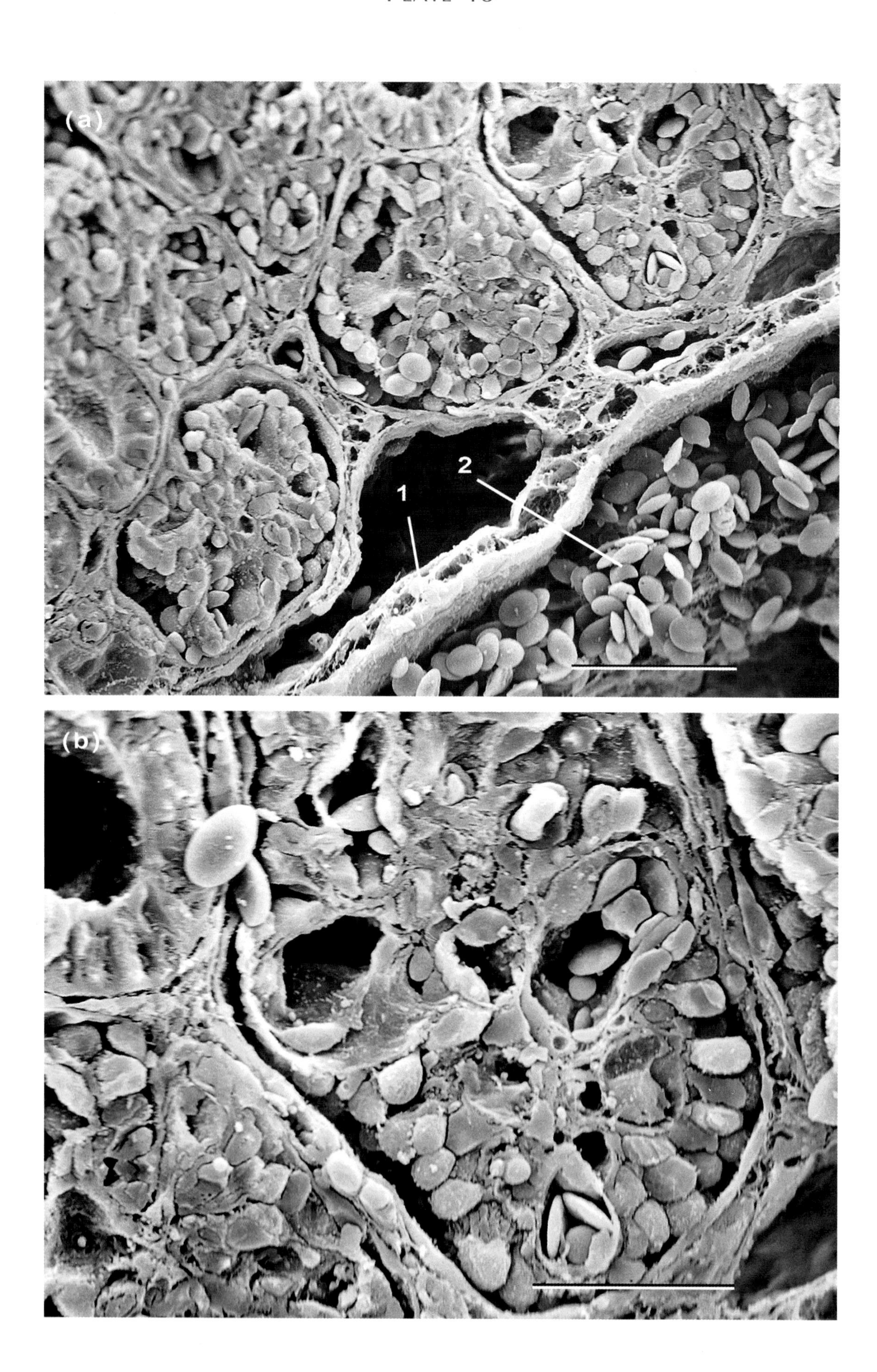
(a)
1
2
(b)

Plate 46 Scanning electron micrographs of the mesonephros

(a) Section across the mesonephros of a 5-day embryo. Bar = 25 μm.
(b) Enlargement of part of (a) to show a glomerulus. Bar = 15 μm.

1. nephric duct
2. red blood corpuscles in subcardinal vein

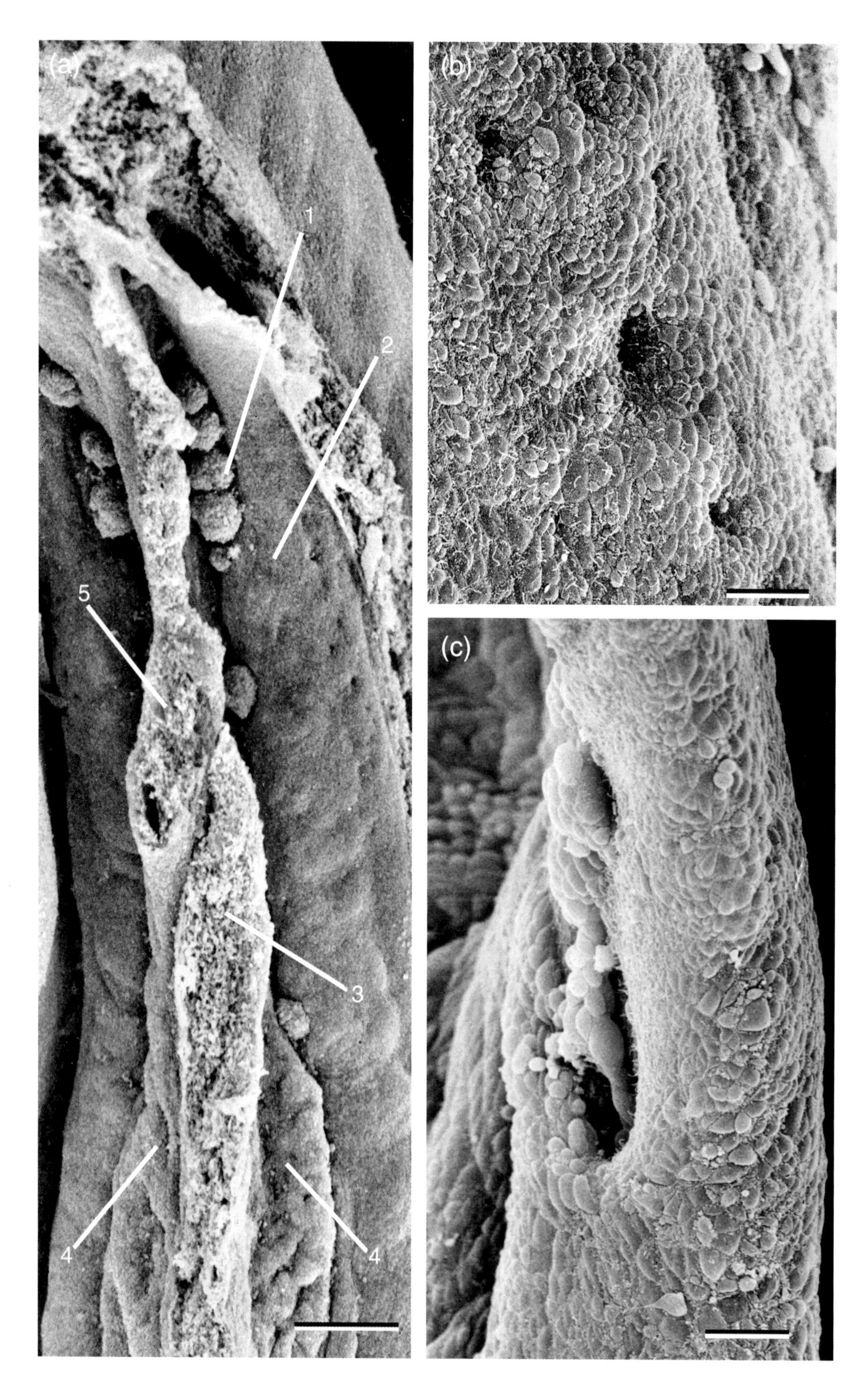
(a)
1
2
5
3
4
4
(b)
(c)

Plate 47 Scanning electron micrographs of the Mullerian duct and mesonephros

(a) Mullerian duct at stage 24 after removal of thoracic and abdominal organs. A row of nephrostomes is visible at the lateral side of the mesonephros. Bar = 100 μm.
(b) Enlargement of one of the nephrostomes in (a). Bar = 10 μm.
(c) Mullerian duct at stage 30 to show the funnel opening. Bar = 10 μm.

1. external glomeruli
2. mesonephros
3. dorsal mesentery
4. gonad
5. Mullerian duct

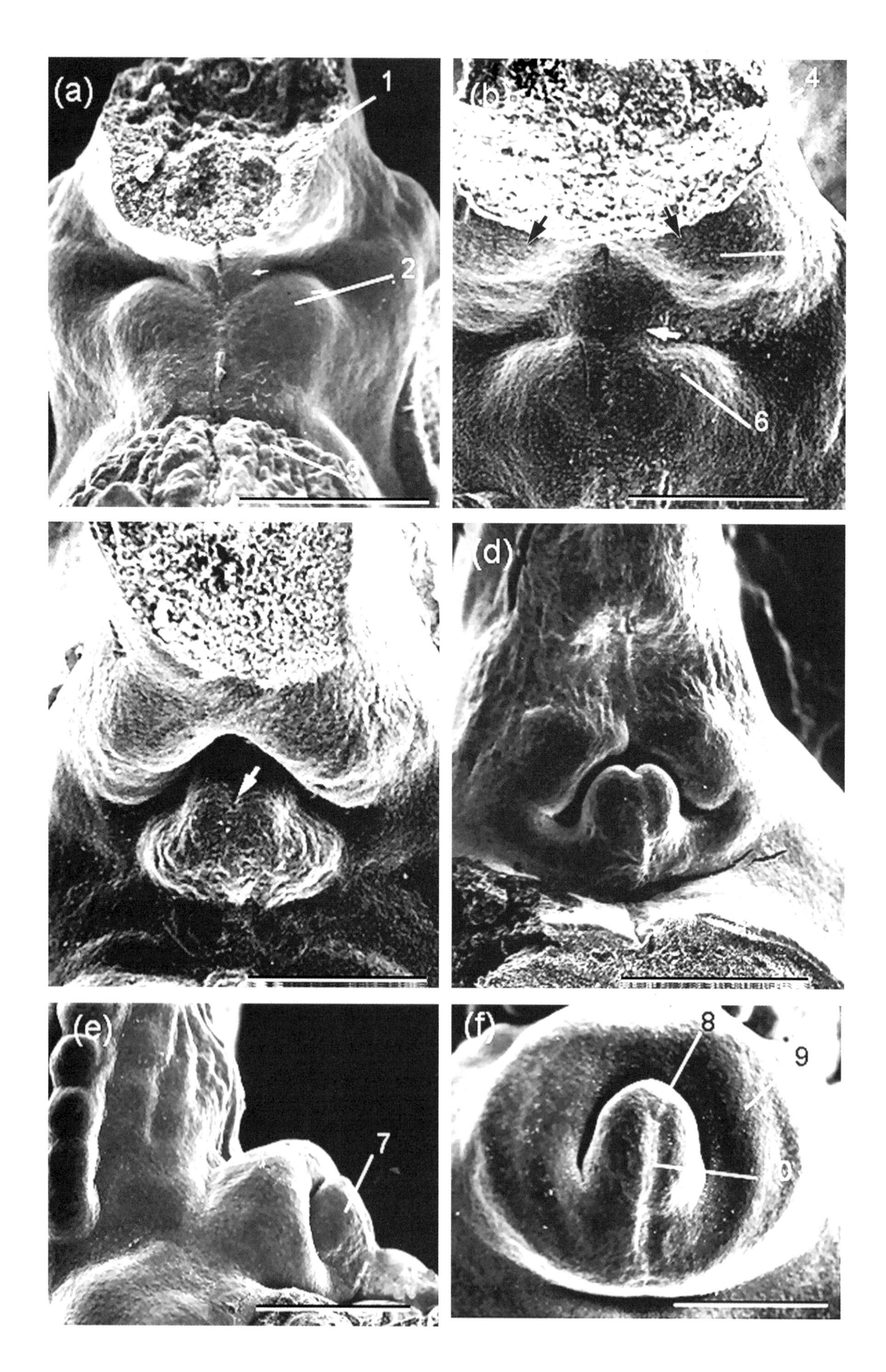
(a)
1
2
3
4
6
(d)
(e)
7
(f)
8
9

Plate 48 Scanning electron micrographs of a sequence of stages in the development of the external genitalia

Apart from (e) these are all seen in frontal view. The tail, or its cut edge, lies at the top of each image, the amnion at the bottom.

(a) 5 days of incubation, showing the genital eminences.
(b) 5.5 days of incubation. The genital eminences have merged in the midline to become the genital tubercle. The genital crest lies between the genital tubercle and the tail bud. Divergent arrows indicate genital swelling; horizontal arrow points to genital crest.
(c) 6 days of incubation. The genital crest has merged with the genital tubercle. Genital swellings overhang the genital tubercle. Arrow indicates phallic sulcus.
(d) 7 days of incubation. The apex of the genital tubercle is now bilobed and its apical diameter is 350 μm.
(e) 8 days of incubation, shown in lateral view. The genital swellings are now confluent with one another and have formed a collar around the genital tubercle.
(f) 9 days of incubation. The collar now totally encloses the genital tubercle.

1. section across tail bud
2. left genital eminence
3. amnion
4. wall of tail bud
5. left genital swelling
6. genital tubercle
7. median phallic body (genital tubercle)
8. tip of genital tubercle
9. inner border of collar
10. phallic sulcus

All bars = 500 μm

Reprinted from Bakst (1986), with permission of *Scanning Microscopy*.

PLATE 49

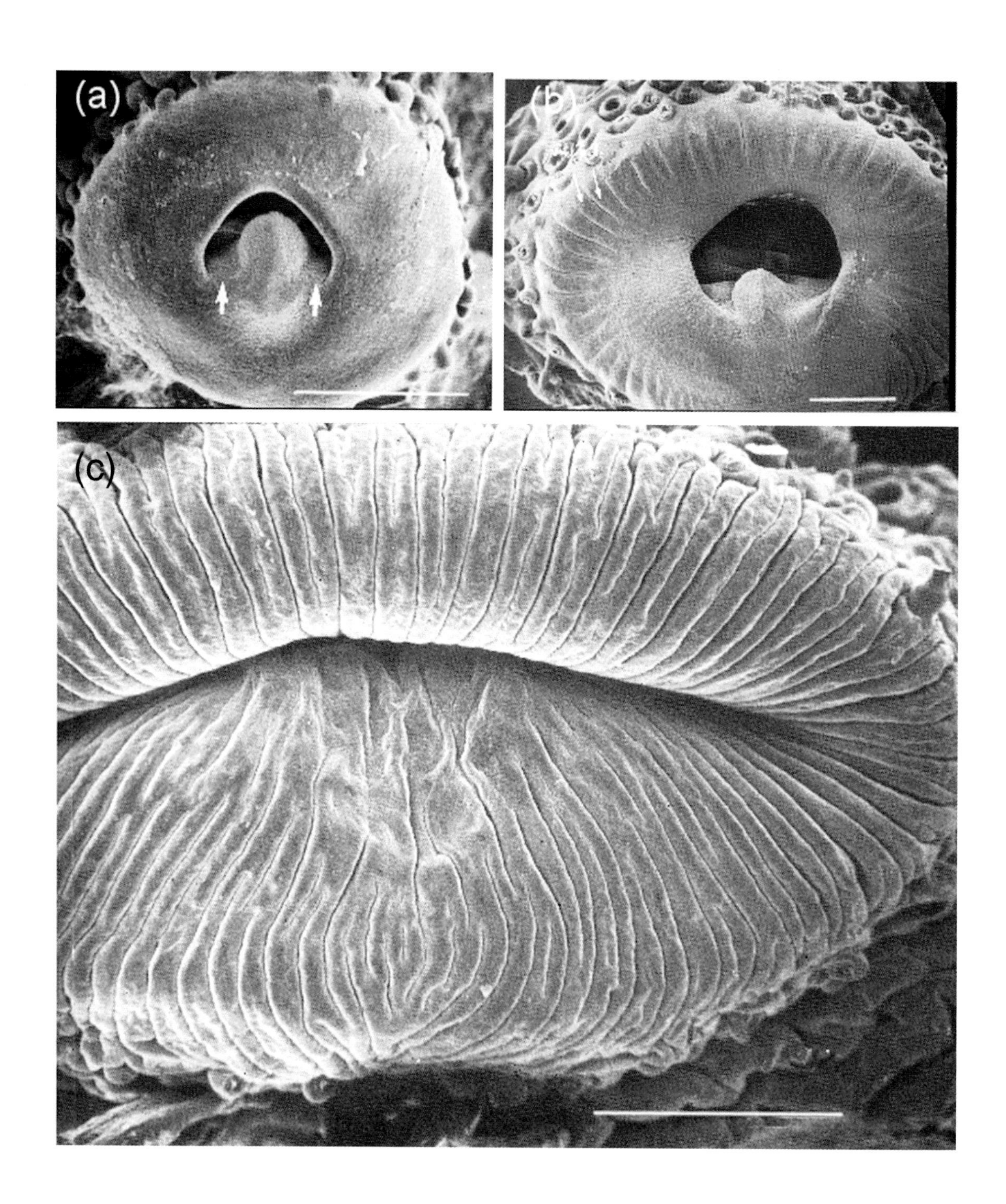

Plate 49 Sequences of stages in the development of the external genitalia (continued)

(a) 12 days of incubation: arrows indicate where the primordia for the lateral phallic bodies will arise.
(b) 15 days of incubation: the external cloaca has now developed folds (small arrows) around its periphery.
(c) 16 days of incubation. Radially orientated folds are now present on both upper and lower cloacal lips. The dorsal lip now projects over the lower lip, giving the vent a slit-like appearance. The phallus is no longer visible externally.

All bars = 500 μm

Reprinted from Bakst (1986), with permission of *Scanning Microscopy.*

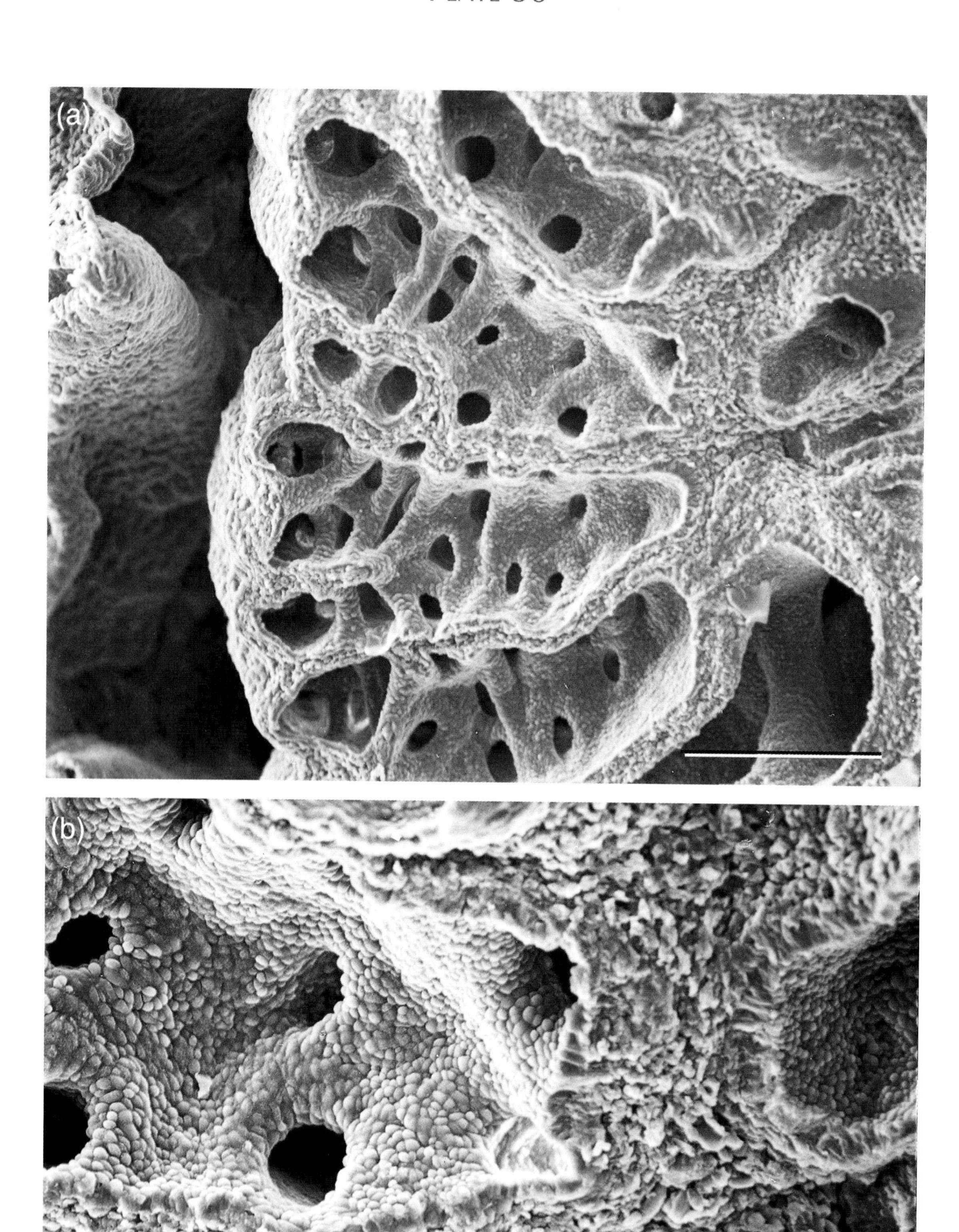
(a)
(b)

Plate 50 Scanning electron micrographs of the lung

(a) Fracture through the lung of an 8-day (stage 34) embryo. Bar = 200 μm.
(b) Enlargement of a portion of (a). Bar = 100 μm.

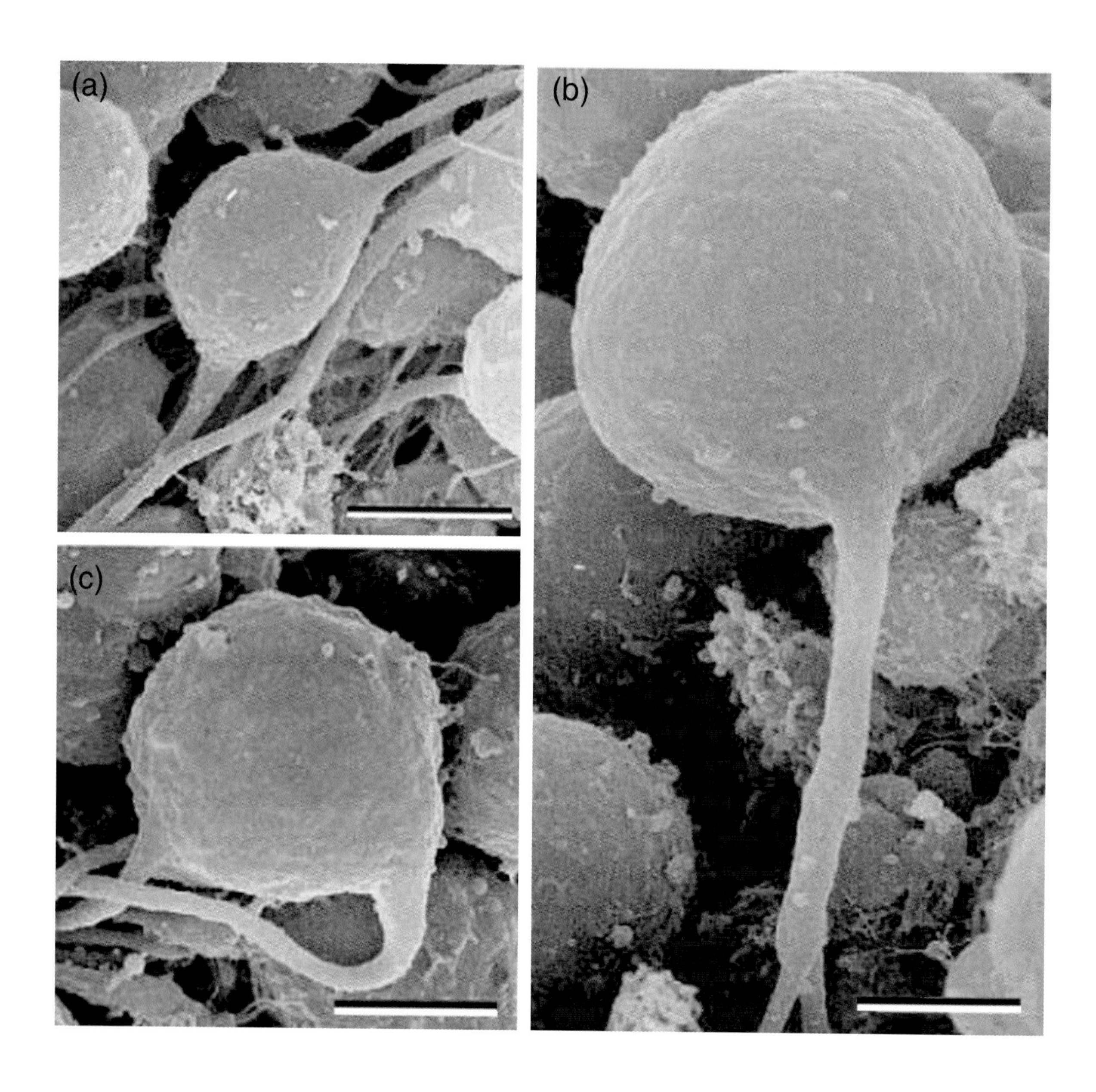
(a)
(b)
(c)

Plate 51 Scanning electron micrographs of neurons and dorsal root ganglion

(a) Bipolar neuron at 8 days of incubation.
(b) Bell-shaped neuron at 13 days of incubation.
(c) Unipolar, long-stemmed dorsal root ganglion at 13 days of incubation.

All bars = 5 μm.

Reprinted from Matsuda *et al.* (1996), with permission from Springer-Verlag.

Plate 52

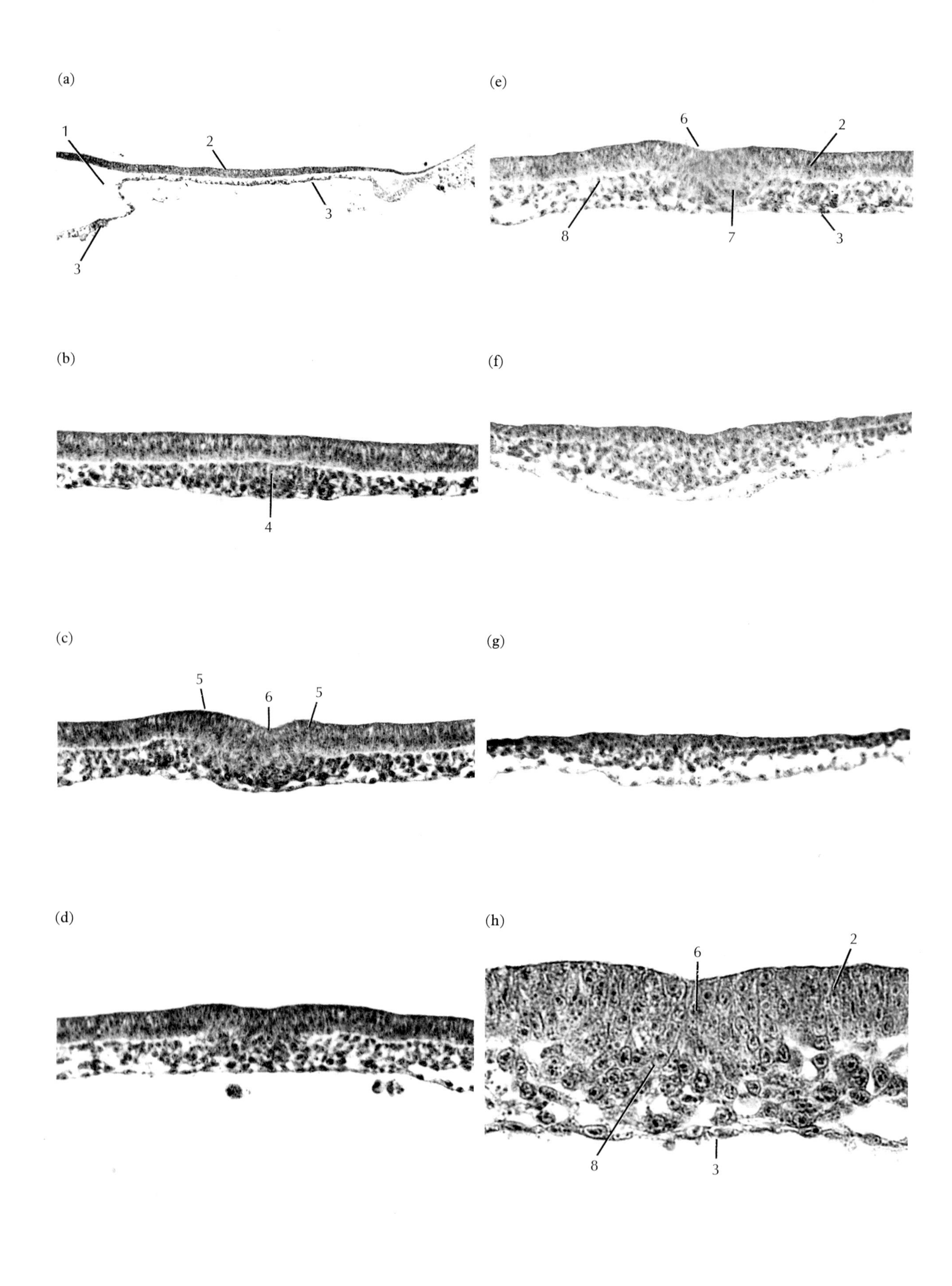

Plate 52 Stage 4 (18 h) TS (transverse sections)

(a) ×35 (b–g) × 150 (h) Enlargement through Hensen's (primitive) node, ×400

1. amnio-cardiac vesicle
2. ectoderm
3. endoderm
4. head process
5. primitive folds
6. primitive groove
7. primitive streak mesoderm
8. mesoderm

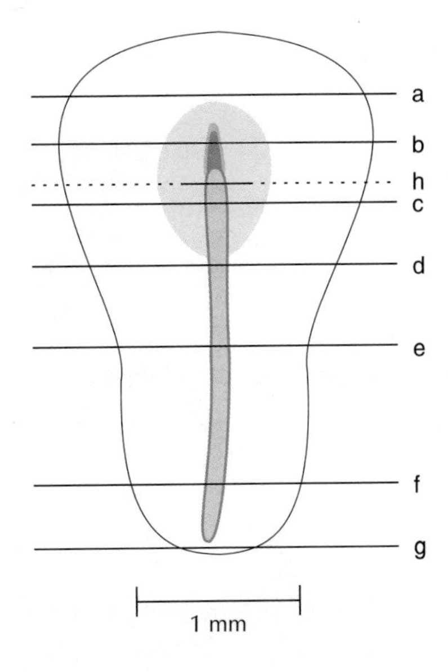

Plate 53

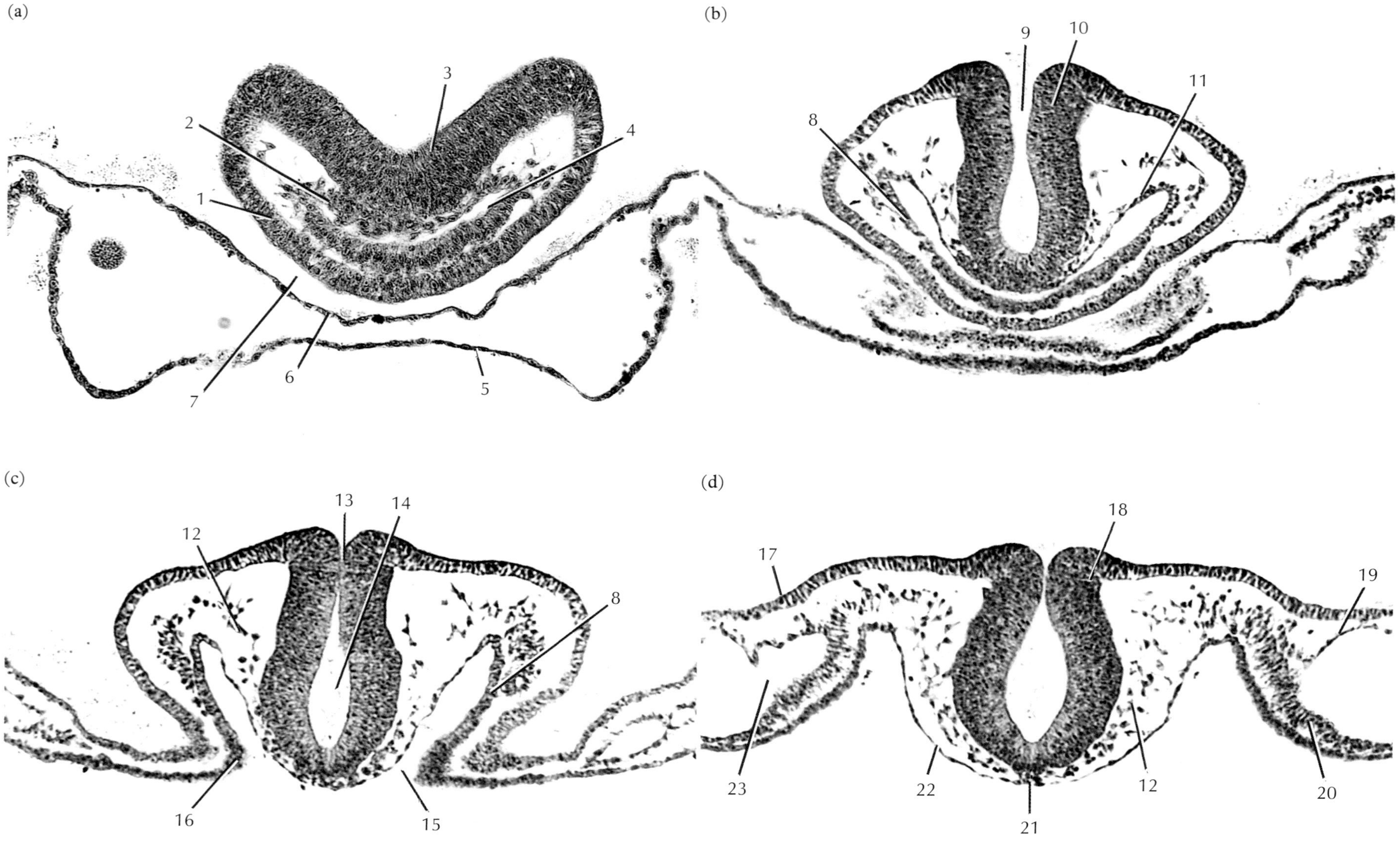

Plate 53 Stage 8 (26–29 h) TS

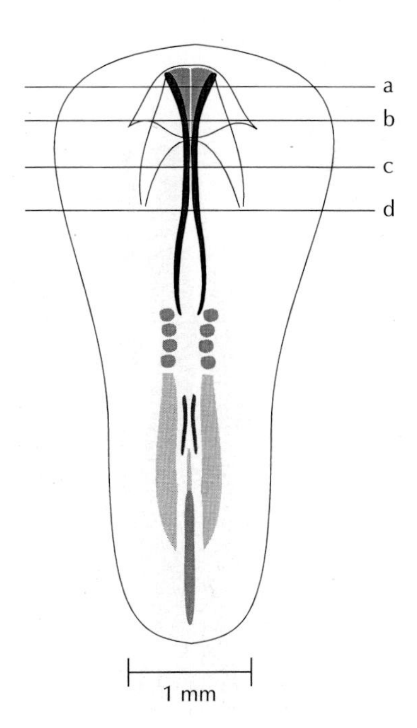

(a–d) × 160

1. head ectoderm
2. prechordal plate mesoderm
3. neural plate of future forebrain
4. anterior end of foregut
5. extra-embryonic endoderm
6. extra-embryonic ectoderm
7. space created by head fold
8. floor of pharynx
9. anterior neuropore
10. neural fold
11. thin roof of pharynx
12. head mesenchyme
13. point of fusion of neural folds
14. lumen of neural tube
15. anterior intestinal portal
16. lateral border of anterior intestinal portal
17. ectoderm
18. future neural crest
19. somatic lateral plate mesoderm
20. splanchnic lateral plate mesoderm
21. anterior tip of notochord
22. endoderm
23. embryonic coelom

Plate 54

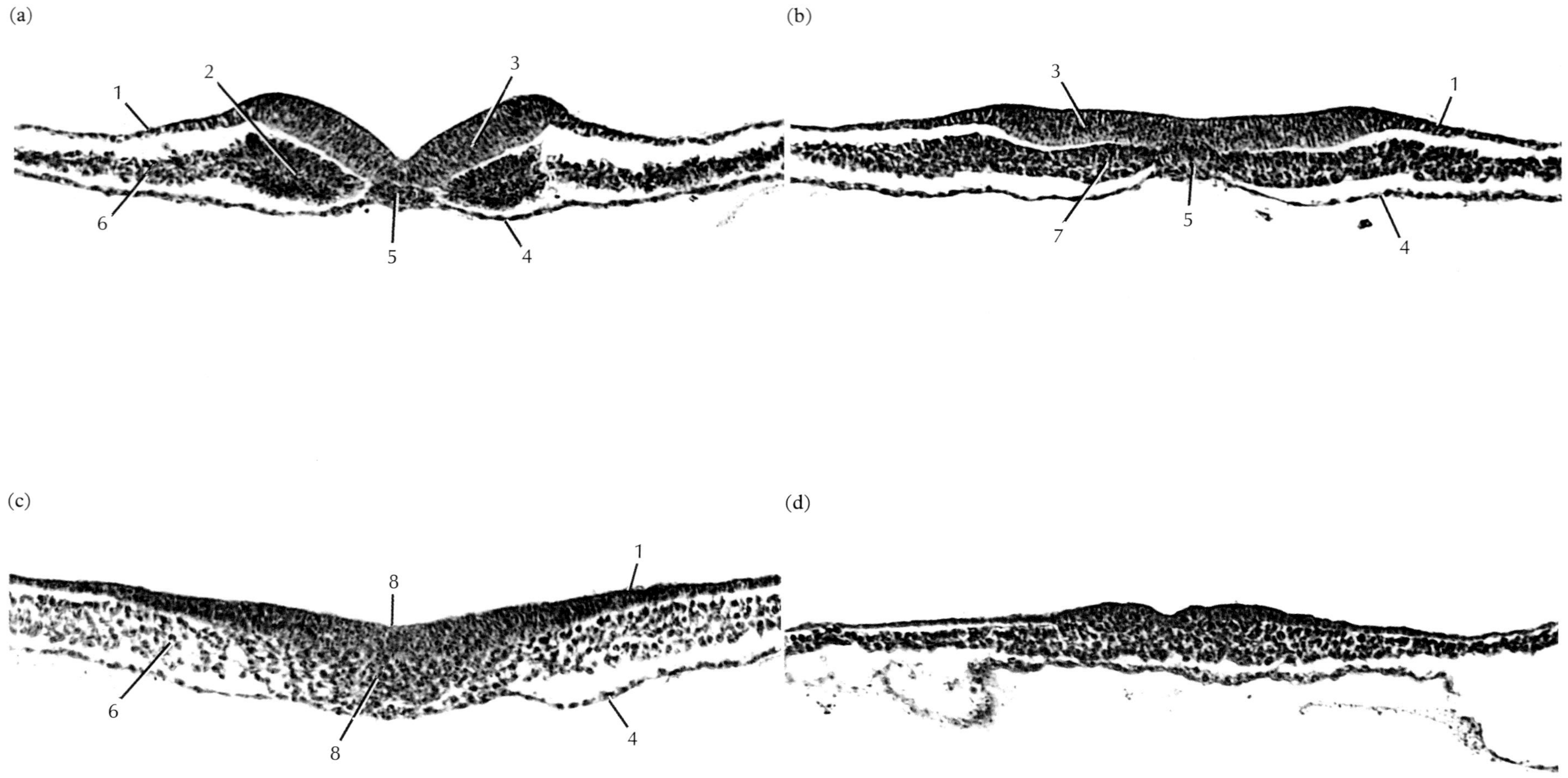

(a)

1 2 3 6 5 4

(b)

3 1 7 5 4

(c)

1 8 6 8 4

(d)

Plate 54 Stage 8 (26–29 h) TS

(a–d) × 160

1. ectoderm
2. second somite
3. open neural plate
4. endoderm
5. notochord
6. lateral plate mesoderm
7. segmental plate mesoderm
8. primitive streak

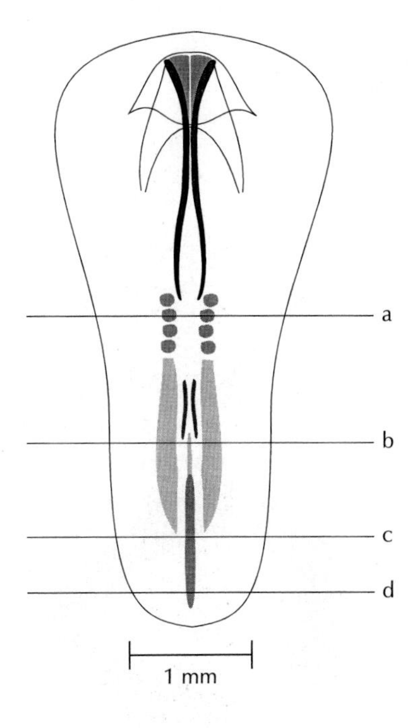

PLATE 55

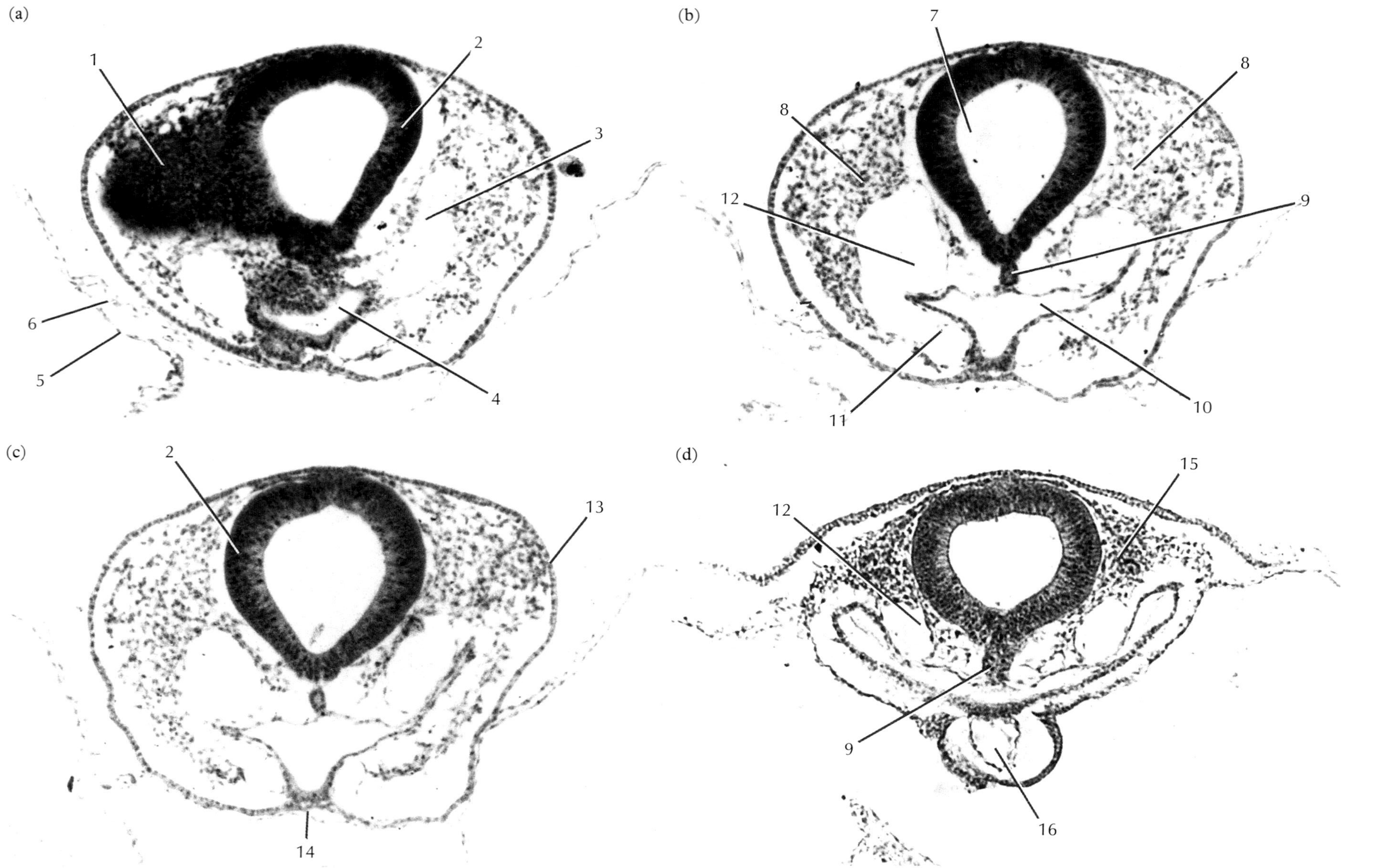

(a)

(b)

(c)

(d)

Plate 55 Stage 9 (29–33 h) TS

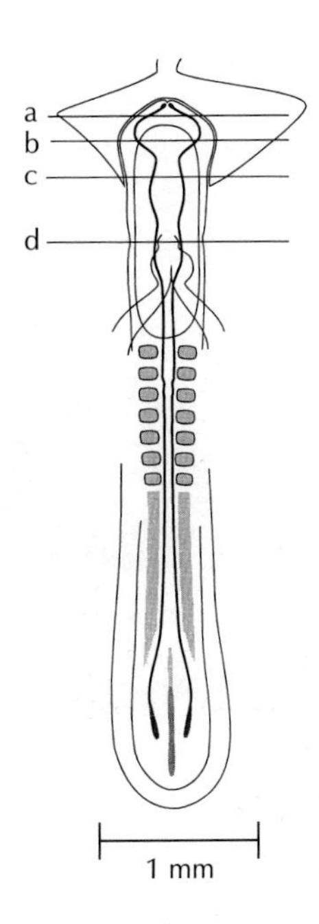

(a–d) × 100

1. edge of right optic vesicle
2. forebrain
3. left dorsal aorta
4. oral cavity
5. splanchnopleure
6. somatopleure
7. lumen of forebrain
8. head mesenchyme
9. notochord
10. pharynx
11. right ventral aorta
12. right dorsal aorta
13. ectoderm of head
14. stomatodaeum (stomatodaeal plate)
15. mesoderm
16. bulbus cordis

Plate 56

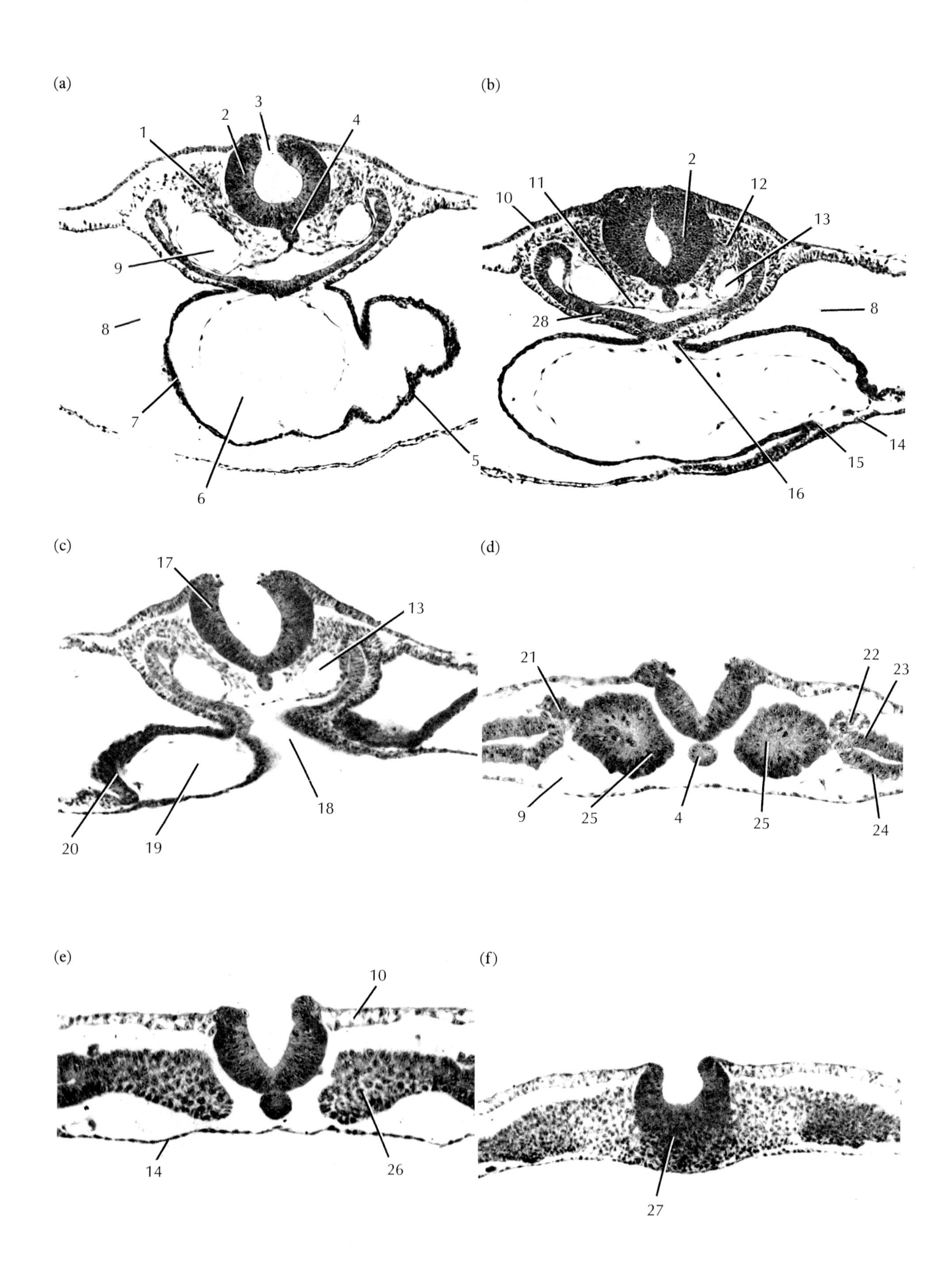

(a)
1
2
3
4
9
8
7
6
5
(b)
10
11
2
12
13
8
28
16
15
14
(c)
17
13
18
19
20
(d)
21
22
23
9
25
4
25
24
(e)
10
14
26
(f)
27

Plate 56 Stage 9 (29–33 h) TS

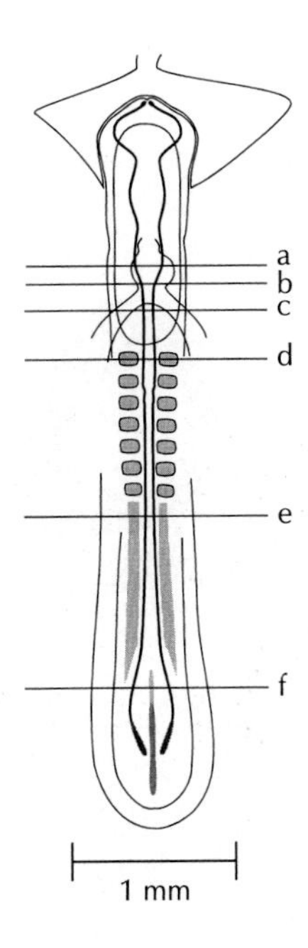

(a–f) × 100

1. mesenchyme
2. hindbrain
3. unclosed region of neural tube
4. notochord
5. wall of atrium
6. lumen of ventricle
7. wall of ventricle
8. coelom
9. right dorsal aorta
10. ectoderm
11. thin roof of pharynx
12. parachordal mesoderm
13. left dorsal aorta
14. endoderm
15. ventral mesentery (ventral mesocardium)
16. dorsal mesentery (dorsal mesocardium)
17. neural plate
18. anterior intestinal portal
19. right sinu-atrial region
20. wall of sinus venosus
21. right intermediate mesoderm
22. left intermediate mesoderm
23. somatic lateral plate mesoderm
24. splanchnic mesoderm of lateral plate
25. somite
26. segmental plate mesoderm
27. Hensen's (primitive) node
28. Floor of pharynx

Plate 57

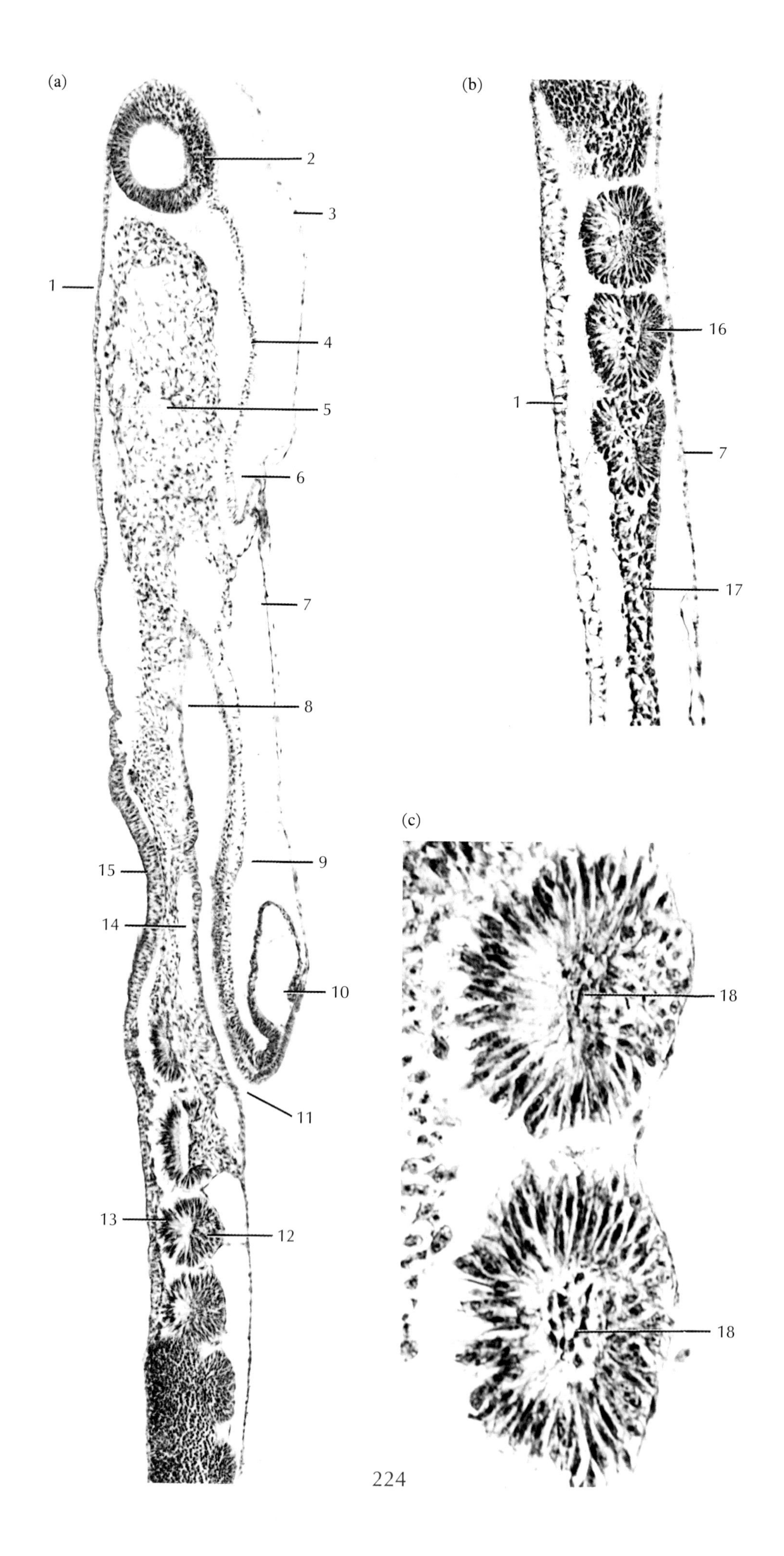

Plate 57 Stage 10 (33–38 h) LS (longitudinal sections: sagittal)

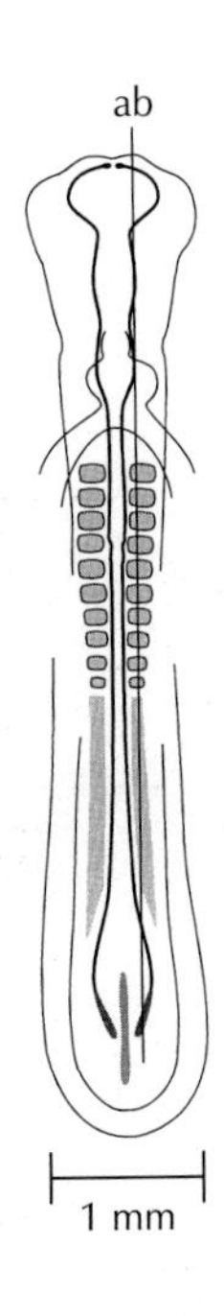

(a) Anterior end of embryo, ×120
(b) Posterior end of same embryo, ×120
(c) Enlargement of two newly formed somites, ×360

1. embryonic dorsal ectoderm
2. prosencephalon (forebrain)
3. extra-embryonic ectoderm and endoderm
4. ventral ectoderm of head
5. head mesenchyme
6. head fold
7. embryonic endoderm
8. foregut
9. pericardial cavity
10. cardiac primordium
11. anterior intestinal portal
12. sclerotome of an anterior somite
13. derma-myotome
14. dorsal aorta
15. neural plate
16. newly formed (posterior) somite
17. segmental plate mesoderm
18. cells in somitocoele of somite

Plate 58

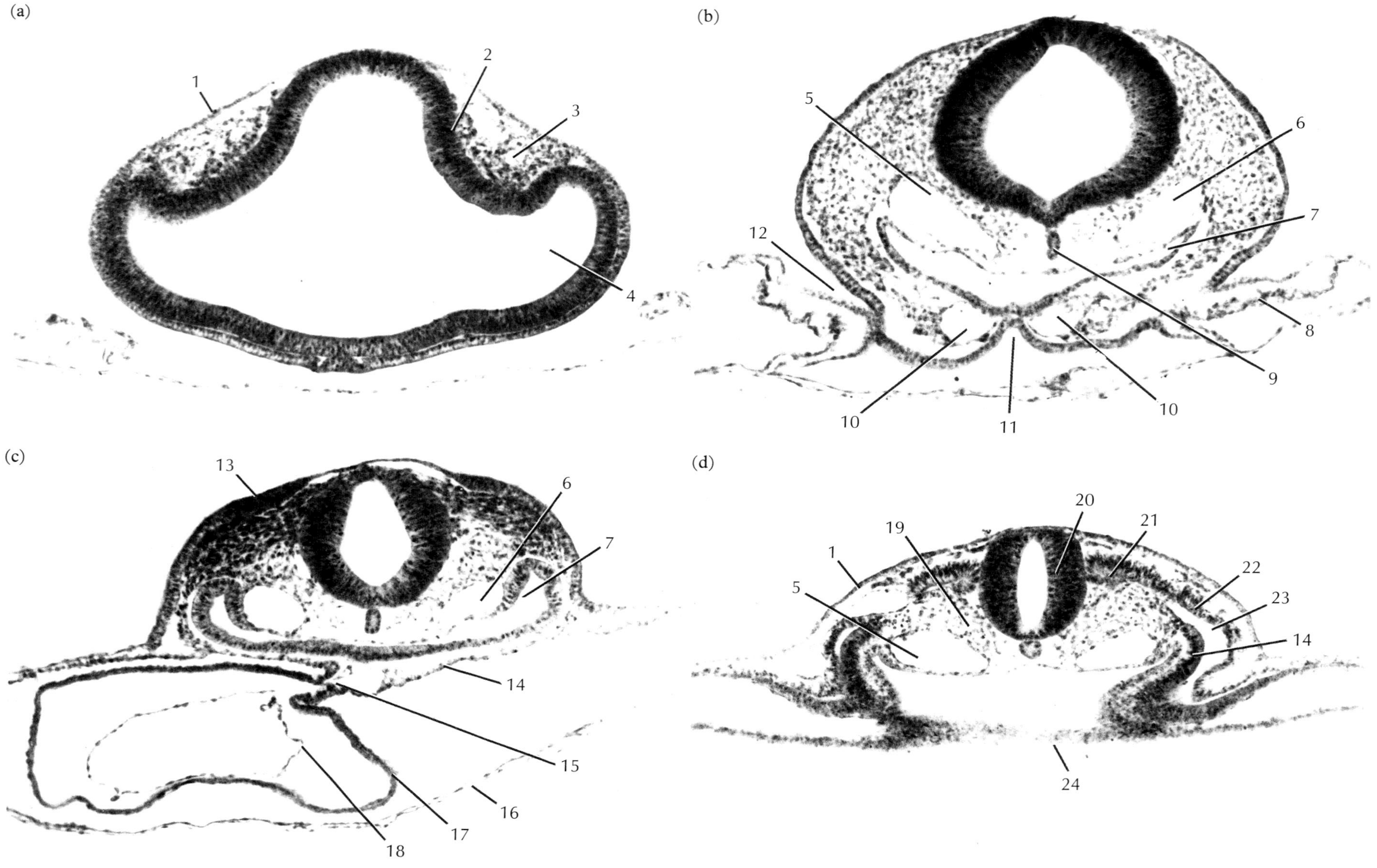

(a)

1
2
3
4

(b)

5
6
7
12
9
8
10
11
10

(c)

13
6
7
14
15
16
17
18

(d)

20
19
21
1
22
5
23
14
24

Plate 58 Stage 10 (33–38 h) TS

(a–d) × 150

1. ectoderm of head
2. prosencephalon
3. head mesenchyme
4. lumen of optic vesicle
5. right dorsal aorta
6. left dorsal aorta
7. pharynx
8. extra-embryonic somatic mesoderm
9. notochord
10. cardiac primordia (left and right)
11. stomatodaeum (stomatodaeal plate)
12. lateral head fold
13. otic placode
14. embryonic splanchnic (lateral plate) mesoderm
15. dorsal mesocardium
16. endoderm
17. epimyocardium
18. endocardium
19. sclerotome
20. neural tube
21. derma-myotome
22. embryonic somatic (lateral plate) mesoderm
23. embryonic coelom
24. lip of anterior intestinal portal

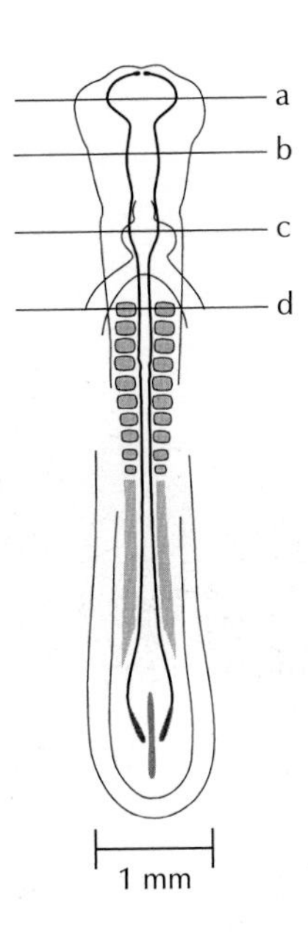

PLATE 59

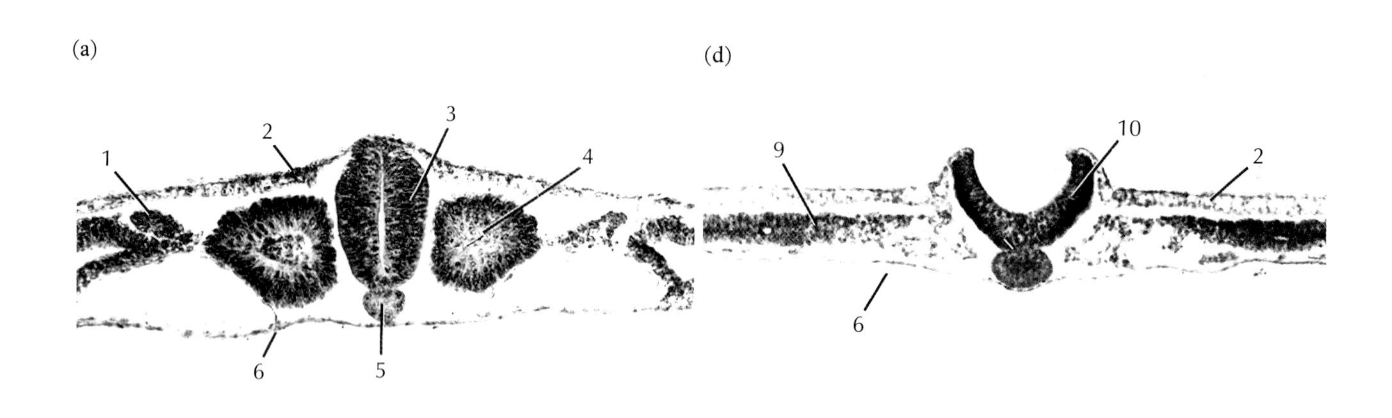

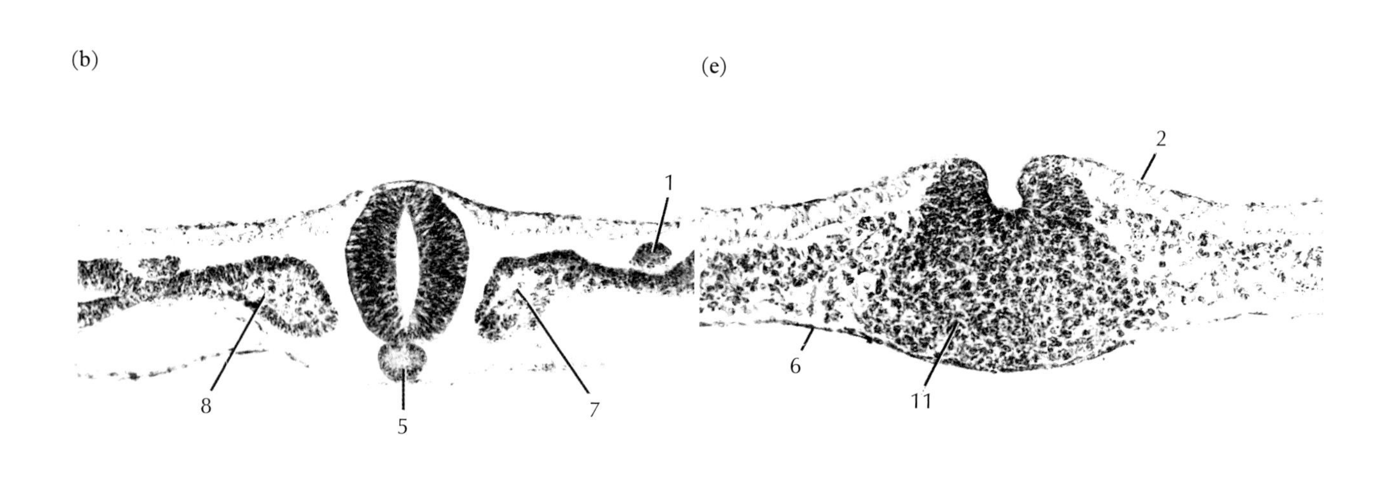

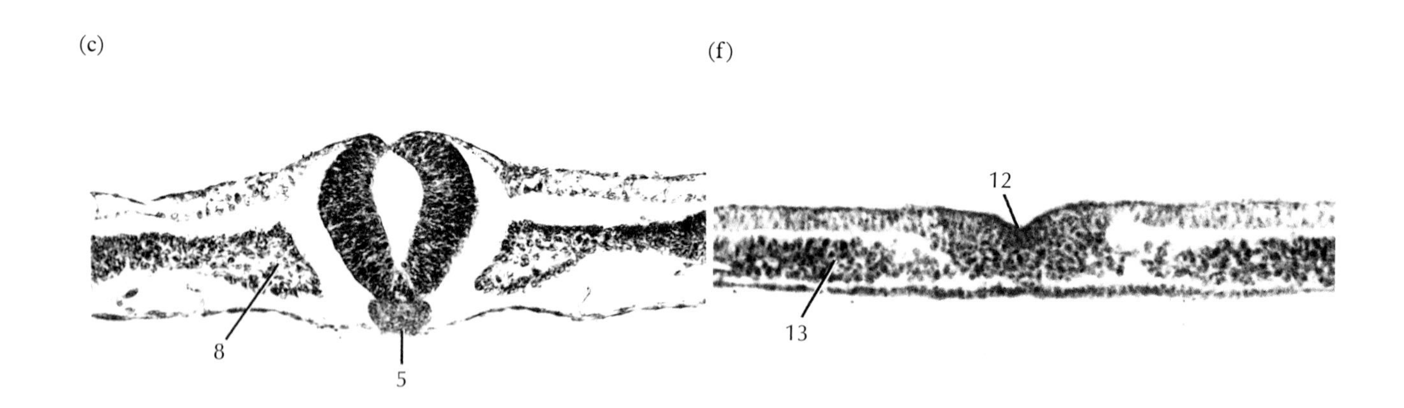

Plate 59 Stage 10 (33–38 h) TS

(a–f) ×170

1. intermediate mesoderm (nephric rudiment)
2. embryonic ectoderm
3. neural tube
4. left somite
5. notochord
6. endoderm
7. left segmental plate mesoderm
8. right segmental plate mesoderm
9. lateral plate mesoderm
10. unclosed neural plate
11. Hensen's node
12. primitive streak
13. mesoderm lateral to primitive streak

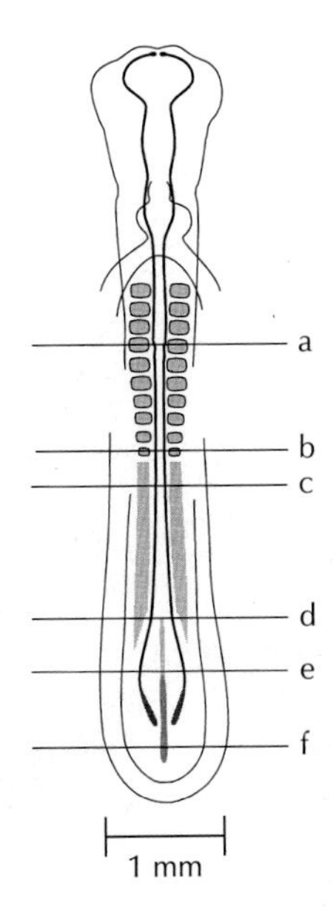

PLATE 60

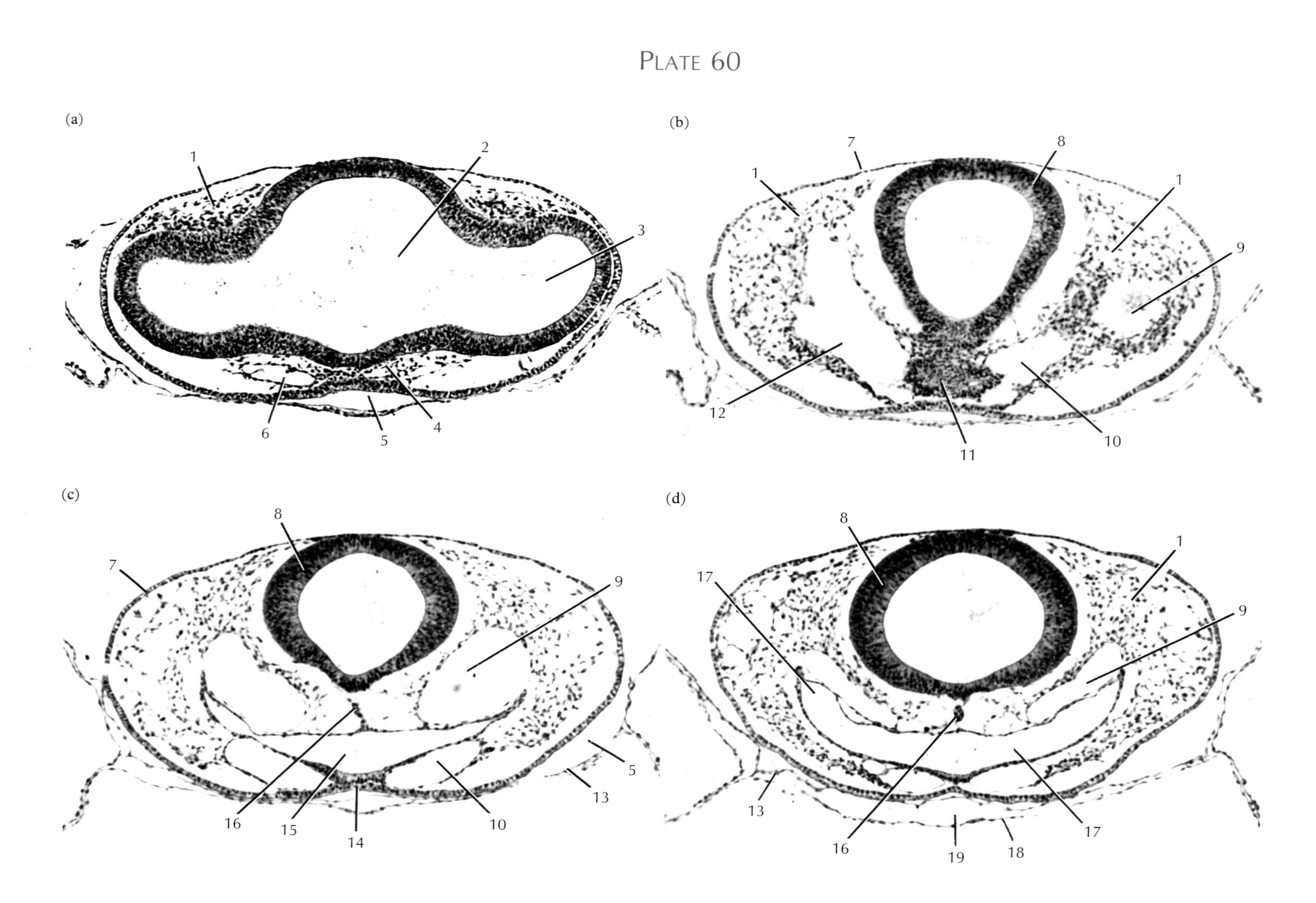

Plate 60 Stage 11 (40–45 h) TS

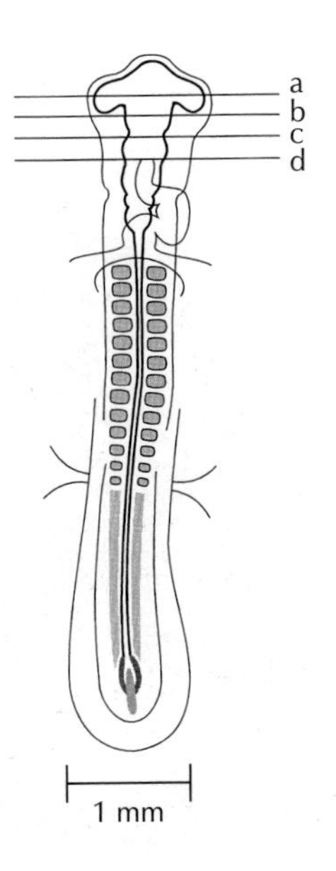

(a–d) × 133

1. head mesenchyme
2. lumen of prosencephalon
3. left optic vesicle
4. pre-notochordal mesenchyme
5. sub-cephalic pocket formed by head fold
6. right ventral aorta
7. ectoderm of head
8. mesencephalon
9. left dorsal aorta
10. left ventral aorta
11. anterior tip of foregut
12. communication of right ventral and dorsal aortae
13. extra-embryonic ectoderm
14. stomatodaeum (stomatodaeal plate)
15. oral cavity
16. tip of notochord
17. lumen of pharynx
18. extra-embryonic endoderm
19. pro-amnion

Plate 61

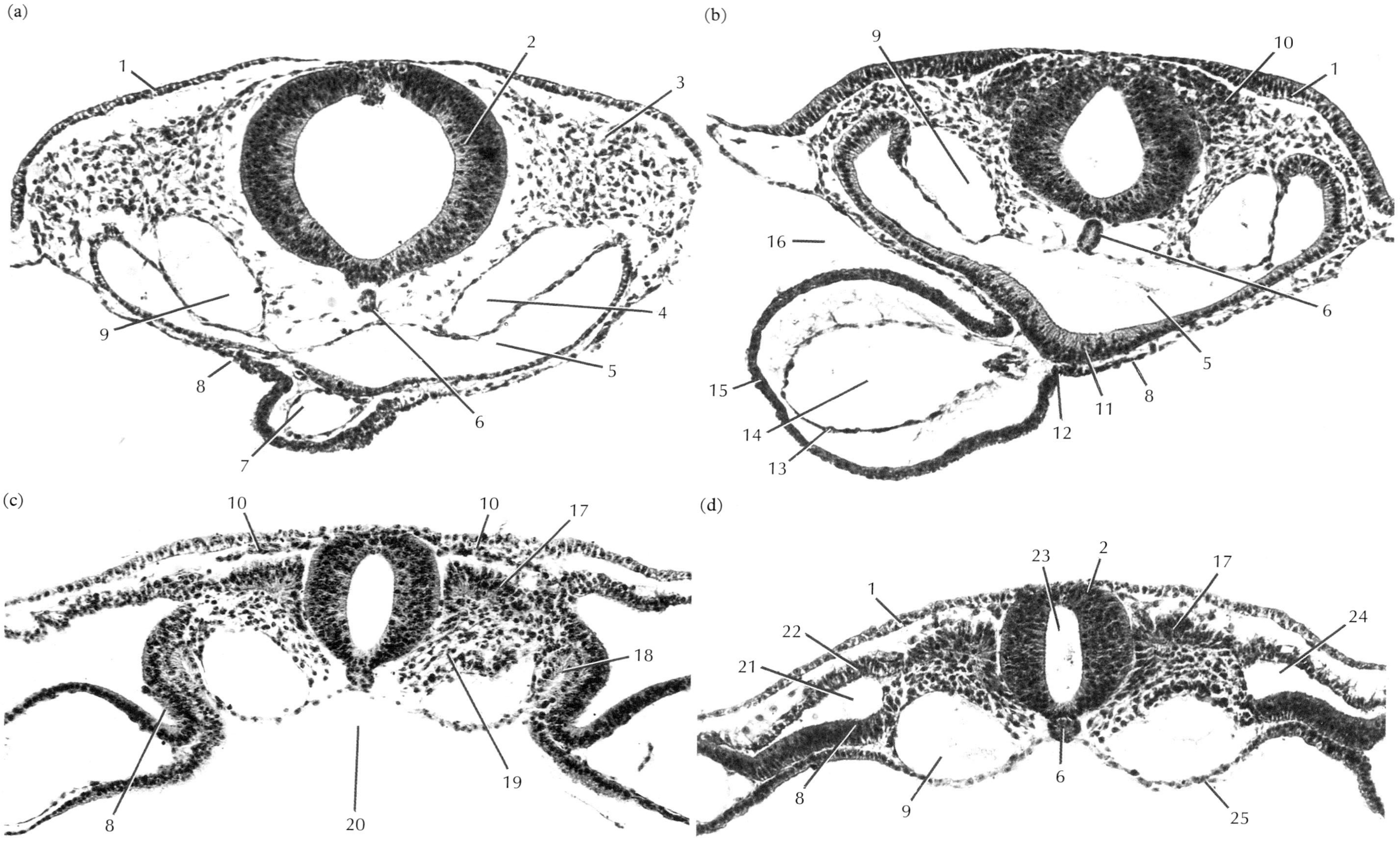

(a)

(b)

(c)

(d)

Plate 61 Stage 11 (40–45 h) TS

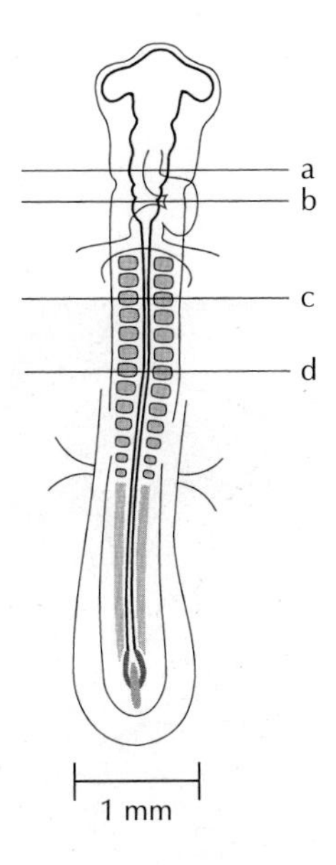

(a–d) × 175

1. ectoderm
2. mesencephalon
3. head mesenchyme
4. left dorsal aorta
5. lumen of pharynx
6. notochord
7. truncus arteriosus
8. splanchnic lateral plate mesoderm
9. right dorsal aorta
10. neural crest
11. keel of pharynx
12. dorsal mesocardium
13. endocardium
14. lumen of ventricle
15. epimyocardium
16. pericardial cavity
17. derma-myotome
18. left edge of midgut
19. sclerotome
20. midgut
21. right coelom
22. right somatic lateral plate mesoderm
23. lumen of neural tube
24. left coelom
25. endoderm

Plate 62

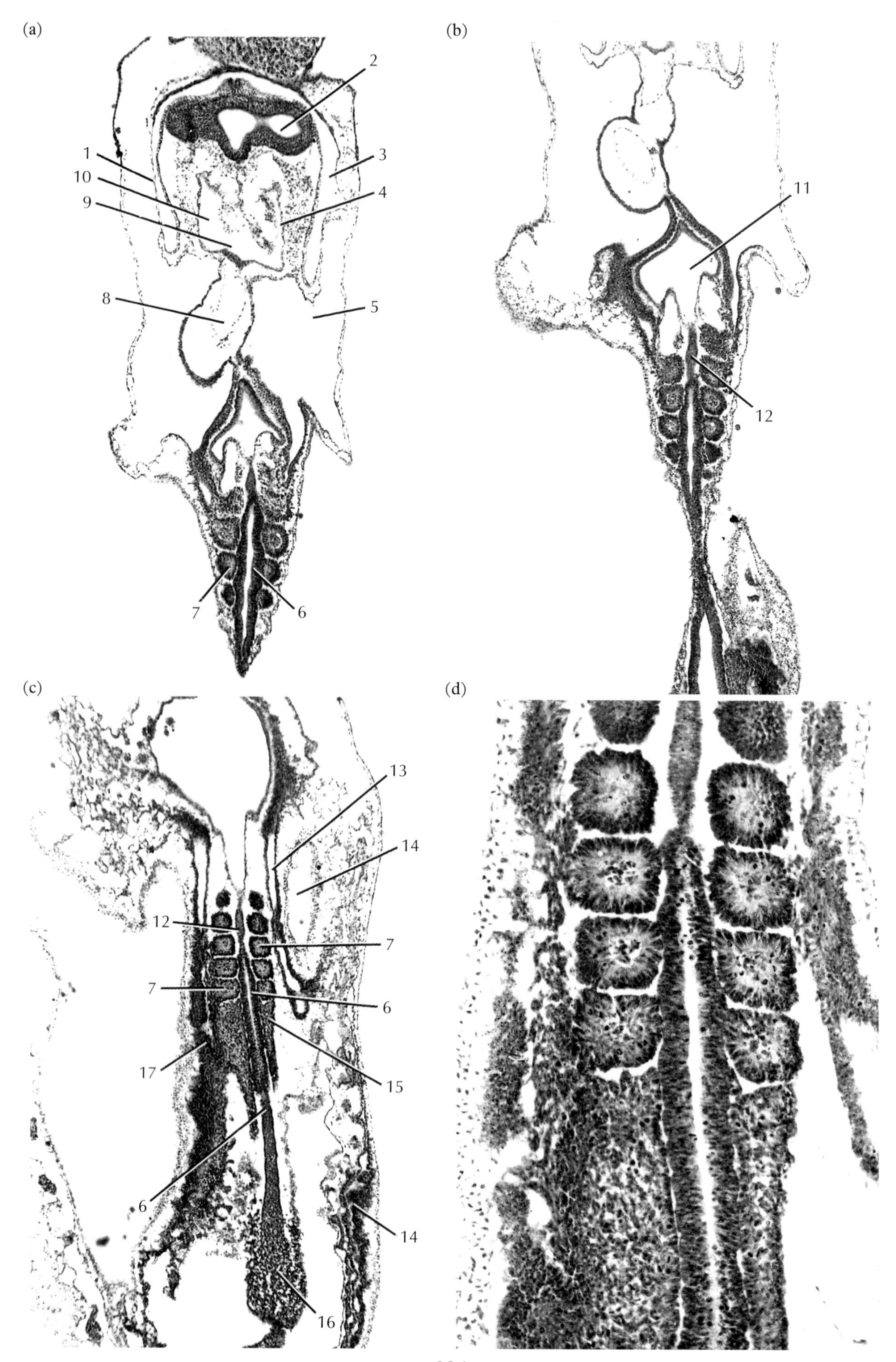

Plate 62 Stage 12 (45–49 h) LS (longitudinal sections: coronal)

(a–c) ×50 (d) ×200

(a) Enlarged on *Plate 63*
(d) Enlargement of part of (c)

Diagram shows view of the embryo from the ventral side. Sections taken through plane of page.

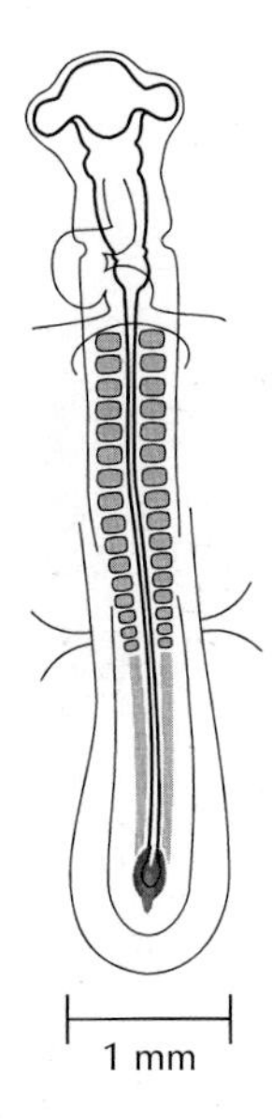

1. amniotic ectoderm
2. left optic vesicle
3. amniotic cavity
4. left anterior cardinal vein
5. pericardial cavity
6. wall of neural tube
7. somite
8. ventricle
9. common cardinal vein
10. right anterior cardinal vein
11. gut in region of anterior intestinal portal
12. notochord
13. left omphalo-mesenteric (vitelline) vein
14. yolk sac
15. segmental plate mesoderm
16. Hensen's node
17. intermediate mesoderm

PLATE 63

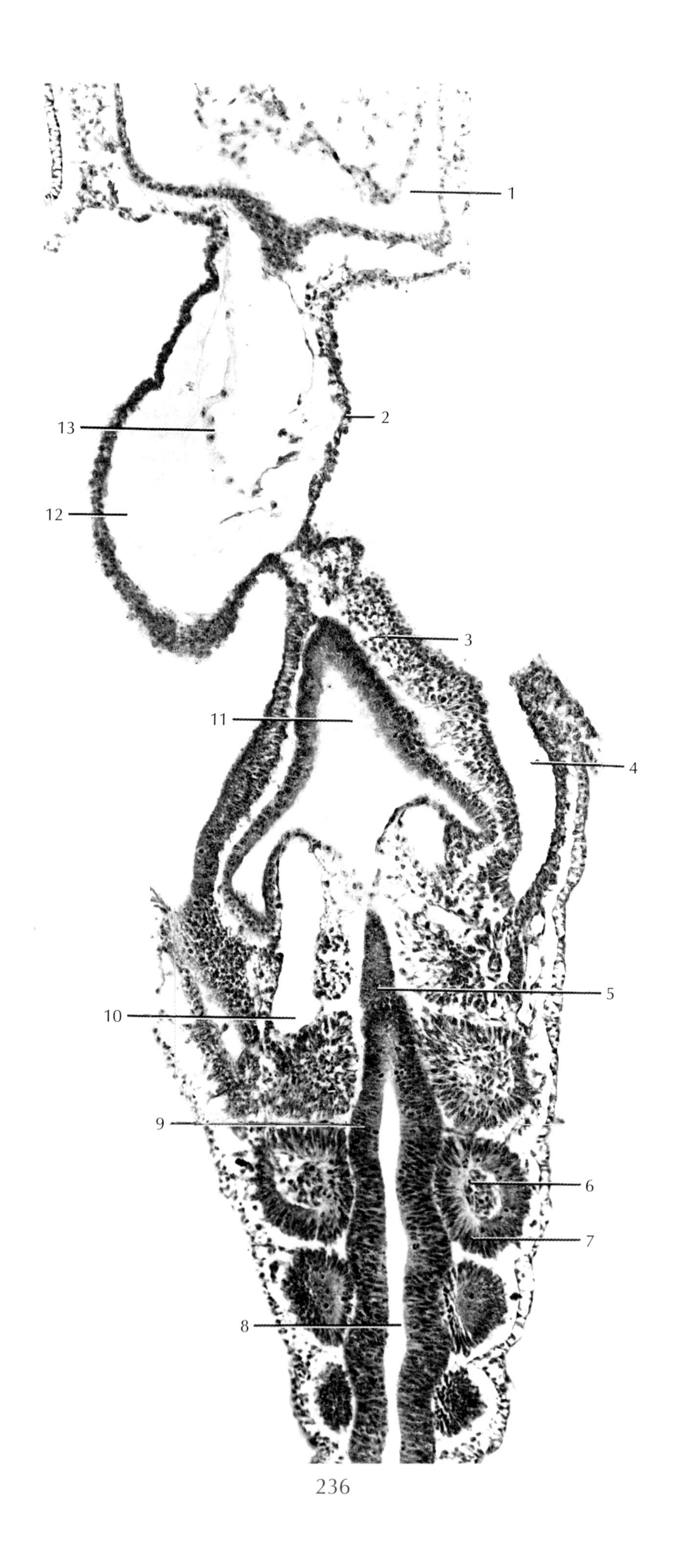

Plate 63 Stage 12 (45–49 h) LS (longitudinal sections: coronal)

Enlargement of part of *Plate 62a*, ×200

Diagram shows view of embryo from ventral side. Sections taken through plane of page.

1. left anterior cardinal vein
2. ventricle
3. splanchnic lateral plate mesoderm
4. embryonic coelom
5. notochord
6. central cells of somite
7. epithelial wall of somite
8. lumen of neural tube
9. wall of neural tube
10. right dorsal aorta
11. gut in region of anterior intestinal portal
12. cardiac jelly
13. endocardium

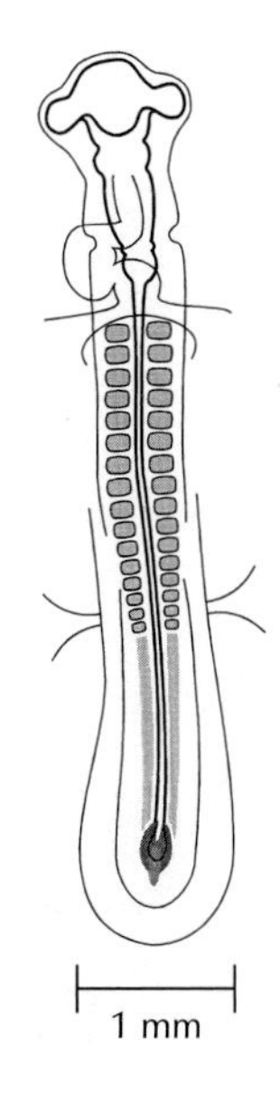

Plate 64

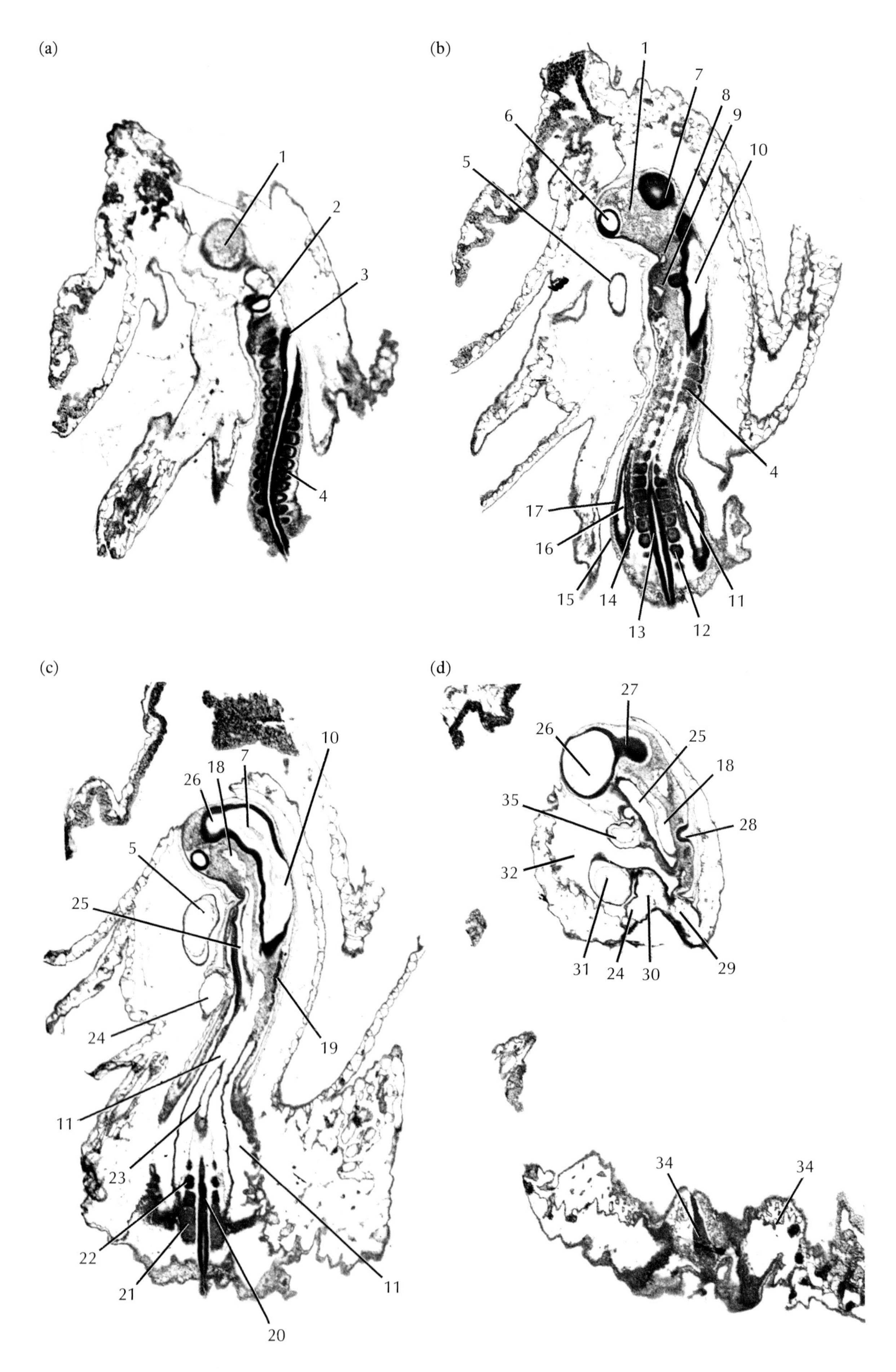

(a)
1
2
3
4
(b)
1
7
8
9
6
10
5
4
17
16
15
14
13
12
11
(c)
10
26
18
7
5
25
24
19
11
23
22
21
20
11
(d)
27
26
25
18
35
28
32
31
24
30
29
34
34

Plate 64 Stage 13 (48–52 h) LS (longitudinal sections: coronal)

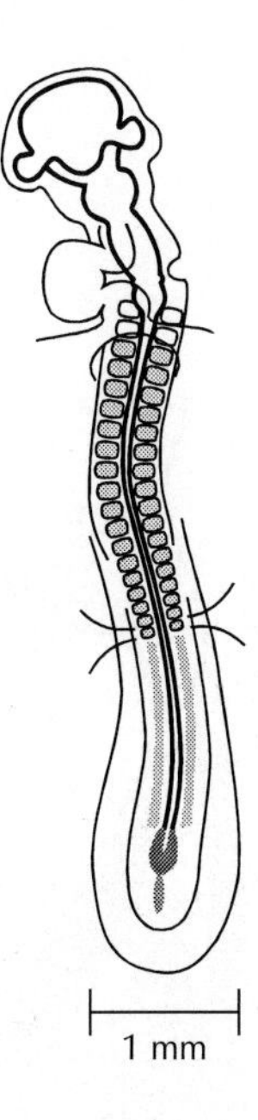

(a–d) ×25

Diagram shows view of embryo from ventral side. Sections taken through plane of page.

1. head mesenchyme
2. left otic vesicle
3. myelencephalon
4. somite
5. lumen of ventricle
6. right eye
7. mesocoele
8. first pharyngeal pouch
9. second pharyngeal pouch
10. myelocoele
11. embryonic coelom
12. 17th somite
13. wall of neural tube
14. nephric duct
15. lateral body wall
16. splanchnic lateral plate mesoderm
17. somatic lateral plate mesoderm
18. anterior cardinal vein
19. myotome
20. roof of neural tube
21. segmental plate mesoderm
22. last formed somite
23. gut
24. right common cardinal vein
25. pharynx
26. prosencephalon
27. left eye
28. left otic placode
29. left common cardinal vein
30. sinus venosus
31. atrium
32. pericardial cavity
33. ventral aorta
34. blood vessels in yolk sac

PLATE 65

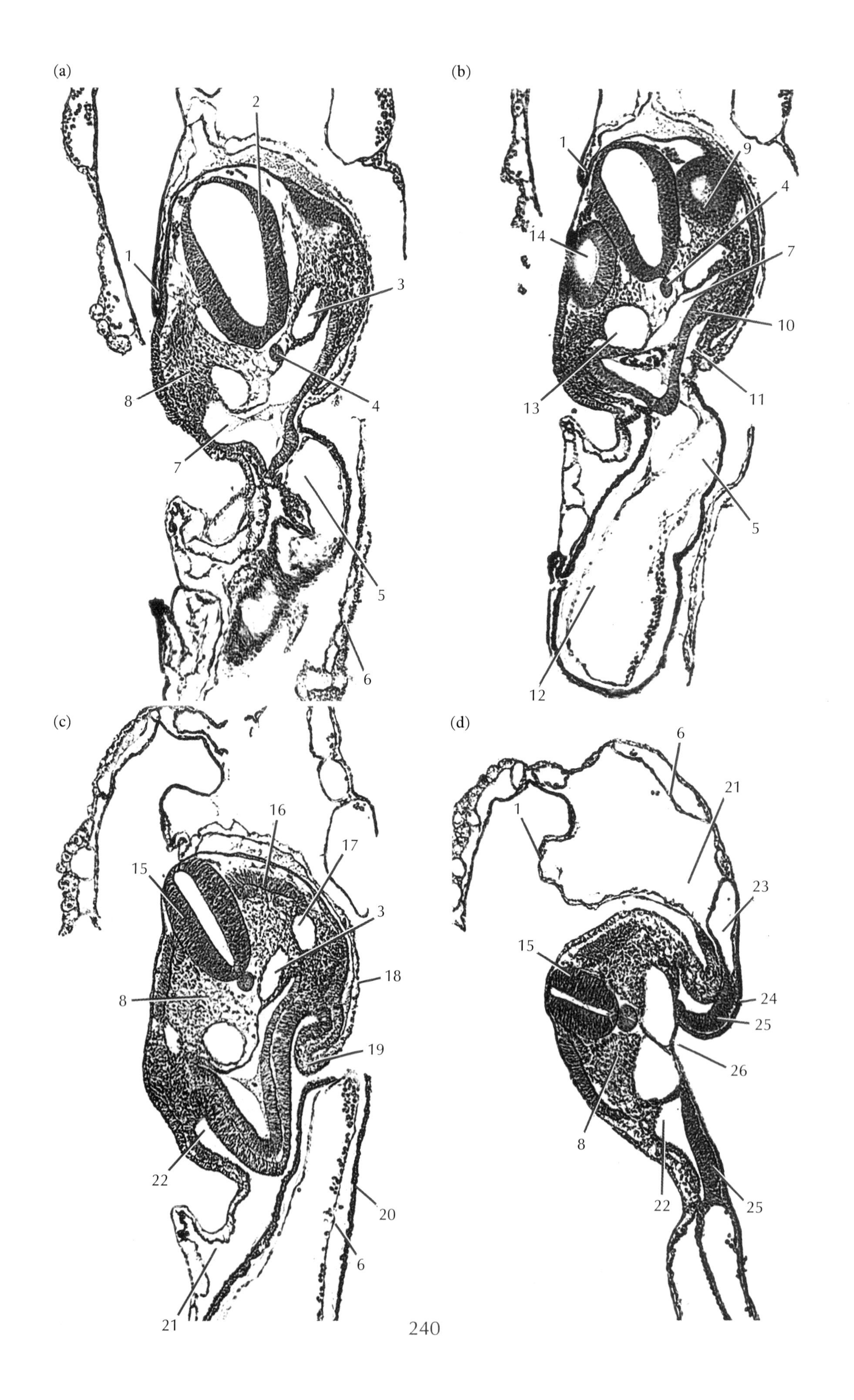

(a)
2
1
3
8
4
7
5
6
(b)
1
9
4
14
7
10
13
11
5
12
(c)
16
15
17
3
18
8
19
22
20
6
21
(d)
6
21
1
23
15
24
25
26
8
22
25

Plate 65 Stage 13 (48–52 h) TS

(a–d) × 100

Diagram shows view of embryo from ventral side.

1. amniotic fold
2. myelencephalon
3. left dorsal aorta
4. notochord
5. truncus arteriosus
6. splanchnic lateral plate mesoderm of yolk sac
7. pharynx
8. sclerotome
9. left otic vesicle
10. thick-walled floor of pharynx
11. left ventral aorta
12. ventricle
13. right dorsal aorta
14. right otic vesicle
15. neural tube
16. derma-myotome
17. left anterior cardinal vein
18. amnion
19. left lateral body fold
20. endodermal layer of yolk sac
21. extra-embryonic coelom
22. embryonic coelom
23. left omphalo-mesenteric vein
24. endoderm
25. splanchnic lateral plate mesoderm of gut
26. midgut

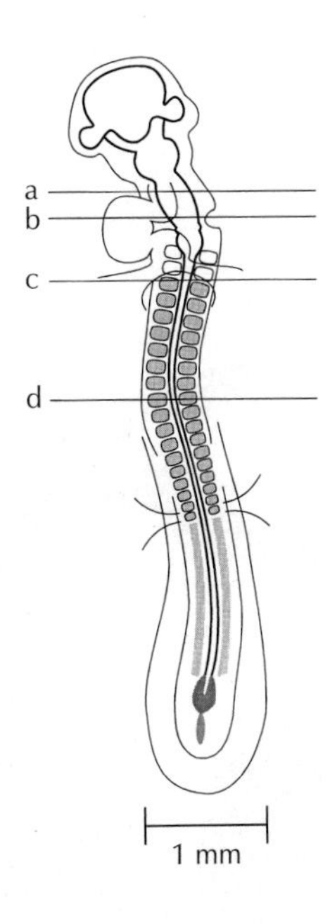

PLATE 66

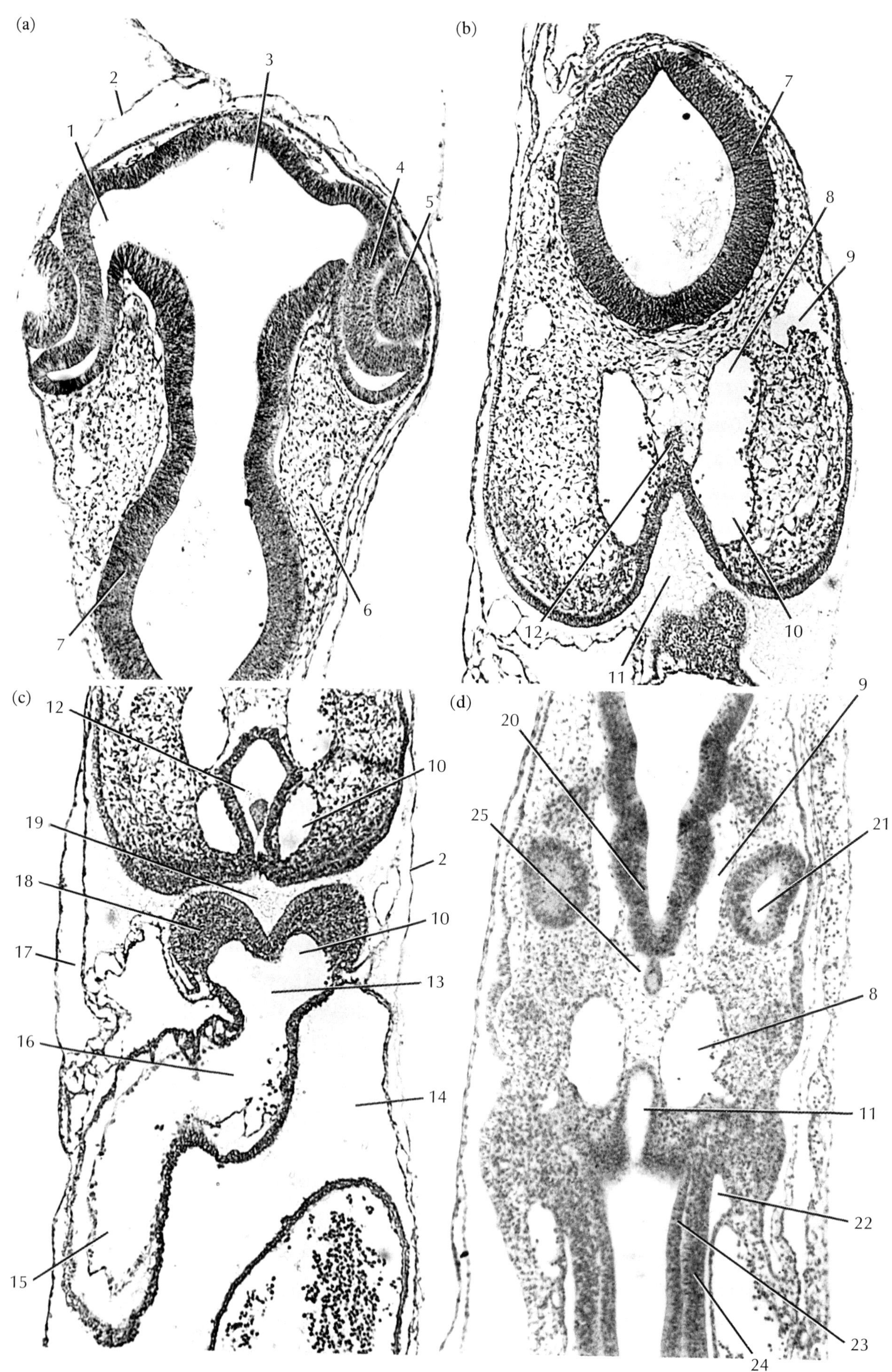

(a)
1
2
3
4
5
6
7
(b)
7
8
9
10
11
12
(c)
12
10
19
2
18
10
17
13
16
14
15
(d)
20
9
25
21
8
11
22
23
24

Plate 66 Stage 14 (50–53 h) LS (longitudinal sections: coronal)

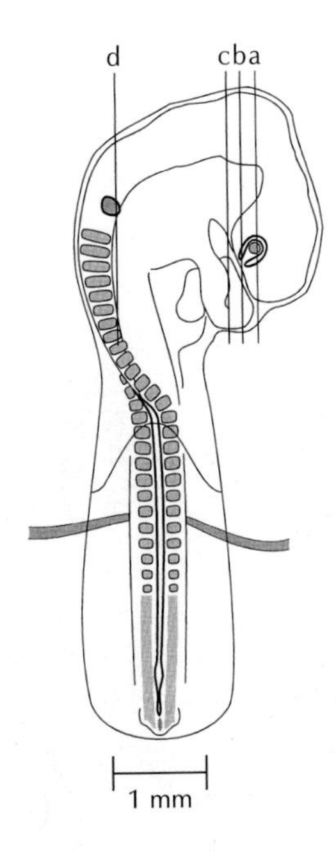

(a–d) × 135

1. optic stalk
2. amnion
3. diocoele (third ventricle)
4. optic cup
5. lens vesicle
6. head mesenchyme
7. diencephalon
8. left dorsal aorta
9. left anterior cardinal vein
10. first aortic arch
11. Rathke's pouch
12. infundibulum
13. aortic sac
14. pericardial cavity
15. ventricle
16. truncus arteriosus
17. amniotic cavity
18. right first (mandibular) pharyngeal arch
19. oral cavity
20. myelencephalon
21. otic vesicle
22. coelom
23. endoderm of gut
24. splanchnic mesoderm of gut
25. notochord

Plate 67

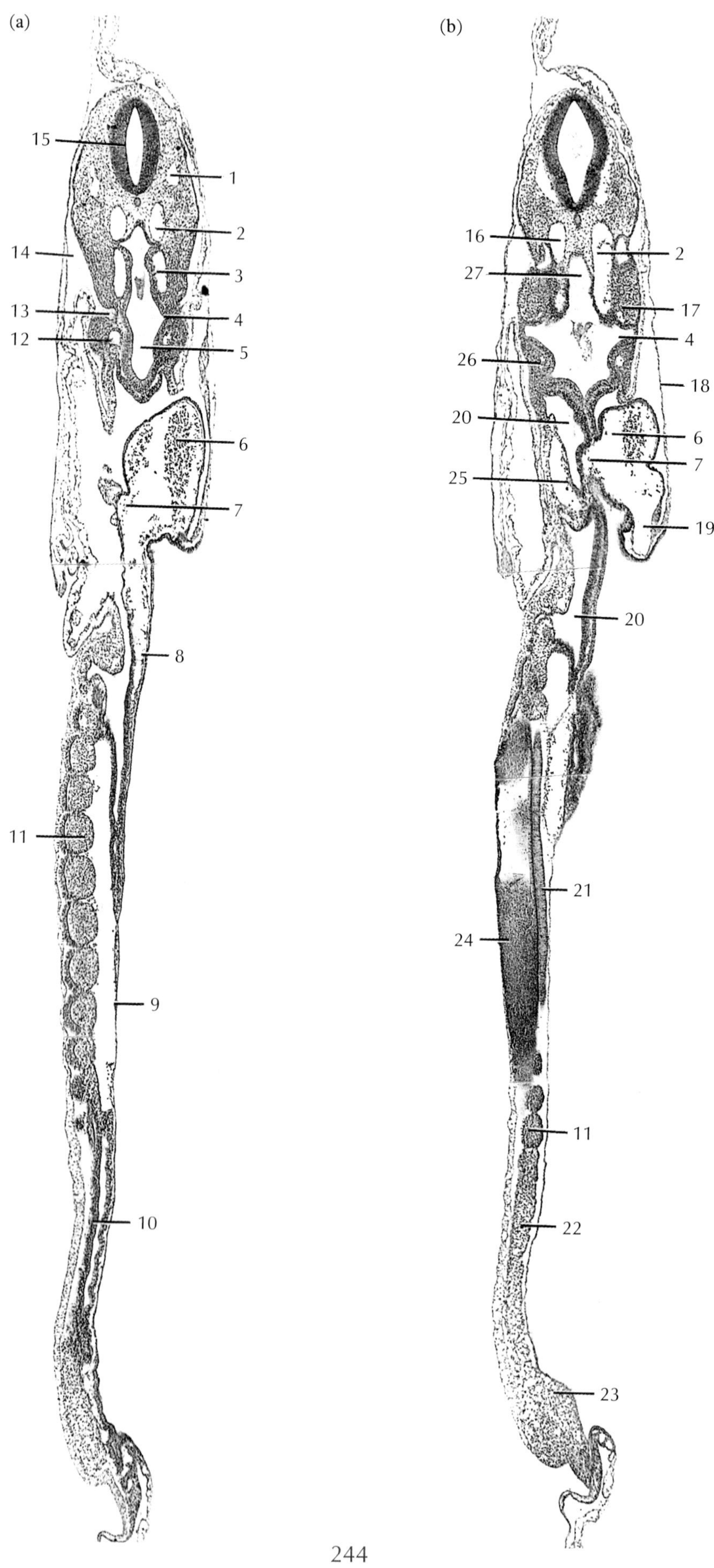

(a)
15
1
2
14
3
13
4
12
5
6
7
8
11
9
10
(b)
16
2
27
17
4
26
18
20
6
25
7
19
20
21
24
11
22
23

Plate 67 Stage 14 (50–53 h) LS (longitudinal sections: coronal and sagittal)

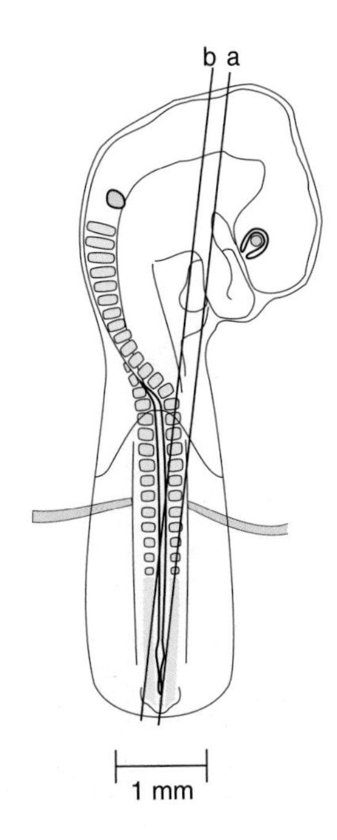

The cervical flexure (see p. 39) is incomplete, so the sections pass coronally through the anterior end of the embryo and sagittally through the posterior end.

(a, b) ×45

1. left anterior cardinal vein
2. left dorsal aorta
3. second aortic arch
4. second pharyngeal pouch
5. pharynx
6. atrium
7. sinus venosus
8. left posterior cardinal vein
9. endoderm
10. lateral plate mesoderm
11. somite
12. third aortic arch
13. second pharyngeal groove
14. amniotic cavity
15. diencephalon
16. right dorsal aorta
17. left first (mandibular) pharyngeal arch
18. amnion
19. ventricle
20. coelom
21. notochord
22. segmental plate mesoderm
23. Hensen's (primitive) node
24. neural tube
25. right anterior cardinal vein
26. third pharyngeal arch
27. Rathke's pouch

Plate 68

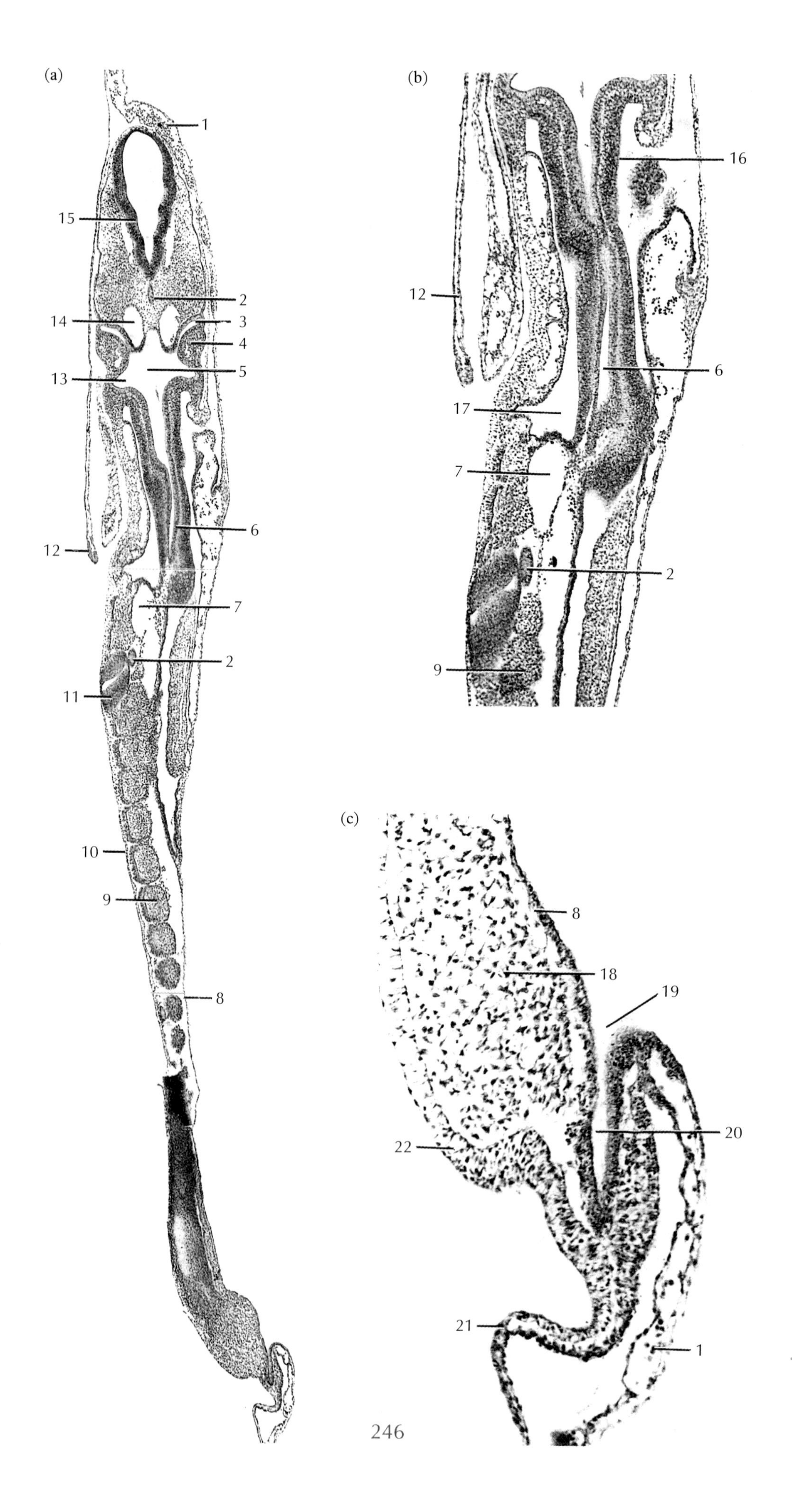

Plate 68 Stage 14 (50–53 h) LS (longitudinal sections: coronal and sagittal)

The cervical flexure (see p. 39) is incomplete, so that the sections pass coronally through the anterior end and sagittally through the posterior end.

(a) ×45 (b) ×85 (c) ×135

1. yolk sac
2. notochord
3. third pharyngeal pouch
4. fourth pharyngeal arch
5. lumen of pharynx
6. oesophagus
7. right dorsal aorta
8. endoderm
9. somite
10. ectoderm
11. neural tube
12. posterior tip of amniotic fold
13. fourth pharyngeal pouch
14. right dorsal aorta
15. myelencephalon
16. wall of pharynx
17. coelom
18. tail bud mesoderm
19. posterior intestinal portal
20. hindgut
21. posterior amniotic fold
22. ectoderm of tail bud

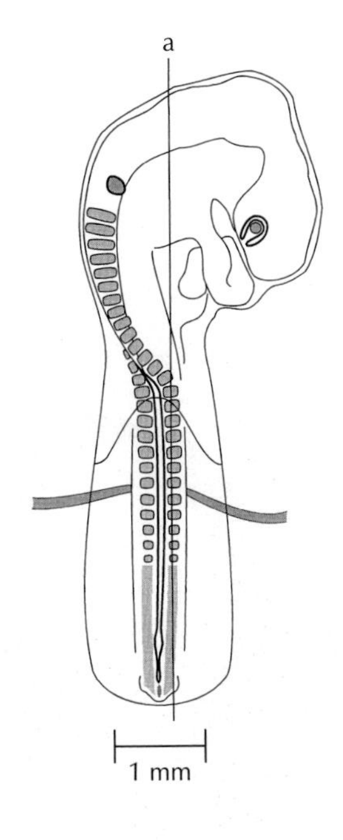

PLATE 69

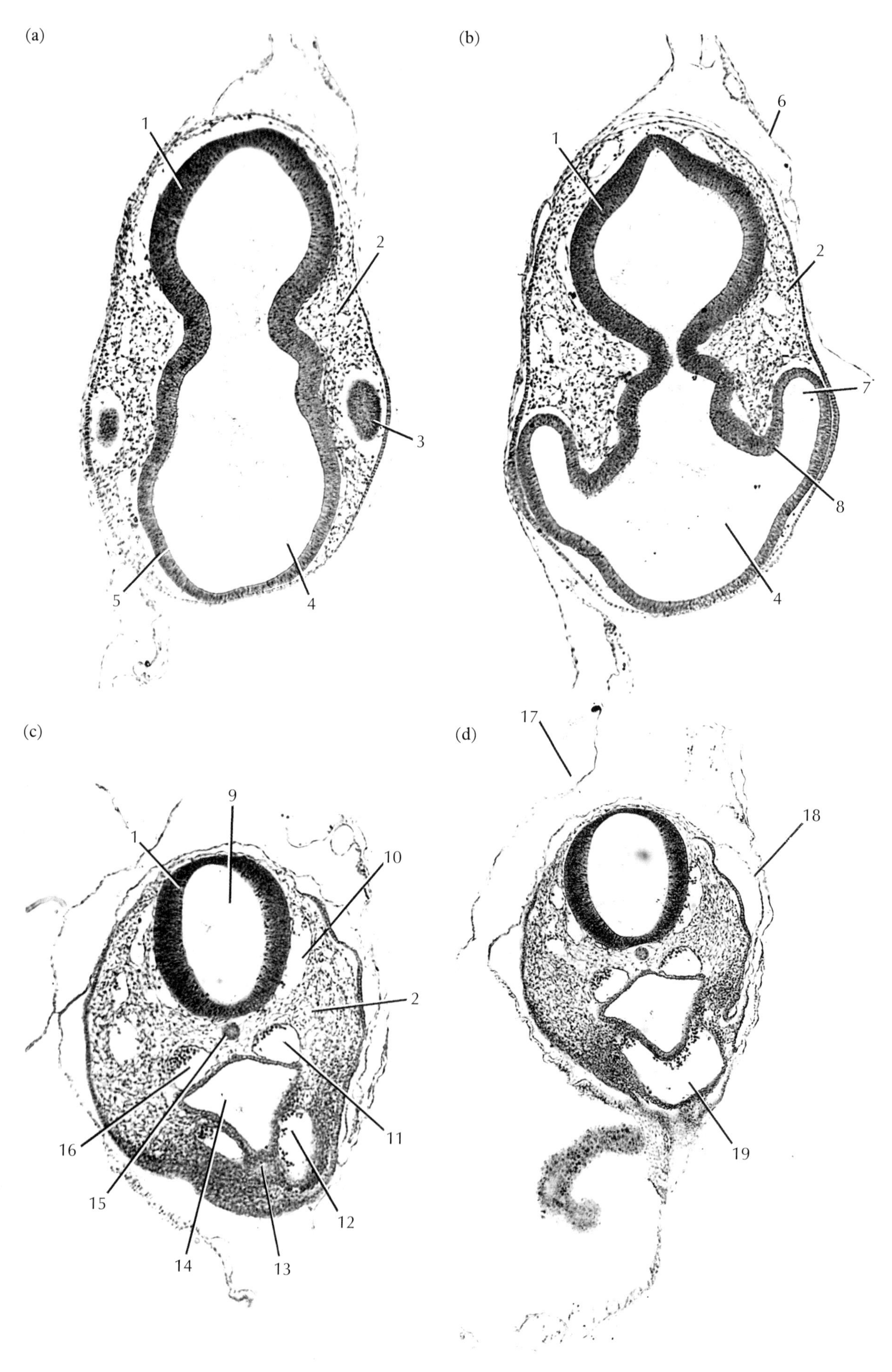

Plate 69 Stage 14 (50–53 h) TS

(a–d) ×85

1. mesencephalon
2. head mesenchyme
3. tip of optic evagination
4. diocoele
5. wall of diencephalon
6. yolk sac
7. left optic evagination
8. left optic stalk
9. mesocoele
10. left anterior cardinal vein
11. left dorsal aorta
12. left ventral aorta
13. stomatodaeum (stomatodaeal plate)
14. pharynx
15. notochord
16. right dorsal aorta
17. chorion
18. amnion
19. aortic sac

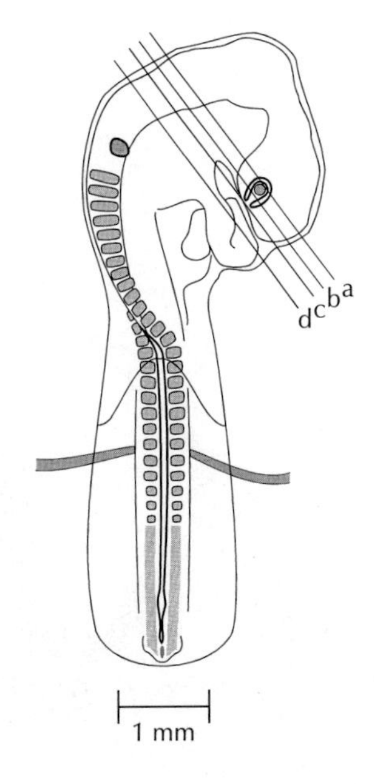

Plate 70

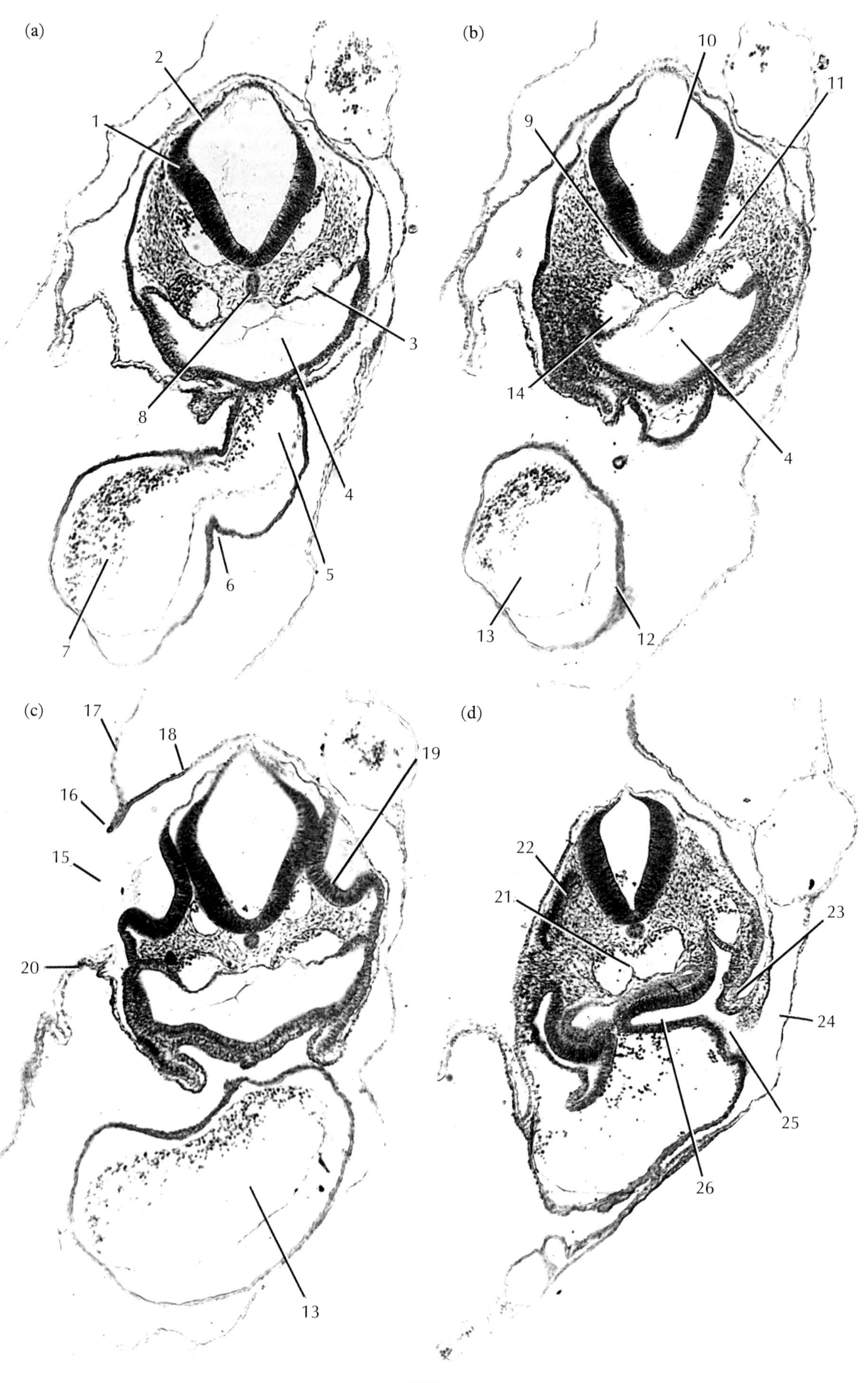

(a)
2
1
3
8
4
5
6
7
(b)
10
11
9
14
4
13
12
(c)
17
18
19
16
15
20
13
(d)
22
21
23
24
25
26

Plate 70 Stage 14 (50–53 h) TS

(a–d) ×85

1. thick wall of myelencephalon
2. thin roof of myelencephalon
3. left dorsal aorta
4. pharynx
5. lumen of truncus arteriosus
6. junction of truncus arteriosus and ventricle
7. ventricle
8. notochord
9. right anterior cardinal vein
10. fourth ventricle
11. left anterior cardinal vein
12. epimyocardium
13. atrium
14. right dorsal aorta
15. gap between the amniotic folds posterior to their fusion
16. left amniotic fold
17. chorion
18. amnion
19. left otic placode
20. right amniotic fold
21. fused dorsal aortae
22. derma-myotome
23. lateral body fold
24. extra-embryonic coelom
25. communication between embryonic and extra-embryonic coeloms
26. embryonic coelom

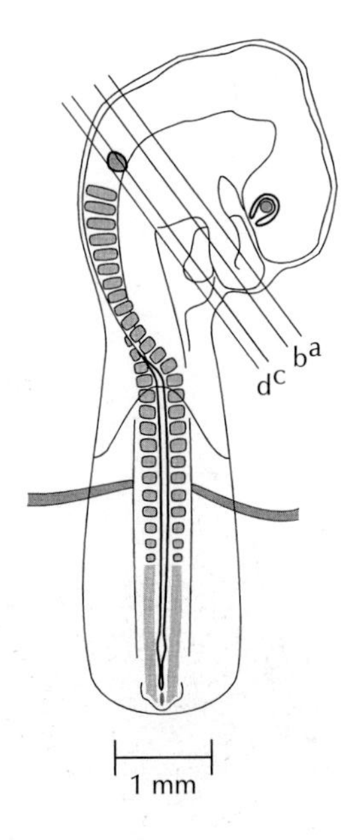

Plate 71

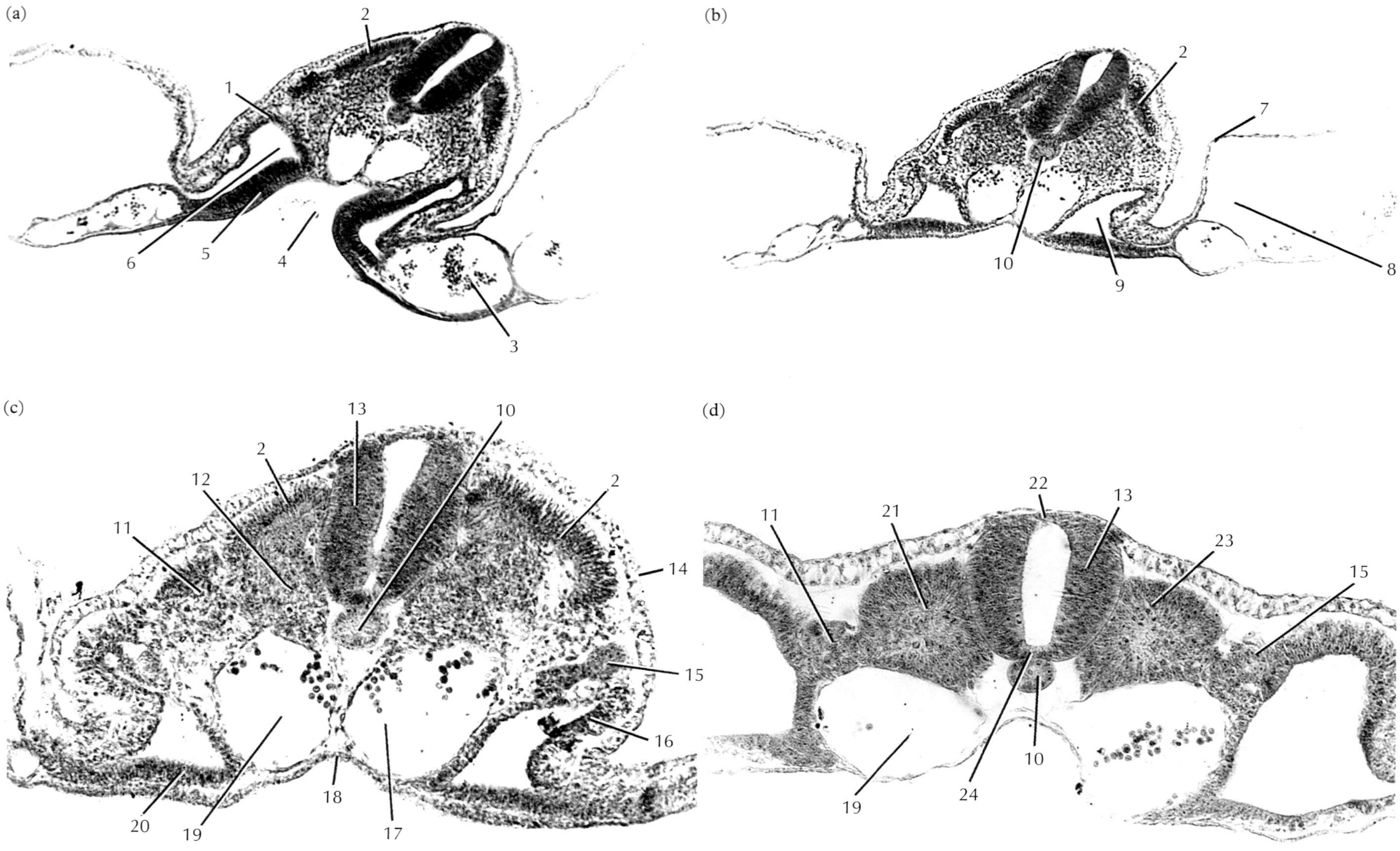

(a)
2
1
6
5
4
3
(b)
2
7
10
9
8
(c)
13
10
2
12
2
11
14
15
16
20
19
18
17
(d)
22
13
11
21
23
15
19
24
10

Plate 71 Stage 14 (50–53 h) TS

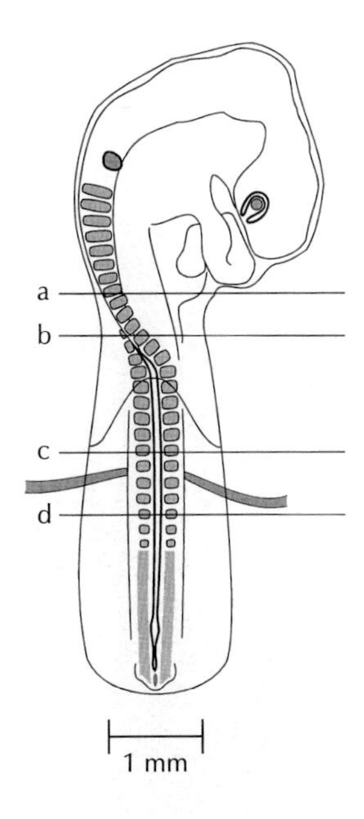

(a, b) ×85 (c, d) ×100

1. lateral plate mesoderm
2. derma-myotome
3. left omphalo-mesenteric vein
4. anterior intestinal portal
5. splanchnopleure
6. right embryonic coelom
7. left amniotic fold
8. left extra-embryonic coelom
9. left embryonic coelom
10. notochord
11. right nephric primordium
12. sclerotome
13. neural tube
14. ectoderm
15. left nephric primordium
16. somatic lateral plate mesoderm
17. left dorsal aorta
18. endoderm of midgut
19. right dorsal aorta
20. splanchnic lateral plate mesoderm
21. right newly formed somite
22. roof of neural tube
23. left newly formed somite
24. floor plate of neural tube

Plate 72

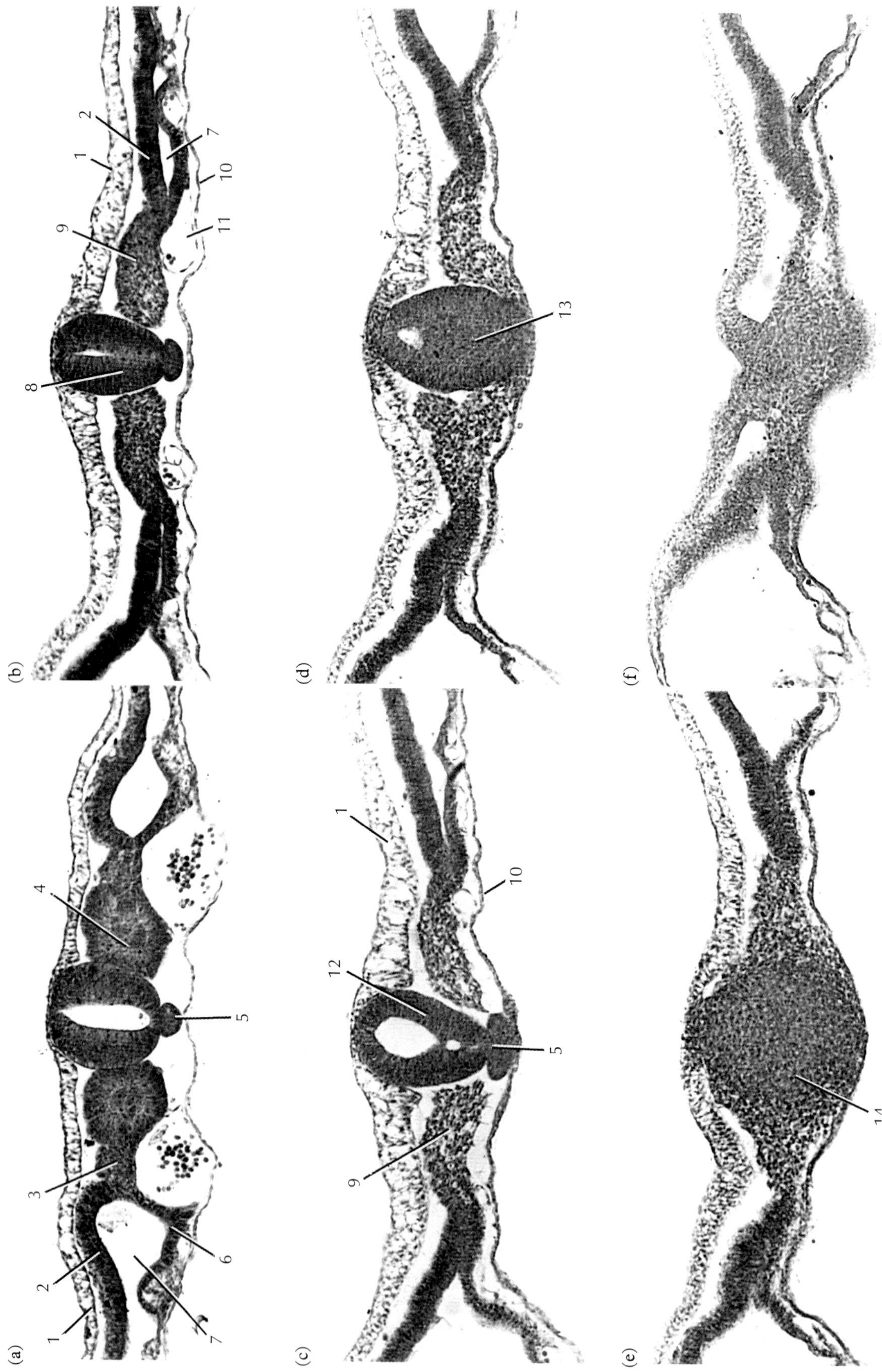

(a) 1 2 3 4 5 6 7

(b) 8 9 1 2 7 10 11

(c) 9 12 1 5 10

(d) 13

(e) 14

(f)

Plate 72 Stage 14 (50–53 h) TS

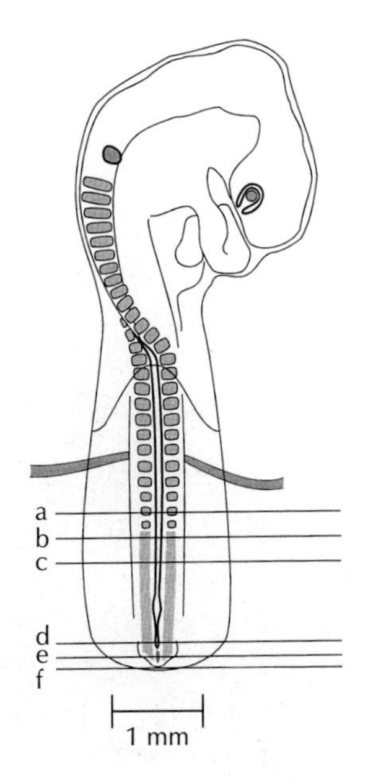

(a–f) × 100

1. ectoderm
2. somatic lateral plate mesoderm
3. nephric primordium
4. somite
5. notochord
6. splanchnic lateral plate mesoderm
7. embryonic coelom
8. posterior end of primary neural tube showing several lumina
9. segmental plate mesoderm
10. endoderm
11. left dorsal aorta
12. region of overlap of primary and secondary neural tube
13. secondary neural tube
14. tail bud mesoderm

Plate 73

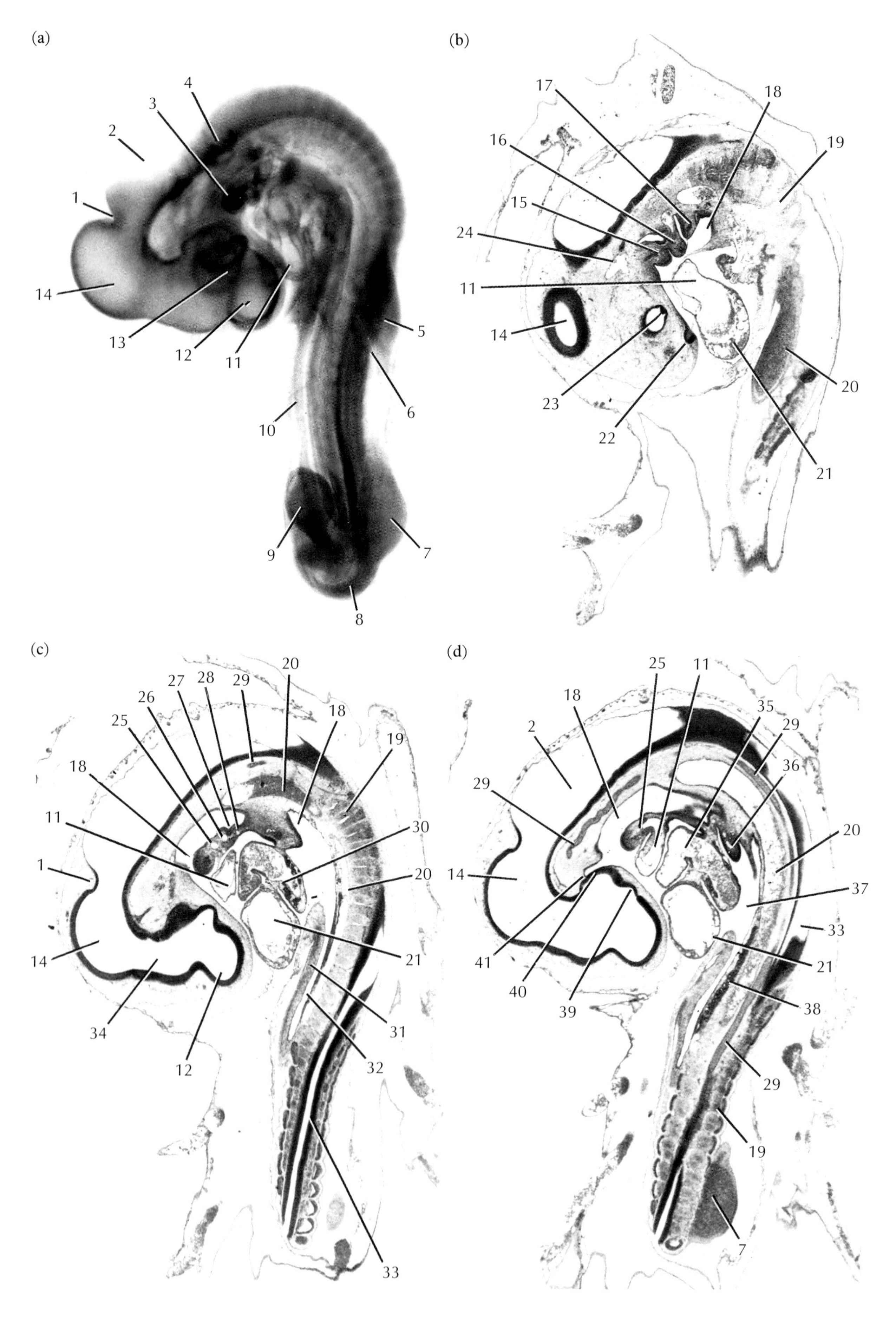

(a)
1
2
3
4
5
6
7
8
9
10
11
12
13
14
(b)
11
14
15
16
17
18
19
20
21
22
23
24
(c)
1
11
12
14
18
18
19
20
20
21
25
26
27
28
29
30
31
32
33
34
(d)
2
7
11
14
18
19
20
21
25
29
29
29
33
35
36
37
38
39
40
41

Plate 73 Stage 18 (65–69 h) LS

(a–d) ×20

(a) Whole-mount.
(b–d) Sagittal sections, taken through plane of page.

Diagram shows view of embryo from ventral side.

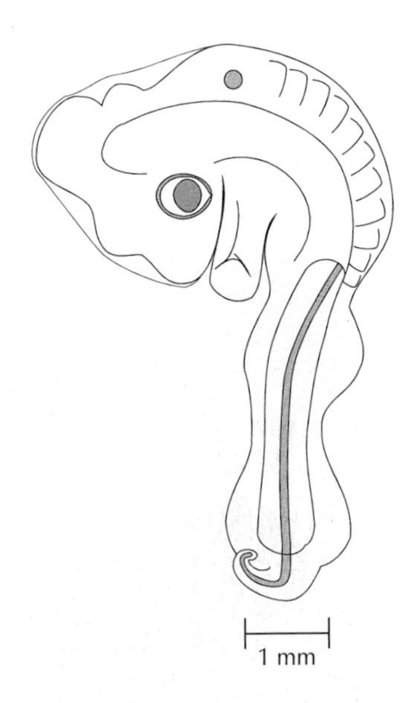

1. meso-metencephalic fold
2. metacoele
3. trigeminal (cr. nerve V) ganglion
4. otic vesicle
5. wing bud
6. anterior intestinal portal
7. leg bud
8. tail
9. allantois
10. edge of amnion
11. conus arteriosus
12. telencephalon
13. eye
14. mesencephalon
15. pharyngeal arch I
16. pharyngeal arch II
17. pharyngeal arch III
18. pharynx
19. myotome
20. dorsal aorta
21. ventricle
22. nasal placode
23. pecten
24. anterior cardinal vein
25. arterial arch II
26. arterial arch III
27. thyroid invagination – foramen caecum
28. arterial arch IV
29. notochord
30. atrium
31. omphalo-mesenteric vein
32. hindgut
33. neural tube
34. diencephalon
35. atrium
36. trachea
37. midgut
38. mesonephros
39. optic recess
40. infundibulum
41. Rathke's pouch

Plate 74

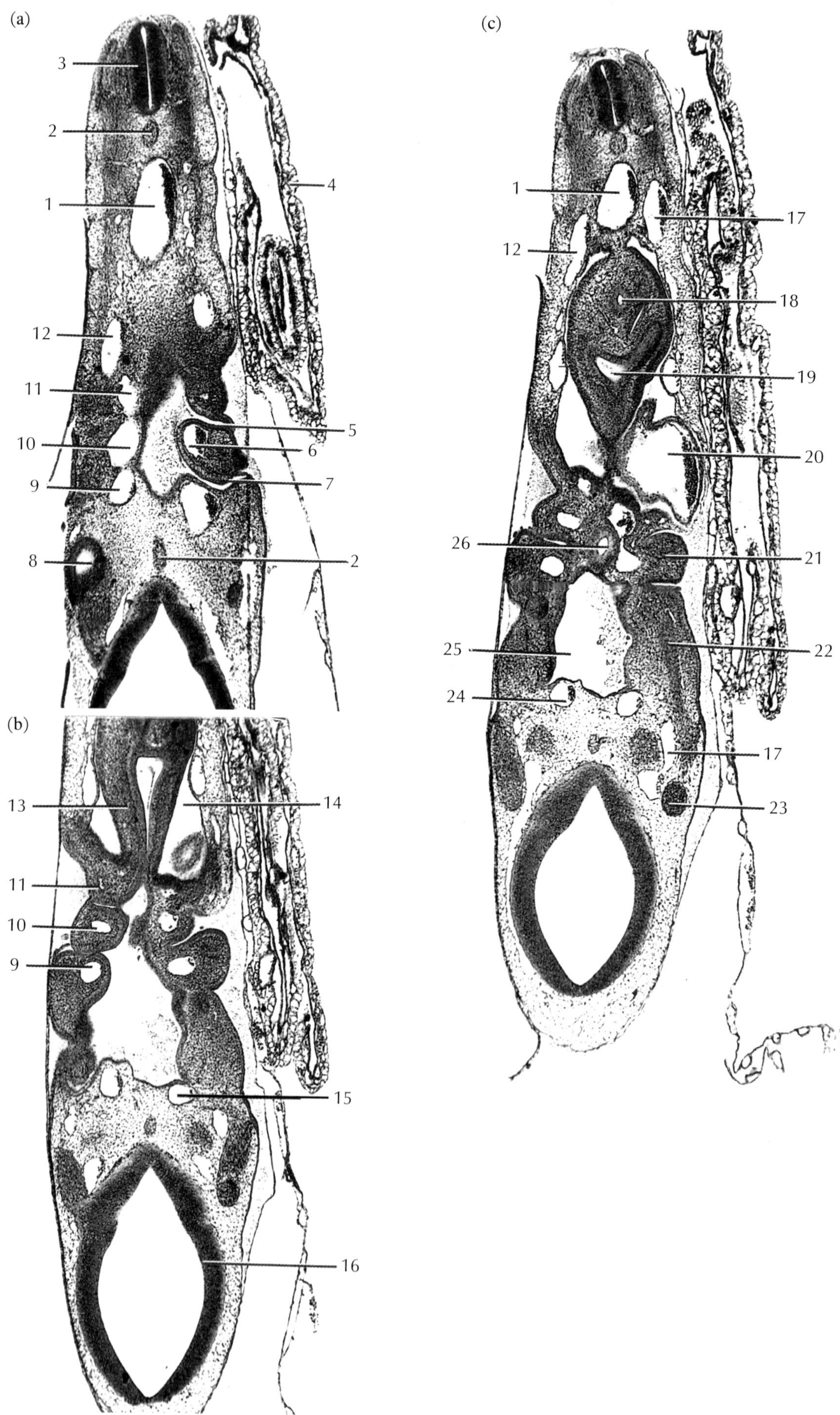

(a)
3
2
1
4
12
11
10
9
8
5
6
7
2
(b)
13
14
11
10
9
15
16
(c)
1
12
17
18
19
20
26
21
25
22
24
17
23

Plate 74 Stage 18 (65–69 h) TS

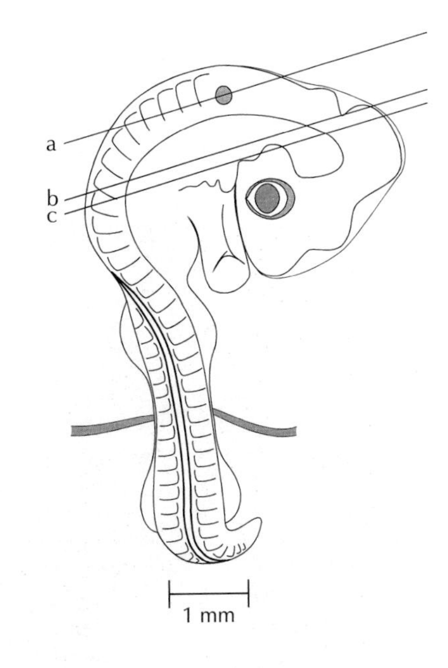

(a–c) ×45

1. dorsal aorta
2. notochord
3. neural tube
4. yolk sac
5. third pharyngeal pouch
6. left third aortic arch
7. second pharyngeal pouch
8. right otic vesicle
9. right second aortic arch
10. right third aortic arch
11. right fourth aortic arch
12. right anterior cardinal vein
13. thick wall of pharynx
14. coelom
15. left internal carotid artery
16. mesencephalon
17. left anterior cardinal vein
18. oesophagus
19. lung bud
20. left atrium
21. second (hyoid) pharyngeal arch
22. myotome
23. ophthalmic branch of trigeminal (cr. V) nerve
24. right internal carotid artery
25. pharynx
26. thyroid invagination – foramen caecum

Plate 75

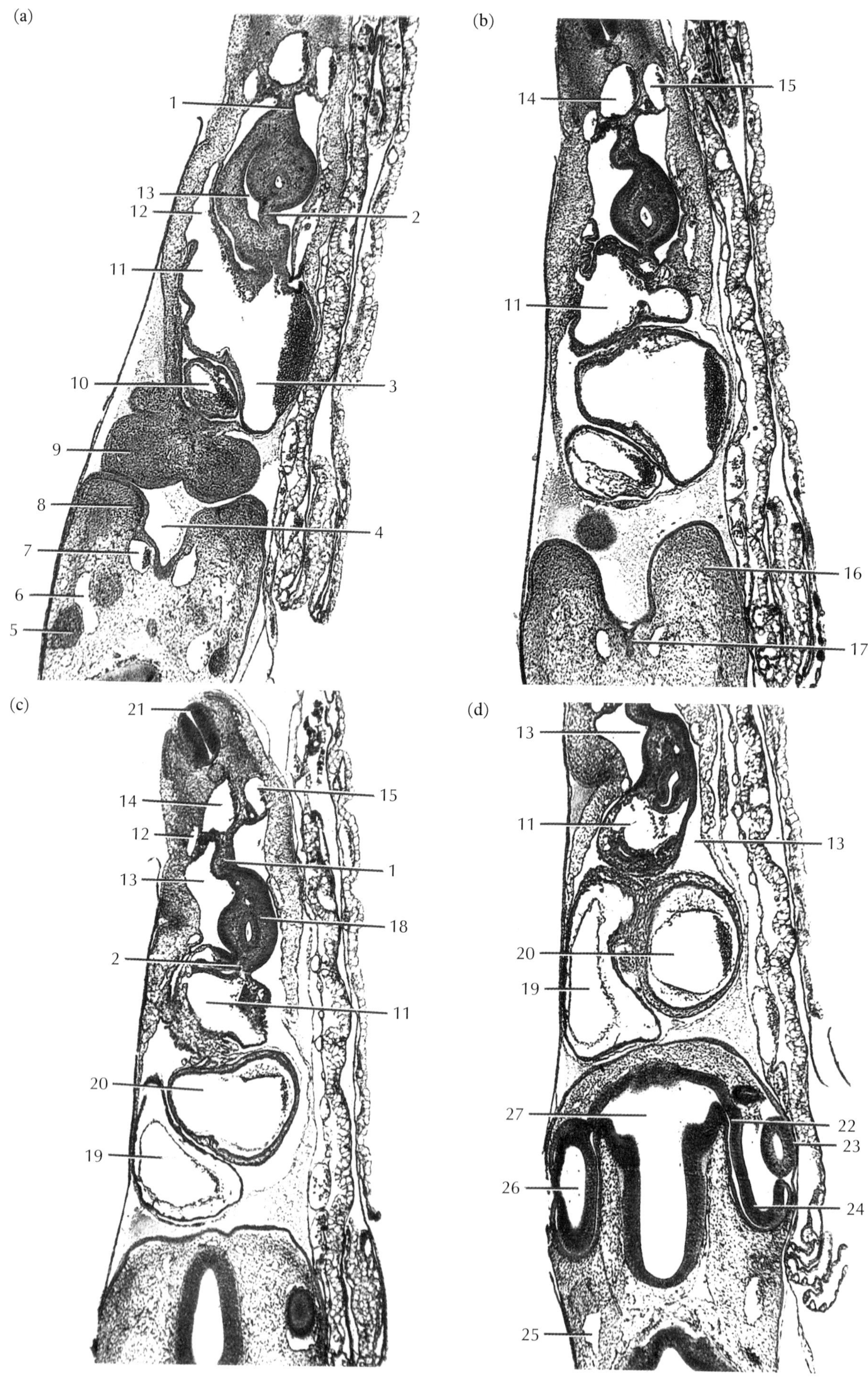
(a)
1
13
12
2
11
10
3
9
8
4
7
6
5
(b)
14
15
11
16
17
(c)
21
14
15
12
1
13
18
2
11
20
19
(d)
13
11
13
20
19
27
22
23
26
24
25

Plate 75 Stage 18 (65–69 h) TS

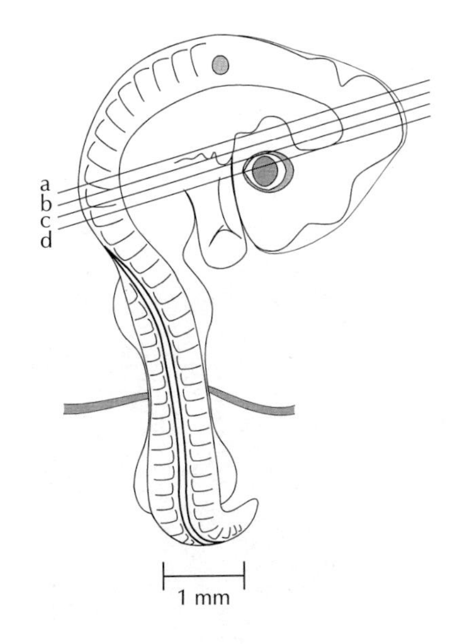

(a–d) ×45

1. dorsal mesentery
2. ventral mesentery
3. left atrium
4. oral cavity
5. ophthalmic branch of trigeminal (cr. V) nerve
6. right anterior cardinal vein
7. right internal carotid artery
8. right first (maxillary) pharyngeal arch
9. right first (mandibular) pharyngeal arch
10. truncus arteriosus
11. sinus venosus
12. right anterior cardinal vein
13. coelom
14. dorsal aorta
15. left anterior cardinal vein
16. left first (maxillary) pharyngeal arch
17. Rathke's pouch
18. oesophagus
19. bulbus cordis
20. ventricle
21. neural tube
22. left optic stalk
23. left lens vesicle
24. left retina
25. head vein
26. right eye
27. prosencephalon

Plate 76

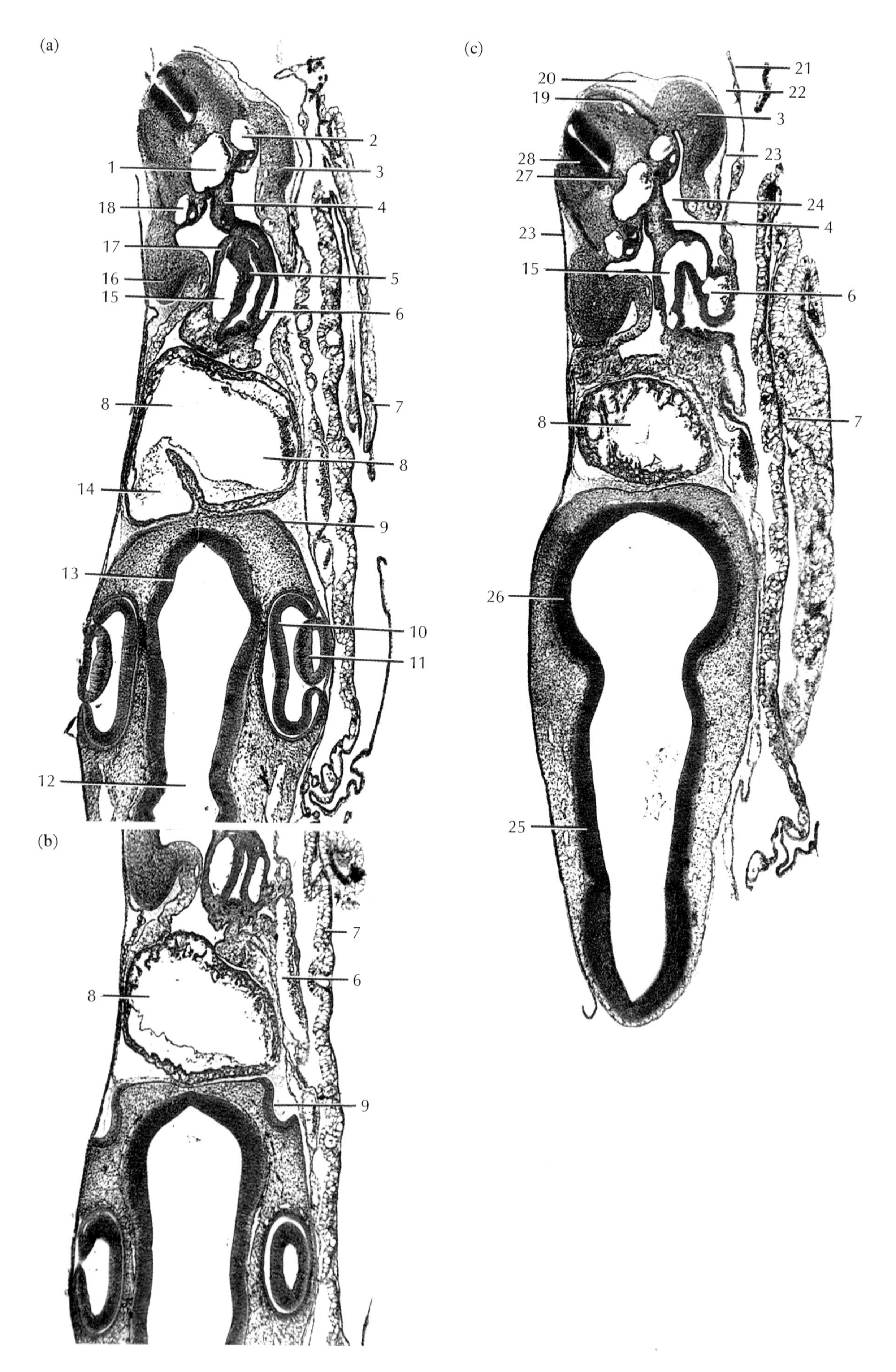

(a)
2
1
3
18
4
17
5
16
15
6
8
7
8
14
9
13
10
11
12
(b)
7
6
8
9
(c)
20
21
22
19
3
23
28
27
24
4
23
15
6
8
7
26
25

Plate 76 Stage 18 (65–69 h) TS

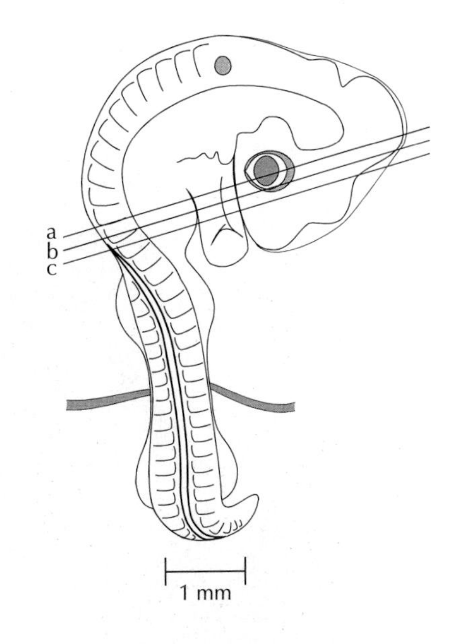

(a–c) ×45

1. dorsal aorta
2. left anterior cardinal vein
3. left forelimb bud
4. dorsal mesentery
5. duodenum
6. left omphalo-mesenteric vein
7. yolk sac
8. ventricle
9. nasal placode
10. optic cup
11. lens vesicle
12. diocoele
13. prosencephalon
14. bulbus cordis
15. right omphalo-mesenteric vein
16. right forelimb bud
17. splanchnopleure
18. right anterior cardinal vein
19. myotome
20. amniotic cavity
21. chorion
22. extra-embryonic coelom
23. amnion
24. left embryonic coelom
25. diencephalon
26. telencephalon
27. notochord
28. neural tube

PLATE 77

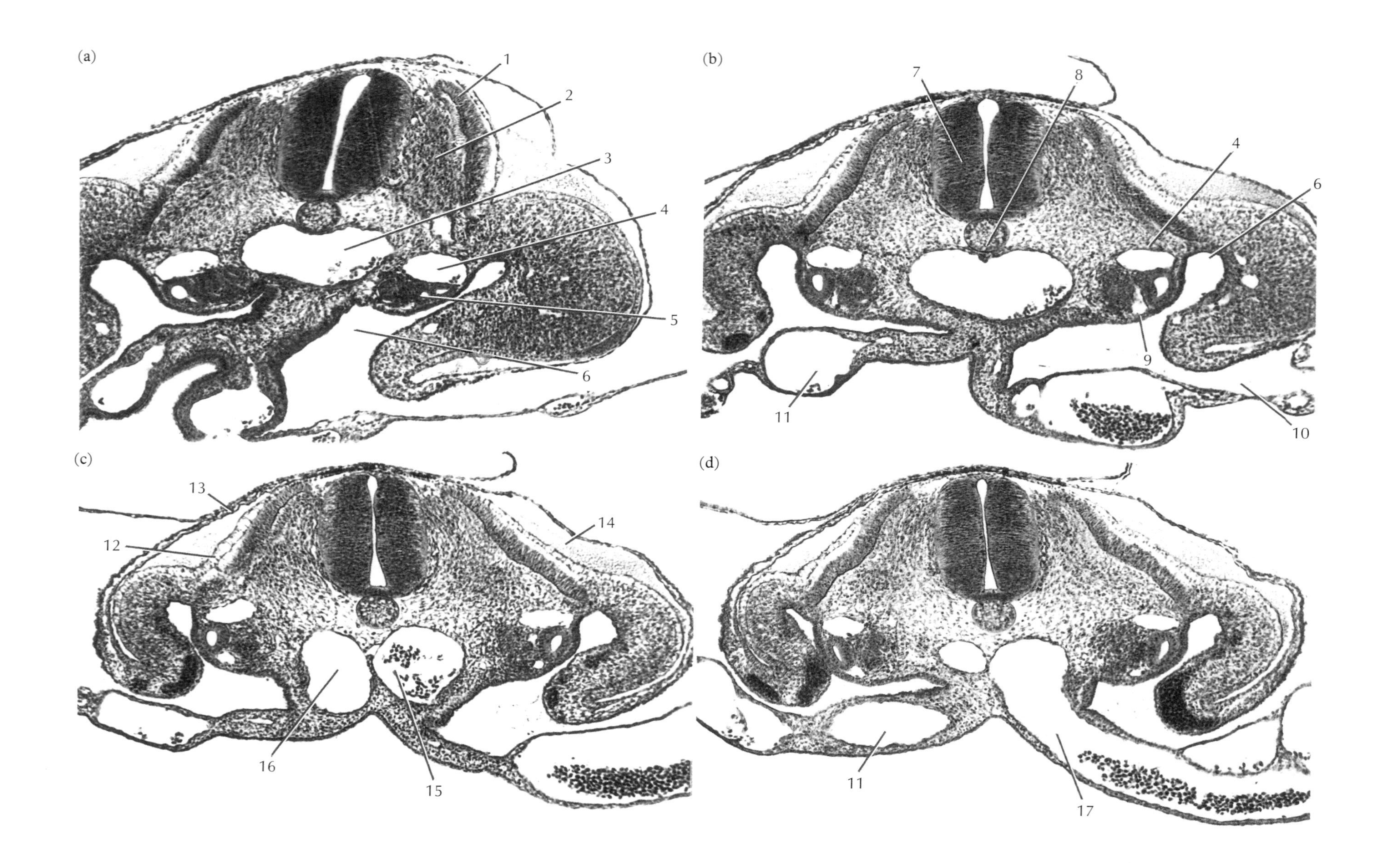

Plate 77 Stage 18 (65–69 h) TS

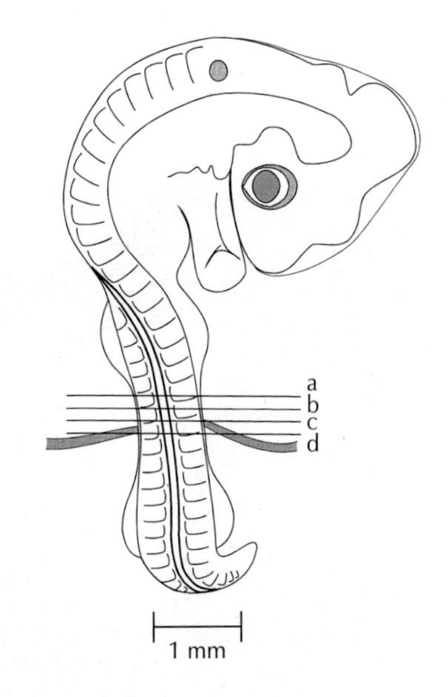

(a–d) ×90

1. dermatome
2. sclerotome
3. single dorsal aorta
4. left posterior cardinal vein
5. left nephric duct
6. left embryonic coelom
7. neural tube
8. notochord
9. left nephrogenic ridge
10. extra-embryonic coelom
11. right omphalo-mesenteric artery
12. embryonic ectoderm
13. amnion
14. amniotic cavity
15. left dorsal aorta
16. right dorsal aorta
17. left omphalo-mesenteric artery leaving dorsal aorta

PLATE 78

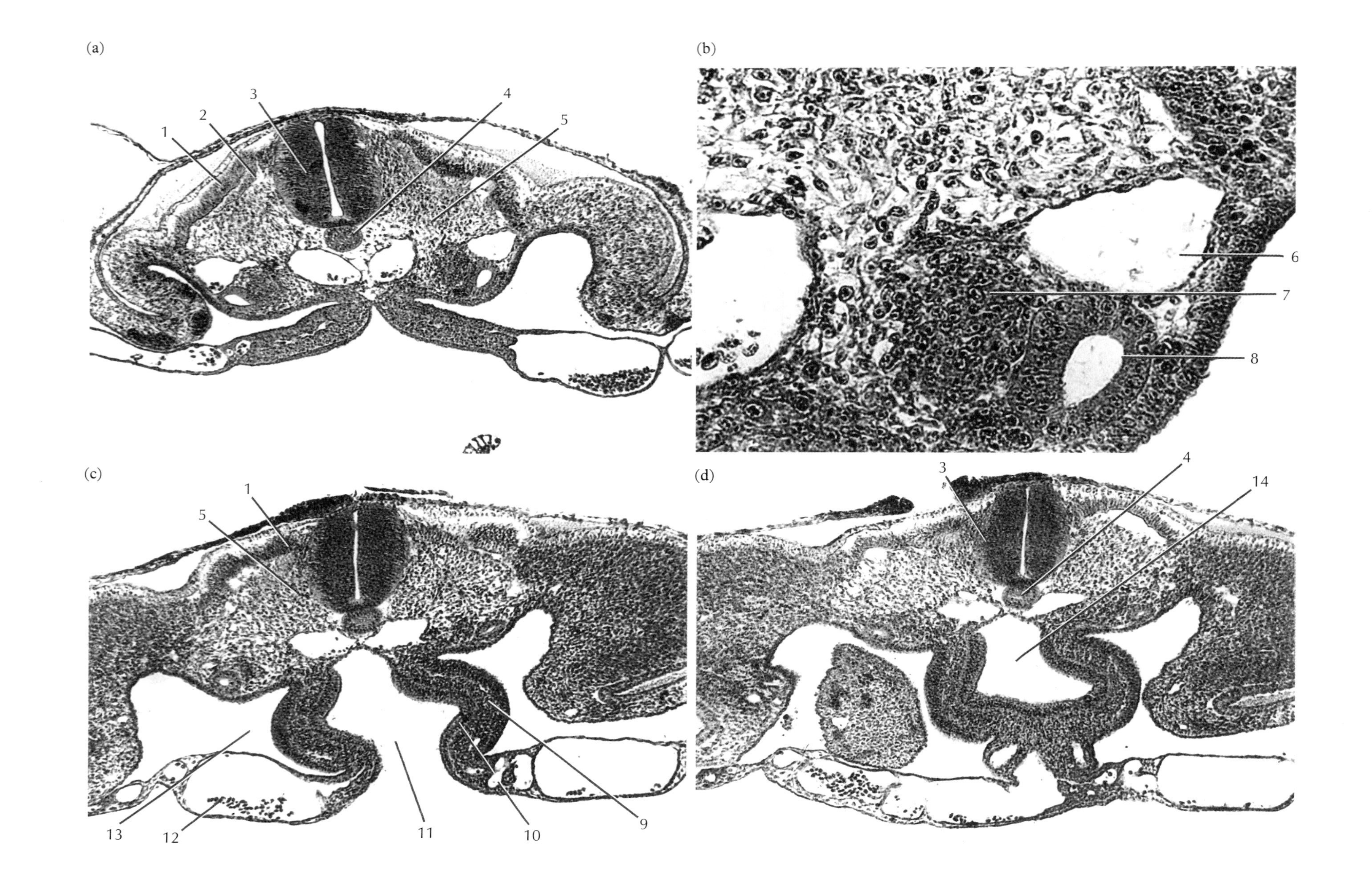

Plate 78 Stage 18 (65–69 h) TS

(a, c, d) ×120

(b) Enlargement of part of (a) ×400

1. dermatome
2. myotome
3. neural tube
4. notochord
5. sclerotome
6. left posterior cardinal vein
7. left nephrogenous mesenchyme
8. left nephric duct
9. splanchnic mesoderm
10. embryonic endoderm
11. posterior intestinal portal
12. right omphalo-mesenteric artery
13. right embryonic coelom
14. hindgut

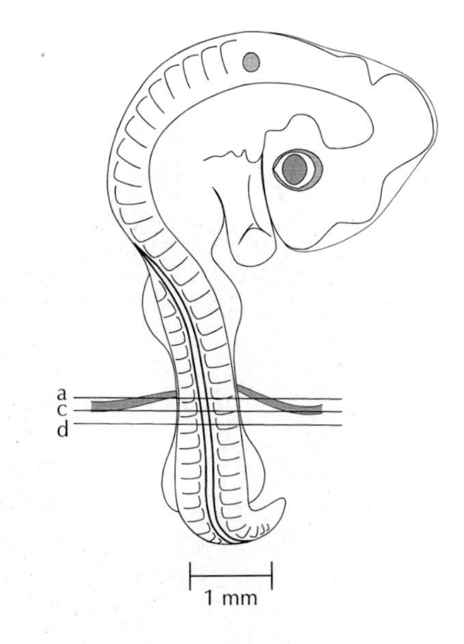

Plate 79

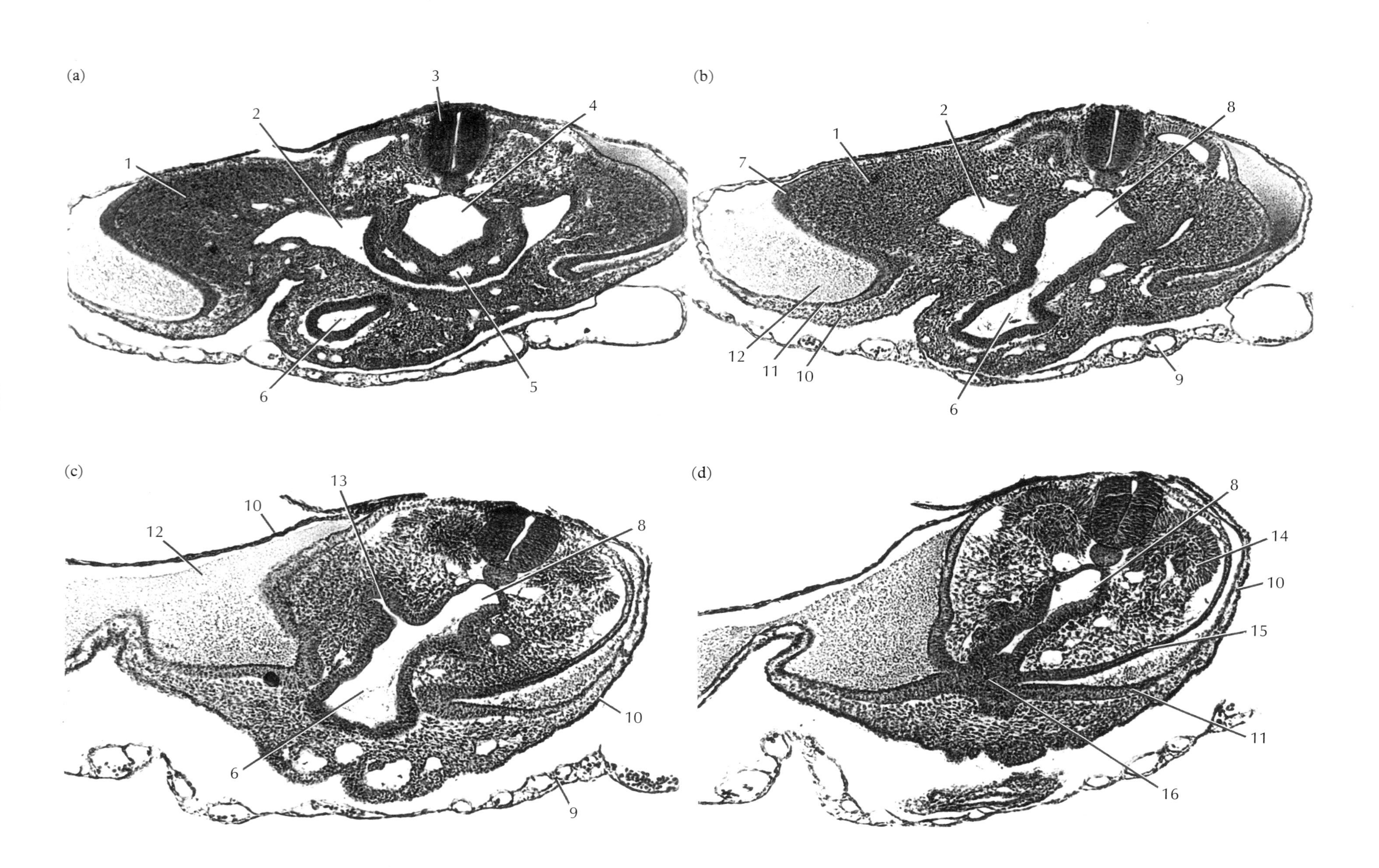

Plate 79 Stage 18 (65–69 h) TS

(a–d) ×120

1. right leg bud
2. embryonic coelom
3. neural tube
4. hindgut
5. subintestinal vein
6. lumen of allantois
7. apical ectodermal ridge of leg bud
8. cloaca
9. yolk sac
10. somatic mesoderm
11. extra-embryonic ectoderm
12. amniotic cavity
13. right nephric duct
14. somite
15. embryonic ectoderm
16. cloacal membrane

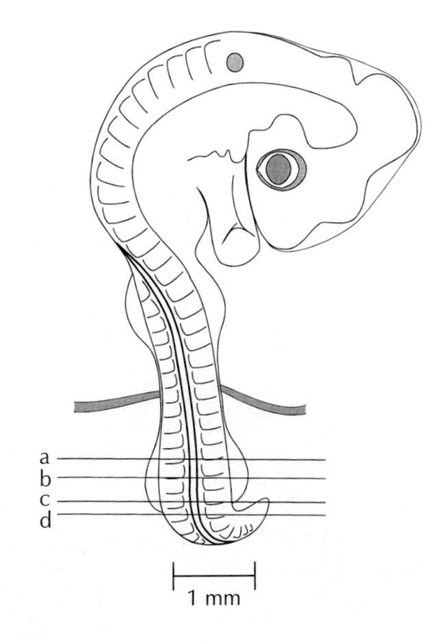

Plate 80

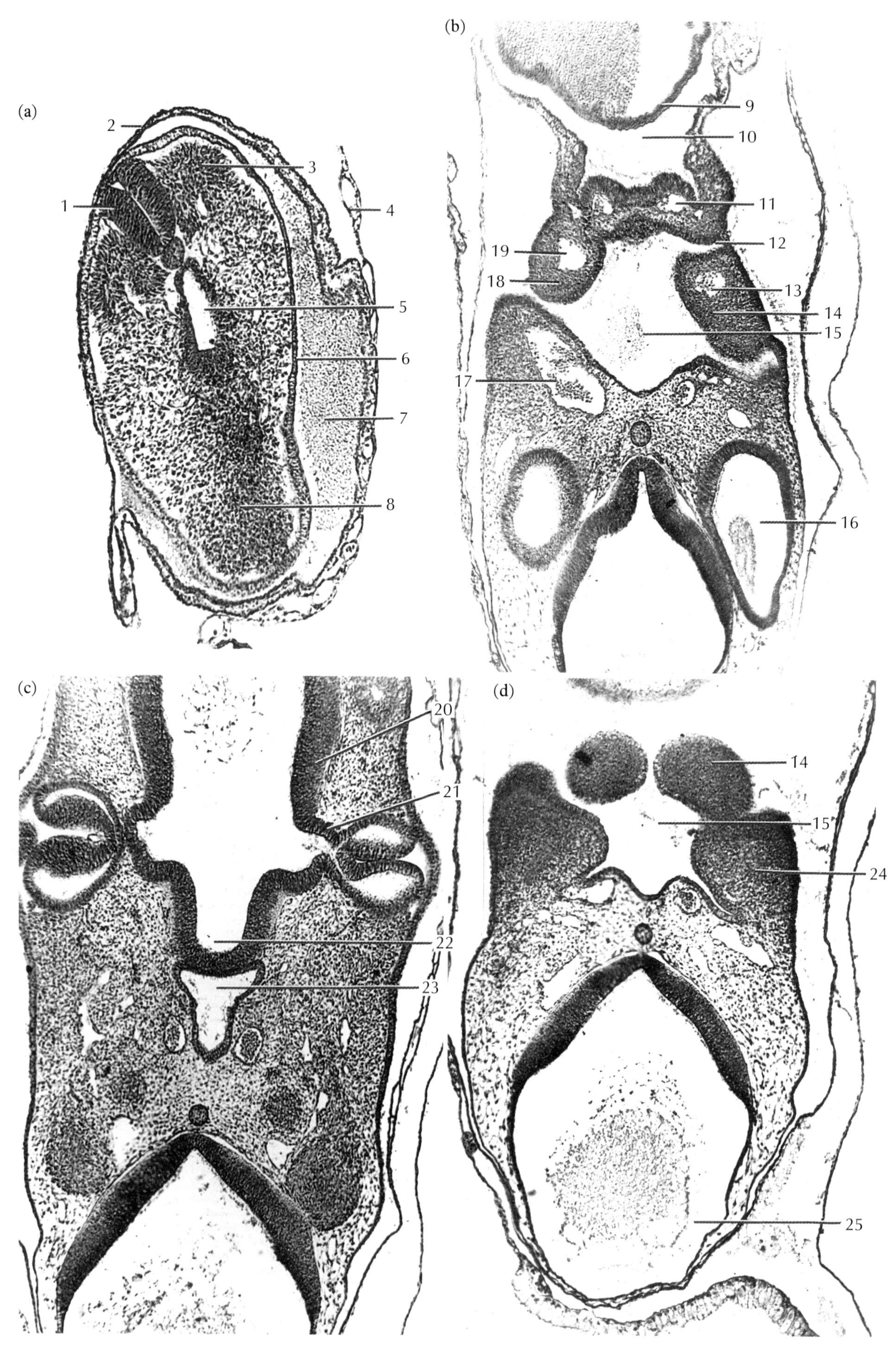

Plate 80 Stage 18 (65–69 h) TS

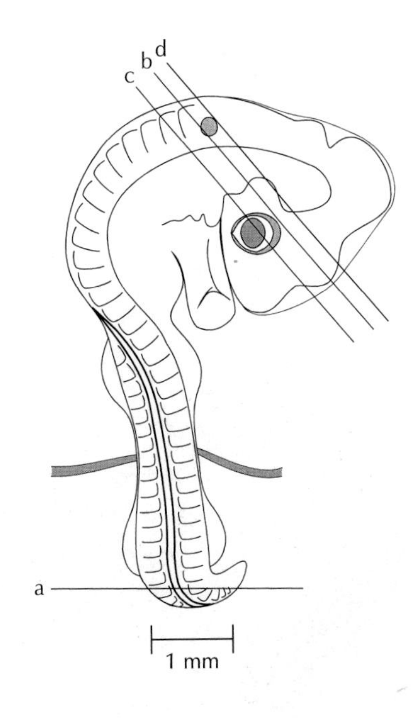

(a) ×120
(b–d) ×70

(a) Same embryo as in *Plates 74–79.*
(b–d) Another embryo to illustrate details of pharyngeal region.

1. neural tube
2. amnion
3. somite
4. yolk sac
5. tail gut
6. embryonic ectoderm
7. amniotic cavity
8. tail mesoderm
9. truncus arteriosus
10. pericardial coelom
11. left second aortic arch
12. left first pharyngeal pouch
13. left first aortic arch
14. left first (mandibular) pharyngeal arch
15. oral cavity
16. left otic vesicle
17. right anterior cardinal vein
18. right first (mandibular) pharyngeal arch
19. right first aortic arch
20. prosencephalon
21. left optic stalk
22. infundibulum
23. Rathke's pouch
24. left first (maxillary) pharyngeal arch
25. myelocoele

Plate 81

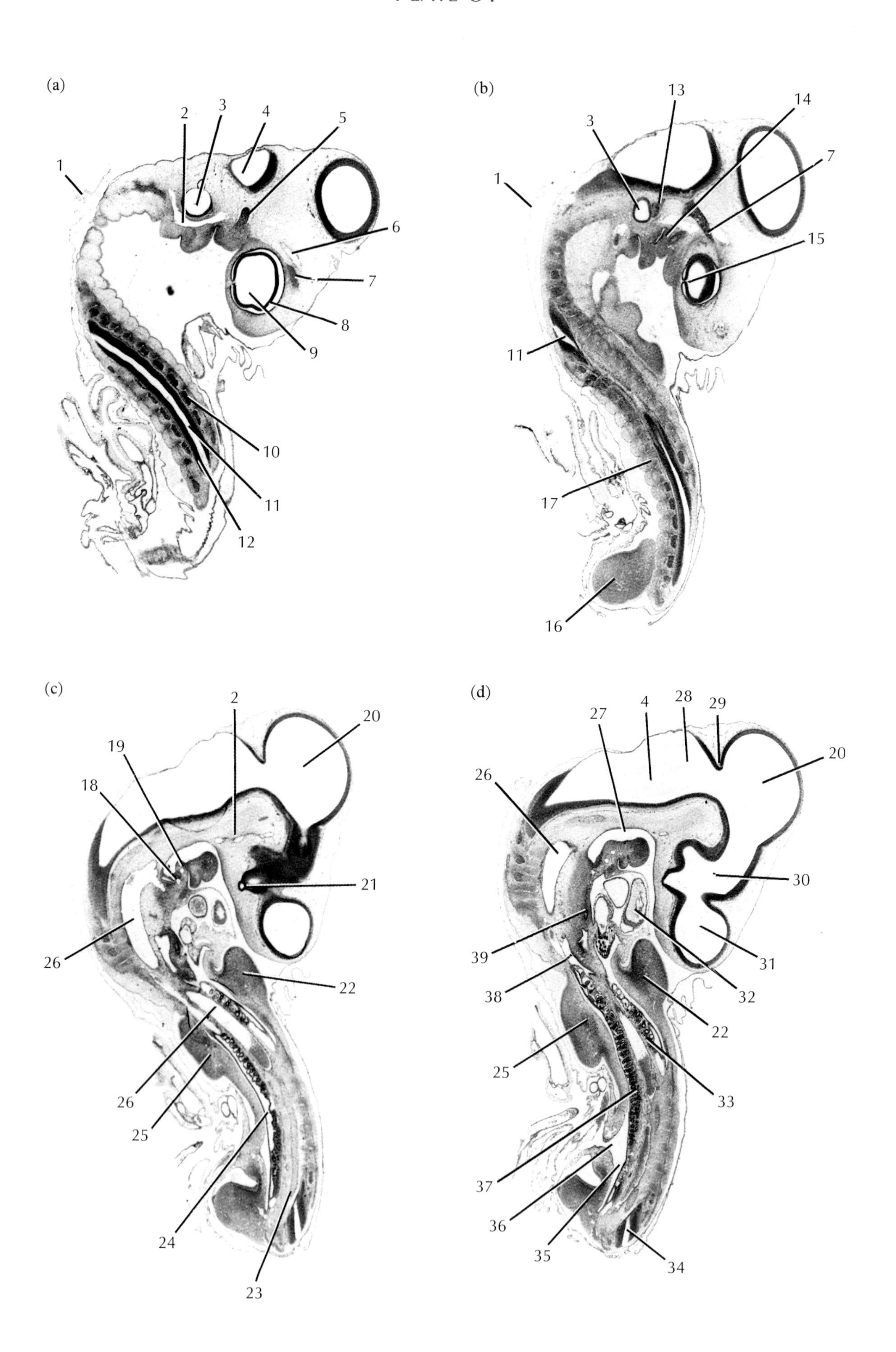

(a)
1
2
3
4
5
6
7
8
9
10
11
12
(b)
13
14
3
7
1
15
11
17
16
(c)
2
20
19
18
21
26
22
26
25
24
23
(d)
27
4
28
29
20
26
30
39
31
38
32
22
25
33
37
36
35
34

Plate 81 Day 4 LS (longitudinal sections: coronal and sagittal)

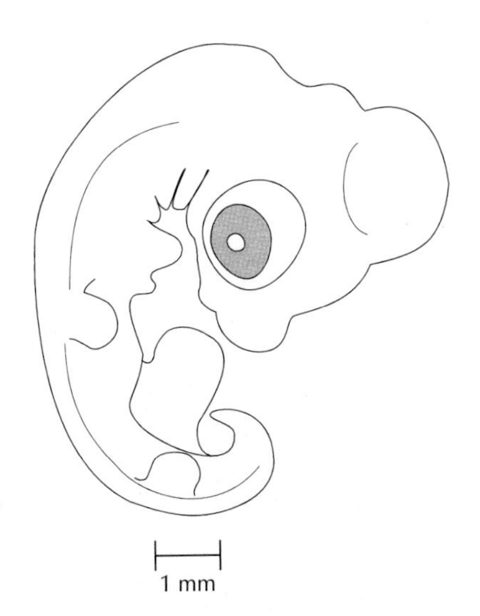

(a–d) × 12

Sections taken through plane of page.

Caudal end of specimen twisted during preparation of material.

1. amnion
2. anterior cardinal vein
3. otic vesicle
4. myelocoele
5. maxillary branch of trigeminal (cr. V) nerve
6. head vein
7. ophthalmic branch of trigeminal (cr. V) nerve
8. pigmented layer of retina
9. lumen of optic vesicle
10. 25th right somite
11. lumen of neural tube
12. wall of neural tube
13. geniculate (cr. nerve VII) ganglion
14. first (mandibular) pharyngeal arch
15. choroid fissure
16. left leg bud
17. 31st left somite
18. fourth aortic arch
19. third aortic arch
20. mesocoele
21. infundibulum
22. right wing bud
23. notochord
24. left nephric (Wolffian) duct
25. left wing bud
26. dorsal aorta
27. oral cavity
28. metacoele
29. meso-metencephalic fold
30. diocoele
31. telocoele
32. truncus arteriosus
33. right mesonephros
34. neural tube of tail region
35. hindgut
36. midgut
37. left mesonephros
38. peritoneal cavity
39. oesophagus

Plate 82

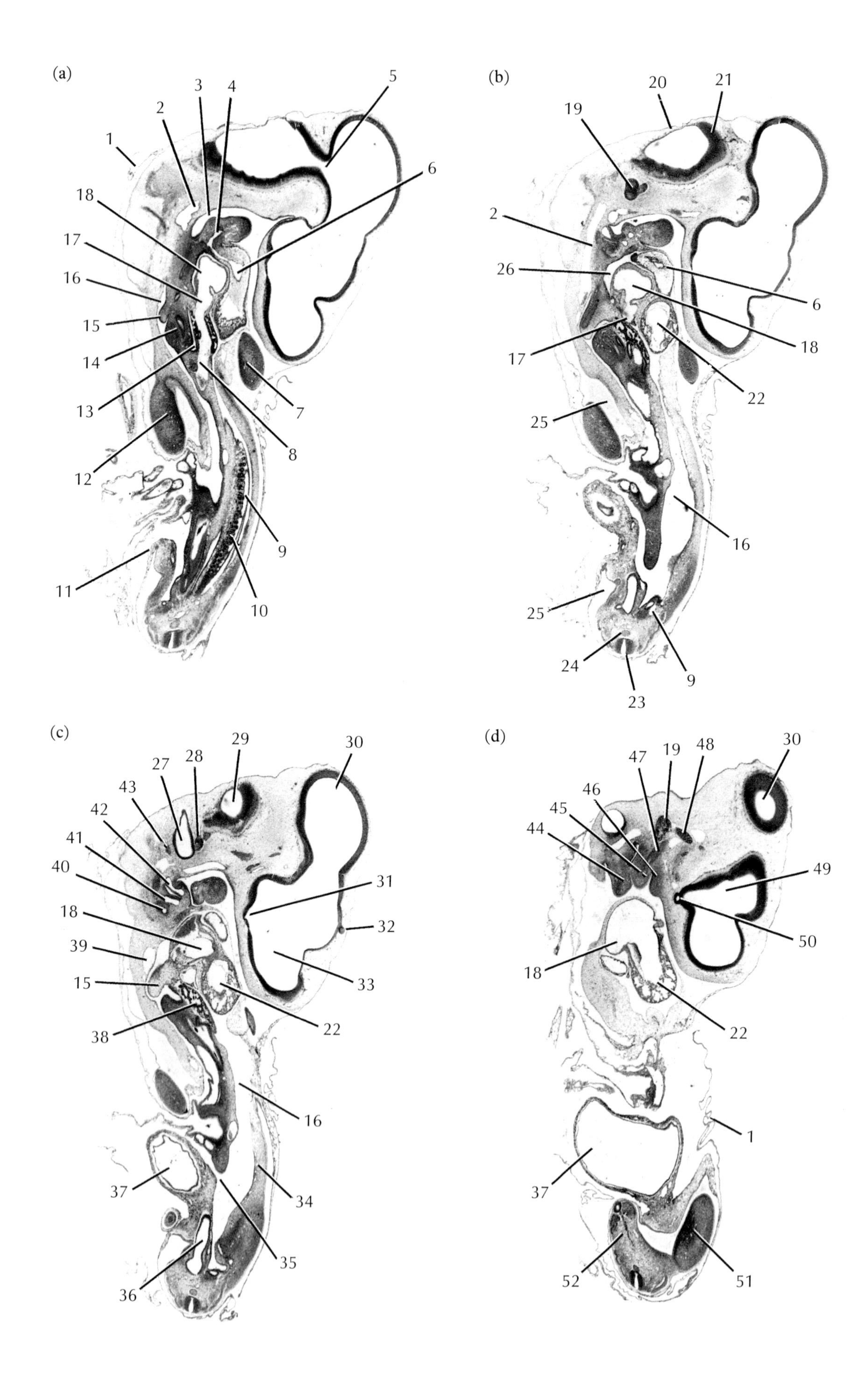

(a)
1
2
3
4
5
6
7
8
9
10
11
12
13
14
15
16
17
18
(b)
19
20
21
2
26
17
25
25
24
23
9
16
22
18
6
(c)
43
42
41
40
18
39
15
38
37
36
27
28
29
30
31
32
33
22
16
34
35
(d)
44
45
46
47
19
48
30
49
50
18
22
1
37
52
51

Plate 82 Day 4 LS (longitudinal sections: sagittal and coronal)

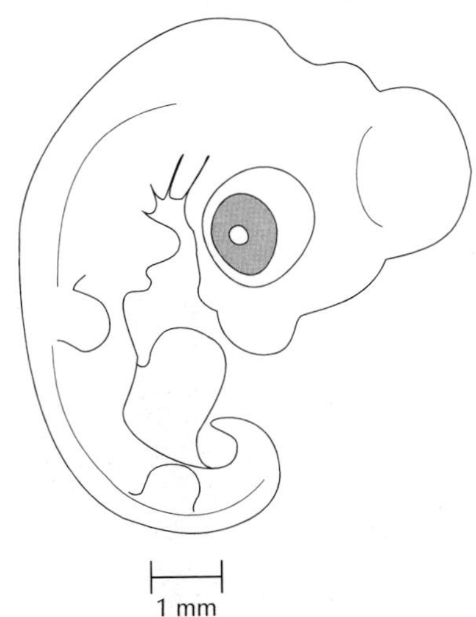

(a–d) × 12

Sections taken through plane of page.

Caudal end twisted in preparation of material.

1. amnion
2. anterior cardinal vein
3. pharynx
4. ventral aorta
5. isthmus in meso-metencephalic fold
6. truncus arteriosus
7. right wing bud
8. posterior cardinal vein
9. nephric duct
10. right mesonephros
11. junction of body wall and amnion
12. left wing bud
13. lesser sac
14. stomach
15. left lung bud
16. coelom
17. sinus venosus
18. atrium
19. trigeminal (cr. nerve V) ganglion
20. myelencephalon
21. metencephalon
22. ventricle
23. neural tube
24. notochord
25. amniotic cavity
26. pericardial cavity
27. otic vesicle
28. geniculate (cr. nerve VII) ganglion
29. myelocoele
30. mesocoele
31. optic recess
32. pineal organ
33. telocoele
34. body wall
35. communication between embryonic and extra-embryonic coeloms
36. cloaca
37. lumen of allantois
38. liver
39. posterior cardinal vein
40. fourth aortic arch
41. third aortic arch
42. second aortic arch
43. superior ganglion of glossopharyngeal (cr. IX) nerve
44. second (hyoid) pharyngeal arch
45. first (mandibular process) pharyngeal arch
46. first (maxillary process) pharyngeal arch
47. trigeminal (cr. V) nerve
48. facial branch of trigeminal (cr. V) nerve
49. diocoele
50. infundibulum
51. right leg bud
52. tail bud

Plate 83

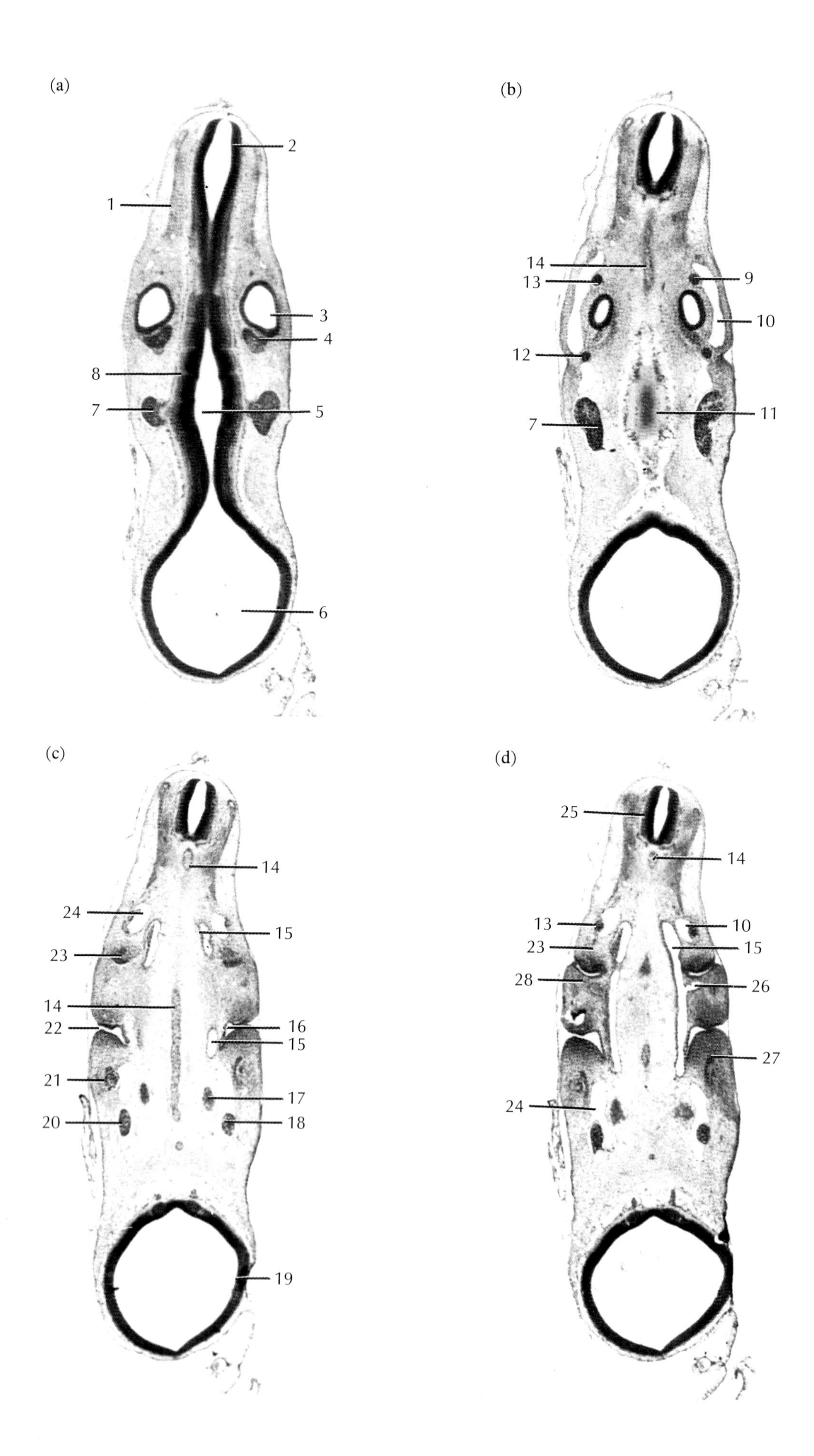
(a)
2
1
3
4
8
7
5
6
(b)
14
9
13
10
12
11
7
(c)
14
24
15
23
14
16
22
15
21
17
20
18
19
(d)
25
14
13
10
23
15
28
26
27
24

Plate 83 Day 4 TS

(a–d) ×22

1. myotome
2. myelencephalon
3. left otic vesicle
4. left acoustico-facialis (cr. nerves VII and VIII) ganglion
5. metacoele
6. diencephalon
7. trigeminal (cr. nerve V) ganglion
8. neuromere
9. left glossopharyngeal (cr. IX) nerve
10. left anterior cardinal vein
11. floor of metencephalon
12. right acoustic (cr. nerve VIII) ganglion
13. right glossopharyngeal (cr. IX) nerve
14. notochord
15. left dorsal aorta
16. left first pharyngeal pouch
17. left trigeminal (cr. V) nerve (maxillary branch)
18. left trigeminal (cr. V) nerve (ophthalmic branch)
19. mesencephalon
20. right trigeminal (cr. V) nerve (ophthalmic branch)
21. right trigeminal (cr. V) nerve (mandibular branch)
22. right first pharyngeal ectodermal groove
23. right second (hyoid) pharyngeal arch
24. right anterior cardinal vein
25. spinal cord
26. left first aortic arch
27. left first (maxillary) pharyngeal arch
28. right first (mandibular) pharyngeal arch

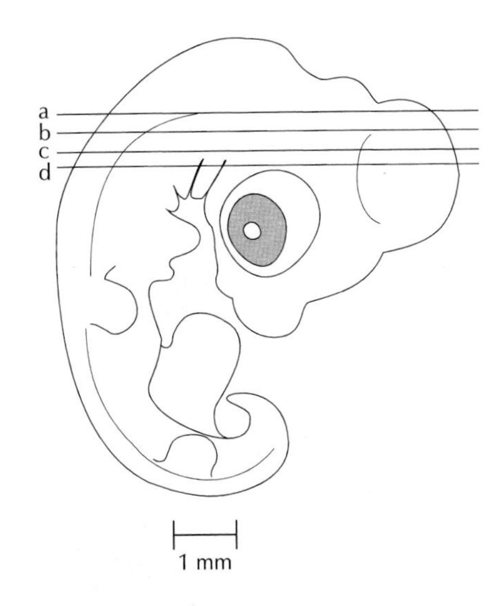

PLATE 84

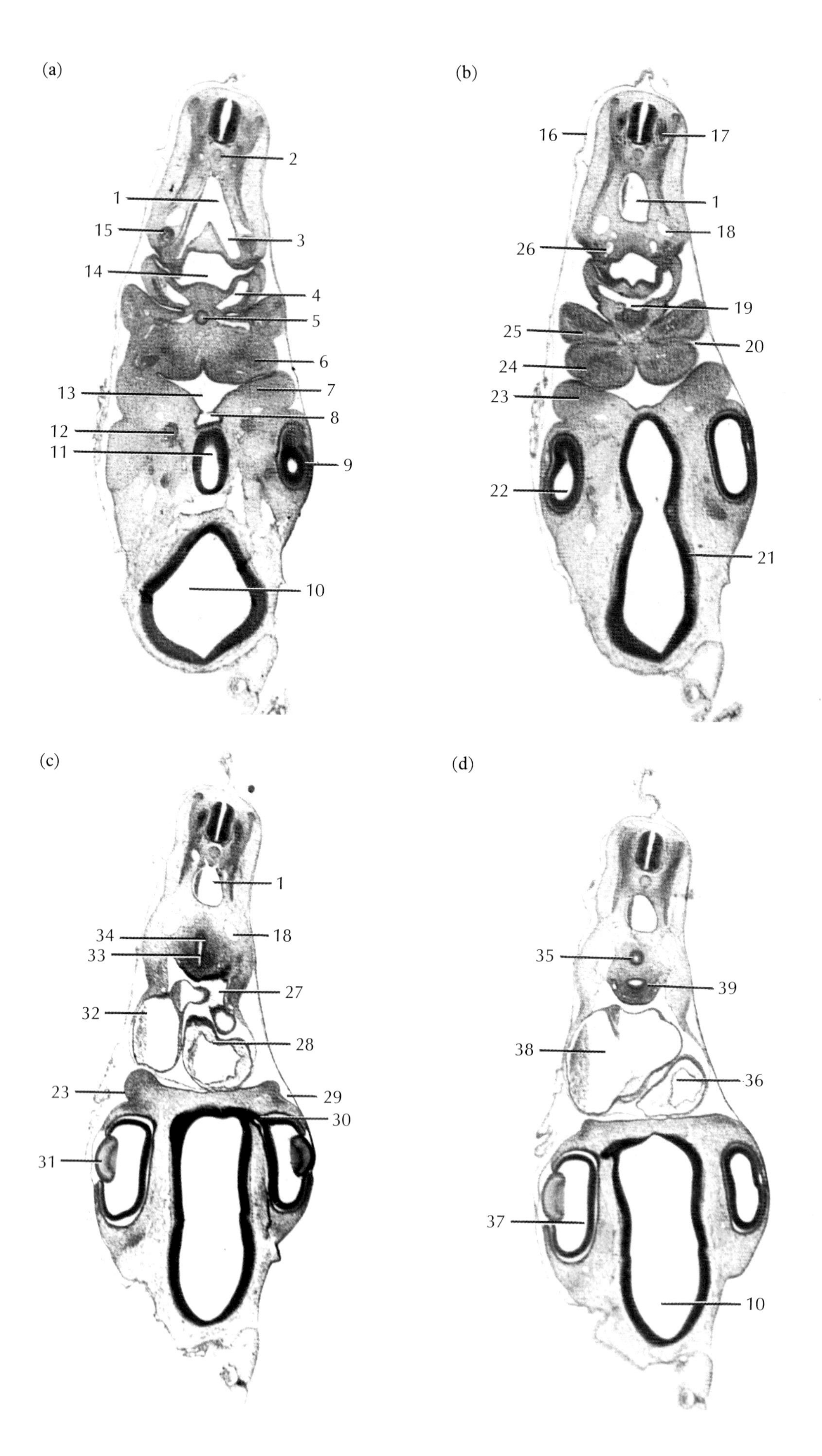
(a)
2
1
15
3
14
4
5
6
13
7
8
12
11
9
10
(b)
16
17
1
18
26
19
25
20
24
23
22
21
(c)
1
34
18
33
27
32
28
23
29
30
31
(d)
35
39
38
36
37
10

Plate 84 Day 4 TS

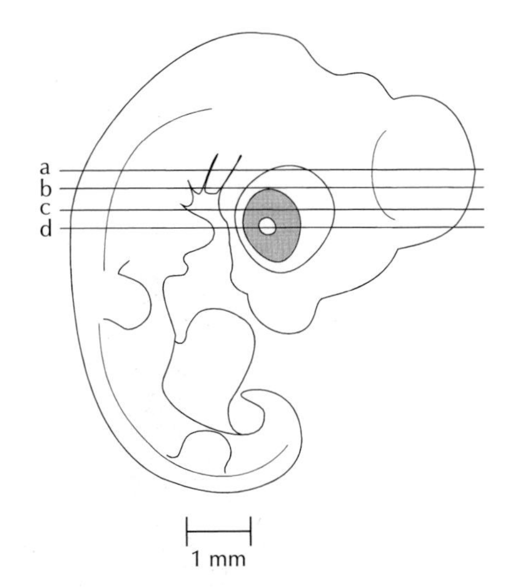

(a–d) ×22

1. dorsal aorta
2. notochord
3. left fourth aortic arch
4. left third aortic arch
5. thyroid
6. left first (mandibular) pharyngeal arch
7. left first (maxillary) pharyngeal arch
8. Rathke's pouch
9. left eye
10. diocoele
11. infundibulum
12. right oculomotor (cr. III) nerve
13. oral cavity
14. pharynx
15. jugular ganglion of right vagus (cr. X) nerve
16. amnion
17. dorsal root ganglion
18. left anterior cardinal vein
19. ventral aorta
20. left first ectodermal groove
21. diencephalon
22. right eye
23. right first (maxillary) pharyngeal arch
24. right first (mandibular) pharyngeal arch
25. right second (hyoid) arch
26. right fourth aortic arch artery
27. pericardial cavity
28. ventricle
29. nasal placode
30. optic stalk
31. lens
32. sinus venosus
33. laryngo-tracheal groove
34. pharynx
35. oesophagus
36. truncus arteriosus
37. retina
38. atrium
39. trachea

PLATE 85

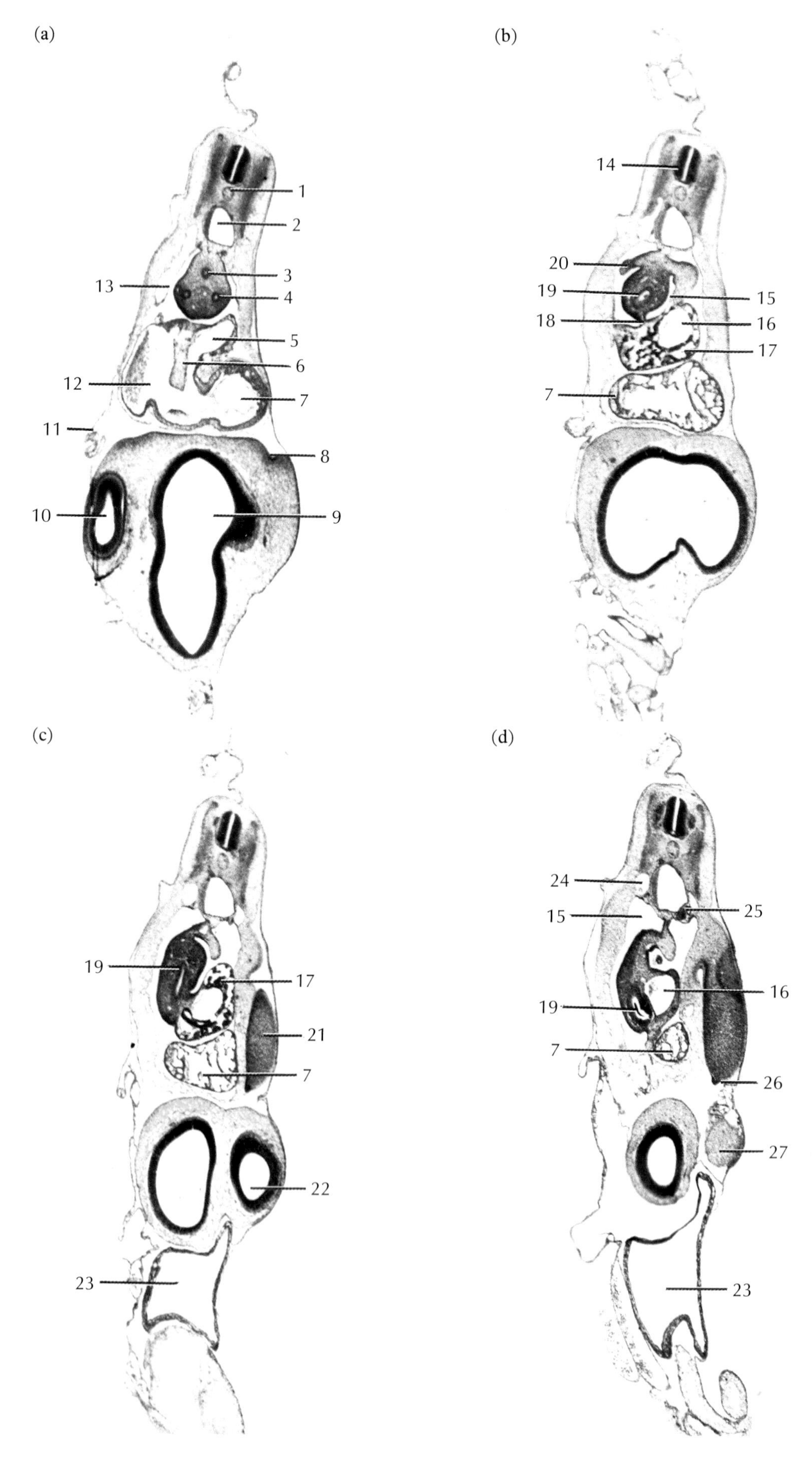
(a)
1
2
3
4
13
5
6
12
7
11
8
10
9
(b)
14
20
19
15
18
16
17
7
(c)
19
17
21
7
22
23
(d)
24
25
15
16
19
7
26
27
23

Plate 85 Day 4 TS

(a–d) ×22

1. notochord
2. dorsal aorta
3. oesophagus
4. trachea
5. left atrium
6. interatrial septum
7. ventricle
8. nasal pit
9. diocoele
10. right eye
11. amnion
12. right atrium
13. right anterior cardinal vein
14. neural tube
15. future pleural cavity
16. ductus venosus
17. liver
18. septum transversum
19. stomach
20. right lung bud
21. left wing bud
22. left lateral telencephalic vesicle
23. lumen of allantois
24. right posterior cardinal vein
25. left nephric duct
26. apical ectodermal ridge of left wing bud
27. left omphalo-mesenteric vein

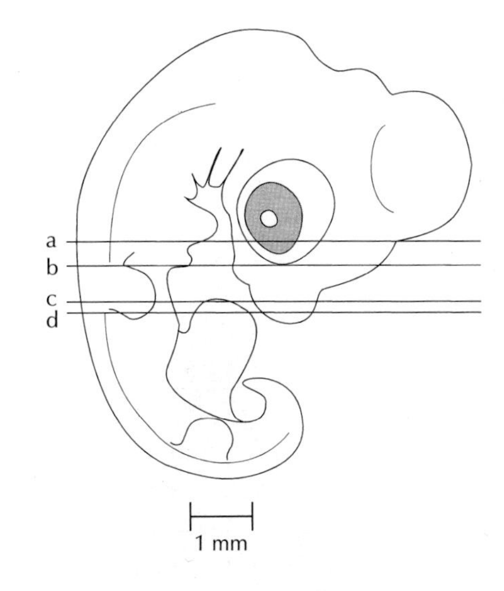

Plate 86

(a)

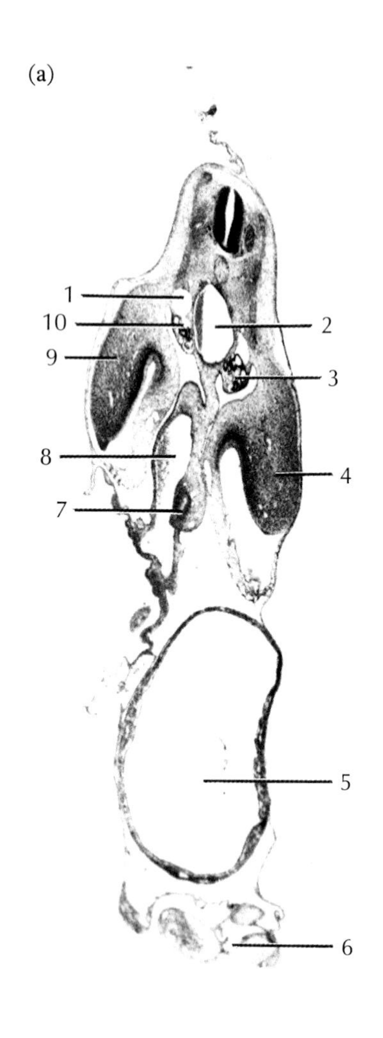

(b)

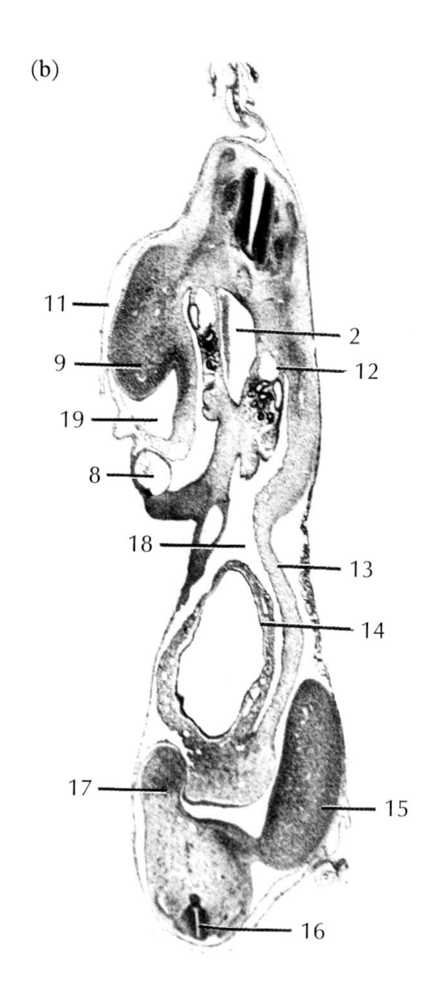

(c)

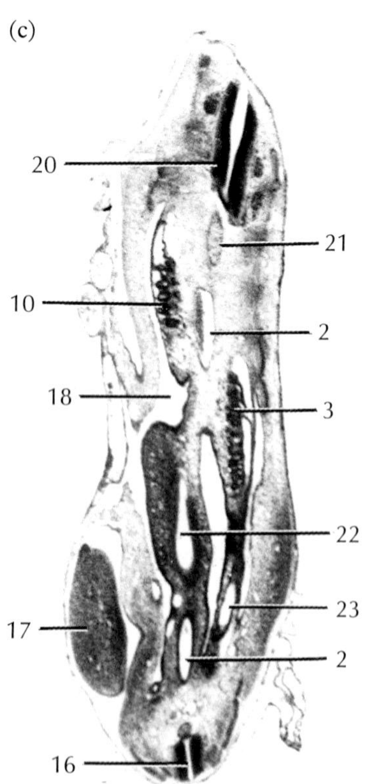

(d)

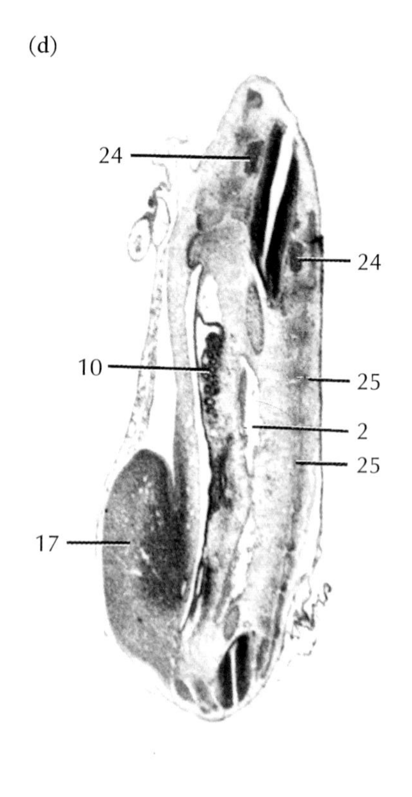

Plate 86 Day 4 TS

(a–d) ×22

1. right posterior cardinal vein
2. dorsal aorta
3. left mesonephros
4. left wing bud
5. lumen of allantois
6. yolk sac
7. duodenum
8. right omphalo-mesenteric vein
9. right wing bud
10. right mesonephros
11. amnion
12. left posterior cardinal vein
13. body wall
14. allantois
15. left leg bud
16. spinal cord of tail
17. right leg bud
18. abdominal coelom
19. amniotic cavity
20. spinal cord
21. notochord
22. hindgut
23. left nephric duct
24. dorsal root ganglion
25. prevertebra

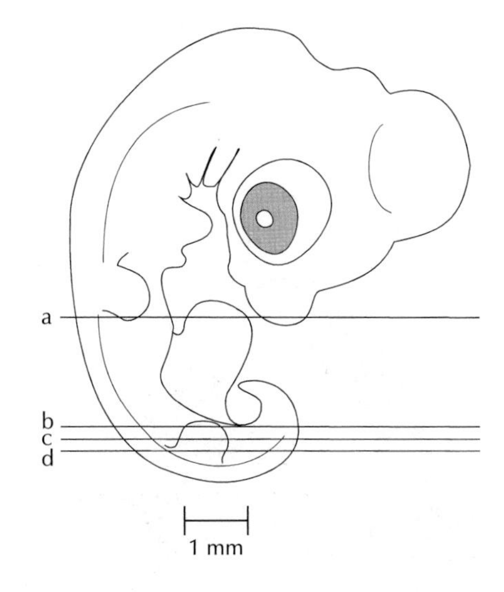

Plate 87

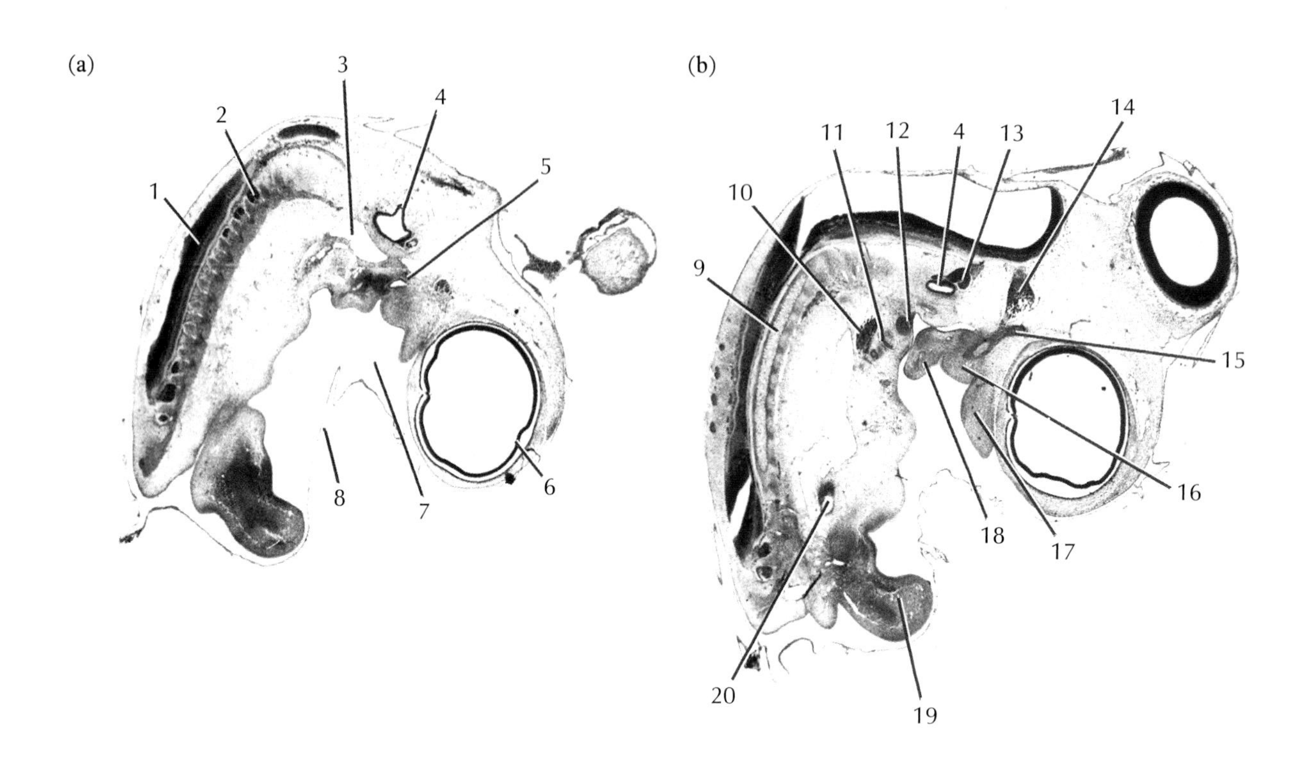

(a)
1
2
3
4
5
6
7
8
(b)
4
9
10
11
12
13
14
15
16
17
18
19
20

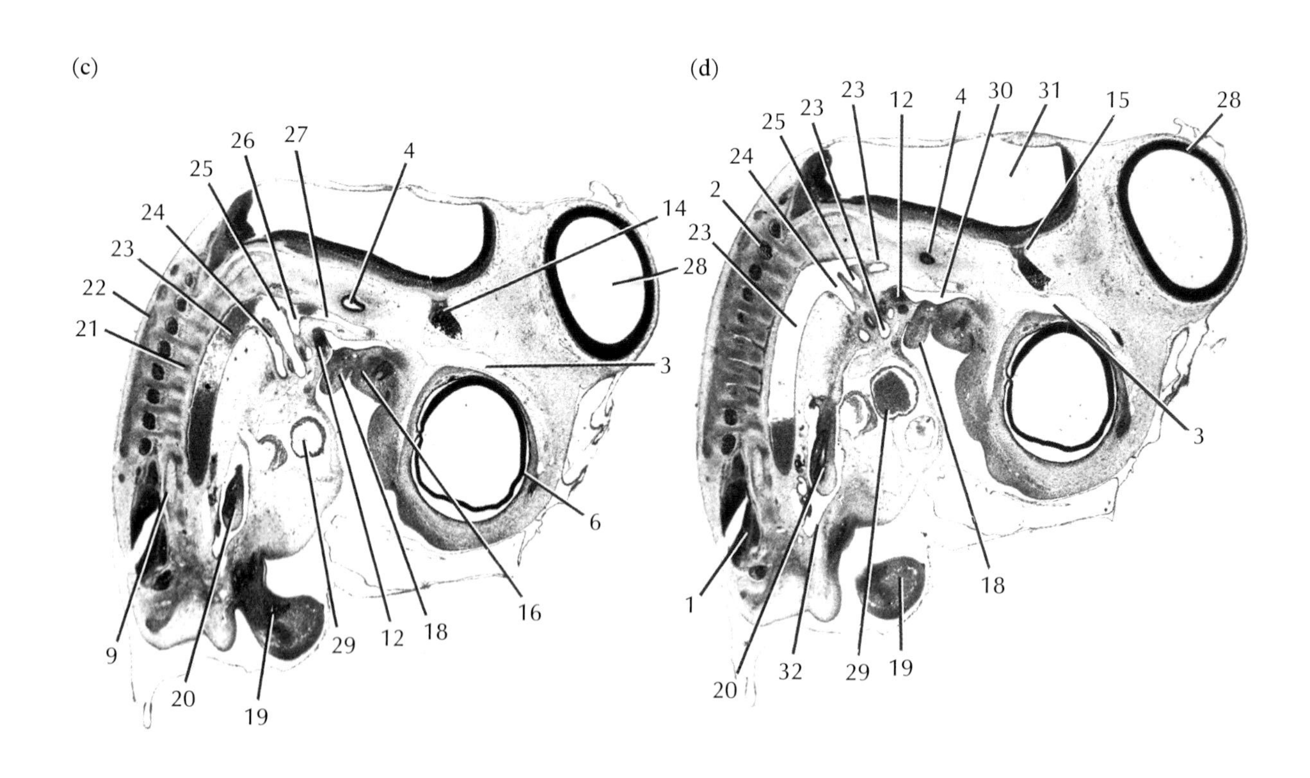

(c)
3
4
6
9
12
14
16
18
19
20
21
22
23
24
25
26
27
28
29
(d)
1
2
3
4
12
15
18
19
20
23
23
23
24
25
28
29
30
31
32

Plate 87 Day 5 LS (longitudinal sections: sagittal)

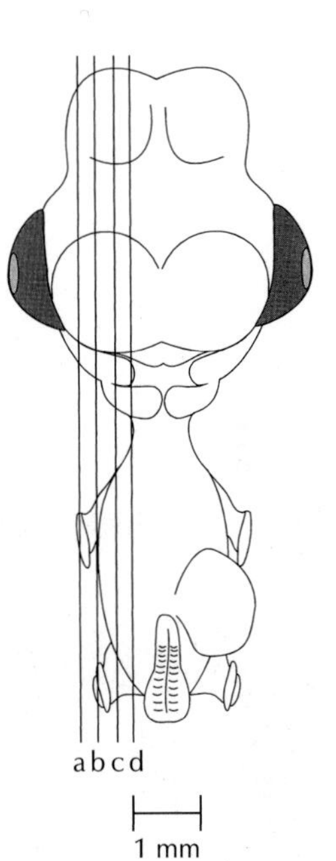

(a–d) ×8

1. spinal cord
2. dorsal root ganglion
3. head vein
4. otic vesicle
5. first pharyngeal pouch
6. pigmented layer of retina
7. amniotic cavity
8. amnion
9. notochord
10. vagus (cr. X) nerve
11. fourth pharyngeal pouch
12. precartilaginous condensation of mesoderm in third pharyngeal arch
13. geniculate (cr. nerve VII) ganglion
14. trigeminal (cr. nerve V) ganglion
15. trigeminal (cr. V) nerve
16. first (mandibular) pharyngeal arch
17. first (maxillary) pharyngeal arch
18. second (hyoid) pharyngeal arch
19. wing bud
20. loop of intestine
21. body of vertebra
22. neural arch of vertebra
23. dorsal aorta
24. fifth aortic arch
25. fourth aortic arch
26. third aortic arch
27. carotid artery
28. mesencephalon (developing optic lobe)
29. atrium
30. pharynx
31. myelocoele
32. peritoneum

Plate 88

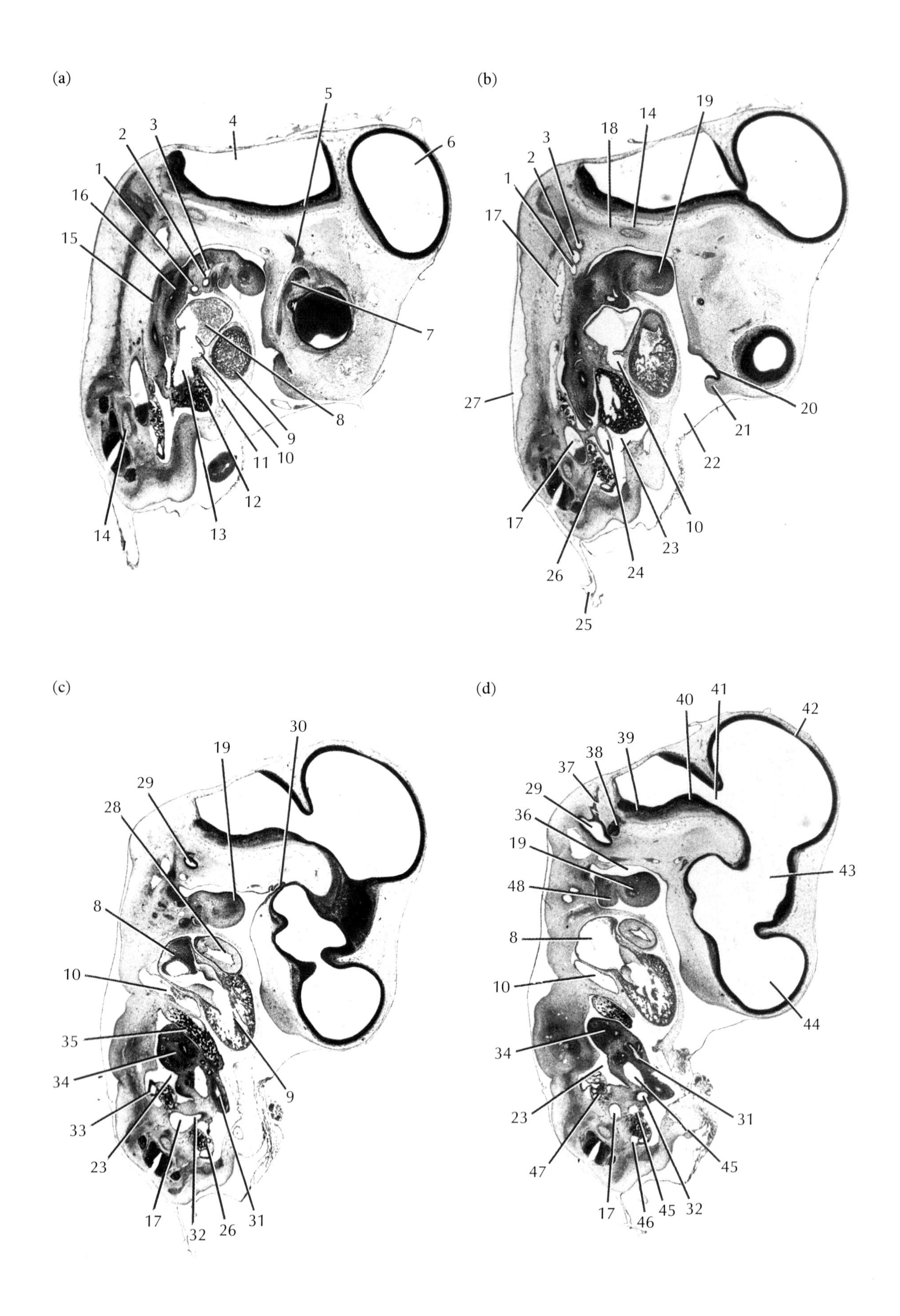
(a)
1
2
3
4
5
6
7
8
9
10
11
12
13
14
15
16
(b)
1
2
3
14
17
18
19
20
21
22
23
24
25
26
27
10
17
(c)
8
9
10
17
19
23
26
28
29
30
31
32
33
34
35
9
(d)
8
10
17
19
23
29
31
32
34
36
37
38
39
40
41
42
43
44
45
45
46
47
48

Plate 88 Day 5 LS (longitudinal sections: sagittal)

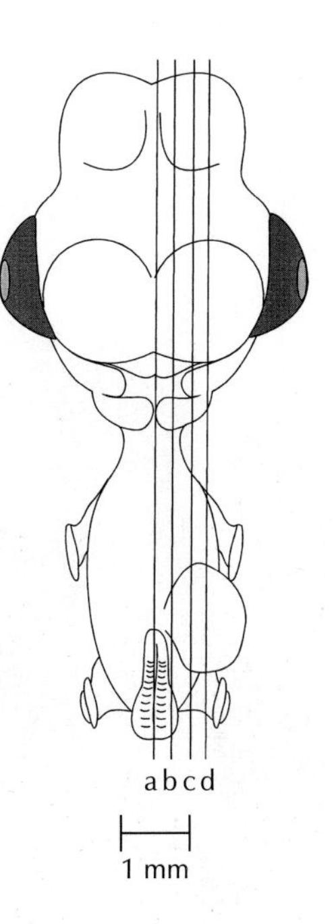

(a–d) ×8

1. fifth aortic arch
2. fourth aortic arch
3. third aortic arch
4. myelocoele
5. trigeminal (cr. V) nerve
6. mesocoele (developing optic lobe)
7. trigeminal (cr. V) nerve (ophthalmic branch)
8. atrium
9. ventricle
10. sinus venosus
11. right allantoic vein
12. liver
13. ductus venosus
14. notochord
15. oesophagus
16. trachea
17. dorsal aorta
18. basal plate of developing chondrocranium
19. first (mandibular) pharyngeal arch
20. nasal pit
21. fronto-nasal process
22. amniotic cavity
23. coelomic cavity
24. left omphalo-mesenteric artery
25. posterior amniotic canal
26. left mesonephros
27. amnion
28. truncus arteriosus
29. otic vesicle
30. Rathke's pouch
31. loop of intestine in mesentery
32. left omphalo-mesenteric artery
33. right oviduct (Mullerian duct)
34. stomach
35. liver in region of falciform ligament
36. pharynx
37. endolymphatic duct
38. acoustic (cr. nerve VIII) ganglion (vestibulo-cochlearis)
39. myelencephalon
40. metencephalon
41. isthmus
42. mesencephalon
43. diocoele
44. telocoele
45. left subcardinal vein
46. posterior cardinal vein
47. right mesonephros
48. second (hyoid) pharyngeal arch

PLATE 89

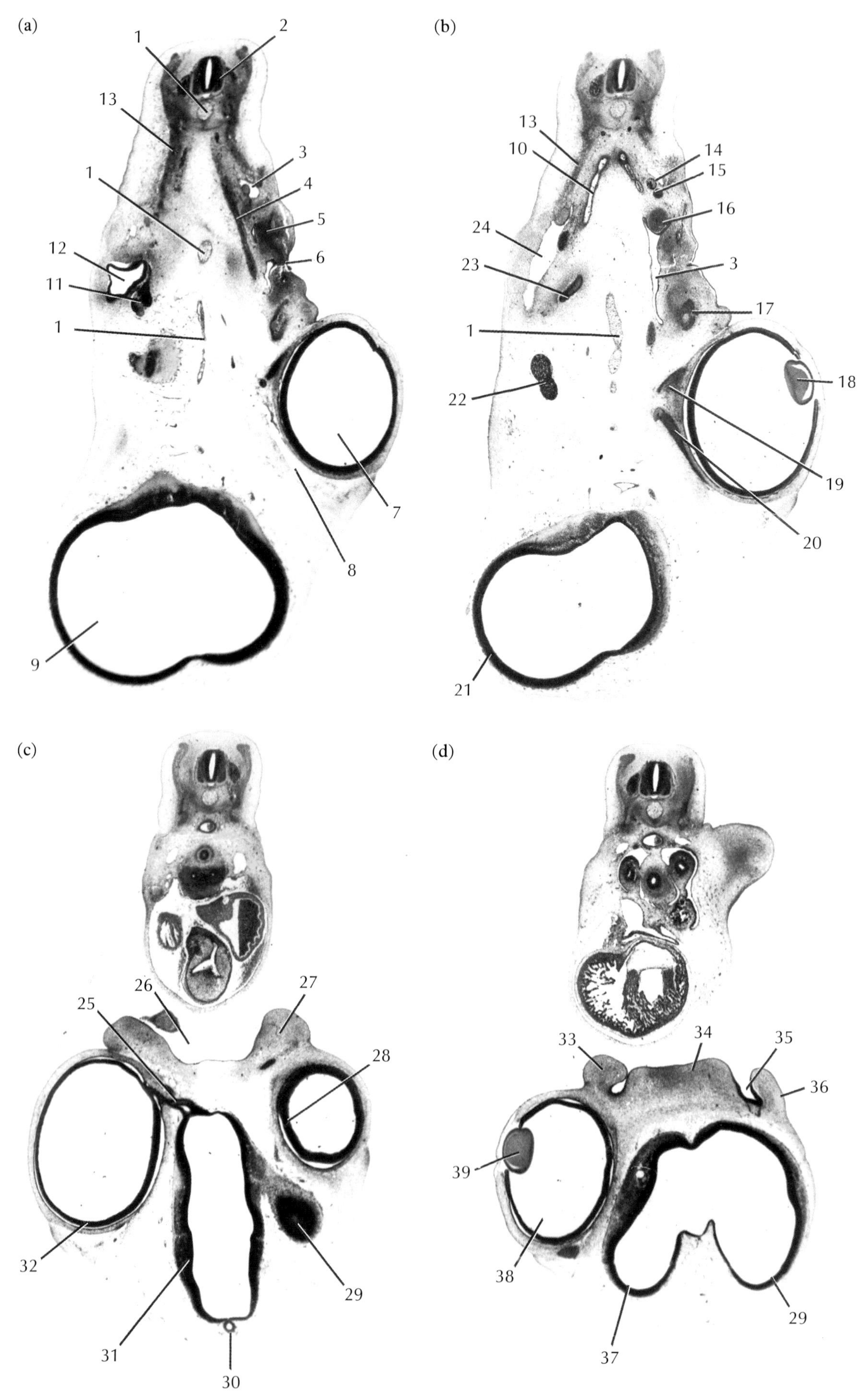

(a)
1
2
13
3
1
4
5
12
6
11
1
7
8
9
(b)
13
10
14
15
16
24
3
23
17
1
18
22
19
20
21
(c)
26
27
25
28
32
29
31
30
(d)
33
34
35
36
39
38
29
37

Plate 89 Day 5 TS

(a–d) × 15

Trunk of (c) enlarged on *Plate 90*

Trunk of (d) enlarged on *Plate 91*

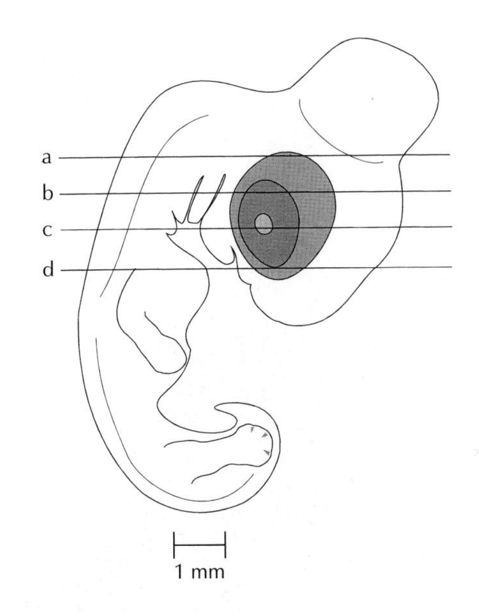

1. notochord
2. spinal cord
3. left anterior cardinal vein
4. left carotid artery
5. left second (hyoid) pharyngeal arch
6. first pharyngeal pouch
7. vitreous humour of left eye
8. left ophthalmic vein
9. mesocoele
10. right carotid artery
11. geniculate (cr. nerves VII and VIII) ganglion
12. otic vesicle
13. myotome
14. quadrate cartilage
15. Meckel's cartilage
16. acoustic (cr. VIII) nerve
17. left maxillary (cr. nerve V) nerve
18. lens of left eye
19. dorsal oblique eye muscle
20. dorsal rectus eye muscle
21. mesencephalon
22. trigeminal (cr. nerve V) ganglion
23. lagena of right ear
24. right anterior vena cava
25. right optic nerve
26. oral cavity
27. median nasal eminence
28. retina of left eye
29. wall of left telencephalic vesicle
30. pineal organ
31. diencephalon
32. retina of right eye
33. right lateral nasal eminence
34. fronto-nasal process
35. left nasal pit (external naris)
36. left lateral nasal eminence
37. wall of right telencephalic vesicle
38. vitreous humour of right eye
39. lens of right eye

PLATE 90

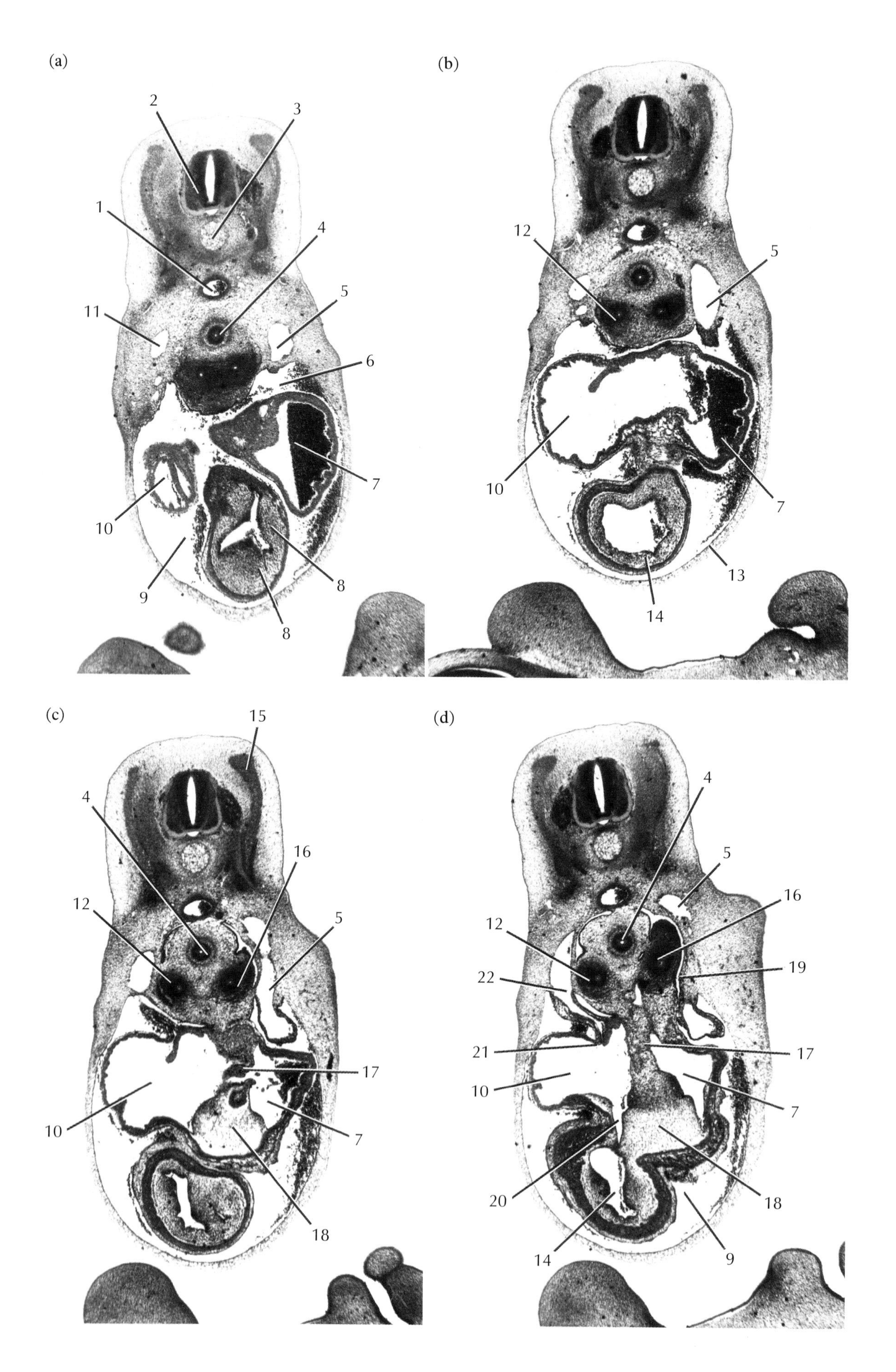
(a)
2
3
1
4
5
11
6
7
10
9
8
8
(b)
12
5
10
7
13
14
(c)
15
4
16
12
5
17
10
7
18
(d)
4
5
16
12
19
22
21
17
10
7
20
18
14
9

Plate 90 Day 5 TS

(a–d) ×30

(a) Enlargement of part of *Plate 89c.*

1. dorsal aorta
2. neural tube
3. notochord
4. oesophagus
5. left anterior cardinal vein
6. extravasated blood in pericardial cavity
7. left atrium
8. aortico-pulmonary ridges in truncus arteriosus
9. pericardial cavity
10. right atrium
11. right anterior cardinal vein
12. right lung bud
13. ventral body wall
14. truncus arteriosus
15. myotome
16. left lung bud
17. interatrial septum
18. atrio-ventricular cushion
19. edge of pleural cavity
20. right atrio-ventricular canal
21. sinu-atrial valve
22. sinus venosus

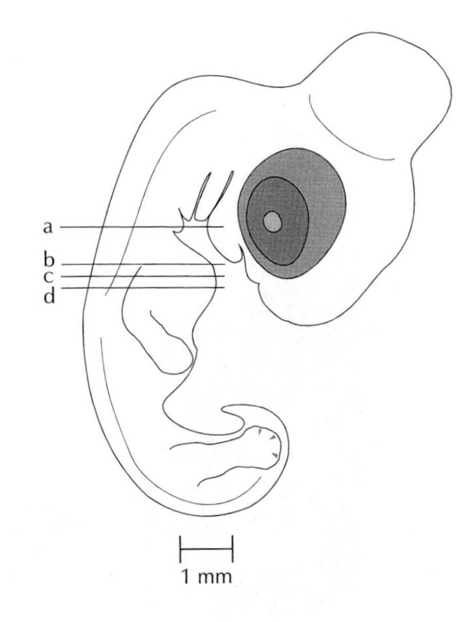

Plate 91

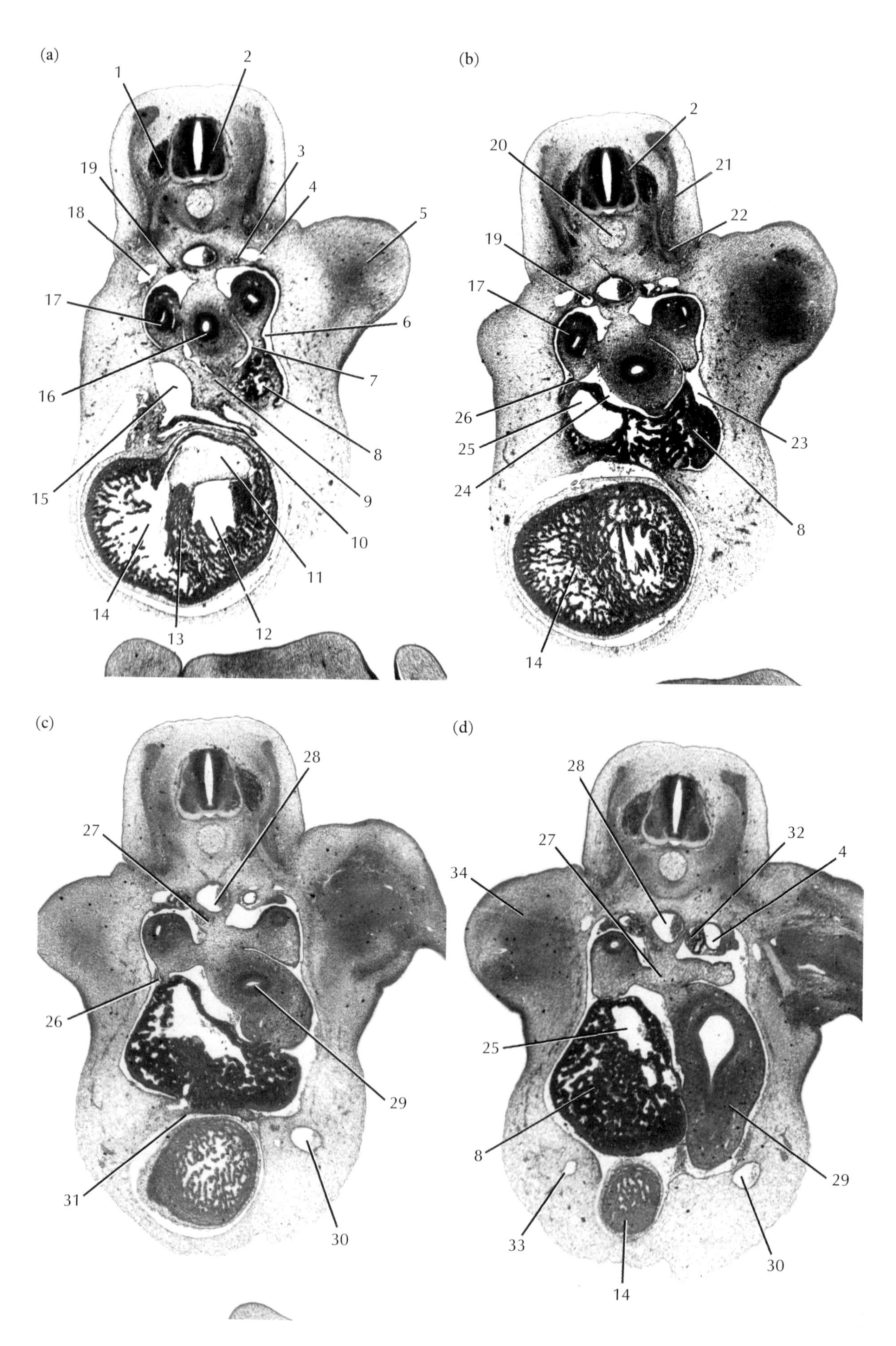

(a)
1
2
3
4
5
6
7
8
9
10
11
12
13
14
15
16
17
18
19
(b)
2
20
21
22
19
17
26
25
24
23
8
14
(c)
28
27
26
29
31
30
(d)
28
27
32
4
34
25
8
29
33
14
30

Plate 91 Day 5 TS

(a–d) ×30

(a) Enlargement of part of *Plate 89d.*

1. dorsal root ganglion
2. spinal cord
3. left nephric (pronephric) duct
4. left anterior cardinal vein
5. left wing bud
6. left pleural cavity
7. lesser sac
8. liver
9. ventral mesentery
10. left common cardinal vein
11. atrio-ventricular cushion
12. left ventricle
13. interventricular septum
14. right ventricle
15. sinus venosus
16. crop
17. right lung bud
18. right anterior cardinal vein
19. right nephric (pronephric) duct
20. notochord
21. spinal nerve
22. ramus communicans nerve
23. peritoneal cavity
24. dorsal hepatic cavity
25. ductus venosus
26. right pulmonary fold
27. dorsal mesentery
28. dorsal aorta
29. gizzard
30. left allantoic vein
31. septum transversum
32. left mesonephros
33. right allantoic vein
34. right wing bud

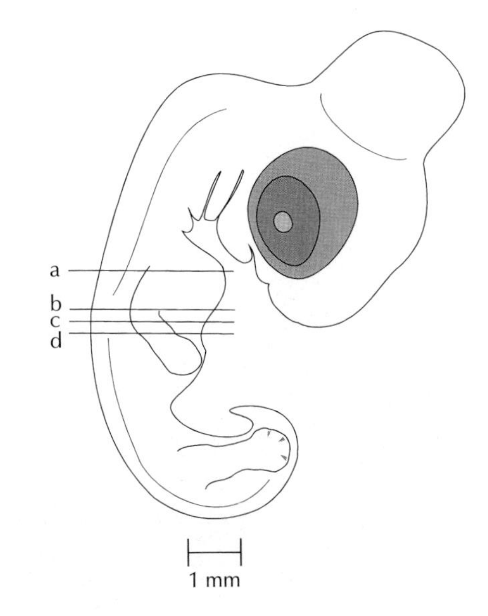

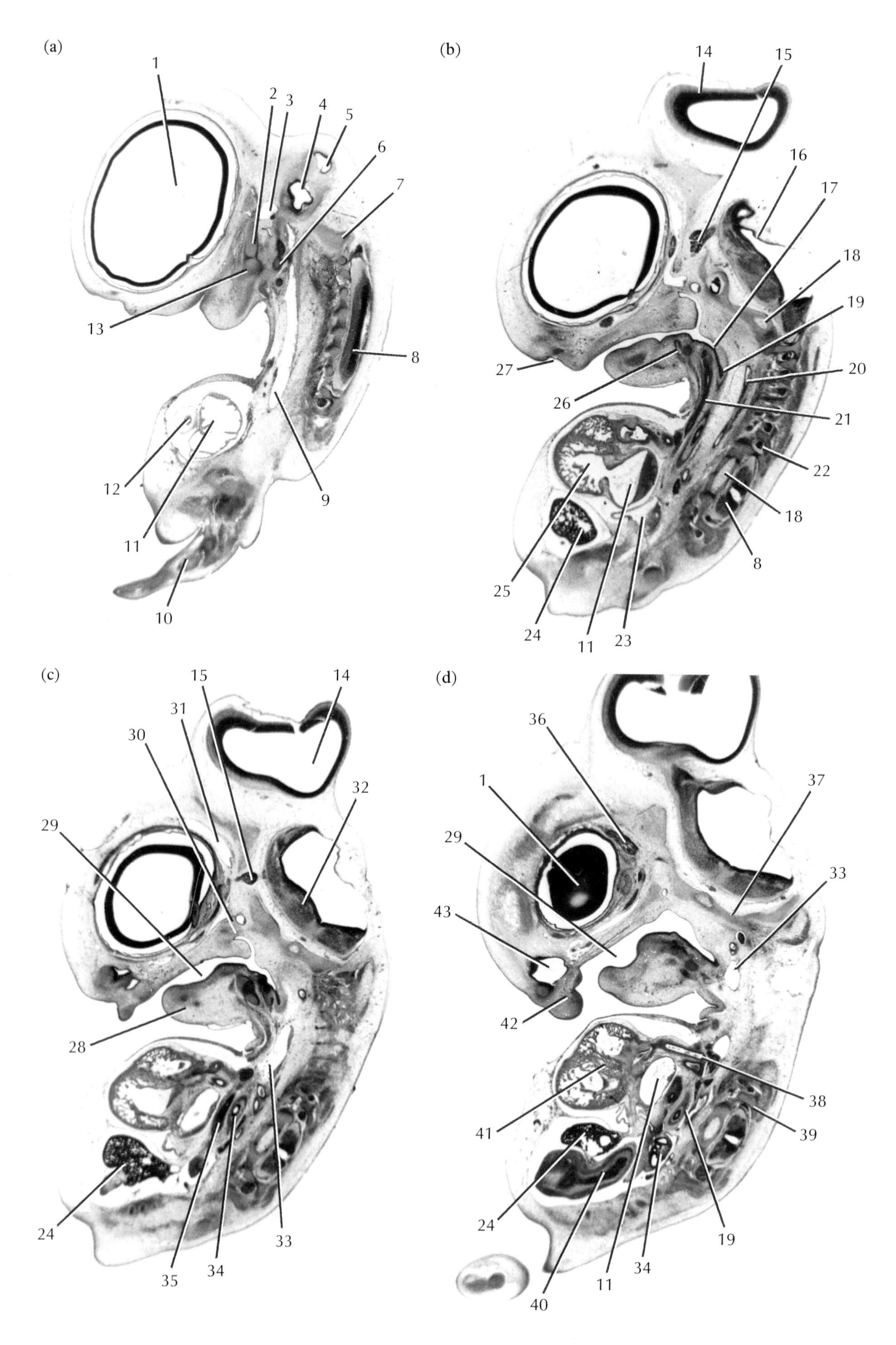
(a)
1
2
3
4
5
6
7
8
9
10
11
12
13
(b)
14
15
16
17
18
19
20
21
22
18
8
23
11
24
25
26
27
(c)
15
14
31
30
32
29
28
24
35
34
33
(d)
36
37
1
29
33
43
42
38
41
39
24
19
34
11
40

Plate 92 Day 6 LS (longitudinal sections: sagittal)

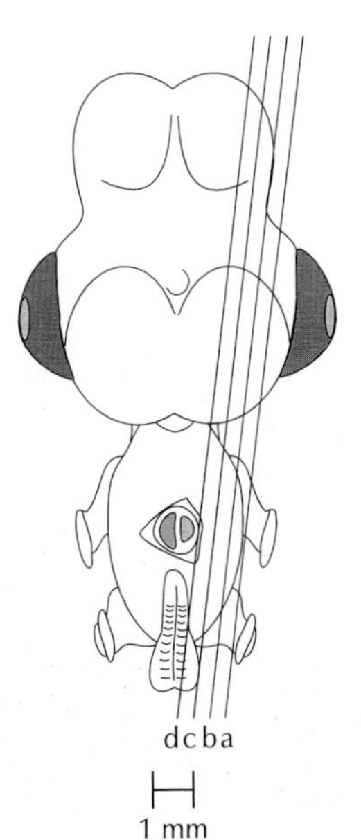

(a–d) ×10

1. left eye
2. quadrate cartilage
3. left jugular vein
4. left otic vesicle
5. semicircular canal
6. left occipital cartilage
7. wall of myelencephalon
8. spinal cord
9. left cranial vena cava
10. left femur
11. left atrium
12. bulbus arteriosus
13. Meckel's cartilage
14. left optic lobe
15. left trigeminal (cr. nerve V) ganglion
16. thin wall of myelencephalon
17. oesophageal opening
18. notochord
19. oesophagus
20. dorsal aorta
21. trachea
22. dorsal root ganglion
23. sinus venosus
24. liver
25. left ventricle
26. thyroglossal duct
27. left external naris
28. lower jaw
29. oral cavity
30. internal naris
31. left head vein
32. floor of myelencephalon
33. right jugular vein
34. left lung bud
35. gut mesentery
36. external eye muscle
37. parachordal cartilage
38. truncus arteriosus
39. transverse process of vertebra
40. gizzard
41. interventricular septum
42. fronto-nasal process
43. left nasal sac

PLATE 93

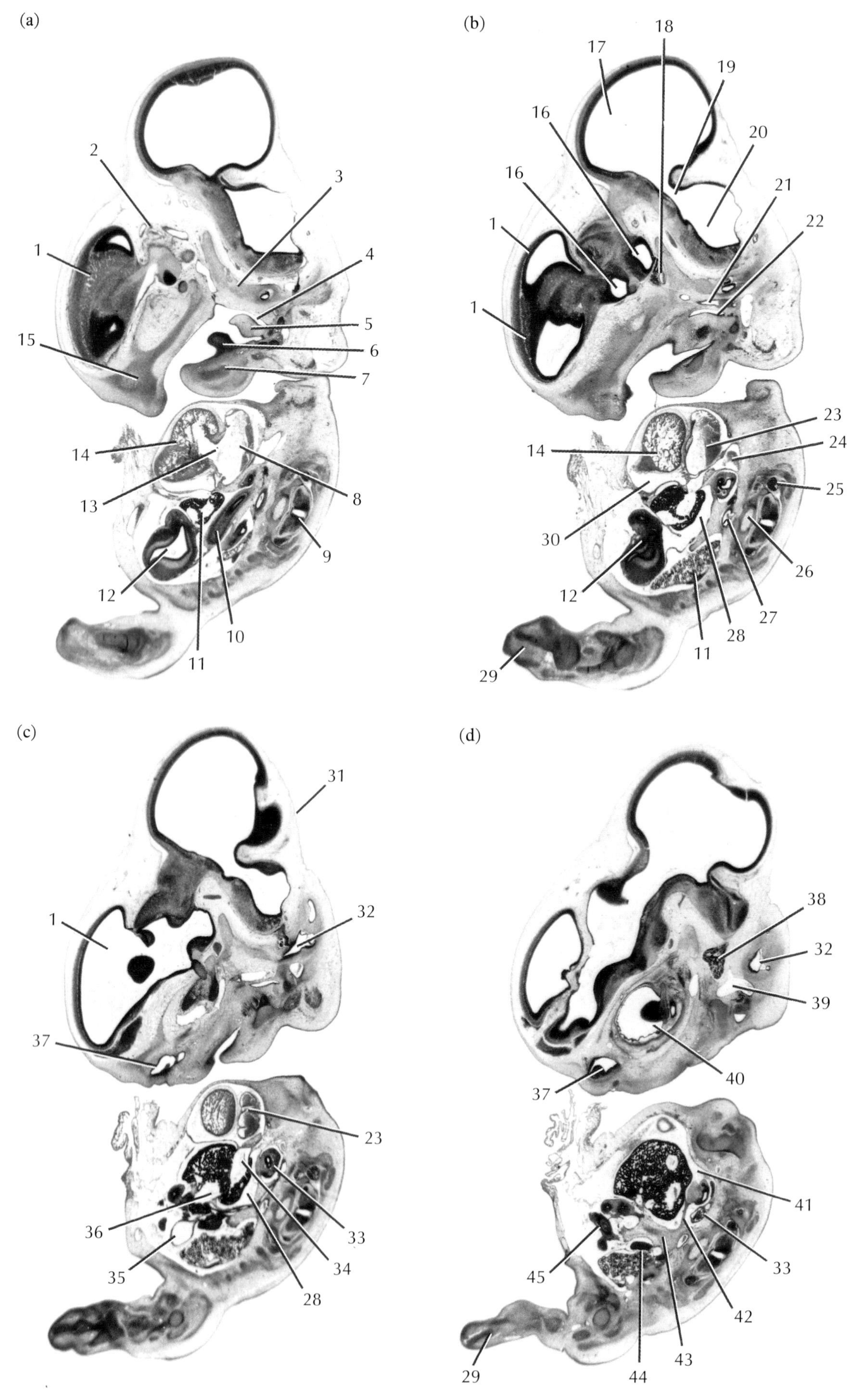

(a)
1
2
3
4
5
6
7
8
9
10
11
12
13
14
15
(b)
1
11
12
14
16
17
18
19
20
21
22
23
24
25
26
27
28
29
30
(c)
1
23
28
31
32
33
34
35
36
37
(d)
29
32
33
37
38
39
40
41
42
43
44
45

Plate 93 Day 6 LS (longitudinal sections: sagittal)

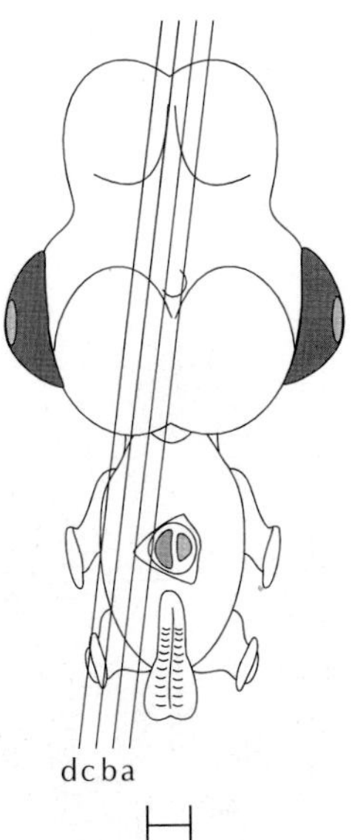

(a–d) × 10

1. left cerebral vesicle
2. branches of head vein
3. basal plate of chondrocranium
4. pharynx
5. wall of pharynx
6. tongue
7. Meckel's cartilage
8. atrium
9. spinal cord
10. oesophagus
11. liver
12. gizzard
13. atrio-ventricular canal
14. right ventricle
15. nasal capsule
16. third ventricle
17. cerebral vesicle
18. infundibulum
19. isthmus
20. myelocoele
21. right carotid artery
22. right auditory tube
23. right atrium
24. sinus venosus
25. dorsal root ganglion
26. notochord
27. dorsal aorta
28. right ventral hepatic cavity
29. right leg
30. pericardial cavity
31. epidermis of head
32. right otic vesicle
33. right lung bud
34. caudal vena cava
35. posterior abdominal air sac
36. omphalo-mesenteric vein
37. right nasal sac
38. right trigeminal (cr. nerve V) ganglion
39. right jugular vein
40. right eye
41. right dorsal hepatic cavity
42. horizontal hepatic ligament
43. dorsal mesentery
44. gonad
45. loop of intestine

Plate 94

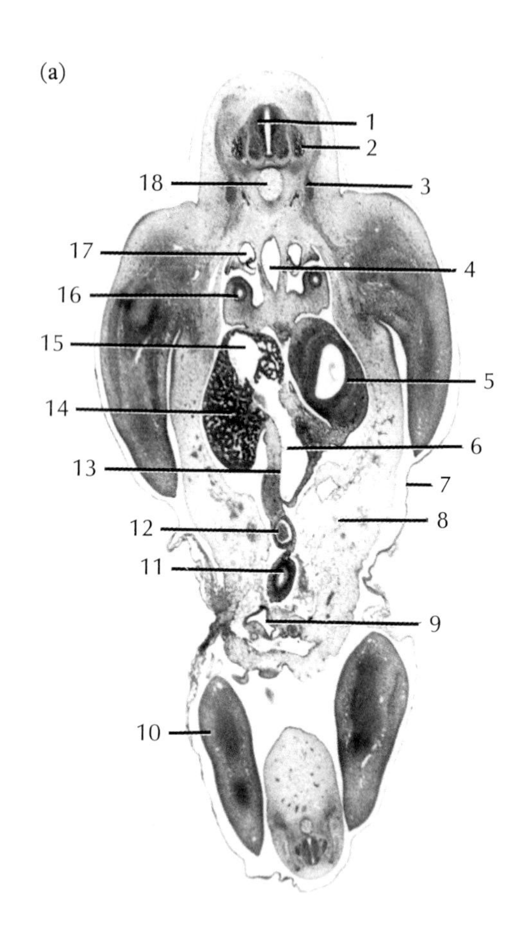
(a)
1
2
18
3
17
4
16
15
5
14
6
13
7
12
8
11
9
10

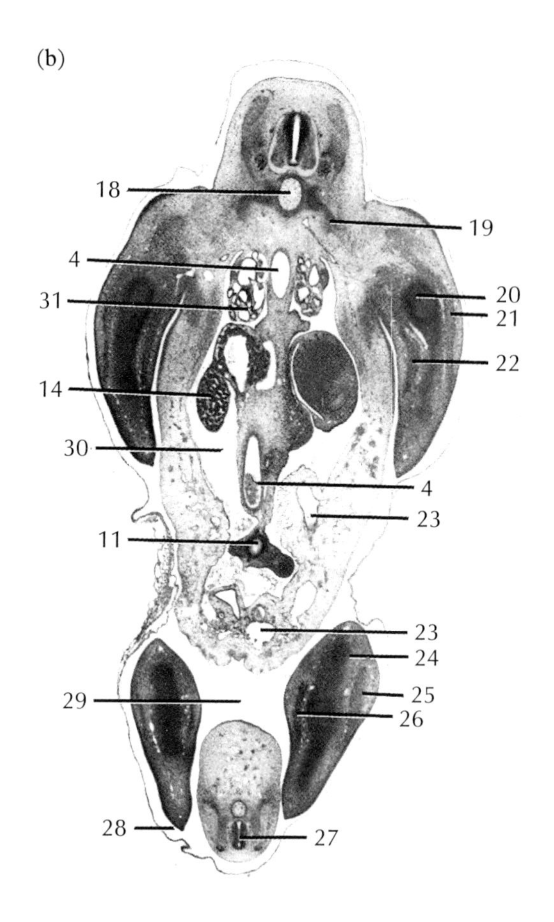
(b)
18
19
4
31
20
21
22
14
30
4
23
11
23
24
29
25
26
28
27

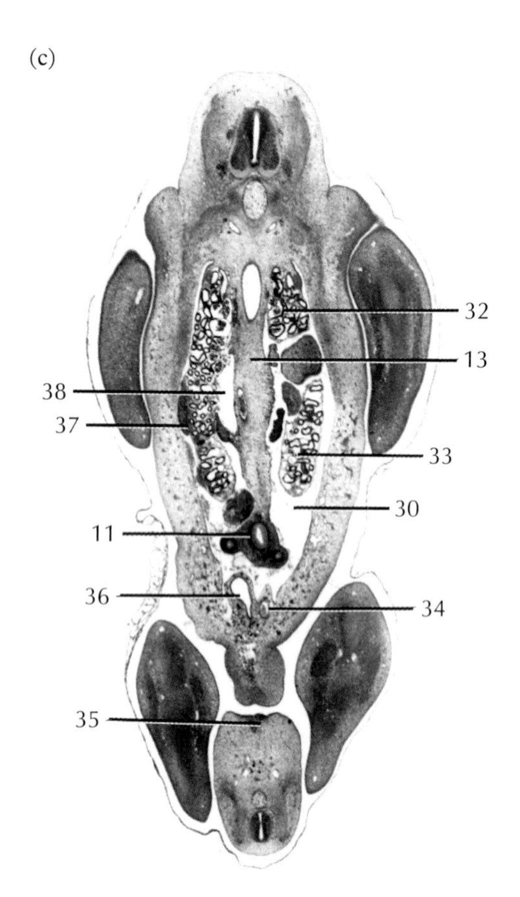
(c)
32
13
38
37
33
30
11
36
34
35

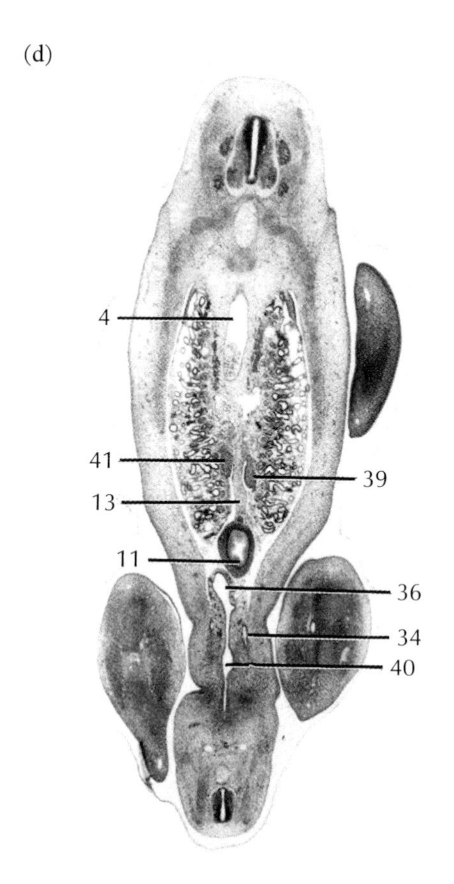
(d)
4
41
39
13
11
36
34
40

Plate 94 Day 6 LS (longitudinal sections: coronal)

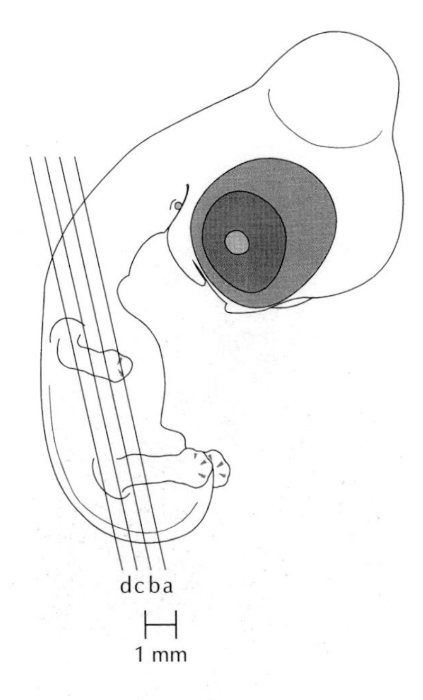

(a–d) × 15

Enlargements of portions of these sections are shown in *Plates 95* and *96*.

1. neural tube of trunk
2. dorsal root ganglion
3. sympathetic ganglion in myotome
4. dorsal aorta
5. gizzard
6. left hepatic vein
7. amnion
8. body wall
9. cloaca
10. right leg bud
11. colon
12. small intestine
13. dorsal mesentery
14. liver
15. caudal (posterior) vena cava
16. right lung bud
17. right anterior cardinal vein
18. notochord
19. rib (not yet cartilaginous)
20. head of humerus
21. dorsal premuscle mass of wing bud
22. ventral premuscle mass of wing bud
23. allantoic vein
24. femur
25. dorsal premuscle mass of thigh
26. ventral premuscle mass of thigh
27. neural tube of tail
28. apical ectodermal ridge of right leg bud
29. amniotic cavity
30. coelomic cavity
31. right metanephros
32. left metanephros
33. left mesonephros
34. nephric (Wolffian) duct
35. cloacal membrane
36. rectum
37. right mesonephros
38. right subcardinal vein
39. left gonad
40. cloaca
41. right gonad

Plate 95

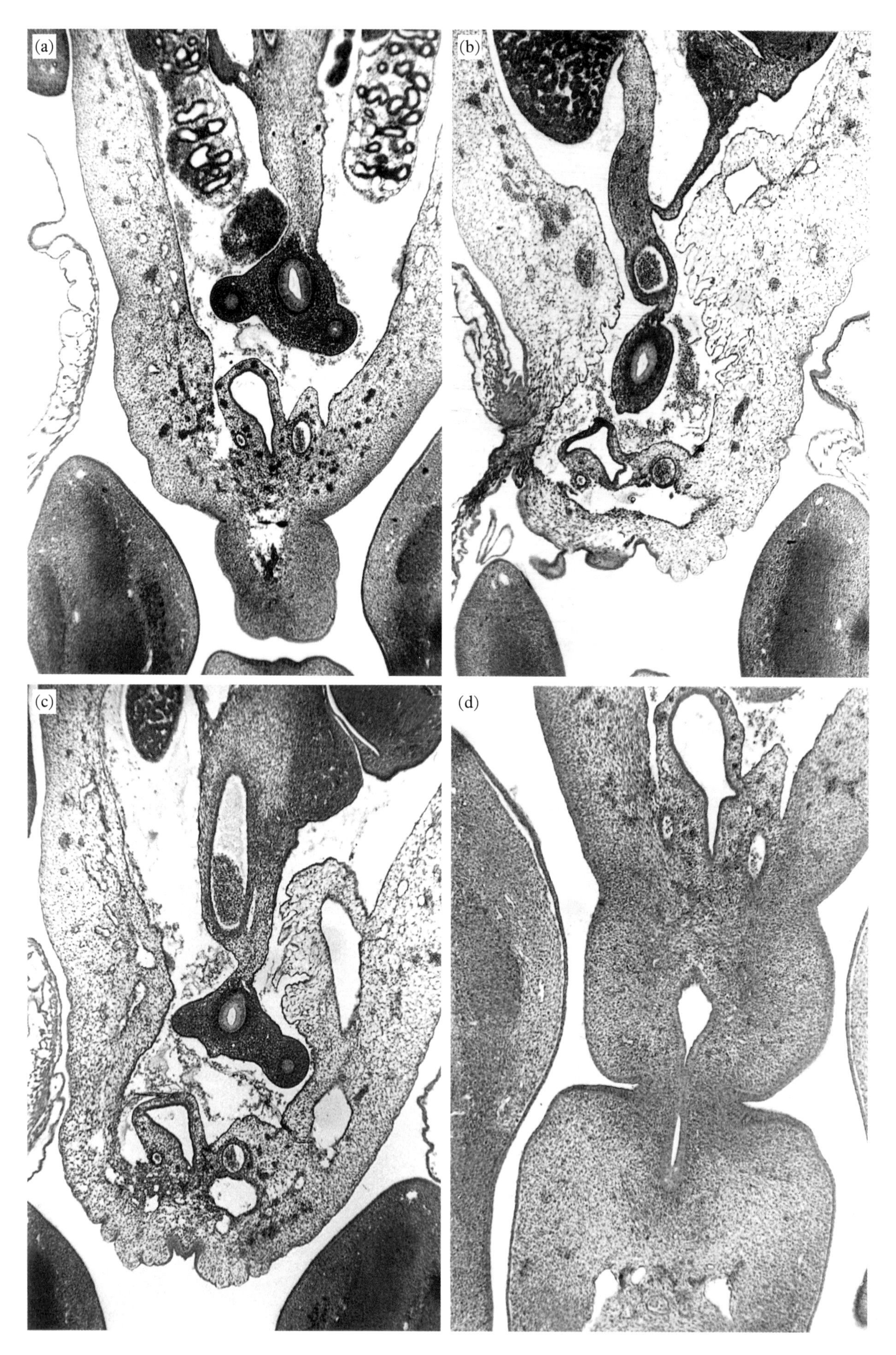

Plate 95 Day 6 LS (longitudinal sections: coronal)

(a–d) ×48

(a) Enlargement of section adjacent to *Plate 94c.*
(b) Enlargement of part of *Plate 94a.*
(c) Enlargement of part of *Plate 94b.*
(d) Enlargement of section adjacent to *Plate 94d.*

Plate 96

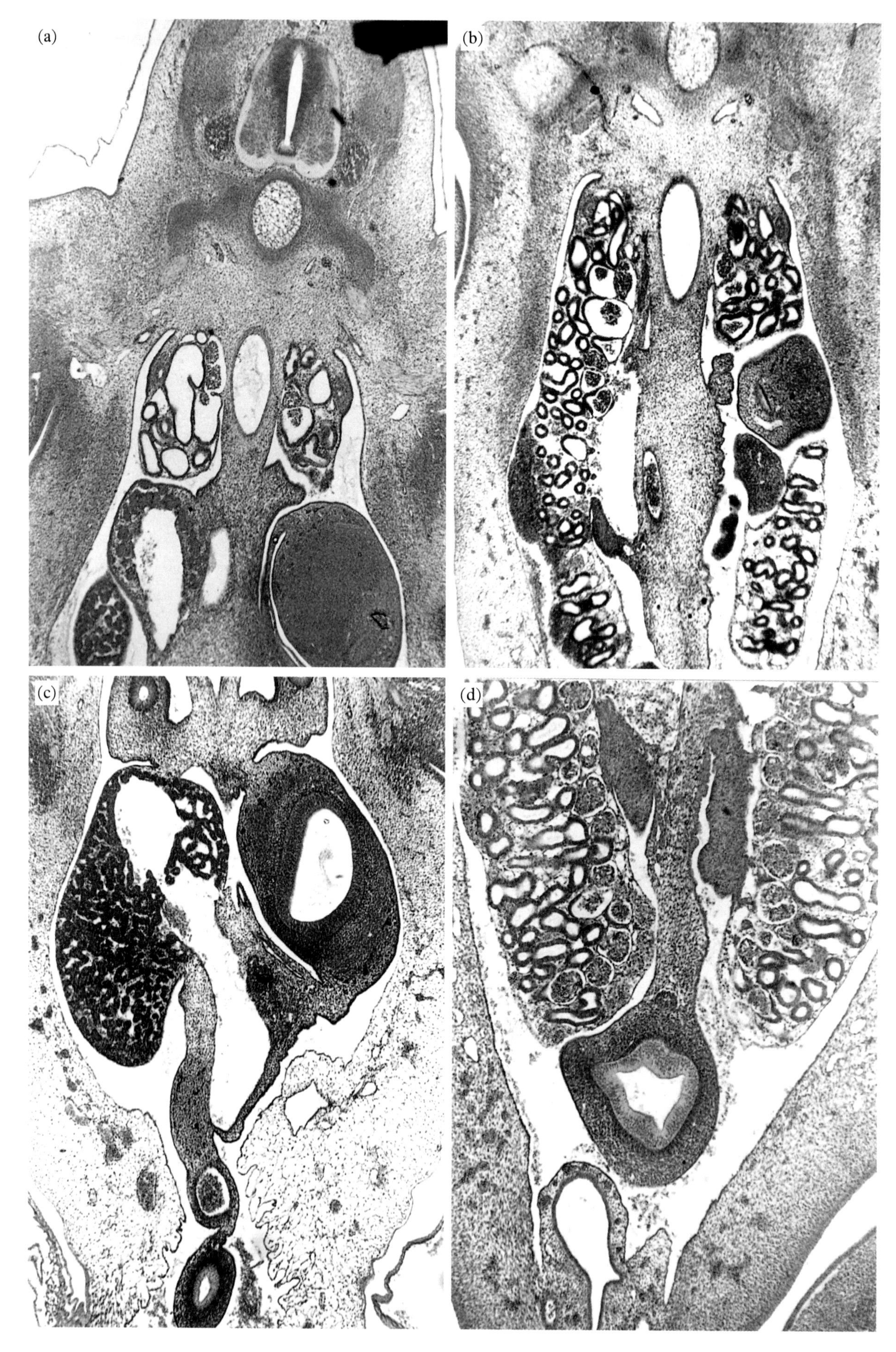

Plate 96 Day 6 LS (longitudinal sections: coronal)

(a–d) ×48

(a) Enlargement of section adjacent to *Plate 94b*.
(b) Enlargement of part of *Plate 94c*.
(c) Enlargement of part of *Plate 94a*.
(d) Enlargement of section adjacent to *Plate 94d*.

PLATE 97

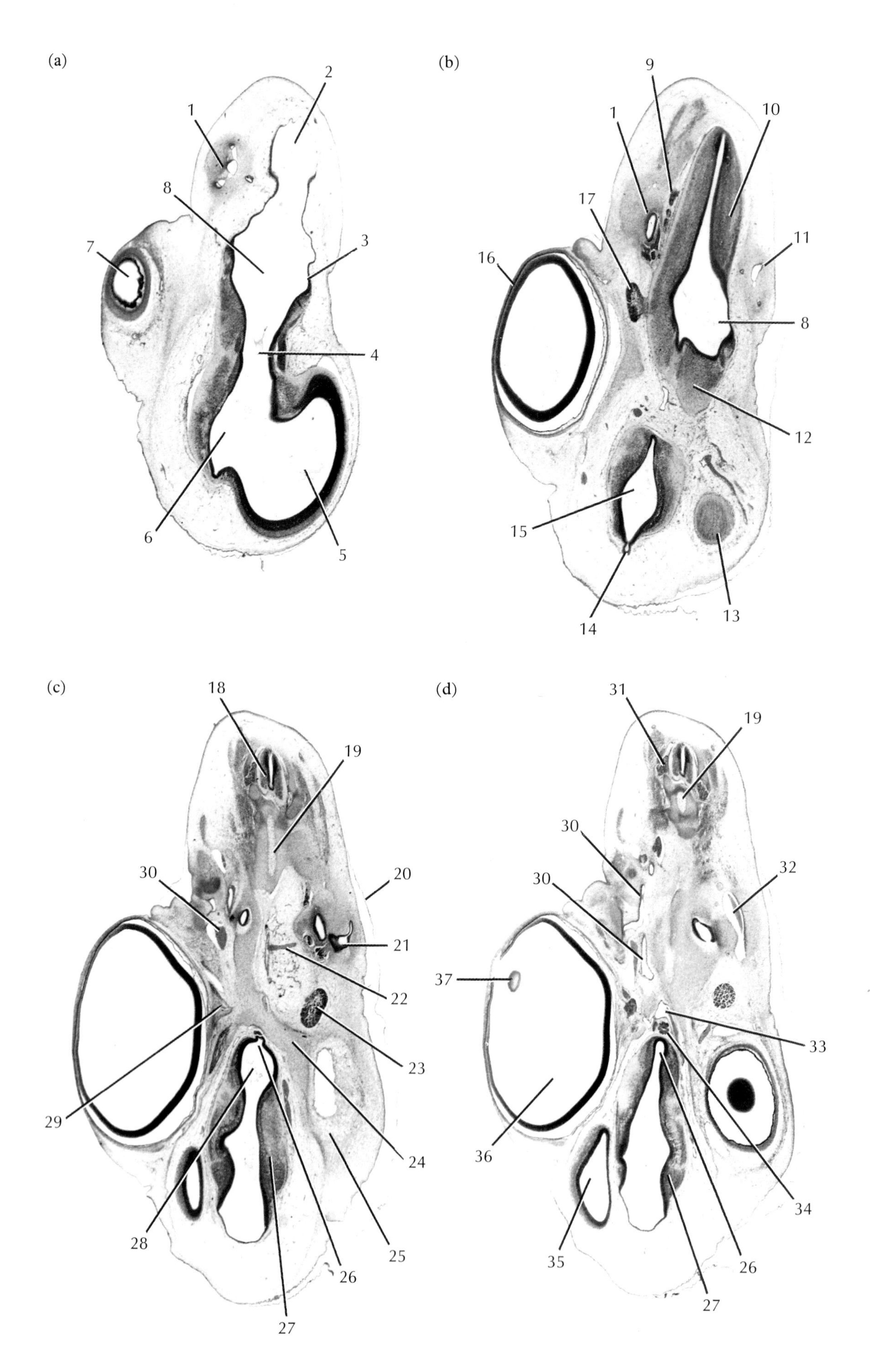

(a)
1
2
8
7
3
4
6
5
(b)
9
1
10
17
11
16
8
12
15
13
14
(c)
18
19
30
20
21
22
23
29
24
28
25
26
27
(d)
31
19
30
30
32
37
33
36
34
35
26
27

Plate 97 Day 6 TS

(a–d) ×14

1. right otic vesicle
2. myelocoele (fourth ventricle)
3. junction between thick-walled metencephalon and thin-walled roof of myelencephalon
4. isthmus
5. lumen of left optic lobe
6. mesocoele (aqueduct of Sylvius)
7. right eye
8. metacoele
9. accessory (cr. XI) nerve
10. thick-walled floor of myelencephalon
11. left otic vesicle
12. floor of diencephalon
13. floor of left optic lobe
14. pineal organ
15. diocoele
16. pigmented retina
17. right trigeminal (cr. nerve V) ganglion
18. spinal cord
19. notochord
20. amnion
21. semicircular canal of left ear
22. occipital artery
23. left trigeminal (cr. nerve V) ganglion
24. cartilage of basal plate of chondrocranium
25. posterior part of left orbital cartilage
26. infundibulum
27. thick lateral wall of diencephalon (optic thalamus)
28. optic recess in diencephalon
29. right optic (cr. II) nerve
30. right jugular vein
31. dorsal root ganglion
32. left jugular vein
33. anastomotic vein (interjugular anastomosis)
34. Rathke's pouch
35. lateral ventricle (telocoele)
36. vitreous chamber
37. lens of right eye

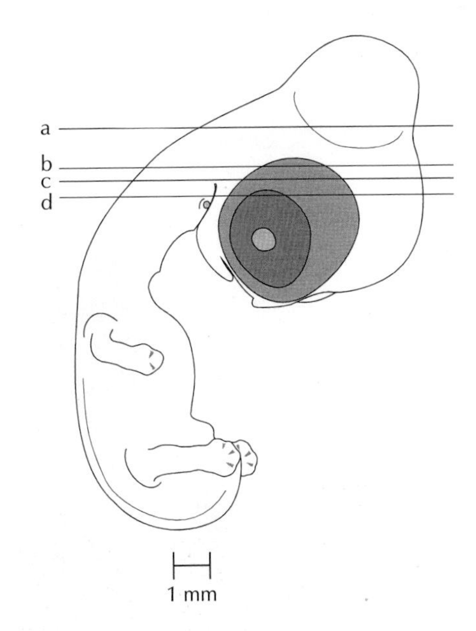

Plate 98

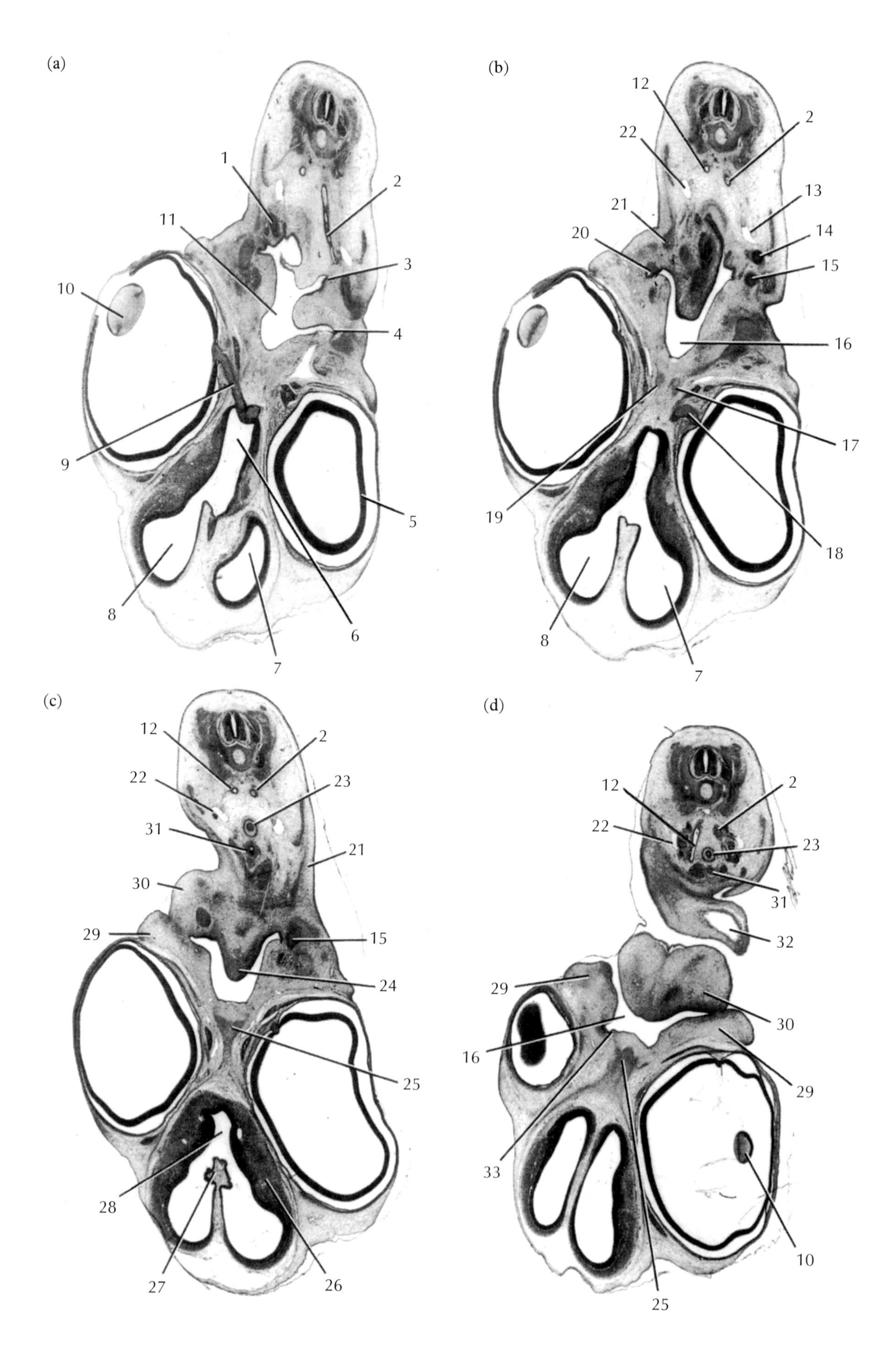

(a)
1
2
11
3
10
4
9
5
8
6
7
(b)
12
2
22
13
21
14
20
15
16
17
19
18
8
7
(c)
12
2
22
23
31
21
30
29
15
24
25
28
27
26
(d)
12
2
22
23
31
32
29
30
16
29
33
10
25

Plate 98 Day 6 TS

(a–d) × 14

1. second (hyoid) pharyngeal arch cartilage
2. left carotid artery
3. second pharyngeal pouch
4. first pharyngeal pouch
5. pigmented layer of left retina
6. diocoele
7. left lateral ventricle
8. right lateral ventricle
9. right optic (cr. II) nerve
10. lens
11. pharynx
12. right carotid artery
13. left jugular vein
14. left second (hyoid) pharyngeal arch cartilage
15. left first (mandibular) pharyngeal arch cartilage
16. oral cavity
17. left trabecula cartilage
18. left optic (cr. II) nerve
19. right trabecula cartilage
20. right first (mandibular) pharyngeal arch cartilage
21. myotome
22. right jugular vein
23. oesophagus
24. tongue
25. interorbital septum
26. corpus striatum (wall of cerebral vesicle)
27. choroid plexus
28. foramen of Monro
29. first (maxillary) pharyngeal arch
30. first (mandibular) pharyngeal arch
31. trachea
32. pericardial cavity
33. internal naris

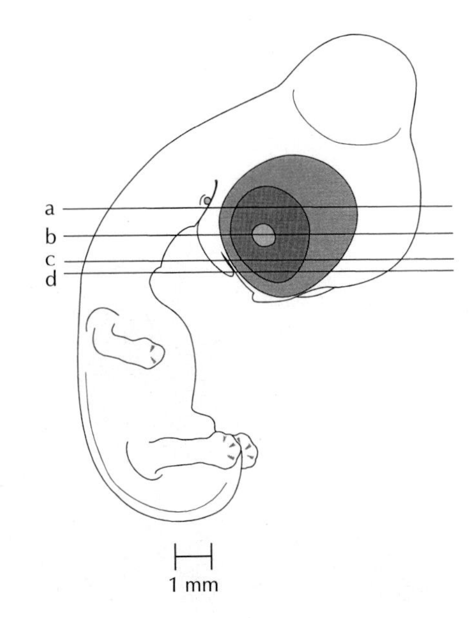

Plate 99

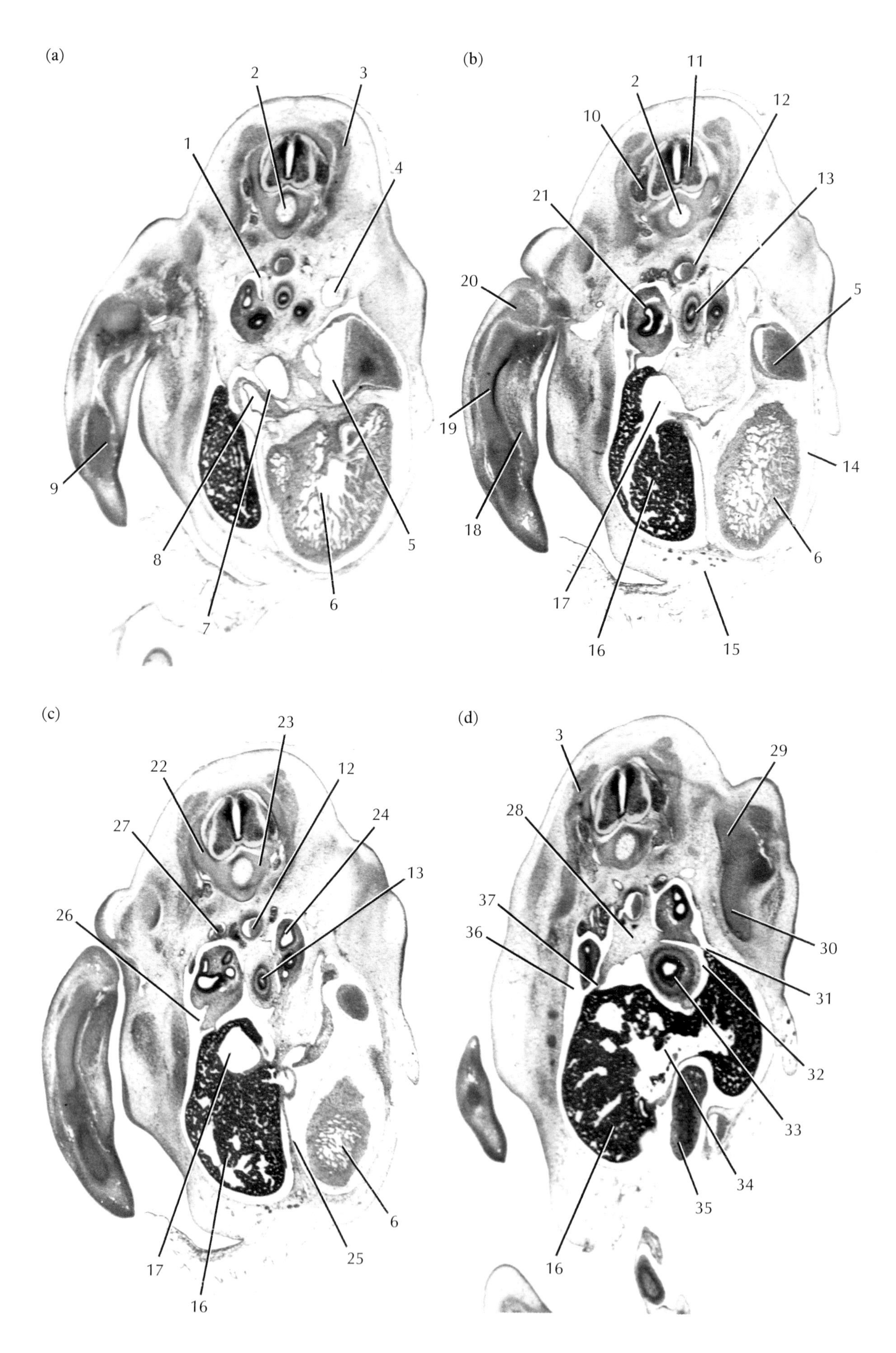

(a)
1
2
3
4
5
6
7
8
9
(b)
2
5
6
10
11
12
13
14
15
16
17
18
19
20
21
(c)
6
12
13
16
17
22
23
24
25
26
27
(d)
3
16
28
29
30
31
32
33
34
35
36
37

Plate 99 Day 6 TS

(a–d) × 18

1. right pleural cavity
2. notochord
3. myotome
4. left anterior cardinal vein
5. left atrium
6. left ventricle
7. sinus venosus
8. right atrium
9. condyle of humerus
10. dorsal root ganglion
11. spinal cord
12. dorsal aorta
13. oesophagus
14. pericardial cavity
15. ventral body wall
16. liver
17. right omphalo-mesenteric vein
18. ventral muscle mass of right wing
19. shaft of humerus
20. dorsal muscle mass of wing
21. right lung bud
22. neural arch
23. body of vertebra
24. left lung bud
25. ventral mesentery
26. pulmonary fold
27. anterior border of right mesonephros
28. dorsal mesentery
29. scapula
30. ischium
31. hepatic ligament of left oblique septum
32. dorsal hepatic cavity
33. crop
34. ductus venosus
35. gizzard
36. right ventral hepatic cavity
37. hepatic ligament of right oblique septum

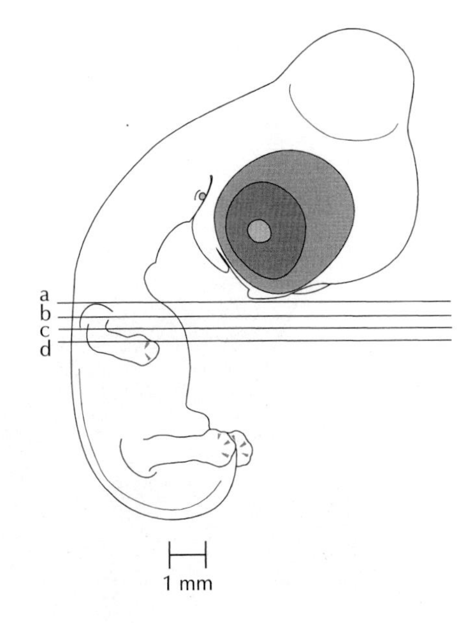

PLATE 100

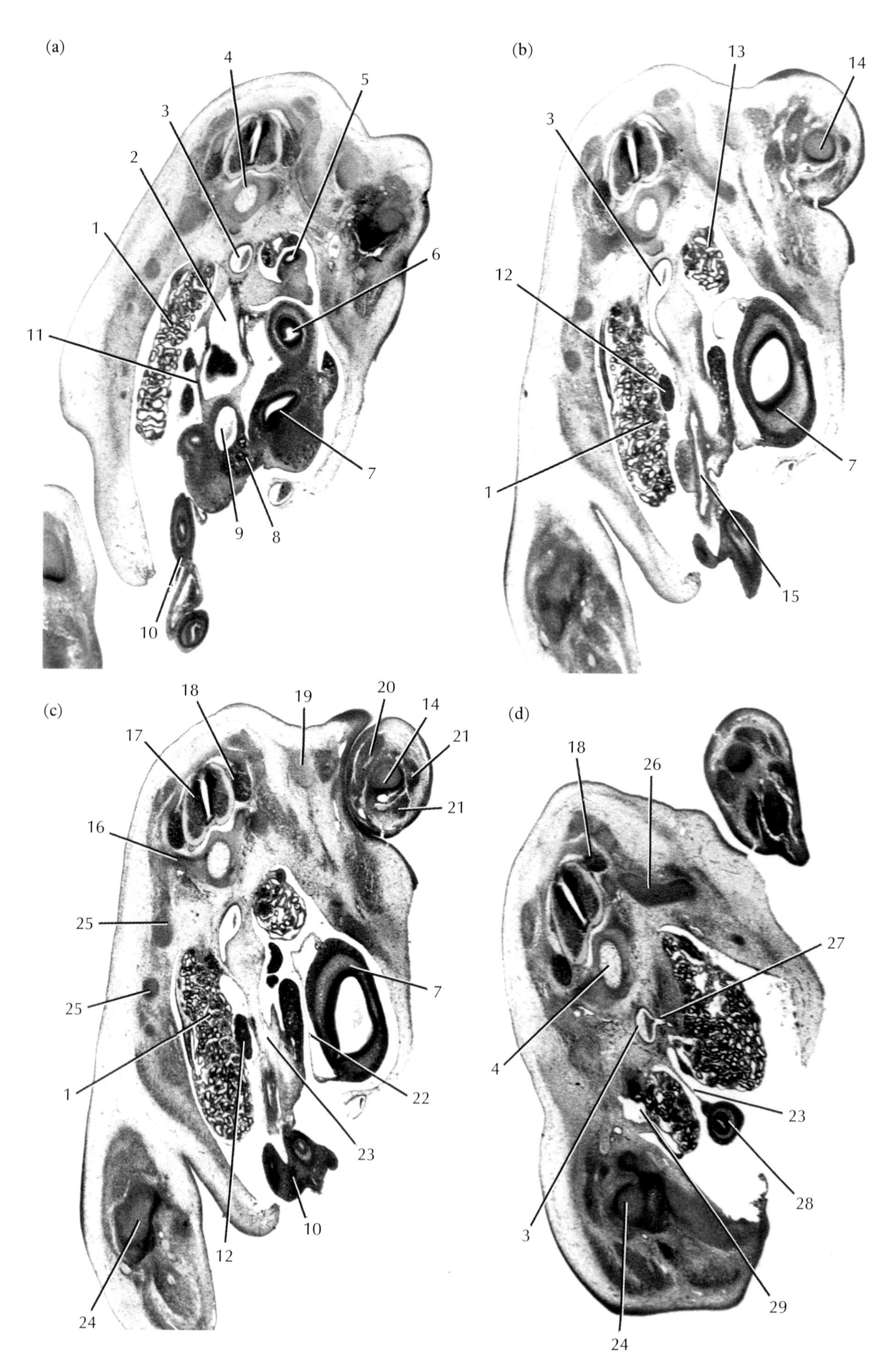

(a)
4
5
3
2
1
6
11
7
9
8
10
(b)
13
14
3
12
7
1
15
(c)
18
19
20
14
21
17
21
16
25
7
25
1
22
23
10
12
24
(d)
18
26
27
4
23
28
3
29
24

Plate 100 Day 6 TS

(a–d) ×18

1. right mesonephros
2. right ventral hepatic cavity
3. dorsal aorta
4. notochord
5. left lung bud
6. junction of crop and gizzard
7. gizzard
8. pancreas
9. hepatic portal vein
10. herniated loop of intestine
11. horizontal hepatic ligament
12. right gonad
13. left mesonephros
14. left femur
15. coeliac artery
16. transverse process of vertebra
17. spinal cord
18. dorsal root ganglion
19. myogenic condensation
20. dorsal muscle mass of left leg
21. ventral muscle mass of left leg
22. post-hepatic septum
23. dorsal mesentery
24. right femur
25. rib
26. cartilaginous precursor of ilium
27. renal artery
28. small intestine
29. ductus venosus

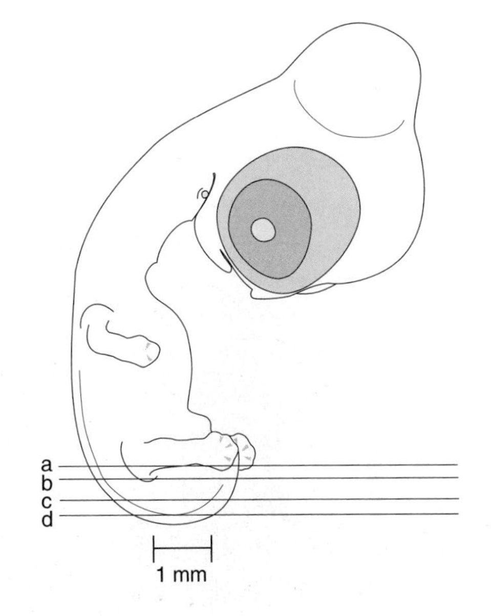

PLATE 101

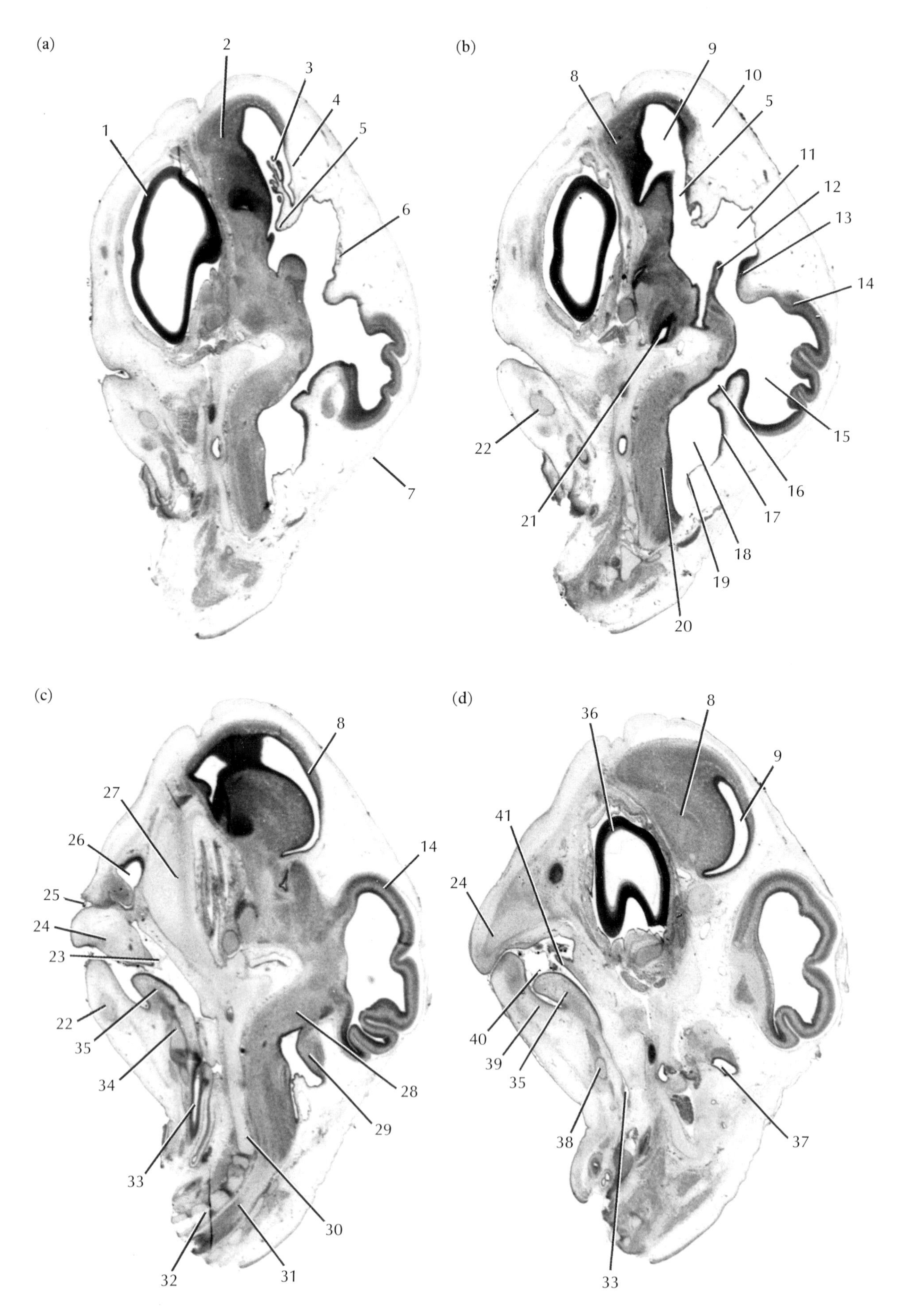
(a)
1
2
3
4
5
6
7
(b)
8
9
10
5
11
12
13
14
15
16
17
18
19
20
21
22
(c)
8
14
27
26
25
24
23
22
35
34
33
32
31
30
29
28
(d)
36
8
9
41
24
40
39
35
38
33
37

Plate 101 Day 7 LS (longitudinal sections: sagittal) of head

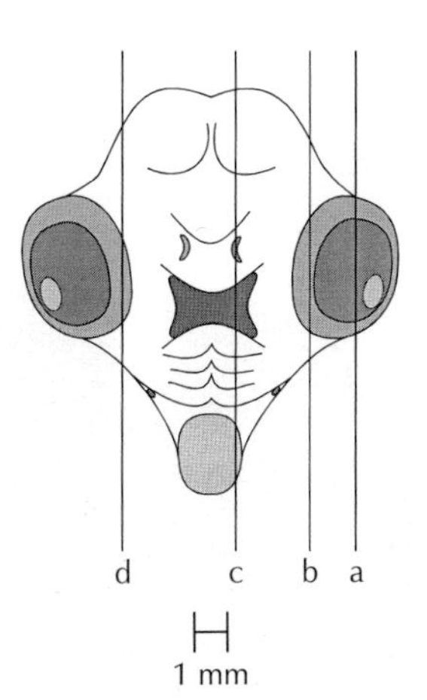

(a–d) × 10

Diagram shows view of embryo from ventral side.

1. retinal pigment of left eye
2. left corpus striatum in wall of cerebral hemisphere
3. choroid plexus
4. pallium
5. foramen of Munro
6. pineal organ
7. skin
8. cerebral hemisphere
9. telocoele (lateral ventricle)
10. head mesenchyme
11. diocoele (third ventricle)
12. sulcus parencephalicus transversus
13. metathalamus in posterior commissure
14. optic lobe
15. mesocoele
16. isthmus
17. metencephalon
18. rhombocoele (fourth ventricle)
19. thin roof of myelencephalon
20. thick floor of myelencephalon
21. infundibulum
22. Meckel's cartilage
23. internal naris
24. first (maxillary) pharyngeal arch
25. external naris
26. nostril
27. interorbital septum
28. thick floor of metencephalon
29. roof of metencephalon
30. basal plate of chondrocranium
31. spinal cord
32. cervical vertebra
33. pharynx
34. paraglossal cartilage
35. tongue
36. retinal pigment of right eye
37. semicircular canal
38. basihyal cartilage (copula I)
39. lower jaw
40. oral cavity
41. palatine process

Plate 102

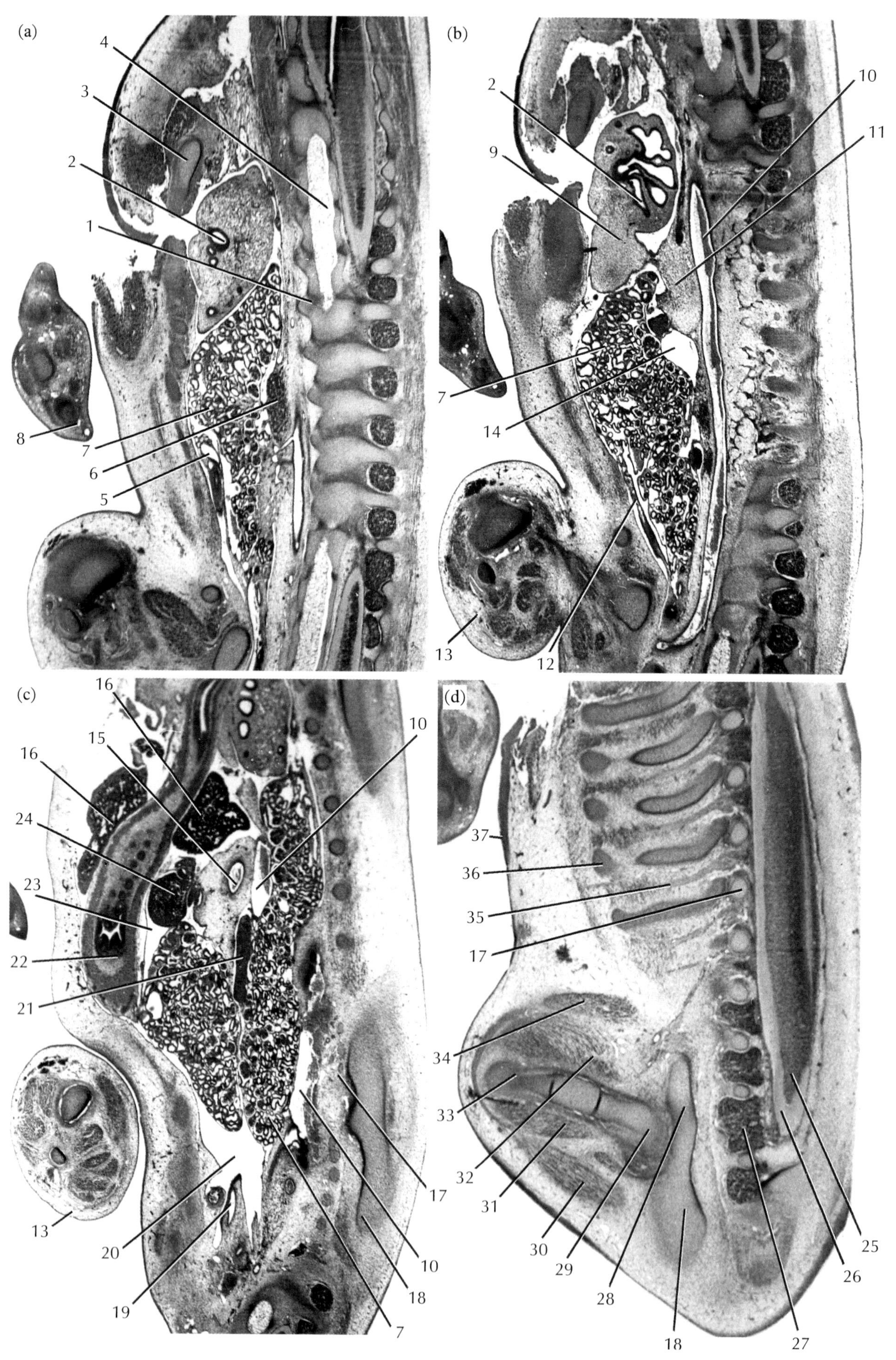

(a)
4
3
2
1
8
7
6
5
(b)
2
9
10
11
7
14
13
12
(c)
16
15
16
24
23
22
21
10
13
20
19
17
10
18
7
(d)
37
36
35
17
34
33
32
31
30
29
28
18
27
26
25

Plate 102 Day 7 LS (longitudinal sections: sagittal) through trunk

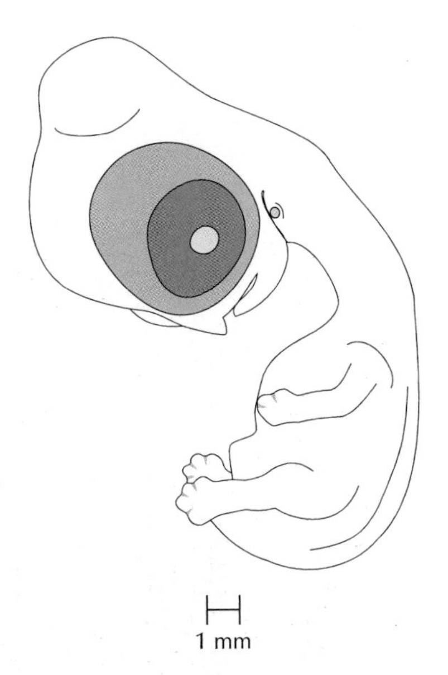

(a–d) ×20

Sections taken through plane of page.

1. body of vertebra
2. lung bud
3. humerus
4. notochord
5. nephric duct
6. right ovary
7. mesonephros
8. wing bud
9. air sac
10. dorsal aorta
11. dorsal mesentery
12. nephric duct
13. leg
14. subcardinal vein
15. coeliac artery
16. liver
17. transverse process of vertebra
18. ischium
19. allantois
20. coelom
21. left ovary
22. colon
23. posterior abdominal coelom
24. spleen
25. spinal cord grey matter
26. spinal cord white matter
27. dorsal root ganglion
28. ilium
29. head of femur
30. flexor cruris muscle
31. puboischiofemoralis muscle
32. femorotibialis muscle
33. shaft of femur
34. iliotibialis cranialis muscle
35. spinal nerve
36. rib
37. abdominal wall

Plate 103

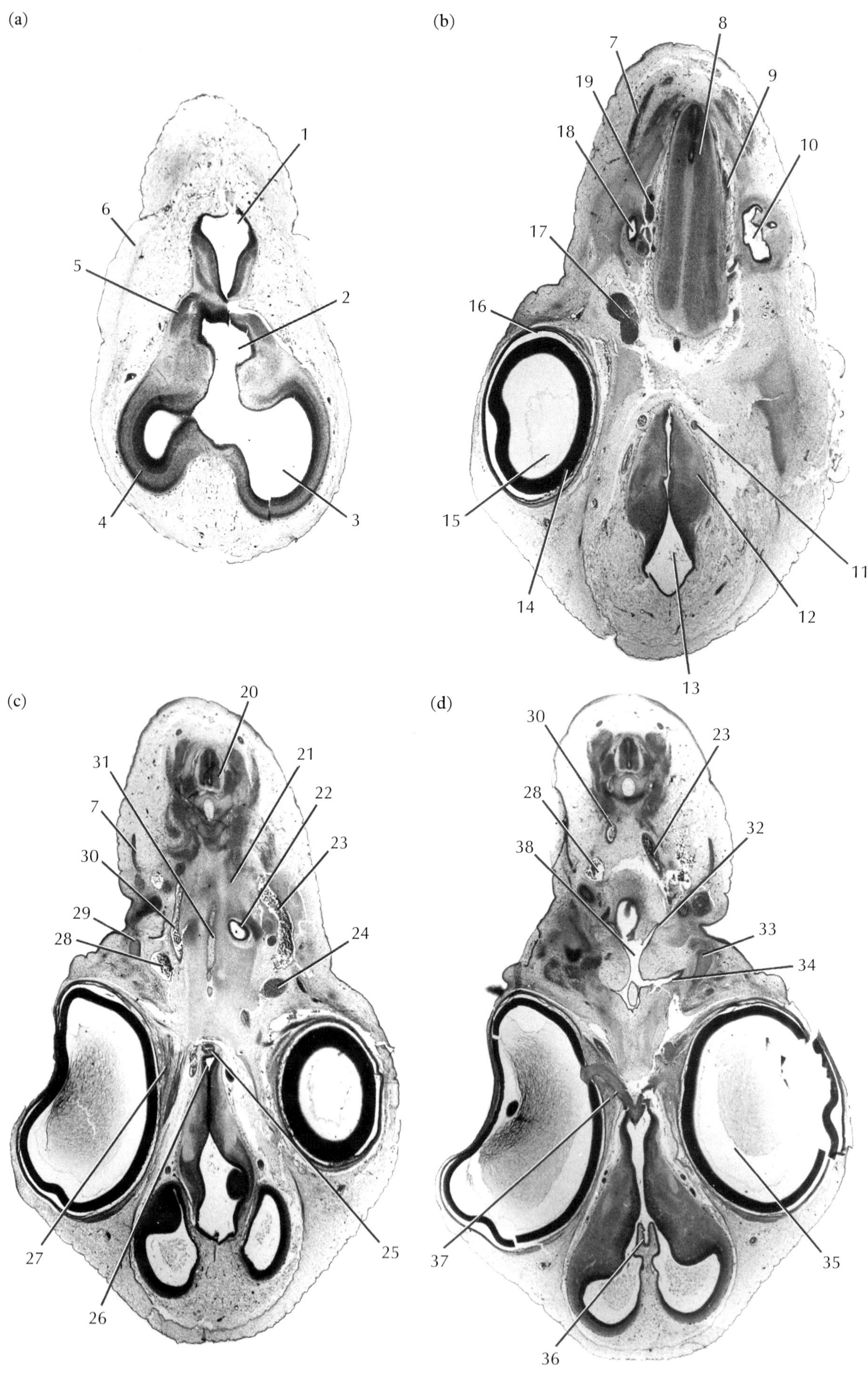
(a)
(b)
(c)
(d)
1
2
3
4
5
6
7
8
9
10
11
12
13
14
15
16
17
18
19
20
21
22
23
24
25
26
27
28
29
30
31
7
30
23
28
32
38
33
34
37
35
36

Plate 103 Day 7 TS

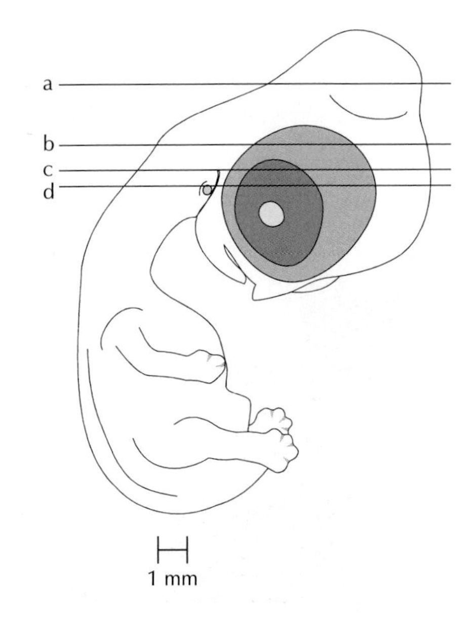

(a–d) × 12

1. thin roof of myelencephalon
2. metencephalon
3. left optic lobe
4. right optic lobe
5. trigemino-cerebellar fibre tract
6. head mesenchyme
7. myotome
8. thick floor of mesencephalon
9. left accessory (cr. XI) nerve
10. left otic vesicle
11. left ethmoidal artery
12. diencephalon
13. diocoele
14. pigmented retina of right eye
15. vitreous humour of right eye
16. sclera of right eye
17. right trigeminal (cr. nerve V) ganglion
18. right otic vesicle
19. right accessory (cr. XI) nerve
20. spinal cord
21. basal plate of chondrocranium
22. pharyngeal diverticulum
23. left carotid artery
24. left trigeminal (cr. nerve V) ganglion
25. Rathke's pouch
26. infundibulum
27. lateral rectus eye muscle
28. right jugular vein
29. right Meckel's cartilage
30. right carotid artery
31. notochord
32. second pharyngeal pouch
33. left Meckel's cartilage
34. first pharyngeal pouch
35. vitreous humour of left eye
36. choroid plexus
37. right optic (cr. II) nerve
38. lumen of pharynx

PLATE 104

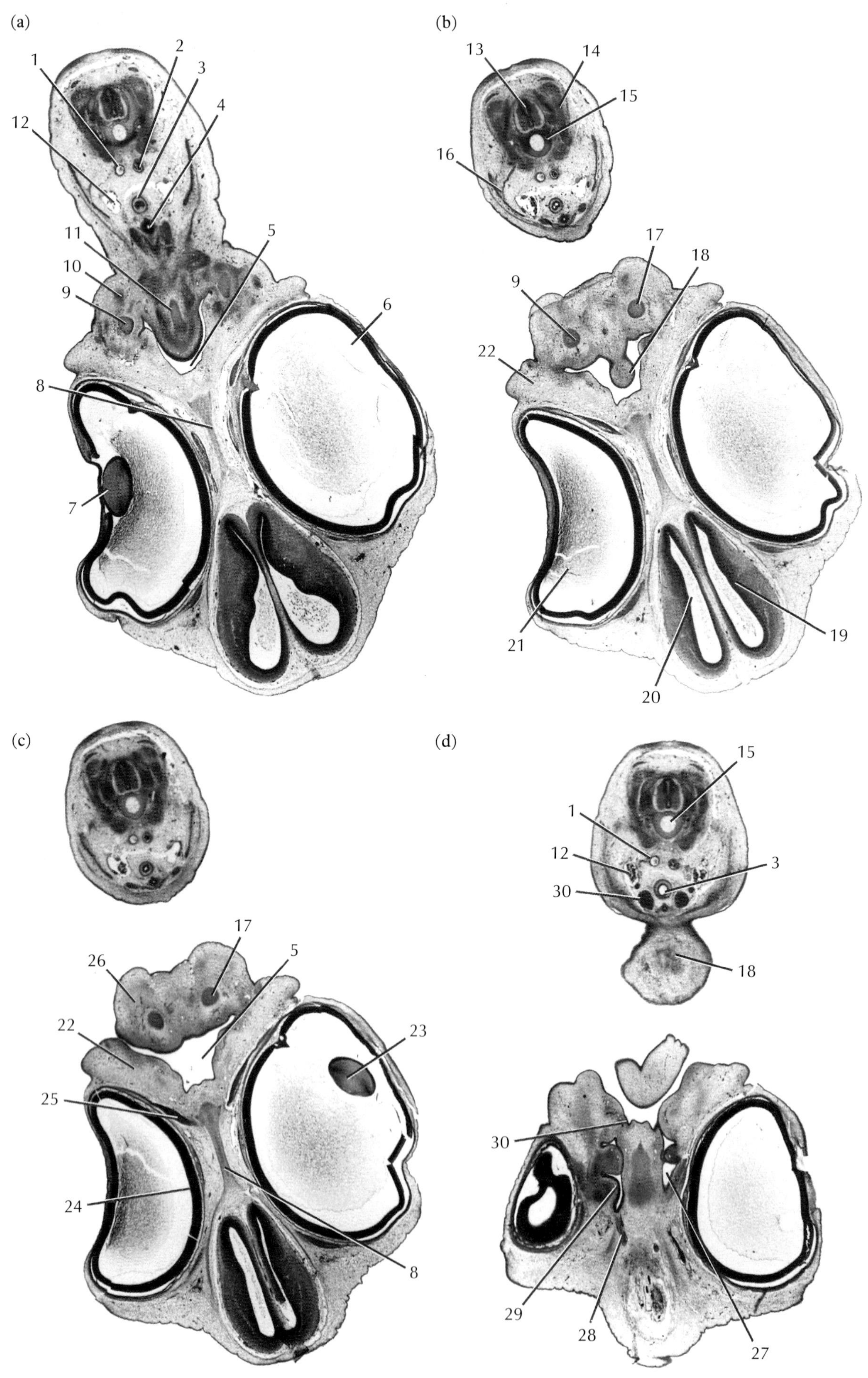

(a)
1
2
3
4
12
11
10
9
5
6
8
7
(b)
13
14
15
16
17
18
9
22
21
20
19
(c)
17
5
26
22
23
25
24
8
(d)
15
1
12
3
30
18
30
29
28
27

Plate 104 Day 7 TS

(a–d) × 12

1. right common carotid artery
2. left common carotid artery
3. oesophagus
4. trachea
5. oral cavity
6. vitreous chamber of left eye
7. lens of right eye
8. interorbital septum
9. right quadrate cartilage
10. right Meckel's cartilage
11. larynx
12. right jugular vein
13. spinal cord
14. primordium of costal process
15. notochord
16. myotome
17. left quadrate cartilage
18. tongue
19. left cerebral vesicle
20. right cerebral vesicle
21. vitreous humour of right eye
22. right first (maxillary) pharyngeal arch
23. lens of left eye
24. pigmented retina of right eye
25. right inferior oblique eye muscle
26. right first (mandibular) pharyngeal arch
27. lumen of left nostril
28. nasal concha
29. lumen of right nostril
30. right internal naris

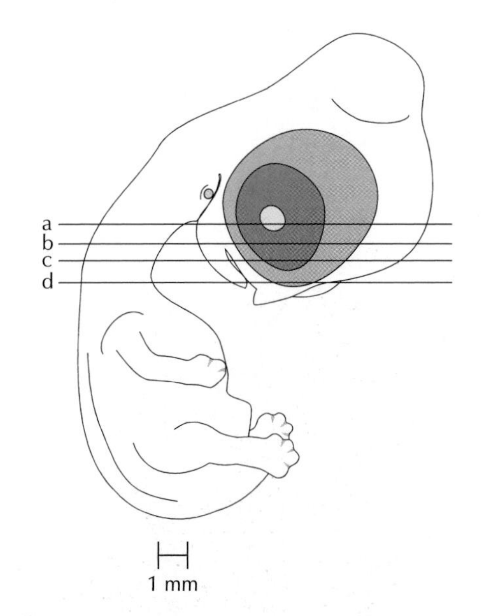

Plate 105

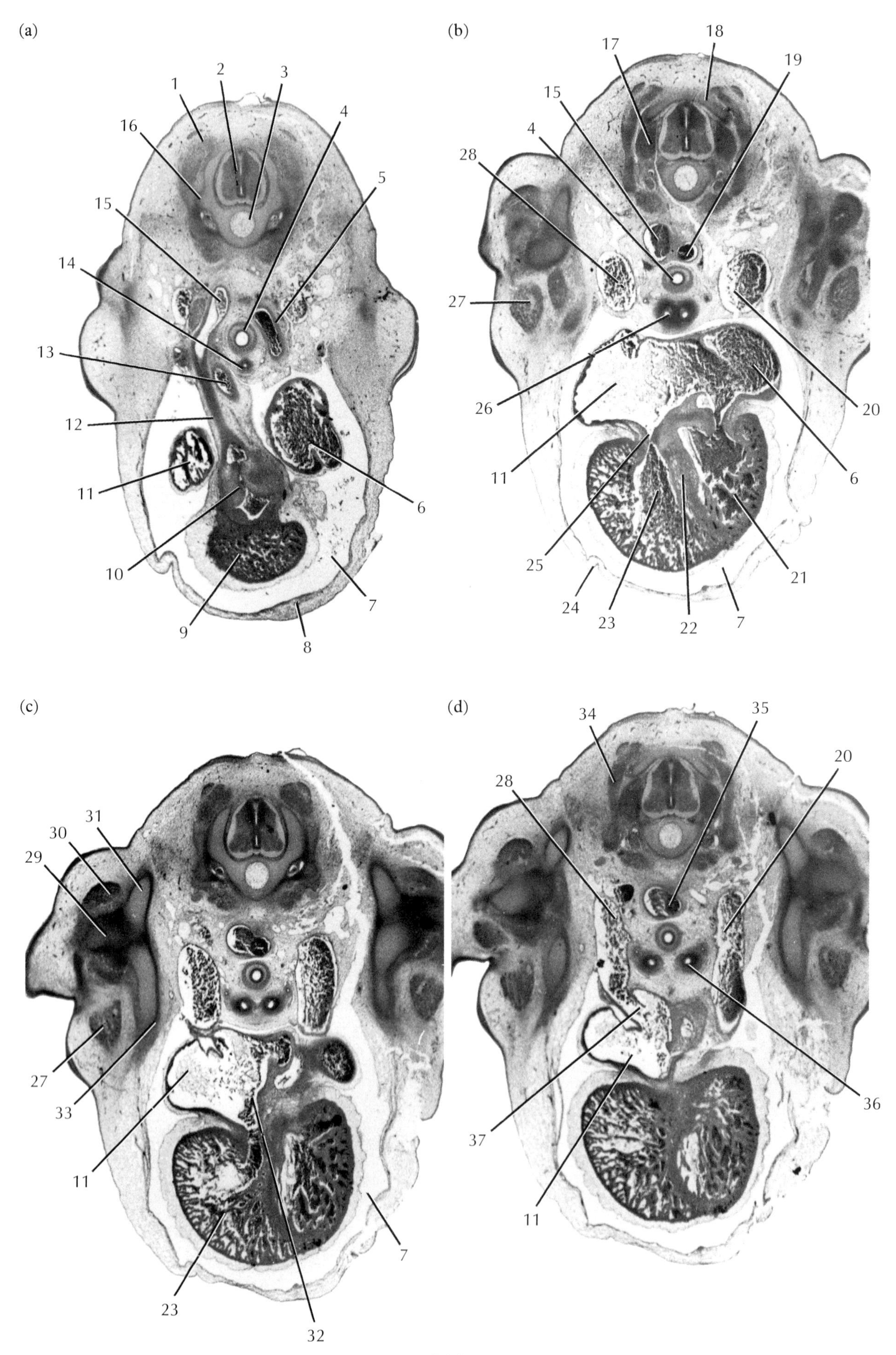

(a)
1
2
3
4
5
6
7
8
9
10
11
12
13
14
15
16
(b)
17
18
19
15
4
28
27
26
11
25
24
23
22
7
21
6
20
(c)
31
30
29
27
33
11
23
32
7
(d)
34
35
20
28
37
11
36

Plate 105 Day 7 TS

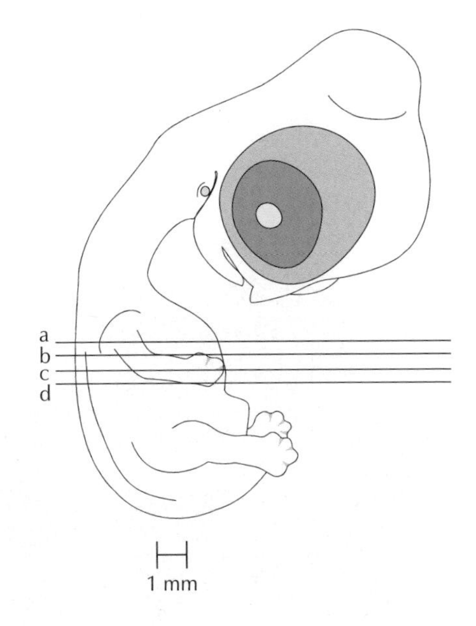

(a–d) ×20

1. myotome
2. spinal cord
3. notochord
4. oesophagus
5. left dorsal aorta
6. left atrium
7. pericardial cavity
8. body wall
9. ventricle
10. aortico-pulmonary ridge
11. right atrium
12. aortico-pulmonary septum
13. pulmonary artery
14. trachea
15. right dorsal aorta
16. transverse process of vertebra
17. dorsal root ganglion
18. neural arch of vertebra
19. left dorsal aorta just cranial to fusion with right dorsal aorta
20. left cranial (anterior) vena cava
21. left ventricle
22. interventricular septum
23. right ventricle
24. pericardium
25. right atrio-ventricular canal
26. right bronchus
27. ventral muscle mass of wing
28. right cranial (anterior) vena cava
29. humerus
30. dorsal muscle mass of wing
31. scapula
32. perforated interatrial septum (interatrial foramina)
33. coracoid
34. developing muscle
35. median dorsal aorta
36. left bronchus
37. sinus venosus

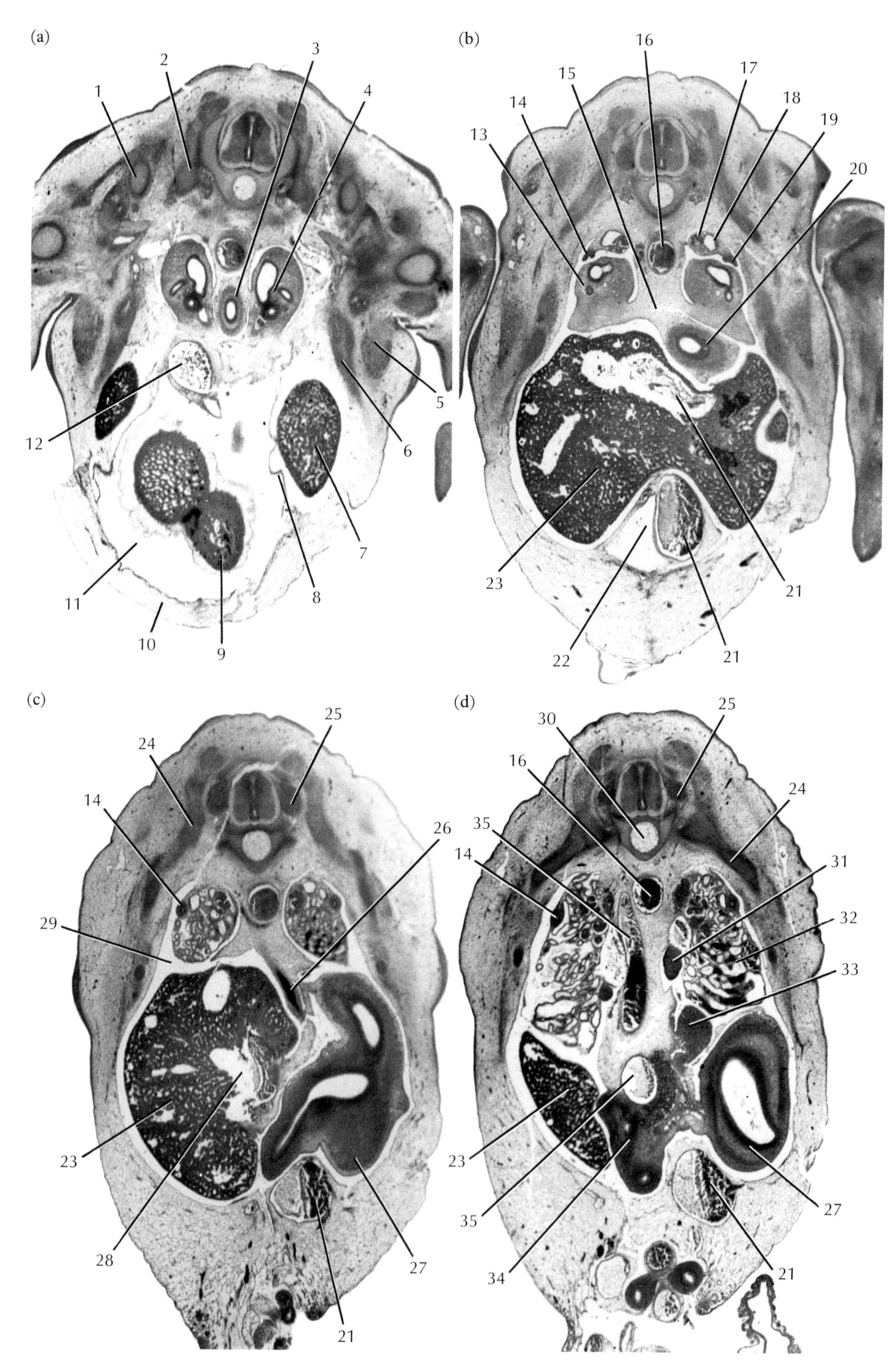
(a)
1
2
3
4
5
6
7
8
9
10
11
12
(b)
13
14
15
16
17
18
19
20
21
22
23
21
(c)
14
24
25
26
27
28
29
23
21
(d)
30
16
25
24
35
14
31
32
33
23
35
34
27
21

Plate 106 Day 7 TS

(a–d) ×20

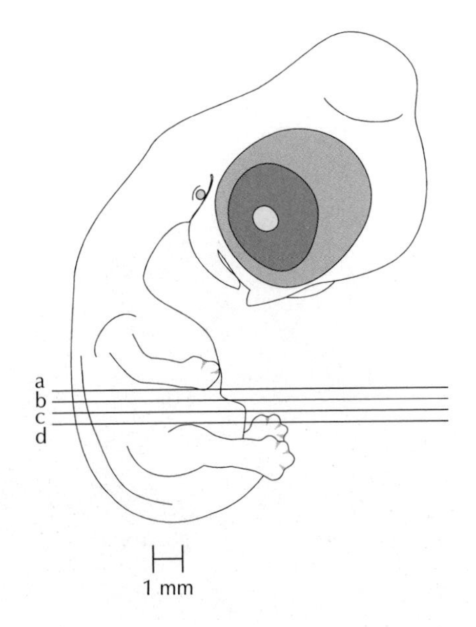

1. right scapula
2. head of rib
3. crop
4. left lung bud
5. intercostalis externus muscle
6. intercostalis internus muscle
7. left lobe of liver
8. oblique septum
9. ventricle
10. body wall
11. pericardial cavity
12. right cranial (anterior) vena cava
13. right lung bud
14. right oviduct (Mullerian duct)
15. dorsal mesentery
16. dorsal aorta
17. cortical substance of left adrenal gland
18. left subcardinal vein
19. left oviduct (Mullerian duct)
20. proventriculus
21. left allantoic vein
22. ventral mesentery
23. right lobe of liver
24. mesenchyme condensing to form rib
25. dorsal root ganglion
26. coeliac artery
27. gizzard
28. right hepatic vein
29. right dorsal hepatic cavity
30. notochord
31. left gonad
32. left mesonephros
33. spleen
34. intestinal loop
35. right allantoic artery

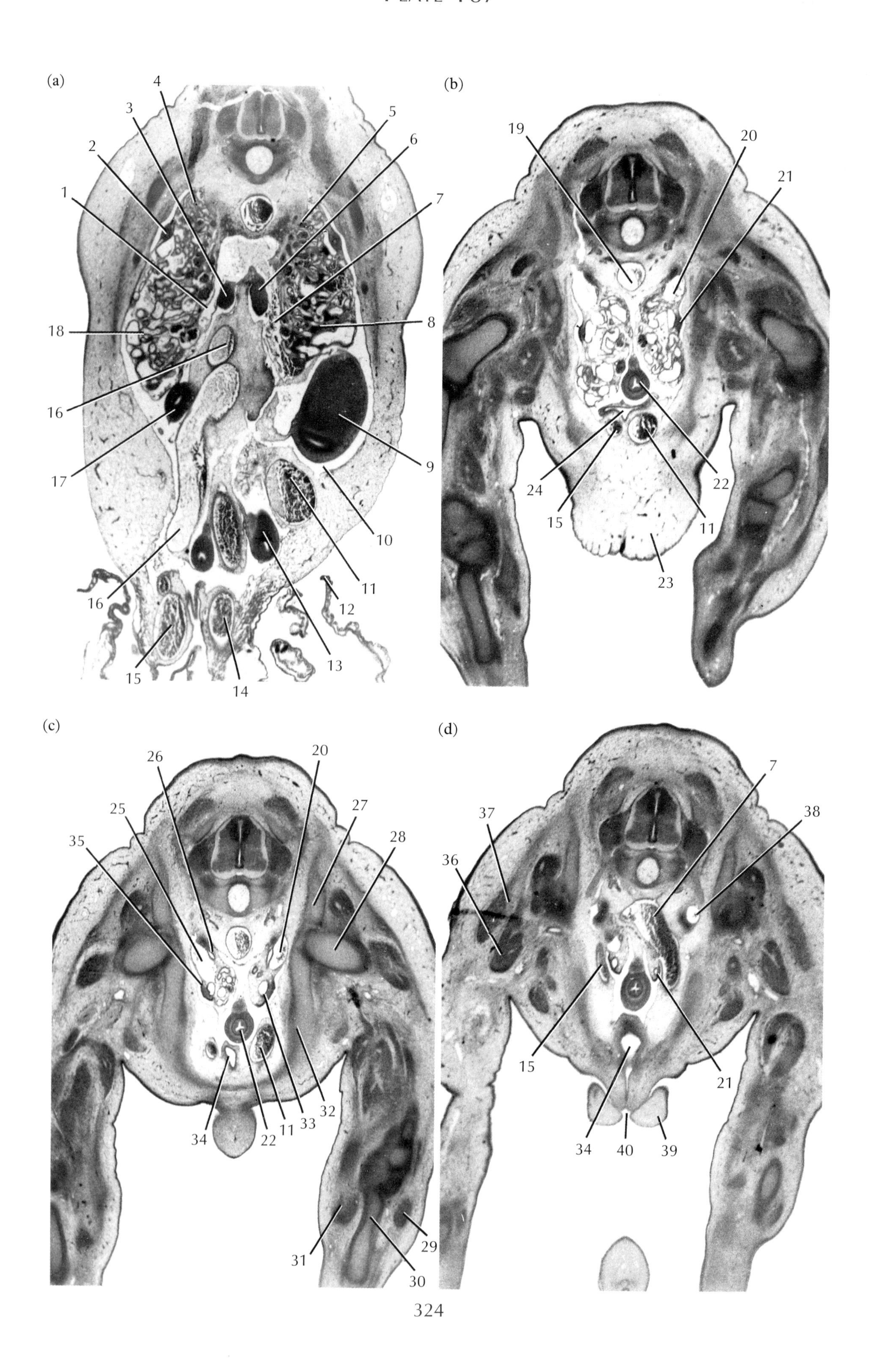
(a)
1
2
3
4
5
6
7
8
9
10
11
12
13
14
15
16
17
18
(b)
19
20
21
22
23
24
15
11
(c)
20
25
26
27
28
29
30
31
32
33
34
35
22
11
(d)
7
15
21
34
36
37
38
39
40

Plate 107 Day 7 TS

(a–d) ×20

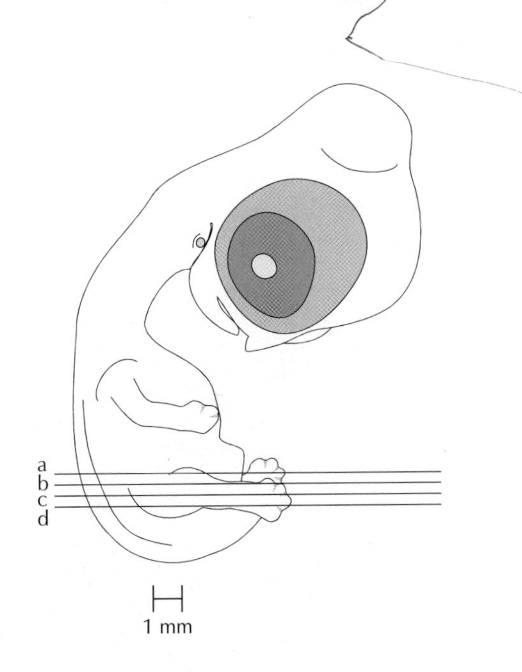

1. right omphalo-mesenteric artery
2. right oviduct (Mullerian duct)
3. right gonad
4. right metanephros
5. left metanephros
6. left gonad
7. left omphalo-mesenteric (vitelline) artery
8. left mesonephros
9. gizzard
10. peritoneal cavity
11. left allantoic vein
12. yolk sac
13. loop of intestine
14. left omphalo-mesenteric (vitelline) vein
15. right omphalo-mesenteric (vitelline) vein
16. right allantoic vein
17. jejunum
18. right mesonephros
19. dorsal aorta
20. left subcardinal vein
21. left oviduct (Mullerian duct)
22. intestine
23. cloacal fold
24. base of allantois
25. right subcardinal vein
26. right metanephric duct
27. ilium
28. femur
29. dorsal muscle mass of lower leg
30. tibiotarsus
31. ventral muscle mass of lower leg
32. pubis
33. left nephric duct
34. cloaca
35. right nephric duct
36. ventral muscle mass of right thigh
37. dorsal muscle mass of right thigh
38. left metanephric duct
39. genital fold
40. cloacal membrane

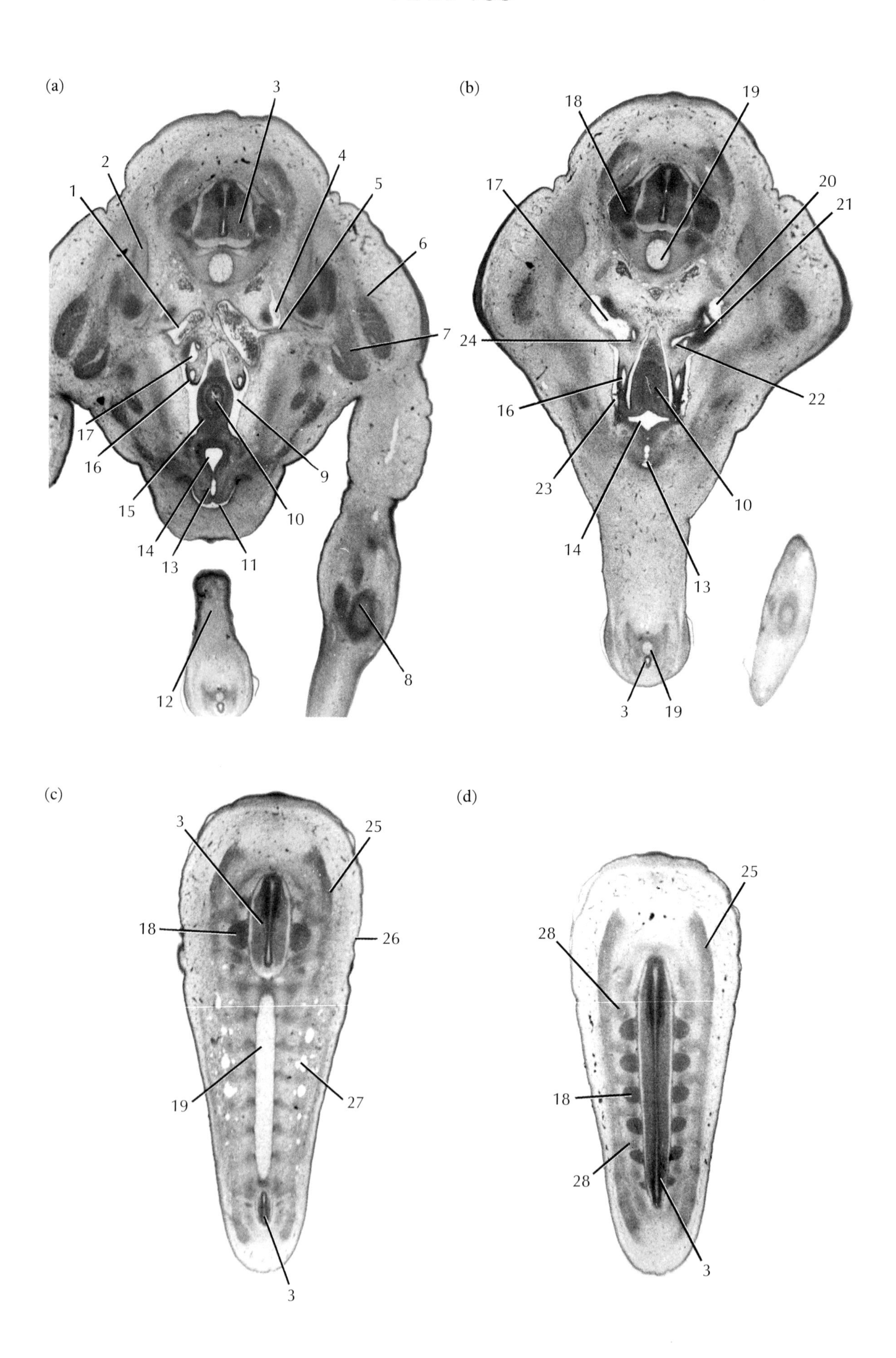
(a)
1
2
3
4
5
6
7
8
9
10
11
12
13
14
15
16
17
(b)
3
10
13
14
16
17
18
19
20
21
22
23
24
(c)
3
18
19
25
26
27
(d)
3
18
25
28

Plate 108 Day 7 TS

(a–d) ×20

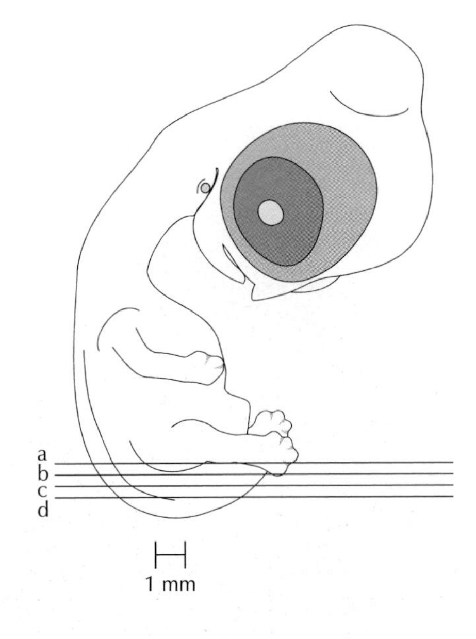

1. right dorsal aorta
2. ilium
3. spinal cord
4. vertebral artery
5. sciatic artery
6. dorsal muscle mass of thigh
7. ventral muscle mass of thigh
8. femur
9. intestinal coelomic cavity
10. colon
11. cloacal fold
12. tail
13. cloacal bursa
14. cloaca
15. mesentery
16. right oviduct (Mullerian duct)
17. right posterior cardinal vein
18. dorsal root ganglion
19. notochord
20. left posterior cardinal vein
21. left metanephric tubule
22. left metanephric (ureteric) duct
23. coelom
24. right metanephric (ureteric) duct
25. derma-myotomal condensation
26. epidermis of tail
27. intersegmental artery
28. sclerotomal condensation for vertebral body

PLATE 109

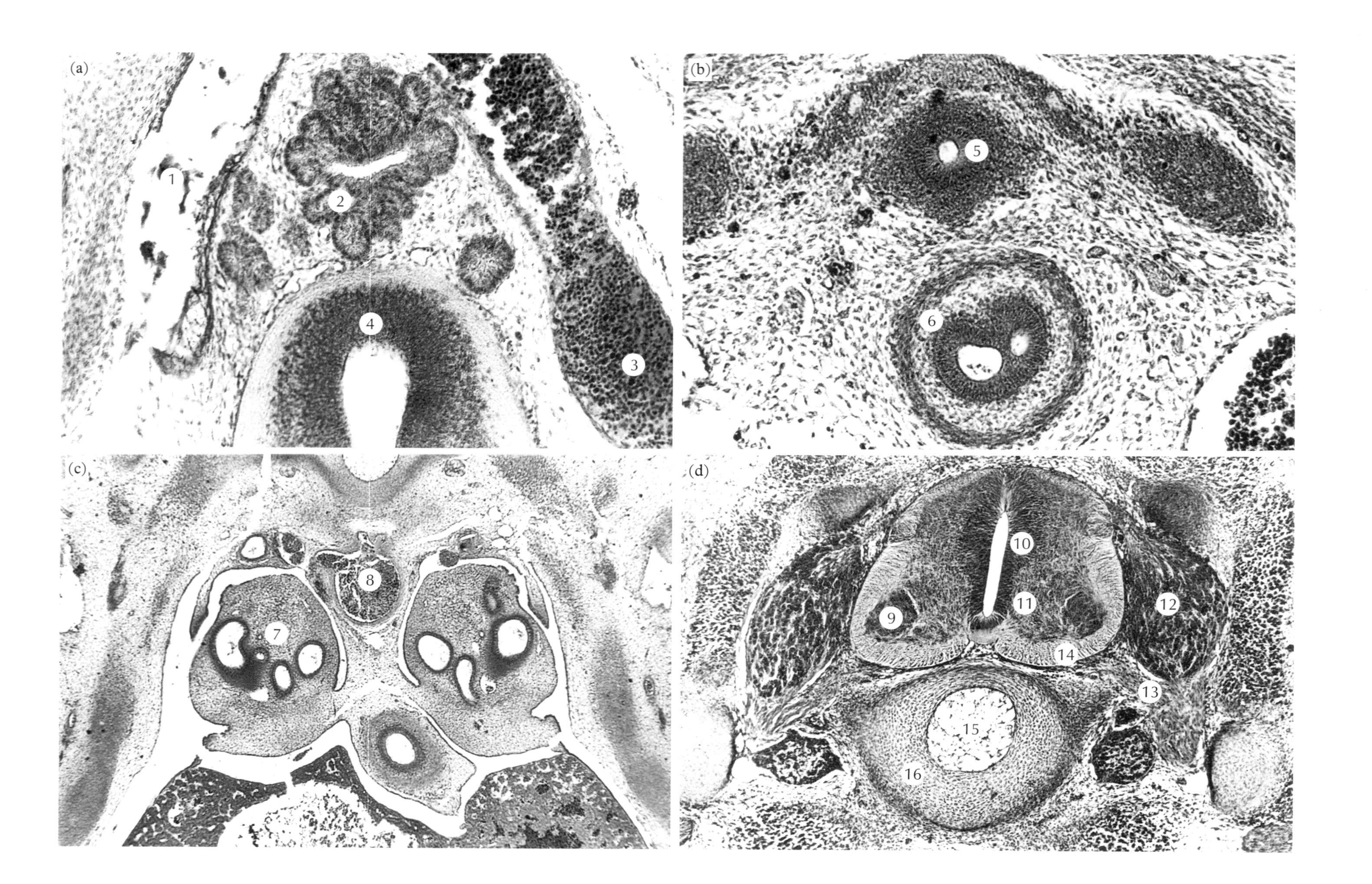

Plate 109 Day 7 TS Enlargements

(a) Rathke's pouch (centre top) and infundibulum (centre bottom) from section adjacent to *Plate 103c*, ×130

(b) Oesophagus (top) and trachea (bottom) from section adjacent to *Plate 105b*, ×120

(c) Lung buds and oviduct from section adjacent to *Plate 106a*, ×50

(d) Section through spinal cord region in mid trunk. ×80

1. right internal carotid artery
2. Rathke's pouch
3. left internal carotid artery
4. infundibulum
5. oesophagus
6. trachea at bifurcation into bronchae
7. right lung bud
8. dorsal aorta
9. ventral horn (motor column)
10. ependyma
11. mantle layer
12. dorsal root ganglion
13. ventral root
14. white matter
15. notochord
16. notochordal sheath

Plate 110

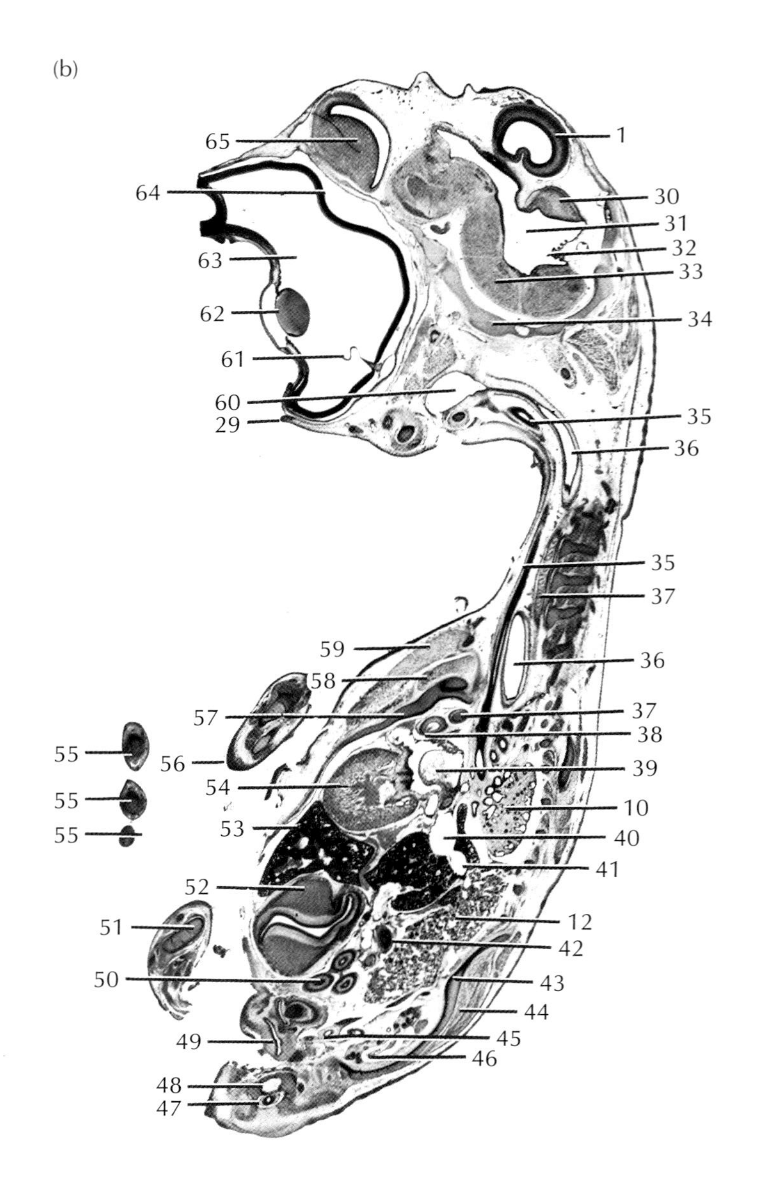

(a)
1
2
3
4
5
6
7
8
9
10
11
12
13
14
15
16
17
18
19
20
21
22
23
24
25
26
27
28
29

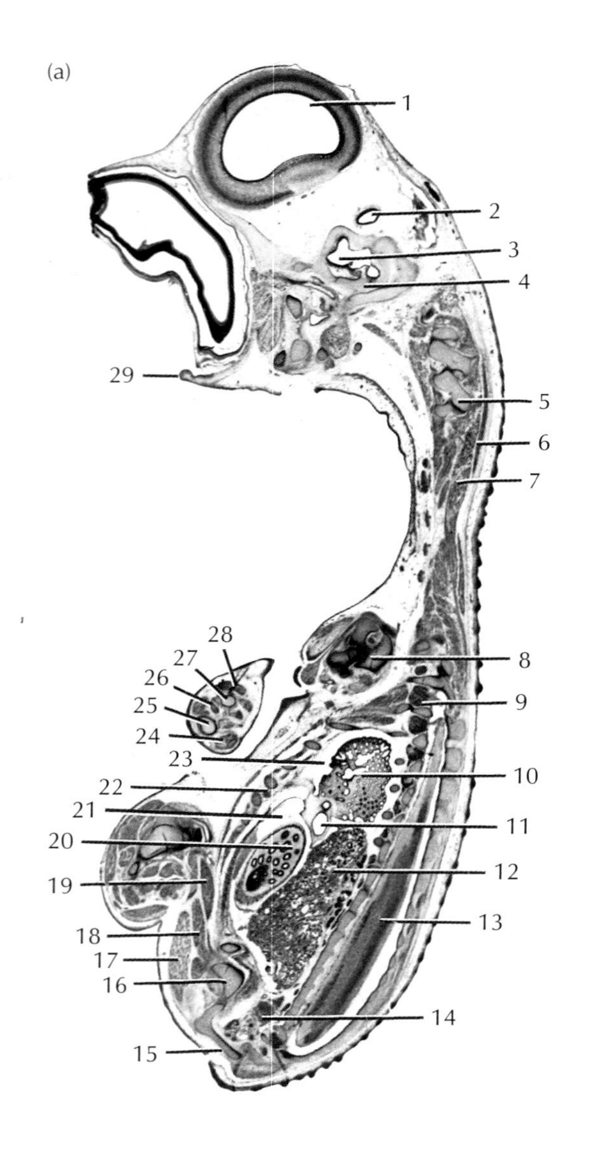

(b)
1
30
31
32
33
34
35
36
35
37
36
37
38
39
10
40
41
12
42
43
44
45
46
47
48
49
50
51
52
53
54
55
55
55
56
57
58
59
60
29
61
62
63
64
65

Plate 110 Day 9 LS (longitudinal sections: sagittal)

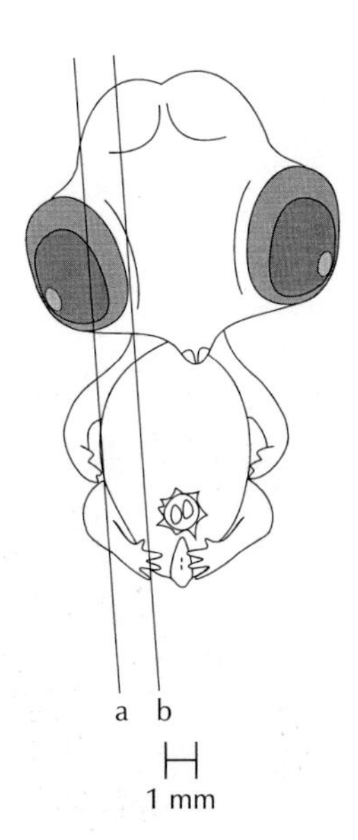

(a, b) ×6

1. optic lobe
2. endolymphatic sac
3. inner ear
4. otic capsule
5. cervical vertebra
6. latissimus dorsi muscle
7. rhomboideus superficialis muscle
8. coracoid
9. dorsal root ganglion
10. lung bud
11. abdominal air sac
12. mesonephros
13. spinal cord
14. metanephros
15. ilium
16. trochanter of femur
17. flexor cruris medialis muscle
18. biceps femoralis muscle
19. iliotibialis muscle
20. proventriculus
21. thoracic air sac
22. sternal rib
23. pleural cavity
24. flexor carpi ulnaris muscle
25. ulna
26. interosseus dorsalis muscle
27. radius
28. extensor metacarpi radialis muscle
29. eyelid
30. cerebellum
31. metencephalon
32. choroid plexus
33. floor of myelencephalon
34. basal plate of chondrocranium
35. trachea
36. oesophagus
37. carotid artery
38. dorsal aorta
39. left atrium
40. caudal (posterior) vena cava
41. right hepatic vein
42. spleen
43. synsacrum
44. ischiofemoralis muscle
45. metanephric duct
46. renal vein
47. tail spinal cord
48. notochord
49. cloaca
50. small intestine
51. foot
52. gizzard
53. liver
54. ventricle
55. toes
56. wing
57. sternum
58. supracoracoideus muscle
59. pectoralis muscle
60. oral cavity
61. pecten
62. lens
63. collapsed eye
64. pigmented retina
65. cerebral hemisphere

Plate 111

(a)

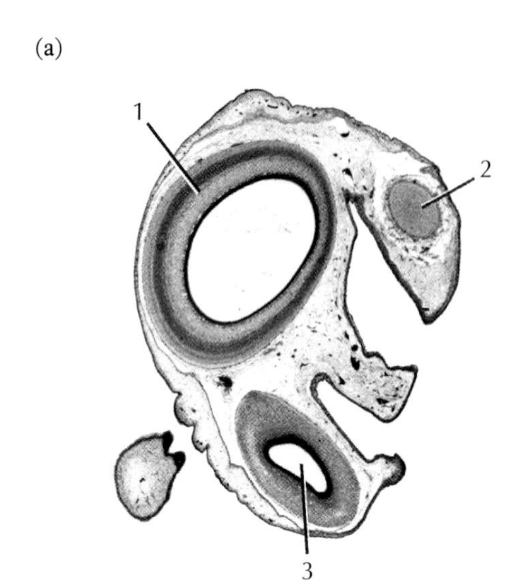

(b)

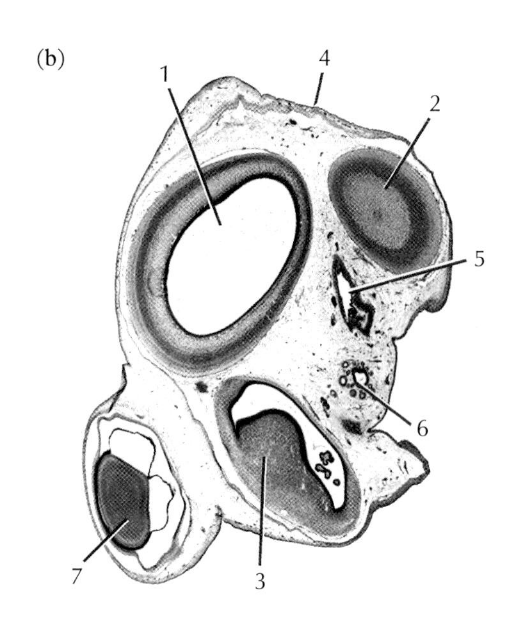

(c)

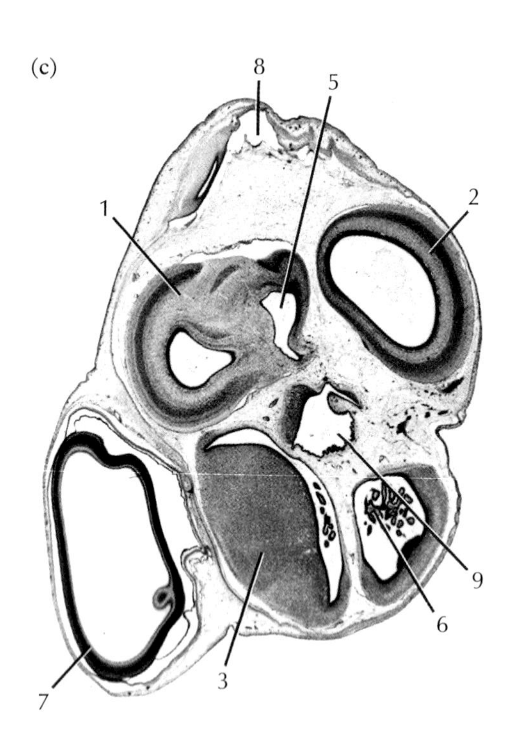

(d)

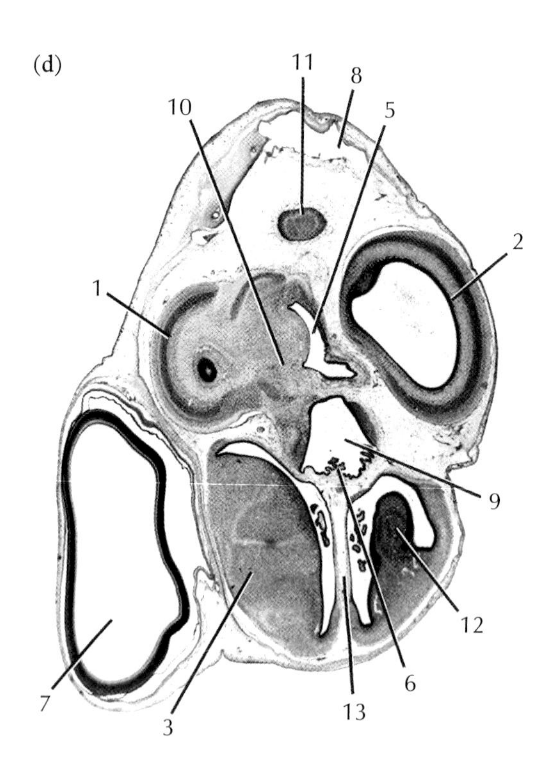

Plate 111 Day 9 TS

(a–d) ×9

1. right optic lobe
2. left optic lobe
3. right telencephalon
4. ectoderm of head
5. mesocoele
6. choroid plexus
7. right eye
8. cranial cavity
9. third ventricle
10. thalamus (diencephalon)
11. edge of spinal cord
12. left telencephalon
13. region between the two telencephalic vesicles

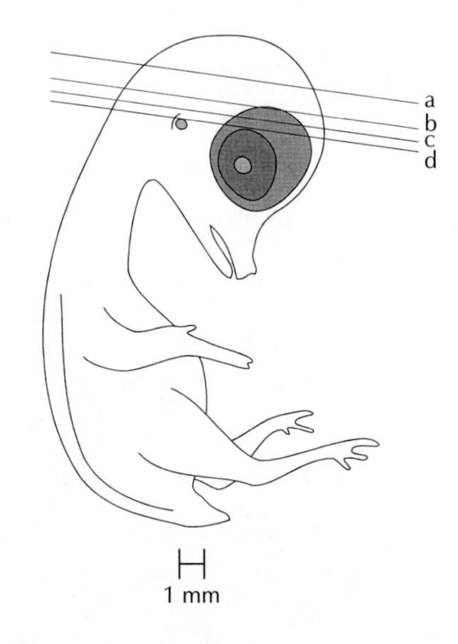

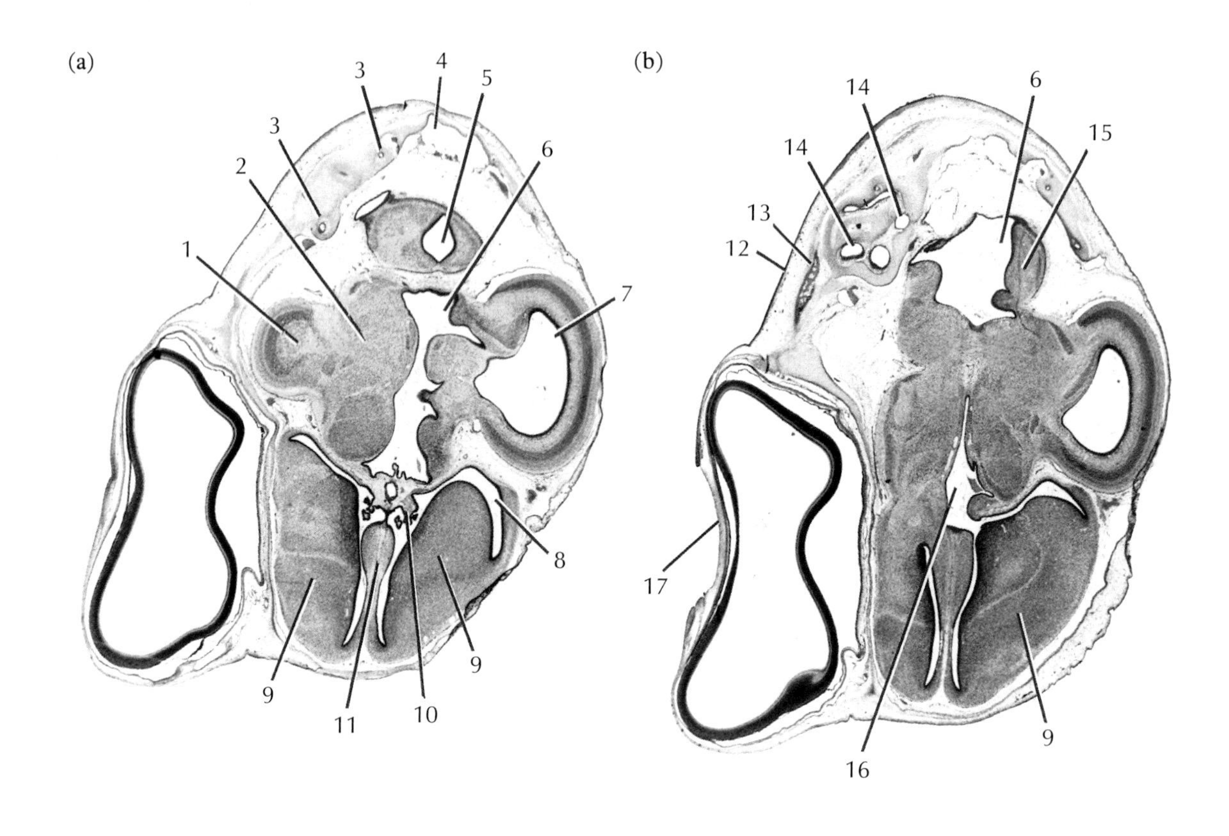
(a)
3
4
5
3
6
2
7
1
8
9
9
11
10
(b)
14
6
14
15
13
12
17
16
9

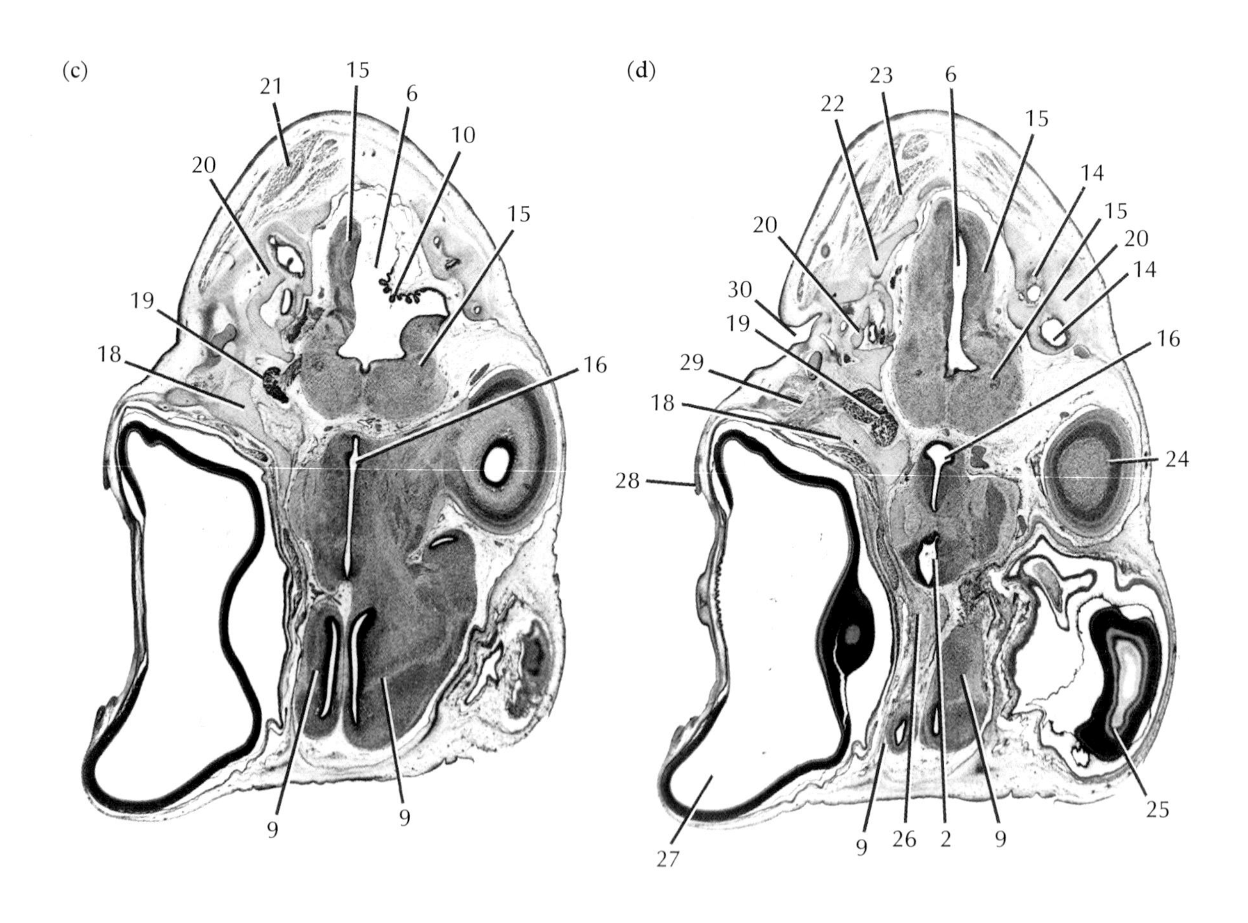
(c)
21
15
6
10
20
15
19
18
16
9
9
(d)
23
6
22
15
14
15
20
20
14
30
19
16
29
18
24
28
27
9
26
2
9
25

Plate 112 Day 9 TS

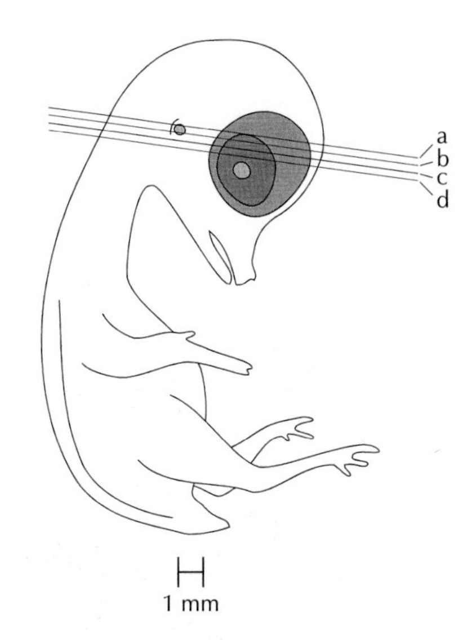

(a–d) × 9

1. right optic lobe
2. thalamus (diencephalon)
3. foramina of emissaria veins
4. cranial cavity
5. cerebellum
6. fourth ventricle
7. lumen of left optic lobe
8. lateral ventricle
9. telencephalon (cerebral hemisphere)
10. choroid plexus
11. interhemispherical fold
12. head ectoderm
13. right squamosal cartilage
14. semicircular canals
15. wall of medulla oblongata
16. third ventricle
17. cornea
18. postorbital cartilage
19. trigeminal (cr. nerve V) ganglion
20. otic capsule
21. splenius capitis muscle
22. right metotic cartilage
23. right biventer cervicis muscle
24. left optic lobe
25. left eye
26. dorsal oblique eye muscle
27. vitreous chamber of eye
28. lower eyelid
29. adductor mandibulae externus muscle
30. external auditory meatus

Plate 113

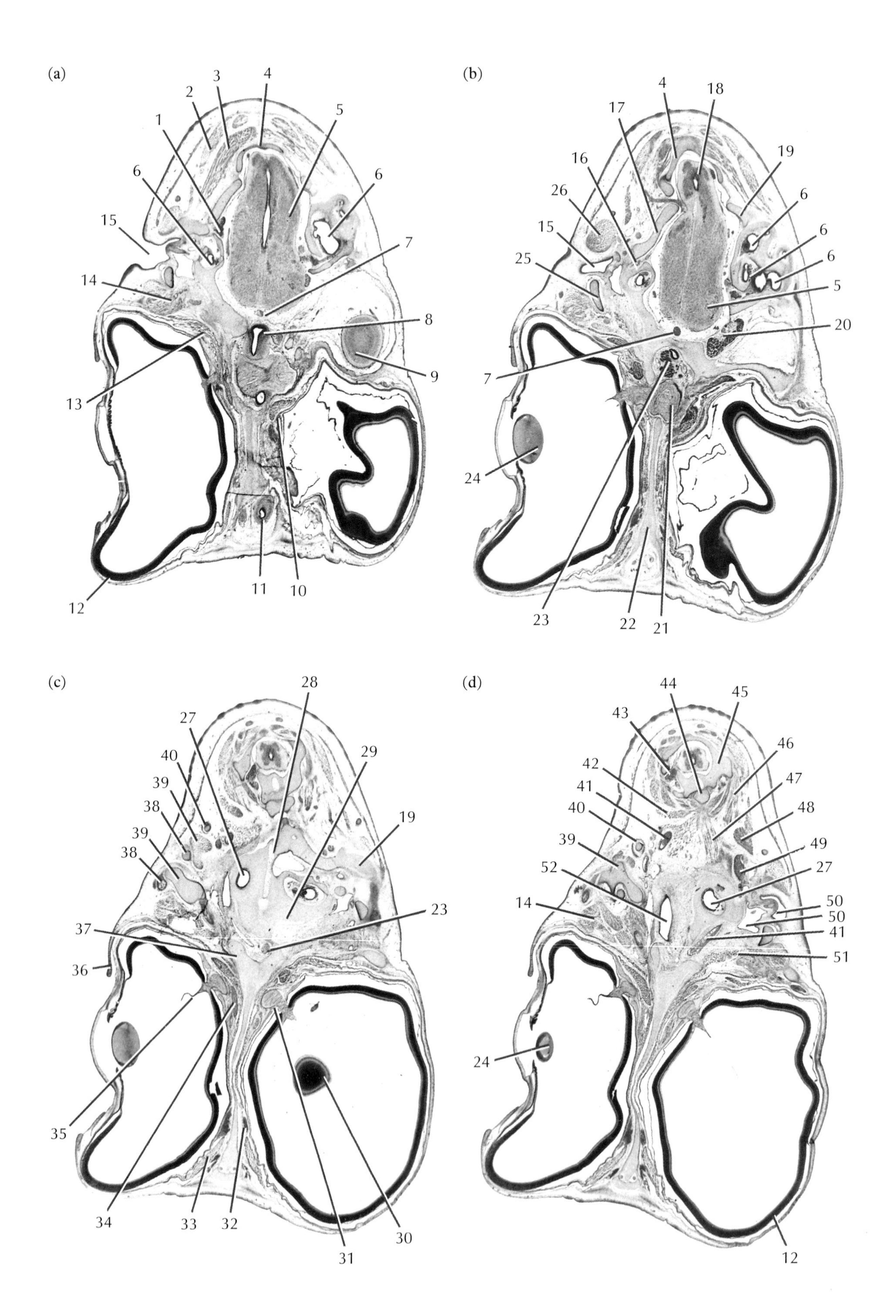

(a)
1
2
3
4
5
6
6
7
8
9
10
11
12
13
14
15
(b)
4
18
17
16
19
26
6
6
15
6
25
5
20
7
24
23
22
21
(c)
28
27
29
40
39
19
38
39
38
23
37
36
35
34
33
32
31
30
(d)
44
45
43
46
42
47
41
48
40
49
39
52
27
14
50
50
41
51
24
12

Plate 113 Day 9 TS

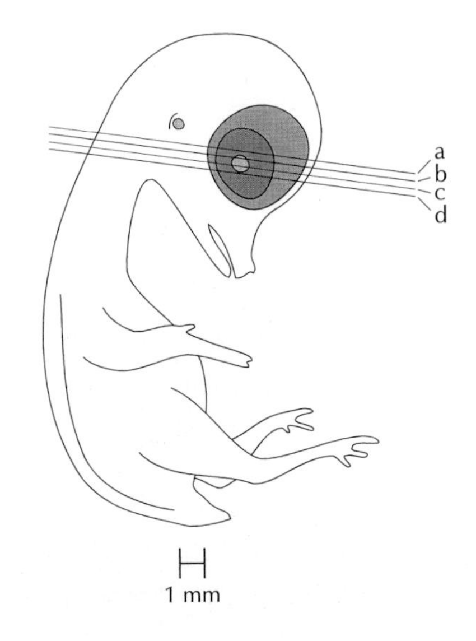

(a–d) ×9

1. glossopharyngeal (cr. IX) nerve
2. complexus muscle
3. biventer cervis muscle
4. neural arch
5. medulla oblongata
6. semicircular canal
7. basilar artery
8. diencephalon (and third ventricle)
9. left optic lobe
10. medial rectus muscle of eye
11. telencephalon
12. pigmented retina
13. lateral rectus muscle of eye
14. adductor mandibulae externus muscle
15. external auditory meatus
16. otic capsule
17. right metotic cartilage
18. spinal cord
19. left metotic cartilage
20. trigeminal (cr. nerve V) ganglion
21. optic chiasma
22. interorbital (nasal) septum
23. hypophysis
24. lens
25. epibranchial cartilage
26. depressor mandibulae muscle
27. cochlea canal
28. notochord
29. basal plate
30. artifactual bulge in retina of left eye
31. optic nerve
32. olfactory (cr. I) nerve
33. ventral oblique muscle of eye
34. external ophthalmic vein
35. pecten
36. lower eyelid
37. trabecular cartilage
38. Meckel's cartilage
39. quadrate cartilage
40. columella
41. common carotid artery
42. rectus capitis medialis muscle
43. dorsal root ganglion
44. odontoid process (body of axis)
45. transverse process
46. splenius capitis muscle
47. rectus capitis ventralis muscle
48. rectus capitis lateralis muscle
49. vagus (cr. X) nerve
50. middle ear
51. pterygoid cartilage
52. cranial (anterior) vena cava

PLATE 114

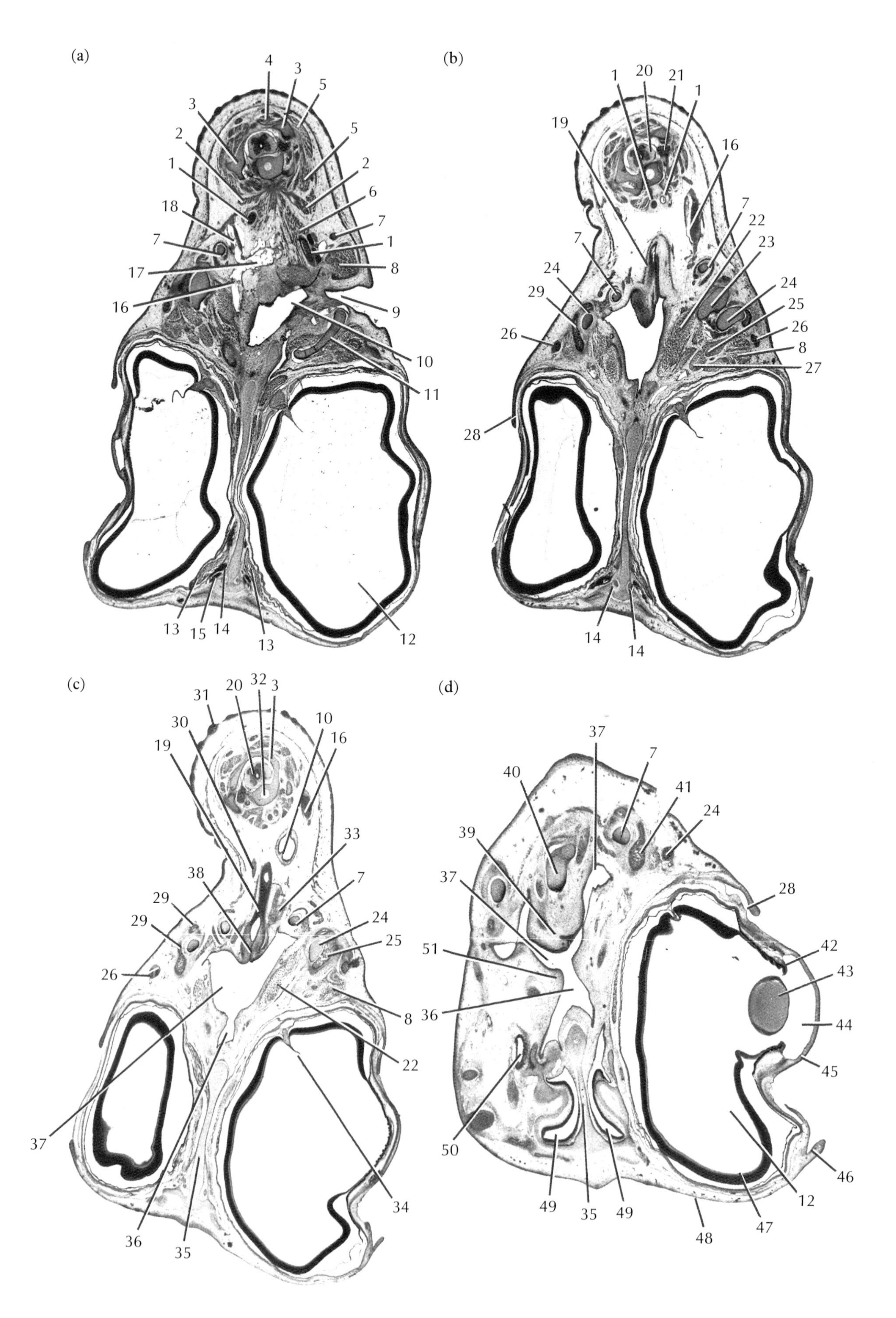
(a)
4
3
5
3
5
2
2
1
6
18
7
7
1
17
8
16
9
10
11
13
15
14
13
12
(b)
1
20
21
1
19
16
7
22
7
23
24
24
29
25
26
26
8
27
28
14
14
(c)
31
20
32
3
30
10
16
19
33
38
7
29
24
29
25
26
8
22
37
34
36
35
(d)
37
7
40
41
24
39
37
28
51
42
43
36
44
45
50
46
49
35
49
12
47
48

Plate 114 Day 9 TS

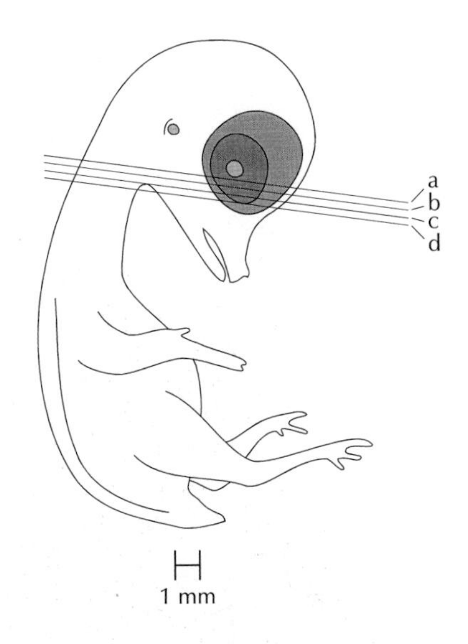

(a–d) ×9

1. carotid artery
2. rectus capitis lateralis muscle
3. neural arch
4. biventer cervicis muscle
5. splenius capitis muscle
6. rectus capitis ventralis muscle
7. epibranchial cartilage
8. adductor mandibulae externus muscle
9. external auditory meatus
10. pharynx
11. otic process of quadrate cartilage
12. vitreous chamber of eye
13. dorsal oblique eye muscle
14. olfactory (cr. I) nerve
15. palatine artery
16. jugular vein
17. anastomotic vein (interjugular anastomosis)
18. occipital vein
19. trachea
20. spinal cord
21. dorsal root ganglion
22. pterygoid muscle
23. Meckel's cartilage
24. quadrato-jugal cartilage
25. pseudotemporalis superficialis muscle
26. facial vein
27. pseudotemporalis profundus muscle
28. lower eyelid
29. pterygoid condyle of quadrate cartilage
30. extrinsic tracheal muscle
31. feather papilla
32. notochord in body of vertebra
33. intrinsic tracheal muscle
34. pecten
35. interorbital (nasal) process
36. internal naris
37. oral cavity
38. glottis
39. tongue
40. paraglossal cartilage
41. lateral condyle of quadrate cartilage
42. iris
43. lens
44. anterior chamber of eye
45. cornea
46. upper eyelid
47. pigmented retina
48. ectoderm of head
49. nasal cavity
50. nasal concha
51. palatine process

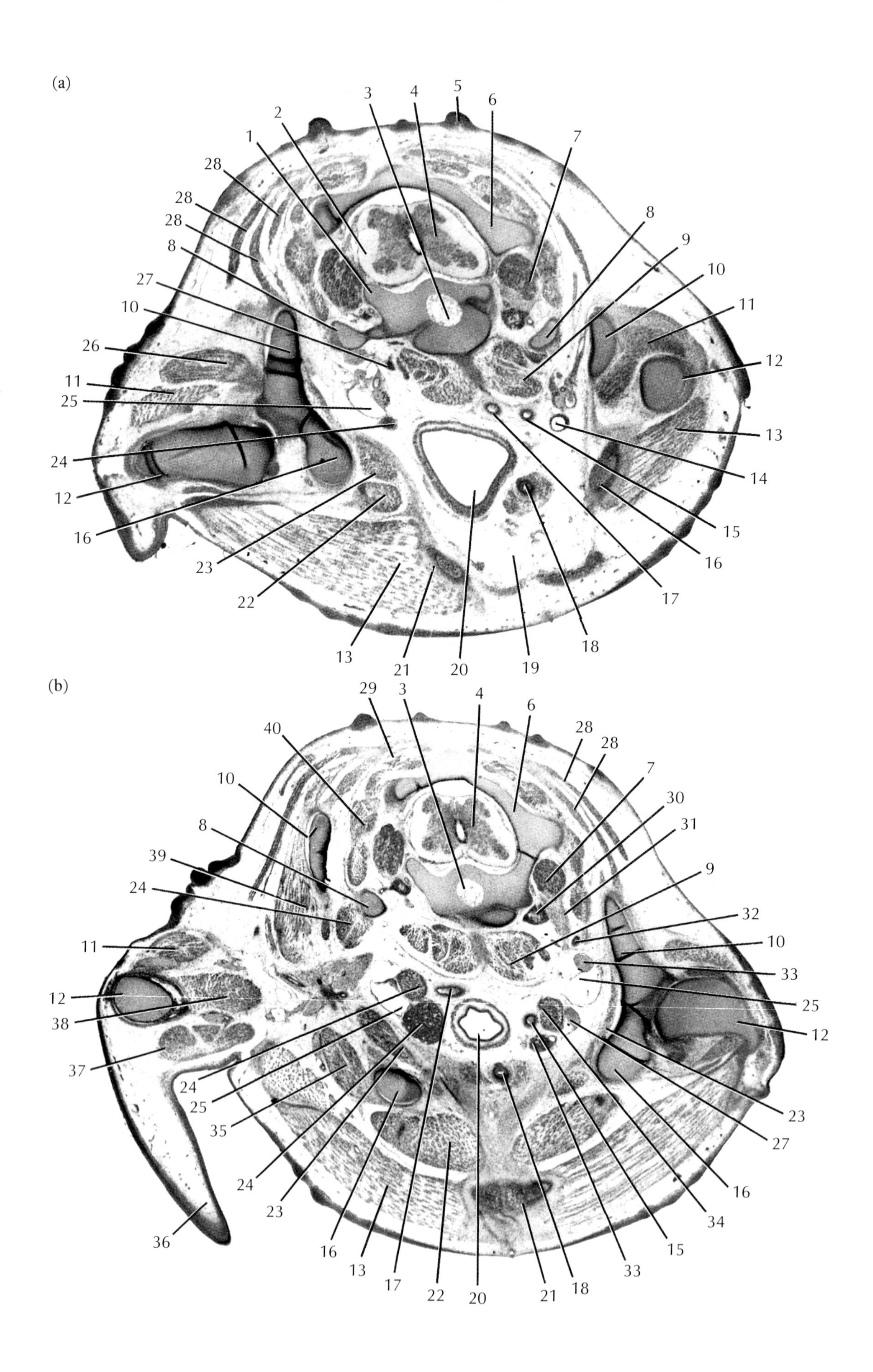
(a)
1
2
3
4
5
6
7
8
9
10
11
12
13
14
15
16
17
18
19
20
21
22
23
24
25
26
27
28
(b)
29
30
31
32
33
34
35
36
37
38
39
40

Plate 115 Day 9 TS

(a, b) ×20

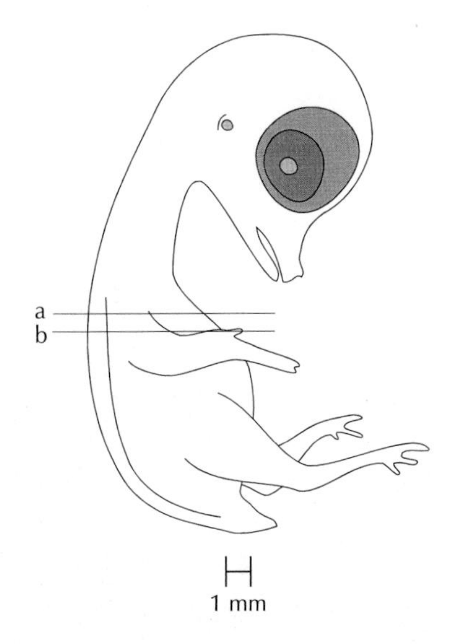

1. cervical vertebra
2. spinal cord, white matter
3. notochord
4. spinal cord, grey matter
5. feather papilla
6. neural arch of cervical vertebra
7. dorsal root ganglion
8. cervical rib
9. longus colli ventralis muscle
10. scapula
11. deltoid major muscle
12. humerus
13. pectoralis muscle
14. oesophago-tracheo-bronchial artery
15. left common carotid artery
16. coracoid
17. right common carotid artery
18. trachea
19. connective tissue
20. crop
21. apex of sternum
22. costo-sternalis muscle
23. subcoracoideus muscle, minor portion
24. thymus of right side
25. internal jugular vein
26. tensor propatagialis longus muscle
27. spinal nerve
28. branches of latissimus dorsi muscle
29. rhomboideus superficialis muscle
30. sympathetic ganglion
31. part of left brachial plexus
32. thoracic rib
33. thymus of left side
34. cervical air sac
35. supracoracoideus muscle, major portion
36. rim of right wing
37. triceps brachii, pars humerotriceps muscle
38. medial head of triceps muscle
39. serratus superficialis muscle
40. rhomboideus profundus muscle

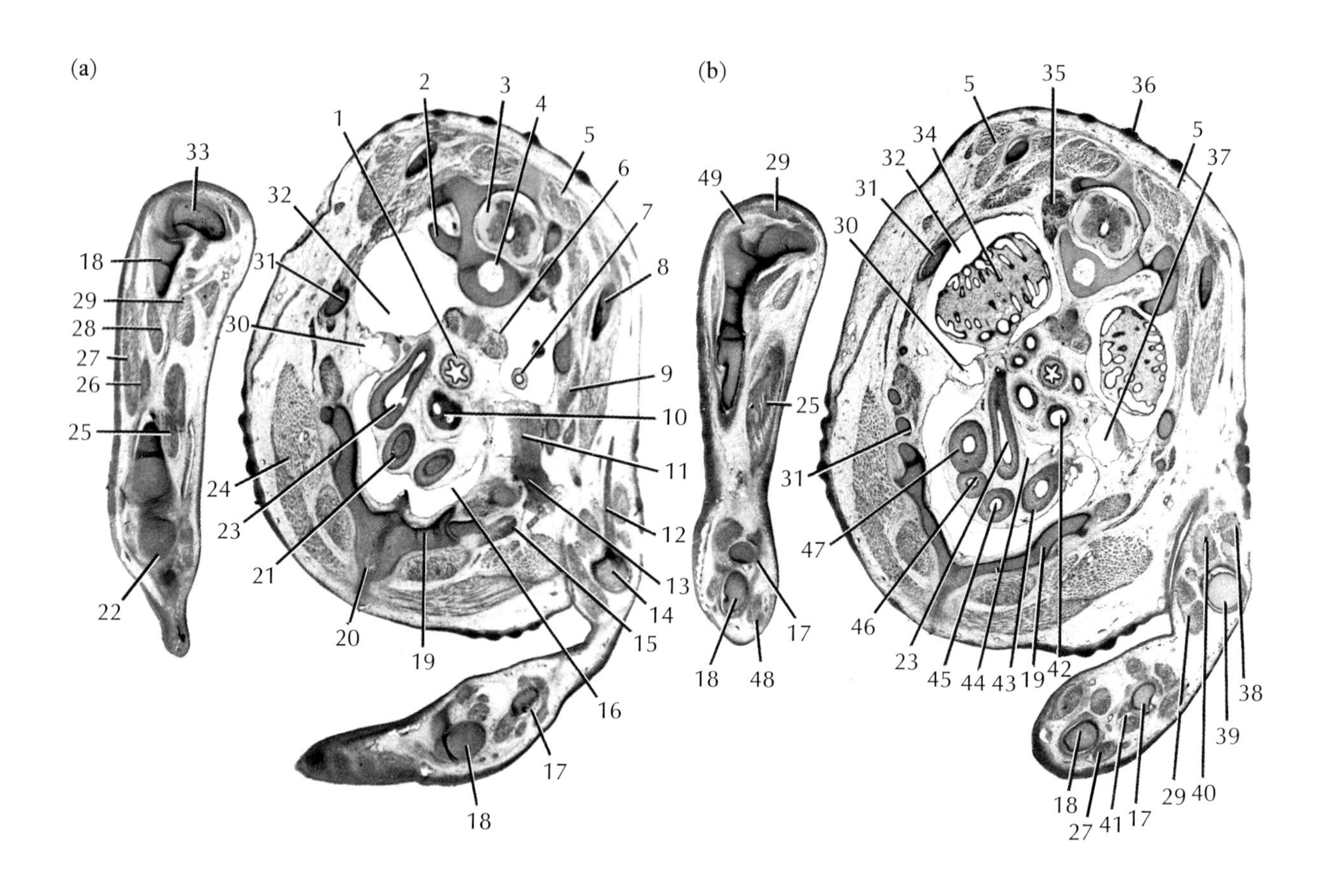
(a)
33
18
29
28
27
26
25
22
32
31
30
24
23
21
20
19
1
2
3
4
5
6
7
8
9
10
11
12
13
14
15
16
17
18
(b)
49
29
25
17
18
48
5
35
36
34
32
31
30
37
31
47
46
23
45
44
43
19
42
38
39
40
29
17
41
27
18

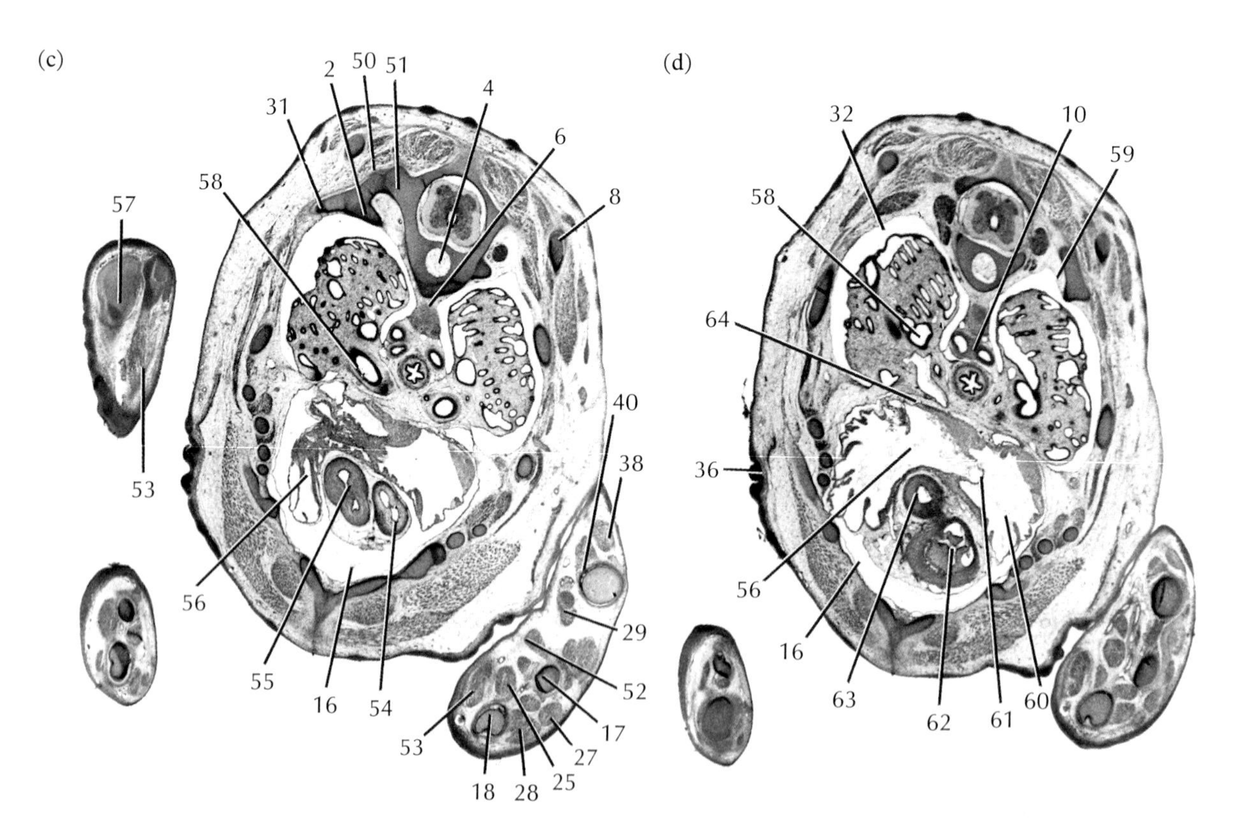
(c)
57
53
31
2
50
51
4
6
8
58
40
38
29
52
17
27
25
28
18
53
54
16
55
56
(d)
32
10
59
58
64
36
56
16
63
62
61
60

Plate 116 Day 9 TS

(a–d) × 14

1. oesophagus
2. head of rib (capitulum)
3. spinal cord, white matter
4. notochord
5. rhomboideus superficialis muscle
6. longus colli ventralis muscle
7. anterior tip of lung
8. scapula
9. intercostal muscle
10. syrinx
11. jugular vein
12. shaft of humerus
13. subclavian vein
14. ventral condyle of humerus
15. coracoid
16. pericardial cavity
17. radius
18. ulna
19. sternum
20. keel of sternum
21. right common carotid artery
22. ulna carpal
23. right pulmonary artery
24. coracobrachialis muscle
25. flexor digitorum profundus muscle
26. extensor digitorum communis muscle
27. extensor metacarpi ulnaris muscle
28. triceps brachii pars humerotriceps muscle
29. pronator profundus muscle
30. right cranial (anterior) vena cava in pleuro-pericardial septum
31. thoracic rib
32. right pleural cavity
33. dorsal condyle of humerus
34. right lung
35. dorsal root ganglion
36. feather papilla
37. left cranial (anterior) vena cava in pleuro-pericardial septum
38. biceps brachii muscle
39. humerus
40. deltoid major muscle
41. extensor indicis longus muscle
42. bronchus
43. left pulmonary artery
44. cervical air sac
45. left branchiocephalic artery
46. right brachiocephalic artery
47. dorsal aorta
48. flexor carpi ulnaris muscle
49. pronator muscle
50. head of rib (tuberculum)

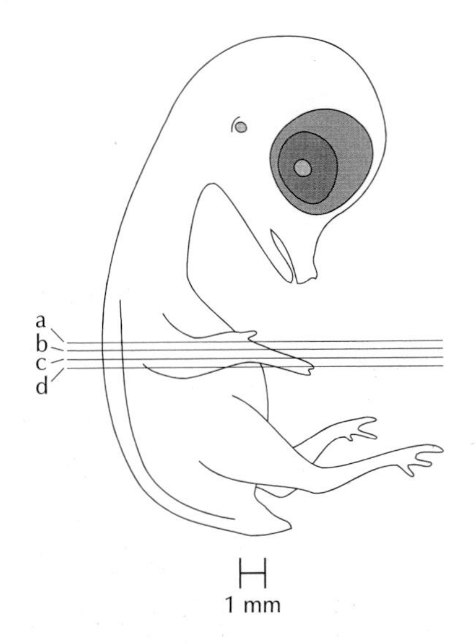

51. dispophysis of vertebra
52. pronator superficialis muscle
53. flexor carpi ulnaris muscle
54. trunk of pulmonary arteries
55. trunk of dorsal aorta and brachiocephalic arteries
56. right atrium
57. ulnimetacarpalis ventralis muscle
58. bronchus
59. left pleural cavity
60. left atrium
61. interatrial septum (with perforations)
62. semilunar valves of pulmonary trunk
63. aortic arch
64. pleuro-pericardial septum

Plate 117

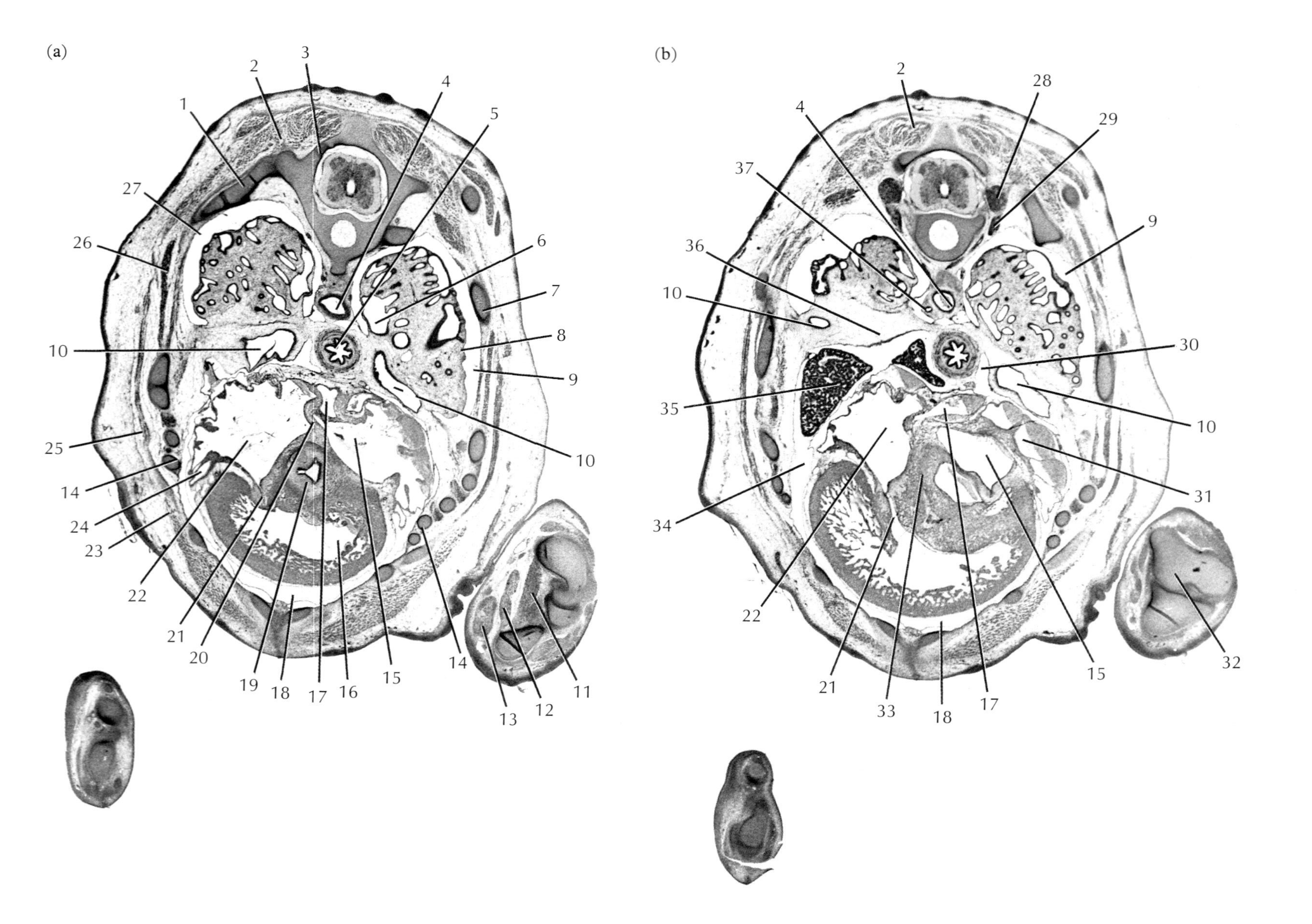
(a)
1
2
3
4
5
6
7
8
9
10
10
10
11
12
13
14
14
15
16
17
18
19
20
21
22
23
24
25
26
27
(b)
2
4
9
10
10
15
17
18
21
22
28
29
30
31
32
33
34
35
36
37

Plate 117 Day 9 TS

(a, b) × 18

1. head of vertebral rib
2. rhomboideus superficialis muscle
3. neural arch of vertebra
4. dorsal aorta
5. oesophagus
6. bronchus
7. vertebral rib
8. lung
9. left pleural cavity
10. junction of anterior thoracic air sac and mesobronchus
11. flexor digitorum profundus muscle
12. ulnametacarpalis dorsalis muscle
13. pronator muscle
14. sternal rib
15. left atrium
16. left ventricle
17. sinus venosus
18. pericardial cavity
19. bulbus arteriosus
20. interatrial septum
21. right atrio-ventricular canal
22. right atrium
23. supracoracoideus muscle
24. lobe of right atrium
25. oblique externus abdominis muscle
26. intercostal muscle
27. right pleural cavity
28. dorsal root ganglion
29. sympathetic ganglion
30. pneumo-enteric recess
31. lobulation of left atrium
32. articular surface of ulna
33. interventricular septum
34. right ventral hepatic cavity
35. right lobe of liver
36. pleuro-peritoneal septum
37. omphalo-mesenteric artery

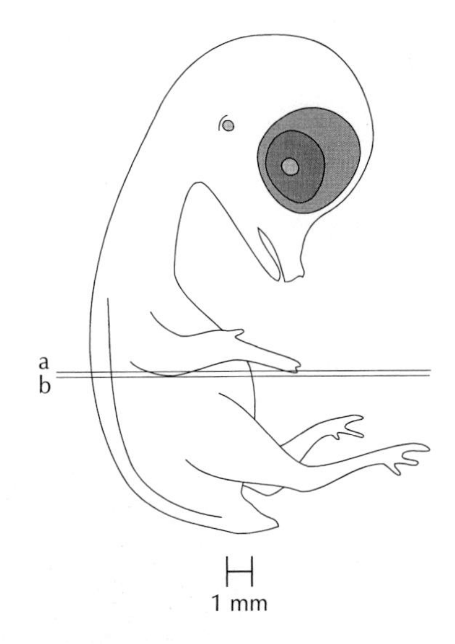

Plate 118

(a)

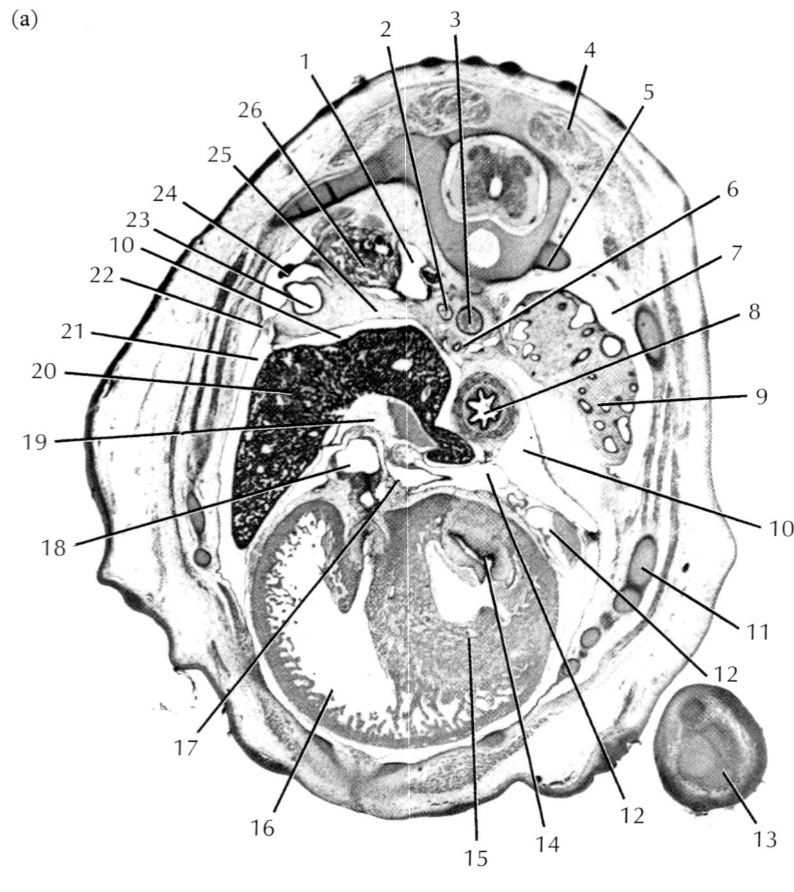
1
2
3
4
5
6
7
8
9
10
11
12
13
14
15
16
17
18
19
20
21
22
23
24
25
26

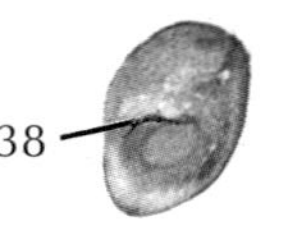
38

(b)

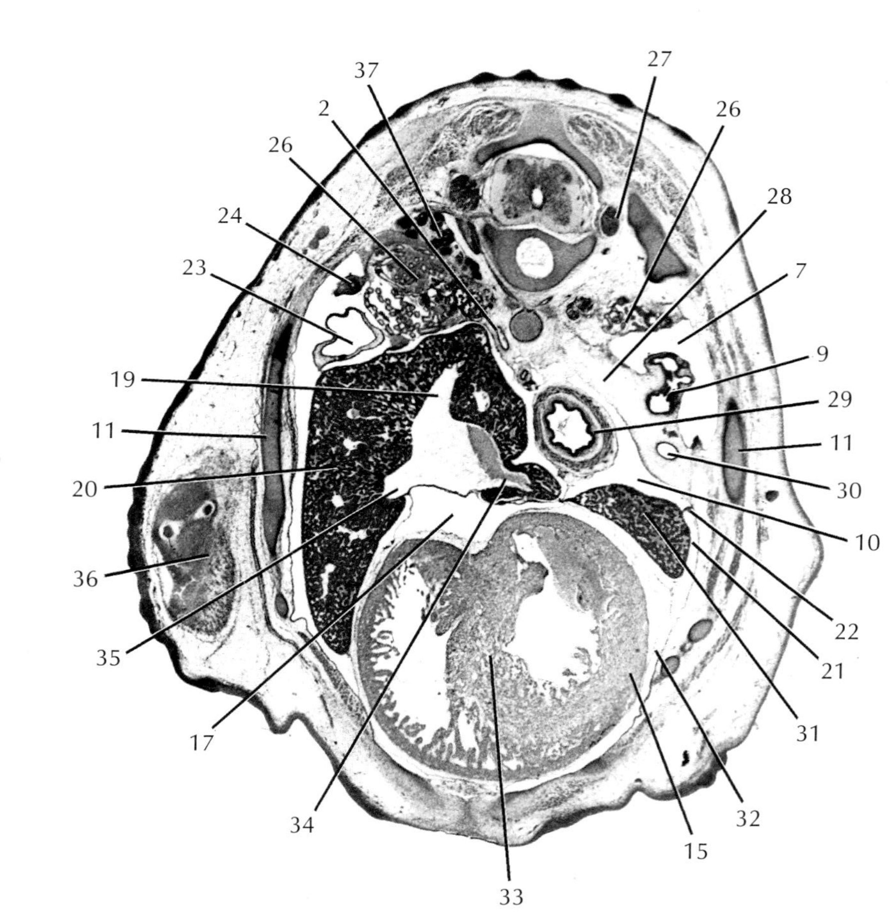
2
7
9
10
11
15
17
19
20
21
22
23
24
26
27
28
29
30
31
32
33
34
35
36
37

Plate 118 Day 9 TS

(a, b) × 18

1. right subcardinal vein
2. omphalo-mesenteric artery
3. dorsal aorta
4. rhomboideus superficialis muscle
5. transverse process of rib
6. coeliac artery
7. pleural cavity
8. oesophagus
9. left lung
10. dorsal hepatic cavity
11. sternal rib
12. lobes of left atrium
13. wing
14. left atrio-ventricular canal
15. left ventricle
16. right ventricle
17. sinus venosus
18. right atrium
19. caudal (posterior) vena cava
20. right lobe of liver
21. ventral hepatic cavity
22. horizontal hepatic ligament
23. right anterior thoracic air sac
24. right oviduct (Mullerian duct)
25. right pleuro-peritoneal septum
26. mesonephros
27. dorsal root ganglion
28. left pleuro-peritoneal septum
29. proventriculus
30. anterior thoracic air sac
31. anterior left lobe of liver
32. pericardium
33. interventricular septum
34. left hepatic vein
35. right hepatic vein
36. femur
37. metanephros
38. right foot

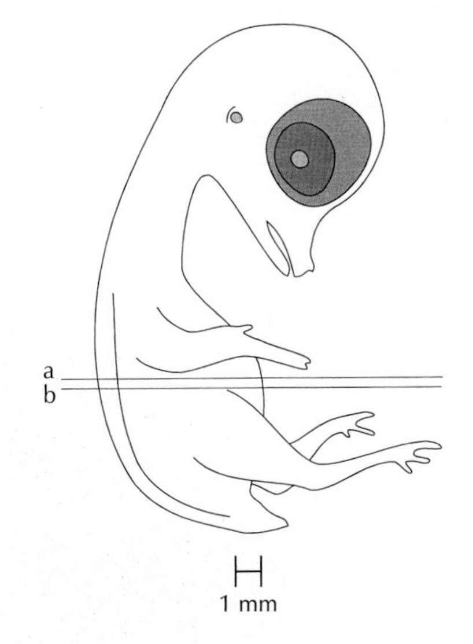

PLATE 119

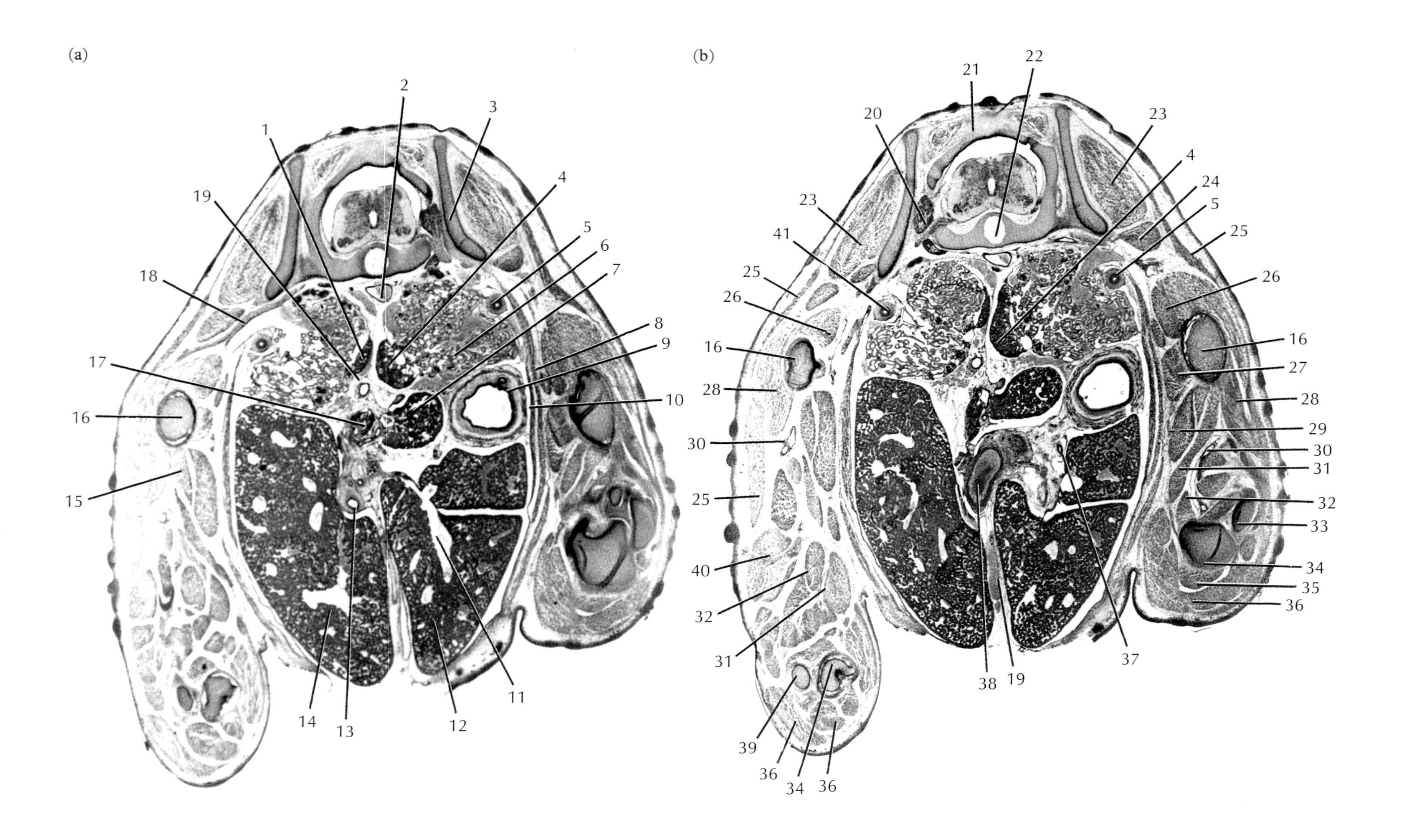

(a)
1
2
3
4
5
6
7
8
9
10
11
12
13
14
15
16
17
18
19
(b)
4
5
16
19
20
21
22
23
24
25
26
27
28
29
30
31
32
33
34
35
36
37
38
39
40
41

Plate 119 Day 9 TS

(a, b) × 18

1. right ovary
2. dorsal aorta
3. ilium
4. left ovary
5. left oviduct (Mullerian duct)
6. mesonephros
7. spleen
8. oblique externus abdominis muscle
9. proventriculus
10. oblique internus abdominis muscle
11. hepatic vein
12. left posterior lobe of liver
13. common bile duct
14. right lobe of liver
15. caudofemoralis muscle
16. femur
17. pancreas
18. trunk of brachial plexus
19. omphalo-mesenteric artery
20. dorsal root ganglion
21. neural arch of vertebra
22. notochord
23. iliotrochantericus cranialis muscle
24. ischiofemoralis muscle
25. iliotibialis lateralis muscle
26. iliotrochantericus caudalis muscle
27. ambiens muscle
28. femorotibialis lateralis muscle
29. puboischiofemoralis muscle
30. femoral artery
31. femorotibialis internus muscle
32. flexor cruris medialis muscle
33. condyle of femur
34. tibiotarsus
35. extensor digitorum longus muscle
36. gastrocnemius muscle
37. coeliac artery
38. duodenum
39. fibula
40. iliofibularis muscle
41. right oviduct (Mullerian duct)

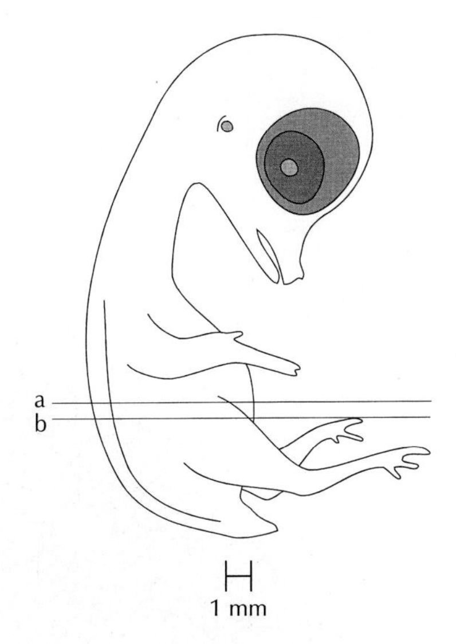

Plate 120

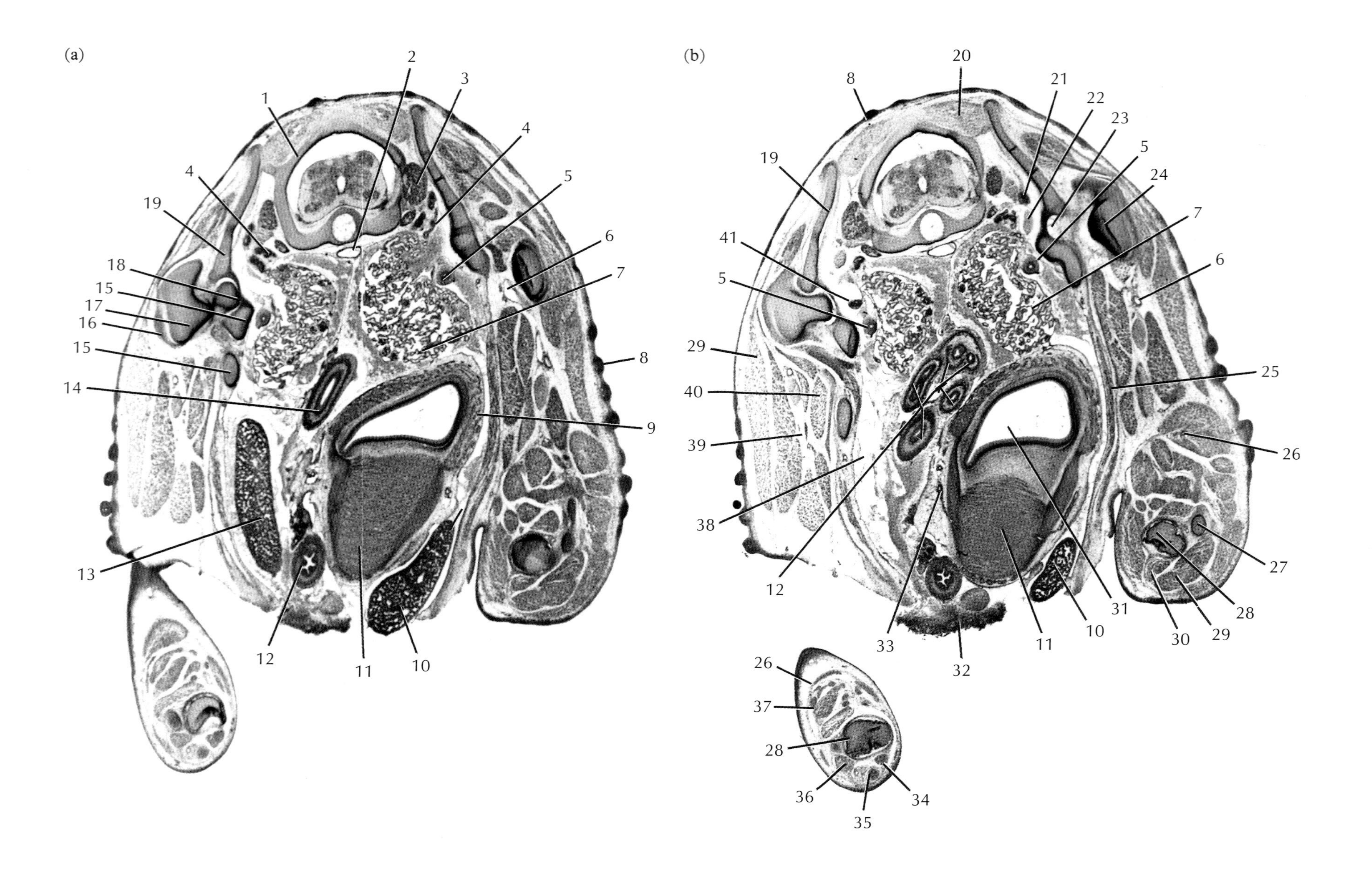

(a)
1
2
3
4
5
6
7
8
9
10
11
12
13
14
15
16
17
18
19
(b)
20
21
22
23
24
25
26
27
28
29
30
31
32
33
34
35
36
37
38
39
40
41

Plate 120 Day 9 TS

(a, b) × 18

1. neural arch of vertebra
2. dorsal aorta
3. dorsal root ganglion
4. metanephros
5. oviduct (Mullerian duct)
6. femoral artery
7. mesonephros
8. feather bud
9. cranio-dorsal muscles of gizzard
10. left posterior lobe of liver
11. ventral lateral muscles of gizzard
12. loops of small intestine
13. right lobe of liver
14. duodenum
15. ischium
16. aponeurosis of iliotibialis lateralis muscle
17. femur
18. head of femur
19. ilium
20. iliotrochantericus muscle
21. spinal nerve, part of sacral plexus
22. left renal vein
23. acetabulum
24. trochanter of femur
25. oblique internus abdominus muscle
26. gastrocnemius muscle
27. fibula
28. tibiotarsus
29. tibialis cranialis muscle
30. extensor digitorum longus muscle
31. lumen of gizzard
32. blood extravasated from umbilicus during preparation
33. gastric branch of coeliac artery
34. flexor hallucis brevis muscle
35. peroneus muscle
36. extensor brevis digitorum IV muscle
37. extensor hallucis longus muscle
38. blood extravasated into coelom
39. fibularis nerve
40. ischiofemoralis muscle
41. right renal vein

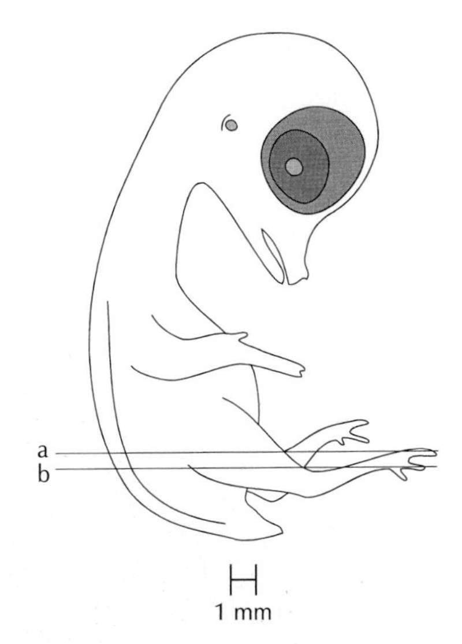

Plate 121

(a)

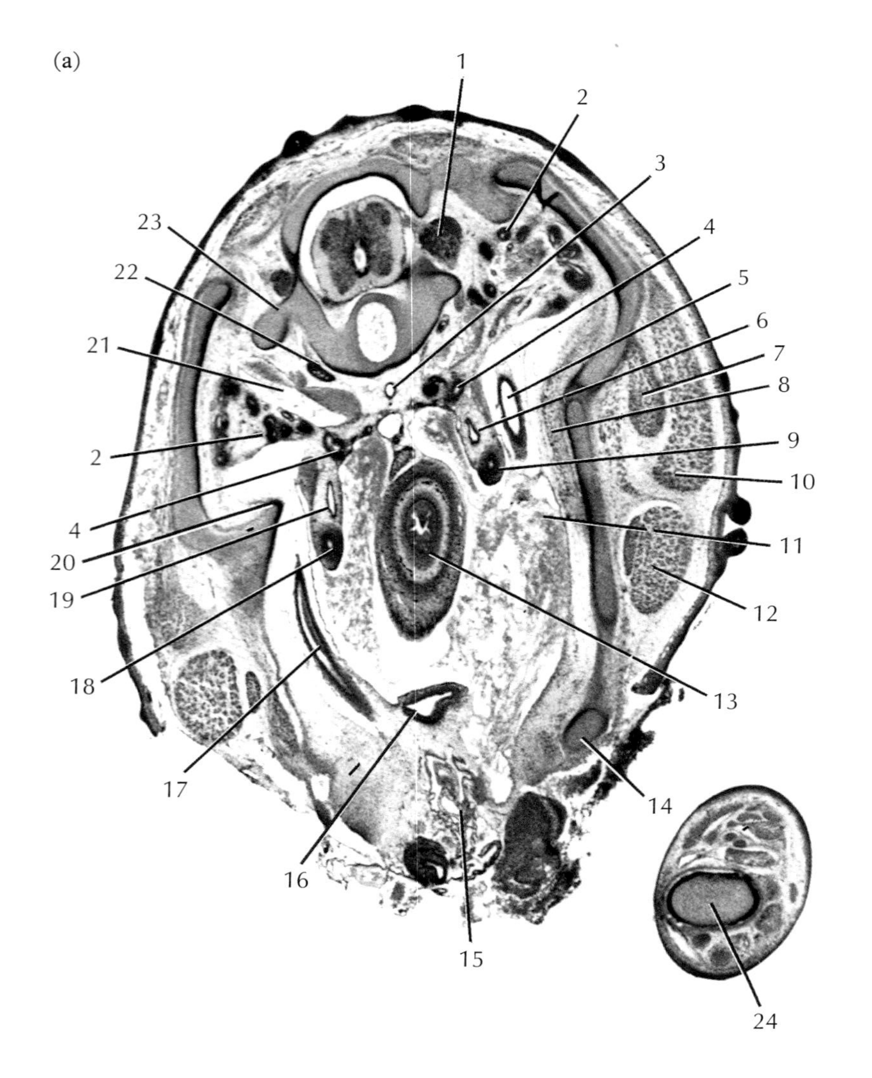

1
2
3
4
5
6
7
8
9
10
11
12
13
14
15
16
17
18
19
20
21
22
23
24

(b)

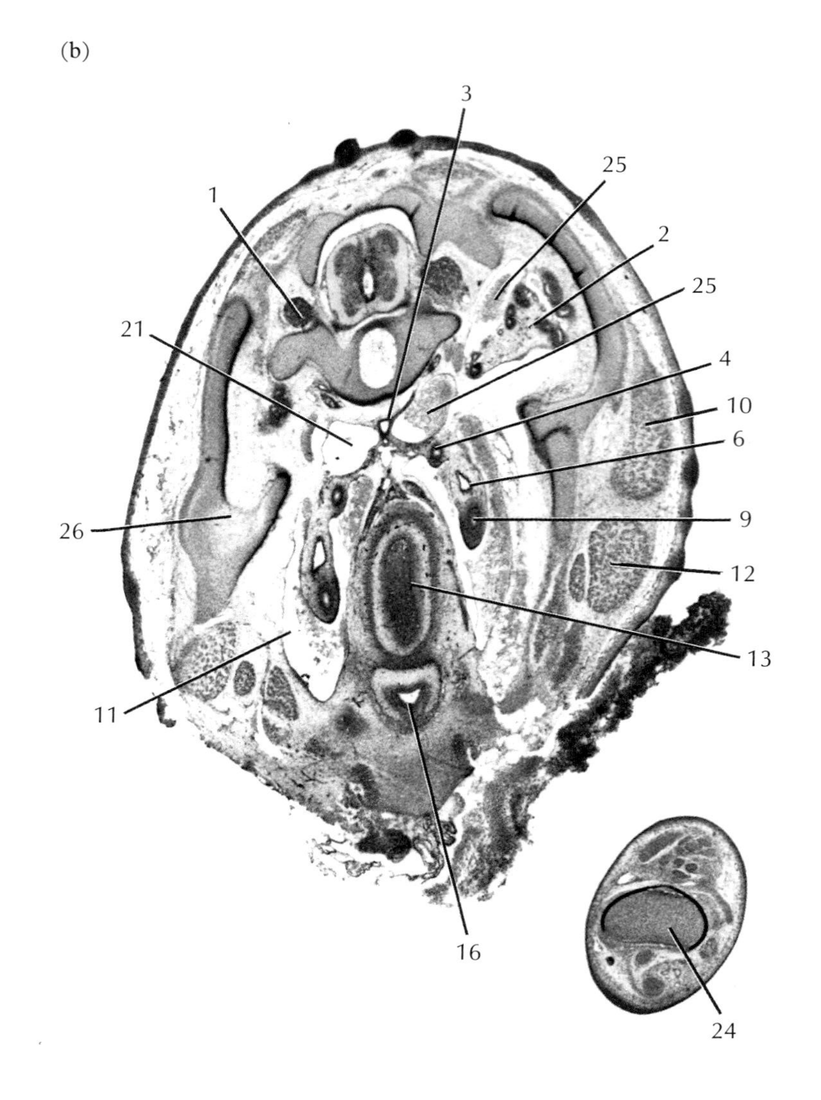

1
2
3
4
6
9
10
11
12
13
16
21
24
25
26

Plate 121 Day 9 TS

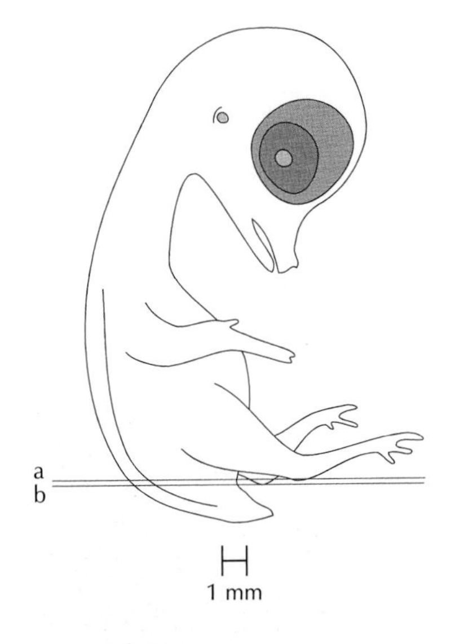

(a, b) ×20

1. dorsal root ganglion
2. metanephros
3. dorsal aorta
4. metanephric duct
5. left omphalo-mesenteric artery
6. left nephric (mesonephric, Wolffian) duct
7. iliofibularis muscle
8. obturator muscle
9. left oviduct (Mullerian duct)
10. left iliotibialis muscle
11. extravasated blood
12. flexor cruris lateralis muscle
13. rectum
14. pubis
15. yolk sac stalk
16. cloaca
17. right omphalo-mesenteric artery
18. right oviduct (Mullerian duct)
19. right nephric (mesonephric, Wolffian) duct
20. pectineal process
21. right renal vein
22. spinal nerve
23. transverse process of rib
24. tibiotarsus
25. left renal vein
26. ischium

PLATE 122

(a)

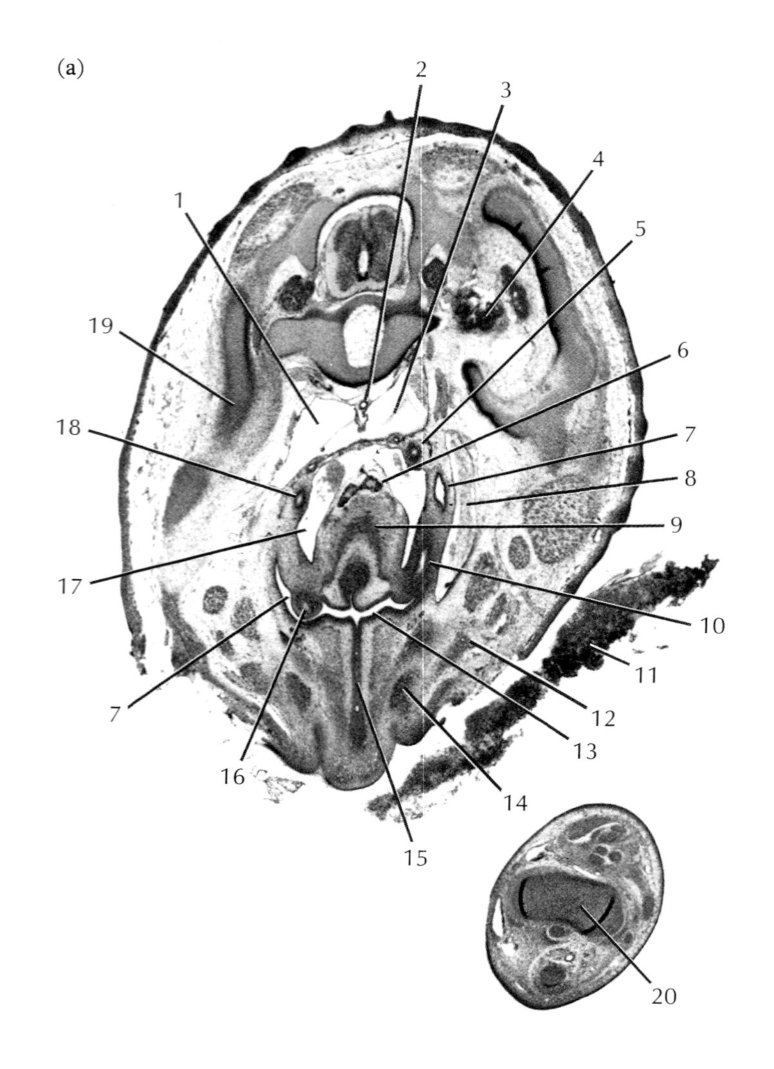

1
2
3
4
5
6
7
8
9
10
11
12
13
14
15
16
7
17
18
19
20

(b)

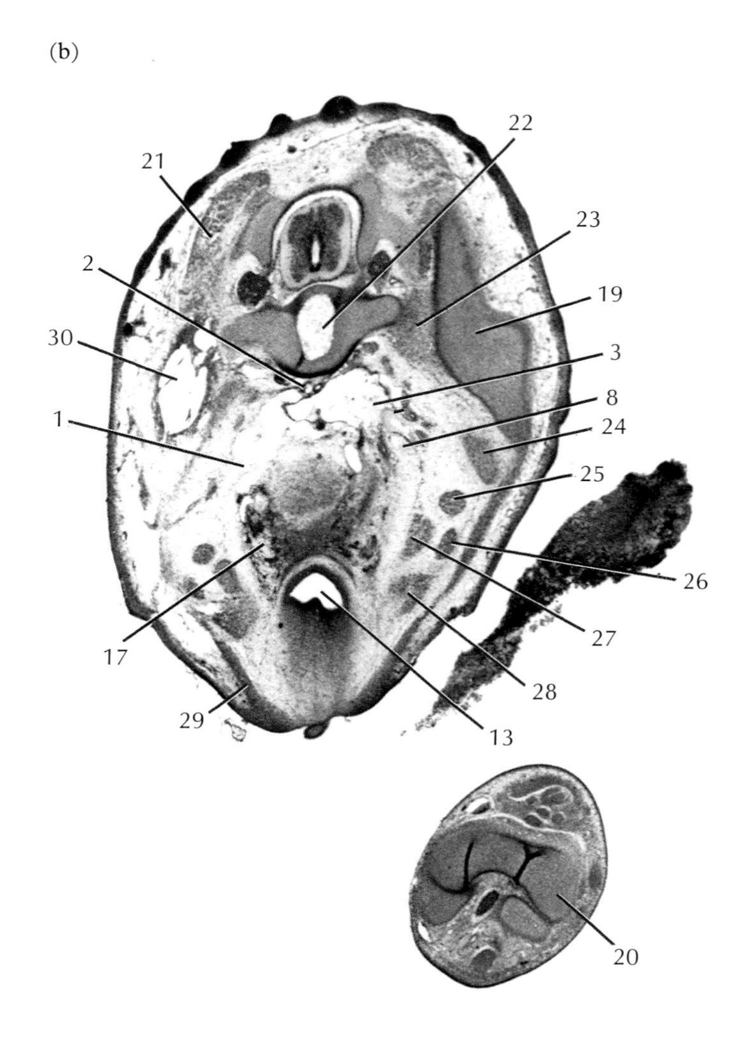

21
22
23
19
2
30
3
8
1
24
25
26
27
28
13
17
29
20

Plate 122 Day 9 TS

(a, b) ×20

1. right omphalo-mesenteric artery
2. dorsal aorta
3. left omphalo-mesenteric artery
4. metanephros
5. left metanephric duct
6. renal artery
7. nephric (mesonephric, Wolffian) duct
8. extravasated blood in posterior coelom
9. rectum
10. left oviduct (Mullerian duct)
11. extravasated blood
12. levator cloacae muscle
13. cloaca
14. sphincter cloacae
15. cloacal membrane
16. right oviduct (Mullerian duct)
17. interiliac vein anastomosis
18. right metanephric duct
19. ischium
20. hindlimb
21. iliotibialis lateralis muscle
22. remains of notochord
23. mesenchymal mass connecting vertebra to girdle

24–28. tail muscles

29. lateralis caudae muscle
30. remnant of right posterior cardinal vein

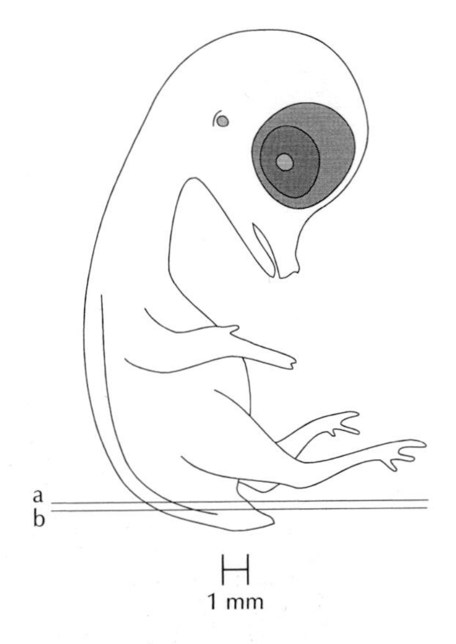

Plate 123

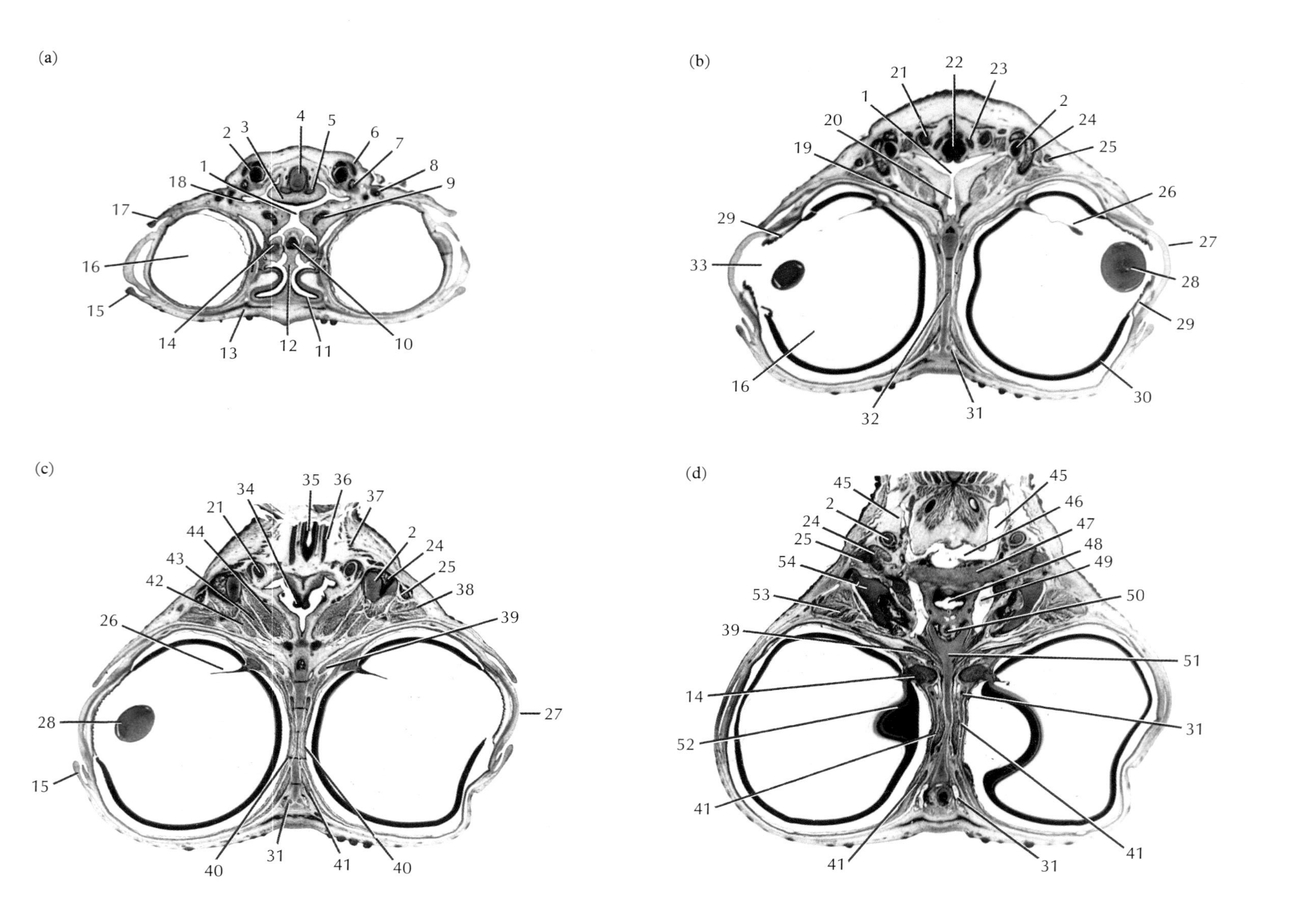

(a)
1
2
3
4
5
6
7
8
9
10
11
12
13
14
15
16
17
18
(b)
1
2
16
19
20
21
22
23
24
25
26
27
28
29
30
31
32
33
(c)
2
15
21
24
25
26
27
28
31
34
35
36
37
38
39
40
41
42
43
44
(d)
2
14
24
25
31
39
41
45
46
47
48
49
50
51
52
53
54

Plate 123 Day 10 Head TS

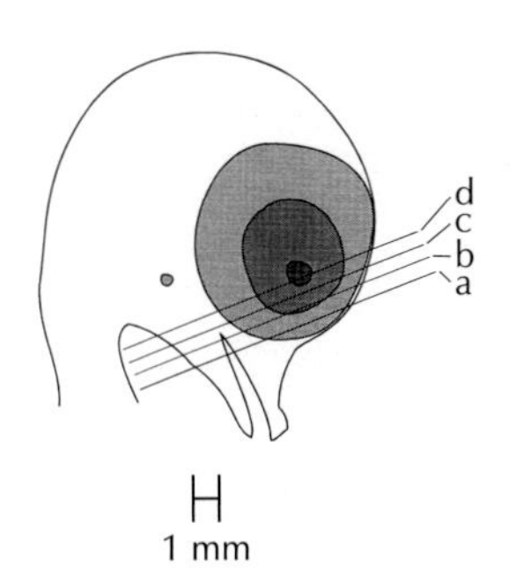

(a–d) ×7

1. median rostral choana
2. Meckel's cartilage
3. tongue
4. paraglossal cartilage (entoglossal)
5. basibranchialis muscle
6. biventer maxilla muscle
7. quadrato-jugal cartilage
8. maxillary cartilage
9. palatine bone
10. base of nasal septum
11. nasal cavity
12. nasal septum
13. parieto-tectal cartilage
14. optic (cr. II) nerve
15. upper eyelid
16. vitreous chamber
17. lower eyelid
18. fold in surface of oral cavity
19. facial nerve (cr. VII, hyomandibular branch)
20. common internal naris
21. cerato-branchial cartilage
22. copulo-paraglossal cartilage
23. hypoglossal muscle
24. dentary
25. quadrate cartilage
26. pecten
27. cornea
28. lens
29. iris
30. pigmented retina
31. olfactory (cr. I) nerve
32. interorbital septum
33. anterior chamber of eye
34. tip of copulo-paraglossal cartilage
35. trachea
36. sterno-tracheal muscle
37. splenius capitis muscle
38. left adductor mandibulae externus muscle
39. ventral oblique eye muscle
40. sclera
41. dorsal oblique eye muscle
42. pseudotemporalis superficialis muscle
43. pseudotemporalis profundus muscle
44. mandibular depressor muscle
45. jugular vein
46. interjugular anastomosis (anastomotic vein)
47. pila otica cartilage
48. larynx
49. cranial (anterior) vena cava
50. hypophyseal fenestra
51. trabecular cartilage
52. collapsed medial wall of eye
53. right adductor mandibulae externus muscle
54. epibranchial cartilage

Plate 124

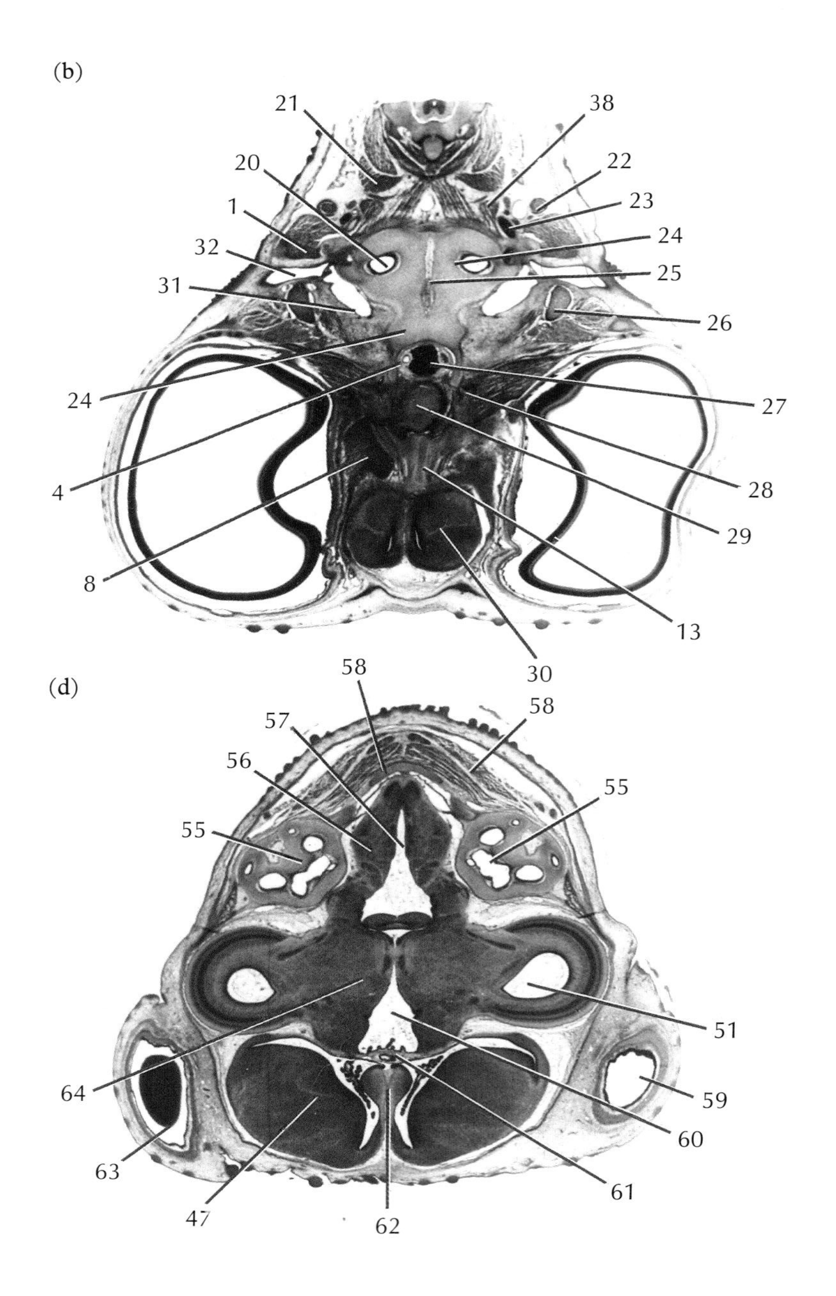
(b)
21
38
20
22
1
23
32
24
31
25
26
24
27
4
28
29
8
13
30
(d)
58
57
58
56
55
55
51
64
59
60
63
61
47
62

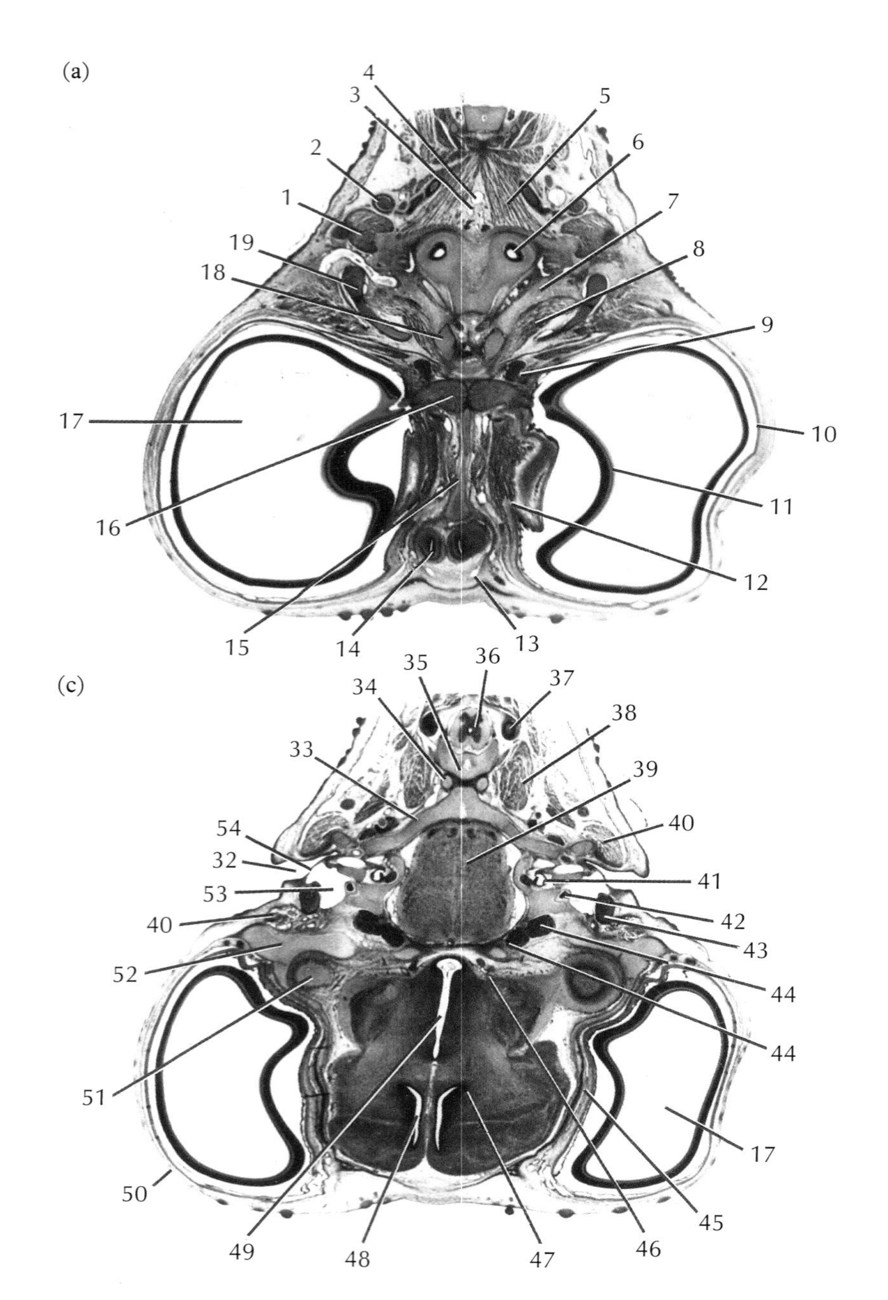
(a)
4
3
5
2
6
1
7
19
8
18
9
17
10
16
11
12
15
14
13
(c)
35
36
34
37
38
33
39
54
40
32
41
53
40
42
43
52
44
44
51
17
50
45
49
48
47
46

Plate 124 Day 10 Head TS

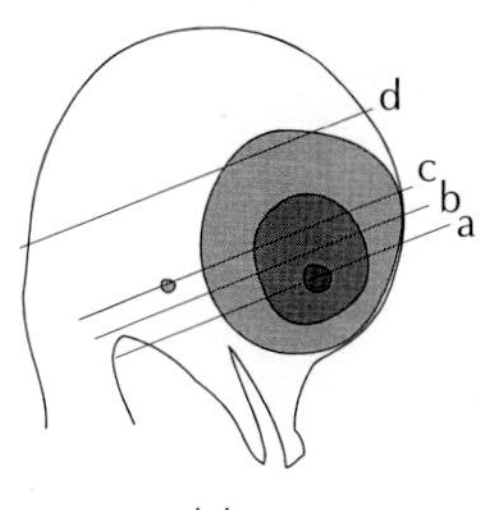

(a–d) ×7

1. quadrate cartilage
2. cerato-branchial cartilage (copula II)
3. right internal carotid artery
4. left internal carotid artery
5. rectus capitis ventralis muscle
6. cochlea canal
7. planum antorbitale cartilage
8. ventral oblique eye muscle
9. oculomotor (cr. III) nerve
10. sclera
11. retina
12. pyramidalis membranae nictitansis muscle
13. olfactory (cr. I) nerve
14. olfactory bulb
15. interorbital septum
16. optic chiasma
17. vitreous chamber
18. infrapolar cartilage
19. orbital process of quadrate cartilage
20. perilymphatic duct of auditory capsule
21. rectus capitis lateralis muscle
22. cerato-branchial cartilage
23. stapedial artery
24. parachordal cartilage (future parasphenoid)
25. remains of notochord
26. pterygoid
27. hypophysis in basicranial fenestra
28. internal oblique eye muscle
29. intratrabecula cartilage
30. left cerebral hemisphere
31. auditory tube
32. external auditory meatus
33. neural arch of axis, forming occipital arch
34. occipital condyle
35. body of atlas
36. spinal cord
37. left petrosal ganglion of cranial nerve IX
38. complexus muscle
39. medulla oblongata
40. adductor mandibulae externus muscle
41. inner ear
42. vestibulo-cochlearis nerve
43. columella
44. trigeminal (cr. nerve V) ganglion (bilobed)
45. superior oblique eye muscle
46. ophthalmic branch of trigeminal (cr. V) nerve
47. cerebral hemisphere
48. lateral ventricle
49. sulcus of hypothalamus
50. cornea
51. optic lobe
52. postorbital cartilage
53. middle ear cavity
54. tympanic membrane
55. columella in otic capsule
56. thick wall of myelencephalon
57. fourth ventricle
58. rectus capitis dorsalis muscle
59. edge of eye in orbit
60. third ventricle
61. pineal organ
62. encephalon
63. edge of sclera and retina
64. wall of mesencephalon (thalamus)

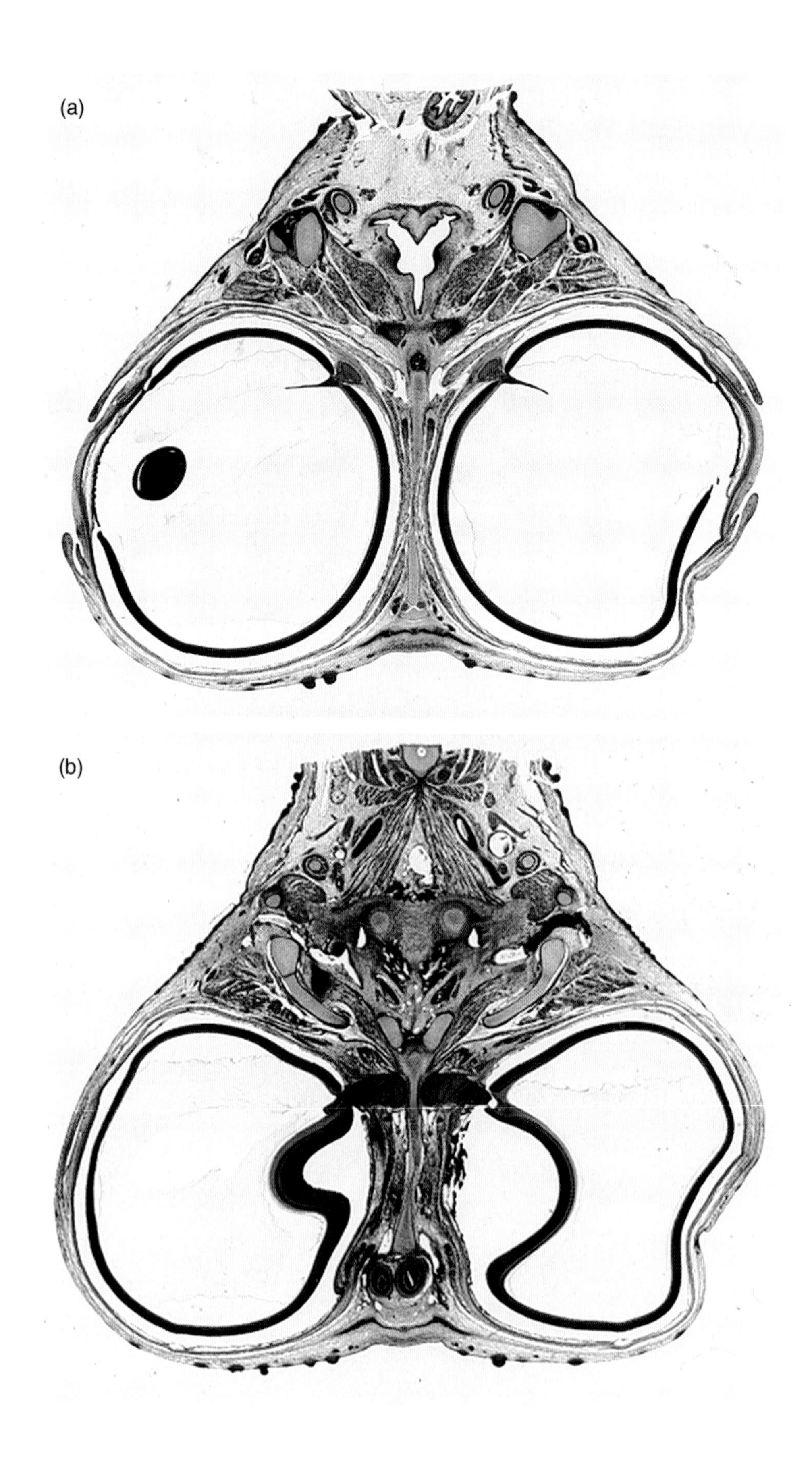
(a)
(b)

Plate 125 Day 10 TS

(a) Enlargement of section adjacent to *Plate 123c*, ×13.5.
(b) Enlargement of section adjacent to *Plate 123d*, ×13.5.

Both sections stained with Masson's trichrome. Cartilages in blue.

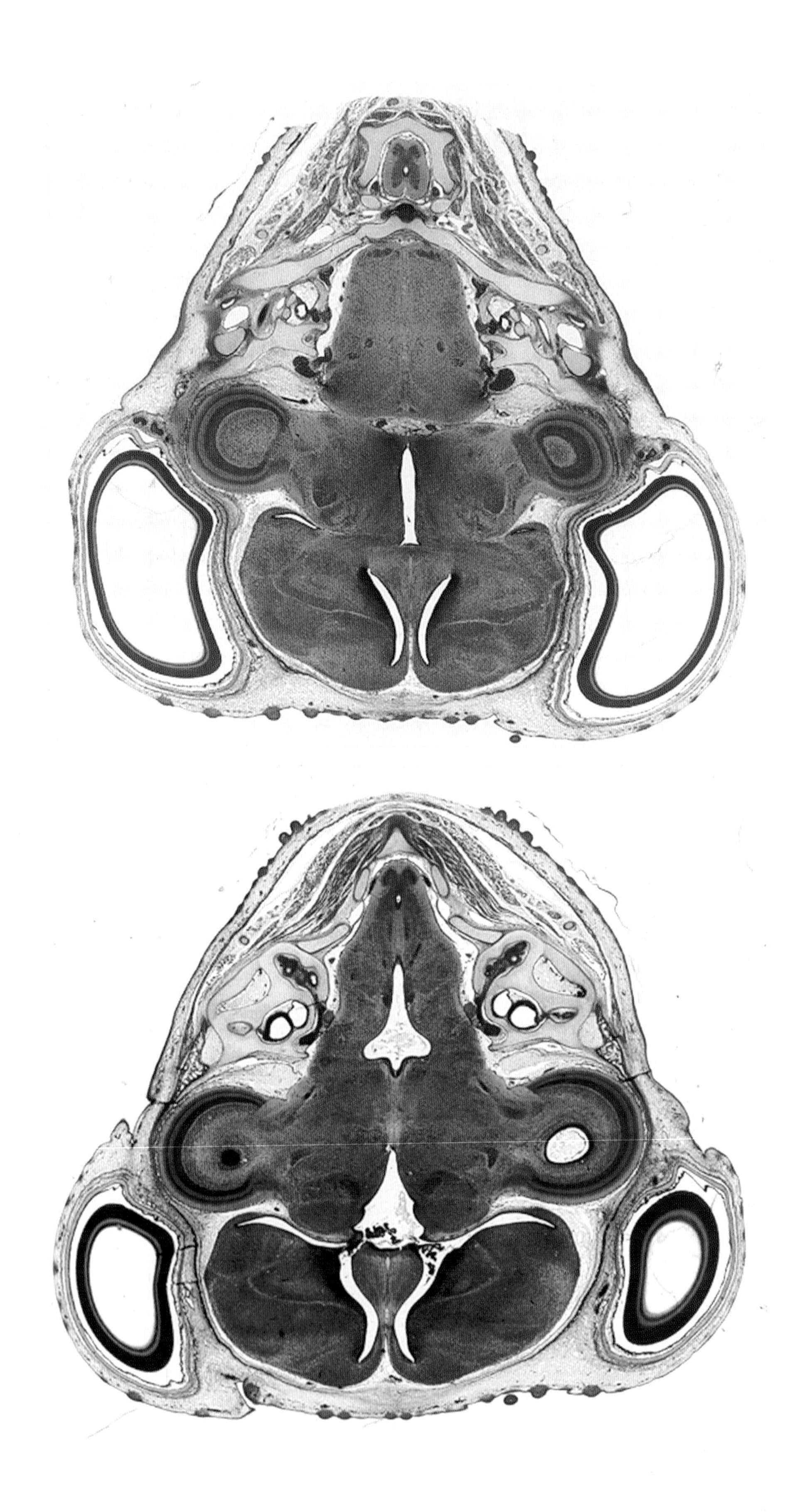

Plate 126 Day 10 TS

(a) Enlargement of section adjacent to *Plate 124c*, ×13.5.
(b) Enlargement of section adjacent to *Plate 124d*, ×13.5.

Both sections stained with Masson's trichrome. Cartilages in blue.

Plate 127

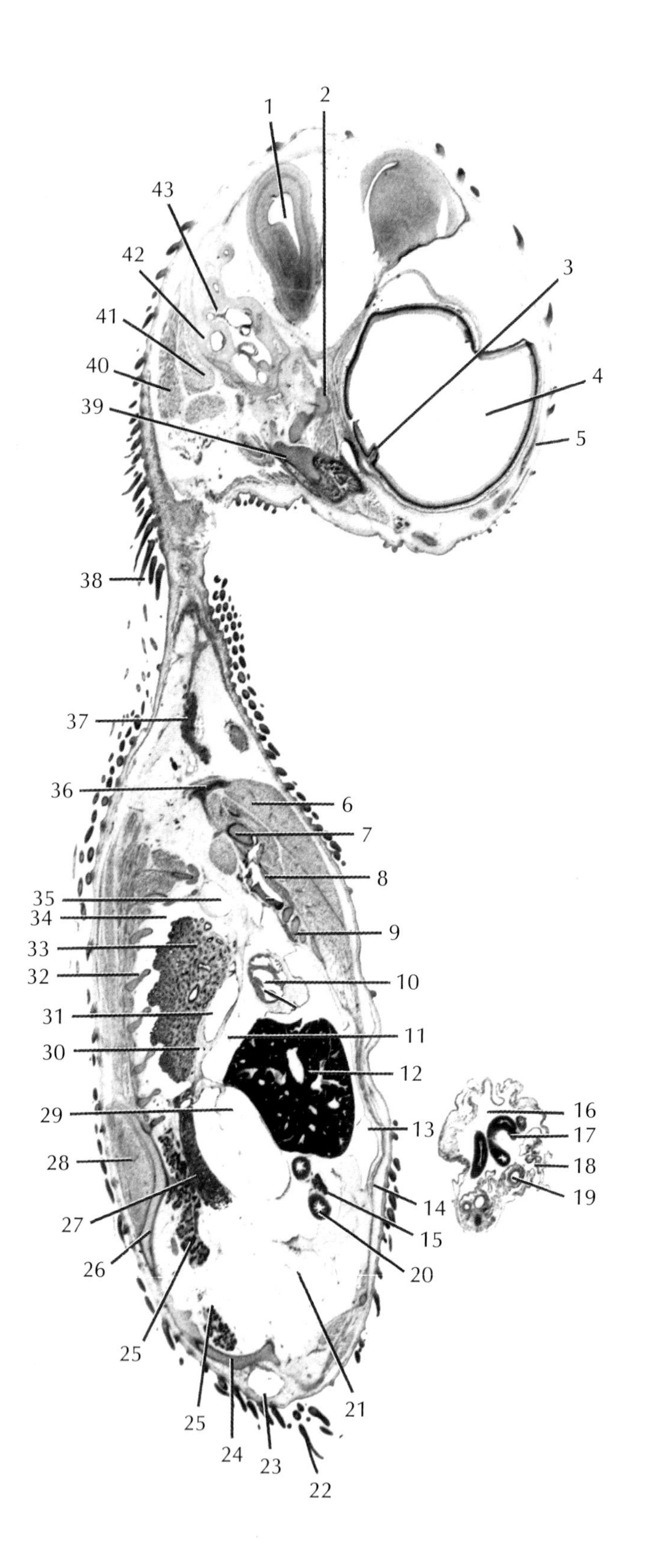

1
2
43
3
42
41
4
40
39
5
38
37
36
6
7
8
35
34
33
9
32
10
31
11
30
12
29
16
13
17
28
18
19
14
27
15
26
20
25
21
25
24
23
22

Plate 127 Day 11 LS

Sagittal section in plane of label 127 on diagram (for sagittal sections 128 and 129 see *Plates 128* and *129*, respectively), ×5.

1. optic tectum
2. epibranchial cartilage
3. pecten
4. vitreous chamber
5. scleral ossicle
6. pectoralis muscle
7. coracoid
8. sternum
9. sternal rib
10. ventricle
11. pulmonary recess
12. liver
13. ventral hepatic cavity
14. oblique internus abdominus muscle
15. liver edge
16. extra-embryonic coelom
17. herniated loop of intestine
18. omphalo-mesenteric vessels
19. allantoic (umbilical) artery
20. intestine
21. mesentery
22. tail feather
23. abdominal air sac
24. pubis
25. metanephros
26. ischium
27. mesonephros
28. iliotrochantericus muscle
29. left umbilical vein
30. oblique septum
31. anterior thoracic air sac
32. rib
33. lung
34. pleural cavity
35. interclavicular air sac
36. scapula
37. thyroid
38. neck feather
39. quadrate cartilage
40. biventer cervicis muscle
41. complexus muscle
42. otic capsule
43. semicircular canal

PLATE 128

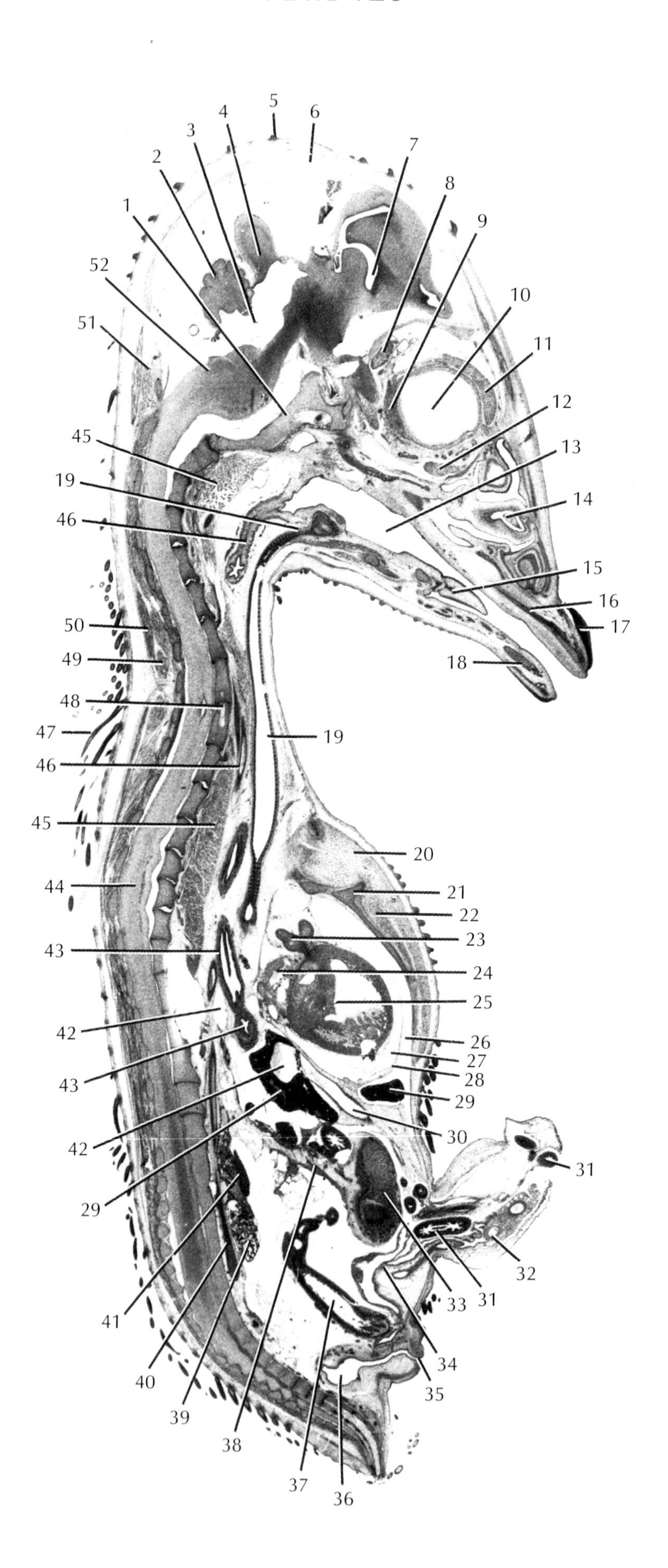

1
2
3
4
5
6
7
8
9
10
11
12
13
14
15
16
17
18
19
20
21
22
23
24
25
26
27
28
29
30
31
32
33
34
35
36
37
38
39
40
41
42
43
44
45
46
47
48
49
50
51
52

Plate 128 Day 11 LS

Sagittal section in plane of label 128 on diagram (for sagittal sections 127 and 129 see *Plates 127* and *129*, respectively), ×5.

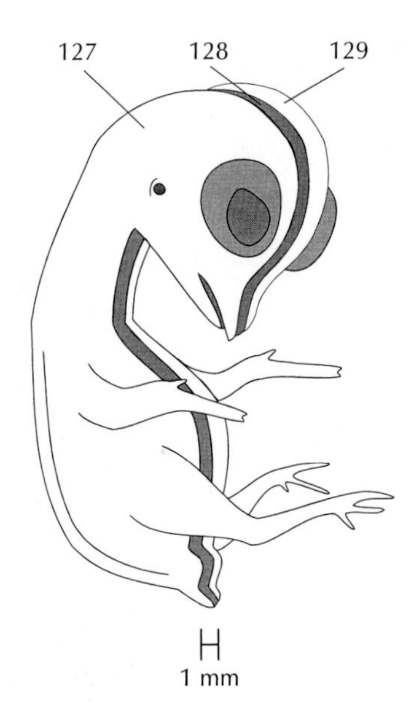

1. basal plate
2. arbor vitae of cerebellum
3. mesencephalon
4. edge of optic lobe
5. feather bud
6. frontal bone
7. cerebral hemisphere (ventricle)
8. dorsal rectus eye muscle
9. ventral rectus eye muscle
10. vitreous chamber of eye
11. dorsal oblique (superior) muscle
12. ventral oblique (inferior) muscle
13. oral cavity
14. nasal conchae
15. tongue
16. maxilla
17. egg tooth
18. dentary
19. trachea
20. pectoralis muscle
21. sternum
22. supracoracoideus muscle
23. truncus arteriosus
24. right atrium
25. right ventricle
26. sterno-cardiac diverticulum of interclavicular air sac
27. pericardial cavity
28. pericardium
29. liver
30. omphalo-mesenteric vein
31. herniated loop of intestine
32. right allantoic vein
33. gizzard wall
34. allantoic stalk
35. genital tubercle
36. cloaca
37. rectum
38. mesentery
39. metanephric kidney
40. dorsal aorta
41. testis
42. caudal (posterior) vena cava
43. intestine
44. spinal cord
45. longus colli ventralis muscle
46. oesophagus
47. feather
48. ninth cervical vertebra
49. rhomboideus supericialis muscle
50. latissimus dorsi muscle
51. adductor mandibulae externus muscle, pars temporalis
52. pontine flexure of medulla oblongata

Plate 129

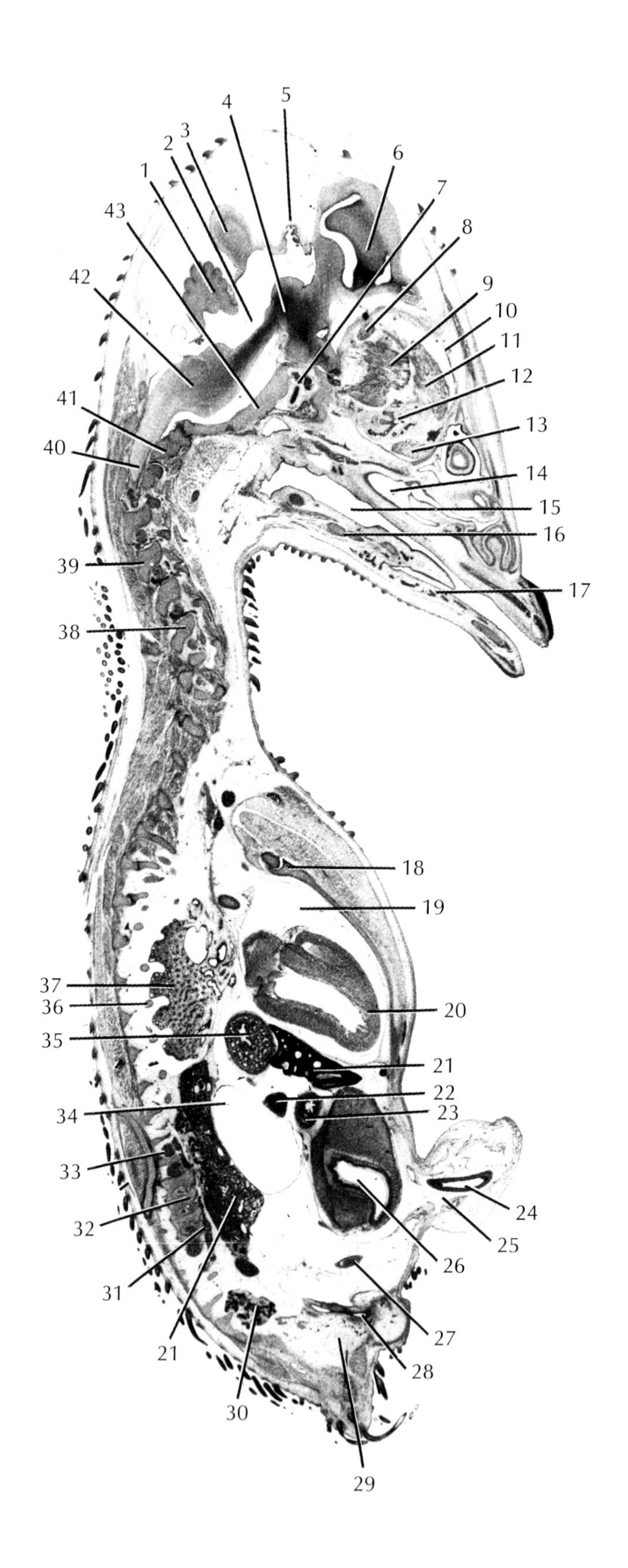

1
2
3
4
5
6
7
8
9
10
11
12
13
14
15
16
17
18
19
20
21
22
23
24
25
26
27
28
29
30
21
31
32
33
34
35
36
37
38
39
40
41
42
43

Plate 129 Day 11 LS

Sagittal section in plane of label 129 on diagram (for sagittal sections 127 and 128 see *Plates 127* and *128*, respectively), ×5.

1. cerebellum
2. mesencephalon
3. edge of optic lobe
4. floor of mesencephalon
5. pineal organ
6. cerebral hemisphere
7. hypophysis and hypophysial artery (branch of carotid)
8. dorsal rectus eye muscle
9. quadratus eye muscle
10. edge of orbit
11. dorsal oblique eye muscle
12. medial rectus eye muscle
13. ventral rectus eye muscle
14. nasal passage
15. oral cavity
16. Meckel's cartilage
17. sublingual glands
18. sternum
19. pericardial cavity
20. ventricle
21. liver
22. spleen
23. duodenum
24. herniated loop of intestine
25. umbilical cord
26. gizzard
27. intestine
28. rectum
29. wall of cloacal bursa (bursa of Fabricius)
30. metanephros
31. spinal nerve
32. lumbar vertebra
33. dorsal root ganglion
34. ventral hepatic cavity
35. proventriculus
36. thoracic rib
37. lung
38. sixth cervical vertebra
39. fourth cervical vertebra
40. cervical spinal cord
41. atlas vertebra
42. floor of medulla oblongata
43. basal plate of chondrocranium

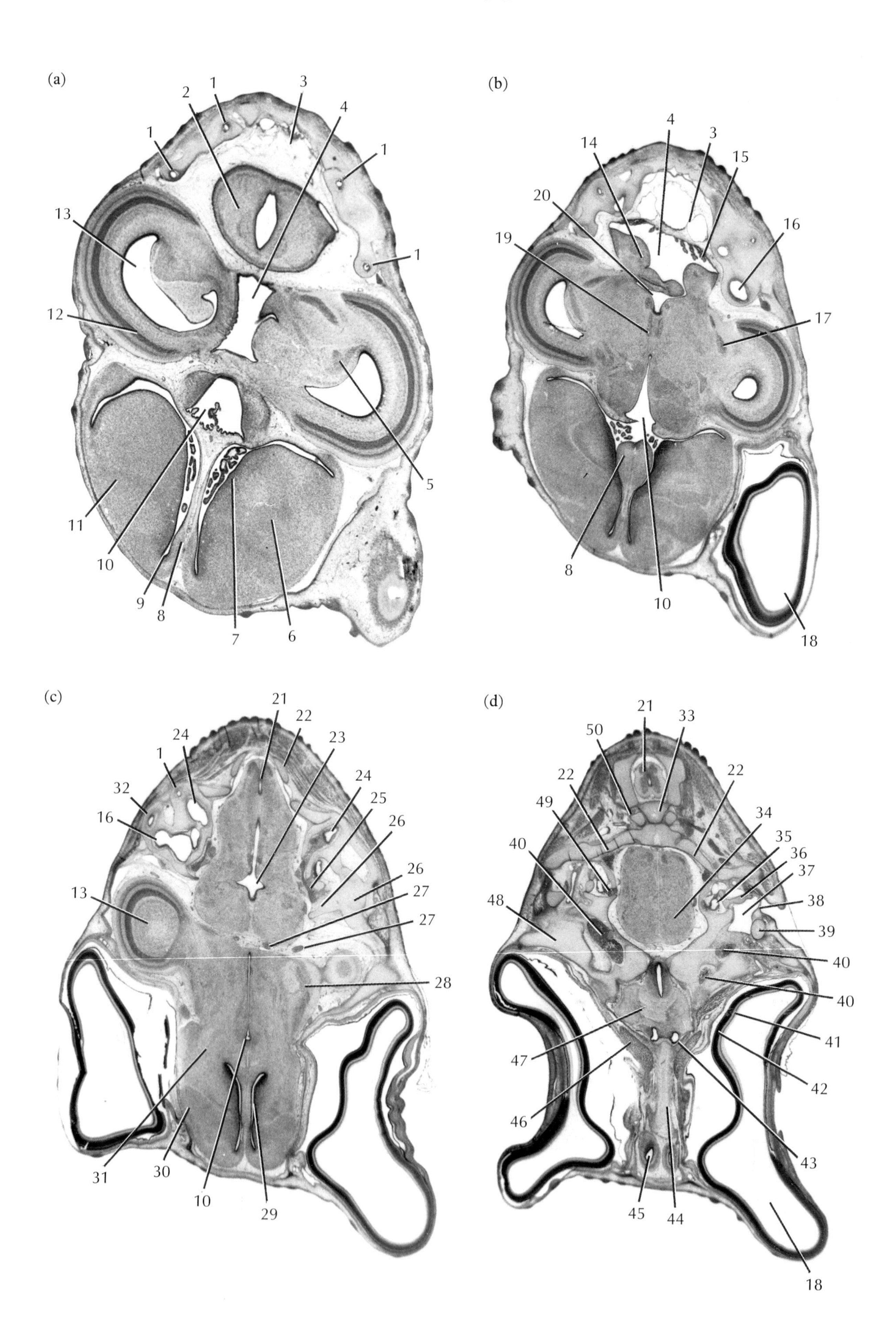
(a)
1
2
1
3
4
1
1
13
1
12
5
11
10
9
8
7
6
(b)
4
14
3
15
20
16
19
17
8
10
18
(c)
21
22
24
23
1
24
32
25
16
26
26
13
27
27
28
31
30
10
29
(d)
21
33
50
22
22
49
34
40
35
36
37
48
38
39
40
40
41
47
42
46
43
45
44
18

Plate 130 Day 11 TS

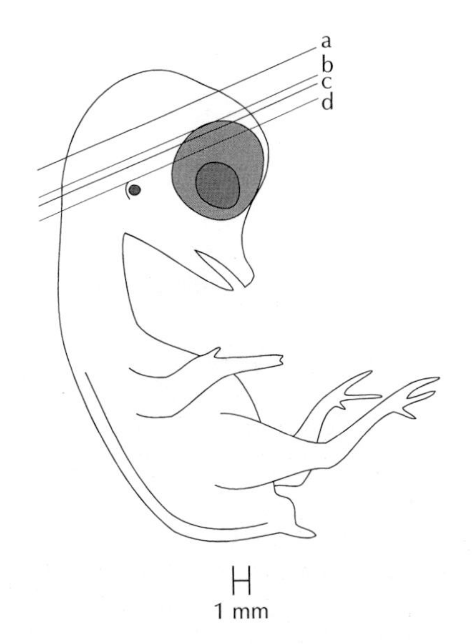

(a–d) × 12

1. foramina for vena emissaria occipitalis
2. cerebellum
3. cranial cavity
4. fourth ventricle
5. nucleus mesencephalicus lateralis
6. corpus striatum
7. left cerebral hemisphere
8. region of contact of left and right hemispheres
9. anterior horn of right lateral ventricle
10. third ventricle
11. right cerebral hemisphere
12. laminated wall of right optic lobe
13. lumen of optic lobe
14. metencephalon (pons)
15. lateral recess of fourth ventricle
16. utriculus
17. left optic (cr. II) nerve
18. left eye (collapsed in preparation)
19. floor of mesencephalon
20. mesencephalic aqueduct (aqueduct of Sylvius)
21. spinal cord
22. tectum synoticum
23. central canal
24. endolymphatic duct
25. vestibular nerve
26. auditory capsule
27. branches of oculomotor (cr. III) nerve
28. optic thalamus
29. lateral ventricle
30. archistriatum
31. palaeostriatum
32. semicircular canal
33. odontoid process (body of axis)
34. medulla oblongata
35. membranous labyrinth
36. columella
37. middle ear cavity
38. tympanic membrane
39. epibranchial cartilage
40. lobes of trigeminal (cr. nerve V) ganglion
41. retina
42. pigmented layer of retina
43. hypophysial fenestra
44. interorbital septum
45. olfactory bulb
46. dorsal (superior) oblique eye muscle
47. hypophysis
48. postorbital cartilage
49. vestibular ganglion of acoustic (cr. VIII) nerve
50. occipital condyle

Plate 131

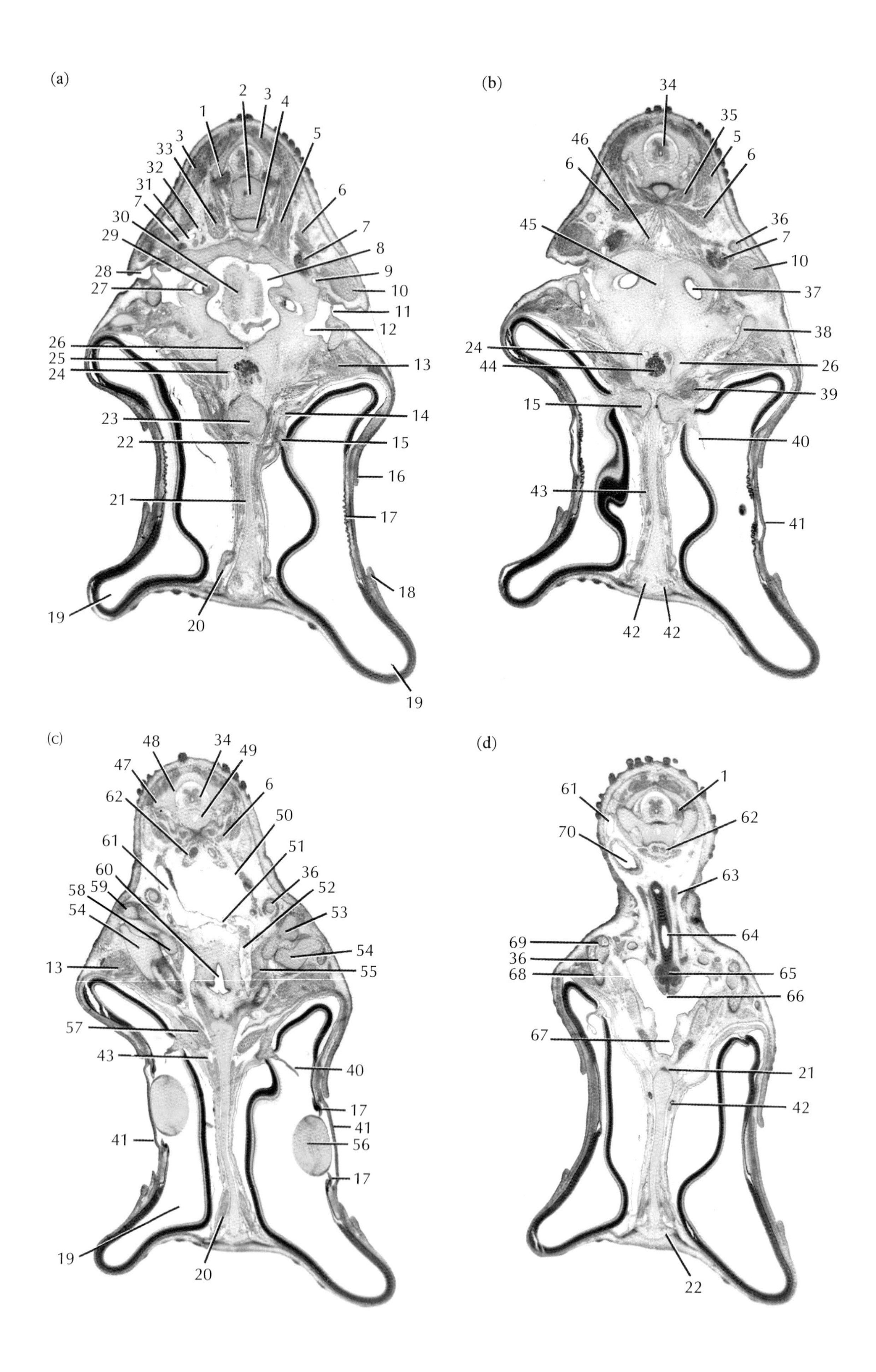

(a)
1
2
3
4
5
6
7
8
9
10
11
12
13
14
15
16
17
18
19
20
21
22
23
24
25
26
27
28
29
30
31
32
33
(b)
34
35
36
37
38
39
40
41
42
43
44
45
46
(c)
47
48
49
50
51
52
53
54
55
56
57
58
59
60
61
62
(d)
63
64
65
66
67
68
69
70

Plate 131 Day 11 TS

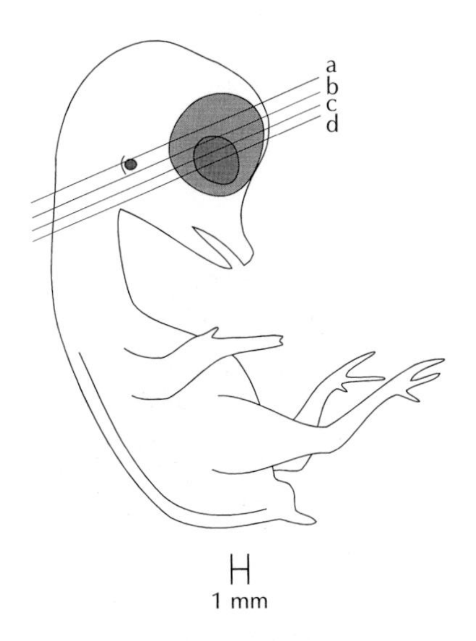

(a–d) × 10

1. dorsal root ganglion
2. axis
3. complexus muscle
4. odontoid process (body of axis)
5. splenius capitis muscle
6. rectus capitis lateralis muscle
7. vagus (cr. X) nerve
8. foramen magnum
9. stapedial artery
10. depressor mandibulae muscle
11. tympanic membrane
12. middle ear
13. adductor mandibulae externus muscle
14. left dorsal oblique muscle of eye
15. optic (cr. II) nerve
16. lower eyelid
17. ciliary muscle
18. upper eyelid
19. vitreous humour of the eye
20. right dorsal oblique muscle of eye
21. interorbital septum
22. tectal process
23. optic chiasma
24. right internal carotid artery
25. abducens (cr. VI) nerve
26. acrochordal cartilage
27. lagena
28. external auditory meatus
29. acoustic ganglion (cr. VIII nerve) (vestibulocochlearis)
30. medulla oblongata
31. occipital vein
32. occipital artery
33. flexor colli lateralis muscle
34. spinal cord
35. flexor colli medialis muscle
36. epibranchial cartilage of hyoid
37. cochlea canal
38. otic process of quadrate cartilage
39. ciliary ganglion
40. pecten
41. cornea
42. olfactory (cr. I) nerve
43. quadratus muscle
44. hypophysis
45. remains of notochord
46. right rectus capitis ventralis muscle
47. transverse process of cervical vertebra
48. neural arch of cervical vertebra
49. body of vertebra with remnant of notochord
50. connective tissue
51. interjugular anastomosis (anastomotic vein)
52. left head vein
53. Meckel's cartilage
54. quadrate cartilage
55. brachio-mandibular muscle
56. lens
57. ventral rectus eye muscle
58. retro-articular process of Meckel's cartilage
59. angular process of Meckel's cartilage
60. pharynx
61. head vein (jugular)
62. common carotid artery
63. sternotracheal muscle
64. trachea
65. larynx
66. glottis
67. base of choana
68. dentary
69. splenial bone
70. crop

Plate 132

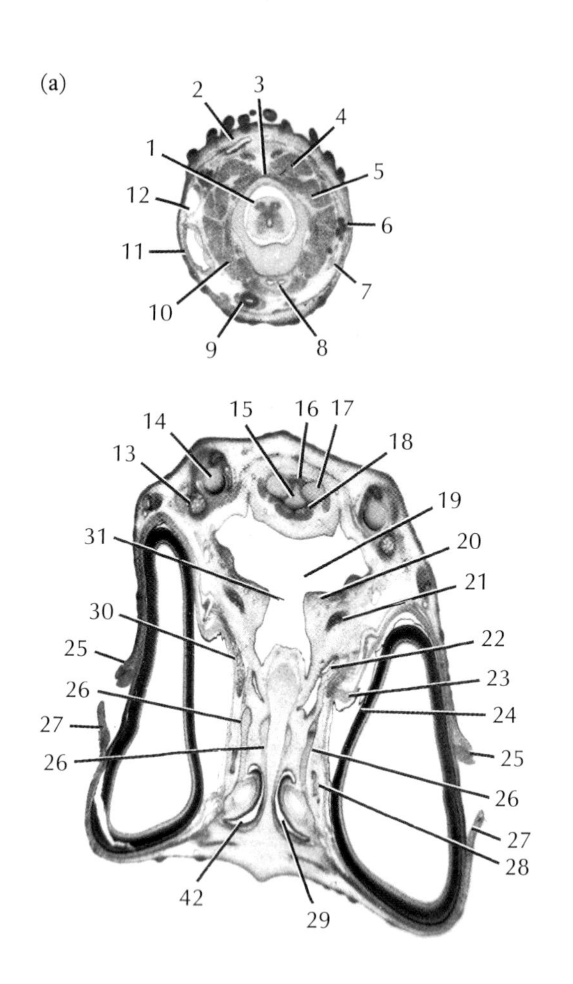

(a)
2
3
4
1
5
12
6
11
7
10
9
8
14
15
16
17
18
13
19
31
20
21
30
22
25
23
26
24
27
25
26
26
27
28
42
29

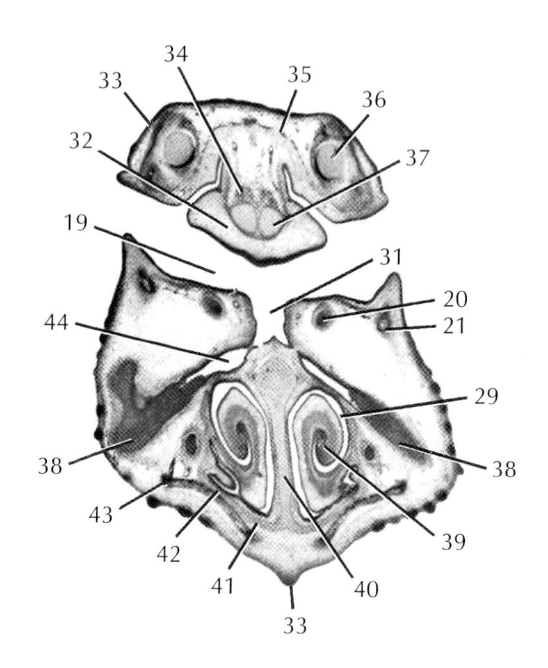

(b)
34
35
33
36
32
37
19
31
20
44
21
29
38
38
43
39
42
41
40
33

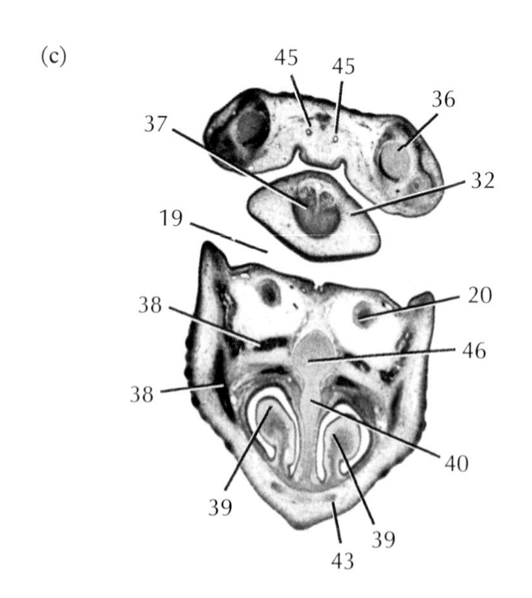

(c)
45
45
36
37
32
19
20
38
46
38
40
39
39
43

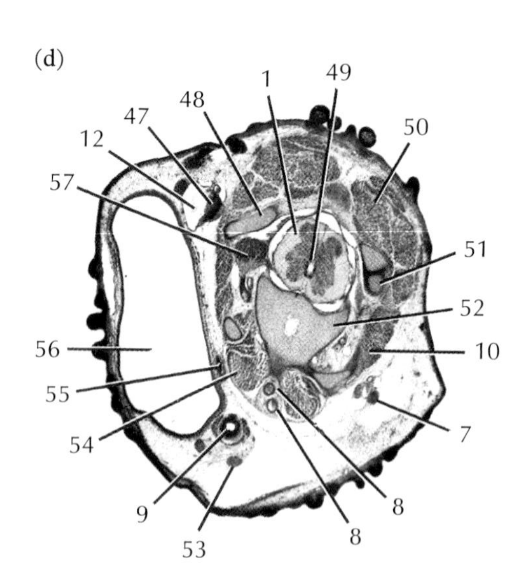

(d)
1
49
47
48
50
12
57
51
52
56
10
55
54
7
9
8
53
8

Plate 132 Day 11 TS

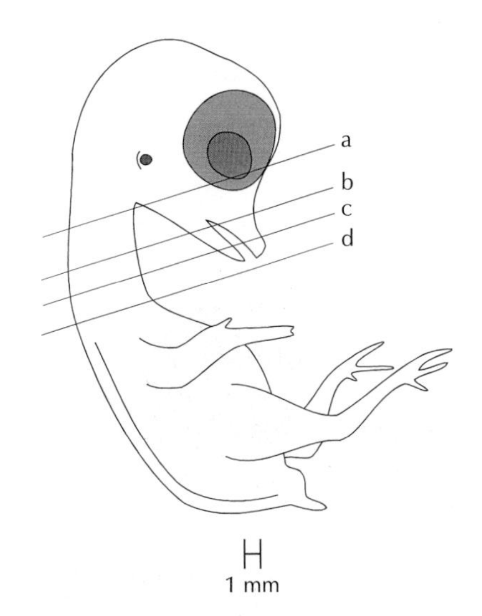

(a–d) ×12

1. spinal cord, cervical region (white matter)
2. branch of right jugular vein
3. neural arch of cervical vertebra
4. splenius capitis lateralis muscle
5. rectus capitis dorsalis muscle
6. lobe of left thymus
7. vagus (cr. X) nerve
8. left and right common carotid arteries
9. trachea
10. intertransversarii muscle
11. crop
12. right jugular vein
13. quadrate cartilage (otic process)
14. articular cartilage (posterior end of Meckel's cartilage)
15. basihyal (copula I) cartilage
16. paraglossoceratobranchial muscle
17. ceratobranchial cartilage
18. hypobranchial muscles
19. oral cavity
20. palatine bone
21. maxillary cartilage
22. palatine sinus
23. optic (cr. II) nerve
24. sclera
25. lower eyelid
26. wall of nasal capsule
27. upper eyelid
28. olfactory nerve
29. left nasal meatus
30. orbital cartilage
31. median palatine fissure
32. tongue
33. epidermis
34. hypoglossus oblique lateralis muscle
35. adductor mandibulae muscle
36. Meckel's cartilage
37. paraglossal cartilage
38. wall of lacrimal duct
39. nasal concha
40. interorbital (nasal) septum
41. roof of nasal capsule
42. right nasal meatus
43. premaxilla
44. lumen of lacrimal duct
45. mandibular artery
46. base of nasal septum
47. right vagus nerve and artery
48. cervical rib
49. neural canal
50. ascendens muscle
51. zygapophysis of adjacent vertebra
52. body of cervical vertebra
53. tracheal branch of vagus nerve
54. longus colli ventralis muscle
55. crop branch of vagus nerve
56. crop
57. dorsal root ganglion

Plate 133

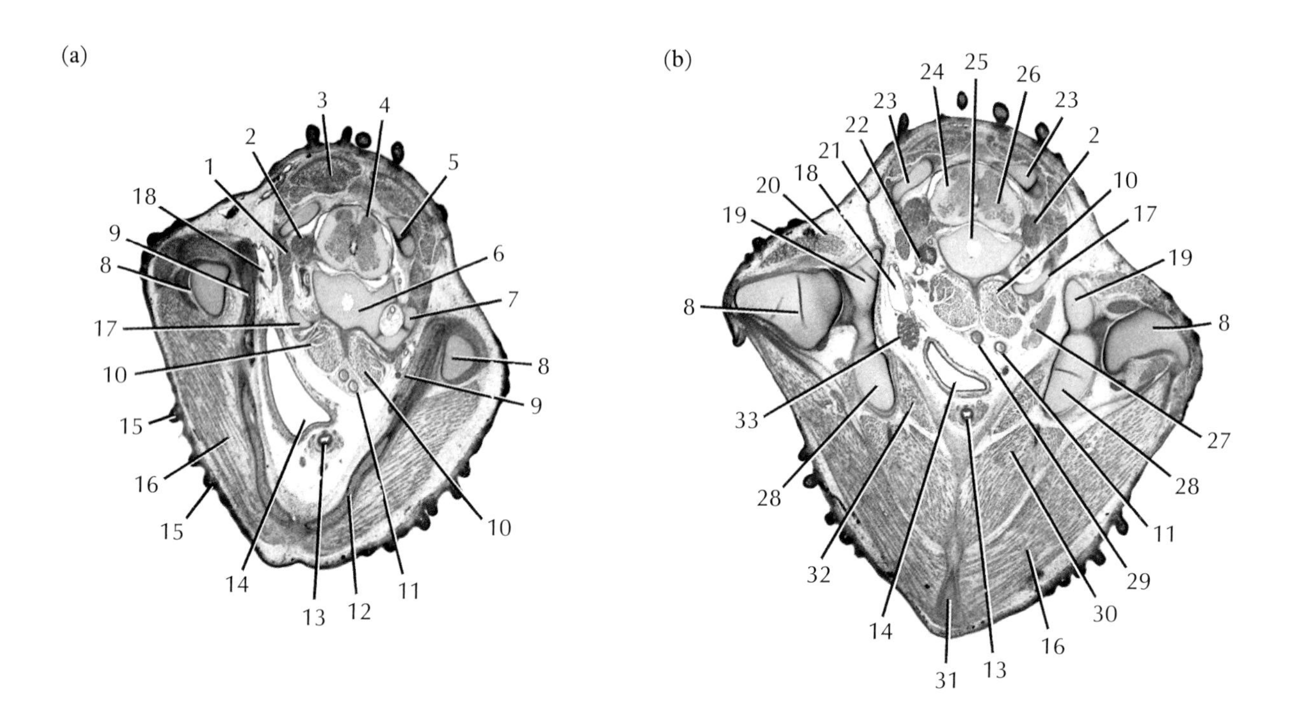
(a)
3
4
2
5
1
18
9
8
6
7
17
8
10
9
15
16
15
10
14
13
12
11
(b)
24
25
26
23
23
22
2
21
18
10
20
17
19
19
8
8
33
27
28
28
32
11
29
14
30
16
31
13

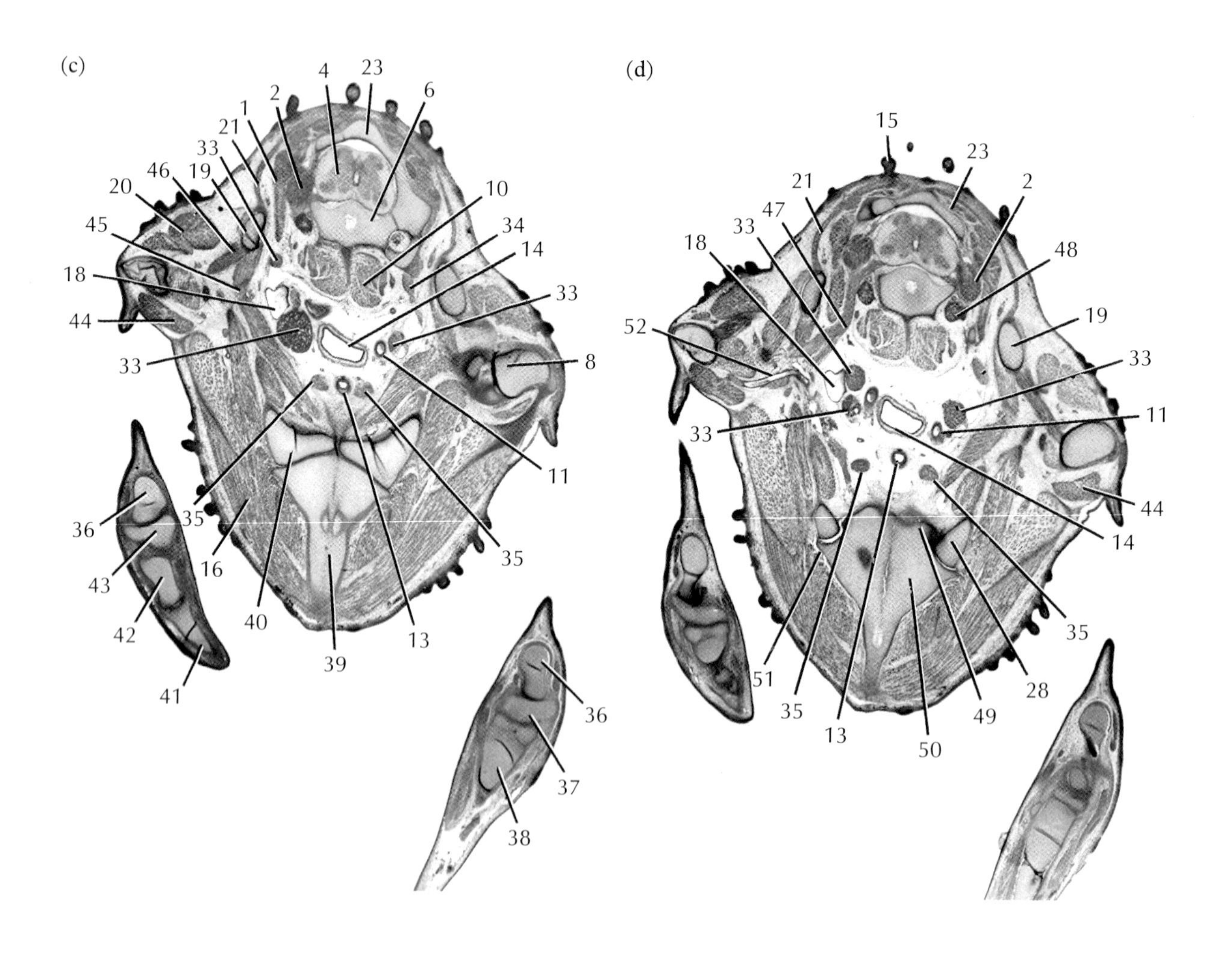
(c)
4
23
6
2
1
21
33
46
19
20
10
34
14
45
33
18
44
8
33
11
36
35
16
35
43
42
40
39
13
41
36
37
38
(d)
15
23
21
2
47
33
48
18
19
52
33
33
11
44
14
35
51
35
13
50
49
28

Plate 133 Day 11 TS

(a–d) × 12

1. intertransversarii muscle
2. dorsal root ganglion
3. biventer cervicis muscle
4. spinal cord
5. neural arch of cervical vertebra
6. body of vertebra
7. cervical rib
8. humerus
9. coracobrachialis caudalis muscle
10. longus colli ventralis muscle
11. left common carotid artery
12. clavicle
13. trachea
14. oesophagus
15. feather bud
16. pectoralis muscle pars thoracicus
17. transverse process of last cervical vertebra
18. right cranial (anterior) cardinal vein
19. scapula
20. triceps brachii muscle, pars humerotriceps
21. rhomboideus superficialis muscle
22. vertebral artery
23. neural arch of thoracic vertebra
24. spinal cord (white matter)
25. notochord in vertebral body
26. spinal cord (grey matter)
27. left anterior vena cava
28. coracoid
29. right common carotid artery
30. supracoracoid muscle
31. costal process of sternum
32. sternocoracoideus muscle
33. lobes of thymus
34. spinal nerve component of brachial plexus
35. thyroid
36. ulna
37. left ulnacarpal
38. carpo-metacarpus
39. keel of sternum
40. coracoid abutting sternum
41. tip of right phalanx III
42. metacarpal III
43. right ulnacarpal
44. biceps brachii muscle
45. junction of subscapularis muscle and subcoracoideus muscle
46. subscapularis externa muscle
47. ventral ramus of spinal nerve contributing to brachial plexus
48. sympathetic ganglion
49. rostrum of sternum
50. sternum

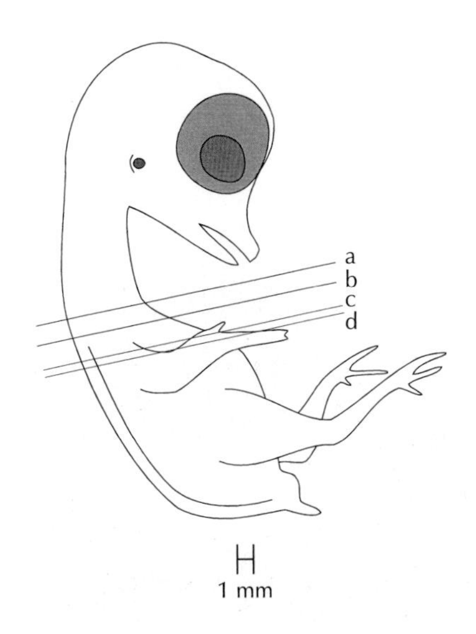

51. costal process of sternum
52. subclavian artery

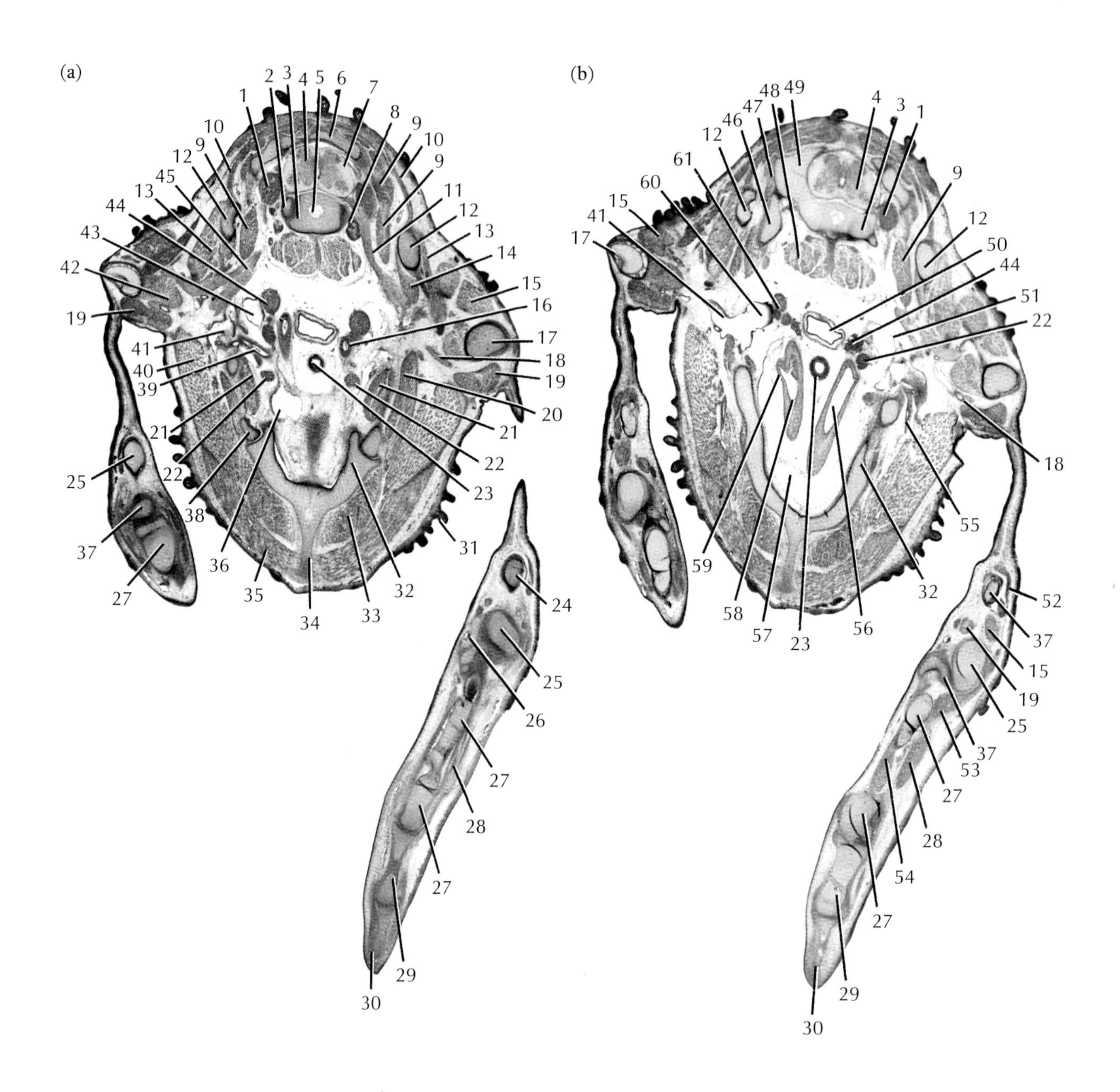
(a)
1
2
3
4
5
6
7
8
9
10
11
12
13
14
15
16
17
18
19
20
21
22
23
24
25
26
27
28
29
30
31
32
33
34
35
36
37
38
39
40
41
42
43
44
45
(b)
46
47
48
49
50
51
52
53
54
55
56
57
58
59
60
61

Plate 134 Day 11 TS

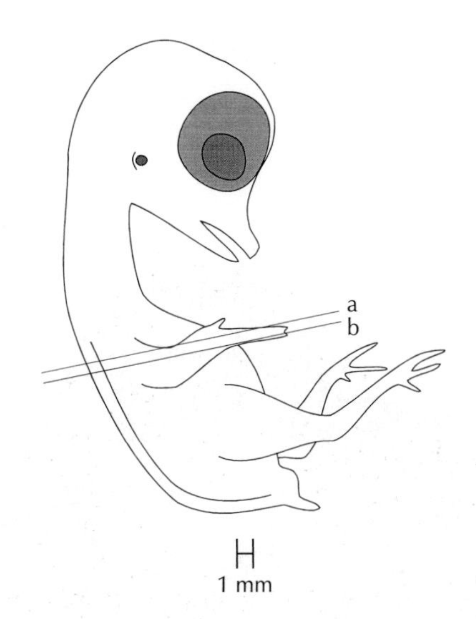

(a, b) × 12

1. dorsal root ganglion
2. capitulum of costa
3. body of vertebra
4. spinal cord (grey matter)
5. notochord
6. neural arch of vertebra
7. spinal cord (white matter)
8. sympathetic ganglion
9. spinal thoracic muscle complex
10. rhomboideus superficialis muscle
11. ventral ramus of spinal nerve 16, contributing to left brachial plexus
12. scapula
13. scapulohumeralis lateralis muscle
14. subscapularis muscle
15. triceps brachii pars humerotriceps muscle
16. left common carotid artery
17. humerus
18. subclavian artery
19. biceps brachii muscle
20. intercostalis externus muscle
21. intercostalis internus muscle
22. thyroid
23. syrinx
24. radius
25. ulna
26. ulnar artery
27. metacarpus
28. extensor medius brevis muscle
29. phalanx
30. digit III
31. feather bud
32. sternum
33. supracoracoideus muscle
34. keel of sternum
35. pectoralis muscle pars thoracicus
36. anterior tip of pericardial cavity
37. radial carpal
38. coracoid
39. subclavian artery
40. coracobrachialis muscle
41. subclavian vein
42. deltoid muscle
43. right anterior cardinal vein
44. thymus
45. part of brachial plexus
46. tuberculum of costa (rib)
47. intertransversarii muscle
48. longus colli ventralis muscle
49. thoracic vertebra
50. oesophagus
51. left anterior cardinal vein
52. left tensor propatagialis brevis muscle
53. brachialis muscle
54. flexor digitorum profundus muscle
55. pectoral artery
56. pulmonary artery
57. pericardial cavity
58. right brachiocephalic artery
59. root of subclavian artery
60. jugular vein
61. lobules of thymus

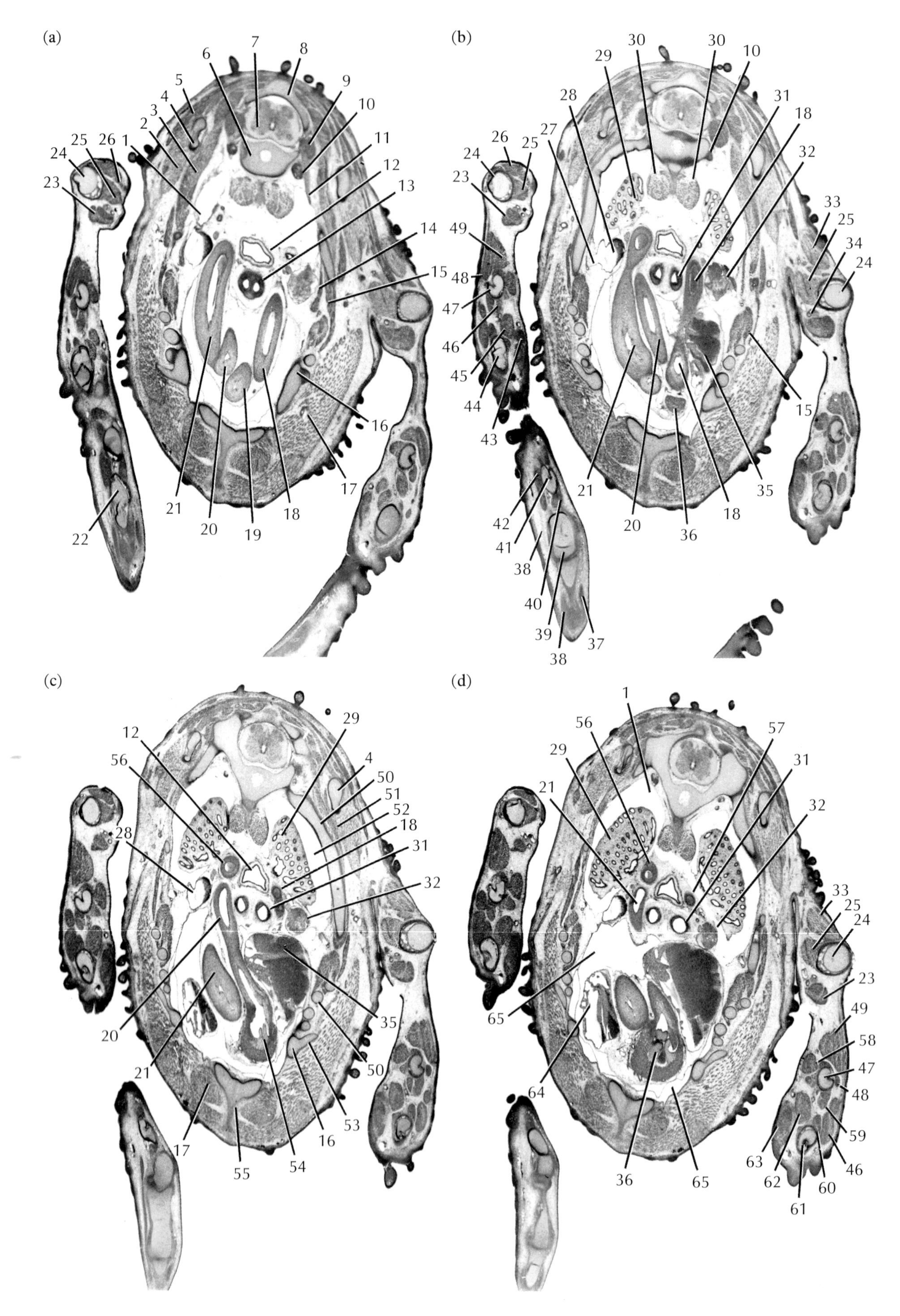
(a)
1
2
3
4
5
6
7
8
9
10
11
12
13
14
15
16
17
18
19
20
21
22
23
24
25
26
(b)
10
15
18
20
21
23
24
25
26
27
28
29
30
31
32
33
34
35
36
37
38
39
40
41
42
43
44
45
46
47
48
49
(c)
4
12
16
17
18
20
21
28
29
31
32
35
50
51
52
53
54
55
56
(d)
1
21
23
24
25
29
31
32
33
36
46
47
48
49
56
57
58
59
60
61
62
63
64
65

Plate 135 Day 11 TS

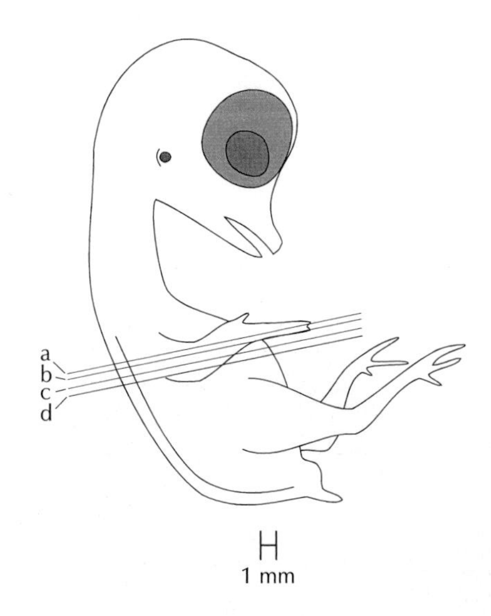

(a–d) × 12

1. right pleural cavity
2. scapulohumeralis muscle
3. serratus profundus muscle
4. scapula
5. latissimus dorsi muscle
6. body of vertebra
7. spinal cord
8. vertebral arch
9. dorsal root ganglion
10. sympathetic nerve ganglion
11. mixed nerve
12. oesophagus
13. bronchus, immediately posterior to syrinx
14. intercostalis internus muscle
15. intercostalis externus muscle
16. sternum
17. supracoracoideus muscle
18. pulmonary artery
19. left brachiocephalic artery
20. right brachiocephalic artery
21. aortic arch
22. carpo-metacarpus
23. biceps brachii muscle
24. humerus
25. triceps brachialis (anconeus) muscle, pars humerotriceps
26. deltoid muscle
27. pectoral vein
28. right cranial (anterior) vena cava
29. lung
30. longus colli ventralis muscle
31. bronchus
32. left cranial (anterior) vena cava
33. triceps brachialis (anconeus) muscle, pars scapulotriceps
34. brachial artery and median ulna nerve
35. left atrium
36. truncus arteriosus
37. flexor digitorum muscle, profundus tendon
38. superficialis muscle tendon
39. carpo-metacarpus
40. right flexor carpi ulnaris muscle
41. ulnar carpal
42. interosseus dorsalis muscle
43. metacarpiradialis muscle
44. radius
45. ectepicondylo-ulnaris muscle
46. extensor metacarpi ulnaris muscle (attachment to ulna)
47. ulna
48. extensor digitorum communis muscle
49. extensor metacarpi ulnaris muscle
50. rib
51. serratus profundus muscle
52. left pleural cavity
53. xiphisternal process of sternum
54. left ventricle
55. keel of sternum
56. right common carotid artery
57. left common carotid artery
58. flexor carpi ulnaris muscle
59. ulnimetacarpalis ventralis muscle
60. extensor indicis longus muscle
61. radius
62. flexor digitorum profundus muscle
63. pronator profundus muscle
64. right atrium
65. pericardial cavity

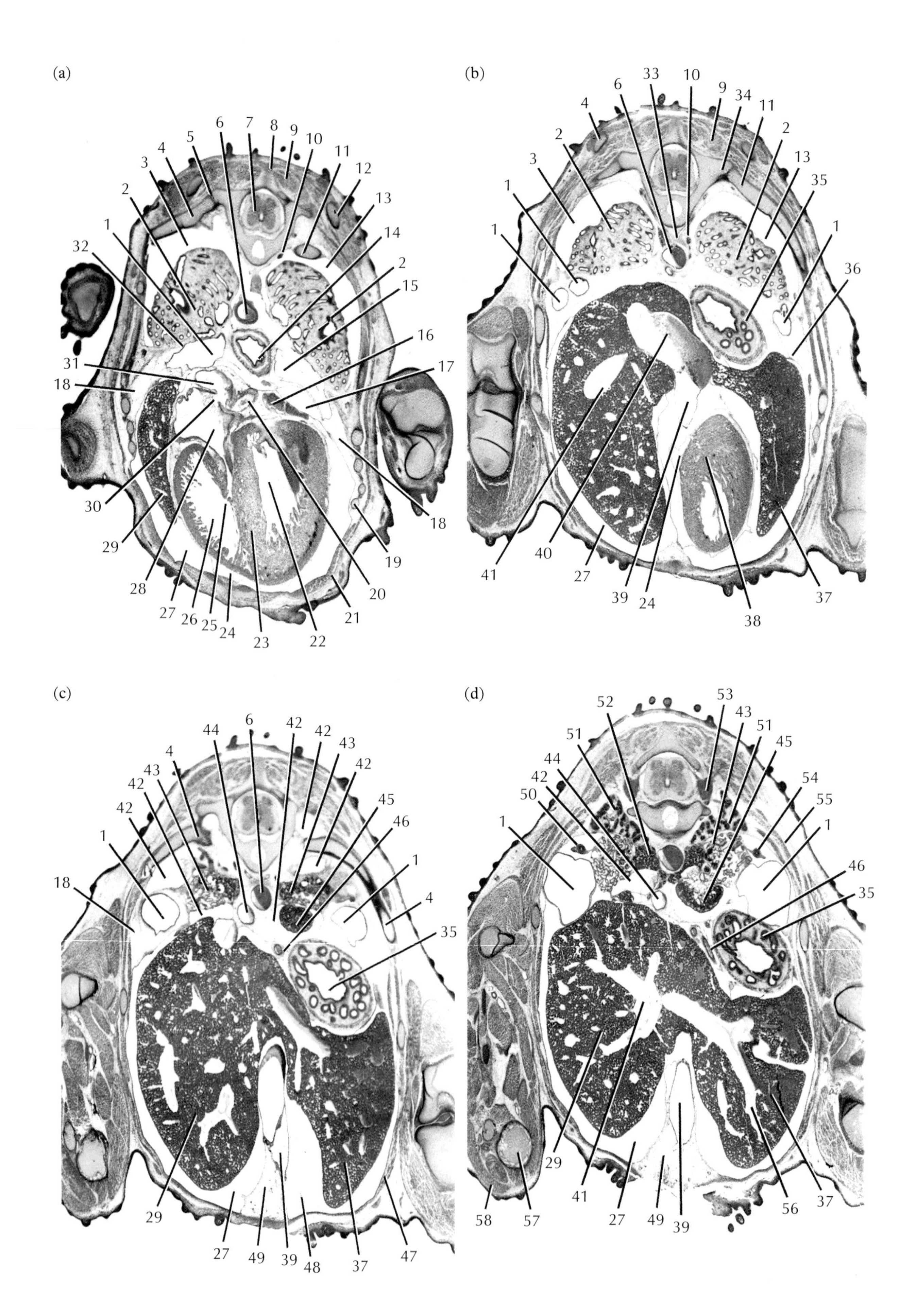
(a)
1 2 3 4 5 6 7 8 9 10 11 12 13 14 2 15 16 17 18 19 20 21 22 23 24 25 26 27 28 29 30 31 18 32
(b)
4 2 3 1 1 6 33 10 9 34 11 2 13 35 1 36 37 38 24 39 27 40 41
(c)
44 6 42 42 43 42 45 46 1 4 35 4 43 42 42 1 18 29 27 49 39 48 37 47
(d)
52 53 43 51 45 54 55 1 51 44 42 50 1 46 35 37 56 39 49 27 41 29 57 58

Plate 136 Day 11 TS

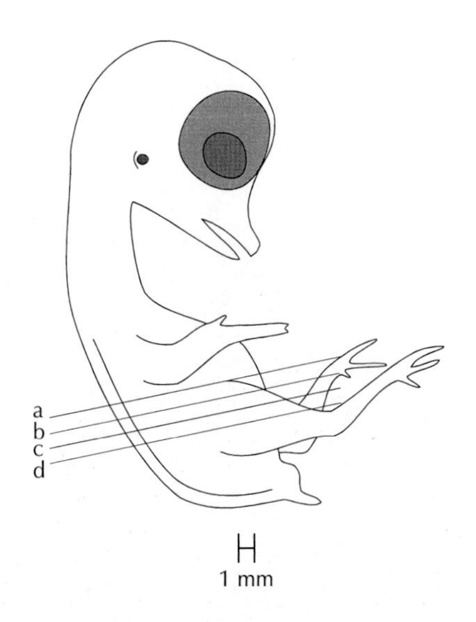

(a–d) × 12

1. abdominal air sac
2. lung
3. right pleural cavity
4. rib
5. articulation of tuberculum of rib with transverse process
6. dorsal aorta
7. spinal cord (grey matter)
8. spinal process of vertebra
9. rhomboideus superficialis muscle
10. sympathetic nerve trunk
11. capitulum of costa (articulates with transverse process)
12. posterior end of scapula
13. left pleural cavity
14. oesophagus
15. left lateral thoracic air sac
16. left atrium
17. dorsal hepatic cavity
18. oblique septum
19. sternal rib
20. cranial (anterior) vena cava
21. posterior end of pectoralis muscle
22. lumen of left ventricle
23. interventricular septum
24. pericardial cavity
25. atrio-ventricular valve
26. lumen of right ventricle
27. right ventral hepatic cavity
28. lumen of right atrium
29. right lobe of liver
30. sinu-atrial canal
31. sinus venosus
32. horizontal septum
33. dorsal mesentery
34. transverse process of thoracic vertebra
35. proventriculus
36. left horizontal hepatic ligament
37. left lobe of liver
38. wall of left ventricle
39. left umbilical vein
40. caudal (posterior) vena cava
41. right hepatic vein
42. intestinal coelomic cavity
43. mesonephros
44. omphalo-mesenteric artery
45. left ovary
46. coeliac artery
47. rectus abdominis muscle
48. left ventral hepatic cavity
49. ventral mesentery (falciform ligament)
50. right oviduct (Mullerian duct)
51. metanephros
52. adrenal gland
53. dorsal root ganglion
54. left oviduct (Mullerian duct)
55. intercostalis externus muscle
56. hepatic vein
57. tibiotarsus of right leg
58. gastrocnemius muscle

Plate 137

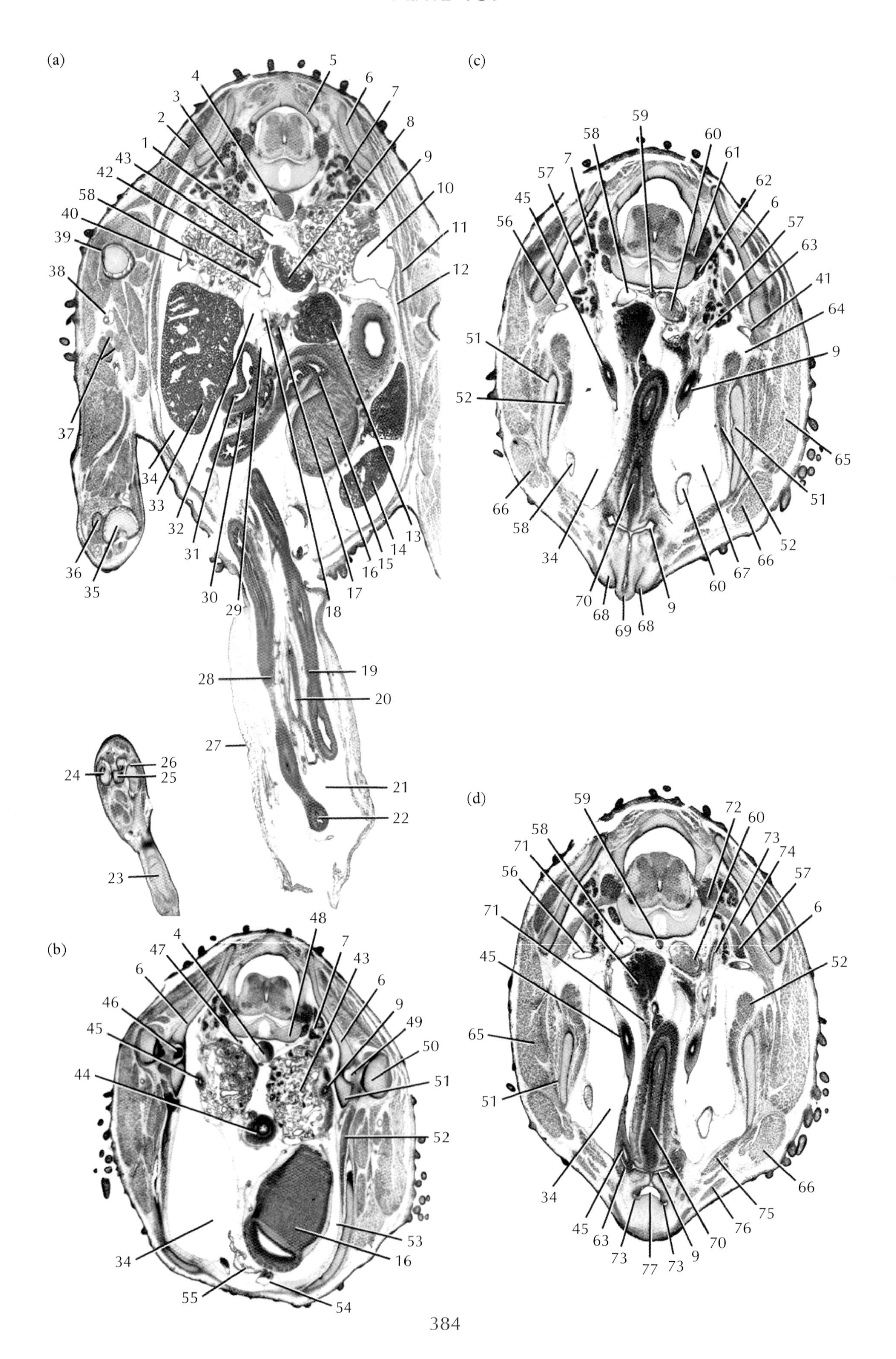

Plate 137 Day 11 TS

(a) × 12; (b) × 10; (c, d) × 15

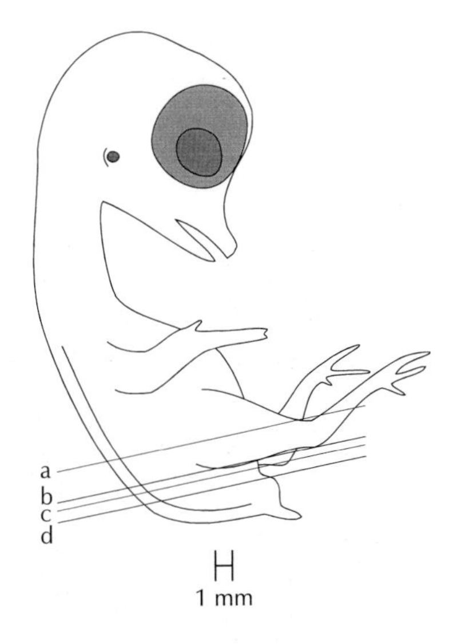

1. caudal vena cava
2. iliotibialis muscle
3. renal vein
4. dorsal aorta
5. neural arch
6. ilium
7. metanephros
8. left ovary
9. left oviduct (Mullerian duct)
10. abdominal air sac
11. oblique externus abdominis muscle
12. oblique internus abdominis muscle
13. spleen
14. left posterior lobe of liver
15. descending duodenum
16. gizzard
17. dorsal pancreas
18. mesenteric artery
19. descending limb of intestine
20. omphalo-mesenteric vein
21. extraembryonic coelom
22. herniated loop of intestine
23. metatarsus
24. digit IV of right foot
25. digit III of right foot
26. digit II of right foot
27. amniotic covering of umbilical cord
28. ascending limb of intestine
29. common mesenteric vein
30. ventral pancreas
31. ascending duodenum
32. gastropancreaticduodenal vein
33. right lobe of liver
34. right ventral hepatic cavity
35. tibiotarsus
36. fibula
37. branches of ischiatic nerve
38. femoral artery and iliac vein
39. femur
40. posterior tip of right abdominal air sac
41. left metanephric artery
42. remnant of right ovary
43. mesonephros
44. intestine
45. right oviduct (Mullerian duct)
46. head of femur in acetabulum
47. sympathetic nerve
48. body of lumbar vertebra
49. acetabulum
50. trochanter of femur
51. ischium
52. obturator muscle
53. wall of ventral hepatic cavity
54. allantoic stalk
55. yolk sac stalk
56. right metanephric artery
57. renal portal vein
58. right omphalo-mesenteric artery
59. caudal artery (caudal end of dorsal aorta)
60. left omphalo-mesenteric artery
61. spinal nerve
62. sympathetic nerve ganglion
63. nephric (mesonephric, Wolffian) duct
64. foramen of ilioischiadicum
65. iliotibialis muscle
66. flexor cruris lateralis (semitendinosus) muscle
67. left ventral hepatic cavity
68. genital fold
69. cloacal ridge
70. rectum
71. clot of extravasated blood (artefact of preparation)
72. dorsal root ganglion
73. metanephric duct
74. obturator nerve
75. pubocaudalis internus muscle
76. pubocaudalis externus muscle
77. cloaca

PLATE 138

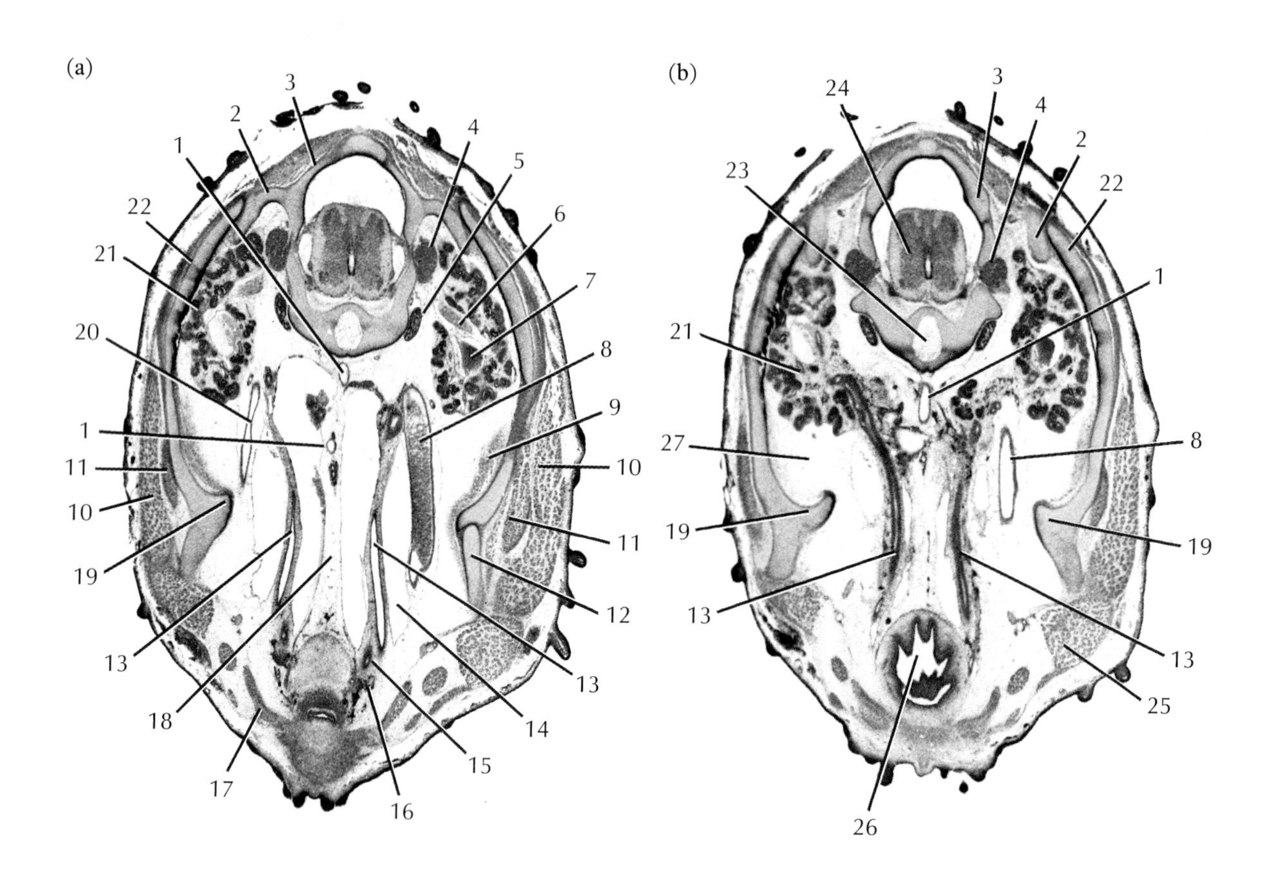
(a)
1
2
3
4
5
6
7
8
9
10
11
12
13
14
15
16
17
18
19
20
21
22
(b)
1
2
3
4
8
13
19
21
22
23
24
25
26
27

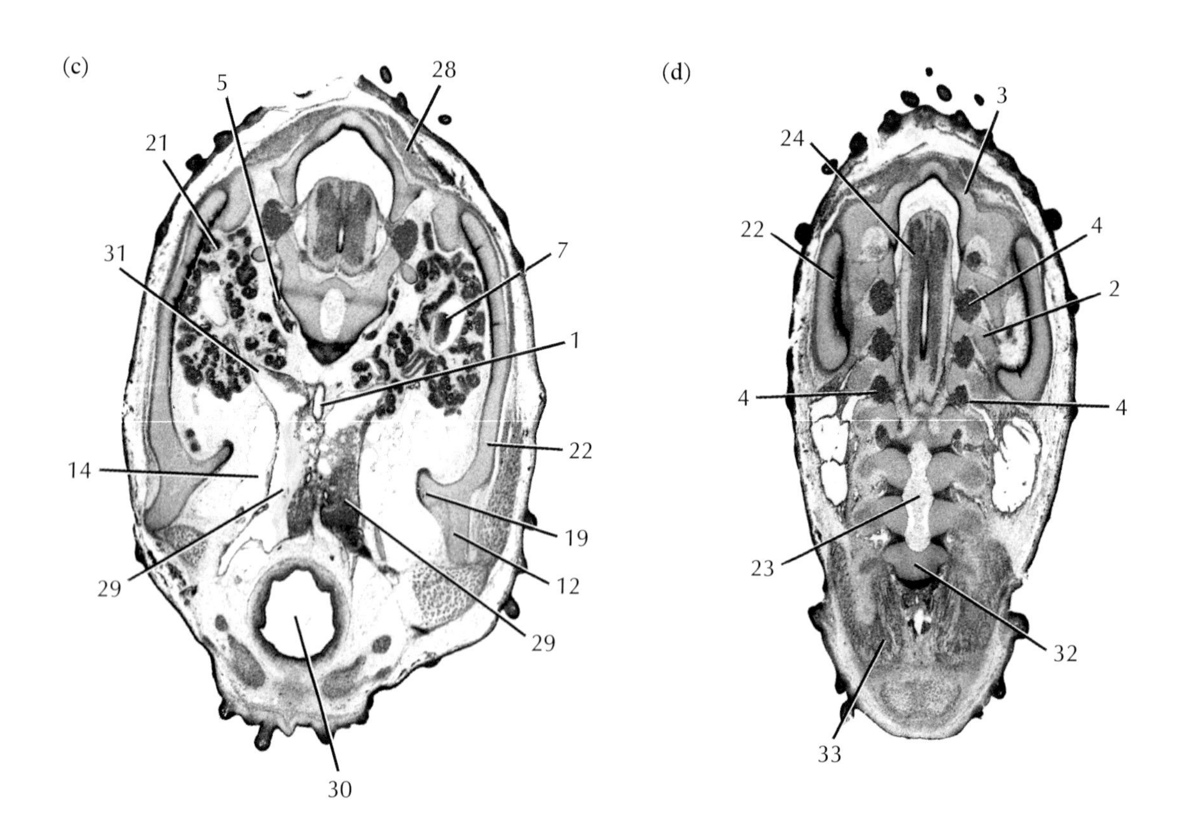
(c)
1
5
7
12
14
19
21
22
28
29
30
31
(d)
2
3
4
22
23
24
32
33

Plate 138 Day 11 TS

(a–d) × 15

1. caudal artery
2. transverse process of sacral vertebra fused to ischium, contributing to synsacrum
3. neural arch
4. dorsal root ganglion
5. sympathetic nerve ganglion
6. spinal nerve, part of pudendal plexus
7. caudal renal portal vein
8. left omphalo-mesenteric artery
9. obturator muscle
10. iliotibialis muscle
11. caudofemoralis muscle, pars iliofemoralis
12. pubis
13. metanephric duct (ureter)
14. abdominal air sac
15. left oviduct (Mullerian duct)
16. nephric (mesonephric, Wolffian) duct
17. levator cloacae muscle
18. dorsal mesentery
19. pectineal process
20. right omphalo-mesenteric artery
21. metanephros
22. ilium
23. notochord
24. spinal cord
25. flexor cruris lateralis muscle
26. cloaca
27. peritoneum
28. lateralis caudae muscle
29. coccygeo-mesenteric vein (fused left and right coccygeal veins)
30. cloacal bursa (bursa of Fabricius)
31. caudal renal portal vein
32. body of caudal vertebra
33. levator caudae muscle

PLATE 139

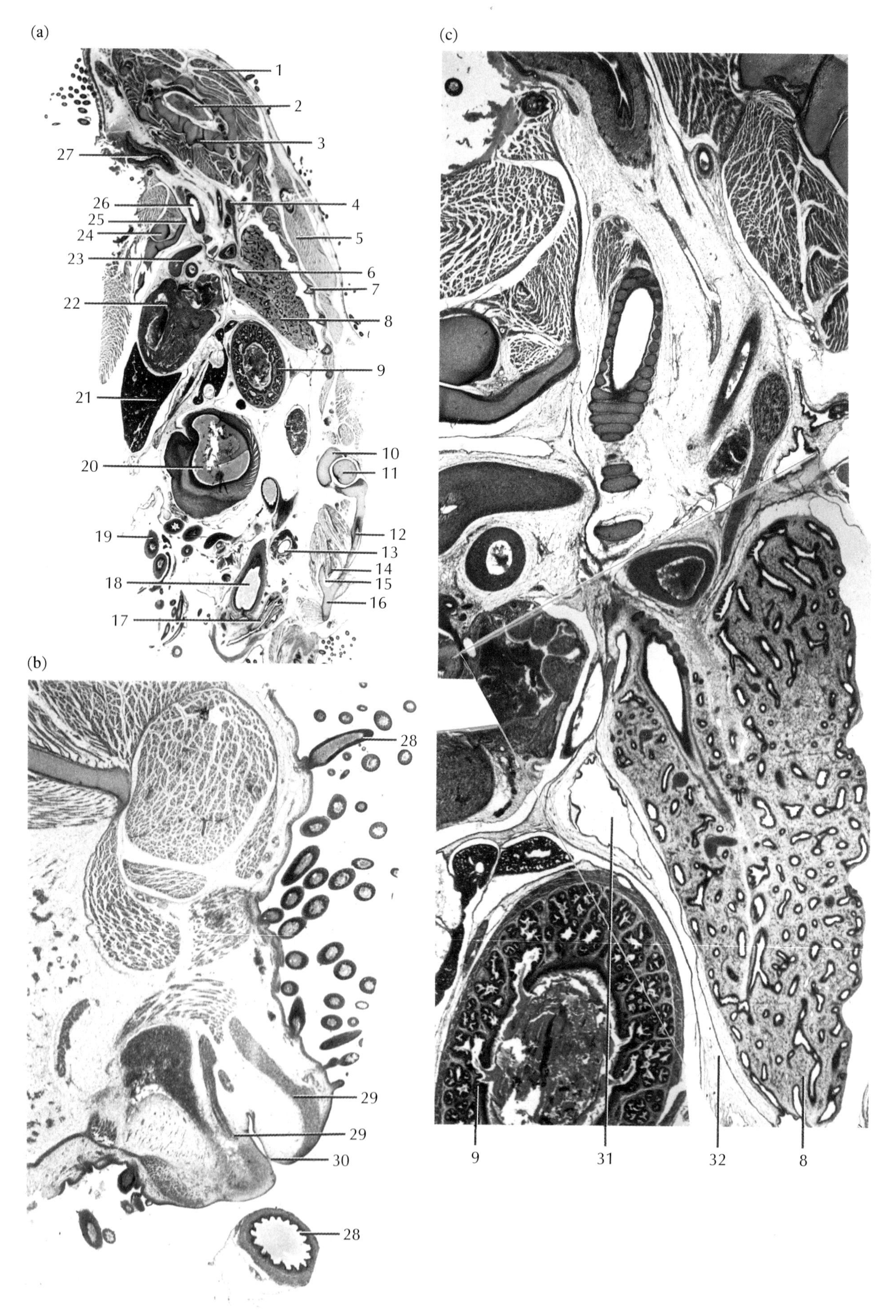
(a)
1
2
3
27
26
4
25
24
5
23
6
7
22
8
9
21
10
20
11
19
12
13
14
18
15
16
17
(b)
28
29
29
30
28
(c)
9
31
32
8

Plate 139 Day 13 LS

Sagittal sections taken through plane of page.

(a) ×4
(b) Cloacal region ×20
(c) Enlargement of part of (a) ×10

1. cervical muscles
2. spinal cord
3. cervical vertebra
4. spinal nerve
5. left rhomboideus muscle
6. left primary bronchus
7. rib
8. left lung
9. proventriculus
10. left ilium
11. head of left femur
12. ischium
13. oviduct (Mullerian duct)
14. obturator muscle
15. pectineal process
16. pubis
17. cloacal bursa (bursa of Fabricius)
18. coprodaeum
19. loop of intestine
20. gizzard
21. liver
22. ventricle
23. pulmonary artery
24. head of left humerus
25. clavicle
26. trachea
27. oesophagus
28. feather bud
29. cloacal sphincter
30. cloaca
31. left anterior thoracic air sac
32. oblique septum

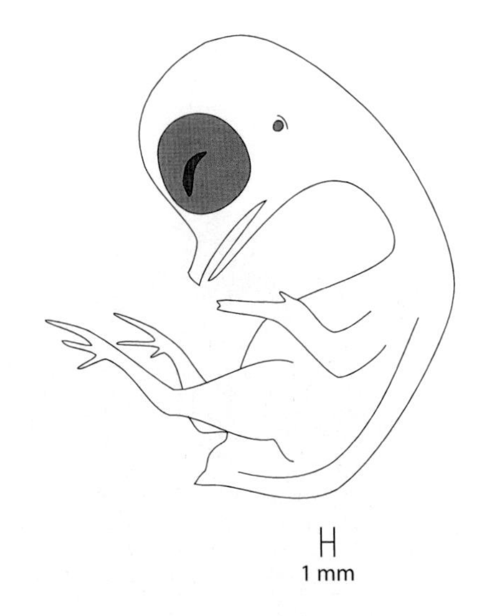

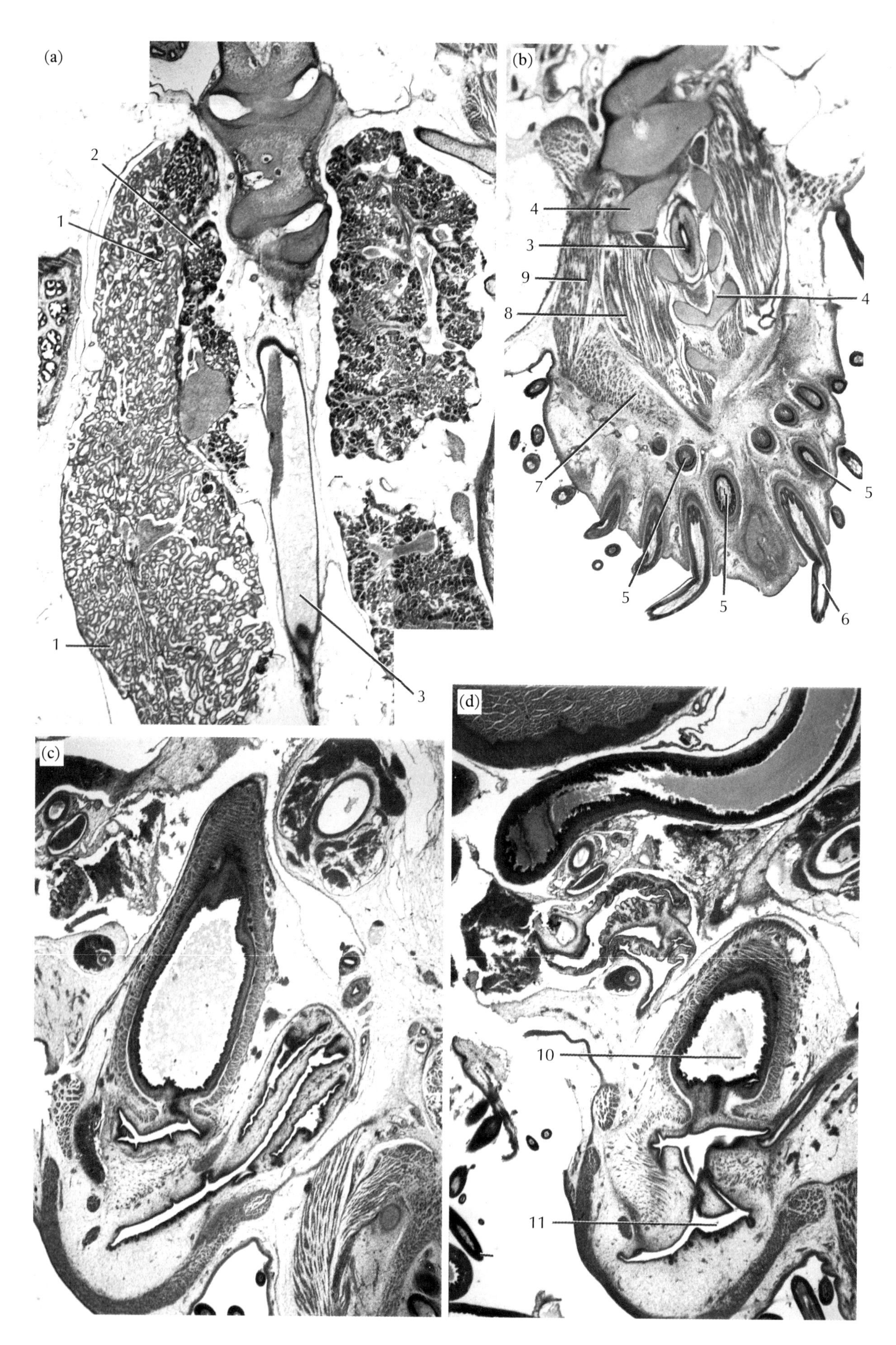
(a)
2
1
1
3
(b)
4
3
9
8
4
7
5
5
5
6
(c)
(d)
10
11

Plate 140 Day13 LS (longitudinal sections: coronal and sagittal)

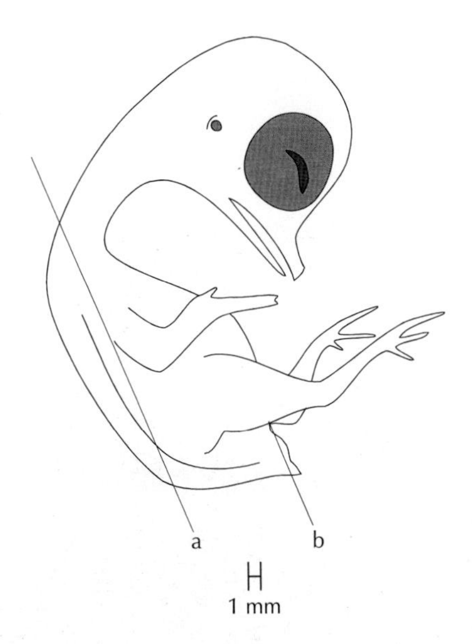

(a) Coronal section, ×18.
(b) Coronal section, ×15.
(c) Sagittal section through plane of page. Enlargement of part of *Plate 139a*, ×18.
(d) Section adjacent to (c), ×15.

1. right mesonephros
2. right metanephros
3. spinal cord
4. caudal vertebra
5. feather papilla
6. feather
7. adductor retricium muscle
8. pubocaudalis internus muscle
9. pubocaudalis lateralis muscle
10. coprodaeum
11. cloacal bursa (bursa of Fabricius)

Plate 141

(a)

(b)

(c)

(d)

Plate 141 Day 13 TS

(a–d) ×7.5

1. thymus
2. spinalis muscle
3. biventer cervicis muscle
4. spinous process of vertebra
5. spinal cord
6. transverse process of vertebra
7. intertraversarii muscle
8. left and right carotid arteries
9. trachea
10. body of vertebra
11. cervical oesophagus (i.e. anterior to crop)
12. epithelium
13. dorsal root ganglion
14. feathers
15. descending vertebral artery
16. longus colli ventralis muscle
17. neural arch
18. triceps brachii pars humerotriceps muscle
19. biceps brachii muscle pars propatagialis
20. deltoid muscle
21. humerus
22. coracoid
23. triceps muscle
24. brachial plexus
25. crop
26. shaft of humerus
27. scapula
28. scapulohumeralis muscle
29. latissimus dorsi muscle
30. rhomboideus superficialis muscle
31. costosternalis muscle
32. sternocoracoid process
33. thyroid
34. supracoracoid muscle
35. pectoralis major muscle
36. thoracic oesophagus (i.e. posterior to crop)
37. coracobrachialis muscle
38. right carotid artery
39. right internal jugular vein

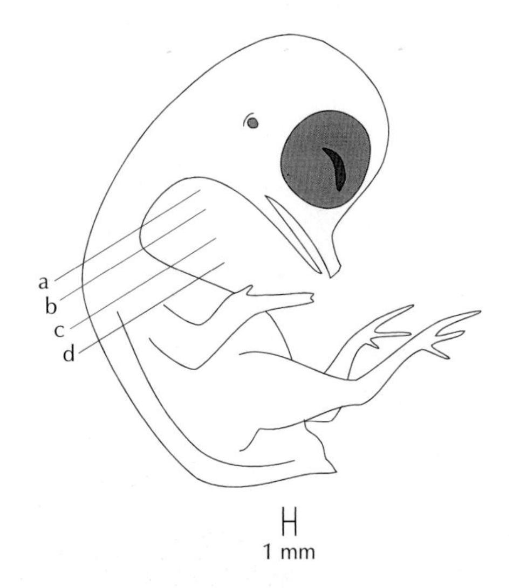

PLATE 142

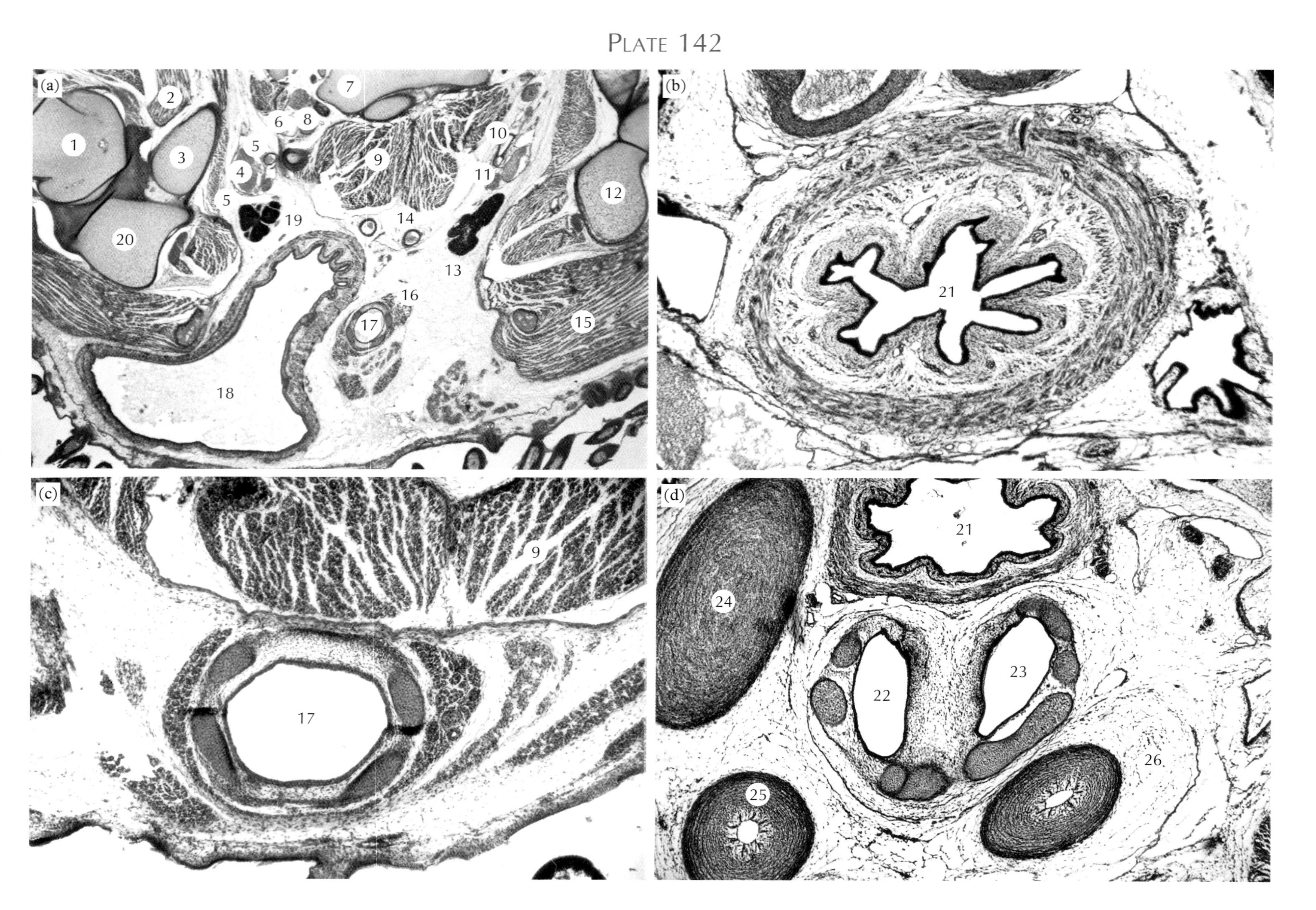

(a)
1
2
3
4
5
5
6
7
8
9
10
11
12
13
14
15
16
17
18
19
20
(b)
21
(c)
9
17
(d)
21
22
23
24
25
26

Plate 142 Day 13 TS Enlargements

(a–d) ×75

(a) Transverse section through crop.
(b) Transverse section through thoracic oesophagus.
(c) Transverse section through trachea.
(d) Transverse section through left and right bronchi immediately posterior to bifurcation of trachea. Enlargement of part of Plate 143c.

1. right humerus
2. right scapulo-humeralis muscle
3. right scapula
4. right internal jugular vein
5. fascial sheath enclosing internal jugular vein, vagus nerve, vagus artery and a lymphatic vessel
6. right vertebral vein
7. body of vertebra
8. vertebral artery
9. longus colli ventralis muscle
10. left internal jugular vein
11. left brachial plexus
12. left humerus
13. lobe of left thymus
14. left carotid artery
15. left pectoralis muscle
16. sternotrachealis muscle
17. lumen of trachea
18. lumen of crop
19. lobe of right thymus
20. right coracoid
21. lumen of thoracic oesophagus
22. right bronchus
23. left bronchus
24. aortic arch
25. right brachiocephalic artery
26. left brachiocephalic artery

PLATE 143

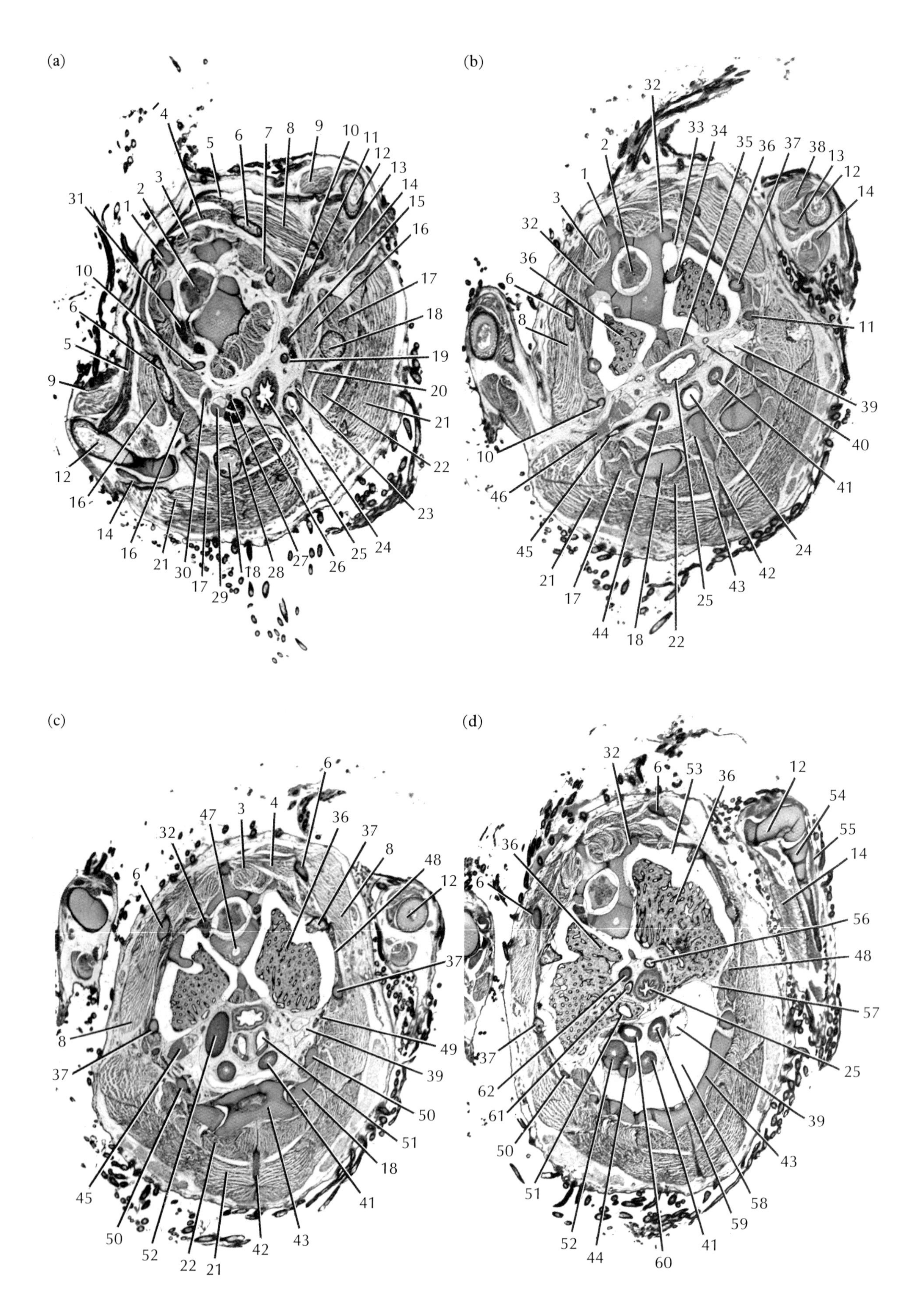

(a)
4 5 6 7 8 9 10 11 12 13 14 15 16 17 18 19 20 21 22 23 24 25 26 27 28 29 30 31
1 2 3 10 6 5 9 12 16 14 16 21 30 17 29 18 28 27 26 25 24
(b)
32 2 1 3 32 36 6 8 33 34 35 36 37 38 13 12 14 11 39 40 41 24 42 43 25 22 18 44 17 21 45 46 10
(c)
47 3 4 6 32 6 36 37 8 48 12 37 8 37 49 39 50 51 18 41 43 42 21 22 52 50 45
(d)
32 6 53 36 12 54 55 14 36 6 56 48 57 25 39 43 58 59 41 60 44 52 51 50 61 62 37

Plate 143 Day 13 TS

(a–d) ×6

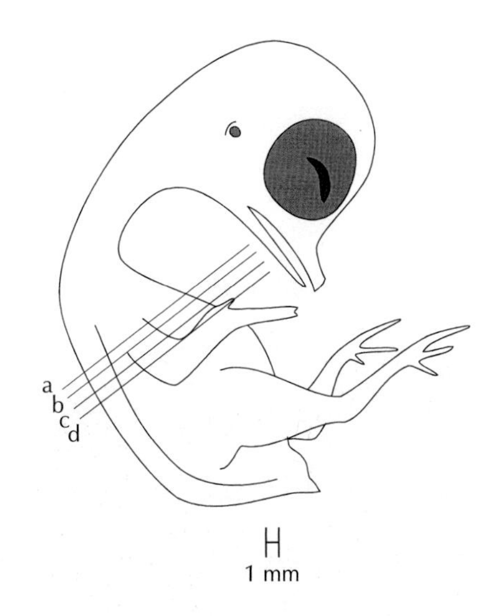

1. spinous process of vertebra
2. spinal cord
3. longus colli dorsalis muscle
4. rhomboideus superficialis muscle
5. latissimus dorsi muscle
6. scapula
7. long head of left triceps muscle
8. scapulohumeralis caudalis muscle
9. deltoid muscle
10. spinal nerve
11. part of left brachial plexus
12. humerus
13. triceps brachialis (anconeus) muscle pars humerotriceps
14. biceps brachii muscle
15. internal jugular vein and part of brachial plexus
16. coracobrachialis muscle
17. subcoracoideus muscle
18. coracoid
19. left carotid artery
20. sternocoracoideus muscle
21. pectoralis muscle
22. supracoracoideus (pectoralis secundum) muscle
23. tracheolaryngeal muscle
24. trachea
25. thoracic oesophagus
26. right carotid artery
27. cervical nerve
28. lobe of right thymus
29. right internal jugular vein
30. part of brachial plexus
31. dorsal root ganglion
32. transverse process of vertebra
33. vertebrarterial canal
34. parapophysis of vertebra
35. longus colli ventralis muscle
36. lung
37. vertebral rib
38. triceps brachialis (anconeus) muscle, pars scapulotriceps
39. abdominal air sac
40. extrapulmonary bronchus
41. left brachiocephalic artery
42. hypocledial ligament
43. sternum
44. right brachiocephalic artery
45. subclavian artery
46. cranial (anterior) vena cava
47. notochord in body of vertebra
48. intercostalis muscle
49. intercostal nerve
50. sternal rib
51. bronchus
52. aortic arch
53. left pleural cavity
54. radius
55. tensor propatagialis longus muscle
56. left ductus arteriosus
57. peritoneum
58. cranial thoracic air sac
59. left pulmonary artery
60. right pulmonary artery
61. right ductus arteriosus
62. dorsal aorta

Plate 144

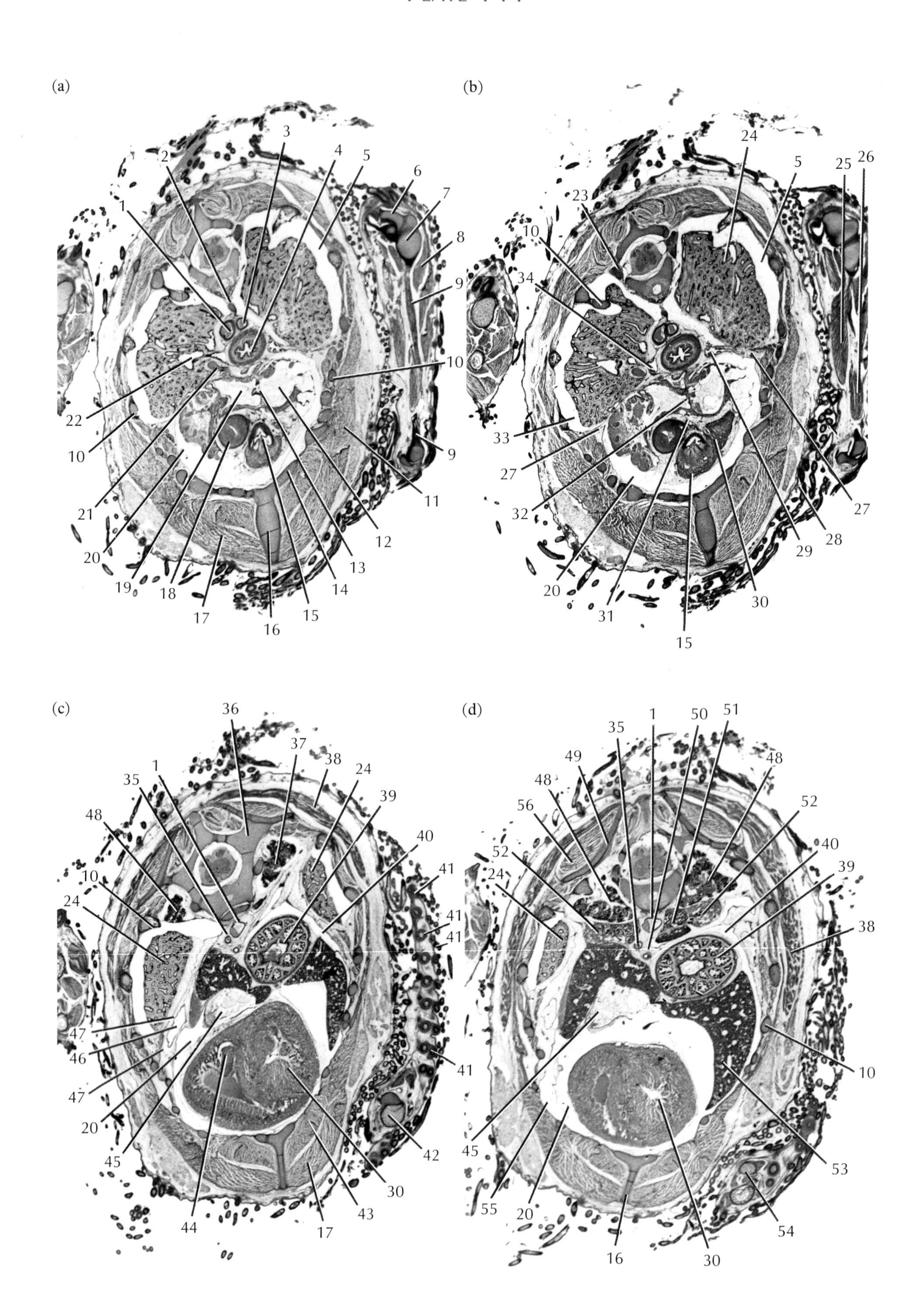

(a)
1
2
3
4
5
6
7
8
9
10
9
11
12
13
14
15
16
17
18
19
20
21
10
22
(b)
24
5
25
26
23
10
34
33
27
32
20
31
15
30
29
28
27
(c)
36
37
38
24
39
40
41
41
41
41
35
1
48
10
24
47
46
47
20
45
44
17
43
30
42
(d)
35
1
50
51
48
49
48
56
52
52
24
40
39
38
10
45
55
20
16
30
53
54

Plate 144 Day 13 TS

(a–d) × 6

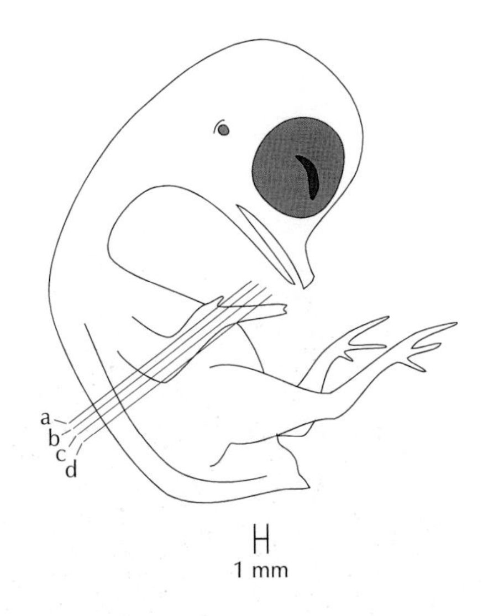

1. dorsal aorta
2. dorsal mesentery
3. left brachiocephalic artery
4. oesophagus
5. left pleural cavity
6. humerus
7. head of radius
8. biceps brachii muscle
9. radius
10. rib
11. coracobrachialis muscle
12. left atrium
13. perforations in interatrial septum
14. sternum
15. truncus arteriosus
16. keel of sternum
17. pectoralis muscle
18. conus arteriosus
19. right atrium
20. pericardial cavity
21. right pulmonary vein
22. bronchus in right lung
23. dorsal root ganglion
24. lung
25. pronator profundus muscle
26. pronator superficialis muscle
27. oblique septum
28. feathers
29. left common cardinal vein
30. left ventricle
31. sinus venosus
32. interatrial septum
33. right pleural cavity
34. right primary bronchus
35. mesenteric artery
36. transverse process of vertebra
37. left metanephros
38. latissimus dorsi muscle
39. proventriculus
40. dorsal hepatic cavity
41. feather papilla
42. ulna carpal
43. supracoracoid
44. right ventricle
45. caudal vena cava
46. thoracic air sac
47. horizontal septum
48. right metanephros
49. ilium
50. coeliac artery
51. left ovary
52. mesonephros
53. left lobe of liver
54. carpo-metacarpal
55. ventral hepatic cavity
56. right ischiofemoralis muscle

Plate 145

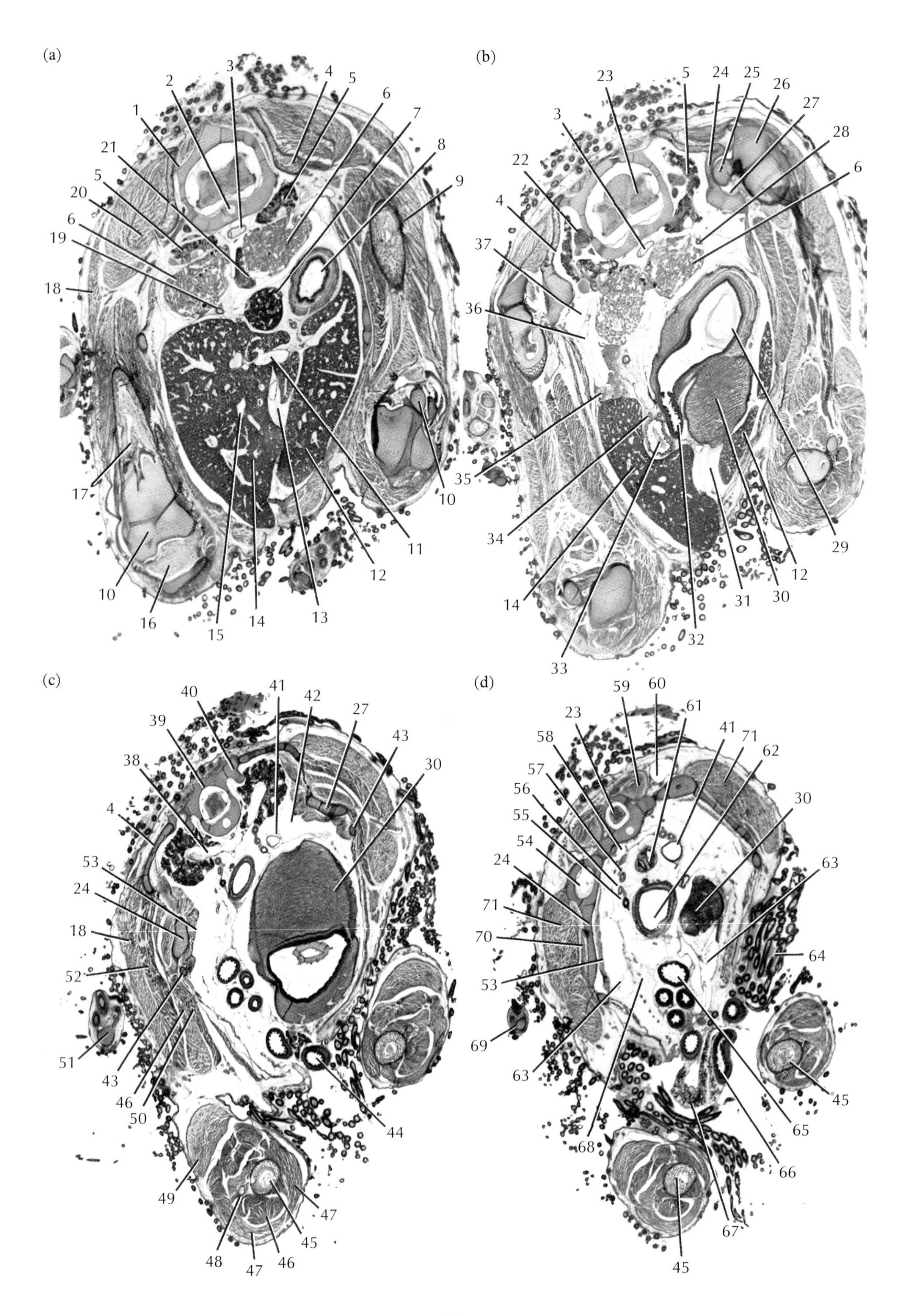

(a)
1
2
3
4
5
6
7
8
9
10
11
12
13
14
15
16
17
18
19
20
21
(b)
22
23
24
25
26
27
28
29
30
31
32
33
34
35
36
37
(c)
38
39
40
41
42
43
44
45
46
47
48
49
50
51
52
53
(d)
54
55
56
57
58
59
60
61
62
63
64
65
66
67
68
69
70
71

Plate 145 Day 13 TS

(a–d) ×6

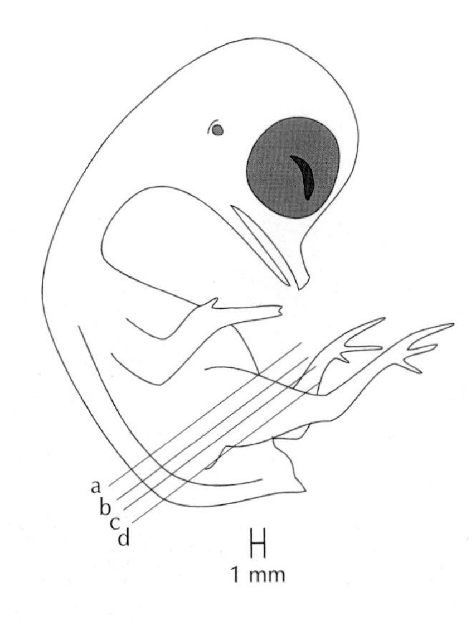

1. neural arch of sacral vertebra
2. notochord
3. dorsal aorta
4. ilium
5. metanephros
6. mesonephros
7. spleen
8. proventriculus
9. left femur
10. condyle of femur
11. caudal vena cava
12. left lobe of liver
13. ventral mesentery
14. right lobe of liver
15. right hepatic vein
16. patella
17. right femur
18. iliotibialis cranialis muscle
19. anterior mesenteric artery
20. ischiofemoralis (obturator externus) muscle
21. left ovary
22. dorsal root ganglion
23. spinal cord
24. ischium
25. head of femur
26. trochanter of femur
27. acetabulum
28. left oviduct (Mullerian duct)
29. lumen of gizzard
30. muscular wall of gizzard
31. ventral hepatic cavity
32. duodenum
33. right allantoic artery
34. branches of coeliac artery
35. horizontal hepatic ligament
36. oblique septum
37. lateral thoracic air sac
38. cranial renal artery
39. synsacrum
40. parapophysis of vertebra
41. ventral gastric artery
42. abdominal air sac
43. pubis
44. herniated loop of intestine
45. tibiotarsus
46. flexor cruris lateralis muscle
47. gastrocnemius muscle
48. fibula
49. fibularis longus muscle
50. flexor cruris medialis muscle
51. right foot
52. iliofibularis muscle
53. obturator medialis muscle
54. sciatic foramen
55. right oviduct (Mullerian duct)
56. right metanephric duct
57. right nephric (mesonephric, Wolffian) duct
58. internal iliac vein
59. longissimus dorsi muscle
60. vertebro-medullary vein
61. rectum
62. cloaca
63. abdominal air sac
64. feathers
65. loop of intestine
66. duodenum
67. pancreas
68. dorsal mesentery
69. toe
70. ischiofemoralis muscle
71. iliofemoralis muscle

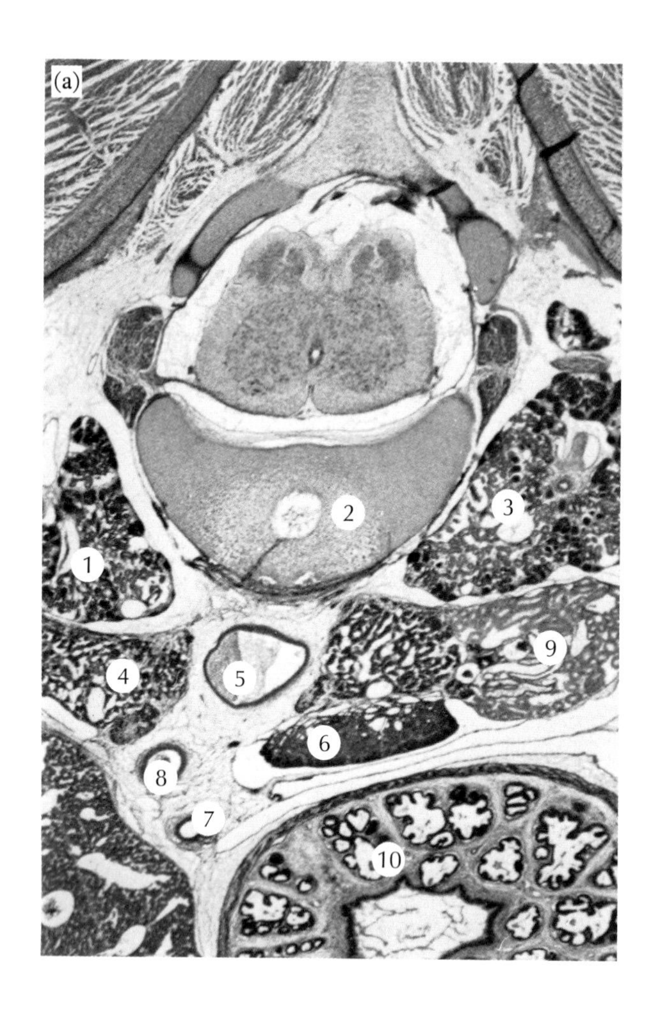
(a)
1
2
3
4
5
6
7
8
9
10

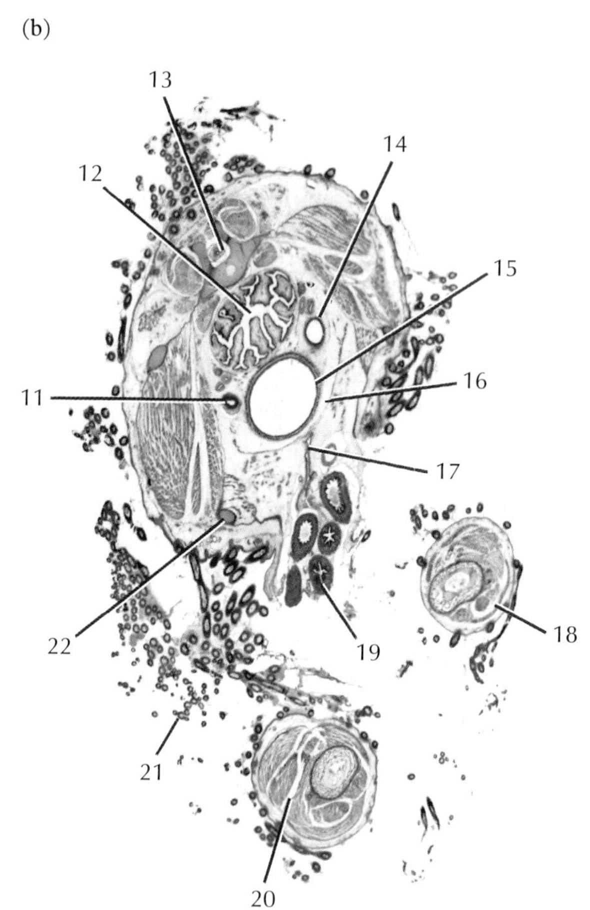
(b)
11
12
13
14
15
16
17
18
19
20
21
22

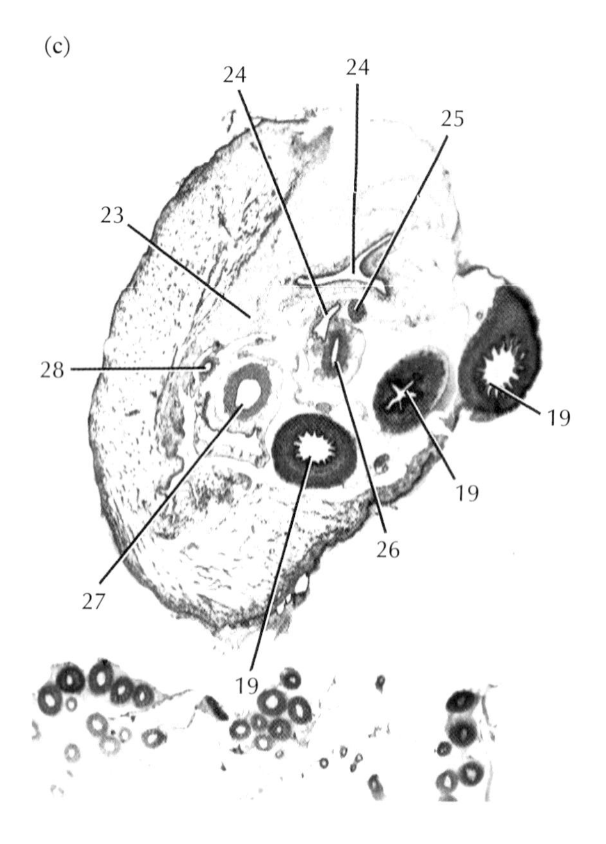
(c)
19
19
19
19
23
24
24
25
26
27
28

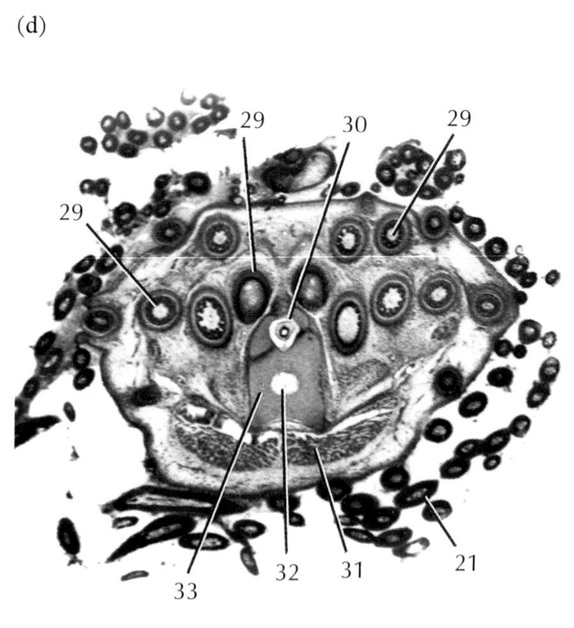
(d)
21
29
29
29
30
31
32
33

Plate 146 Day 13 TS

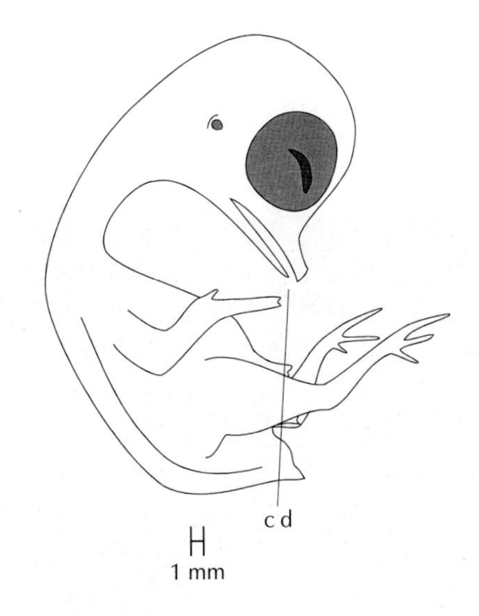

(a) Enlargement of part of *Plate 144d*, ×35.
(b) Section across cloacal bursa, adjacent section to *Plate 145d*, ×6.
(c) Section through base of umbilical cord (same section as (d)), ×18.
(d) Section through tail (same section as (c)), ×18.

1. right metanephros
2. body of vertebra
3. left metanephros
4. right mesonephros
5. dorsal aorta
6. left ovary
7. coeliac artery
8. mesenteric artery
9. left mesonephros
10. proventriculus
11. right oviduct (Mullerian duct)
12. cloacal bursa (bursa of Fabricius)
13. spinal cord
14. left oviduct (Mullerian duct)
15. cloaca
16. embryonic coelom
17. neck of allantois
18. left leg
19. herniated loop of intestine
20. right leg
21. feathers
22. pubis
23. extra-embryonic coelom
24. allantoic stalk
25. vitelline vessel
26. left allantoic artery
27. allantoic vein
28. right allantoic artery
29. feather papillae
30. vertebral foramen
31. left depressor cauda muscle
32. remnant of notochord
33. body of caudal vertebra

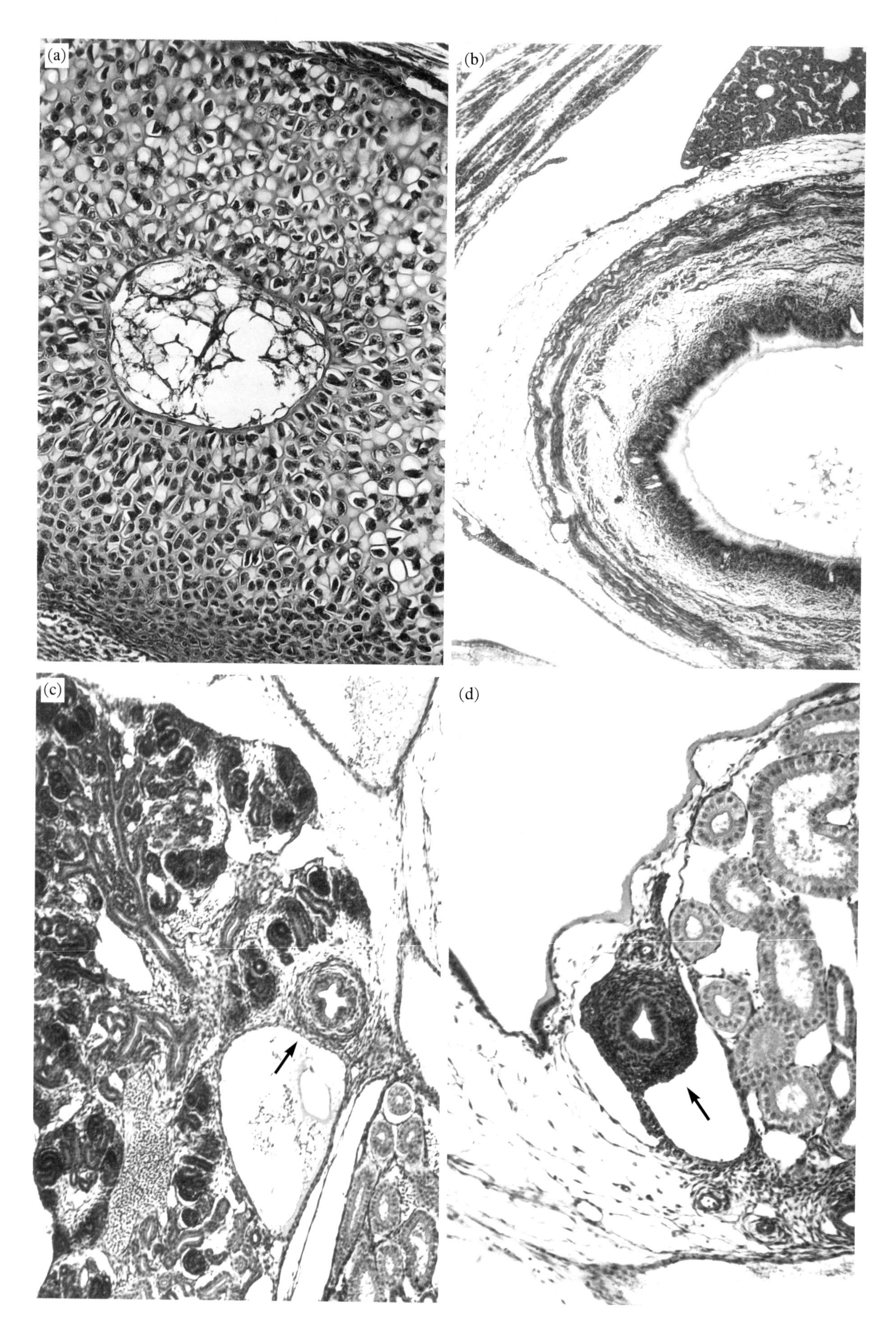
(a)
(b)
(c)
(d)

Plate 147 Day 13 TS Enlargements

(a) Section through remnant of notochord of *Plate 144* to show cartilaginous body of vertebra, ×250.
(b) Section through wall of proventriculus, ×100.
(c) Section through metanephros and metanephric duct (arrow), ×100.
(d) Section through mesonephros and left oviduct (arrow), ×250.

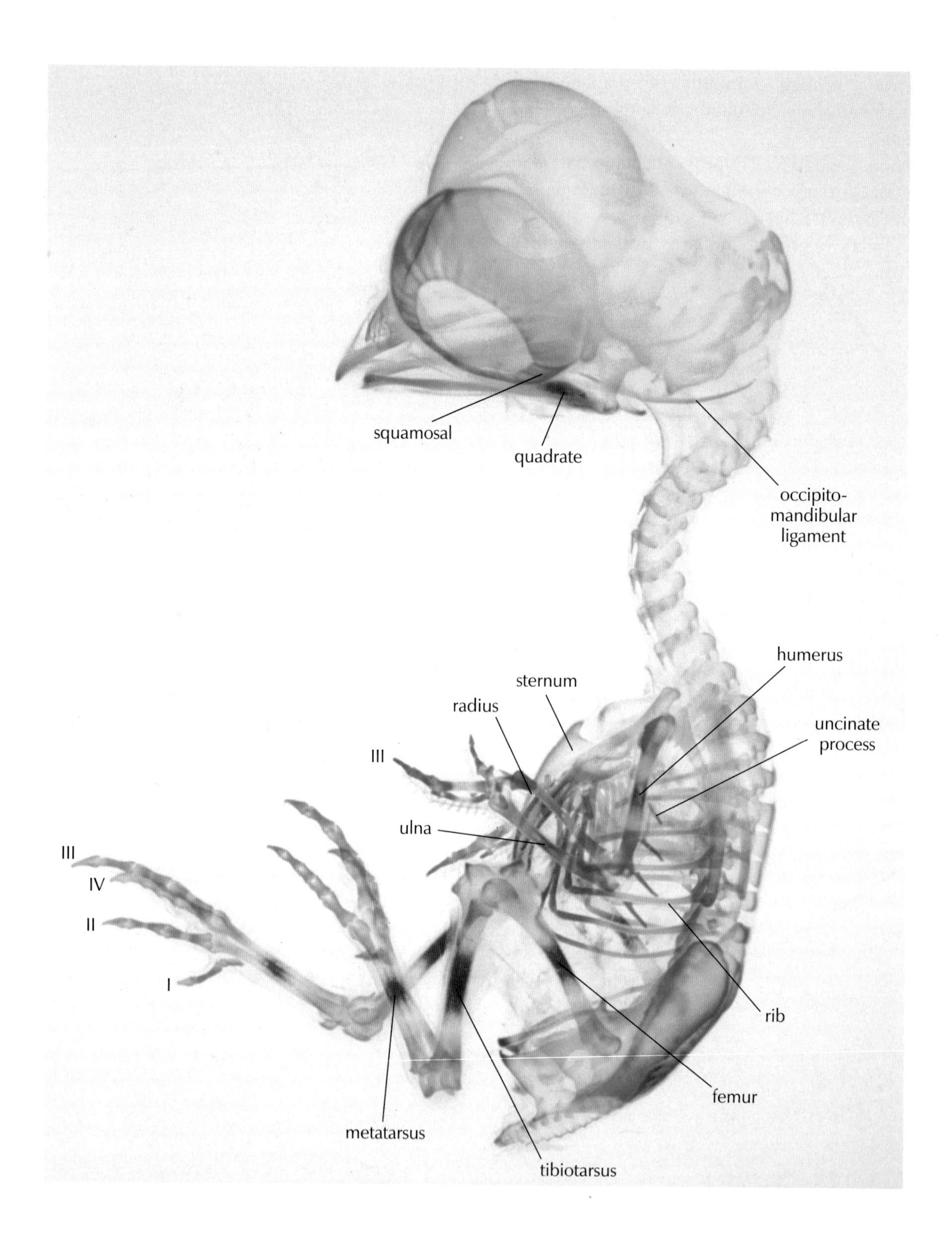

Plate 148 Day 11

Whole-mount photograph of embryo stained with alizarin red and alcian blue to show skeletal elements. Membranes and cartilages: blue; ossified regions: red.

Ossification has occurred in the quadrate and in the middle of the long bones. Most regions remain unossified. Roman numerals indicate digits, ×5.5.

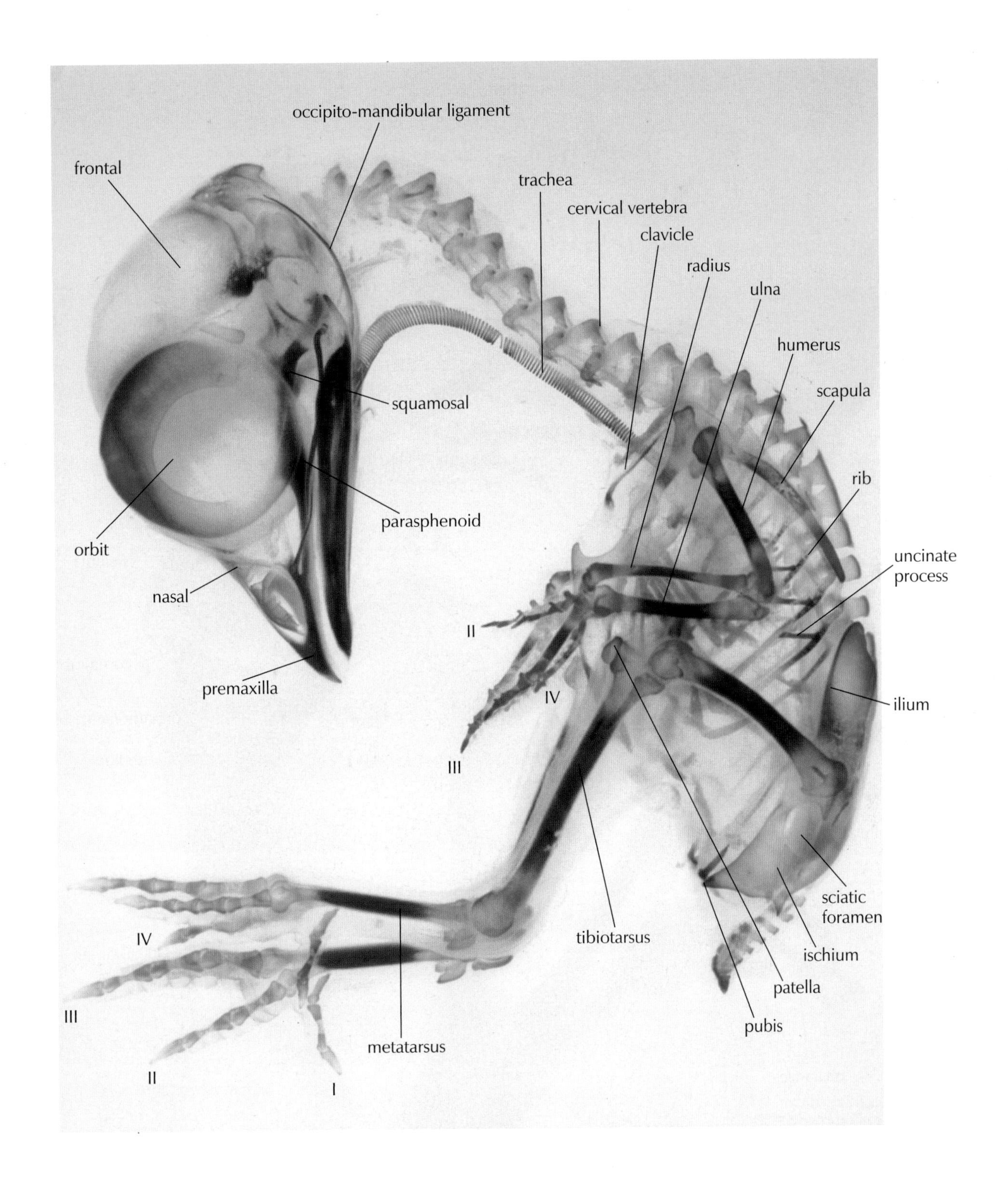

Plate 149 Day 13

Whole-mount photograph of embryo stained with alizarin red and alcian blue to show skeletal elements. Membranes and cartilages: blue; ossified regions: red.

Ossification has occurred in the upper and lower jaws, has increased in the long bones, and is visible in the clavicle, the coracoid, the thoracic ribs and, slightly, in the ilium, ×5.

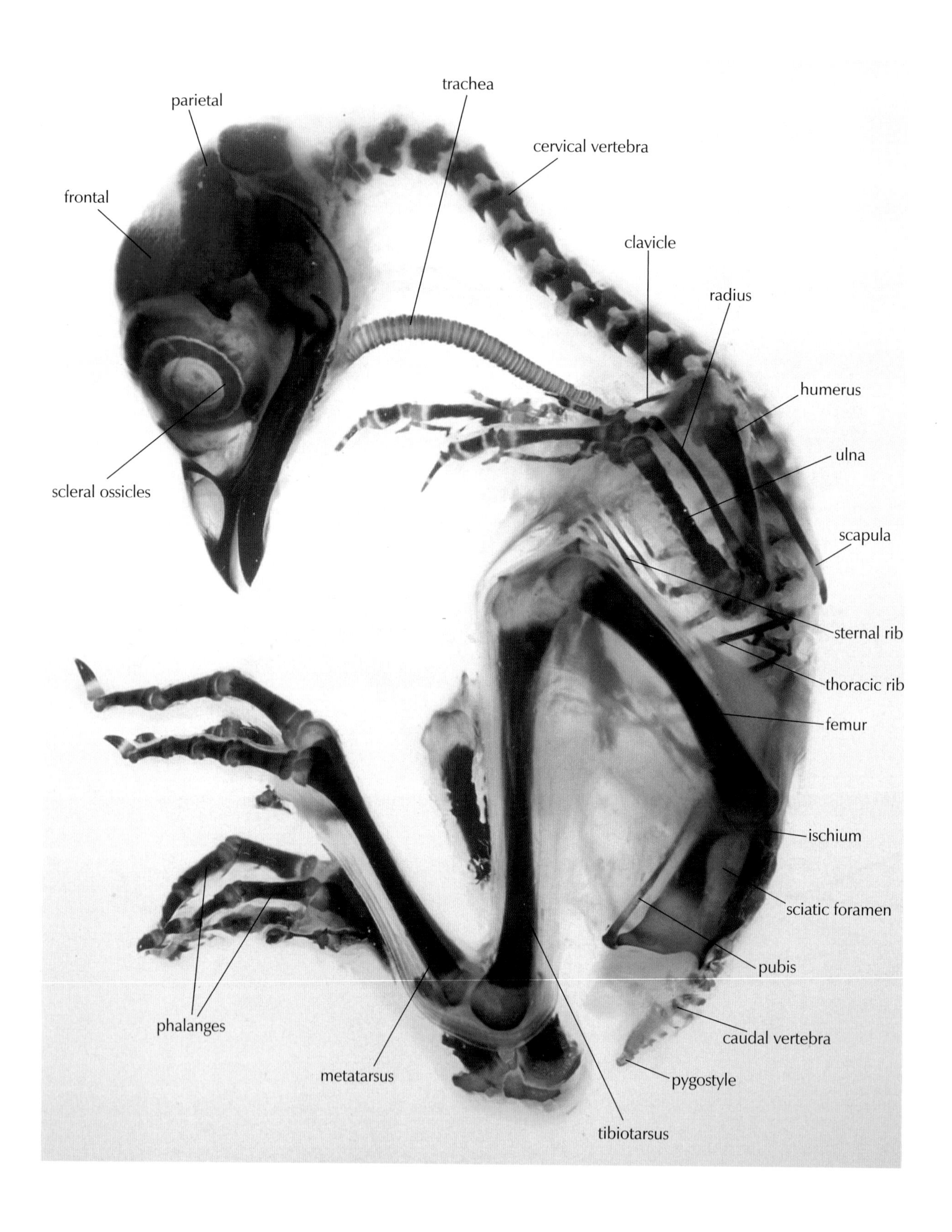
parietal
trachea
cervical vertebra
frontal
clavicle
radius
humerus
ulna
scleral ossicles
scapula
sternal rib
thoracic rib
femur
ischium
sciatic foramen
pubis
phalanges
caudal vertebra
metatarsus
pygostyle
tibiotarsus

Plate 150 Day 19.5

Whole-mount photograph of embryo stained with alizarin red and alcian blue to show skeletal elements. Membrane and cartilages: blue; ossified regions: red.

The skull shows extensive ossification, especially of the frontal and parietal bones and of the scleral ossicles. Ossification is well advanced in the digits and the vertebrae, though the caudal vertebrae remain largely cartilaginous, ×4.5.

Appendices
Normal Tables

Embryologists have long found it convenient to classify normal embryos in a series of developmental stages. Such a classification is called a **normal table**. The most widely used table for the domestic fowl is by Hamburger and Hamilton (1951, reprinted 1992) (Appendix II). It is based exclusively on external features and each stage is illustrated by a photograph and accompanied by a brief description. Tables of this sort are of particular value to experimentalists in that they enable the developmental state of an embryo to be assessed and communicated relatively accurately. There are, however, some problems involved in the use of Hamburger and Hamilton's table. The stages from 7 to 14 are based largely on the numbers of somites that have developed, but the rate of differentiation of other structures does not always keep pace with the somites, e.g. stage 8 of the normal table possesses four pairs of somites and a brain that has scarcely begun to bulge, whereas in practice embryos with four pairs of somites may possess a less or more developed head, comparable to those of stages 7 or 9. In later stages the shape of the limb buds is used as the most distinguishing feature for each stage, but again this does not always bear the same relationship to the development of other external features.

Another problem with the Hamburger and Hamilton table is that not enough stages have been allocated for early incubation when development is taking place very rapidly. Stages 1–4 cover the period until the primitive streak is fully formed at about 18 h and the problem is compounded by the fact that marked differences occur not only between different batches of eggs but often within the same batch, even though incubation conditions are identical. A supplementary normal table (Appendix I) has therefore been published (Eyal-Giladi and Kochav, 1975) to cover the period from the earliest stages normally to be found in unincubated eggs to the stage when the primitive streak is about half formed. A further table, an elaboration of the stages in wing development (Murray and Wilson, 1994), is reproduced in Appendix III.

Because the purpose of a normal table is for quick recognition of the stage as indicated by external features, there is no guarantee of a precise correlation with the differentiation of internal organs. For that reason it is often preferable when embryos are older than about 3 days to describe them by the length of incubation time, a practice that has been followed in this book. Nevertheless, it is possible to make some approximations with internal differentiation and these, extracted from the monograph of Freeman and Vince (1974) are inserted into the normal table of Hamburger and Hamilton (Appendix II).

Some references providing normal tables for other domestic birds are listed in Appendix IV.

Appendix I

Normal Table of Eyal-Giladi and Kochav (1975)

The following description is slightly modified and abbreviated from the original publication. Reproduced by permission of Academic Press (Elsevier).

Cleavage Stages

Stage I (0–1 h in shell gland, *Figures 1* and *2*). Each egg was surrounded by viscous albumen and a very flaccid shell membrane. Early cleavage. Germinal disc 3.5–4 mm diameter. (c.fur = cleavage furrows; p.b. = polar body; su.sp. = supernumerary sperm; vac. = large vacuoles.)

Stage II (about 2 h in shell gland, *Figures 3* and *4*). About 14–16 laterally closed cells raised above the rest of the germinal disc. Cleavage furrows spread out from the centre.

Stage III (about 3–4 h in shell gland, *Figures 5* and *6*). 80–90 laterally closed blastomeres in the centre. (cl.bl. = closed blastomeres.)

Stage IV (about 5 h in shell gland, *Figures 7* and *8*). About 250–300 closed cells in upper layer. About 80–90 closed cells now visible in lower layer. (s.c. = subgerminal cavity.)

Stage V (about 8–9 h in shell gland, *Figures 9* and *10*). Cells still bead-like. Subgerminal cavity enlarged.

Stage VI (about 10–11 h in shell gland, *Figures 11* and *12*). Cells now small and flat and have lost bead-like appearance. All cytoplasm now cleaved into true cells (blastomeres).

Formation of Area Pellucida

Stage VII (about 12–14 h in shell gland, *Figures 13* and *14*). Upper layer cells have become progressively smaller. Lower layer cells have remained large. The first indication of the area pellucida as a thinner region in the embryo where cells have been shed from the underside of the upper layer. (a.p. = area pellucida).

Stage VIII (about 15–17 h in shell gland, *Figures 15* and *16*). The area pellucida has spread but the peripheral region has remained thick and is forming the area opaca. (a.o. = area opaca; a.p. = area pellucida; y.l.c. = yolk-laden cells.)

Stage IX (about 17–19 h in the shell gland, *Figures 17* and *18*). Further spreading of the area pellucida. Border between area opaca and area pellucida not yet sharply defined. (a.o. = area opaca; a.p. = area pellucida.)

Stage X (freshly laid egg, *Figures 19* and *20*). The area pellucida and area opaca are now clearly demarcated from one another. Clusters of small cells lie beneath posterior part of area pellucida except for the extreme posterior region (i.ag. = isolated cell aggregates.)

Hypoblast Formation (Early Stages of Incubation)

Stage XI (*Figures 21* and *22*). A transparent belt-like area (t.b.) at the poster border of the area pellucida and area opaca. Anterior to this region clusters of cells lie in a horseshoe-shaped arrangement known as Koller's sickle (k.s.).

Stage XII (*Figures 23* and *24*). The hypoblast (hyp) is beginning to form [from Koller's sickle] but is not yet a continuous sheet.

Stage XIII (*Figures 25* and *26*). The hypoblast is now completely formed (as a sheet).

Stage XIV (*Figures 27* and *28*). The anterior border of the hypoblast is now clearly defined. A cellular bridge (pos.br.) begins to link the posterior end of the hypoblast to the area opaca. This bridge is also visible in *Figures 29* and *30*, which correspond with stage 2 of Hamburger and Hamilton (see Appendix II). (p.s. = earliest indication of primitive streak.)

Appendix 1 Plates

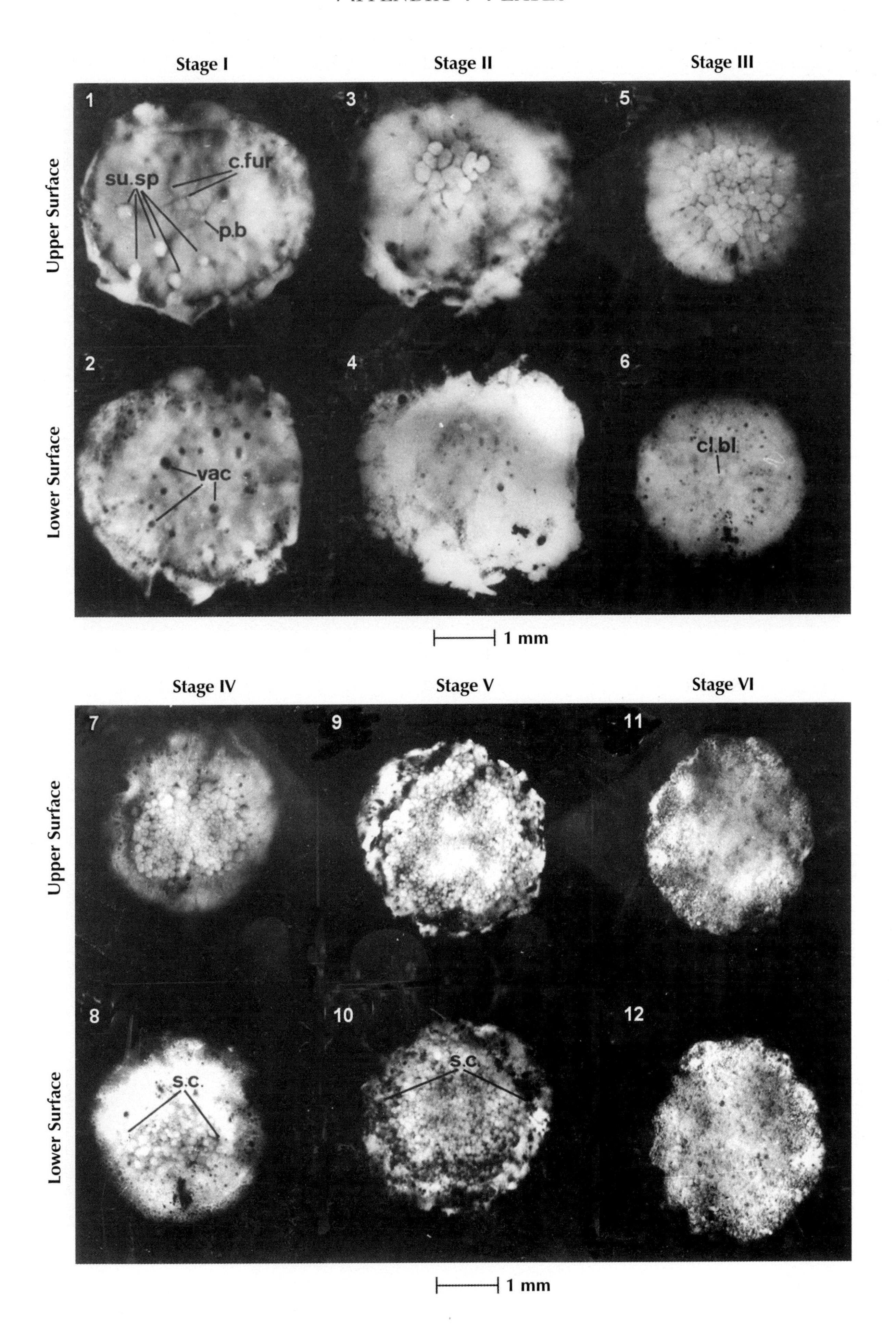

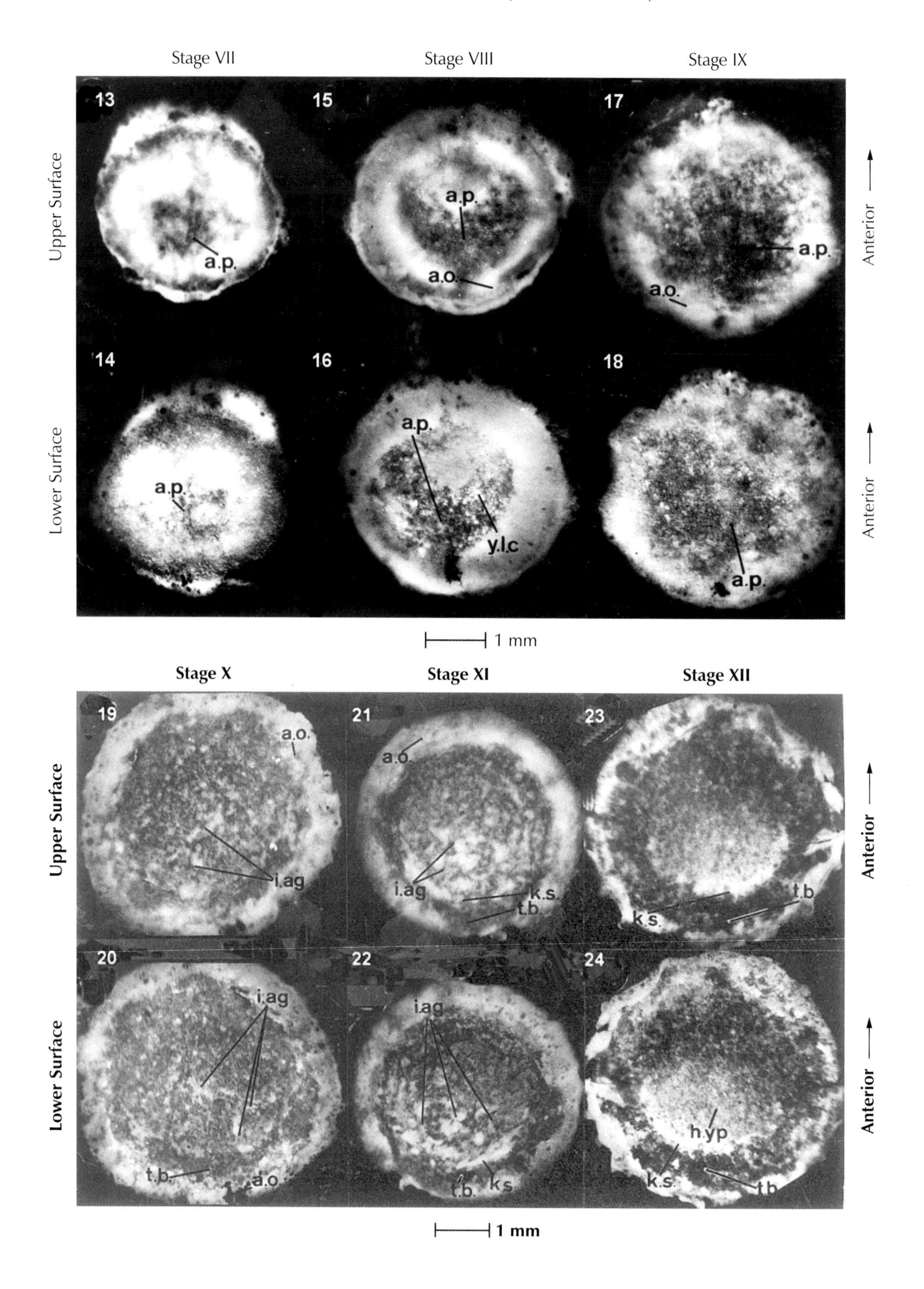
Stage VII
Stage VIII
Stage IX
13
15
17
14
16
18
Upper Surface
Lower Surface
Anterior
a.p.
a.o.
y.l.c
1 mm
Stage X
Stage XI
Stage XII
19
21
23
20
22
24
i.ag
k.s.
t.b.
h.yp
1 mm

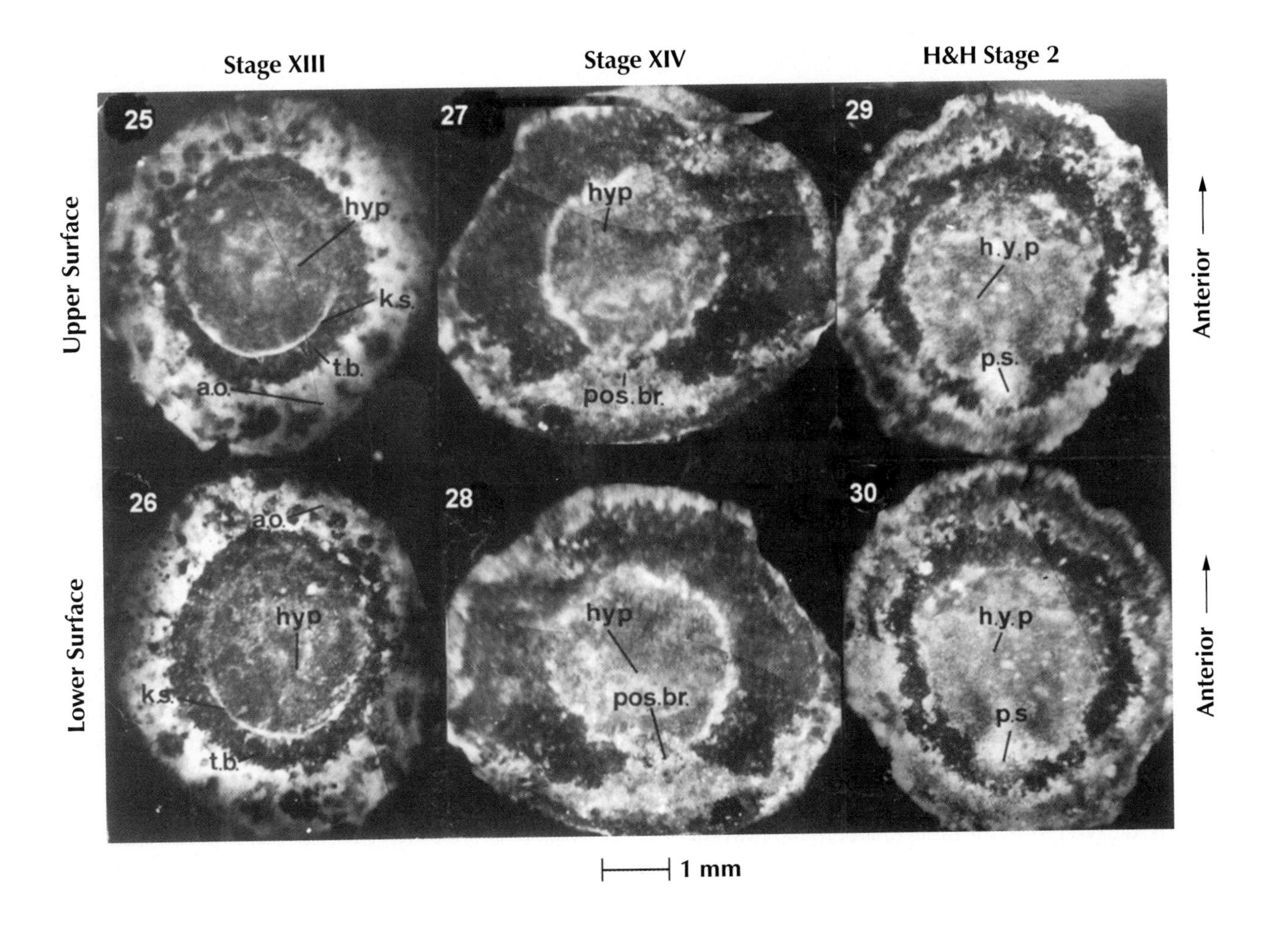
Stage XIII
Stage XIV
H&H Stage 2
Upper Surface
Lower Surface
25
27
29
26
28
30
hyp
k.s.
t.b.
a.o.
pos.br.
h.y.p
p.s.
Anterior
Anterior
1 mm

Appendix II

Normal Table of Hamburger and Hamilton (1951; 1992)

A fuller description of each stage is given in the original publications (1951, reprinted 1992), the following being abbreviated and with minor modifications. (Reproduced by permission of Wiley-Liss, Inc., a subsidiary of John Wiley and Sons, Inc.)

Stages 1 and 2 probably correspond to stages XII–XIV of Eyal-Giladi and Kochav (Appendix I). Approximate incubation times are indicated for each stage in parentheses. Notes in italics have been added by the present authors; those in square brackets, derived from the table of Freeman and Vince (1974), describe internal features.

References to 'HH Plates' refer to plates in this normal table.

Stage 1. ×20 Pre-streak. An 'embryonic shield' may be visible towards the posterior half of the blastoderm.

Stage 2. ×20 Initial streak (about 6–7 h). The primitive streak is short and conical (0.3–0.5 mm in length).

Stage 3. ×20 Intermediate streak (12–13 h). The primitive streak extends to about the centre of the area pellucida.

Stage 4. ×20 Definitive streak (18–19 h). The primitive streak has reached its maximum length (about 1.88 mm). The primitive groove, primitive pit and Hensen's node are present. The area pellucida has become pear-shaped and the primitive streak extends over about two-thirds to third-quarters of its length. (*NB: The beginning of the head process is already visible in the photograph as a dark region at the anterior tip of the primitive streak.*)

Stage 5. ×20 Head process (19–22 h). The head process extends anteriorly from Hensen's node.

Stage 6. ×20 Head fold (23–25 h). No somites yet visible.

Stages 7–14 are based primarily on the numbers of pairs of somites visible.

Stage 7. ×20 One pair of somites (23–26 h). This is actually the second somite pair, the first not yet clearly defined. Neural folds are visible in the head region.

Stage 8. ×20 Four pairs of somites (26–29 h). The neural folds meet at the level of the midbrain. Blood islands are present in the posterior half of the blastoderm.
[*Nephric ducts starting to develop.*]

Stage 9. ×20 Seven pairs of somites (29–33 h). Primary optic vesicles are present. Paired primordia of heart begin to fuse.
[*Inner ear beginning to form.*]

Stage 10. ×12 Ten pairs of somites (33–38 h). First indication of cranial flexure. Three primary brain vesicles are clearly visible. Heart primordia fused, bent slightly to right.
[*Haemoglobin synthesis begins; thyroid anlage differentiates; pronephros forming.*]

Stage 11. ×12 Thirteen pairs of somites (40–45 h). Slight cranial flexure. Five neuromeres of hindbrain are distinct. Anterior neuropore is closing. Optic vesicles constricted at bases. Heart bent to right.
[*Rathke's pouch forming; pronephros degenerates; mesonephros differentiates; amniotic head fold appears; neurohypophysis begins to form.*]

Stage 12. ×12 Sixteen pairs of somites (45–49 h). Head is turning onto left side. Anterior neuropore is closed. Telencephalon indicated. Primary optic vesicles and optic stalk well established. Otic pit is deep

but wide open. Heart is slightly S-shaped. Head fold of amnion covers entire region of forebrain.
[*Circulatory system developing; ductus venosus and liver primordia appear; telencephalon differentiating; amniotic tail fold.*]

Stage 13. ×12 Nineteen pairs of somites (48–52 h). Head is partly to fully turned onto the left side. Cranial and cervical flexures make broad curves. Enlargement of telencephalon. Atrio-ventricular canal indicated by constriction. Head fold of amnion covers forebrain, midbrain and anterior part of hindbrain.
[*Proventriculus and gizzard differentiate; interatrial septum forms; allantoic bud and hindgut develop.*]

Stage 14. ×12 Twenty-two pairs of somites (50–53 h). Cranial flexure: axes of forebrain and hindbrain form almost a right angle. Cervical flexure is a broad curve. Rotation of body as far back as somites 7–9.
Visceral (*pharyngeal*) arches 1 and 2, and clefts (*grooves*) 1 and 2 are distinct. Optic vesicle begins to invaginate and lens placode is formed. Opening of otic pit constricted. Rathke's pouch recognizable. Ventricular loop of heart now ventral to atrio-ventricular canal. Amnion extends to somites 7–10.
[*Pancreatic anlage differentiate.*]

Beyond stage 14 the number of somites becomes increasingly difficult to determine with accuracy.

Stage 15. ×12 (*c.* 50–55 h).
Lateral body folds extend to level of somites 15–17.
Limb primordia: inconspicuous condensations of mesoderm for wing bud. 24–27 pairs of somites.
Amnion extends to level of somites 7–14.
Flexures and rotation: cranial flexure: axes of forebrain and hindbrain form acute angle. Cervical flexure is a broad curve. Rotation extends to level of somites 11–13.
Optic cup completely formed.
Visceral (pharyngeal) arches: arch 3 and cleft 1 are distinct.
[*Lung primordia differentiate from pharynx.*]

Stage 16. ×12 (*c.* 51–56 h).
Lateral body folds extend to level of somites 17–20, between levels of wings and legs.
Limbs: wings are lifted off blastoderm. Primordia of legs flat.
Somites: 26–28 pairs.
Amnion: extends to level of somites 10–18.
Flexures and rotation: all flexures more accentuated. Rotation extends to somites 14–15.
Tail bud: a short, straight cone.
[*Epiphysis develops.*]

Stage 17. ×12 (*c.* 52–64 h).
Lateral body folds extend around entire circumference of body.
Limb buds: both wing and leg buds lifted off blastoderm.
Somites: 29–32 pairs.
Amnion: variable in extent.
Rotation: extends to level of somites 17–18.
Tail bud: bent ventrally. Its mesoderm unsegmented.
Pineal: a distinct knob.
Allantois: not yet formed.

Stage 18. ×12 (*c.* 65–69 h).
Limb buds enlarged (see *HH Plates*, pp. 426–7).
Somites: 30–36 pairs.
Amnion: usually closed.
Rotation: now extends to posterior part of body.
Visceral (*pharyngeal*) arches: maxillary process absent or inconspicuous. Fourth cleft (groove) indistinct or absent.
Tail bud: turned to right.
Allantois: short, thick-walled pocket.
[*First passive movements can be detected.*]

Stage 19. ×12 (*c.* 68–72 h).
Limb buds: leg buds slightly larger than wing buds (see *HH Plates*, p. 427).
Somites: 37–40 pairs, extend into tail.
The contour of the posterior part of the trunk is straight to the tail.
Tail bud curved, tip pointing forward (*anteriorly*).
Visceral (*pharyngeal*) arches: maxillary process about same length as mandibular process. The second (*hyoid*) projects slightly over the surface.
Allantois: small pocket of variable size.
Eyes: unpigmented.
[*Aortic arches form, the left soon to degenerate; nephric ducts unite with cloaca.*]

Stage 20. ×12 (*c.* 70–72 h).
Limb buds: wing buds approximately symmetrical, leg buds slightly asymmetrical (see *HH Plates*, p. 427).
Somites: 40–43 pairs.
Rotation: completed.
Visceral (*pharyngeal*) arches: maxillary equals or exceeds mandibular in length. Second (*hyoid*) arch projects over surface. Fourth arch smaller than third arch.
Allantois: vesicular, variable size.
Eye pigment: faint grey.
[*Pulmonary arch forms; ductus arteriosus; embryo surrounded by amnion.*]

Stage 21. ×12 (*c.* 3.5 days).
Limbs: both wing and leg buds are slightly asymmetrical with proximo-distal axes directed caudally.

Somites: 43–44 pairs.
Visceral (*pharyngeal*) arches: maxillary process longer than mandibular. Second (*hyoid*) overlaps third arch ventrally. Fourth arch is distinct.
Allantois: variable in size, may extend to head.
Eye pigmentation: faint.
[*First active movements of head and neck; earliest motor fibres contact anterior trunk muscles.*]

Stage 22. ×8 (*c.* 3.5 days).
Eye pigmentation: distinct.
Somites: somites extend to tip of tail. (*This is incorrect since the mesoderm at the tip of the tail never becomes fully segmented.*)
[*Adrenal cortical and medullary cells differentiate; insulin synthesis begins; gonadal anlagen differentiate; amniotic contractions begin; crop differentiates; erythropoiesis in yolk sac begins; oestrogen and oestrodiol-17β synthesis begin.*]

Stage 23. ×8 (*c.* 4 days).
Limbs: both wing buds and leg buds approximately as long as they are wide.
Visceral (*pharyngeal*) arches: see *HH Plates*, p. 429.
[*Pronephros disappears.*]

Stage 24. ×8 (*c.* 4.5 days).
Limbs: both wing and leg buds longer than wide. Toe plate in leg bud distinct.
Visceral (*pharyngeal*) arches: see *HH Plates*, pp. 429–430.
[*Chorion and allantois fuse to give chorio-allantois; metanephros begins to differentiate; cochlear nucleus becomes visible.*]

Stage 25. ×8 (*c.* 4.5–5 days).
Limbs: elbow and knee joints distinct. Digital plate in wing distinct but no demarcation of digits.
Visceral (*pharyngeal*) arches: see *HH Plates*, pp. 429–430.
[*Spleen differentiates.*]

Stage 26. ×8 (*c.* 5 days).
Limbs: longer. Contour of digital plate rounded. Demarcation of first three toes distinct.
Visceral (*pharyngeal*) arches: see *HH plates*, pp. 430–431.
[*Mesonephros becomes functional; production of definitive erythrocytes begins; adult haemoglobin synthesized; first active movements of trunk; corticosteroid synthesis begins; duodenum begins to differentiate; first mouth movements; bursa of Fabricius begins to differentiate; four-chambered heart formed; nerves from retina reach optic lobe; thymus anlagen differentiate.*]

Stage 27. ×5 (*c.* 5–5.5 days).
Limbs: grooves between first, second and third digits indicated. Distinct grooves between toes.
Visceral (*pharyngeal*) arches: see *HH Plates*, pp. 430–431.
[*Beak forming; amnion begins to contract rhythmically.*]

Stage 28. ×5 (*c.* 5.5–6 days).
Limbs: second digit and third toe longer than others. Three digits and four toes distinct.
Visceral (*pharyngeal*) arches: see *HH Plates*, pp. 430–431.
Beak: a distinct outgrowth is visible in profile.
[*First reflexes are established.*]

Stage 29. ×5 (*c.* 6–6.5 days).
Limbs: wing bent at elbow. Shallow grooves between first, second and third digits. Second to fourth toes stand out as ridges separated by grooves and with indications of webs. Rudiments of fifth toes visible.
Visceral (*pharyngeal*) arches: mandibular process and second arch are broadly fused. Auditory meatus distinct at dorsal end of fusion.
Neck has lengthened: see *HH plates*, p. 432.
[*Parathyroids differentiate; air sacs begin to differentiate; first eyelid and independent limb movements; ductus venosus is lost.*]

Stage 30. ×5 (*c.* 6.5–7 days).
Limbs: the three major segments of limbs are well demarcated. Wing bent at elbow and leg bent at knee. Distinct grooves between first and second digits.
Neck: lengthened.
Feather germs: two dorsal rows on either side of spinal cord at brachial level, three rows at level of legs.
Scleral papillae: one on either side of choroid fissure.
Egg tooth: distinct.
[*Thyroid concentrates iodine; sexual differentiation begins; testosterone production begins in male.*]

Stage 31. ×4 (*c.* 7–7.5 days).
Limbs: rudiment of fifth toe still distinct.
Feather germs: on dorsal surface, continuous from brachial to lumbo-sacral level. Approximately seven rows at lumbo-sacral level. Distinct feather papillae on thigh.
Scleral papillae: usually six, four on the dorsal side, two on the ventral.
[*Left allantoic blood vessel is lost; first eyeball movements begin; thyroid able to synthesize monoiodotyrosine; ACTH secretion begins.*]

Stage 32. ×4 (*c.* 7.5 days).
Limbs: all digits and four toes have lengthened conspicuously. Webs between digits and toes are thin and their contours concave. Rudiment of fifth toe has disappeared. Differences in size of individual digits and toes conspicuous.
Feather germs: eleven rows or more on dorsal surface at level of legs. One row on tail distinct.
Scleral papillae: 6–8 in groups.

Stage 33. ×4 (*c.* 7.5–8 days).
Limbs: web on radial margin of arm and first digit.
Feather germs: three distinct rows in tail, the middle one considerably larger than the other ones.

Scleral papillae: 13 present, forming almost complete circle.
[*Mineralization of bone begins; development of right Mullerian duct ceases in female and of both left and right in male.*]

Stage 34. ×4 (*c.* 8 days).
Limbs: differential growth of second digit. Third toe conspicuous.
Feather germs: visible with good illumination over scapula, ventral side of neck, procoracoid and posterior edge of wing. One row on inner side (*medial*) of each eye. None around umbilical cord.
Scleral papillae: 13 or 14.
Nictitating membrane: extends halfway between outer rim of eye and scleral papilla.

Stage 35. ×4 (*c.* 8–9 days).
Limbs: webs between digits and toes inconspicuous.
Beak: lengthened.
Feather germs: mid-dorsal line stands out distinctly in profile. At least four rows on inner side of each eye. New feather germs near mid-ventral line and extending to both sides of umbilical cord.
Nictitating membrane: approaches the outer scleral papillae.
Eyelids (external to nictitating membrane) have extended towards beak and begun to overgrow eyeball.
[*Thyroid able to synthesize diiodotyrosine; male Mullerian ducts regress; chorioallantois becomes fixed in relation to shell; haemopoietic activity in bone marrow; fibres of acoustic ganglion enter cochlear nucleus.*]

Stage 36. ×1.2 (*c.* 10 days).
Limbs: tapering primordia of claws just visible on termini of toes.
Comb: primordium visible as prominent ridge with slightly serrated edge along dorsal midline of beak. Nostril has narrowed to a slit.
Feather germs: flight feathers conspicuous. Feather germs now cover the tibio-fibular portion of leg. At least nine rows of feather germs between upper eyelid and dorsal midline. Sternal tracts prominent.
Eyelids: lower lid has grown upward to level of cornea.
[*Whole body movements become jerky and random; thyroid secretes thyroxine; TSH secretion begins; parathormone secretion begins; embryo's position fixed at right angles to long axis of egg; first local proprioceptive muscle reflexes.*]

Stage 37. ×1.2 (*c.* 11 days).
Limbs: claws of toes flattened laterally and curved ventrally. Pads on plantar surface of foot. Length of third toe = 7.4 ± 0.3 mm.
Comb more prominent and clearly serrated.
Feather germs: elongated into long, tapering cones along back and tail.
Eyelids: lower lid covers one-third to one half of cornea. Upper lid has reached dorsal edge of cornea.
[*Metanephros begins to function; first auditory evoked responses from cochlear nuclei.*]

Stage 38. ×1.2 (*c.* 12 days).
Limbs: primordia of scales visible as ridges over entire surface of leg. Length of third toe = 8.4 ± 0.3 mm.
Feather germs: coverts of web of wing becoming conical. Sternum covered with feather germs except along midline. Upper eyelid covered with newly formed feather germs.
Lower eyelid covers two-thirds to three-quarters of cornea.
[*Absorption of albumen begins; mesonephros degenerates; Mullerian ducts lost in male; calcium absorption from shell begins; inbibition of amniotic fluid begins.*]

Stage 39. ×1.2 (*c.* 13 days).
Limbs: scales overlapping on superior surface of leg. Major pads of phalanges covered with papillae. Length of third toe = 9.8 ± 0.3 mm.
Mandible and maxillae cornified (opaque) as far back as proximal level of egg tooth.
Feather germs: coverts of web of wing are long, tapering cones.
Eyelids: opening between lids reduced to a thin crescent.
[*Increased transport of lipids by yolk sac; amniotic contractions cease; sero-amniotic connection ruptures; pituitary–gonad axis definitely established; neurohypophysis becomes active; gland cells of proventriculus begin secretion.*]

Stages 40–44 are based mainly on the length of the beak and on the length of the third (longest) toe, since other external features have lost their diagnostic value.

Stage 40. ×1 (*c.* 14 days).
Length of beak from anterior edge of nostril to tip of bill = 4.0 mm.
Length of third toe = 12.7 ± 0.5 mm.
[*Exocrine pancreas begins maturation; oestriol synthesis begins in female; maximum sensitivity to sound at 400 Hz; sporadic electrical activity in cerebrum.*]

Stage 41. ×1 (*c.* 15 days).
Length of beak from anterior angle of nostril to tip of upper bill = 4.5 mm.
Length of third toe = 14.9 ± 0.8 mm.
[*Stomach begins to contract; electrical activity in optic lobes.*]

Stage 42. ×1 (*c.* 16 days).
Length of beak from anterior angle of nostril to tip of upper bill = 4.8 mm.
Length of third toe = 16.7 ± 0.8 mm.

[*Embryo capable of respiratory movements; first electrical activity in cerebellum.*]

Stage 43. ×1 (*c.* 17 days).
Length of beak from anterior angle of nostril to tip of upper bill = 5.0 mm.
Length of third toe = 18.6 ± 0.8 mm.
[*Coordinated and stereotyped movements begin.*]

Stage 44. ×1 (*c.* 18 days).
Length of beak from anterior angle of nostril to tip of upper bill = 5.7 mm.
Length of third toe = 20.4 ± 0.8 mm.
[*Beak becomes tucked under right wing; maximum sensitivity to sound rises to 800 Hz. Duodenum begins maturation; calcitonin secretion begins; first behavioural responses to light.*]

Stage 45. ×1 (*c.* 19–20 days).
Beak length no longer diagnostic because of reduction in length due to sloughing off of periderm, but is now shiny.
Length of third toe unchanged (except for certain specialized breeds with a slightly longer incubation period).
Extra-embryonic membranes: yolk sac half enclosed in body.
Chorioallantoic membrane 'sticky' in living embryo.
[*Absorption of allantoic fluid completed; hatching muscle matures; hepatic glycogen stores mobilized; postural reflexes fully developed; yolk sac withdrawal begins; cerebral auditory waves attain regularity of those of hatched chick.*]
Day 20: [*Begins to breathe and vocalize; inner shell membrane pierced; shell 'pipped'; ductus arteriosus closed; interatrial foramina closed; chorioallantoic ectoderm degenerates; chorioallantoic circulation reduced; withdrawal of yolk sac completed.*]

Stage 46. ×1 (*c.* 20–21 days). Newly hatched chick.

Appendix 2 Plates

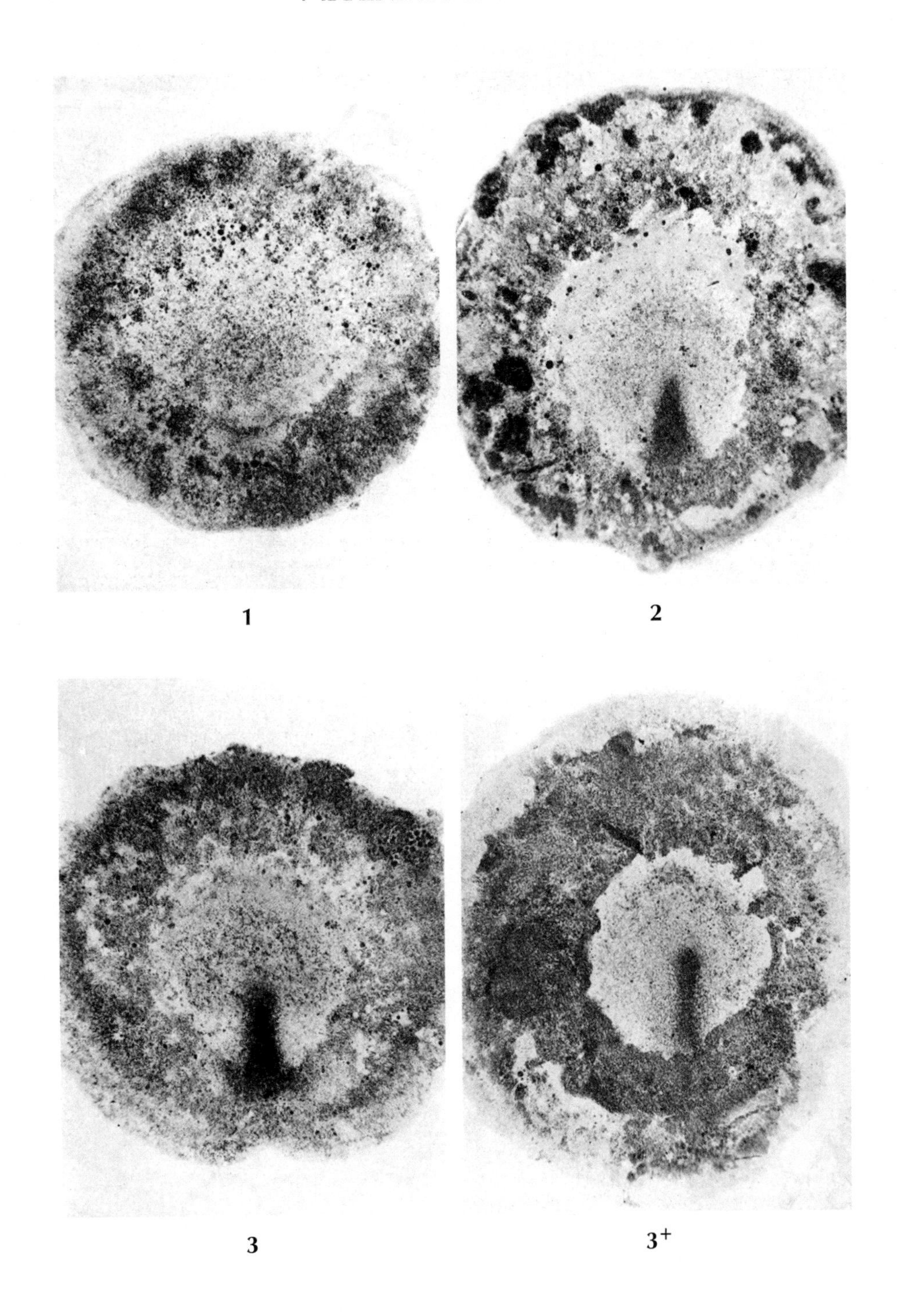

Appendix 2 Plates (Continued)

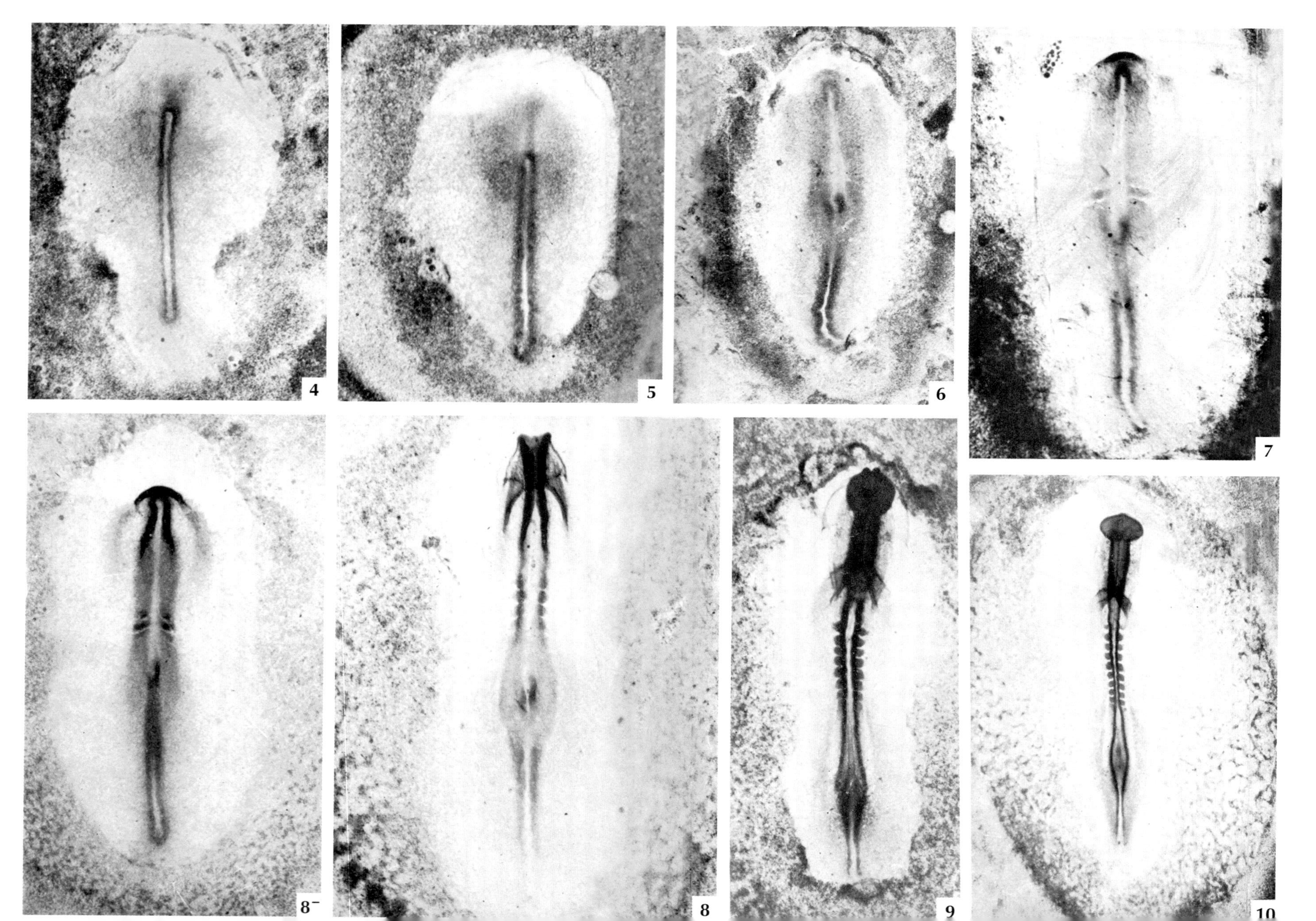

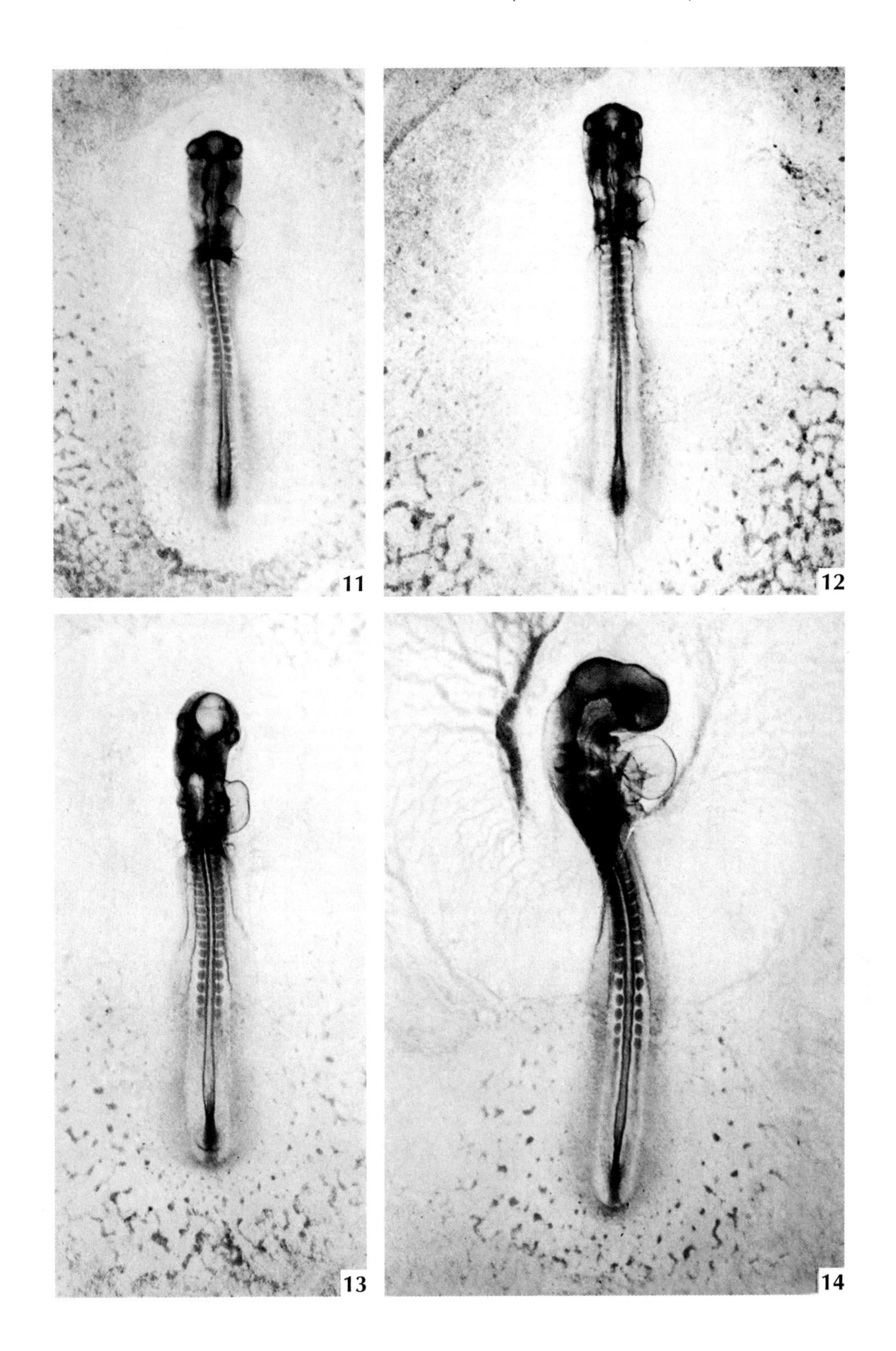
11
12
13
14

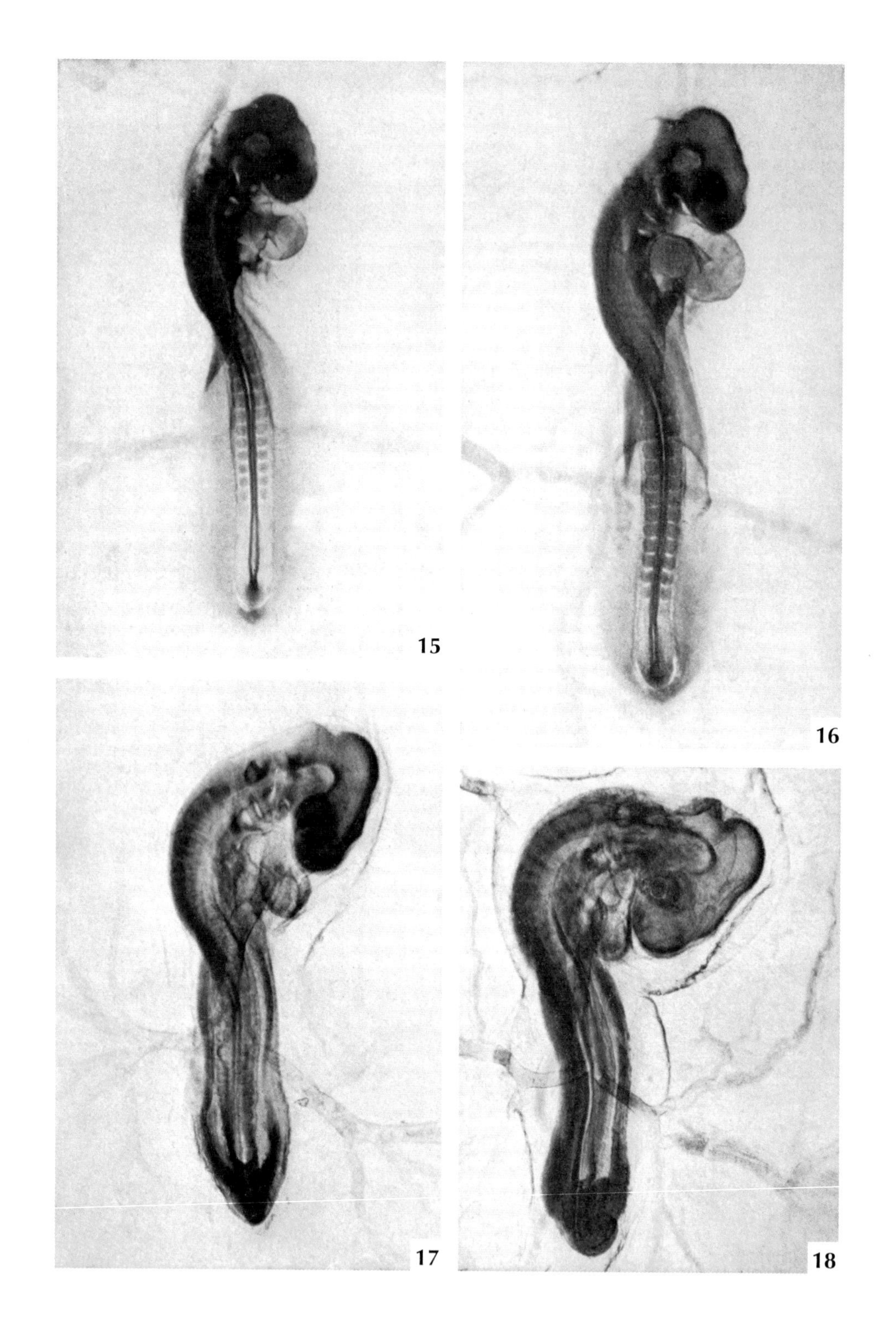
15
16
17
18

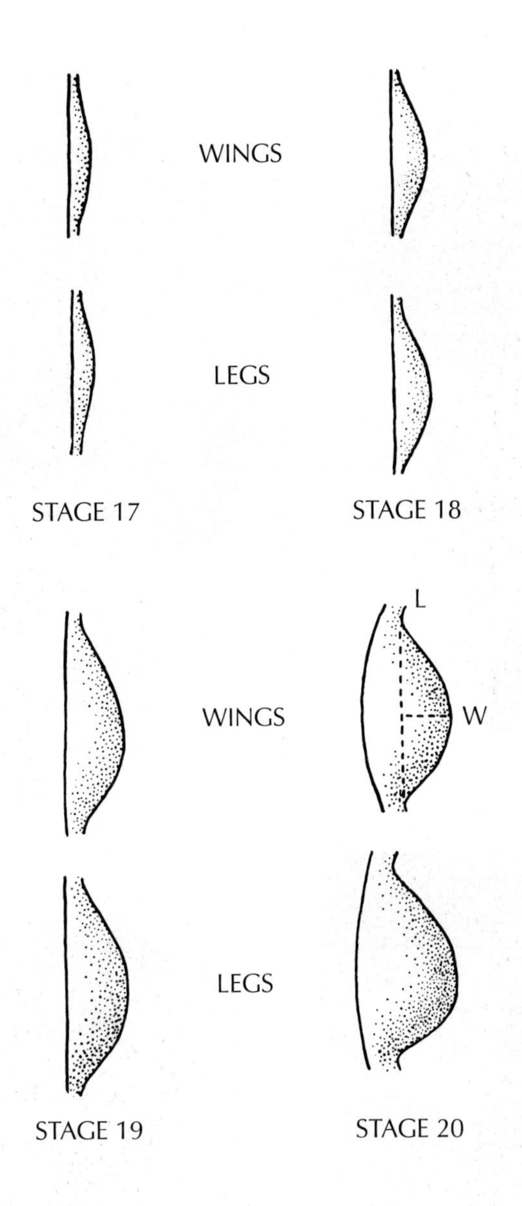
WINGS
LEGS
STAGE 17
STAGE 18
L
WINGS
W
LEGS
STAGE 19
STAGE 20

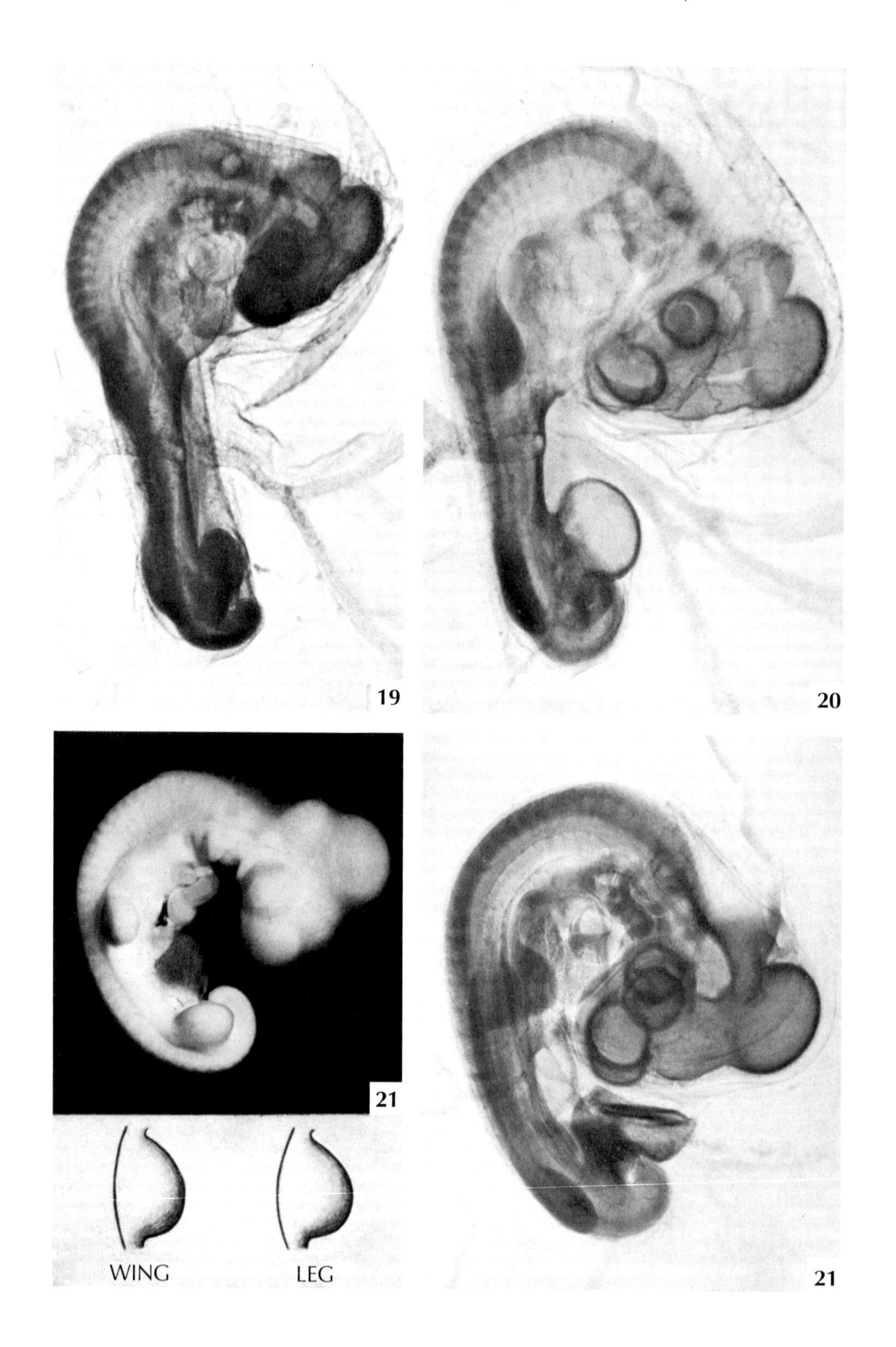
19
20
21
WING
LEG
21

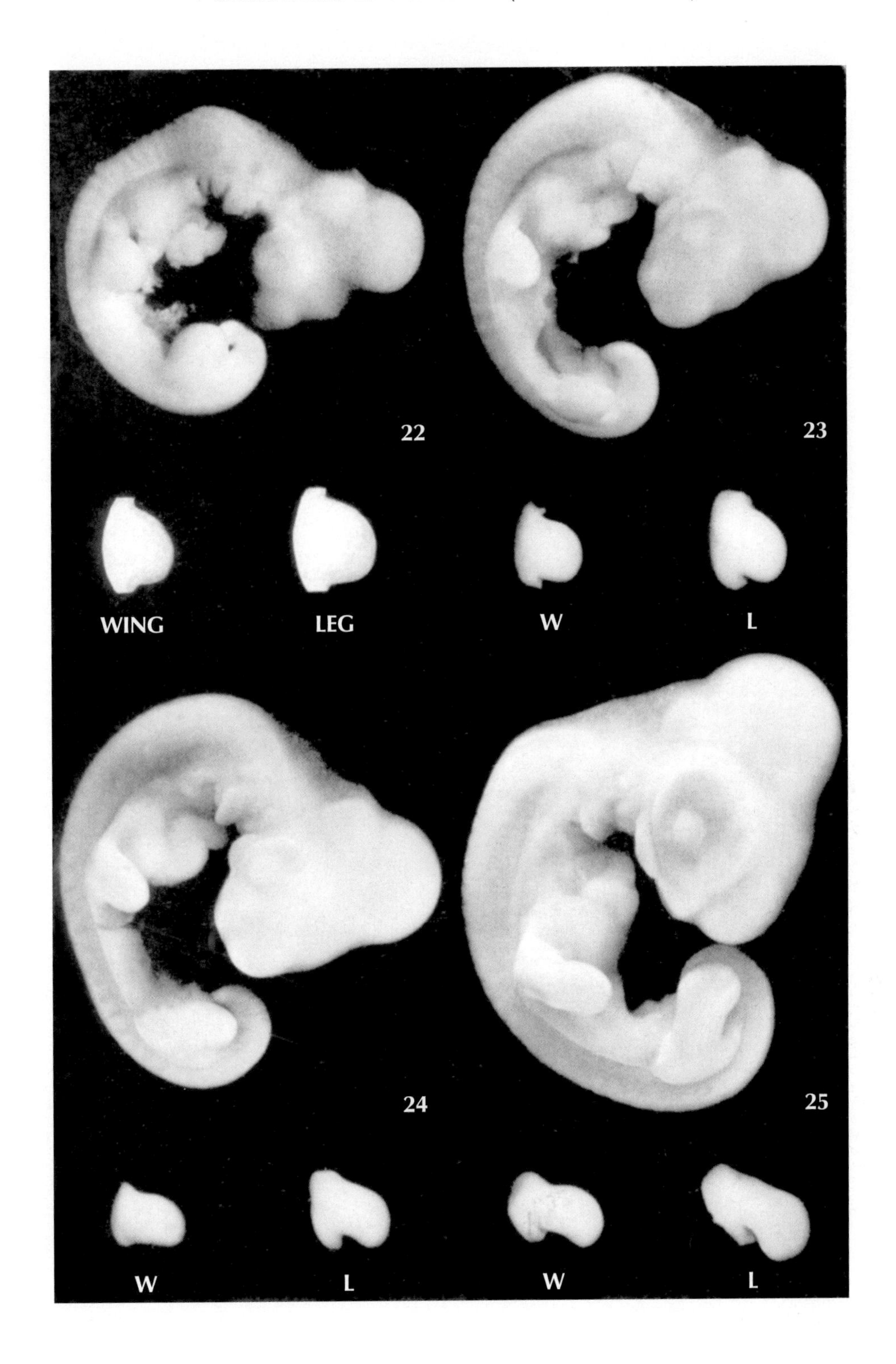
22
23
WING
LEG
W
L
24
25
W
L
W
L

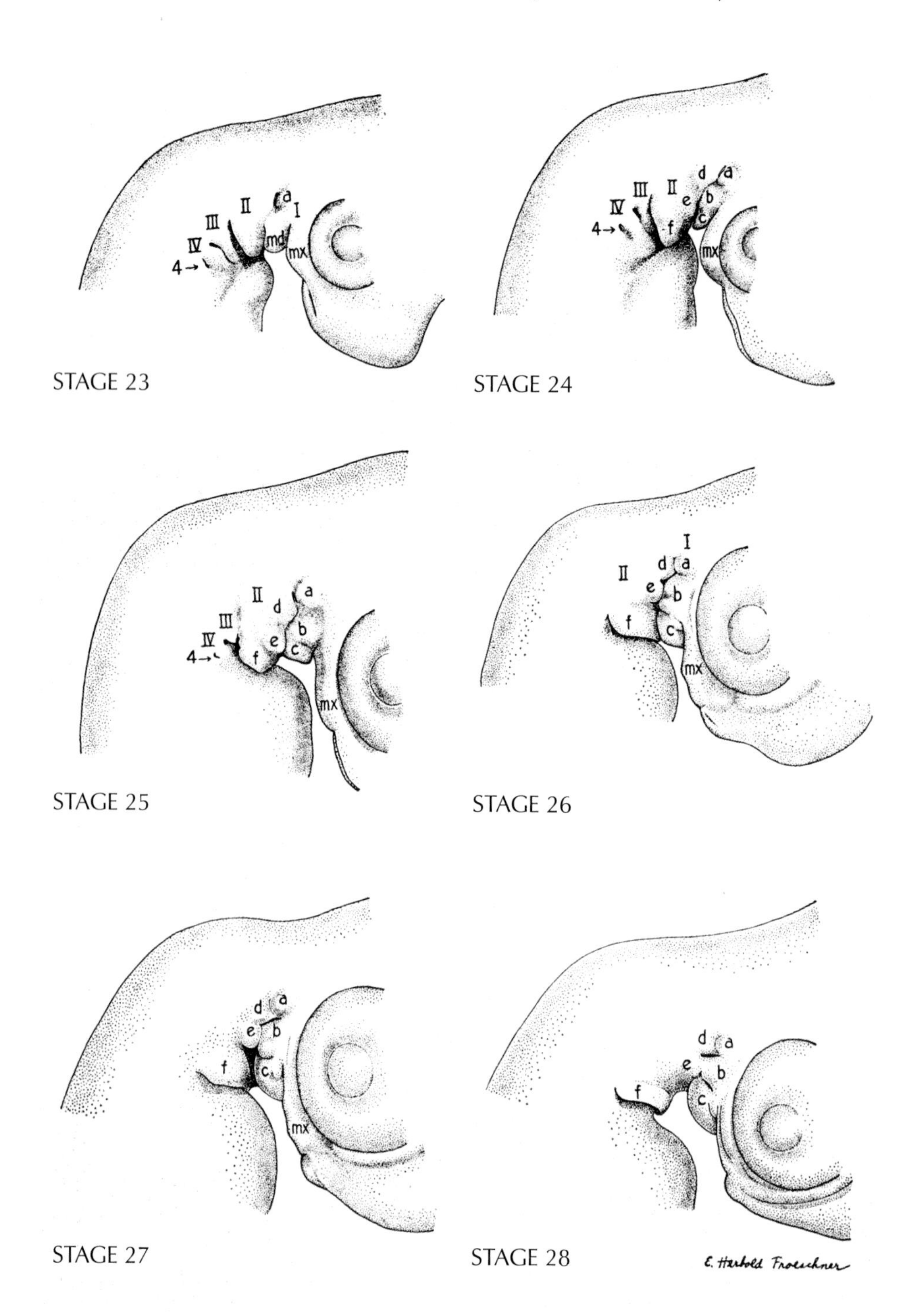
STAGE 23
STAGE 24
STAGE 25
STAGE 26
STAGE 27
STAGE 28
a
b
c
d
e
f
md
mx
I
II
III
IV
4→
E. Harbold Froeschner

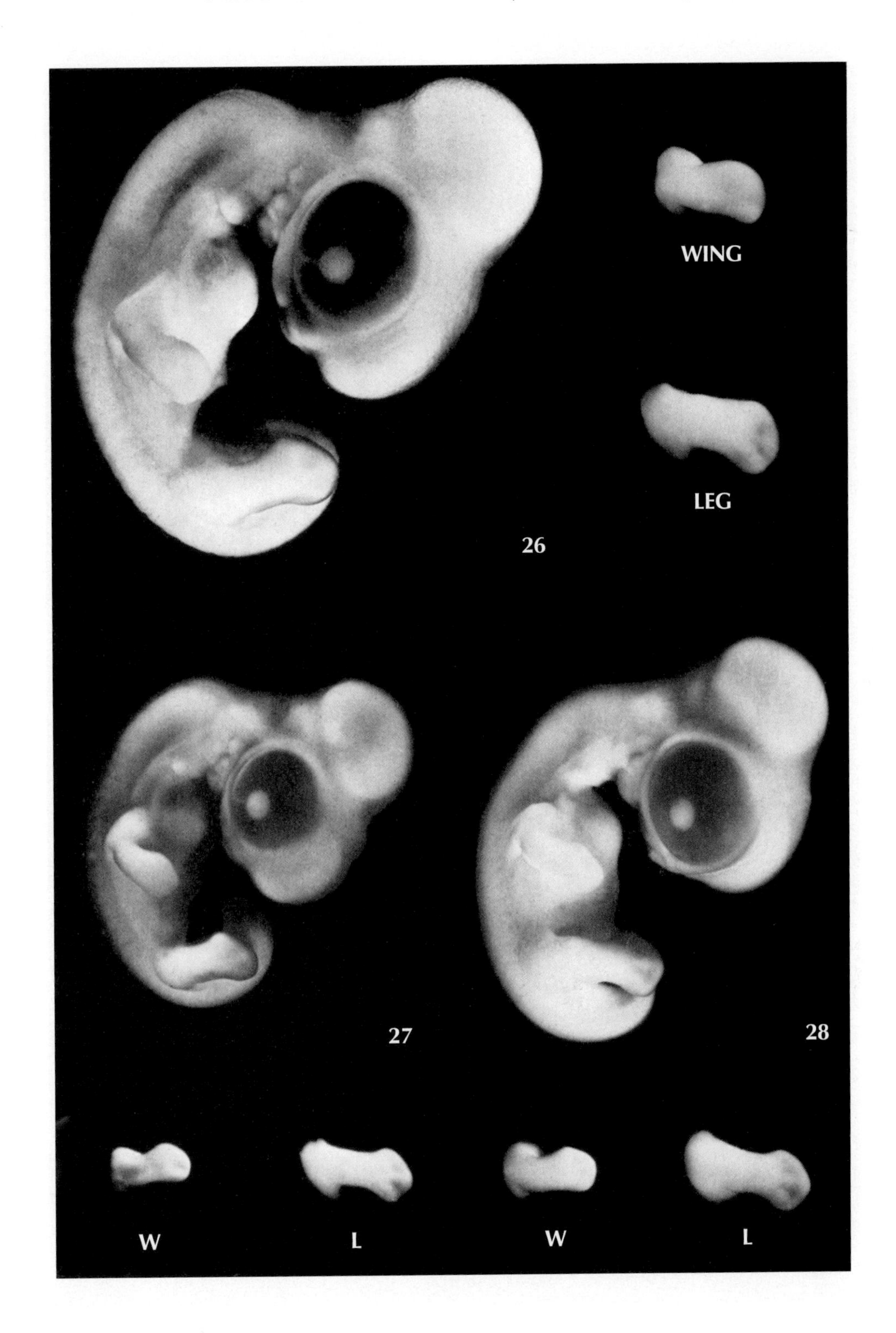
WING
LEG
26
27
28
W
L
W
L

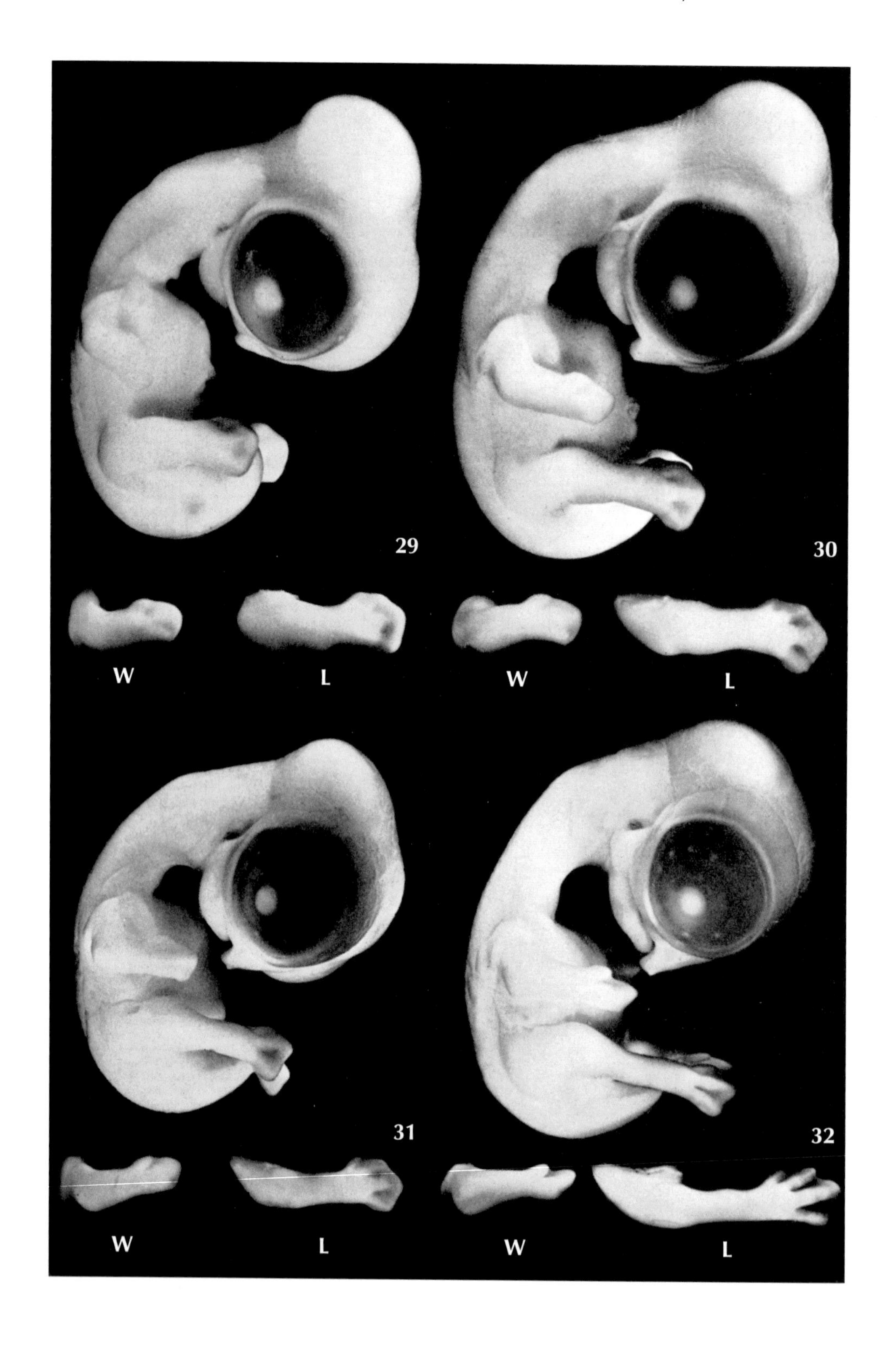
29
W
L
30
W
L
31
W
L
32
W
L

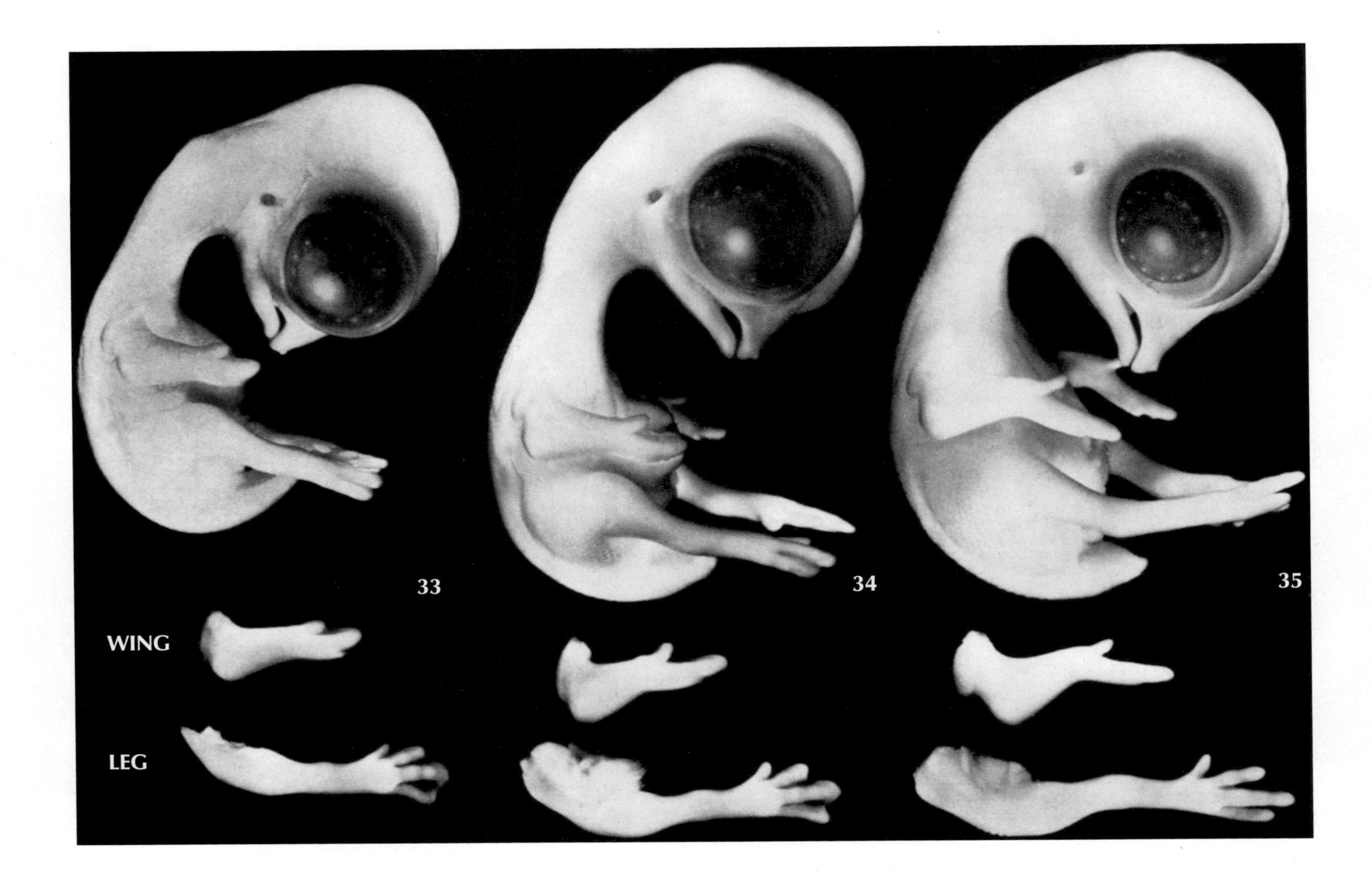
33
34
35
WING
LEG

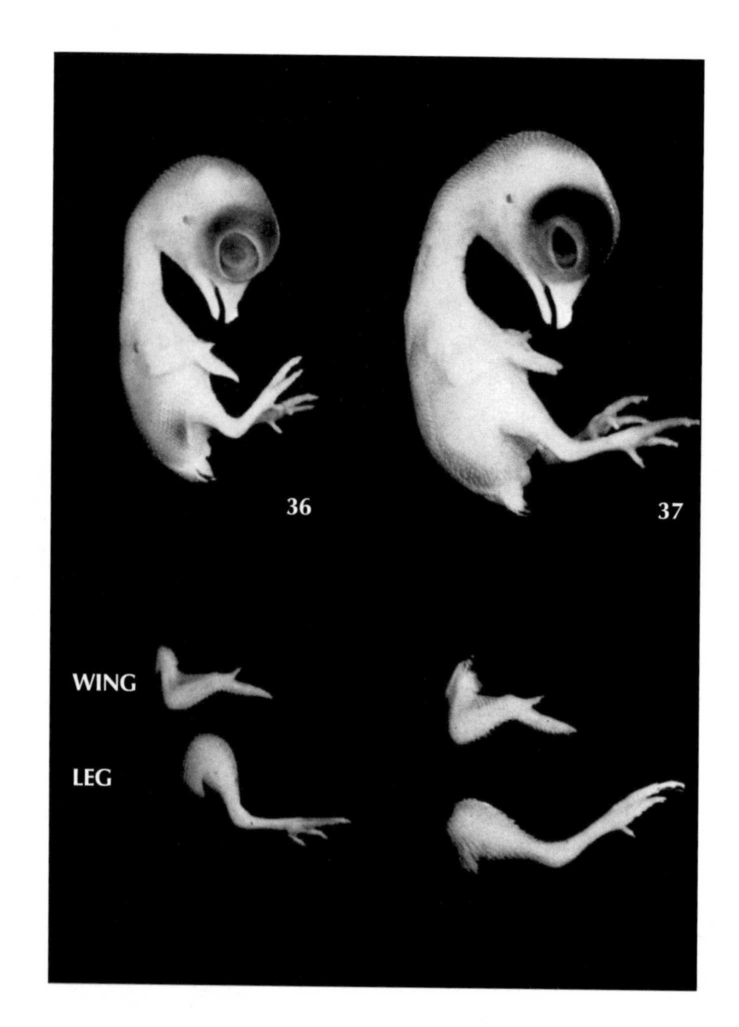
36
37
WING
LEG

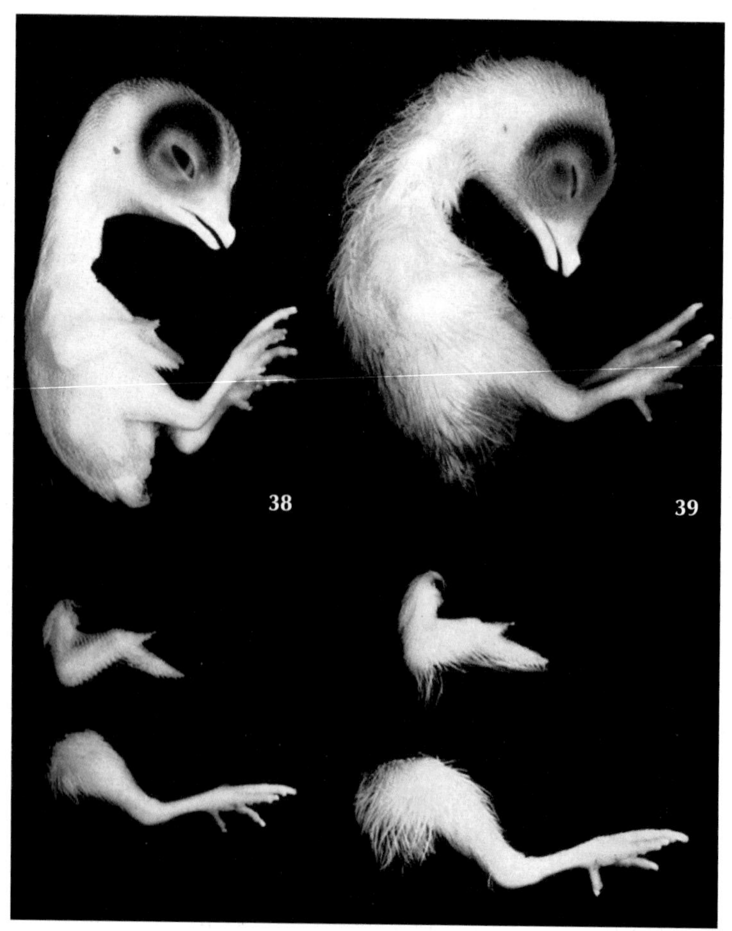
38
39

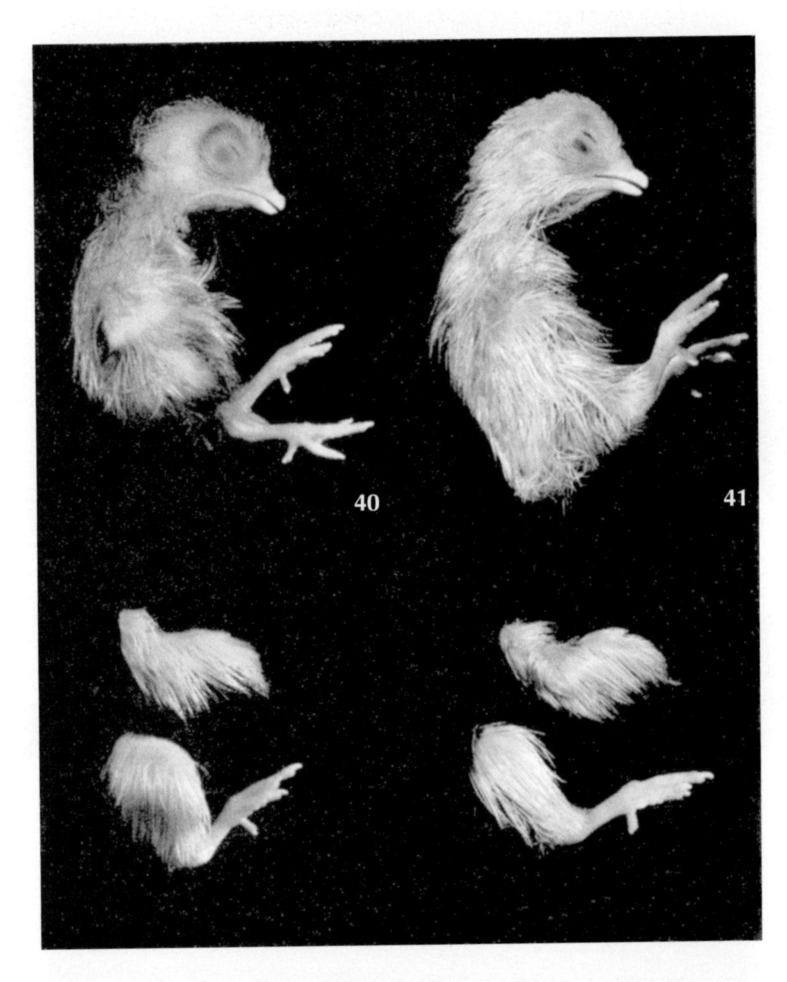
40
41

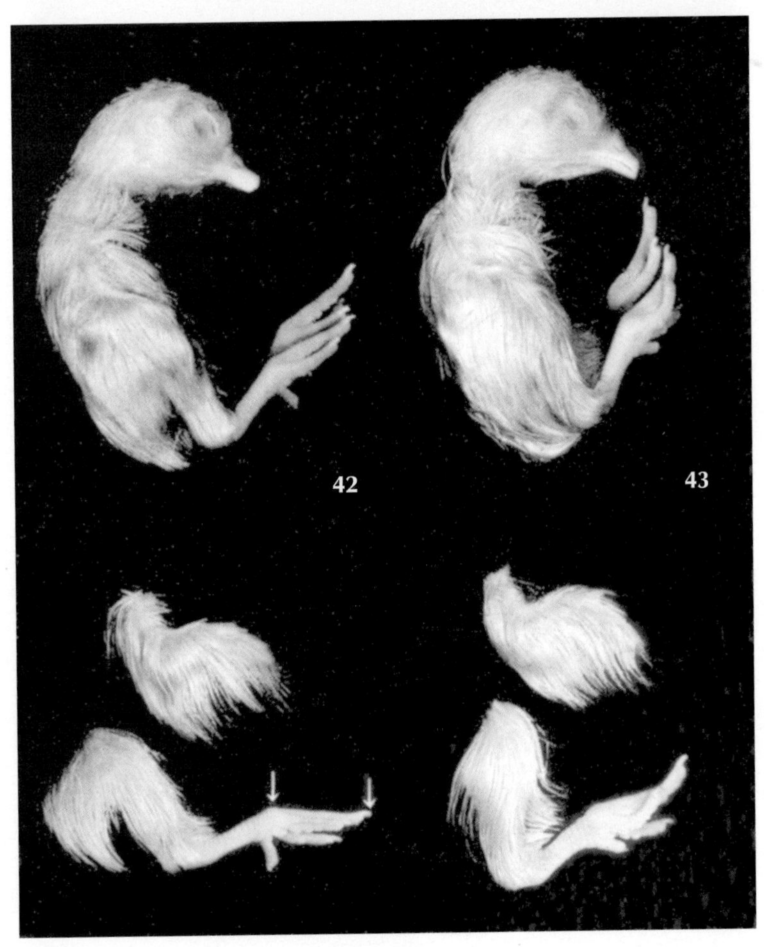
42
43

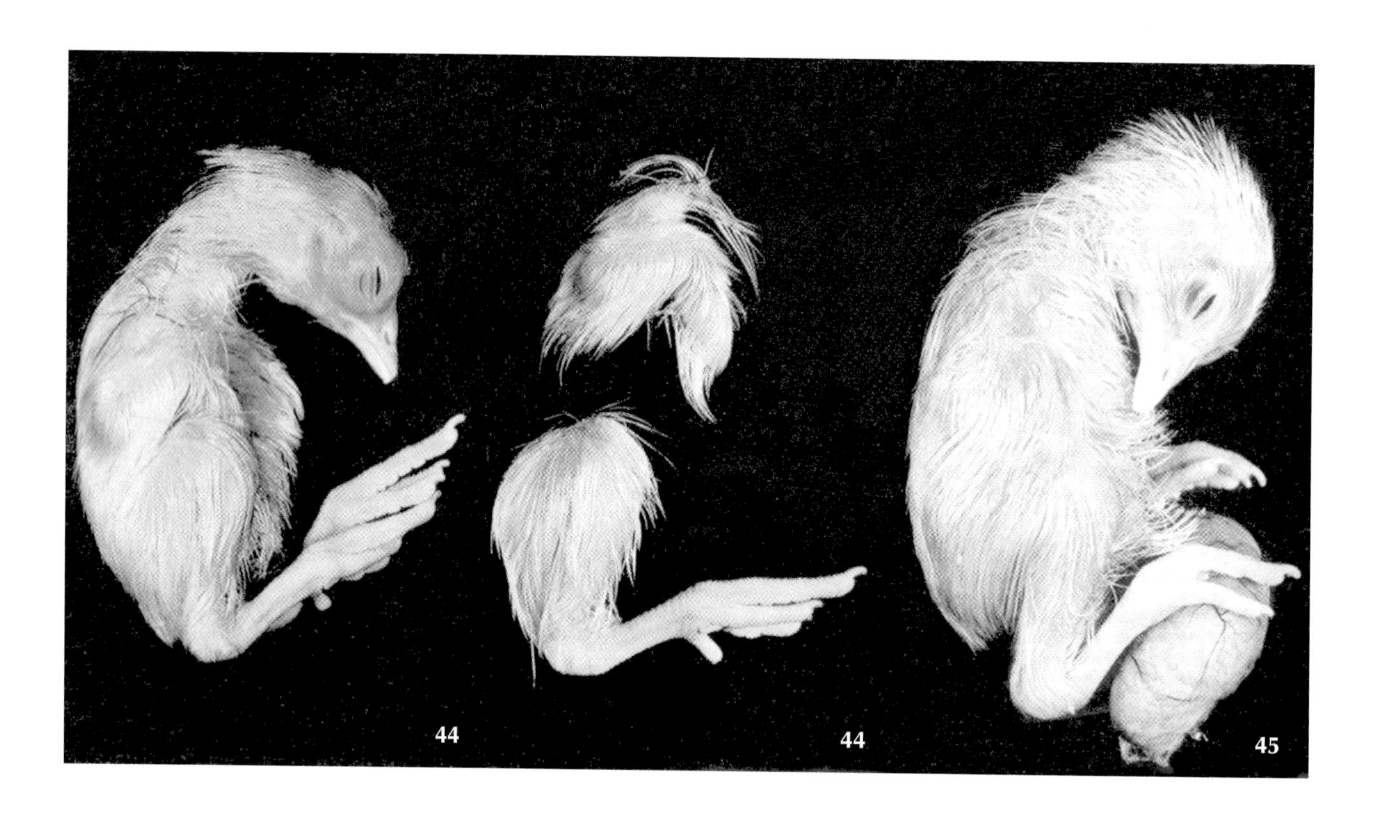
44
44
45

Appendix III

Supplement to Appendix II by Murray and Wilson (1994)

This table is an elaboration of the wing stages (19–36) of Hamburger and Hamilton (Appendix II). The text has been slightly abbreviated from that of the original publication. Reprinted by permission of Springer-Verlag.

Plate IIIa

(a) **Stage 19** (68–72 h). The wing bud is symmetrical about the midline of the antero-posterior axis, but is distinctly wider than it is long. The wing bud extends across the width of six somites. The apical ectodermal ridge (AER) is indicated by the arrow. Bar = 250 μm.

(b) **Stage 20** (70–72 h). The wing bud is still approximately symmetrical about the antero-posterior axis but now extends across the width of only five somites. There is an increase in the proximo-distal length of the bud. The length : width ratio of the limb bud is 1 : 3.5. Bar = 200 μm.

(c) **Stage 21** (3.5 days). The wing bud is asymmetrical with the proximo-distal axis lying at right angles to the main body axis, but with an increased proportion of the limb posterior to the midline. Post-axially the wing bud meets the flank at a right angle. The wing bud is still wider than it is long. The length : width ratio is 1 : 2.5. Bar = 140 μm.

(d) **Stage 22** (3.5 days). The wing bud extends across the width of only four somites. The proximo-distal axis continues to be directed caudally and the AER is becoming located more ventrally on the apex of the wing. The length : width ratio is 1 : 1. Bar = 100 μm.

(e) **Stage 23** (3.5–4 days). The wing bud extends across the width of three and a half somites. The anterior and posterior margins of the limb bud are approximately parallel at their bases. Post-axially, the limb forms an angle of approximately 45° with the flank. There is marked elongation of the limb in the proximo-distal axis. Bar = 140 μm.

(f) **Stage 24** (4 days). The wing bud extends across the width of three somites. It is now distinctly longer than wide. There is no demarcation of the digital plate but a central bulge (arrow) at the base of the limb is apparent (coinciding with onset of development of the proximal cartilaginous humerus). Bar = 150 μm.

Plate IIIb

(a) **Stage 25** (4.5 days). The wing bud continues to elongate in the proximo-distal axis, but there is movement of the proximal anterior part of the wing bud away from the base of the somites, which is the first indication of the formation of the shoulder. There is a distinct bend in the wing bud where the direction of growth changes from lateral to ventral, which indicates the position of the prospective elbow. Compared with the preceding stage, the wing bud has become narrower around the humerus and broader in the posterior distal region, forming the paddle-shaped area corresponding to the future digital plate. A depression is visible in the posterior distal region of the wing bud, which may represent the space between the developing digits III and IV (arrow). Bar = 250 μm.

(b) **Stage 26** (4.5–5 days). The shoulder appears more prominent and the elbow is more clearly defined. The anterior and posterior margins of the paddle-shaped region have become rounded. The groove between the developing digits III and IV is still visible. There is a marked change in the posterior digital region of the limb (arrow) which may be due to cell death in the posterior necrotic zone. Bar = 225 μm.

(c) **Stage 27** (5 days). The three future sections of the wing are now apparent: the **stylopod** (humerus), the **zeugopod** (radius and ulna) and the **autopod** digital plate). The digital plate is beginning to assume blunt triangular arrow-head shape (which coincides with the first cartilaginous elements of the metacarpals). The digital plate is slightly longer apically in the region of the future digit III. There is a change in the contours of the wing in the region of the future wrist (where the anterior and posterior cartilaginous elements of the wrist have begun to form). Bar = 300 μm.

(d) **Stage 28** (5.5 days). The stylopod, zeugopod and autopod regions of the wing are more clearly defined and the wing is more bent at the elbow. The digital plate is now distinct from the proximal parts of the wing bud, being broader and irregular in appearance. It is distinctly longer in the region of digit III (the cartilaginous element of metacarpal III having appeared) and interdigital grooves are just visible (arrows). Bar = 300 μm.

(e) **Stage 29** (6 days). The wing remains bent at the elbow and is narrowed at the wrist. The anterior border of the digital plate is deeply angled at the wrist (overlying the region in which the proximal parts of digit II will form). There is a decrease in the curvature of the posterior contours of the digital plate (perhaps due to regression of the cartilaginous elements of digit V). The digital plate remains pointed in the region of digit III but the angulation at the apex is sharper. Interdigital grooves are clearly visible (arrows) with the groove between digits II and III being deeper than those between digits III and IV. Bar = 300 μm.

(f) **Stage 30** (6.5 days). The shoulder region has narrowed significantly and the anterior contours of the wing are straighter at the level of the elbow joint. There is a distinct depression along the anterior distal border of the wing between digits II and III (arrow). The digits have considerably increased in length and the grooves between them have deepened, with that between II and III remaining deeper than that between III and IV. There is an indication of a web forming between digits II and III. Bar = 600 μm.

Plate IIIc

(a) **Stage 31** (7 days). The apex of digit III appears to be more pointed than the other two digits, and there is an increase in the depth of the groove between digits III and IV. Bar = 750 μm.

(b) **Stage 32** (7.5 days). The wing now forms a right angle at the elbow. External contouring of the underlying digital elements, particularly those of digits II and III, is now obvious with distinct webs having formed between the digits. The differences in size between the digits has now increased, with digit III being the longest element (the terminal cartilaginous elements having now formed). Feather germs are not yet visible on the wing but are discernible in the shoulder region as low swellings (arrow). Bar = 750 μm.

(c) **Stage 33** (7.5–8 days). The wing/body wall boundary is becoming increasingly distinct, and the wing remains angulated at the elbow. The interdigital webs have continued to regress, in particular digit II has become separated from digit III. The pattern of feather germs of the shoulder region is similar to that observed in the previous stage, but a double row of feather germs is now visible on the posterior margin of the wing (arrows). Bar = 700 μm.

(d) **Stage 34** (8 days). The contours of most of the underlying cartilaginous elements are now distinct. There is little overall shape change in the zeugopod and stylopod regions. The most marked difference from previous stages is that only the proximal part of digit II is connected to digit III by a web. This freeing of digit II coincides with an increase in the angle between the two digits. The angle at the wrist joint has changed also so that digits III and IV now point posteriorly. A small protuberance is now visible at the posterior margin of digit III (white arrow) which according to Hamburger and Hamilton (1951) is a remnant of the web joining the two digits. Feather germs are now becoming apparent on the proximal anterior margin of the wing (black arrow). Bar = 800 μm.

(e) **Stage 35** (8–9 days). The wing forms an acute angle at the elbow, rather than a right angle as in the previous stage. Webs between the digits are not evident and the protruberance on the posterior margin of digit III has now disappeared. Slight tapering of the distal tip of digit

II is becoming apparent, indicating the site of the development of the future claw. Feather germs in the shoulder region are now distinct swellings. Bar = 700 μm.

(f) **Stage 36** (10 days). There is some angulation of the elbow. The angle at the wrist has increased so that digits III and IV are directed more posteriorly than at stage 35. The most marked difference in the digital region at this stage is the appearance of the claw on digit II (arrows). The number of rows of feather germs along the posterior border of the zeugopod has now increased and they appear as distinct swellings. Bar = 750 μm.

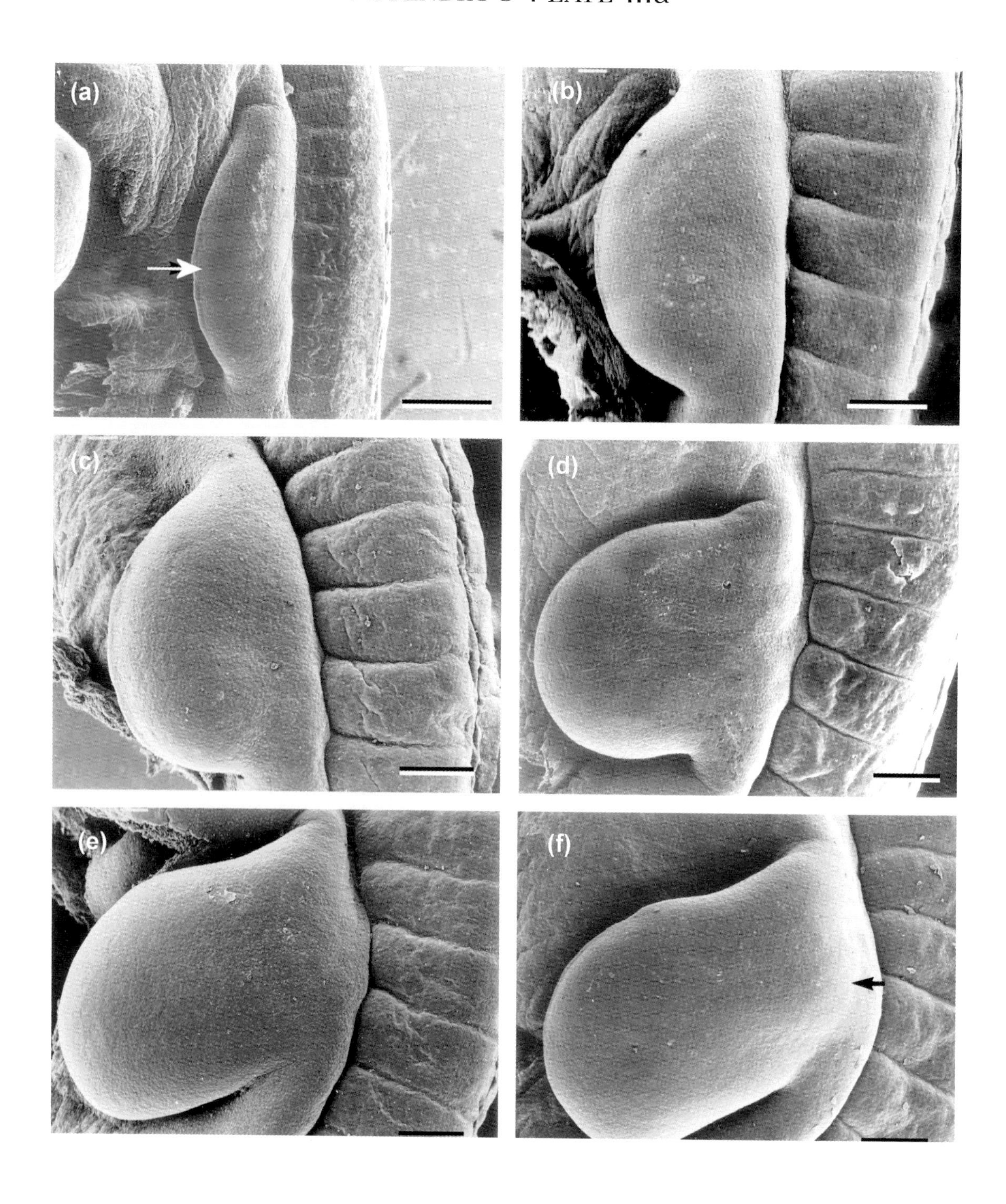
(a)
(b)
(c)
(d)
(e)
(f)

APPENDIX 3 PLATE IIIb

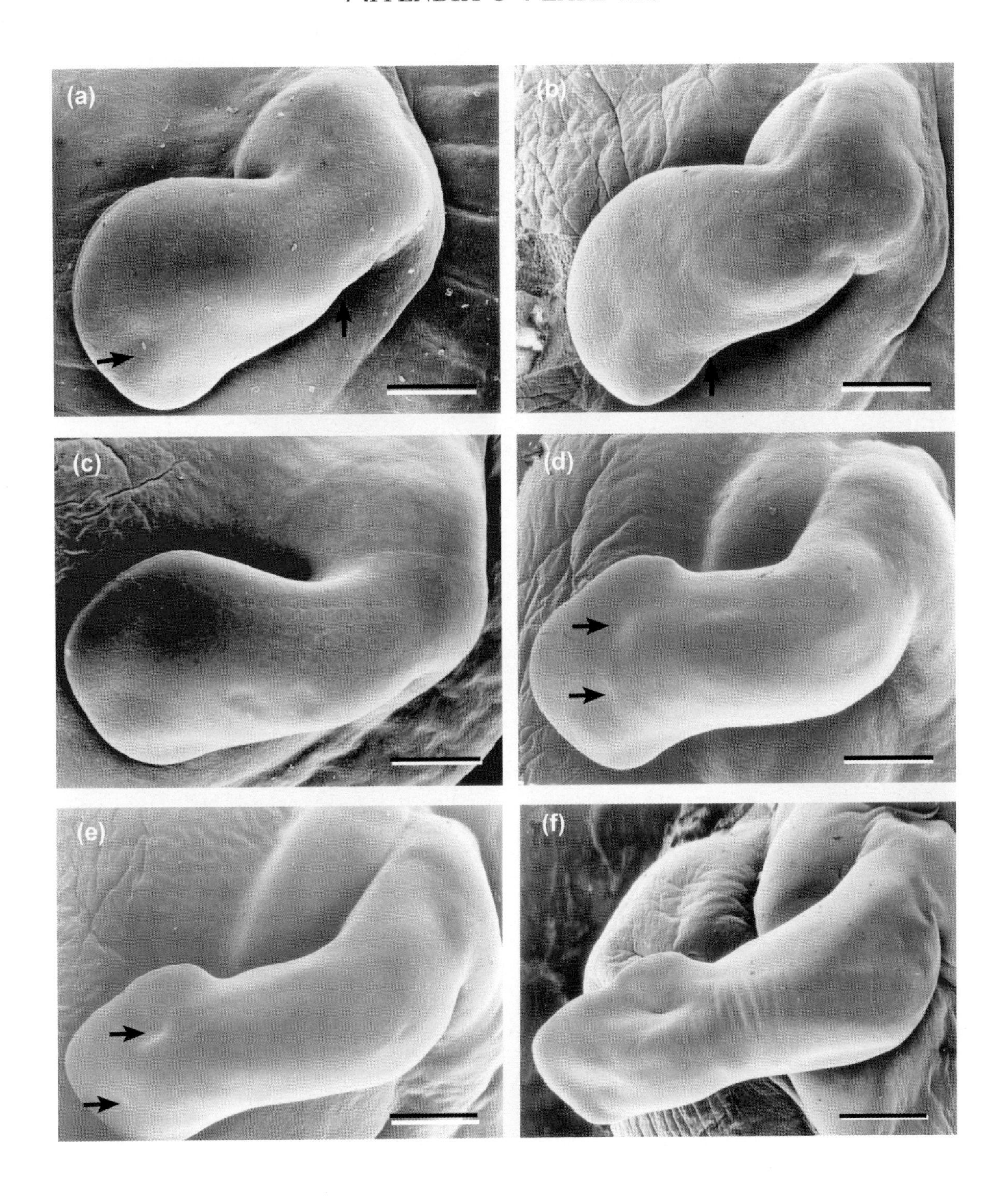

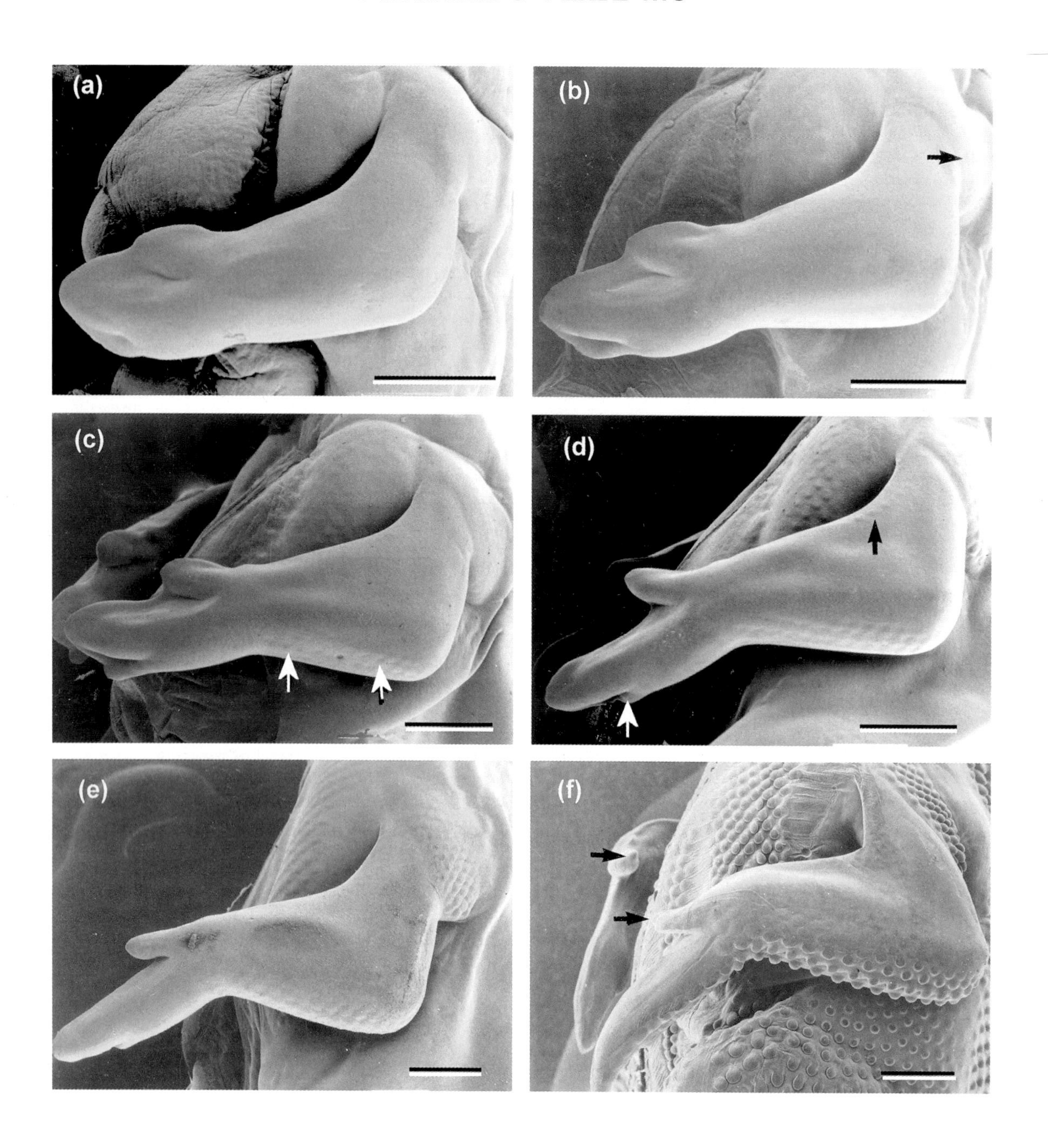
(a)
(b)
(c)
(d)
(e)
(f)

Appendix IV
Additional Normal Tables

Chick

Tolhurst, B.E. (1974) Development of the chick embryo in relation to the shell, yolk, albumen and extra-embryonic membranes. *Development of the Avian Embryo,* In (eds. B.M. Freeman, and M.A. Vince) *Chapman and Hall, pp. 277–91.*

Duck

Dupuy, V., Nersessian, B. and Bakst, M.R. (2002) Embryonic development from first cleavage through seventy-two hours incubation in two strains of pekin duck (*Anas platyrhynchos*). *Poult. Sci.* **81,** 860–8.

Koecke, H.-U. (1958) Normalstadien der Embryonalentwicklung bei der Hausente (*Anas Boschas domestica*). *Embryologia* **4,** 55–78.

Pheasant

Fant, R.J. (1957) Criteria for aging pheasant embryos. *J. Wildlife Management* **21,** 324–8.

Quail

Padgett, C.S. and Ivey, W.D. (1960) The normal embryology of the *Coturnix* Quail. *Anat. Rec.* **137,** 1–11.

Zacchei, A.M. (1961) Lo sviluppo embrionale della quaglia giaponese (*Coturnix coturnix japonica*). *Archo Ital. Anat. Embriol.* **66,** 36–62.

Turkey

Bakst, M.R., Gupta, S.K. and Akuffo, V. (1997) Comparative development of the turkey and chicken embryo from cleavage through hypoblast formation. *Poult. Sci.* **76,** 83–90.

Phillips, R.E. and Williams, C.S. (1964) External morphology of the turkey during the incubation period. *Poult. Sci.* **23,** 270–77.

References

Abbott, U.K. and Pisenti, J.M. (1993) Making the connection: exploring classical concepts in normal and abnormal limb development using contemporary approaches. *Prog. Clin. Biol. Res.* **383,** 99–112.

Abdel-Malek, E.T. (1950) Early development of the urinogenital system in the chick. *J. Morph.* **86,** 599–626.

Adler, R. and Belecky-Adams, T.L. (2002) The role of bone morphogenetic proteins in the differentiation of the optic cup. *Development* **129,** 3161–71.

Allen, S.P., Bogardi, J.P., Barlow, A.J., Mir, S.A., Qayyum, S.R., Verbeek, F.J., Anderson, R.H., Francis-West, P.H., Brown, N.A. and Richardson, M.K. (2001) Misexpression of noggin leads to septal defects in the outflow tract of the chick heart. *Dev. Biol.* **235,** 98–109.

Andrews, J.E., Smith, C.A. and Sinclair, A.H. (1997) Sites of oestrogen receptor and aromatase expression in the chicken embryo. *Gen. Comp. Endocrinol.* **108,** 182–90.

Aoyama, H. and Asamoto, K. (2000) The developmental fate of the rostral/caudal half of a somite for vertebra and rib formation: experimental confirmation of the re-segmentation theory using chick-quail chimeras. *Mech. Develop.* **99,** 71–82.

Arcalis, T., Carretero, A., Navarro, M., Ayuso, E. and Ruberte, J. (2002) Vasculogenesis and angiogenesis in the subcardinal venous plexus of quail mesonephros: spatial and temporal morphological analysis. *Anat. Embryol.* **205,** 19–28.

Arguello, C., Alanis, J. and Valenzuela, B. (1988) The early development of the atrioventricular node and bundle of His in the embryonic chick heart. An electrophysiological and morphological study. *Development* **102,** 623–37.

Atkins, R.L., Wang, D. and Burke, R.D. (2000) Localized electroporation: a method for targeting expression of genes in avian embryos. *Biotechniques* **28,** 94–6.

Bachvarova, R.F., Skromme, I. and Stern, C.D. (1998) Induction of primitive streak and Hensen's node by the posterior marginal zone in the chick embryo. *Development* **125,** 3521–34.

Bagnall, K.M. (1988) A method to increase the survival rate of early chick embryos in experiments involving surgical intervention. *Teratology* **38,** 75–77.

Bagnall, K.M., Higgins, S.J. and Sanders, E.J. (1988) The contribution made by a single somite to the vertebral column: experimental evidence in support of resegmentation using the chick-quail chimaera model. *Development* **103,** 69–85.

Bakst, M.R. (1986) Embryonic development of the chicken external cloaca and phallus. *Scanning Electron Micr.* **II,** 653–9.

Bakst, M.R. and Howarth, B. Jr. (1975) The head, neck and midpiece of cock spermatozoa examined with the transmission electron microscope. *Biol. Reprod.* **12,** 632–40.

Bakst, M.R. and Howarth, B. Jr. (1977) The fine structure of the hen's egg at ovulation. *Biol. Reprod.* **17,** 361–9.

Bakst, M.R., Brillard, J.P. and Wishart, G.J. (1994) Sperm selection, transport and storage in the avian oviduct. *Poultry Sci. Revs* **5,** 143–77.

Baldwin, H.S. and Solursh, M. (1989) Degradation of hyaluronic acid does not prevent looping of the mammalian heart *in situ*. *Dev. Biol.* **136,** 555–9.

Bancroft, M. and Bellairs, R. (1974) The onset of differentiation in the epiblast of the chick blastoderm. *Cell Tiss. Res.* **155,** 399–418.

Bancroft, M. and Bellairs, R. (1975) Differentiation of the neural plate and neural tube in the young chick embryo. *Anat. Embryol.* **147,** 309–35.

Bancroft, M. and Bellairs, R. (1976) The neural crest cells of the trunk region of the chick embryo studied by SEM and TEM. *Zoon* **4,** 73–85.

Bancroft, M. and Bellairs, R. (1977) Placodes of the chick embryo studied by SEM. *Anat. Embryol.* **151,** 97–109.

Barnes, G.L. and Sawyer, P.H. (1995) Histidine rich protein B of embryonic feathers is present in the transient embryonic layers of scutate scales. *J. Exp. Zool.* **271,** 307–14.

Basch, M.L., Selleck, M.A. and Bronner-Fraser, M. (2000) Timing and competence of neural crest formation. *Dev. Neurosci.* **22,** 217–27.

Baumel, J.J., King, A.S., Breuzile, J.E., Evans, H.E. and Van den Berge, J.C. (1993) Handbook of Avian Anatomy, Nomina Anatomica Avium, 2nd ed. Nuttal Orthological Club, Cambridge, Mass.

Beebe, D.C. and Coates, J.M. (2000) The lens organizes the anterior segment: specification of neural crest cell differentiation in the avian eye. *Dev. Biol.* **220,** 424–31.

Bellairs, R. (1961) Cell death in the chick embryo as studied by electron microscopy. *J. Anat.* **95,** 54–60.

Bellairs, R. (1964) Biological aspects of the yolk of the hen's egg. *Adv. Morphogen.* **4,** 217–72.

Bellairs, R. (1967) Aspects of the development of the yolk spheres in the hen's oocyte, studied by electron microscopy. *J. Embryol. Exp. Morph.* **17,** 267–81.

Bellairs, R. (1971) *Developmental processes in Higher Vertebrates.* Logos Press, London.

Bellairs, R. (1979) The mechanism of somite segmentation in the chick embryo. *J. Embryol. Exp. Morph.* **51,** 227–43.

Bellairs, R. and Sanders, E.J. (1986) Somitomeres in the chick tail bud: an SEM study *Anat. Embryol.* **175,** 235–40.

Bellairs, R. (1987) The primitive streak and the neural crest: comparable regions of migration? In *Developmental and Evolutionary Aspects of the Neural Crest* (ed. P. Maderson), John Wiley, New York, pp. 123–45.

Bellairs, R. (1993) Fertilization and early development in poultry. *Poultry Sci.* **72,** 874–81.

Bellairs, R. and Boyde, A. (1969) Scanning electron microscopy of the shell membranes of the hen's egg. *Z. Zellforsch.* **96,** 237–49.

Bellairs, R., Harkness, M. and Harkness, R.D. (1963) The vitelline membrane of the hen's egg: a chemical and electron microscopical study. *J. Ultra. Res.* **8,** 339–59.

Bellairs, R., Bromham, D.R. and Wylie, C.C. (1967) The influence of the area opaca on the development of the young chick embryo. *J. Embryol. Exp. Morph.* **27,** 195–212.

Bellairs, R., Boyde, A. and Heaysman, J.E.M. (1969) The relationship between the edge of the chick blastoderm and the vitelline membrane. *Wilhelm Roux' ArchEntwMech. Org.* **163,** 113–21.

Bellairs, R., Lorenz, F.W. and Dunlap, T. (1978) Cleavage in the chick embryo. *J. Embryol. Exp. Morph.* **43,** 55–69.

Bellairs, R., Lear, P., Yamada, K.M., Rutishauser, U. and Lash, J.W. (1995) Posterior extension of the chick nephric (Wolffian) duct: the role of fibronectin and NCAM polysialic acid. *Develop. Dynamics* **202,** 333–42.

Bereiter-Hahn, J., Matoltsty, A.G. and Richards, K.S. (eds.) (1986) *Biology of the Integument,* Vol. 2, Vertebrates. Springer-Verlag, Berlin.

Beresford, B. (1983) Brachial muscles in the chick embryo: the fate of individual somites. *J. Embryol. Exp. Morph.* **77,** 99–116.

Berger, A.J. (1960) The musculature. In *Biology and Comparative Physiology of Birds* (ed. A.J. Marshall), Vol. 1, Academic Press, New York and London, pp. 301–44.

Bishop-Calame, S. (1965) Nouvelle recherches concernant le role du canal de Wolff dans la différenciation du mésonéphros de l'embryon de poulet. *J. Embryol. Exp. Morph.* **14,** 239–45.

Bissonette, J.P. and Fekete, D.M. (1996) Standard atlas of the gross anatomy of the developing inner ear of the chicken. *J. Comp. Neurol.* **368,** 620–30.

Board, R.G. and Sparks, N.H.C. (1991) Shell structure and formation in avian eggs. In *Egg Incubation: Its Effects on Embryonic Development in Birds and Reptiles* (eds. D.C. Deeming and M.W.J. Ferguson), Cambridge University Press, pp. 371–83.

Boettger, T., Knoetgen, H., Wittler, L., and Kessel, M. (2001) The avian organizer. *Int. J. Dev. Biol.* **45,** 281–7.

Borghjid, S. and Siddiqui, M.A. (2000) Chick homeobox gene cDix expression demarcates the forebrain anlage, indicating the onset of forebrain regional specification at gastrulation. *Dev. Neurosci.* **22,** 183–96.

Bortier, H. and Vakaet, L.C.A. (1992) Fate mapping the neural plate and the intraembryonic mesoblast in the upper layer of the chicken blastoderm with xenografting and time-lapse videography. *Development, Suppl.*, 93–7.

Bortier, H., Callebaut, M. and Vakaet, L.C.A. (1996) Time-lapse cinephotomicrography, videography, and videomicrography of the avian blastoderm. *Meth. Cell Biol.* **51,** 331–54.

Borycki, A. and Emerson, A.G. (2000) Multiple tissue interactions and signal transduction pathways control somite myogenesis. *Curr. Topics Develop. Biol.* **48,** 165–224.

Boyden, E.A. (1927) Experimental obstruction of the mesonephric ducts. *Proc. Soc. Exp. Biol. Med.* **24,** 572–76.

Boyer, A.S. and Runyan, R.B. (2001) TGFβ Type III and TGFβ Type II receptors have distinct activities during epithelial-mesenchymal cell transformation in the embryonic heart. *Dev. Dyn.* **221,** 454–9.

Brand-Saberi, B. and Christ, B.B. (1999) Genetic and epigenetic control of muscle development in vertebrates. *Cell Tissue Res.* **296,** 199–212.

Brockman, D.E., Redmond, M.E. and Kirby, M.L. (1990) Altered development of pharyngeal arch vessels after neural crest ablation. *Ann. N.Y. Acad. Sci.* **588,** 296–304.

Brown, A.J. and Sanders, E.J. (1991) Interactions between mesoderm cells and the extracellular matrix following gastrulation in the chick embryo. *J. Cell Sci.* **99,** 431–41.

Brown, J.M., Robertson, K.E., Wedden, S.E. and Tickle, C. (1997) Alterations in Msx1 and Msx2 expression correlate with inhibition of outgrowth of chick facial primordia induced by retinoic acid. *Anat. Embryol.* **195,** 203–7.

Burke, A.C. (2000) *Hox* genes and the global patterning of the somitic mesoderm. *Curr. Topics Develop. Biol.* **47,** 155–81.

Burke, A.C., Nelson, C.E., Morgan, B.A. and Tabin, C. (1995) Hox genes and the evolution of vertebrate axial morphology. *Development* **121,** 333–46.

Burke, W.H. and Henry, M.H. (1999) Gonadal development and growth of chickens and turkeys hatched from eggs injected with an aromatase inhibitor. *Poultry Sci.* **78,** 1019–33.

Burley, R.W. and Vadehra, D.V. (1989) *The Avian Egg.* John Wiley and Sons, New York.

Burns, A.S. and Le Douarin, N.M. (2001) Enteric nervous system development: analysis of the selected developmental potentialities of vagal and sacral neural crest cells using quail-chick chimeras. *Anat. Rec.* **262,** 16–28.

Burns, R.K. (1956) Urino-genital system. In *Analysis of Development* (eds. B.H. Willier, P.A. Weiss and V. Hamburger), W.B. Saunders, Philadelphia, pp. 462–91.

Butler, J.K. (1952) An experimental analysis of cardiac loop formation of the heart tube in the chick. MA Thesis, University of Texas (not seen).

Callebaut, M. (1983) The constituent oocytal layers of the avian germ and the origin of the primordial germ cell yolk. *Arch. Anat. Microsc. Morphol. Exp.* **72,** 199–214.

Callebaut, M. (1987) Ooplasmic localization and segregation in quail germs: fate of the four ooplasms. *Arch. Biol. (Bruxelles)* **98,** 441–73.

Callebaut, M. and Van Neuten, E. (1994) Rauber's (Koller's) sickle: the early gastrulation organizer of the avian blastoderm. *Eur. J. Morph.* **32,** 35–48.

Callebaut, M., Van Neuten, E., Van Nassauw, L., Bortier, H. and Harrisson, F. (1998) Only the endophyll-Rauber's sickle complex and not cells derived from the caudal margin zone induce a primitive streak in the upper layer of avian blastoderms. *Reprod. Nutr. Dev.* **38,** 449–63.

Callebaut, M., Van Neuten, E., Harrisson, F., Van Nassauw, L. and Bortier, H. (2000a) Avian junctional endoblast has strong embryo-inducing and dominating potencies. *Euro. J. Morphol.* **38,** 3–16.

Callebaut, M., Van Neuten, E., Harrisson, F. and Bortier, H. (2000b) Development of the sickle canal, an unrecognized formation in the avian blastoderm, and its spatial relationship with the first appearing blood islands, induced by Rauber's sickle. *Belg. J. Zool.* **130,** 143–56.

Callebaut, M., Harrisson, F. and Bortier, H. (2001) Effect of gravity on the interaction between the avian germ and neighboring ooplasm in inverted egg yolk balls. *Eur. J. Morphol.* **39,** 27–38.

Calvo, J. and Boya, J. (1979) Ultrastructural study of the embryonic development of the pineal gland of the chicken (*Gallus gallus*). *Acta Anat. Basel,* **103,** 39–73.

Canning, D.R. and Stern, C.D. (1988) Changes in the expression of the carbohydrate epitope HNK-1 associated with mesoderm induction in the chick embryo. *Development* **104,** 643–55.

Caprioli, A., Minko, K., Drevon, C., Eichmann, A., Dieterlen-Lièvre, F. and Jaffredo, T. (2001) Hemangioblast commitment in the avian allantois: cellular and molecular aspects. *Dev. Biol.* **238,** 64–78.

Carretero, A., Ditrich, H., Navarro, M., Splechtna, H. and Ruberte, J. (1993) Technical improvements in corrosion casting of small specimens: a study of mesonephric tubules and vessels of chick embryos. *Scanning Microsc.* **7,** 1333–8.

Carretero, A., Ditrich, H., Perez-Aparicio, F.J., Splechtna, H. and Ruberte, J. (1995) Development and degeneration of the arterial system in the mesonephros and metanephros of chicken embryos. *Anat. Rec.* **243,** 120–8.

Carretero, A., Ditrich, H., Navarro, M. and Ruberte, J. (1997) Afferent portal venous system in the mesonephros and metanephros of chick embryos: development and degeneration. *Anat. Rec.* **247,** 63–70.

Castro-Quezada, A., Nadal-Ginard, B. and De La Cruz, M. (1972) Experimental study of the formation of the bulbo-ventricular loop in the chick. *J. Embryol. Exp. Morph.* **27,** 623–37.

Catala, M., Teillet, M.A., De Robertis, E.M. and Le Douarin, N.M. (1996) A spinal cord fate map in the avian embryo: while regressing, Hensen's node lays down the notochord and floor plate thus joining the spinal cord lateral walls. *Development* **122,** 2599–610.

Catala, M., Teillet, M.-A., and Le Douarin, N.M. (1995) Organization and development of the tail bud analyzed with the quail-chick chimaera system. *Mech. Develop.* **51,** 51–65.

Chang, W., ten Dijke, P. and Wu, D.J. (2002) BMP pathways are involved in otic capsule formation and epithelial-mesenchymal signalling in the developing chicken inner ear. *Dev. Biol.* **251,** 380–94.

Chapman, S.C., Collignon, J., Schoenwolf, G.C. and Lumsden, A. (2001) Improved method for chick whole-embryo culture using a filter paper carrier. *Dev. Dyn.* **220,** 284–9.

Charrier, J.B., Teillet, M.A., Lapointe, F. and Le Douarin, N.M. (1999) Defining subregions of Hensen's node essential for caudalward movement, midline development and survival. *Development* **126,** 4771–83.

Charrier, J.B., Lapointe, F., Le Douarin, N.M. and Teillet, M.-A. (2001) Anti-apoptotic role of sonic hedgehog protein at the early stages of nervous system organogenesis. *Development* **128,** 4011–20.

Chernoff, A.E.G. and Overton, J. (1979) Scanning electron microscopy of chick epiblast expansion on the vitelline membrane. *Dev. Biol.* **57,** 33–46.

Chernoff, E.A., Clarke, D.O., Wallace-Evers, J.L., Hungate-Muegge, L.P. and Smith, R.C. (2001) The effects of collagen synthesis inhibitory drugs on somitogenesis and myogenein expression in cultured chick and mouse embryos. *Tissue Cell* **23,** 97–110.

Chevallier, A. (1975) Role du mésoderme somitique dans la développement de la cage thoracique de l'embryon d'oiseau I. Origine du segment sternal et mécanismes de la différenciation des côtes. *JEEM* **33,** 291–311.

Chevallier, A, (1977) Origines des ceintures scapulaires et pelviennes chez l'embryon d'oiseau. *JEEM* **423,** 275–92.

Chevallier, A., Kieny, M. and Mauger, A. (1978) Limb-somite relationship: effect of removal of somitic mesoderm on the wing musculature. *J. Embryol. Exp. Morph.* **43,** 263–78.

Chiba, Y., Khandoker, A.H., Nobuta, M., Moriya, K., Abiyama, R. and Tazawa, H. (2002) Development of respiratory rhythms in perinatal chick embryos. *Comp. Biochem. Physiol.* **131,** 817–24.

Chin, K., Kurian, R. and Saunders, J.C. (1997) Maturation of tympanic membrane layers and collagen in the embryonic and post-hatch chick (Gallus domesticus). *J. Morphol.* **223,** 257–66.

Chodankar, R., Chang, C.H., Yue, Z., Jiang, T.X., Suksa, S., Burrus, L., Chuong, C.M. and Widelitx, R. (2003) Shift of localized growth zones contributes to appendage morphogenesis: role of Wnt/beta catenin pathway. *J. Invest. Dermatol.* **120,** 20–6.

Christ, B. and Ordahl, C.P. (1995) Early stages of chick somite development. *Anat. Embryol.* **191,** 381–96.

Christ, B. and Wilting, J. (1992) From somites to vertebral column. *Ann. Anat.* **174,** 23–32.

Christ, B., Jacob, H.J. and Jacob. M. (1977) Experimental analysis of the origin of the wing musculature in avian embryos. *Anat. Embryol.* **150,** 171–86.

Christ, B., Jacob, M. and Jacob, H.J. (1983) On the origin and development of the ventro-lateral trunk musculature in the avian embryo. An experimental and ultrastructural study. *Anat. Embryol.* **166,** 87–101.

Christ, B., Huang, R. and Wilting, J. (2000) The development of the avian vertebral column. *Anat. Embryol.* **202,** 179–94.

Chuong, C.M. (1993) The making of a feather – homeoproteins, retinoids and adhesion molecules. *BioEssays* **15,** 513–22.

Climent, S., Sarasa, M., Villar, J.M. and Murillo-Ferrol, N. (1995) Neurogenic cells inhibit the differentiation of cardiogenic cells. *Dev. Biol.* **171,** 130–48.

Cobos, I., Shimamura, K., Rubenstein, J.L.R., Martinez, S. and Puelles, L. (2001). Fate map of the avian anterior forebrain at the four-somite stage, based on the analysis of quail-chick chimeras. *Dev. Biol.* **239,** 46–67.

Coco, C.M., Hargis, B.M. and Hargis, P.S. (1992) Research Note: effect of *in ovo* 17β-estradiol or tamoxifen administration on sexual differentiation of the external genitalia. *Poultry Sci.* **71,** 1947–51.

Coffin, J.D. and Poole, T.J. (1988) Embryonic vascular development: immunohistochemical identification of the origin and subsequent morphogenesis of the major vessel primordia in quail embryos. *Development* **102,** 735–48.

Cohen, G.M. and Hersing, W. (1993) Development of the chick's auditory ossicle, the columella. *Physiologist* **36S,** 75–6.

Cohen, Y.E., Hernandez, H.N. and Saunders, J.C. (1992) Middle-ear development: II Morphometric changes in the conducting apparatus of the chick. *J. Morphol.* **212,** 257–67.

Colas, J.F., Lawson, A. and Schoenwolf, G.C. (2000) Evidence that the translation of smooth muscle alpha-actin mRNA is delayed in the chick promyocardium until fusion of the bilateral heart-forming regions. *Dev. Dyn.* **218,** 316–30.

Cook, G.M.W., Bellairs, R., Rutherford, N.G., Stafford, C.A. and Alderson, T. (1985) Isolation, characterization and localization of a lectin within the vitelline membrane of the hen's egg. *J. Embryol. Exp. Morph.* **90,** 389–407.

Cooke, J. and Zeeman, E.C. (1976) A clock and wavefront model for control of the number of repeated structures during animal morphogenesis. *J. Theor. Biol.* **58,** 455–76.

Cornish, J.A. and Etkin, L.D. (1993) The formation of the pronephric duct in Xenopus involves recruitment of posterior cells by migrating pronephric duct cells. *Dev. Biol.* **159,** 338–45.

Coulombre, A.J., Coulombre, J.L. and Mehta, H. (1962) The skeleton of the eye. I. Conjuctival papillae and scleral ossicles. *Dev. Biol.* **5,** 382–401.

Couly, G. and Le Douarin, N.M. (1988) The fate map of the cephalic neural primordium at the presomitic to the 3-somite stage in the avian embryo. *Development* **103,** Suppl., 101–13.

Couly, G.F., Coltey, P.M. and Le Douarin, N.M. (1992) The developmental fate of the cephalic mesoderm in quail-chick chimeras. *Development* **114,** 1–15.

Couly, G.F., Coltey, P.M. and Le Douarin, N.M. (1993) The triple origin of skull in higher vertebrates: a study in quail-chick chimeras. *Development* **117,** 409–29.

Croisille, Y. and Le Douarin, N. (1964) Development and regeneration of the liver. In *Organogenesis* (eds. R.L. DeHaan and H.L. Ursprung), Holt, Rhinehart and Winston, New York, pp. 421–66.

Croisille, Y., Gumpel-Pinot, M. and Martin, C. (1976) Embryologie expérimentale. La différenciation des tubes sécréteurs du rein chez les oiseaux; effets des inducteurs hétérogènes. *C. R Acad. Sci. Paris* **D282,** 1987–90.

D'Amico-Martel, A. and Noden, D.M. (1983) Contributions of placodal and neural crest cells to avian cranial peripheral ganglia. *Am. J. Anat.* **166,** 445–68.

Daniel, J.C. Jr. (1955) A simplified method for the preparation of avian embryos. *Turtox News* **33,** 142–3.

Darland, C. and D'Amore (2001) Cell-cell interactions in vascular development. *Curr. Top. Dev. Biol.* **52,** 107–49.

Darnell, D.K. and Schoenwolf, G.C. (1997) Modern techniques for cell labelling in avian and murine embryos. In *Molecular and Cellular Methods in Developmental Toxicology* (ed. G.P. Daston) CRC Press, New York. pp. 231–72.

Dathe, V., Gamel, A., Männer, J., Brand-Saberi, B. and Christ, B. (2003) Morphological left-right asymmetry of Hensen's node precedes the asymmetric expression of Shh and Fgf8 in the chick embryo. *Anat. Embryol.* **205,** 342–54.

D'Costa, S., Pardue, S.L. and Petitte, J.N. (2001) Comparative development of avian primordial germ cells and production of germ line chimeras. *Avian Poultry Biol. Rev.* **12,** 151–68.

Debby-Brafman, A., Burstyn-Cohen, T., Klar, A. and Kalcheim, C. (1999) F-Spondin, expressed in somite regions avoided by neural crest cells, mediates inhibition of distinct somite domains to neural crest migration. *Neuron* **22,** 475–88.

Deeming, D.C. (1989) Characteristics of unturned eggs: critical period, retarded embryonic growth and poor albumen utilisation. *Br. Poultry Sci.* **30,** 239–49.

DeFouw, D.O., Rizzo, V.J., Steinfeld, R.N. and Feinberg, R.N. (1989) Mapping of the microcirculation in the chick chorioallantoic membrane during normal angiogenesis. *Microvasc. Res.* **38,** 136–47.

DeHaan, R.L. (1959) Cardia bifida and the development of pacemaker function in the early chick heart. *Dev. Biol.* **1,** 586–602.

DeHaan, R.L. (1965) Morphogenesis of the vertebrate heart. In *Organogenesis* (eds. R.L. DeHaan and H. Ursprung), Holt, Rinehart and Winston, New York, pp. 377–419.

DeHaan, R.L. (1990) The embryonic origin of the heartbeat. In *The Heart* (ed. J.W. Hurst) 7th ed. McGraw-Hill, New York, pp. 72–77.

Denetclaw, W.F. Jr., Berdougo, E., Venters, S.J. and Ordahl, C.P. (2001) Morphogenetic cell movements in the middle region of the dermomyotome dorsomedial lip associated with patterning and growth of the primary epaxial myotome. *Development* **128,** 1745–55.

DeRuiter, M.C., Gittenberger-de-Groot, A.C., Poelmann, R.E., VanIperen, L. and Mentink, M.M. (1993) Development of the pharyngeal arch system related to the pulmonary and bronchial vessels in the avian embryo. With a concept on system-pulmonary collateral artery formation. *Circulation* **87,** 1306–9.

De Santa Barbara, P. and Roberts, D.J. (2002) Tail gut endoderm and gut/genitourinary/tail development: a new tissue-specific role for Hoxa13. *Development* **129,** 551–61.

Diaz, C., Glover, J.C., Puelles, L. and Bjaalic, J.G. (2003) The relationship between hodological and cytoarchitectonic organization in the vestibular complex of the 11-day chicken embryo. *J. Comp. Neurol.* **457,** 87–105.

Diaz-Ruiz, C., Perez-Tomas, R., Cullere, X. and Domingo, J. (1993) Immunohistochemical localization of transforming growth factor-alpha and epidermal growth factor-receptor in the mesonephros and metanephros of the chicken. *Cell Tissue Res.* **271,** 3–8.

Dickinson, M.E., Selleck, M.A., McMahon, A.P. and Bronner-Fraser, M. (1995) Dorsalization of the neural tube by the non-neural ectoderm. *Development* **121,** 2099–106.

Didier, E., Didier, P., Fargeix, N., Guillot, J. and Thiery, J.P. (1990) Expression and distribution of carbohydrate sequences in chick germ cells; a comparative study with lectins and the NC-1/HNK-1 monoclonal antibody. *Int. J. Dev. Biol.* **34,** 421–31.

Dieterlen-Lièvre, F. (1984) Emergence of intra-embryonic blood stem cells in the avian chimeras by means of monoclonal antibodies. *Dev. Comp. Immunol.* **3,** 75–80.

Dieterlen-Lièvre, F. (1997) Intraembryonic hematopoietic stem cells. *Hematol. Oncol. Clin. North Am.* **11,** 1149–71.

Dietrich, S., Schubert, F.R., Healy, C., Sharpe, P.T. and Lumsden, A. (1998) Specification of the hypaxial musculature. *Development* **125,** 2235–49.

Dingerkus, G. and Uhler, L. (1977) Enzyme clearing of alcian blue stained whole small vertebrates for demonstration of cartilage, *Stain Technol.* **52,** 229–232.

Djonov, V.G., Galli, A.B. and Burri, P.H. (2000) Intussusceptic arborization contributes to vascular tree formation in the chick chorio-allantoic membrane. *Anat. Embryol.* **202,** 347–57.

Dockter, J.L. (2000) Sclerotome induction and differentiation. *Curr. Topics in Develop. Biol.* **48,** 77–127.

Dockter, J. and Ordahl, C.P. (2000) Dorsoventral axis determination in the somites: a re-examination. *Development* **127,** 2201–6.

Drushel, R.F. and Caplan, A.I. (1991) Three-dimensional reconstruction and cross-sectional anatomy of the thigh musculature of the developing chick embryo (*Gallus gallus*). *J. Morph.* **208,** 293–309.

Dudley, J. (1942) The development of the ultimobranchial body of the fowl, *Gallus domesticus*. *Am. J. Anat.* **71,** 65–97.

Dunn, B.E. (1991) Methods for shell-less and semi-shell-less culture of avian and reptilian embryos. In *Egg Incubation: its effects on embryonic development in birds and reptiles* (eds. D.C. Deeming and M.W.J. Ferguson), Cambridge University Press, pp. 409–18.

Dupin, E., Glavieux, C., Vaigot, P. and Le Douarin, N.M. (2000) Endothelin3 induces the reversion of melanocytes to glia through a neural crest-derived glial-melanocytic progenitor. *Proc. Natl. Acad. Sci. USA* **97,** 7882–7.

Duval, M. (1889) *Atlas d'Embryologie.* Masson et Cie, Paris.

Easton, H.S., Bellairs, R. and Lash, J.W. (1990) Is chemotaxis a factor in the migration of precardiac mesoderm in the chick? *Anat. Embryol.* **181,** 461–8.

Ebensperger, C., Wilting, J., Brand-Saberi, B., Mizutani, Y., Christ, B., Balling, R. and Koseki, H. (1995) *Pax-1*, a regulator of sclerotome development is induced by notochord and floorplate signals in avian embryos. *Anat. Emb.* **191,** 297–310.

Ede, D. A. (1964) *Bird Structure: an Approach Through Evolution, Development and Function in the Fowl.* Hutchinson Educational.

Egawa, C. and Kameda, Y. (1995) Innervation of the chicken parathyroid glands; Immunohistochemical study with the TuJ1, galanin, VIP, substance P, CGRP and tyrosine hydroxylase antibodies. *Anat. Embryol.* **191,** 445–50.

Erickson, C.A. and Goins, T.L. (1995) Avian neural crest cells can migrate in the dorsolateral path only if they are specified as melanocytes. *Development* **121,** 915–24.

Etches, R.J. (1996) *Reproduction in Poultry.* CAB International, Oxon.

Etches, R.J., Carsience, R.S., Clark, M.E., Fraser, R.A., Toner, A. and Verrinder Gibbins, A.M. (1993) Chimeric chickens and their use in manipulation of the chicken genome. *Poultry Science* **72,** 882–9.

Eyal-Giladi, H. (1991) The early development of the chick as an epigenetic process. *Crit. Rev. Poultry Biol.* **3,** 143–6.

Eyal-Giladi, H. and Fabian, B. (1980) Axis determination in uterine chick blastodiscs under changing spatial positions during the sensitive period of polarity. *Dev. Biol.* **77,** 228–32.

Eyal-Giladi, H. and Kochav, S. (1975) From cleavage to primitive streak formation; a complementary Normal table and a new look at the first stages of the development of the chick. I. General morphology. *Dev. Biol.* **49,** 321–37.

Fabian, B. and Eyal-Giladi, H. (1981) A SEM study of cell shedding during the formation of the area pellucida in the chicken embryo. *J. Embryol. Exp. Morph.* **64,** 11–22.

Fairman, C.L., Clagett-Dame, M., Lennon, V.A. and Epstein, M.L (1995) Appearance of neurons in the chick gut. *Dev. Dyn.* **204,** 192–201.

Faraco, C.D., Vaz, S.A., Pastor, M.V. and Erickson, C.A. (2001) Hyperpigmentation in the Silkie fowl correlates with abnormal migration of fate-restricted melanoblasts and loss of environmental barrier molecules. *Dev. Dyn.* **220,** 215–25.

Fasenko, G.M., Robinson, F.E., Whelan, A.I., Kremeniuk, K.M. and Walker, J.A. (2001) Prestorage incubation of long-term stored broiler breeder eggs: 1. Effects on hatchability. *Poultry Sci.* **80,** 1406–11.

Faucounau, N., Ichas, F., Stoll, R. and Maraud, R. (1995) Action of testosterone on the estradiol-induced feminization of the male chick embryo. *Anat. Embryol.* **191,** 377–9.

Fisher, J.R. and Eakin, R.E. (1957) Nitrogen excretion in developing chick embryos. *J. Embryol. Exp. Morph.* **5**, 215–24.

Flamme, I. (1987) Edge cell migration in the extra-embryonic mesoderm of the chick embryo. *Anat. Embryol.* **176,** 477–91.

Flynn, M.E., Pikalow, A.S., Kimmelman, R.S. and Searls, R.L. (1991) The mechanism of cervical flexure formation in the chick. *Anat. Embryol.* **184,** 411–20.

Foley, A., Skromme, I. and Stern, C.D. (2000) Reconciling different models of forebrain induction and patterning: a dual role for the hypoblast. *Development* **127,** 3839–54.

Francis-West, P.H., Tatla, T. and Brickell, P.M. (1994) Expression patterns of the bone morphogenetic proteins Bmp-4 and Bmp-2 in the developing chick face suggest a role in outgrowth of the primordia. *Dev. Dyn.* **201,** 168–78.

Freedman, S.L., Akuffo, V.G. and Bakst, M.R. (2001) Evidence for the innervation of sperm storage tubules in the oviduct of the turkey (*Melleagri gallopavo*). *Reproduction* **121,** 809–14.

Freeman, B.M. and Vince, M.A. (1974) *Development of the Avian Embryo.* Chapman and Hall, London.

Freund, R., Döfler, D., Popp, W. and Wachtler, F. (1996). The metameric pattern of the head – does it exist? *Anat. Embryol.* **193,** 73–80.

Fukuda, K. and Yasugi, S. (2002) Versatile roles for sonic hedgehog in gut development. *J. Gastroenterol.* **37,** 239–46.

Fukui, Y. (1988) Cell proliferation during bud formation of the quail uropygial gland. *Dev. Growth Diff.* **30,** 501–14.

Fukui, Y. (1989) Cell proliferation in the quail uropygial gland during placode stage to lumen formation. *Anat. Embryol.* **179,** 347–51.

Gambaryan, S.P. (1992) Development of the metanephros in the chick: maturation of glomerular size and nephron size. *Anat. Embryol.* **185,** 291–7.

Garcia-Martinez, V. and Schoenwolf, G. (1992) Positional control of mesoderm movement and fate during avian gastrulation and neurulation. *Dev. Dyn.* **193,** 249–56.

Garcia-Martinez, V. and Schoenwolf, G.C. (1993) Primitive-streak origin of the cardiovascular system in avian embryos. *Dev. Biol.* **159,** 706–19.

Garcia-Martinez, V., Alvarez, I.S. and Schoenwolf, G.C. (1993) Locations of the ectodermal and neurectodermal subdivisions of the epiblast at stages 3 and 4 of avian gastrulation and neurulation. *J. Exp. Zool.* **267,** 431–46.

Garda, A.L., Puelles, L., Rubenstein, J.L. and Medina, L. (2002) Expression patterns of Wnt8b and Wnt7b in chicken embryogenic brain suggest a correlation with chicken forebrain patterning centers and morphogenesis. *Neuroscience* **113,** 689–98.

Gasc, J.M. (1991) Distribution and regulation of progesterone receptor in the urinogenital tract of the chick embryo. An immunohistochemical study. *Anat. Embryol.* **183,** 415–26.

Gerlach, L.M., Hutson, M.R., Germiller, J.A, Nguyen-L., Luu, D., Victor, J.C. and Barald, K.F. (2000) Addition of BMP4 antagonist, noggin, disrupts avian inner ear development. *Development* **27,** 45–54.

Gheri, G. and Gheri-Bryk, S. (1987) Computerised morphometric analysis of the chick embryo ileum organogenesis. *Z. Mikrosk. Anat. Forsch.* **101,** 1011–22.

Gheri, G., Bryk, S.G., Gloria, L. and Audino, S. (1992) The development of the caeca in the chick embryo ileum organogenesis. *Arch. Ital. Anat. Embryol.* **97,** 89–107.

Gheri-Bryk, S., Gheri, G. and Pacini, P. (1990). The development of the chick embryo gall bladder studied by scanning electron microscope. *Anat. Anz.* **171,** 297–305.

Gheri-Bryk, S., Gheri, G. and Sgambati, E. (1993) Computerised morphometric analysis of the chick embryo oesophagus. *Arch. Ital. Anat. Embryol.* **98,** 1–11.

Gilbert, A.B. (1971) The egg: its physical and chemical aspects. In *The Physiology of the Domestic Fowl*, Vol. **3** (eds. D.J. Bell and D.M. Freeman), Academic Press, London, pp. 1379–1409.

Gilbert, S. F. (2000) *Developmental Biology.* 6th ed. Sinauer Associates, Sunderland, Mass, p. 749.

Ginsburg, M. (1997) Primordial germ cell development in avians, *Poultry Sci.* **76,** 91–5.

Gleiberman, A.S., Fedtsova, N.G. and Rosenfeld, M.G. (1999) Tissue interactions in the induction of anterior pituitary: role of the ventral diencephalons, mesenchyme, and notochord. *Dev. Biol.* **213,** 340–53.

Glick, B. and Olah, I. (1993) Bursal secretory dendritic-like cell: a microenvironment issue. *Poultry Sci.* **72,** 1262–6.

Goetinck, P.F. and Sekellic, M.J. (1972) Observations on collagen synthesis, lattice formation and morphology of scaleless and normal embryonic skin. *Dev. Biol.* **28,** 636–48.

Goldschneider, I. and Barton, R.W. (1976) Development and differentiation of lymphocytes. In *The Cell Surface in Animal Embryogenesis and Development* (eds. G. Poste and G.L. Nicolson), North Holland, Amsterdam, pp. 599–695.

Goldstein, R.S., Avivi, C. and Geffen, R. (1995) Initial axial level-dependent differences in size of avian dorsal root ganglia are imposed by the sclerotome. *Dev. Biol.* **168,** 214–22.

Gomperts, M., Wylie, C. and Heasman, J. (1994) Primordial germ cell migration. *Ciba Found. Symp.*, **182,** 121–34.

Gonzalez-Sanchez, A. and Bader, D. (1990) *In vitro* analysis of cardiac progenitor cell differentiation. *Dev. Biol.* **139,** 197–209.

Goodrum, G.R. and Jacobson, A.G. (1981) Cephalic flexure formation in the chick embryo. *J. Exp. Zool.* **216,** 399–408.

Gräper, L. (1929) Die Methodik der stereokinematographischen Untersuchung der lebenden vitalfärbten Hühnerembryos. *Arch. Entw. Mech. Org.* **115,** 523–43.

Grapin-Botton, A., Majithia, A.R. and Melton, D.A. (2001) Key events of pancreas formation are triggered in gut endoderm by ectopic expression of pancreatic regulatory genes. *Genes Dev.* **15,** 444–54.

Griffith, C.M., Wiley, M.J. and Sanders, E.J. (1992). The vertebrate tail bud: three germ layers from one tissue. *Anat. Embryol.* **185,** 101–13.

Grim, M. and Halata, Z. (2000) Developmental origin of avian Merkel cells. *Anat. Embryol.* **202,** 401–10.

Groenendijk-Huijbers, M.M. (1967) Experimental studies concerning the compensatory growth of the rudimantary right ovary of sinistrally castrated chick embryos. *Verh. Anat. Ges.* **62,** 321–25.

Gruenwald, P. (1941) The relation of the growing tip of the Mullerian duct to the Wolffian duct and its importance for the genesis of malformations. *Anat. Rec.* **81,** 1–19.

Gumpel-Pinot, M. (1984) Muscle and skeleton of limbs and body wall. In *Chimeras in Developmental Biology* (eds. N.L. Le Douarin and A. McClaren), Academic Press, London, pp. 281–310.

Gunhaga, L., Jessell, T.M. and Edlund, T. (2000) Sonic hedgehog signalling at gastrula stages specifies ventral telencephalic cells in the chick embryo. *Development* **127,** 3283–93.

Hacker, A. and Guthrie, S. (1998) A distinct developmental programme for the cranial paraxial mesoderm in the chick embryo. *Development* **126,** 3461–72.

Halata, Z., and Grim, M. (1993) Sensory nerve ending in the back skin of the Japanese quail. *Anat. Embryol.* **187,** 131–8.

Halata, Z., Grim, M. and Christ, B. (1990) Origin of spinal meninges, sheaths of peripheral nerves, and cutaneous receptors including Merkel cells. *Anat. Embryol.* **182,** 529–37.

Hall, B.K. (1981) Specificity in the differentiation and morphogenesis of neural crest-derived scleral ossicles and of epithelial scleral papillae in the eye of the embryonic chick. *J. Embryol. Exp. Morph.* **66,** 175–90.

Hall, B.K. (1999) *The Neural Crest Development*, Springer-Verlag, New York.

Hall, B.K. (2001) Development of the clavicles in birds and mammals. *J. Exp. Zool.* **289,** 153–61.

Hall, B.K. and Hörstadius, S. (1988) *The Neural Crest*, Oxford University Press, London.

Hamburger, V. (1947) *A Manual of Experimental Embryology.* University of Chicago Press, Chicago, Ill.

Hamburger, V. and Hamilton, H.L. (1951) A series of normal stages in the development of the chick embryo. *J. Morph.* **88,** 49–92. (Reprinted 1992, *Dev. Dyn.* **195,** 229–30.)

Harris, M.P., Fallon, J.F. and Prum, R.O. (2002) Shh-Bmp2 signalling module and the evolutionary origin and diversification of feathers. *J. Exp. Zool.* **294,** 160–76.

Harrison, T.A., Stadt, H.A., Kumiski, D. and Kirby, M.L. (1995) Compensatory responses and development of nodose ganglion following ablation of placode precursors in the embryonic chick (Gallus domesticus). *Cell Tissue Res.* **281,** 379–85.

Harrisson, F., Andries, L., and Vakaet, L. (1988) The chicken blastoderm: current views on cell biological events guiding intercellular communication. *Cell Differ.* **22,** 1–23.

Harvey, S., Johnson, C.D. and Sanders, E.J. (2000) Extra-pituitary growth hormone in peripheral tissues of early chick embryos. *J. Endocrinol.* **166,** 489–502.

Hatada, Y. and Stern, C.D. (1994) A fate map of the epiblast of the early chick embryo. *Development,* **120,** 2879–90.

Hebrock, M., Kim, S.K. and Melton, D.A. (1998) Notochord repression of endodermal Sonic hedgehog permits pancreas development. *Genes Dev.* **12,** 1705–13.

Herzog, Y., Kalcheim, C., Kahane, N., Reshef, R. and Neufeld, G. (2001) Differential expression of neuropilin-2 in arteries and veins. *Mech. Dev.* **109,** 115–9.

Hilfer, S.R. (1964) Follicle formation in the embryonic chick thyroid. I. Early morphogenesis. *J. Morph.* **115,** 135–51.

Hirakow, R. and Hiruma, T. (1981) Scanning electron microscopic study on the development of primitive blood vessels in chick embryos at the early somite stage. *Anat. Embryol.* **163,** 290–306.

Hirata, M. and Hall, B.K. (2000) Temporospatial patterns of apoptosis in chick embryos during the morphogenetic period of development. *Int. J. Dev. Biol.* **44,** 757–68.

Hirota, A., Kamino, K., Komuro, H. (1987) Mapping of early development of electrical activity in the embryonic chick heart using multiple-site optical recording. *J. Physiol.* **383,** 711–28.

Hirsinger, E., Jouve, C., Malapert, P. and Pourquié, O. (1998) Role of growth factors in shaping the developing somite. *J. Mol. Cell Endocrinol.* **25,** 83–7.

Hiruma, T. and Hirakow, R. (1985) An ultrastructural topographic study on myofibrillogenesis in the heart of the chick embryo during pulsation onset period. *Anat. Embryol.* **172,** 325–9.

Hiruma, T. and Hirakow, R. (1995) Formation of the pharyngeal arch arteries in the chick embryo. Observations of corrosion casts by scanning electron microscopy. *Anat. Embryol.* **191,** 415–23.

Hiruma, T. and Nakamura, H. (2003) Origin and development of the pronephros in the chick embryo. *J. Anat.* **203,** 539–52.

Hofmann, C., Drossopoulou, G., McMahon, A., Balling, R. and Tickle, C. (1998) Inhibitory action of BMPs on *Pax1* expression and on shoulder girdle formation during limb development. *Dev. Dyn.* **213,** 199–206.

Höhn, E.O. (1961) Endocrine glands, thymus and pineal body. In *Biology and Comparative Physiology of Birds* (ed. A.J. Marshall) Academic Press, London, pp. 87–114.

Holm, L., Ekwall, H., Wishart, G.J. and Ridderstrale, Y. (2000) Localization of calcium and zinc in the sperm storage tubules of chicken and turkey using X-ray microanalysis. *J. Reprod. Fertil.* **118,** 331–6.

Horder, T.J., Witowski, J.A. and Wylie, C.C. (eds.) (1986) *A History of Embryology.* Cambridge University Press.

Horrocks, A.J., Stewart, S., Jackson, L. and Wishart, G.J. (2000) Induction of acrosomal exocytosis in chicken spermatozoa by inner perivitelline-derived N-linked glycans. *Biochem. Biophys. Res.* **278,** 84–9.

Huang, R., Zhi, Q., Wilting, J. and Christ, B. (1994) The fate of somitocoele cells in avian embryos. *Anat. Embryol.* **190,** 243–50.

Huang, R., Zhi, Q., Izpisua-Belmonte, J.-C., Christ, B. and Patel, K.(1999) Origin and development of the avian tongue muscles. *Anat. Embryol.* **200,** 137–52.

Huang, R., Lang, E.R., Otto, W.R., Christ, B. and Patel, K. (2001) Molecular and cellular analysis of embryonic avian tongue development. *Anat. Embryol.* **204,** 179–87.

Huang, R., Brand-Saberi, B. and Christ, B. (2000a) New experimental evidence for somite resegmentation. *Anat. Embryol.* **202,** 195–200.

Huang, R., Zhi, Q., Patel, K., Wilting, J. and Christ, B. (2000b) Dual origin and segmental organisation of the avian scapula. *Development* **127,** 3789–94.

Huang, R., Zhi, Q., Patel, K., Wilting, J. and Christ, B. (2000c) Contribution of single somites to the skeleton and muscles of the occipital and cervical regions in avian embryos. *Anat. Embryol.* **202,** 375–83.

Huang, R., Zhi, Q., Schmidt, C., Wilting, J., Brand-Saberi, B. and Christ, B. (2000d) Sclerotomal origin of the ribs. *Development* **127,** 527–32.

Huang, R., Zhi, Q. and Christ, B. (2003) The relationship between limb muscle and endothelial cells migrating from a single somite. *Anat. Embryol.* **206,** 283–9.

Hughes, A.F.W. (1934) On the development of the blood vessels in the head of the chick. *Phil. Trans. Roy. Soc. Lond.* Ser. B **224,** 75–129.

Hughes, A.F.W. (1937) Studies on the area vasculosa of the embryo chick. II The influence of the circulation on the diameter of the vessels. *J. Anat.* **72,** 1–17.

Hume, C.R. and Dodd, J. (1993) A novel Wnt gene with a potential role in primitive streak formation and hindbrain organization. *Development* **119,** 1147–60.

Hutson, J.M and Donahoe, P.K. (1984) Improved histology for chick-quail chimeras. *Stain Technol.* **59,** 105–11.

Icardo, J.M. (1989) Changes in the endocardial cushion morphology during the development of the endocardial cushions. *Anat. Embryol.* **179,** 443–8.

Icardo, J.M. (1990) Development of the outflow tract. A study of hearts with *situs solitus* and *situs inversus*. *Ann. N.Y. Acad. Sci.* **588,** 26–40.

Inagaki, T., Garcia-Martinez, V. and Schoenwolf, G.C. (1993) Regulative ability of the prospective cardiogenic and vasculogenic areas of the primitive streak during avian gastrulation. *Dev. Dyn.* **197,** 57–68.

Itasaki, N., Nakamura, H., Sumida, H. and Yasuda, M. (1991) Actin bundles on the right side in the caudal part of the heart play a role in dextro-looping in the embryonic chick heart. *Anat. Embryol.* **183,** 29–39.

Izpisúa-Belmonte, J.C., De Robertis, E.M., Storey, K.G. and Stern, C.D. (1993) The homeobox gene *goosecoid* and the origin of organizer cells in the early chick blastoderm. *Cell* **74,** 645–59.

Jacob, H.J. and Christ, B. (1978) Experimental investigations on the excretion apparatus of young chick embryos. In *XIXth Morphological Congress Symposia, Charles University, Prague*, pp. 219–35.

Jacob, M., Christ, B. and Jacob, H.J. (1979) The migration of myogenic cells from the somites into the leg region of avian embryos. *Anat. Embryol.* **157,** 291–309.

Jacob, M., Jacob, H.J., Wachtler, F. and Christ, B. (1984) Ontogeny of avian extrinsic ocular muscles. I Light and electronmicroscopic study. *Cell Tiss. Res.* **234,** 549–57.

Jacob, M., Konrad, K. and Jacob, H.J. (1999) Early development of the Mullerian duct in avian embryos with reference to the human. An ultrastructural and immunohistochemical study. *Cells Tissues Organs* **164,** 63–81.

Jacobson, A.G., Miyamoto, D.M. and Mai, S.H. (1979) Rathke's pouch morphogenesis in the chick embryo. *J. Exp. Zool.* **207,** 351–66.

Jarzem, J. and Meier, S.P. (1987) A scanning electron microscope survey of the origin of the primordial pronephric duct cells in the avian embryo. *Anat. Rec.* **218,** 175–81.

Jaskoll, T.F. and Maderson, P.F. (1978) A histological study of the development of the avium middle ear and tympanum. *Anat. Rec.* **190,** 177–99.

Jeffs, P. and Osmond, M. (1992) A segmented pattern of cell death during development of the chick embryo. *Anat. Embryol.* **185,** 589–98.

Jin, E.J., Erickson, C.A., Takada, S. and Burrus, L.W. (2001) Wnt and Bmp signalling govern lineage segregation of melanocytes in the avian embryo. *Dev. Biol.* **223,** 22–37.

Johnston, S.D., Orgeig, S., Lopatko, O.V. and Daniels, C.B. (2000) Development of the pulmonary surfactant system in two oviparous vertebrates. *Am. J. Physiol. Regul. Integr. Comp.Physiol.* **278,** 486–93.

Jordanov, J., Georgiev, I. and Boyadjieva-Mihailova, A. (1966) Physicochemical and electron microscopical investigations on the vitelline membrane of the hen's egg with a view to its permeability to macromolecules. *C. R. Acad. Bulg. Sci.* **19,** 153–6.

Kagami, H., Clark, M.E., Verrinder Gibbins, A.M. and Etches, R.J. (1995) Sexual differentiation of chimeric chickens containing ZZ and ZW cells in the germline. *Mol. Reprod. Dev.* **42,** 379–87.

Kameda, Y. (1984) Ontogeny of chicken ultimobranchial glands studied by an immunoperoxidase method using calcitonin, somatostatin and 19S-thyroglobulin antisera. *Anat. Embryol.* **170,** 139–44.

Kameda, Y. (1990) Ontogeny of the carotid body and glomus cells distributed in the wall of the common carotid artery and its branches in the chicken. *Cell Tissue Res.* **261,** 525–37.

Kameda, Y. (1993) Electron microscopic study on the development of the chicken ultimobranchial glands, with special reference to the innervation of C-cells. *Anat. Embryol.* **188,** 561–70.

Kameda, Y., Miura, M. and Ohno, S. (2000) Expression of the common alpha-subunit mRNA of glycoprotein hormones during the chick pituitary organogenesis, with special reference to the pars tuberalis. *Cell Tissue Res.*, **299,** 71–80.

Karagenc, L, Cinnamon, Y., Ginsburg, M. and Petitte, J.N. (1996) Origin of primordial germ cells in the prestreak chick embryo. *Dev. Genet.* **19,** 290–301.

Kato, N, and Ayoyama, H. (1998) Dermomyotomal origin of the ribs as revealed by extirpation and transplantation experiments in chick and quail embryos. *Development* **125,** 3437–43.

Kawakami, Y., Capadevilla, J., Buscher, D., Itoh, T., Esteban, C. and Izpisua Belmonte, J.C. (2001) Wnt signals control FGF-dependent limb initiation and ART induction in the chick embryo. *Cell* **23,** 891–900.

Kelly, R.G. and Buckingham, M.E. (2002) The anterior heart- forming field: voyage to the arterial pole of the heart. *Trends Genet.* **18,** 210–6.

Kent, J., Wheatley, S.C., Andrews, J.E., Sinclair, A.H. and Koopman, P. (1996) A male-specific role for SOX9 in vertebrate sex determination. *Development* **122,** 2813–22.

Kerr, J.G. (1919) *Text-Book of Embryology, with the exception of mammals.* Vol. 2 Vertebrata, Macmillan, London.

Kessel, J. and Fabian, B.C. (1985) Graded morphogenetic patterns during the development of the extraembryonic blood system and coelom of the chick blastoderm: a scanning electron microscope and light microscope study. *Amer. J. Anat.* **173,** 99–112.

Keyes, W.M. and Sanders E.J. (1999) Cell death in the endocardial cushions of the developing heart. *J. Mol. Cell Cardiol.* **31,** 1015–23.

Keyes, W.M., Logan, C., Parker, E. and Sanders, E.J. (2003) Expression and function of bone morphogenetic proteins in the development of the embryonic endocardial cushions. *Anat. Emb.* **207,** 135–47.

Keynes, R.J. and Stern, C.D. (1984) Segmentation in the vertebrate nervous system. *Nature, Lond.* **310,** 786–9.

Keynes, R.J. and Stern, C.D. (1988) Mechanisms of vertebrate segmentation. *Development* **103,** 413–29.

Khaner, O. (1995) The rotated hypoblast of the chicken embryo does not initiate an ectopic axis in the epiblast. *Proc. Natl. Acad. Sci.* **92,** 10733–7.

Khaner, O. (1998) The ability to initiate an axis in the avian blastula is concentrated mainly at a posterior site. *Dev. Biol.* **194,** 256–66.

Kido, S., Janado, M. and Nunoura, H. (1976) Macromolecular components of the vitelline membrane of hen's eggs. II Physicochemical properties of glycoprotein II. *J. Biochem.* **79,** 1351–6.

Kieny, M., Mauger, A. and Sengel, P. (1972) Early regionalization of the somitic mesoderm as studied by the development of the axial skeleton of the chick embryo. *Dev. Biol.* **28,** 142–61.

King, D.B. and May, J.D. (1984) Thyroidal influence on body growth. *J. Exp. Zool.* **232,** 453–60.

King, A.S. and McLelland, J. (1979) *Form and Function in Birds,* Vol. 1. Academic Press, London.

King, A.S. and McLelland, J. (1979–89) *Form and Function in Birds,* 4 vols. Academic Press, London.

Kirby, M.L. (1990) Alteration of cardiogenesis after neural crest ablation. *Ann. N.Y. Acad. Sci.* **588,** 289–95.

Kirunda, D.F. and McKee, S.R. (2000) Relating quality characteristics of aged eggs and fresh eggs to vitelline membrane strength as determined by a texture analyser. *Poultry Sci.* **79,** 1189–93.

Knospe, C., Seichert, V. and Rychter, Z. (1991) The topogenesis of the thyroid in chick embryos. *Eur. J. Morphol.* **29,** 291–6.

Kosaka, Y., Akimoto, Y., Obinata, A. and Hirano, H. (2000) Localization of HB9 homeobox gene mRNA and protein during the early stages of chick feather development. *Biochem. Biophys. Res.* **276,** 1112–7.

Kucera, P. and Burnand, M.-B. (1987a) Routine teratogenicity test that uses chick embryos *in vitro. Teratogen. Carcinogen. Mutagen.* **7,** 427–47.

Kucera, P. and Burnand, M.B. (1987b) Mechanical tension and movement in the chick blastoderm as studied by real-time image analysis. *J. Exp. Zool.* Suppl. 1, 329–39.

Kucera, P., Abriel, H. and Katz, U. (1994) Ion transport across the early chick embryo: Electrical measurements, ionic fluxes and regional heterogeneity. *J. Membr. Biol.* **141,** 149–57.

Kuratani, S. and Kirby, M.L ((1991) Initial migration and distribution of the cardiac neural crest in the avian embryo: an introduction to the concept of the circumpharyngeal crest. *Am. J. Anat.* **191,** 215–27.

Kuratani, S. and Tanaka, S. (1990) Peripheral development of the avian vagus nerve with special reference to the morphological innervation of heart and lung. *Anat. Embryol.* **182,** 435–45.

Kuwana, T., Maeda-Suga, H. and Fujimoto, T. (1986) Attraction of chick primordial germ cells by gonadal anlage in vitro. *Anat. Rec.* **215,** 403–6.

Lambson, R. (1970) An electron microscopical study of the endodermal cells of the yolk sac of the chick during incubation and after hatching. *Am. J. Anat.* **129,** 1–20.

Lance-Jones, C. (1988) The somitic level of origin of embryonic chick hindlimb muscles. *Dev. Biol.* **126,** 394–407.

Lance-Jones, C. (1990) Pathfinding by motoneurons in the limb of the avian embryo. In *The Avian Model in Developmental Biology: from Organism to Genes* (eds. N. Le Douarin, F. Dierterlen-Lièvre and J. Smith), Editions du CNRS, Paris, pp. 129–38.

Landauer, W. (1967) *The Hatchability of Chicken Eggs as influenced by Environment and Heredity.* Monograph 1 (revised) of Storrs Agricultural Development Station, Storrs, Conn.

Lapao, C., Gama, L.T. and Soares, M.C. (1999) Effects of broiler breeder age and length of early storage on albumen characteristics and hatchability. *Poultry Sci.* **78,** 640–5.

Lash, J.W., Seitz, A.W., Cheney, C.M. and Ostrovsky, D. (1984) On the role of fibronectin during the compaction stage of somitogenesis in the chick embryo. *J. Exp. Zool.* **232,** 197–206.

Lash, J.W., Linask, K.K. and Yamada, K.M. (1987) Synthetic peptides that mimic the adhesion recognition signal of fibronectin. Differential effects on cell-cell and cell-substratum adhesion in embryonic chick cells. *Dev. Biol.* **123,** 411–20.

Lash, J.W., Gosfield, E.III, Ostrovsky, D. and Bellairs, R. (1990) Migration of chick blastoderm under the vitelline membrane: the role of fibronectin. *Dev. Biol.* **139,** 407–16.

Latter, G.V. and Baggott, G.K. (1996) Effect of egg turning and fertility upon the sodium concentration of albumen of the Japanese quail. *Br. Poult. Sci.* **37,** 301–8.

Lawson, A. and Schoenwolf, G.C. (2001) New insights into critical events of avian gastrulation. *Anat. Embryol.* **262,** 238–52.

Le, A.C. and Musil, L.S. (2001) FGF signalling in chick lens development. *Dev. Biol.* **233,** 394–411.

Le, S.H., Fu, K.K., Hui, J.N. and Richman, J.M. (2001) Noggin and retinoic acid transform the identity of avian facial prominences. *Nature, Lond.* **414,** 909–12.

Le Douarin, N. (1969) Particularités du noyaux interphasique chez la caille japonaise (*Coturnic coturnix* japonica). Utilisation de ses particularités comme 'marquage biologique' dans les recherches sur les interactions tissulaires et les migrations cellulaires au cours de l'ontogenèse. *Bull. Biol. Fr. Belg.* **103,** 435–52.

Le Douarin, N.M. (1988) On the origin of pancreatic endocrine cells. *Cell* **53,** 169–71.

Le Douarin, N.M. (2001) Early neurogenesis in amniote vertebrates. *Int. J. Dev. Biol.* **45,** 373–8.

Le Douarin, N.M. and Kalcheim, C. (1999) *The Neural Crest*, 2nd ed. Cambridge University Press, Cambridge.

Le Douarin, N.M., Jotereau, F.V., Houssaint, E. and Thiery, J.-P. (1984a) Primary lymphoid organ ontogeny in birds. In *Chimeras in Developmental Biology* (eds. N. Le Douarin and A. McLaren), Academic Press, London, pp. 179–216.

Le Douarin, N.M., Teillet, M.A. and Fontaine-Perus, J. (1984b) Chimeras in the study of the peripheral nervous system in birds. In: *Chimeras in Developmental Biology* (eds. N.M. Le Douarin and A. McClaren), Academic Press, London, pp. 313–52.

Lear, P.V. (1993) Nephrogenesis in the chick embryo. PhD Thesis, University of London.

Li, Z., Song, Y., Ma, Y., Wei, H., Liu, C., Huang, J., Wan, N., Sha, J. and Sakurai, F. (2002) Influence of simulated microgravity on avian primordial germ cell migration and reproductive capacity. *J. Exp. Zool.* **292,** 672–6.

Liem, K.F., Jessel, T.M. and Briscoe, J. (2000) Regulation of the neural patterning activity of sonic hedgehog by secreted BMP inhibitors expressed by notochord and somites. *Development* **127,** 4855–66.

Lillie, F.R. (1952) *Development of the Chick: An Introduction to Embryology*, 3rd ed. (Revised by H.L. Hamilton), Henry Holt, New York.

Linask, K.K. and Lash, J.W. (1986) Precardiac cell migration: fibronectin localization at mesoderm:endoderm interface during directional migration. *Dev. Biol.* **114,** 87–101.

Locy, W. A. and Larsell, O. (1916) The embryology of the bird's lung based on observations of the domestic fowl. *Am. J. Anat.* **19,** 447–504, and **20,** 1–44.

Lopez-Sanchez, C., Garcia-Martinez, V. and Schoenwolf, G.C. (2001) Localization of cells of the prospective neural plate, heart and somites within the primitive streak and epiblast of avian embryos at intermediate primitive-streak stages. *Cells Tissues Organs* **169,** 334–46.

Lough, J. and Sugi, Y. (2000) Endoderm and heart development. *Dev. Dyn.* **217,** 327–42.

Loveless, W., Bellairs, R., Thorpe, S.J., Page, M. and Feizi, T. (1990) Developmental patterning of the carbohydrate antigen FC10.2 during early embryogenesis in the chick. *Development* **108,** 97–106.

Lucas, A.M. and Stettenheim, P.R. (1972) *Avian Anatomy – Integument*. 2 vols. U.S. Department of Agriculture.

Maderson, P.F.A. (1985) Some developmental problems of the reptilian integument. In *Biology of the Reptilia*, Vol. 14 (eds. C. Gans, F. Billett and P.F.A. Maderson), Wiley, New York, pp. 523–98.

Mahowald, A.P. (1971) Polar granules of *Drosophila*. III The continuity of polar granules during the life cycle of *Drosophila*. *J. Exp. Zool.* **176,** 329–43.

Maina, J.N. (2003) Developmental dynamics of the bronchial (airway) and air/sac systems of the avian respiratory system from day 3 to day 26 of life: a scanning electron microscopic study of the domestic fowl, *Gallus gallus* variant *domesticus*. *Anat. Embryol.* **2007,** 119–34.

Malago, W. Jr., Franco, H.M., Matheucci, E. Jr., Medaglia, A. and Henrique-Silva, F. (2002) Large scale sex typing of ostriches using DNA extracted from feathers. *BMC Biotechnol.* **2,** 19.

Männer, J. (1993) Experimental study on the formation of the epicardium in chick embryos. *Anat. Embryol.* **187,** 281–9.

Männer, J., Seidl, W. and Steding, G. (1989) The role of extracardiac factors in normal and abnormal development of the chick embryo heart: cranial flexures and ventral thoracic wall. *Anat. Embryol.* **191,** 61–72.

Männer, J., Seidl, W. and Steding, G. (1993) Correlation between embryonic head flexures and cardiac development. An experimental study in chick embryos. *Anat. Embryol.* **188,** 269–285.

Männer, J., Perez-Pomares, J.M., Macias, D. and Muñoz-Chápuli, R. (2001) The origin, formation and developmental significance of the epicardium: a review. *Cells Tissues Organs* **169,** 69–103.

Mareel, M., Bellairs, R., De Bruyne, G. and Van Peteghem, M.C. (1984) Effect of microtubule inhibitors on the expansion of hypoblast and margin of overgrowth of chick blastoderms. *J. Embryol. Exp. Morph.* **81,** 273–86.

Martin, C. (1971) Contribution du canal de Wolff et de ses dérivés à l'édification des tubed urinaires du mésonéphros chez l'embryon d'oiseaux. *C. R. Acad. Sc. Paris*, série D **272,** 1305–7.

Martin, C. (1990) Quail chick chimeras, a tool for developmental immunology. In *The Avian Model in Developmental Biology: from organism to genes* (eds. N. Le Douarin, F. Dieterlen-Lièvre and J. Smith) (Editions du CNRS, Paris, pp. 207–17.

Marvin, M.J., Di Rocco, G., Gardiner, A., Bush, S.M. and Lassar, A.B. (2001) Inhibition of Wnt activity induces heart formation from posterior mesoderm. *Genes Dev.* **15,** 316–27.

Matsuda, S., Baluk, P., Shimizu, D. and Fujiwara, F. (1996) Dorsal root ganglion neuron development in chick and rat. *Anat. Embryol.* **193,** 475–80.

Matsushita, S. (1995) Fate mapping study of the splanchnopleural mesoderm of the 1.5 day-old chick embryo. *Roux's Arch. Dev. Biol.* **204,** 392–99.

Matsushita, S. (1996) Fate mapping study of the endoderm of the 1.5 day-old chick embryo. *Roux's Arch. Dev. Biol.* **205,** 225–31.

Matsushita, S. (1998) Developmental changes in mucosubstances revealed by immunostaining with antimucus monoclonal antibodies and lectin staining in the epithelium lining the segment from gizzard to duodenum of the chick embryo. *J. Anat.* **193,** 587–97.

Matsushita, S. (1999) Fate mapping study of the endoderm in the posterior part of the 1.5-day-old chick embryo. *Develop. Growth Differ.* **41,** 313–9.

Matsushita, S., Ishii, Y., Scotting, P.J., Kuroiwa, A. and Yasugi, S. (2002) Pre-gut endoderm of chick embryos is regionalised by 1.5 days of development. *Dev. Dyn.* **223,** 33–47.

Matulionis, D.H. (1970) Morphology of the down feathers of chick embryos. A descriptive study at the ultrastructural level of differentiation and keratinization. *Z. Anat. Entw.* **132,** 107–57.

Mauch, T.J., Yang, G., Wright, M., Smith, D. and Schoenwolf, G.C. (2000) Signals from trunk paraxial mesoderm induce pronephros formation in chick imtermediate mesoderm. *Dev. Biol.* **220,** 62–75.

Mayer, A.W. (1939) *The Rise of Embryology.* Stamford University Press.

Mayerson, P.L. and Fallon, J.F. (1985) The spatial pattern and temporal sequence in which feather germs arise in the White Leghorn chick embryo. *Dev. Biol.* **109,** 259–67.

McCann, J.P., Owens, P.D.A. and Wilson, D.J. (1991) Chick frontonasal process excision significantly affects midfacial development. *Anat. Embryol.* **184,** 171–8.

McGonnell, I.M., McKay, I.J. and Graham, A. (2001) A population of caudally migrating cranial neural crest cells: functional and evolutionary implications. *Dev. Biol.* **236,** 354–63.

McLaren, A. (1981) Germ cell lineages. In *Chimeras in Developmental Biology.* Academic Press, London. pp 111–29.

McLelland, J. (1979) Digestive system. In *Form and Function in Birds* (eds. A.S. King and J. McLelland), Vol. 1 Academic Press, London, pp. 69–181.

McLelland, J. and King, A.S. (1970). The gross anatomy of the peritoneal coelomic cavities of *Gallus domesticus. Anat. Anz.* **127,** 480–90.

McLelland, J and King, A.S. (1975) Aves coelomic cavities and mesenteries. In *Sisson and Grossman's 'The Anatomy of the Domestic Animals'* (ed. R. Getty), W.B. Saunders, Philadelphia, London, Toronto, pp. 1849–56.

Meier, S. (1979) Development of the chick mesoblast. Formation of the embryonic axis and establishment of the metameric pattern. *Dev. Biol.* **73,** 25–45.

Meier, S. (1981) Development of the chick embryo mesoblast: morphogenesis of the prechordal plate and cranial segments. *Dev. Biol.* **83,** 49–61.

Mekki-Dauriac, S., Agius, E., Kan, P. and Cochard, P. (2002) Bone morphogenetic proteins negatively coat oligodendrocyte precursor specification in the chick spinal cord. *Development* **129,** 5117–30.

Menna, T.M. and Mortola, J.P. (2002) Metabolic control of pulmonary ventilation in developing chick embryo. *Respir. Physiol. Neurobiol.* **130,** 43–55.

Menon, G.K., Aggarwal, S.K. and Lukas, A.M. (1981) Evidence for the holocrine nature of lipoid secretion by avian epidermal cells: a histochemical and ultrastructural study of rictus and the uropygial gland. *J. Morphol.* **167,** 185–99.

Merido-Velasco, J.A., Sanchez-Montesinos, I., Espin, J., Garcia-Garcia, J.D. and Roldan-Schilling, V. (1996) Grafts of the third branchial arch in chick embryo. *Acta Anat.* (Basel) **155,** 73–80.

Merino, R., Gavan, Y., Macias, D., Economides, A.N., Sampath, K.T. and Hurle, J.M. (1999a) Morphogenesis of digits in the avian limb is controlled by FGFs, TGFβs and noggin through BMP signalling. *Dev. Biol.* **200,** 35–45.

Merino, R., Macias, D., Ganan, Y., Rodriguez-Leon, J., Economides, A.N., Rodriguez-Esteban, C., Izpisua-Belmonte, J.C. and Hurle, J.M. (1999b) Control of digit formation by activin signalling. *Development* **126,** 2161–70.

Meyer, D.B. (1964) The migration of the primordial germ cells in the chick embryo. *Dev. Biol.* **10,** 154–90.

Michaille, J.J., Kanzler, B., Blanchet, S., Garnier, J.M. and Dhouailly, D. (1995) Characterization of cDNAs encoding two chick retinoic acid receptor alpha isoforms and distribution of retinoic acid receptor alpha, beta and gamma transcripts during chick skin development. *Int. J. Dev. Biol.* **39,** 587–96.

Miller, S.A. (1982) Differential proliferation in morphogenesis of lateral body folds. *J. Exp. Zool.* **221,** 205–11.

Miller, S.A. and Briglin, A. (1996) Apoptosis removes chick embryo tail gut and remnant of the primitive streak. *Dev. Dyn.* **206,** 212–8.

Miller, S.A., Bresee, K.L., Michaelson, C.L. and Tyrell, D.A. (1994) Domains of differential cell proliferation and formation of amnion folds in chick embryo ectoderm. *Anat. Rec.* **238,** 225–36.

Mjaatvedt, C.H., Nakoaka, T., Moreno-Rodriguez, R., Norris, R.A., Kern, M.J., Eisenberg, C.A., Turner, D. and Markwald, R.R. (2001) The outflow tract of the heart is recruited from a novel heart-forming area. *Dev. Biol.* **238,** 97–109.

Mobbs, I.G. and McMillan, D.B. (1981) Transport across endodermal cells of the chick yolk sac during early stages of development. *Am. J. Anat.* **160,** 285–308.

Monnet-Tschudi, F. and Kucera, P. (1988) Myosin, tubulin and laminin immunoreactivity in the ectoderm of the growing area opaca of the chick embryo. *Anat. Embryol.* **179,** 157–64.

Monsoro-Burq, A.H., Bontoux, M., Teillet, M.A. and Le Douarin, N.M. (1994) Heterogeneity in the development of the vertebra. *Proc. Natl. Acad. Sci.* **91,** 10435–9.

Morais da Silva, S., Hacker, A., Harley, V., Goodfellow, P., Swain, A. and Lovell-Badge, R. (1996) *Sox9* expression during gonadal development implies a conserved role for the gene in testis differentiation in mammals and birds. *Nat. Genet.* **14,** 62–8.

Moro-Balbás, J.A., Gato, A., Alonso, M.I., Martin, P. and de la Mano, A. (2000) Basal heparan sulphate proteoglycan is involved in otic invagination in chick embryos. *Anat. Embryol.* **202,** 333–43.

Moury, J.D. and Schoenwolf, G.C. (1995) Cooperative model of epithelial shaping and bending during avian neurulation: autonomous movements of the neural plate, autonomous movements of the epidermis, and interactions in the neural plate/epidermis transition zone. *Dev. Dyn.* **204,** 323–37.

Müller, E., Schneider, M. and Rickenbacher, J. (1983) Metabolic pathways in chick amnion muscle. *Acta Anat.* **115,** 272–81.

Muroyama, Y., Fujihara, M., Ikeya, M., Kondoh and Takada, S. (2002) Wnt signalling plays an essential role in neuronal specification of the dorsal spinal cord. *Genes Dev.* **16,** 548–53.

Murray, B.M. and Wilson, D.J. (1994) A scanning electron microscopic study of the normal development of the chick wing from stages 19–36. *Anat. Embryol.* **89,** 147–58.

Nafstad, P.H.J. (1986) On the avian Merkel cells. *J. Anat.* **145,** 25–33.

Naito, M., Nirasawa, K. and Oishi, T. (1991) Development in culture of the chick embryo from fertilized ovum to hatching. *J. Exp. Zool.* **254,** 322–6.

Nakabayashi, O., Kikuchi, H., Kikuchi, T. and Mizuno, S. (1998) Differential expression of genes for aromatase and oestrogen receptor during gonadal development in chicken embryos. *J. Mol. Endocrinol.* **20,** 193–202.

Nakamura, A., Kulikowski, R.R., Lacktis, J.W. and Manasek, D.M.D. (1980) Heart looping: a regulated response to deforming forces. In *Etiology and Morphogenesis of Congenital Heart Disease* (ed. R. Van Pragh), Futura, New York, pp. 81–98.

Narita, T., Saitoh, K., Kameda, T., Kuroiwa, A., Mizutani, M., Koike, C., Iba, H. and Yasugi, S. (2000) BMPs are necessary for stomach gland formation in the chick embryo: a study using virally induced NMP-2 and noggin expression. *Development* **127,** 981–8.

Nechaeva, M.V. and Turpaev, T.M. (2002) Rhythmic contractions in chick amnio-yolk sac and snake amnion during embryogenesis. *Comp. Biochem. Physiol. A. Mol Integr. Physiol.* **131,** 861–70.

Needham, J. (1934) *A History of Embryology,* 2nd ed. Cambridge University Press.

Nellemann, C., de Bellard, M.E., Barembaum, M., Laufer, E. and Bronner-Fraser, M. (2001) Excess lunatic fringe causes cranial neural crest over-proliferation. *Dev. Biol.* **235,** 121–30.

New, D.A.T. (1955) A new technique for the cultivation of the chick embryo *in vitro. J. Embryol. Exp. Morph.* **3,** 326–31.

New, D.A.T. (1956) The formation of sub-blastodermic fluid in hens' eggs. *J. Embryol. Exp. Morph.* **4,** 221–7.

New, D.A.T. (1959) The adhesive properties and expansion of the chick blastoderm. *J. Embryol. Exp. Morph.* **7,** 146–64.

New, D.A.T. (1966) *The Culture of Avian Embryos.* Logos Press, London.

Newgreen, D.F., Scheel, M. and Kaster, V. (1986) Morphogenesis of sclerotome and neural crest cells in avian embryos. In vivo and in vitro studies on the role of notochordal extracellular material. *Cell Tiss. Res.* **244,** 299–313.

Newman, H.H. (1917) *The Biology of Twins.* University of Chicago Press, Ill.

Nicolet, G. and Gallera, J. (1963) Dans quelles conditions l'amnios de l'embryon de poulet peut-il se former en culture *in vitro? Experientia* **19,** 165–6.

Niedorf, H.R. and Wolters, B. (1978) Development of the Harderian gland in the chicken: light and electron microscopic investigations. *Invest. Cell Pathol.* **1,** 205–15.

Nieuwkoop, P.D. and Satasurya, L.A. (1979) *Primordial Germ Cells in the Chordates.* Cambridge University Press.

Noden, D.M. (1983a) The embryonic origins of avian cephalic and cervical muscles and associated connective tissues. *Am. J. Anat.* **168,** 257–76.

Noden, D.M. (1983b) The role of the neural crest in patterning of avian cranial, skeletal, connective and muscle tissue. *Dev. Biol.* **96,** 144–65.

Noden, D.M. (1986) Patterning of avian craniofacial muscles. *Dev. Biol.* **116,** 347–56.

Noden, D.M. (1987) Interactions between cephalic neural crest and mesodermal populations. In *Developmental and Evolutionary Aspects of the Neural Crest* (ed. P.F.A. Maderson) Wiley, New York, pp. 89–119.

Noden, D.W. (1988) Interactions and fates of craniofacial mesenchyme. *Development* **103,** 121–40.

Noden, D.W. (1990) Origin and assembly of avian embryonic blood vessels. *Ann. N.Y. Acad. Sci.* 588, 236–49.

Noden, D.M. (1992) Morphogenetic movements of avian prechordal mesoderm. *Anat. Rec.* **232,** 65a.

Noden, D.M. and DeLahunta, A. (1985) *The Embryology of Domestic Animals: Devlopmental Mechanisms and Malformations.* Williams and Wilkins, Baltimore.

Noveen, A., Jiang, T.X., Ting-Bereth, S.A. and Chuong, C.M. (1995) Homeobox genes Msx-l and Msx-2 are associated with the induction and growth of skin appendages. *J. Invest. Dermatol.* **104,** 711–19.

Nowicki, J.L. and Burke, A.C. (1999) Testing *Hox* genes by surgical manipulation. *Dev. Biol.* **210,** 228.

Noy, Y. and Sklan, D. (1998) Yolk utilisation in the newly hatched poult. *Br. Poult. Sci.* **39,** 446–51.

Ohashi, H., Takewaki, T., Uno, T. and Komori, S. (1993) Development of vagal innervation to the muscles of the avian gizzard. *J. Auton. Nerv. Syst.* **42,** 233–40.

Oldfield, S.F. and Evans, D.J.R. (2003) Tendon morphogenesis in the developing avian limb: plasticity of fetal tendon fibroblasts. *J. Anat.* **202,** 153–64.

Olivera-Martinez, I., Coltey, M., Dhouailly, D. and Pourquie, O. (2000) Mediolateral somitic origin of ribs and dermis determined by quail-chick chimeras. *Development* **127,** 4611–7.

Olivera-Martinez, I., Thelu, J., Teillet, M.A. and Dhouailly, D. (2001) Dorsal dermis development depends on a signal from the dorsal neural tube which can be substituted by Wnt-l, *Mech Dev.* **100,** 233–44.

Olsen, M.W. (1962) Polyembryony in unfertilized turkey eggs. *J. Hered.* **53,** 125–9.

Olsson, J.E., Kamachi, Y., Penning, S., Muscat, G.E., Kondoh, H. and Koopman, P. (2001) Sox18 expression in blood vessels and feathers during chicken embryogenesis. *Gene* **271,** 151–8.

Ooi, V.E.C., Sanders, E.J. and Bellairs, R. (1986) The contribution of the primitive streak to the somites in the avian embryo. *J. Embryol. exp. Morph.* **92,** 192–206.

Oppenheim, R.W., Homma, S., Marti, E., Prevette, D., Wang, S., Yaginuma, H. and McMahon, A.P. (1999) Modulation of early but not later stages of programmed cell death in embryonic avian spinal cord by sonic hedgehog. *Mol. Cell Neurosci.* **13,** 348–61.

Oppenheimer, J. (1955) Problems, concepts and history. In *Analysis of Development* (eds. B.H. Willier, P.A. Weiss and V. Hamburger), W.B. Saunders, Philadelphia, pp. 1–24.

Ordahl, C.P. (1993) Myogenic lineages within the developing somite: In *Molecular Basis of Morphogenesis.* Liss, New York, pp. 165–76.

Ordahl, C.P. and Le Douarin, N.M. (1992) Two myogenic lineages within the developing somite. *Development* **114,** 339–53.

Ordahl, C.P., Berdougo, E., Venters, S.J. and Denetclaw, W. Jr. (2001) The dermomyotome dorsomedial lip drives growth and morphogenesis of both the primary myotome and dermomyotome epithelium. *Development* **128,** 1731–44.

Oreal, E., Pieau, C., Mattei, M.G., Josso, N., Picard, J.Y., Carre-Eusebe, D. and Magre, S. (1998) Early expression of AMH in chicken embryonic gonads precedes testicular *SOX9* expression. *Dev. Dyn.* **212,** 522–32.

Orosz, S.E., Ensley, P.K. and Haynes, C.J. (1992) *Avian Surgical Anatomy. Thoracic and Pelvic Limbs.* W.R. Saunders Co., Phladelphia.

Orts-Llorca, F. and Collado, J.J. (1970) The development of heterologous grafts of the cardiac area (labelled with thymidine-3H) to the caudal area of the chick blastoderm. *Arch. Anat. Histol. Embryol.* **53,** 113–24.

Osmond, M.K., Butler, A.J., Voon, F.C.T. and Bellairs, R. (1991) The effects of retinoic acid on heart formation in the early chick embryo. *Development* **113,** 1405–17.

Overton, J. (1989) Fusion of epithelial sheets as seen in formation of the chick amnion. *Cell Tissue Res.* **257,** 141–47.

Packard, D.S. and Meier, S. (1983) An experimental study of the somitomeric organization of the avian segmental plate. *Dev. Biol.* **97,** 191–202.

Paganelli, C.V. (1991) The avian eggshell as a mediating barrier: respiratory gas fluxes and pressure during development. In *Egg Incubation: Its Effects on Embryonic Development in Birds and Reptiles* (eds. D.C. Deeming and M.W.J. Ferguson), Cambridge University Press, Cambridge, pp. 261–76.

Panman, L. and Zeller, R. (2003) Patterning the limb before and after SHH signalling. *J. Anat.* **202,** 3–12.

Pannett, C.A. and Compton, A. (1924) The cultivation of tissues in saline-embryonic juice. *Lancet* **1,** 381–4

Pardenaud, L. and Dieterlen-Lièvre, F. (1993) Emergence of endothelial and hemopoietic cells in the avian embryo. *Anat. Embryol.* **187,** 107–14.

Pardenaud, L. and Dieterlen-Lièvre, F. (1999) Manipulation of the angiopoietic/hemangiopoietic commitment in the avian embryo. *Development* **126,** 617–27.

Pardenaud, L., Buck, C. and Dieterlen-Lièvre, F. (1987) Early germ cell segregation and distribution in the quail blastodisc. *Cell Diff.* **22,** 47–60.

Parker, W.K. and Bettany, G.T. (1877) *The Morphology of the Skull.* Macmillan, London.

Pasteels, J. (1937) Etudes sur la gastrulation du vertebrés meroblastiques. III Oiseaux. *Arch. Biol.* **48,** 381–488.

Patel, K., Makarenkova, H. and Jung, H.S. (1999) The role of long range, local and direct signal molecules during chick feather bud development involving the BMPs, follistatin and the Eph receptor tyrosine kinase Eph-A4. *Mech. Dev.* **86,** 51–62; 399–415.

Patten, B.M. (1922) The formation of the cardiac loop in the chick. *Am. J. Anat.* **30,** 373–97.

Patten, B.M. (1950) *Early Embryology of the Chick*, 4th ed. H.K. Lewis, London.

Patten, I. and Placzek, M. (2002) Opponent activities of Shh and BMP signaling during floor plate induction in vivo. *Curr. Biol.* **12,** 47–52.

Patterson, S.B. and Minkoff, R. (1985) Morphometric and autoradiographic analysis of frontonasal development in the chick embryo. *Anat. Rec.* **212,** 90–9.

Paul, E.R., Ngai, P.K., Walsh, M.P. and Groschel-Stewart, U. (1995) Embryonic chicken gizzard: expression of the smooth muscle regulatory proteins caldesmon and myosin light chain kinase. *Cell Tissue Res.* **279,** 331–7.

Paulsen, D.F. (1994) Retinoic acid in limb-bud outgrowth: review and hypothesis. *Anat. Embryol.* **190,** 399–415.

Pearse, R.V. II and Tabin, C.J. (1998) The molecular ZPA. *J. Exp. Zool.* **282,** 677–90.

Pedernera, E., Solis, L., Peralta, I. and Vel Markazquez, P.N. (1999) Proliferative and steroidogenic effects of follicle-stimulating hormone during chick embryo gonadal hormone. *Gen. Comp. Endocrinol.* **116,** 213–20.

Peebles, F. (1898) Some experiments on the primitive streak of the chick. *Arch. Entw. Mech Org.* **7,** 405–29.

Perry, M.M. (1988) A complete culture system for the chick embryo. *Nature, Lond.* **331,** 70–2.

Perriton, C.L., Powles, N., Chiang, C., Maconochie, M.I. and Cohn, M.J. (2002) Sonic hedgehog signalling from the urethral epithelium controls external genital development. *Dev. Biol.* **247,** 26–46.

Petitte, J.N., Clark, M.E., Liu, G., Verrinder Gibbins, A.M. and Etches, R.J. (1990) Production of somatic and germline chimeras in the chicken by transfer of early blastodermal cells. *Development* **108,** 185–9.

Pfeffer, P.L., De Robertis, E.M. and Izpisua-Belmonte, J.C. (1997) Crescent, a novel chick gene encoding a frizzled-like cysteine-rich domain, is expressed in anterior regions during early embryogenesis. *Int. J. Dev. Biol.* **41,** 449–58.

Pikalow, A.S., Flynn, M.E. and Searls, R.L. (1994) Development of the cranial flexure and Rathke's pouch in the chick embryo. *Anat. Rec.* **238,** 407–14.

Pinot, M. (1969) Etude expérimentale de la morphogenèse de la cage thoracique chez l'embryon de poulet: mécanismes et origine du matérial. *J. Embryol. Exp. Morph.* **21,** 149–64.

Poole, T.J. and Steinberg, M.S. (1982) Evidence for the guidance of pronephric duct migration by a craniocaudally travelling adhesion gradient. *Dev. Biol.* **92,** 144–58.

Poole, T.J., Finkelstein, E.B. and Cox, C.M. (2001) The role of FGF and VEGF in angioblast induction and migration during vascular development. *Dev. Dynamics* **220,** 1–17.

Pourquié, O. (2001) Vertebrate somitogenesis. *Annu. Rev. Cell Dev. Biol.* **17,** 311–50.

Pownall, M.E., Strunk, K.E. and Emerson, C.J. Jr. (1996) Notochord signals control the transcriptional cascade of myogenic bHLH genes in somites of quail embryos. *Development* **122,** 1475–88.

Prin, F. and Dhouailly, D. (2004) How and when the regional competence of chick epidermis is established: feathers *vs.* scutate and reticulate scales, a problem *en route* to a solution. *Int. J. Dev. Biol.* **48,** 137–48.

Psychoyos, J. and Stern, C.D. (1996) Fates and migratory routes of primitive streak cells in the chick embryo. *Development* **122,** 1523–34.

Quay, W.B. (1986) The skin of birds – uropygial gland. In *Biology of the Integument,* (eds. J. Bereiter-Hahn, A. Matoltsty, and K.S. Richards), Vol. 2, Vertebrates Springer-Verlag, Berlin and New York, pp. 248–54.

Quiring, D.P. (1933) The development of the sino-atrial region of the chick heart. *J. Morph.* **55,** 81–118.

Rautenfeld, D.B.V. (1993) Systema lymphaticum et spleen. In *Handbook of Avian Anatomy: Nomina Anatomia Avium,* 2nd ed. (eds. J.J. Baumel and others) Nuttal Ornithological Club, Cambridge, Mass., pp. 477–491.

Rawdon, B.B. (1998) Morphogenesis and differentiation of the avian endocrine pancreas, with particular reference to experimental studies on the chick embryo. *Microsc. Res. Tech.* **15,** 292–305.

Rawdon, B.B. and Andrew, A. (1999) Gut endocrine cells in birds: an overview, with particular reference to the chemistry of gut peptides and the distribution, ontogeny, embryonic origin and differentiation of the endocrine cells. *Prog. Histochem. Cytochem.* **34,** 3–82.

Rawdon, B.B. and Larsson, L.I. (2000) Development of hormonal peptides and processing enzymes in the embryonic avian pancreas with special reference to co-localisation. *Histochem. Cell Biol.* **114,** 105–12.

Rawles, M. (1943) The heart-forming areas of the early chick blastoderm. *Physiol. Zool.* **16,** 22–4.

Richman, J.M. and Tickle, C. (1989) Epithelia are interchangeable between facial primordia of chick embryos and morphogenesis is controlled by the mesenchyme. *Dev. Biol.* **136,** 201–10.

Richman, J.M., Herbert, M., Matinovic, E. and Walin, J. (1997) Effect of fibroblast growth factors on outgrowth of facial mesenchyme. *Dev. Biol.* **189,** 135–47.

Riddle, R.D., Johnson, R.L., Lauter, E. and Tabin, C. (1993) *Sonic hedgehog* mediates the polarizing activity of the ZPA. *Cell* **75,** 1401–16.

Risau, W. and Flamme, I. (1995) Vasculogenesis. *Annu. Rev. Cell Biol.* **11,** 73–91.

Roberts, D.J., Johnson, R.L., Burke, A.C., Nelson, C.E., Morgan, B.A. and Tabin, C. (1995) Sonic hedgehog is an endodermal signal inducing Bmp-4 and *Hox* genes during induction and regionalization of the chick hindgut. *Development* **121,** 3163–74.

Robertson, C.P., Gibbs, S.M. and Roelink, H. (2001) cGMP enhances the sonic hedgehog response and neural plate cells. *Dev. Biol.* **238,** 157–67.

Robertson, L., Wishart, G.J. and Horrocks, A.J. (2000) Identification of perivitelline N-linked glycan as mediators of sperm-egg interaction in chickens. *J. Reprod. Fertil.* **120,** 397–403.

Rodriguez-Estaban, C., Capdevila, J., Kawakami, Y. and Belmonte, J.C.I. (2001) Wnt signalling and PKA control Nodal expression and left-right determination in the chick embryo. *Development* **128,** 3189–96.

Romano, L.A. and Runyan, R.B. (2000) Slug is an essential target of TGFβ2 signaling in the developing chicken heart. *Dev. Biol.* **223,** 91–102.

Romanoff, A.L. (1960) *The Avian Embryo. Structure and Functional Development.* MacMillan, New York.

Romanoff, A.L. and Romanoff, A.J. (1949) *The Avian Egg.* Wiley, New York.

Romer, A.S. (1927) The development of the thigh musculature of the chick. *J. Morph.* **43,** 347–86.

Ros, M.A., Rivero, F.B., Hinchliffe, J.R. and Hurle, J.M. (1995) Immunohistochemical and ultrastructural study of the developing tendons of the avian foot. *Anat. Embryol.* **192,** 483–96.

Rosenquist, G.C. (1966) A radioautographic study of labelled grafts in the chick blastoderm: development from primitive streak to stage 12. *Contr. Embryol. Carneg. Inst.* **38,** 71–110.

Rosenquist, G.C. (1971) The location of pregut endoderm in the chick embryo at the primitive streak stage as determined by radioautographic mapping. *Dev. Biol.* **26,** 323–35.

Rothenberg, F., Hitomi, M., Fisher, S. and Watanabe, M. (2003) Initiation of apoptosis in the developing avian outflow tract myocardium. *Dev. Dyn.* **223,** 469–82.

Rowlett, K. and Simkiss, K. (1987) Explanted embryo culture: in vitro and in ovo techniques for domestic fowl. *Br. Poultry Sci.* **28,** 91–101.

Ruffins, S. and Bronner-Fraser, M. (2000) A critical period for conversion of ectodermal cells to a neural crest fate. *Dev. Biol.* **218,** 13–20.

Ruiz I Altaba, A., Placzek, M., Baldassare, M., Dodd, J. and Jessell, T.M. (1995) Early stages of notochord and floor plate development in the chick embryo defined by normal and induced expression of HNF-3 beta. *Dev. Biol.* **170,** 299–313.

Saber, G.M., Parker, S.B. and Mintoff, R. (1989) Influence of epithelial-mesenchymal interaction on the viability of facial mesenchyme in vitro. *Anat. Rec.* **225,** 56–66.

Sakiyama, J., Yokouchi, Y. and Kuroiwa, A. (2000) Coordinated expression of Hoxb genes and signalling molecules during development of the chick respiratory tract. *Dev. Biol.* **227,** 12–27.

Sanders, E.J. (1989) *The Cell Surface in Embryogenesis and Carcinogenesis.* The Telford Press, New Jersey.

Sanders, E.J. (1991) Embryonic cell invasiveness: an *in vitro* study of chick gastrulation. *J. Cell Sci.* **98,** 403–7.

Sanders, E.J. (1997) Cell death in the avian sclerotome. *Dev. Biol.* **192,** 551–63.

Sanders, E.J. and Parker, E. (2001) Ablation of axial structures activates apoptotic pathways in somite cells of the chick embryo. *Anat. Embryol.* **204,** 389–98.

Sanders, E.J. and Prasad, S. (1991) Possible roles for TGFβ1 in the gastrulating chick embryo. *J. Cell Sci.* **99,** 617–26.

Sanders, E.J., Varedi, M. and French, A.S. (1993) Cell proliferation in the gastrulating chick embryo: a study using BrdU incorporation and PCNA localization. *Development* **118,** 389–99.

Sanders, E.J., Bellairs, R. and Portch, P.A. (1978) *In vivo* and *in vitro* studies on the hypoblast and definitive endoblast of avian embryos. *J. Embryol. Exp. Morph.* **46,** 187–205.

Sanders, E.J., Khare, M.K., Ooi, V.C. and Bellairs, R. (1986) An experimental and morphological analysis of tail bud mesenchyme of the chick embryo. *Anat. Embryol.* **174,** 179–85.

Sanders, E.J., Torkkeli, P.H. and French, A.S. (1997) Patterns of cell death during gastrulation in chick and mouse embryos. *Anat. Embryol.* **195,** 147–154.

Sanz-Ezquerro, J.J. and Tickle, C. (2003) Digital development and morphogenesis. *J. Anat.* **202,** 51–8.

Sasaki, F., Doshita, A., Matsumoto, Y., Kuwahara, S., Tsukamoto, Y, and Ogawa, K. (2003) Embryonic development of the pituitary gland in the chick. *Cells Tissues Organs* **173,** 65–74.

Sato, K., Mochida, H., Yazawa, I., Sasaki, S. and Momose-Sato, Y. (2002) Optical approaches to functional organization of glossopharyngeal and vagal motor nuclei in the embryonic chick hindbrain. *J. Neurophysiol.* **88,** 383–93.

Saunders, J.W. Jr., Cairns, J.M. and Gasseling, M.T. (1957) The role of the apical ridge of ectoderm in the differentiation of the morphological structure and inductive specificity of limb parts of the chick. *J. Morphol.* **101,** 57–88.

Sawada, K. and Aoyama, H. (1999) Fate maps of the primitive streak in chick and quail embryo: ingression timing of progenitor cells of each rostro-caudal axial level of somites. *Int. J. Dev. Biol.* **43,** 809–15.

Sawyer, R.H. (1972) Avian scale development. I. Histogenesis and morphogenesis of the epidermis and dermis during formation of the scale ridge. *J. Exp. Zool.* **181,** 365–84.

Sawyer, R.H., Knapp, L.W. and O'Guin, W.M. (1986) The skin of birds: epidermis, dermis and appendages. In *Biology of the Integument,* Vol. 2 Vertebrates. (eds. J. Bereiter-Hahn, A. Matoltsty and K.S. Richards), Springer-Verlag, Berlin and New York, pp. 194–238.

Saxod, R. (1980) Development of Merkel corpuscles and so-called 'transitional' cells in the white leghorn chicken. *Am. J. Anat.* **151,** 453–74.

Scaal, M., Prols, F., Fuchtbauer, E.M., Patel, K., Hornik, C., Kohler, T., Christ, B. and Brand-Saberri, B. (2002) BMPs induce dermal markers and ectopic feather tracts. *Mech. Dev.* **110,** 51–60.

Scanes, C.G., Hart, L.E., Decuypere, E. and Kuhn, E.R. (1987) Endocrinology of the avian embryo: an overview. *J. Exp. Zool.* Suppl. 1, 253–64.

Schepelmann, K. (1990) Erythropoietic bone marrow in the pigeon: development of its distribution and volume during growth and pneumatization of bones. *J. Morph.* **203,** 21–34.

Schmidt, C., Christ, B., Maden, M., Brand-Saberi, B. and Patel, K. (2001) Regulation of Epha-4 expression in paraxial and lateral plate mesoderm by ectoderm-derived signals. *Dev. Dyn.* **220,** 377–86.

Schneider, R.A., Hu, D., Rubenstein, J.L., Maden, M. and Helms, J.A. (2001) Local retinoid signalling coordinates forebrain and facial morphogenesis by maintaining FGF8 and SHH. *Development* **128,** 2755–67.

Schoenwolf, G.C. and Singh, U. (1981) Changes in the surface morphologies of the cells in the bursa cloacalis (bursa of Fabricius) during ontogeny of the chick embryo. *Anat. Rec.* **201,** 303–16.

Schoenwolf, G.C. and Smith, J.L. (1990) Mechanisms of neurulation: traditional viewpoint and recent advances. *Development* **109,** 243–70.

Schoenwolf, G.C. and Watterson, R.L. (1989) *Laboratory Studies of Chick, Pig, and Frog Embryos. Guide and Atlas of Vertebrate Development,* 6th ed. Macmillan, New York.

Schöfer, C., Frei, K., Weipoltshammer, K. and Wachtler, F. (2001) The apical ectodermal ridge, fibroblast growth factors (FGF-2 and FGF-4) and insulin-like growth factor I (IGF-I) control the migration of epidermal melanoblasts in chicken wing buds. *Anat. Embryol.* **203,** 137–46.

Schramm, C. and Solursh, M. (1990) The formation of premuscle masses during chick wing bud development. *Anat. Embryol.* **182,** 235–47.

Schreiner, K.E. (1902) Über die Entwicklung der Amniotenniere. *Zeit. wiss. Zool.* **71,** 1–188.

Schweitzer, R., Chyung, J.H., Murtaugh, L.C., Brent, A., Rosen, V., Olson, E.N., Lassar, A. and Tabin, C.J. (2001) Analysis of the tendon cell fate using Scleraxis, a specific marker for tendons and ligaments. *Development* **128,** 3855–66.

Searls, R.L. (1986) A description of caudal migration leading to the formation of pericardial and pleural coeloms, to caudal migration of the arches and to development of the shoulder. *Am. J. Anat.* **177,** 271–83.

Seidl, W. and Steding, G. (1978) Topogenesis of the anterior intestinal portal. *Anat. Embryol.* **155,** 37–45.

Sekine, K., Ohuchi, H., Fujiwara, M., Yamasaki, M., Yoshizawa, T., Sato, T., Yagishita, N., Matsui, D., Koga, Y., Itoh, N. and Kato, S. (1999) FGF10 is essential for limb and lung formation. *Nat. Genet.* **21,** 138–41.

Sela-Donenfeld, D. and Kalcheim, C. (2000) Inhibition of noggin expression in the dorsal neural tube by somitogenesis: a mechanism for coordinating the timing of neural crest emigration. *Development* **127,** 4845–54.

Selleck, M.A.J. and Bronner-Fraser, M. (1995) Origins of the avian neural crest: the role of neural plate-epidermal interactions. *Development* **121,** 525–38.

Selleck, M.A. and Bronner-Fraser, M. (2000) Avian neural crest cell fate decisions: a diffusible signal mediates inductions of neural crest by the ectoderm. *Int. J. Dev. Neurosci.* **18,** 621–7.

Selleck, M.A.J. and Stern, C.D. (1991) Fate mapping and cell lineage analysis of Hensen's node in the chick embryo. *Development* **112,** 615–6.

Selleck, M.A.J. and Stern, C.D. (1992) Commitment of mesoderm cells in Hensen's node of the chick embryo to notochord and somite. *Development* **114,** 403–15.

Sengel, P. (1976) *Morphogenesis of Skin.* Cambridge University Press, London.

Sengel, P. (1986) Epidermal-dermal interaction. In *Biology of the Integument, Vol. 2 Vertebrates* (eds. J. Bereiter-Hahn, A. Matoltsty and K.S. Richards), Springer-Verlag, Berlin, pp. 374–408.

Shah, S.B., Skromme, I., Hume, C.R., Kessler, D.S., Lee, K.J., Stern, C.D. and Dodd, J. (1997) Misexpression of chick Vg1 in the marginal zone induces primitive streak formation. *Development* **124,** 5127–38.

Shapiro, F. (1992) Vertebral development of the chick embryo during days 3–19 of incubation. *J. Morph.* **213,** 317–33.

Shen, X., Steyrer, E., Retzek, H., Sanders, E.J. and Schneider, W. (1993) Chicken oocyte growth: receptor-mediated yolk deposition. *Cell Tissue Res.* **272,** 459–71.

Sheng, G. and Stern, C.D. (1999) *Gata2* and *Gata3*: novel markers for early embryonic polarity and for non-neural ectoderm in the chick embryo. *Mech. Dev.* **87,** 213–6.

Shimade, Y. and Ho, E. (1980) Scanning electron microscopy of the chick heart: formation of the epicardium and surface structure of the four heterotypic cells that constitute the embryonic heart. In *Etiology and morphogenesis of cardiac heart disease* (ed. R. Van Pragh) Futura, pp. 63–80.

Simkiss, K. (1967) *Calcium in Reproductive Physiology.* Chapman and Hall, London.

Simkiss, K. (1991) Fluxes during embryogenesis. In *Egg Incubation: Its Effects on Embryonic Development in Birds and Reptiles* (eds. D.C. Deeming and M.W.J. Ferguson), Cambridge University Press, Cambridge, pp. 47–52.

Simkiss, K. and Tyler, C. (1971) Shell formation. In *Biochemistry and Physiology of the Domestic Fowl* (eds. R.F. Bell and D.J. Freeman) Vol. 1, Academic Press, London, pp. 1331–43.

Skromme, I. and Stern, C.D. (2001) Interactions between Wnt and Vg1 signalling pathways initiate primitive streak formation in the chick embryo. *Development* **128,** 2915–27.

Smith, C.A., Smith, M.J. and Sinclair, A.H. (1999) Expression of chicken steroidogenic factor-1 during gonadal sex differentiation. *Gen. Comp. Endocrinol.* **113,** 187–96.

Smith, D.M. and Tabin, C.J. (1999) BMP signalling specifies the pyloric sphincter. *Nature* **402,** 748–9.

Smith, D.M.and Tabin, C.J. (2000) Clonally related cells are restricted to organ boundaries early in the development of the chicken gut to form compartment boundaries. *Dev. Biol.* **227,** 422–31.

Smith, D.M., Grasty, R.C., Theodosiou, N.A., Tabin, C.J. and Nascone-Yoder,N.M. (2000a) Evolutionary relationships between the amphibian, avian, and mammalian stomachs. *Evol. Dev.* **2,** 348–59.

Smith, D.M., Nielsen, C., Tabin, C.J. and Roberts, D.J. (2000b). Roles of BMP signalling and Nkx2.5 in patterning at the chick midgut-foregut boundary. *Development* **127,** 3671–81.

Sosic, D., Brand-Saberi, B., Schmidt, C., Christ, B. and Olson, E.N. (1997) Regulation of paraxis expression and somite formation by ectoderm- and neural tube-derived signals. *Dev. Biol.* **185,** 229–43.

Spemann, H. (1938) *Embryonic Development and Induction.* Yale University Press, New Haven, Conn.

Spratt, N.T. Jr. (1963) Role of the substratum, supracellular continuity and differential growth in morphogenetic cell movements. *Dev. Biol.* **7,** 51–63.

Stabellini, G., Locci, P., Calvitti, M., Evangelisti, R., Marinucci, L., Bodo, M., Caruso, A., Canaider, S. and Carinci, P.(2001) Epithelial-mesenchymal interactions and lung branching morphogenesis. Role of polyamine and transforming growth factor beta 1. *Eur. J. Histochem.* **45,** 151–62.

Stalsberg, H. (1969) The origins of heart asymmetry: right and left contributions to the early chick embryo heart. *Dev. Biol.* **19,** 100–27.

Stark, M.R., Rao, M.S., Schoenwolf, G.C., Yang, G., Smith, D. and Mauch, T.J. (2000) *Frizzled-4* expression during chick kidney development. *Mech. Dev.* **98,** 121–5.

Stéphan, F. (1949) Les suppléances obtenue expérimentalement dans le systeme des arcs aortiques de l'embryon d'oiseau. *C. Rend. Assoc. Anat.* **36,** 647–51.

Stephens, T.D., Sanders, D.D. and Yap, C.Y.F. (1992) Visual demonstration of the limb-forming zone in the chick embryo lateral plate. *J. Morph.* **213,** 305–16.

Stepinska, U. and Olszanska, B. (1983) Cell multiplication and blastoderm development in relation to egg envelope formation during uterine development of quail (*Coturnix coturnix*) embryo. *J. Exp. Zool.* **228,** 505–10.

Stern, C.D. (1990) The marginal zone and its contribution to the hypoblast and primitive streak of the chick embryo. *Development* **109,** 667–82.

Stern, C.D. (1993a) Transplantation in avian embryos. In *Essential Developmental Biology, a Practical Approach* (eds. C.D. Stern and P.W.H. Holland), IRL Press at Oxford University Press, pp. 111–118.

Stern, C.D. (1993b) Avian embryos. In *Essential Developmental Biology, a Practical approach* (eds. C.D. Stern and P.W.H. Holland), IRL Press at Oxford University Press, pp. 45–54.

Stern, C.D. and Canning, D.R. (1988) Changes in the expression of the carbohydrate epitope HNK-1 associated with mesoderm induction in the chick embryo. *Development,* **104,** 643–55.

Stern, C.D. and Canning, D.R. (1990) Origin of cells giving rise to mesoderm and endoderm in chick embryo. *Nature, Lond.* **343,** 273–275.

Stern, C.D. and Holland, P.W.H. (1993) *Essential Developmental Biology, a Practical Approach.* IRL Press at Oxford University Press.

Stern, C.D. and Ireland, G.W. (1981) An integrated experimental study of endoderm formation in avian embryos. *Anat. Embryol.* **163,** 245–63.

Stockdale, F.E., Nikovits, J.R. and Christ, B. (2000) Molecular and cellular biology of avian somite development. *Dev. Dyn.* **219,** 304–21.

Stoll, R., Ichas, F., Fancounan, N. and Maraud, R. (1993) Action of estradiol and tamoxifen on the testis-inducing activity of the chick embryonic testis grafted to the female embryo. *Anat. Embryol.* **188,** 587–92.

Storey, K.G., Crossley, J.M., De Robertis, E.M., Norris, W.E. and Stern, C.D. (1992) Neural induction and regionalisation in the chick embryo. *Development* **114,** 729–41.

Storey, K.G., Selleck, M.A.J. and Stern, C.D. (1995) Neural induction and regionalisation by different subpopulations of cells in Hensen's node. *Development* **121,** 417–28.

Streit, A. and Stern, C.D. (1999) Establishment and maintenance of the border of the neural plate in the chick: involvement of FGF and BMP activity. *Mech. Develop.* **82,** 51–66.

Sudo, H., Takahashi, Y., Tonegawa, A., Arase, Y., Ayoyama, H., Mizutani-Koseki, Y., Moriya, H., Wilting, J., Christ, B. and Koseki, H. (2001) Inductive signals from the somatopleure mediated by bone morphogenetic proteins are essential for the formation of the sternal component of avian ribs. *Dev. Biol.* **232,** 284–300.

Sullivan, G.E. (1962) Anatomy and embryology of the wing musculature of the domestic fowl (*Gallus*). *Aust. J. Zool.* **10,** 458–518.

Sullivan, L.C. and Orgeig, S. (2001) Dexamethasone and epinephrine stimulate surfactant secretion in type II cells of embryonic chicks. *Am. J. Physiol. Regul. Integr. Comp. Physiol.* **281,** R770–7.

Sun, D., Baur, S. and Hay, E.D. (2000) Epithelial-mesenchymal transformation is the mechanism for fusion of the craniofacial primordia involved in morphogenesis of the chicken lip. *Dev. Biol.* **228,** 337–49.

Tajima, A., Hayashi, H., Kamizumi, A., Ogura, J., Kuwana, T. and Chikamune, T. (1999) Study on the concentration of circulating primordial germ cells (cPGCs) in early chick embryos. *J. Exp. Zool.* **284,** 759–64.

Tao, H., Yoshimoto, Y., Yoshioka, H., Nohno, T., Noji, S. and Ohuchi, H. (2002) FGF10 is a mesenchymally derived stimulator of epidermal development in the chick embryonic skin. *Mech. Dev.* **116,** 39–49.

Thommes, R.C. (1987) Ontogenesis of thyroid function and regulation in the developing chick embryo. *J. Exp. Zool.* Suppl. 1, 273–80.

Teillet, M., Watanabe, Y., Jeffs, P., Duprez, D., Lapoin, F. and Le Douarin, N.M. (1998) Sonic hedgehog is required for survival of both myogenic and chondrogenic somatic lineages. *Development* **125,** 2019–30.

Thurston, R.J. and Hess, R.A. (1987) Ultrastructure of spermatozoa from domesticated birds: comparative study of turkey, chicken and guinea fowl. *Scanning Microsc.* **1,** 1829–38.

Tickle, C. (1991) Retinoic acid and chick limb bud development. *Dev. Biol.* Suppl. 1, 113–21.

Tickle, C. (1993) Chick limb buds. In *Essential Developmental Biology. A Practical Approach* (eds. C.D. Stern and P.W.H. Holland), IRL Press at Oxford University Press, pp. 119–25.

Tickle, C. and Munsterberg, A. (2001) Vertebrate limb development, *Curr. Opin. Genet.* **11,** 476–81.

Torrey, T.W. (1965) Morphogenesis of the vertebrate kidney. In *Organogenesis* (eds. R.L. DeHaan and H. Ursprung), Holt, Rinehart and Winston, New York, pp. 559–79.

Toyoshima, K., Seta, Y. and Shimamura, A. (1993) Fine structure of Merkel corpuscles in the lingual mucosa of Japanese quail, *Coturnix coturnix japonica. Arch. Oral. Biol.* **38,** 1009–12.

Trousse, F., Esteve, P. and Bovolenta, P. (2001) Bmp4 mediates apoptotic cell death in the developing chick eye. *J. Neurosci.* **21,** 1292–301.

Tsunekawa, N., Naito, M., Sakai, Y., Nishida, T. and Noce, T. (2000) Isolation of chicken vasa homolog gene and tracing the origin of primordial germ cells. *Development* **127,** 2741–50.

Tuan, R.S., Ono, T., Akins, R.E. and Koide, M. (1991) Experimental studies on cultured, shell-less fowl embryos: calcium transport, skeletal development and cardio-vascular functions. In *Egg Incubation: its Effects on Embryonic Development in Birds and Reptiles.* (eds. D.C. Deeming and M.J. Ferguson), Cambridge University Press, pp. 419–33.

Ukeshima, A. and Fujimoto, T. (1991) A fine morphological study of germ cells in asymmetrically developing right and left ovaries of the chick. *Anat. Rec.* **230,** 378–86.

Ukeshima, A., Kudo, M. and Fujimoto, T. (1987) Relationship between genital ridge formation and settlement site of primordial germ cells in chick embryos. *Anat. Rec.* **219,** 311–4.

Vakaet, L.C. (1962) Some new data concerning the formation of the definitive endoblast in the chick embryo. *J. Embryol. Exp. Morph.* **10,** 38–57.

Vakaet, L. (1970) Cinephotomicrographic investigations of gastrulation in the chick blastoderm. *Arch. Biol.* **81,** 387–426.

Vakaet, L. (1984) Early development in birds. In *Chimeras in Developmental Biology* (eds. N. Le Douarin and A. McLaren), Academic Press, London, pp. 71–88.

Vakaet, L., Vanroelen, Chr. and Andries, L. (1980) An embryological model of non-malignant invasion or ingression. In *Cell Movement and Neoplasia* (ed. M. de Brabander), Pergamon Press, Oxford, pp. 65–75.

Vanden Berge, J.C. and Zweers, G.A. (1993) Myologia. In *Handbook of Avian Anatomy: Nomina Anatomica Avium,* 2nd ed. (eds. J.J. Baumel, A.S. King, J.E. Breazile, H.E. Evans and J.C. Vanden Berge), Nuttall Ornithological Club, Cambridge, Mass, pp. 189–250.

Veini, M. and Bellairs, R. (1990) Early mesoderm differentiation in the chick embryo. *Anat. Embryol.* **183,** 143–9.

Vendrell, V., Carnicero, E., Giraldez, F., Alonso, M.T. and Schimmang, T. (2000) Induction of inner ear fate. *Development* **127,** 2011–9.

Verbout, A.J. (1985) The development of the vertebral column. *Adv. Anat. Embryol. Cell Biol.* **89,** 1–122.

Vesque, C., Ellis, S., Lee, A., Szabo, M., Thomas, P., Beddington, R. and Placzek, M. (2000) Development of chick axial mesoderm: specification of prechordal mesoderm by anterior endoderm-derived TGFbeta family signalling. *Development* **127,** 2795–809.

Viallet, J.P., Prin, F., Olovera-Martinez, I., Hirsinger, E., Pourquier, O. and Dhouailly, D. (1998) Chick Delta-1 gene expression and the formation of the feather primordia. *Mech. Dev.* **72,** 159–68.

Villalpando, I., Sanchez-Bringas, G., Sanchez-Varga, I., Pedernera, E. and Villafa-Monroy, H. (2000) The *P450 aromatase (P450arom)* gene is asymmetrically expressed in a critical period for gonadal sexual differentiation in the chick. *Gen. Comp. Endocrinol.* **117,** 325–34.

Virágh, S., Gittenberger-de Groot, A.C., Poelmann, R.E. and Kálmán, F. (1993) Early development of quail heart epicardium and associated vascular and glandular structures. *Anat. Embryol.* **188,** 381–93.

Vogel-Hopker, A., Momose, T., Rohrer, H. Yasuda, K., Ishihara, L. and Rapaport, D.H. (2000) Multiple functions of fibroblast growth factor 8 (FGF-8) in chick eye development. *Mech. Dev.* **94,** 25–36.

Vorster, W. (1989) The development of the chondrocranium of *Gallus gallus. Adv. Anat. Embryol. Cell Biol.* **113,** 1–77.

Waddington, C.H. (1932) Experiments on the development of chick and duck embryos cultivated *in vitro. Phil. Trans.* **B221,** 179–230.

Waddington, C.H. (1952) *The Epigenetics of Birds.* Cambridge University Press, Cambridge.

Waddington, C.H. and Schmidt, G.A. (1933) Induction by heteroplastic grafts of the primitive streak in birds. *Arch. Entw. Mech. Org.* **128,** 522–63.

Waddington, D., Gribbin, C., Sterling, R.J., Sang, H. and Perry, M.M. (1998) Chronology of events in the first cell cycle of the polyspermic egg of the domestic fowl (*Gallus domesticus*). *Int. J. Dev. Biol.* **42,** 625–8.

Wagner, J., Schmidt, C., Nikowits, W. Jr. and Christ, B. (2000) Compartmentalization of the somite and myogenesis in chick embryos are influenced by wnt expression. *Dev. Biol.* **228,** 86–94.

Waldo, K.L., Kumiski, D.H., Wallis, K.T., Stadt, H.A., Hutson, M.R., Platt, D.H. and Kirby, M.L. (2001) Conotruncal myocardium arises from a secondary heart field. *Development* **128,** 3179–88.

Walsh, C. and McLelland, J. (1978) The development of the epithelium and its innervation in the avian extrapulmonary respiratory tract. *J. Anat.* **125,** 171–82.

Watanabe, M., Kinutani, M., Naito, M., Ochi, O. and Takashima, Y. (1992) Distribution analysis of transferred donor cells in avian blastodermal chimeras. *Development* **114,** 331–8.

Watanabe, Y., Duprez, D., Monsoro-Burq, A.H., Vincent, C. and Le Douarin, N.M. (1998) Two domains in vertebrate development: antagonistic regulation by SHH and BMP4 proteins. *Development,* **125,** 2631–9.

Webb, S., Brown, N.A., Anderson, R.H. and Richardson, M.K. (2000) Relationship in the chick of the developing pulmonary vein to the embryonic systemic venous sinus. *Anat. Rec.* **259,** 67–75.

Wei, Y. and Mikawa, T. (2000) Fate diversity of primitive streak cells during heart field formation in ovo. *Dev. Dyn.* **219,** 505–13.

Weismann, A. (1885) *Die Continuität des Keimplasma's als Grundlage einer Theorie des Vererbung.* Fischer Verlag, Jena.

Wenner, P. and O'Donovan, M.J. (2001) Mechanisms that initiate spontaneous network activity in the chick spinal cord. *J. Neurophysiol.* **86,** 1481–98.

West, N.H., Langille, B.L. and Jones, D.R. (1981) In *Form and Function in Birds,* Vol. 2 (eds. A.S. King and J. McLelland), Academic Press, London, pp. 235–339.

Willier, B.H. (1937) Experimentally produced sterile gonads and the problem of the origin of the germ cells in the chick embryo. *Anat. Rec.* **70,** 89–112.

Wilt, F.H. and Wessels, N.K. (1967) *Methods in Developmental Biology.* Thomas Y. Crowell Co., New York.

Wilting, J., Papoutsi, M., Othman-Hassan, K., Rodriguez-Niedenfuhr, M., Prols, F., Tomarev, S.I. and Eichmann, A. (2001) Development of the avian lymphatic system. *Microsc. Res. Tech.* **55,** 81–91.

Wishart, G. and Horrocks, A.J. (2000) Fertilization in birds. In *Fertilization in Protozoa and Metazoan Animals* (eds. J.J. Tarina and A. Cano), Springer-Verlag, Heidelberg, pp. 193–222.

Withington, S., Beddington, R. and Cooke, J. (2001) Foregut endoderm is required at head process stages for anteriormost neural patterning in chick. *Development* **128,** 309–20.

Wolff, E. (1953) Le role des hormones embryonnaires dans la differenciation sexuelle des oiseaux. *Arch. Anat. Microscop. Morphol. Exptl.* **39,** 426–50.

Woods, J.E. (1987) Maturation of the hypothalamo-adenohypophyseal-gonadal (HAG) axes in the chick embryo. *J. Exp. Zool.* Suppl. 1, 265–72.

Wolpert, L. (2002) *Principles of Development,* 2nd ed. Oxford University Press.

Wortham, R.A. (1948) The development of the muscles and tendons in the lower leg and foot of chick embryos. *J. Morph.* **83,** 105–48.

Yander, G. and Searle, R.L. (1980) A scanning electron microscopic study of the development of the shoulder, visceral arches, and the region ventral to the cervical somites of the chick embryo. *Am. J. Anat.* **157,** 27–39.

Yang, A. and Siegel, P.B. (1997) Late embryonic and early posthatch growth of heart and lung in White Leghorn lines of chickens. *Growth Dev. Aging* **61,** 119–26.

Yassine, F., Feddeka-Brunner, B. and Dieterlen-Lièvre, F. (1989) The ontogeny of the chick embryo spleen; a cytological study. *Cell Differ. Dev.* **27,** 29–45.

Yee, G.W. and Abbott, U.K. (1978) Facial development in normal and mutant chick embryos. I Scanning electron microscopy of primary palate formation. *J. Exp. Zool.* **206,** 307–21.

Yew, D.T. (1978) The origin and initial development of the pecten oculi. *Anat. Anz.* **143,** 383–7.

Yu, M., Yue, Z., Wu, D.-Y., Mayer, J.-A., Medina, M., Widelitz, R.B., Jang, T.-X. and Chuong, C.-M. (2004) The developmental biology of feather follicles. *Int. J. Dev. Biol.* **48,** 181–91.

Yu, X., St Amand, T.R., Wang, S., Li, G., Zhang, Y., Hu, Y.P., Nguyen, L., Qiu, M.S. and Chen, Y.P. (2001) Differential expression and functional analysis of *Pitx2* isoforms in regulation of heart looping in the chick. *Development* **128,** 1005–13.

Yutzey, K.E. and Kirby, M.L. (2002) Wherefore heart thou? Embryonic origins of cardiogenesis mesoderm. *Dev. Dyn.* **223,** 307–20.

Zaccanti, F., Vallisneri, M. and Quagia, A. (1990) Early aspects of sex differentiation in the gonads of chick embryos. *Differentiation* **43,** 71–80.

Zehavi, N., Reich, V. and Khaner, O. (1998) High proliferation rate characterizes the site of axis formation in the avian blastula-stage embryo. *Int. J. Dev. Biol.* **42,** 95–8.

Zemanova, Z. and Ujec, E. (2002) Transepithelial potential in mesonephric nephrons of 7-day-old chick embryos in relation to the histochemically detected sodium pump. *Physiol. Res.* **51,** 43–8.

Zhi, Q., Huang, R., Christ, B. and Brand-Sabieri, B. (1996) Participation of individual brachial somites in skeletal muscles of the avian forearm and hand. *Anat. Embryol.* **194,** 327–39.

Zimmermann, A., Haina, A. and Groschel-Stewart, U. (1995) Neural and smooth muscle development in the chicken gizzard. *Roux's Arch. Dev. Biol.* **204,** 271–4.

Zwaan, J. and Ikeda, A. (1966) Studies of the differentiation of the chicken lens with the fluorescent antibody technique. *Anat. Rec.* **154,** 447.

Zwilling, E. (1949) The role of epithelial components in the developmental origin of the "wingless" syndrome of chick embryos. *J. Exp. Zool.* **111,** 175–88.

Note added in proof. A collection of important reviews has recently been published: Stern, C.D. (ed.) (2004) The chick in developmental biology. *Mechanisms of Develop.* **121,** 1011–1186.

Index

The italic numerals indicate plate page numbers